# Methods in Enzymology

Volume 69
PHOTOSYNTHESIS AND NITROGEN FIXATION
Part C

# METHODS IN ENZYMOLOGY

EDITORS-IN-CHIEF

Sidney P. Colowick        Nathan O. Kaplan

*Methods in Enzymology*

*Volume 69*

# Photosynthesis and Nitrogen Fixation

## Fixation

*Part C*

EDITED BY

*Anthony San Pietro*

BIOLOGY DEPARTMENT
INDIANA UNIVERSITY
BLOOMINGTON, INDIANA

1980

## ACADEMIC PRESS

*A Subsidiary of Harcourt Brace Jovanovich, Publishers*

New York   London   Toronto   Sydney   San Francisco

ACADEMIC PRESS, INC.
111 Fifth Avenue, New York, New York 10003


*United Kingdom Edition published by*
ACADEMIC PRESS, INC. (LONDON) LTD.
24/28 Oval Road, London NW1 7DX


**Library of Congress Cataloging in Publication Data**

San Pietro, Anthony Gordan, Date
    Photosynthesis.

    (Methods in enzymology, v. 23–24,    )
    Pt. B–C have title:  Photosynthesis and nitrogen
fixation.
    Includes bibliographical references.
    1.  Photosynthesis--Research.  2.  Enzymes.
3.  Biological chemistry--Technique.  4.  Nitrogen--
Fixation.  I.  Title.  II.  Title: Photosynthesis
and nitrogen fixation.  III.  Series.
QP601.M49  vol. 23, etc.  [QK882]   581.1'3342
ISBN  0–12–181969–8 (v. 69)      72–179811


PRINTED IN THE UNITED STATES OF AMERICA

80 81 82 83    9 8 7 6 5 4 3 2 1

# Table of Contents

## Section I. Mutants

## Section II. Cellular and Subcellular Preparations

## Section III. Components

## Section IV. Methodology

## Section V. Inhibitors

# Section VI. Nitrogen Fixation

## Addendum

# Contributors to Volume 69, Part C

Article numbers are in parentheses following the names of contributors.
Affiliations listed are current.

S. L. ALBRECHT* (68), *Biochemistry Department, University of Wisconsin, Madison, Wisconsin 53706*

CHARLES J. ARNTZEN (50), *USDA/ARS, Department of Botany, University of Illinois, Urbana, Illinois 61801*

MORDHAY AVRON (60), *Biochemistry Department, Weizmann Institute of Science, Rehovot, Israel*

ASSUNTA BACCARINI-MELANDRI (28), *Institute of Botany, University of Bologna, Bologna, Italy*

J. BARBER (56), *Botany Department, Imperial College, London, S. W. 7, United Kingdom*

SHOSHANA BAR-NUN (34), *Department of Biological Chemistry, The Hebrew University of Jerusalem, Jerusalem, Israel*

R. BARR (35), *Department of Biological Sciences, Purdue University, West Lafayette, Indiana 47907*

ALAN J. BEARDEN (21), *Department of Biophysics and Medical Physics, University of California, Berkeley, California 94720*

RICHARD J. BERZBORN (47), *Lehrstuhl für Biochemie der Pflanzen, Abteilung Biologie der Ruhr-Universität, 4630 Bochum, West Germany*

CLANTON C. BLACK, JR. (6), *Department of Biochemistry, University of Georgia, Athens, Georgia 30602*

PETER BÖGER (10), *Lehrstuhl für Physiologie und Biochemie der Pflanzen, Universität Konstanz, D-7750 Konstanz, Germany*

BOB B. BUCHANAN (36), *Department of Cell Physiology, University of California, Berkeley, California 94720*

JOHN J. BURKE (50), *USDA/ARS, Department of Botany, University of Illinois, Urbana, Illinois 61801*

JACOBO CÁRDENAS (23), *Departamento de Bioquimica, Facultad de Ciencias y C.S.I.C., Universidad de Sevilla, Sevilla, Spain*

KEVIN R. CARTER (67), *Laboratory for Nitrogen Fixation Research, Oregon State University, Corvallis, Oregon 97331*

GEORGE M. CHENIAE (33), *Department of Agronomy, University of Kentucky, Lexington, Kentucky 40506*

J. J. CHILD (69), *National Research Council, Prairie Regional Laboratory, University Campus, Saskatoon S7N 0W9, Saskatchewan, Canada*

R. CHOLLET (30), *Laboratory of Agricultural Biochemistry, University of Nebraska, Lincoln, Nebraska 68583*

NAM-HAI CHUA (40), *The Rockefeller University, New York, New York 10021*

W. A. CRAMER (18), *Department of Biological Sciences, Purdue University, West Lafayette, Indiana 47907*

F. L. CRANE (35), *Department of Biological Sciences, Purdue University, West Lafayette, Indiana 47907*

BRUNO CURTI (22), *Biochemistry Unit, Faculty of Sciences, University of Milano, 20133, Milano, Italy*

* Present address: USDA/SEA and Agronomy Department, University of Florida, Gainesville, Florida 32611

R. A. DILLEY (48), *Department of Biological Sciences, Purdue University, West Lafayette, Indiana 47907*

M. J. DILWORTH (74), *School of Environmental and Life Sciences, Murdoch University, Murdoch, Western Australia 6153*

ROLAND DOUCE (26), *DRF/Biologie Vegetale-CENG, and USMG, 85 X-F 38041, Grenoble-Cedex, France*

ROBERT R. EADY (70), *Agricultural Research Council, Unit of Nitrogen Fixation, University of Sussex, Brighton, BN1 9QJ, United Kingdom*

DALE E. EDMONDSON (37), *Molecular Biology Division, Veterans Administration Medical Center, and Department of Biochemistry and Biophysics, University of California, San Francisco, California 94121*

WILLIAM L. ELLEFSON* (19), *Department of Biochemistry, Purdue University, West Lafayette, Indiana 47907*

HAROLD J. EVANS (67), *Laboratory for Nitrogen Fixation Research, Oregon State University, Corvallis, Oregon 97331*

PETER FAY (73), *Department of Botany and Biochemistry, Westfield College, University of London, London, NW3 7ST, United Kingdom*

DONALD R. GEIGER (53), *Department of Biology, University of Dayton, Dayton, Ohio 45469*

C. GIERSCH (63), *Botanisches Institut der Universität Düsseldorf, 4000 Düsseldorf, West Germany*

W. O. GILLUM (71), *Western Electric Company, Princeton, New Jersey 08540*

JOHN H. GOLBECK (12), *Martin Marietta Laboratories, Baltimore, Maryland 21227*

ELIZABETH L. GROSS (45), *Department of Biochemistry, The Ohio State University, Columbus, Ohio 43210*

R. H. HAGEMAN (24), *Department of Agronomy, University of Illinois, Urbana, Illinois 61801*

D. O. HALL (8), *Department of Plant Sciences, King's College, London SE 24 9JF, United Kingdom*

F. JOE HANUS (67), *Laboratory for Nitrogen Fixation Research, Oregon State University, Corvallis, Oregon 97331*

R. W. F. HARDY (54), *E. I. du Pont de Nemours and Company, Central Research and Development Department, Experimental Station, Wilmington, Delaware 19898*

G. HAUSKA (62), *University of Regensburg, FB Biology, Botany, 31, 84 Regensburg, Germany*

U. HEBER† (63), *Botanisches Institut der Universität Düsseldorf, 4000 Düsseldorf, West Germany*

H. W. HELDT‡ (57), *Institut für Physiologische Chemie, Physikalische Biochemie und Zellbiologie, der Universität München, 8 München 2, West Germany*

T. HORIO (32), *Institute for Protein Research, Osaka University, 5311 Yamada-Kami, Suita, Osaka 565, Japan*

S. IZAWA (39), *Department of Biology, Wayne State University, Detroit, Michigan 48202*

* Present address: Department of Microbiology, University of Illinois, Urbana, Illinois 61801
† Present address: Institut für Botanik und Pharmazeutische Biologie der Universität Würzburg, 87 Würzburg, West Germany
‡ Present address: Lehrstuhl für Biochemie der Pflanzen, der Universität Göttingen, Untere Karspüle 2, 3400 Göttingen, West Germany

JACQUES JOYARD (26), *DRF/Biologie Vegetale-CENG, and USMG 85 X-F 38041, Grenoble-Cedex, France*

SAMUEL KAPLAN (3), *Department of Microbiology, University of Illinois, Urbana, Illinois 61801*

F. KOENIG* (41), *Max-Planck-Institut für Züchtungsforschung (Erwin-Baur-Institut), 5 Köln 30, West Germany*

B. KOK† (25), *Martin Marietta Laboratories, Baltimore, Maryland 21227*

RUUD KRAAYENHOF (49), *Biological Laboratory, Free University Amsterdam, de Boelelaan 1087, 1007 MC Amsterdam, The Netherlands*

DAVID W. KROGMANN (19), *Department of Biochemistry, Purdue University, West Lafayette, Indiana 47907*

S. D. KUNG (30), *Department of Biological Sciences, UMBC, Catonsville, Maryland 21228*

W. G. W. KURZ (69), *National Research Council, Prairie Regional Laboratory, University Campus, Saskatoon S7N 0W9, Saskatchewan, Canada*

PAUL A. LOACH (15), *Department of Biochemistry and Molecular Biology, Northwestern University, Evanston, Illinois 60201*

MANUEL LOSADA (23), *Departemento de Bioquimica, Facultad de Ciencias y C.S.I.C., Universidad de Sevilla, Sevilla, Spain*

HARVARD LYMAN (2), *Department of Biology, SUNY, Stony Brook, New York 11794*

RICHARD E. McCARTY (51, 66), *Section of Biochemistry, Molecular and Cell Biology, Cornell University, Ithaca, New York 14853*

GERHARD MADER (64), *Lehrstuhl für Biochemie der Pflanzen, Universität Göttingen Untere Karspüle 2, 3400 Göttingen, West Germany*

RICHARD MALKIN (21), *Department of Cell Physiology, University of California, Berkeley, California 94720*

BARRY MARRS (3), *Edward A. Doisy Department of Biochemistry, Saint Louis University, School of Medicine, Saint Louis, Missouri 63130*

T. V. MARSHO (25, 30), *Department of Biological Sciences, UMBC, Catonsville, Maryland 21228*

HISAKO MATSUMURA (43), *Institute of Applied Microbiology, University of Tokyo, Bunkyo-ku, Tokyo, 113 Japan*

BRUNO ANDREA MELANDRI (28), *Institute of Botany, University of Bologna, Bologna, Italy*

W. MENKE (41), *Max-Planck-Institut für Züchtungsforschung, (Erwin-Baur-Institut), 5 Köln 30, West Germany*

DONALD MILES (1), *Division of Biological Sciences, University of Missouri, Columbia, Missouri 65211*

SHIGETOH MIYACHI (43), *Institute of Applied Microbiology, University of Tokyo, Bunkyo-ku, Tokyo 113, Japan*

L. E. MORTENSON (71), *Department of Biological Sciences, Purdue University, West Lafayette, Indiana 47907*

JACK MYERS (44), *Department of Zoology, University of Texas, Austin, Texas 78712*

H. Y. NAKATANI (56), *Botany Department, Imperial College, London, S. W. 7 United Kingdom*

NATHAN NELSON (27), *Department of Biology, Technion–Israel Institute of Technology, Haifa, Israel*

* Present address: Research Division, Brigham Young University, Provo, Utah 84602
† Deceased

WILLIAM L. OGREN (61), *United State Department of Agriculture, Science and Education Administration, Agricultural Research, Department of Agronomy, University of Illinois, Urbana, Illinois 61801*

ITZHAK OHAD (34), *Department of Biological Chemistry, The Hebrew University of Jerusalem, Jerusalem, Israel*

Y. OKON\* (68), *Biochemistry Department, University of Wisconsin, Madison, Wisconsin 53706*

O. OLLINGER (52), *Martin Marietta Laboratories, Baltimore, Maryland 21227*

JOHN M. OLSON (31), *Biology Department, Brookhaven National Laboratory, Upton, New York 11973*

LESTER PACKER (59), *Membrane Bioenergetics Group, Lawrence Berkeley Laboratory, and The Department of Physiology-Anatomy, University of California, Berkeley, California 94720*

GEORGE C. PAPAGEORGIOU (58), *Nuclear Research Center Demokritos, Department of Biology, Aghia Paraskevi, Athens, Greece*

URI PICK (51), *Section of Biochemistry, Molecular and Cell Biology, Cornell University, Ithaca, New York 14853*

RAYMOND P. POINCELOT (11), *Department of Biology, Fairfield University, Fairfield, Connecticut 06430*

L. J. PROCHASKA† (48), *Department of Biological Sciences, Purdue University, West Lafayette, Indiana 47907*

BRUNO QUEBEDEAUX (54), *E. I. du Pont de Nemours and Company, Central Research and Development Department, Experimental Station, Wilmington, Delaware 19898*

R. RADMER (52), *Martin Marietta Laboratories, Baltimore, Maryland 21227*

A. J. REED (24), *Department of Agronomy, University of Illinois, Urbana, Illinois 61801*

S. G. REEVES (8), *Brewing Research Foundation, Lyttel Hall, Nutfield, Redhill, Surrey, RH1 4HY United Kingdom*

IRAJ ROUHANI (6), *Department of Biochemistry, University of Georgia, Athens, Georgia 30602*

ALBERT RUESINK (7), *Department of Biology, Indiana University, Bloomington, Indiana 47405*

HANS J. RURAINSKI (64), *Lehrstuhl für Biochemie der Pflanzen, Universität Göttingen, Untere Karspüle 2, 3400 Göttingen, West Germany*

GEORGE K. RUSSELL (2), *Department of Biology, Adelphi University, Garden City, New York 11530*

JEROME A. SCHIFF (2), *Institute for Photobiology of Cells and Organelles, Brandeis University, Waltham, Massachusetts 02154*

PETER SCHÜRMANN‡ (36), *Department of Cell Physiology, University of California, Berkeley, California 94720*

JEROME C. SERVAITES§ (61), *Department of Agronomy, University of Wisconsin, Madison, Wisconsin 53706*

---

\* Present address: Department of Plant Pathology and Microbiology, The Hebrew University of Jerusalem, Rehovot, 76100 Israel

† Present address: Institute of Molecular Biology, University of Oregon, Eugene, Oregon 97403

‡ Present address: Laboratoire de Physiologie Végétale et Biochimie, Université de Neuchâtel, 2000 Neuchâtel, Switzerland

§ Present address: Department of Biology, Virginia Polytechnic Institute and State University, Blacksburg, Virginia 24061

VINOD K. SHAH (72), *Department of Bacteriology and Center for Studies of Nitrogen Fixation, College of Agricultural and Life Sciences, University of Wisconsin, Madison, Wisconsin 53706*

YOSEPHA SHAHAK (60), *Biochemistry Department, Weizmann Institute of Science, Rehovot, Israel*

K. T. SHANMUGAM (5), *Plant Growth Laboratory, Department of Agronomy and Range Science, University of California, Davis, California 95616*

N. SHAVIT (29, 46), *Department of Biology, Ben Gurion University of the Negev, Beer Sheva, Israel*

WILLIAM SHEPHERD (3), *Department of Microbiology, University of Illinois, Urbana, Illinois 61801*

JUDITH ANN SHIOZAWA (13), *Lehrstuhl für Mikrobiologie, Universität Freiburg, 7800-Frieburg i, Br., West Germany*

V. SHOSHAN* (46), *Department of Biology, Ben Gurion University of the Negev, Beer Sheva, Israel*

HARTMUT SPILLER † (10), *Faculty of Biology, University of Konstanz, D-7750 Konstanz, Germany*

H. STROTMANN (29), *Botanisches Institut, Abteilung Biochemie der Pflanzen der Tierarztlichen, Hochschule Hannover, Germany*

MASAURU TANAKA(20), *Department of Information of Computer Science, Faculty of Engineering Science, University of Toyonaka, Osaka 560, Japan*

JANET M. THORNBER (14, 16), *School of Public Health, University of California, Los Angeles, California 90024*

J. PHILIP THORNBER (14, 16), *Department of Biology and Molecular Biology Institute, University of California, Los Angeles, California 90024*

GORDON TOLLIN (37), *Department of Biochemistry, University of Arizona, Tucson, Arizona 85721*

A. TREBST (65), *Department of Biology, Ruhr-University, Bochum, Germany*

ELDON A. ULRICH (19), *Department of Chemistry, Purdue University, West Lafayette, Indiana 47907*

R. C. VALENTINE (5), *Plant Growth Laboratory, Department of Agronomy and Range Science, University of California, Davis, California 95616*

CHASE VAN BAALEN (4), *Department of Botany, University of Texas at Austin, Austin, Texas 78712*

JOSÉ M. VEGA (23), *Departmento de Bioquimica, Facultad de Ciencias y C.S.I.C. Universidad de Sevilla, Sevilla, Spain*

D. A. WALKER (9), *Department of Botany, University of Sheffield, Sheffield S10 2TN United Kingdom*

RICHARD T. WANG (38), *Department of Zoology, University of Texas, Austin, Texas 78712*

AARON R. WASSERMAN (17), *Department of Biochemistry, McGill University, Montreal, Quebec, H3G 1Y6 Canada*

J. WHITMARSH‡ (18), *Department of Biological Sciences, Purdue University, West Lafayette, Indiana 47907*

RICARDO A. WOLOSIUK§ (36), *Department of Cell Physiology, University of California, Berkeley, California 94720*

* Present address: College of Agricultural and Life Sciences, Department of Biochemistry, University of Wisconsin–Madison, Madison, Wisconsin 53706
† Present address: Plant Growth Laboratory, Department of Agronomy and Range Science, University of California, Davis, California 95616
‡ Present address: Department of Biology, Queen's University, Kingston, Ontario K76 3NG, Canada
§ Present address: Instituto de Investigaciones Bioquimicas, Fundación Campomar, Buenos Aires, Argentina

J. YAMASHITA (32), *Institute for Protein Research, Osaka University, 5311 Yamada-kami, Suita, Osaka 565, Japan*

KERRY T. YASUNOBU (20), *Department of Biochemistry and Biophysics, University of Hawaii at Manoa, Honolulu, Hawaii 96822*

CHARLES F. YOCUM (55), *Division of Biological Sciences, The University of Michigan, Ann Arbor, Michigan 48109*

GIULIANA ZANETTI (22), *Laboratory of Biochemistry, University of Milano, 20133 Milano, Italy*

ISRAEL ZELITCH (42), *Department of Biochemistry and Genetics, Connecticut Agriculture Experimental Station, New Haven, Connecticut 06504*

# Preface

The first two volumes (Volume XXIII, Part A and Volume XXIV, Part B) on photosynthesis and nitrogen fixation appeared almost a decade ago. The availability of new experimental techniques and sophisticated instrumentation during this intervening period has allowed for more precise and critical evaluation of the major problem areas. The experimental results of these investigations are detailed in this volume under six major headings: Mutants, Cellular and Subcellular Preparations, Components, Methodology, Inhibitors, and Nitrogen Fixation. It is hoped that even more sophisticated and powerful investigative techniques will become available in the near future. The degree to which the present volume stimulates such developments will be a major measure of its usefulness.

My deepest gratitude is extended to all authors for their excellent contributions. The boundless patience, help, and cooperation of the staff of Academic Press are gratefully acknowledged.

ANTHONY SAN PIETRO

# METHODS IN ENZYMOLOGY

EDITED BY

## Sidney P. Colowick and Nathan O. Kaplan

VANDERBILT UNIVERSITY DEPARTMENT OF CHEMISTRY
SCHOOL OF MEDICINE UNIVERSITY OF CALIFORNIA
NASHVILLE, TENNESSEE AT SAN DIEGO
LA JOLLA, CALIFORNIA

# METHODS IN ENZYMOLOGY

EDITORS-IN-CHIEF

Sidney P. Colowick     Nathan O. Kaplan

Section I

# Mutants

# [1] Mutants of Higher Plants: Maize

## By Donald Miles

The selection and use of genetic mutants to study complex biochemical problems is a well-established technique of enzymology. Selected mutants have been valuable in the analysis of individual reactions in photosynthesis. Thus far, the majority of such mutants studied have been isolated from a few species of green algae. Photosynthesis mutants (i.e., those lacking one or more reactions of photosynthesis) have been isolated from *Scenedesmus*,[1] *Chlamydomonas*,[2] *Euglena*,[3] and other species. These mutants have been widely used in studies of primary photochemical reactions, electron-transport pathways, and photophosphorylation. The data gained have been valuable, and in some cases the information could be obtained in no other way. However, in spite of the variety of microorganisms used in genetic studies of photosynthesis, very few good examples exist among higher plants. The main reason is the difficulty of handling the large number of individuals required for such work in higher plants. The purpose of this article is to summarize higher plant work and to outline methods used to select and characterize photosynthesis mutants of *Zea mays*.

### The Use of Higher Plant Mutants

There are a variety of reasons why a higher land plant, specifically a crop plant, is as desirable to use in such work as a green alga. There are definite differences in the photosynthetic apparatus and growth habits of angiosperms and the green algae. Therefore, photosynthesis mutants should be studied from both of these taxonomic groups. In many crop species, detailed genetic information is available, and careful genetic analyses are possible. The genome is well mapped in barley, pea, tomato, maize, and other species. Such genetic information can be well correlated with biochemical data. This correlation is not possible with most algae that have been used for photosynthetic studies; moreover, the difference between nuclear and chloroplast genes are not as clear in algae as in most higher plants. A further advantage is that higher plants are diploid

[1] N. I. Bishop, Vol. 23, p. 131.
[2] R. P. Levine, Vol. 23, p. 119.
[3] G. H. Russell, H. Lyman, and R. L. Heath, *Plant Physiol.* **44**, 929 (1969).

METHODS IN ENZYMOLOGY, VOL. 69

and recessive genes can be carried in the heterozygote, whereas an algal mutant must be able to take up and metabolize an exogenous carbon source. Stocks of higher plant mutants can be easily maintained, since continuous culture is unnecessary and seed stored properly remain viable for many years. This allows one to work with a large number of mutants and reduces the chance of suppressor mutations or reversions.

A problem with using photosynthesis mutants of higher plants was that until recently very few mutants were available. Table I shows the characteristics of some mutants exclusive of those in maize. Most higher plant photosynthesis mutants were not specifically selected, but were detected during studies of chlorophyll mutants. In contrast, photosynthesis mutants of microorganisms were usually specially selected. Since higher plants cannot be provided with an alternate carbon source to replace that lost through mutation of photosynthesis, selection of mutants among higher plants leads to three types of mutations. The most common type are mutants which survive because they contain incomplete (leaky) blocks of photosynthesis and grow slowly. An example is the yellow *Nicotiana tabacum* nuclear mutant.[4] A second common type is the cytoplasmic (plastome) mutants which show variegation of normal and mutant sectors in the leaf. In these plants, the normal sectors carry on sufficient photosynthesis to supply energy to the mutant sectors. They are usable if the sectors are well defined and large but the mutations are non-Mendelian. Some *Oenothera hookeri* mutants of this type exist.[5] A third type of photosynthesis mutant which has been selected primarily in *Zea mays* is the seedling lethal.[6] These can represent complete blocks in photosynthesis in which growth is maintained by respiration of endosperm starch. These are carried as recessive genes in heterozygous stock.

One problem common to nearly all higher plant mutants was that the photosynthetic pigment content or type was altered. Most often there were gross changes in pigment which were associated with changes in chloroplast, membrane, and leaf structure. These pleiotropic mutants are complex and more useful for developmental studies than those in which the mutation has had a limited effect. Such mutants have been used because of the obvious ease of selection by phenotypic color differences from normal wild-type plants.

The methods outlined here are for the selection of the third type of photosynthesis mutant, seedling lethals. Sixty percent of these mutants selected have full, normal pigmentation, normal chloroplast ultrastruc-

[4] G. H. Schmid, Vol. 23, p. 171.
[5] D. C. Fork and U. W. Heber, *Plant Physiol.* **43**, 606 (1968).
[6] C. D. Miles, *Stadler Genet. Symp.* **7**, 135 (1975).

TABLE I

PHOTOSYNTHESIS MUTANTS OF HIGHER PLANTS[a]

| Species | Mutant | Inheritance | Pigmentation | Component or photosystem lost |
|---|---|---|---|---|
| Vicia faba[b] | — | Nuclear | Yellow-green | Fd-NADP+ reductase |
| Glycine max[c] | LG | Nuclear | Yellow-green | Increased PQ oxidation |
| Oenothera hookeri[a] | I alpha, gamma, delta | Plastome | Pale-green | PS-II |
| Oenothera suaveoleus[a] | II gamma | Plastome | Pale-green | PS-II, Q |
| | II alpha | Plastome | Pale-green | PS-I |
| Nicotiana tabacum[e] | NC 95 | Extranuclear | Yellow-green | PS-II |
| Gossypium hirsutum[f] | 50/8 | Nuclear | Pale-green | Between PS-I and PS-II |
| Oryza sativa[g] | M-4, M-5 | Nuclear | Pale-green | PS-II |
| Lycopersicon | albina-2 | Nuclear | Pale-green | PS-I |
| esculentum[h] | sulfures (pura) | Nuclear | Yellow-green | PS-I |
| Antirrhinum majus[h] | en:alba-1 | Extranuclear | Pale-green | PS-I |
| | en:viridis-1 | Extranuclear | Pale-green | PS-II |
| Hordeum vulgare[i] | xantha-b,[12] c,[23] d[31] | Nuclear | Yellow-green | PS-II |
| | xantha-1[35] | Nuclear | Yellow-green | PS-I |
| | viridis-k[23] | Nuclear | Yellow-green | PS-I |
| Arachis hypogaea[j] | 0026 | Nuclear | Virescent | PS-I |

[a] Exclusive of Zea mays mutants.
[b] U. Heber and W. Gottschalk, Z. Naturforsch., Teil B **18**, 36 (1963).
[c] R. W. Keck, R. A. Dilley, and B. Ke, Plant Physiol. **46**, 669 (1970).
[d] D. C. Fork and U. W. Heber, Plant Physiol. **43**, 606 (1968).
[e] G. H. Schmid, Vol. 23, p. 171.
[f] M. M. Yakubova, A. B. Rubin, G. A. Khramova, and D. N. Matorin, in "Genetic Aspects of Photosynthesis" (Y. E. Nasyrov and Z. Šesták, eds.), p. 263. Junk, The Hague, 1975.
[g] Y. Inoue, T. Ogawa, T. Kawai, and K. Shibata, Physiol. Plant. **29**, 390 (1973).
[h] R. Hagemann, F. Herrmann, and T. Börner, in "Genetic Aspects of Photosynthesis" (Y. E. Nasyrov and Z. Šesták, eds.), p. 115. Junk, The Hague, 1975.
[i] K. W. Henningsen, N. C. Nielsen, and R. M. Smillie, Port. Acta Biol. **14**, 323 (1974).
[j] R. S. Alberte, J. D. Hesketh, and J. S. Kirby, Z. Pflanzenphysiol. **77**, 152 (1976).

ture, and normal plant morphology. Mutations of single, recessive nuclear genes can be selected which cause loss of one step in the light or dark reaction. Although the general methods should apply to most species, the methods presented will be for *Zea mays*. Any new program for mutation in other species would be most productive in genetically well-characterized plants.

Detailed methods for determining which electron carrier or enzyme has been altered by mutation will not be presented here, since previous methods in this series can be used,[7] especially as outlined for algae by Bishop[1] and Levine.[2] The emphasis here will be for screening, selection, and initial testing of photosynthesis mutants induced in higher plants.

Induction of Mutants

Common mutagenic agents which are effective in the induction of higher plants mutants are X rays, γ rays, ultraviolet radiation, and a number of chemical mutagens.[8] Chemical mutagens are most convenient to use with higher plants and among them ethyl methane sulfonate (EMS) and *N*-methyl-*N*'-nitro-*N*-nitrosoguanidine (NG) have been very effective for induction of photosynthesis mutants in microorganisms,[1,2] and higher plants.[9] Sodium azide is also quite effective for induction of higher plant chlorophyll mutants, as seen in *Hordeum vulgare*.[10] Although chemical agents usually cause point mutations, they can also cause chromosome mutations such as translocations. Chromosome mutations are less desirable, since they can lead to loss of more than one gene. Chromosome mutations can be detected much more readily in a higher plant such as maize in which the cytology is advanced.

The site of induced mutations could be nuclear or plastid. Both types of photosynthesis mutants have been described.[11] With the procedures described here only nuclear mutations will be observed.

There are two common procedures for chemical mutagen treatment which are convenient and useful. EMS and NG have been applied directly to maize kernels at 10 m$M$ in 20 m$M$ phosphate buffer (pH 7.5) for 10 hr

[7] A. San Pietro, ed., Vol. 23.
[8] E. Amano and H. H. Smith, *Mutat. Res.* 2, 344 (1965).
[9] M. G. Neuffer, *Maize Genet. Coop. Newsl.* 42, 124 (1968).
[10] R. A. Nilan, E. G. Sideris, A. Kleinhofs, C. Sander, and C. F. Konzak, *Mutat. Res.* 17, 142 (1973).
[11] B. Walles, *in* "Structure and Function of Chloroplasts" (M. Gibbs, ed.), p. 51. Springer-Verlag, Berlin and New York, 1971.

at 25°.[12] The kernels were planted immediately or the mutagen had to be rinsed out for later planting. This treatment gave 52% survivors, and of those 50% were chlorophyll mutants. A problem with this procedure is that diploid and triploid tissue made up of thousands of cells is being exposed. We preferred a second technique of pollen treatment.

Chlorophyll mutants were successfully induced with EMS and NG treatment in haploid pollen cells by Neuffer.[9] The mature pollen grain consist of two haploid nuclei, the pollen tube nucleus and the generative nucleus. Mutations of the tube nucleus would not be transmitted, thus allowing mutagen to be directed to one specific nucleus type, the generative nucleus. Only these will be transmitted.

The equivalent of 8.9 m$M$ EMS (Sigma Chemical Co., St. Louis, Missouri) is suspended in light, domestic paraffin oil (Fisher Chemical Co.) by shaking for 1 hr at room temperature. Pollen is collected using standard tassel bags from a vigorous genetic stock and immediately placed in 15 times its volume of the EMS–oil suspension. The mutagen-pollen mixture is shaken in a sealed plastic vial to prevent clumping of the pollen. After 50 min at summer field temperature, the pollen is applied to previously prepared, fresh silks (styles cut back to 2 cm above the ear) by lifting the pollen on a small (No. 10) camel hair brush and spreading the oil-covered pollen on the silk. Plants other than the pollen donor were receptors. The genetic handling of the treated material is outlined in Fig. 1.

A large number of the pollen grains are killed by this treatment, and since excess pollen is applied only the viable pollen will fertilize the egg and yield a kernel. These kernels are termed the $M_1$ generation ($F_1$ after mutagen treatment) and each kernel is planted and self-pollinated. These $M_1$ plants can have a mutation in only one gene of an allelic pair, since only half of the chromosomes came from the treated male parent and the other half came from the untreated female parent. Any mutation appearing in the $M_1$ generation would therefore be a dominant mutation. These occur in about 0.4% of the population.[13] Any gene causing the loss of photosynthesis appearing here would be a dominant lethal and could not be carried on.

When these $M_1$ generation plants are self-pollinated, any newly induced recessive gene could be revealed when homozygous recessive in the $M_2$ generation. The $M_2$ kernels resulting from each $M_1$ self were planted in a greenhouse bench of well-drained, coarse sand. Approxi-

[12] R. W. Briggs, *Maize Genet. Coop. Newsl.* **43**, 23 (1969).
[13] M. G. Neuffer and O. H. Calvert, *J. Hered.* **66**, 265 (1975).

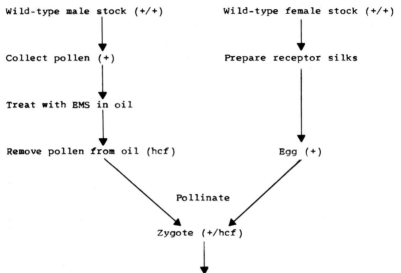

Fig. 1. Mutation and genetic handling of maize stocks. Parentheses indicate genotype when EMS induced a recessive hcf mutation.

mately 100 kernels per $M_1$ plants were planted per row. Environmental conditions for growth of the seedlings were not critical, but day temperature was between 23° and 30° and night temperature (especially of the sand) should not go below 20°. It may be necessary to heat the sand with a standard soil heat tape in cool climates. Supplemental lighting is not necessary except under very cloudy conditions.

Each $M_2$ family will now segregate recessive mutants which were induced by the pollen treatment in Mendelian ratios. In over 4000 such $M_2$ families, all types of pigmentation mutants occur: albino, yellow, yellow-green, pale-green, virescent (slowly and unevenly greening), striped (parallel to veins), banded (across the leaf at approximate 1 day growth intervals), and others were observed in about 20% of the families. Photosynthesis mutants occur in 2% of the $M_2$ families.

Screening for Photosynthesis Mutants

Photosynthesis mutants can be detected in the following four general ways. In the past higher plant mutants were all selected by the first method, but now the third and fourth methods below seem best.

*Alteration of Photosynthetic Pigments.* With the exception of the recent work with maize, this was the only way in which higher plant photosynthesis mutants were selected. One obvious problem with all of these mutants was that in addition to limiting photosynthesis, pigments were altered in quantity or in type. Most often there were gross changes in chlorophyll, alteration of the $a/b$ ratio, and changes in chloroplast structure. Such mutations were most often developmental mutations rather than being specifically associated with individual photosynthetic enzymes. One exception to this may be the *Vicia faba* mutant missing NADP reductase.[14]

*Incorporation of $^{14}CO_2$.* Lack of uptake of $^{14}CO_2$ has been used to detect photosynthetic mutants.[15] Whereas this procedure has been useful for microorganisms, no one has developed a way to use it to screen the large number of individuals necessary for crop plants studies.

*Chlorophyll Fluorescence Yield.* A third technique developed for algal mutants was to examine the total yield of chlorophyll fluorescence.[16,17] Any limitation of electron transport at or beyond the primary acceptor of photosystem II (Q) will increase the level of fluorescence. This is a useful technique, since mutants with normal quantities of pigment can be detected in a large population of plants. Fluorescence of higher plant leaves show a marked increase in visible fluorescence when electron transport is inhibited by DCMU or other electron transport inhibitors.[18]

Plant leaves are illuminated in a darkroom or in a greenhouse on a moonless night. Fluorescence is activated with a long-wave ultraviolet

[14] U. Heber and W. Gottschalk, *Z. Naturforsch., Teil B* **18**, 36 (1963).
[15] R. P. Levine, *Nature (London)* **188**, 339 (1960).
[16] J. Garnier, *C. R. Hebd. Seances Acad. Sci., Ser. D* **265**, 874 (1967).
[17] P. Bennoun and R. P. Levine, *Plant Physiol.* **42**, 1284 (1967).
[18] C. D. Miles and D. J. Daniel, *Plant Sci. Lett.* **1**, 237 (1973).

lamp (Model UVL 56, Ultra Violet Products Inc.). The peak output of this lamp is at 366 nm at an intensity of $7 \times 10^3$ ergs/cm$^2$ sec at 15 cm distance. A similar source of intense blue light could be substituted. Leaf chlorophyll fluorescence which peaks at 683 nm can be viewed directly by eye or through a red cut-off filter (50% cut at 610 to 670 nm) which will exclude the actinic light but pass fluorescence. Almost any type of red filter is suitable, such as the series of Corning filters 2418 to 2030. Any red plastic material which does not pass the light from the UV lamp is good. Best for screening are welder's goggles which have an inner set of fixed lenses and an outer set of hinged "flip-up" lenses. Lenses from UV-protective glasses (UVC 303, Ultra Violet Products, Inc.) are fitted in the inner set and red cut-off filters are placed in the hinged set.

Mutant seedlings are grown in $7 \times 38$ cm plastic trays (Rubbermaid, JB1-2917, Wooster, Ohio) in which a single row of 30 seedlings can be planted. For screening, the UV lamp can be held horizontally behind the row of plants and fluorescence viewed with the goggles.

In order best to see the difference between mutant and normal seedlings the total concentration of chlorophyll is important. Fluorescence activated by this lamp is almost totally reabsorbed in very dark green material, whereas in pale green leaves the chlorophyll concentration can be so low that wild-type plants fluoresce well. The best distinction between mutant and wild-type in maize is with plants grown under $5 \times 10^4$ ergs/cm$^2$ sec from fluorescent and tungsten lamps at 25° in a growth chamber.

This has been a very productive screening procedure since seventy photosynthesis mutants have been isolated in our laboratory and a number of high fluorescence mutants were selected by D.S. Robertson at Iowa State University using the same procedure.[19]

*Resistance to Photodynamic Inhibitors.* A fourth general method to screen for photosynthesis mutants is to select plants which are resistant to various chemical inhibitors. To be effective the selective inhibitor should require functional photosynthetic reactions in order to kill normal tissue and should allow the growth of only those plants which are nonphotosynthetic. Seedlings lacking photosynthesis survive by metabolism of starch reserves. Sodium arsenate at 25 m$M$ has been successfully used to select mutant colonies of *Euglena*.[20] Actinidione at 40 $\mu$g/ml was also useful for selecting *Euglena* mutants, though not as good as arsenate.[20] Arsenate has been applied to *Chlamydomonas* with good results for photosynthesis mutants.[21] The selective mechanism for the function

[19] D. S. Robertson, personal communication (1977).
[20] A. Shneyour and M. Avron, *Plant Physiol.* **55**, 142 (1975).
[21] R. K. Togasaki and M. O. Hudock, *Plant Physiol.* **49**, S-52 (1972).

of arsenate is unknown, though it kills dividing cells 300 times more efficiently than nondividing cells.

Another class of photodynamic inhibitors of photosynthesis are electron acceptors of low redox potential which are reduced by a ferredoxin-requiring photosystem I reaction. These acceptors support a Mehler-type reaction in which reactive radicals are formed with oxygen uptake and hydrogen peroxide production. Bipyridyliums with potentials of $-300$ to $-500$ mV function in this way, producing peroxides in sufficient quantities in the light to have herbicidal activity.[22] Diquat (1,1'-ethylene-2,2'-dipyridylium dibromide) and paraquat or methyl viologen (1,1'-dimethyl-4,4'-bipyridinium dichloride) are two commonly used electron acceptors and herbicides. Both diquat and methyl viologen are effective for selecting photosynthetic mutants in higher plants, but we have found commercial preparations of diquat (35%-diquat from Chevron Chemical Co., Ortho Division, San Francisco, California) to be most useful, since methyl viologen tends to cause excessive chlorosis of leaf tissue.[23]

Diquat solutions (1 m$M$) can be applied as a foliar spray using an insect spraying gun or any device to produce a fine, even spray. Two-week-old maize seedlings which were segregating potential mutants (the $M_2$ generation, Fig. 1) were treated in conditions of bright, uniform light. When plants were treated under conditions of rapid photosynthesis, isolated necrotic spots were seen on the leaf surfaces within 15 to 30 min. Two hours later, as the herbicide is translocated the brown areas increase in size to cover large sections of the leaf. By 8 to 12 hr after treatment the entire leaf is dead. The timing of these events is lengthened if the rate of photosynthesis is slow or if a lower concentration of diquat is used. Screening for resistant plants can take place at any time during the 24 hr following treatment. Typically, resistant plants showing no brown spots are seen 30 min after spraying. Resistant plants are marked (with a toothpick in the soil beside the plant) and checked several hours later to see if they still appear different from wild-type plants. After 24 to 48 hr even resistant plants begin to show some necrosis due to other effects of the herbicide. If wild-type plants are treated with diquat but kept 24 hr in the dark, no necrotic spots are seen. Killing in the first day is therefore dependent on light. Also, treatment of plants with diquat in a nitrogen atmosphere and light does not cause necrosis of tissue. Under best conditions the herbicide treatment will kill only the leaf tissue of wild-type or heterozygous plants and will not be translocated to the stem.

[22] D. E. Moreland and J. L. Hilton, in "Herbicides: Physiology, Biochemistry, Ecology" (L. J. Audus, ed.), 2nd ed., Vol. 1, p. 493. Academic Press, New York, 1976.
[23] C. D. Miles, *Plant Physiol.* **57**, 284 (1976).

Vigorous stocks of treated maize heterozygous seedlings will produce new leaves after about five days and can be grown to mature plants. This selective technique is best used with a family of $M_2$ seedlings that are segregating potential mutants. If a family shows one-quarter resistant seedlings, then more of the original $M_2$ seed can be planted and self-pollinated to obtain a large seed supply.

A similar photosystem I electron acceptor of low redox potential ($-325$ mV) has recently been used for enrichment of photosynthesis mutants in *Chlamydomonas*.[24] Metronidazole (2-methyl-5-nitroimidazole-1-ethanol) at 10 m$M$ concentration effectively kills wild-type cells in the light for 24 hr, while having no effect on photosynthesis mutants. Metronidazole probably kills cells by a mechanism similar to that of the bipyridylium herbicides. Metronidazole also appears to be effective in selecting nonphotosynthetic mutants in species of blue-green algae.[25]

The selection of resistant mutants has an advantage over the fluorescence technique, since it is a selective rather than screening procedure. One mutant seedling can be selected from thousands without screening each individual. Also, resistant mutants are selected under normal lighting conditions. Since these compounds accept electrons from photosystem I, any mutational loss of activity past this point in the electron-transport scheme would not be resistant. Therefore, this class of mutants will be missed. However, mutants of that segment of the electron transport chain could be detected by high fluorescence. Conversely, diquat resistance can detect mutants on the oxidizing side of photosystem II which could not be detected by fluorescence, since they should exhibit lower than normal fluorescence yield.[26] Use of both fluorescence yield and of photosystem I electron-acceptor resistance should allow isolation of all types of photosynthesis mutants.

One additional method which does not fall into the four general classes above could be used on previously selected photosynthesis mutants. When a plant has lost one major source of cellular energy supply (photosynthesis) and is rapidly consuming stored carbon, the total energy supply of the cell is low. Any growth or growth response requiring that energy could be altered. Geotropism requires energy and is not seen in two-week-old photosynthesis mutants which otherwise appear phenotypically no different from wild type.

A 7 × 38 cm tray of 30 two-week-old seedlings segregating mutants

[24] G. W. Schmidt, K. S. Matlin, and N. Chua, *Proc. Natl. Acad. Sci. U.S.A.* **74**, 610 (1977).
[25] L. S. Sherman, personal communication (1977).
[26] B. L. Epel and R. P. Levine, *Biochim. Biophys. Acta* **226**, 154 (1971).

is placed on its side so that the plants axis is now 90° to the field of gravity. After 24 hr, wild-type plants grow upward at an average angle of 40° while mutants remained horizontal. This procedure only tests for energy supply and could not be used for initial screening but could be used for daily selection of known mutants.

## Naming of Mutants

Photosynthetic mutants were named on the basis of the characteristic used for their selection. High chlorophyll fluorescence mutants were named hcf, hcf-2, hcf-3, etc. Diquat-resistant mutants were designated dqr, dqr-2, dqr-3, etc.

## Initial Testing of Selected Mutants

There are a series of tests that can be performed on plants which were selected as mutants to confirm the lack of photosynthetic function. If mutants are shown to be lethal, not to fix $CO_2$, not to evolve oxygen, and to have altered fluorescence kinetics, then a more detailed examination is warranted.

*Lethality.* It is important to know if the mutation causes death at the time the plant has exhausted its starch reserves in the endosperm. If photosynthesis is completely blocked in maize seedlings, lethality occurs at 16 to 20 days after planting. If the kernel is retrieved from the soil at this time and tested for starch with iodine, no starch is seen. At this time the plant has 3 to 5 leaves and is about 15 cm tall. The first sign is browning and necrosis of the leaf tip of the second or third leaves. This necrosis spread down the midrib toward the stem and later begins on the first (oldest) leaf which is small, and the younger fourth and fifth leaves. The leaf blades are brown and dying 12 hr later. After 24 hr the entire plant is shriveled and dry.

*Variable Fluorescence.* When screening the $M_2$ generation for mutants, suspected leaves can be excised and along with sibling control material, whole leaf fluorescence induction kinetics measured in the laboratory. The fluorometer used is illustrated in Fig. 2. Leaf segments are held in a hinged mask to expose a 6 × 20 mm area of the upper leaf surface. When the shutter is opened, it exposes the leaf to an actinic light at $5 \times 10^4$ ergs/cm² sec at a peak wavelength of 465 nm. Fluorescence is recorded at 45° to the leaf surface by a red-sensitive photomultiplier. After a 2 min dark adaption, the induction kinetics of fluorescence, which peaks at 683 nm, was recorded on a fast response chart recorder or with a storage oscilloscope. This variable fluorescence is a good monitor of

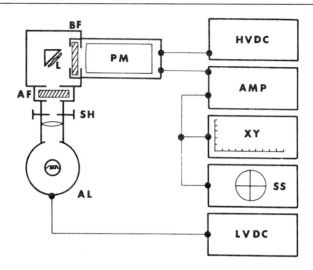

Fig. 2. Diagram of fluorometer. LVDC, low voltage power supply for actinic lamp (Harrison Lab 7.5V-10A); AL, actinic lamp (Bausch and Lomb microscope illuminator); SH, photographic shutter; AF, actinic filter, blue (Corning 4305); L, leaf; BF, blocking filter, red (Corning 2030); PM, S-20 response photomultiplier (EMI 9558); HVDC, high voltage (Keithley 240); AMP, photocurrent amplifier (Keithley 414S); XY, plotter (Houston Inst. 2000); SS, storage oscilloscope (Hughs Memoscope).

the relative function of electron transport in photosystem I and II. It is also very useful, since it is a nondestructive procedure. The usual results are as indicated in Fig. 3.

Upon illumination of a normal green leaf fluorescence rises very rapidly to an initial point (O) and then rises more slowly to a maximum level (P). Under these conditions P is reached after 2 to 5 sec. This slow rise is associated with normal photosystem II activity and intersystem electron transport.[27] If this variable fluorescence is recorded on an oscilloscope, you can distinguish a small rise from O to a point (I) followed by a dip in fluorescence (D) before continuing to P. The O–I–D change occurs in 10 to 50 msec and is due to the primary electron acceptor of photosystem II (Q) being reduced and then reoxidized by intersystem electron carriers.[28] In higher plant leaves fluorescence declines from P to a semi-steady-state level (S). This decline is due to normal electron transport on the reducing side of photosystem I.[29]

[27] L. N. M. Duysens and H. E. Sweers, in "Studies on Microalgae and Photosynthetic Bacteria" (Jpn. Soc. Plant Physiol., eds.), p. 353. Univ. of Tokyo Press, Tokyo, 1963.
[28] P. Joliot, Biochim. Biophys. Acta 102, 135 (1965).
[29] J. S. Munday and Govindjee, Biophys. J. 9, 1 (1969).

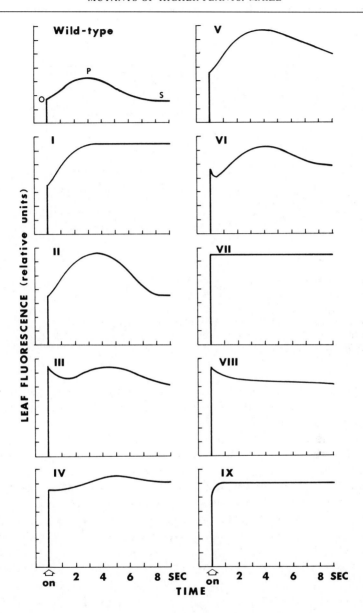

Fig. 3. Classes of fluorescence induction kinetics observed in wild-type and mutant (I to IX) maize leaves.

TABLE II
FLUORESCENCE CLASS VERSUS MUTATION TYPE

| Fluorescence class[a] | Type of Mutation | hcf No. |
|---|---|---|
| I | Electron transport, PS-I | 1, 17, 50, 325 |
| | Photophosphorylation | 18, 23, 323 |
| II | Electron transport | 20 |
| | $CO_2$ fixation | 4, 11, 14, 21 28, 29, 31, 42 |
| III | Electron transport, PS-II | 41 |
| IV | Electron transport | 35, 43 |
| | Photophosphorylation | 7, 15, 30, 34, 48, 317, 324 |
| V | Electron transport | 33, 37 |
| | Cytochrome $f$ | 6 |
| VI | Cytochrome $b_{559}$ | 16, 19 |
| | Cytochrome $b_6$ | 47 |
| | Cytochrome $f$ | 38, 44 |
| VII | Cytochrome $b_{559}$ | 3, 9 |
| | Photophosphorylation | 8 |
| IX | Electron transport, PS-II | 26, 49 |
| | Plastoquinone | 2, 5 |

[a] See Fig. 3 for fluorescence classes.

Almost all mutants show higher than normal fluorescence yield with a variable fluorescence which falls into one of nine classes (Fig. 3). By examining the variable fluorescence, we can make a general statement about the function of one or both photosystems. Table II lists 42 maize mutants and which class of variable kinetics they have. With more data we may be able to make a good prediction of mutant type from fluorescence class.

*Carbon Dioxide Uptake.* To determine the condition of the total photosynthetic process, light-dependent fixation of $CO_2$ was measured by infrared gas analysis. Air was taken into the gas train from outside the laboratory going first into a 20-liter mixing chamber and then through a Drierite drying column. From there it was directed to the analyzer or through one of two plant chambers. The two tightly sealed chambers were made from 1-liter glass beakers covering 1-liter plastic beakers filled with soil and containing five seedling mutants or normal sibs. The chambers were constructed to allow air to enter evenly above the plants and to be drawn out below the leaves. The air leaving the plant chambers was drawn through a second drying column and into a Beckman 215A $CO_2$ analyzer. Flow rate was controlled by a flowmeter in the line be-

TABLE III
RATE OF $CO_2$ FIXATION IN HIGH FLUORESCENCE MAIZE MUTANTS[a]

| Mutant | mg | Percent | Lethality |
|--------|-----|---------|-----------|
| WT | 38 | 100 | − |
| 2 | 0 | 0 | + |
| 3 | 0 | 0 | + |
| 21 | 0 | 0 | + |
| 48 | 0 | 0 | + |
| 317 | 0 | 0 | + |
| 9 | 2 | 5 | + |
| 28 | 9 | 24 | + |
| 26 | 9 | 24 | ? |
| 34 | 10 | 28 | ? |
| 31 | 11 | 30 | + |
| 7 | 12 | 32 | + |
| 18 | 14 | 36 | ? |
| 33 | 15 | 40 | ? |
| 4 | 16 | 46 | + |
| 17 | 19 | 49 | + |
| 324 | 20 | 54 | + |
| 30 | 21 | 55 | ? |
| 1 | 22 | 60 | − |
| 16 | 22 | 60 | + |
| 50 | 22 | 60 | ? |
| 14 | 23 | 62 | − |
| 20 | 27 | 73 | + |

[a] Data in mg $CO_2$ fixed/dm$^2$ of leaf × hr.

tween the outlet of the analyzer and the vacuum pump. The response of the 215 A analyzer was recorded on a slow response strip chart recorder. The results of measuring 22 selected mutants are shown in Table III.

*Oxygen Evolution in vivo.* A second method used to measure total photosynthesis was to record light dependent oxygen evolution of leaf segments with bicarbonate as the ultimate electron acceptor.[30] Leaf material was weighed and washed, and 200 mg was cut into 2-mm squares in 2 ml of 0.4 $M$ sucrose, 10 m$M$ NaCl, and 20 m$M$ Tricine at pH 7.4. This was done with a sharp razor blade on a clean flat glass. The segments were transferred to an oxygen electrode cuvette and the volume brought to 5 ml with 15 m$M$ sodium bicarbonate. The segments were stirred rapidly and illuminated with 5 × 10$^5$ ergs/cm$^2$ sec of red light from a projection lamp. Oxygen was measured with a membrane-covered Clark

[30] H. G. Jones and C. B. Osmond, *Aust. J. Biol. Sci.* **26,** 15 (1973).

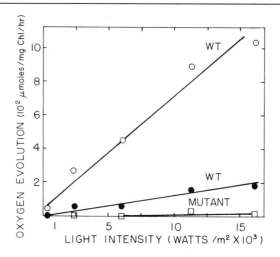

FIG. 4. Oxygen evolution with maize leaf segments as a function of light intensity. Upper wild-type (WT) and mutant responses contained bicarbonate. Middle WT had no bicarbonate. Mutant was hcf-3.

electrode and a Yellow Springs Instruments Model 53 oxygen monitor. The results of one such experiment are shown in Fig. 4. Light- and bicarbonate-dependent oxygen evolution is seen with wild-type leaves over 8-min illumination. The oxygen yield was inhibited by $5 \times 10^{-6}$ $M$ DCMU (data not shown). Mutant hcf-3 leaf segments show little or no oxygen evolution with bicarbonate.

## Photosynthetic Properties of Selected Maize Mutants

When a mutant has been isolated and shown to involve photosynthesis by the above test, it can be further analyzed with a multitude of other standard methods available elsewhere in this series.[7] Details of these procedures will not be given here. Instead, we will summarize the characteristics of the first seventy high chlorophyll fluorescence mutants isolated.

Chloroplasts were isolated with the medium and procedures of Anderson.[31] Mesophyll chloroplasts could be separated from bundle sheath chloroplasts by various techniques.[32] Once chloroplasts were isolated,

[31] J. M. Anderson, N. K. Boardman, and D. Spencer, *Biochim. Biophys. Acta* **245,** 235 (1971).
[32] S. C. Huber and G. E. Edwards, *Physiol. Plant.* **35,** 203 (1975).

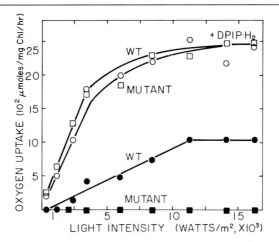

FIG. 5. Oxygen uptake with isolated maize mesophyll chloroplasts. Methyl viologen was the electron acceptor and in the upper two curves (open) ascorbate–DPIP couple was the electron donor. The loss of PS-II electron transport in the muant (hcf-3) is compared to wild type (WT).

uncoupled electron-transport reactions were measured with photosystem I or photosystem II electron acceptors and electron donors. One example of these data for hcf-3 is shown in Fig. 5. Collectively, the measurement of the relative function of the photosystems was possible. Photophosphorylation and light-induced proton pumping was measured with cyclic or noncyclic electron acceptors. With these standard procedures, a general assessment of electron transport and phosphorylation was made. Using these and other data, 42 of the mutants were separated into four classifications. The first two classifications were electron transport mutants (Table IV) and phosphorylation mutants (Table V). For electron transport mutants the rate of phosphorylation reflects the rate of electron transport, but for phosphorylation mutants the rate of ATP synthesis is less than the rate of electron transport. Some of these electron transport mutants (hcf-2,5) have lost significant quantities of plastoquinone (Table VI), which was analyzed according to Barr and Crane.[33] Another mutant (hcf-1) has low NADP reductase activity.[34]

A third class of mutants are electron-transport mutants which do not show normal cytochrome spectra as measured by oxidized–reduced, differential spectroscopy.[35] Examples of such spectra in which cytochrome

[33] R. Barr and F. L. Crane, Vol. 23, p. 372.
[34] C. D. Miles and D. J. Daniel, *Plant Physiol.* **53**, 589 (1974).
[35] D. S. Bendall, H. E. Davenport, and R. Hill, Vol. 23, p. 327.

TABLE IV
ELECTRON-TRANSPORT MUTANTS

| Mutant | Chlorophyll pigments | Fluorescence class[a] | Electron transport[b] | | ATP synthesis[c] | |
|---|---|---|---|---|---|---|
| | | | PS-II | PS-I | Noncyclic | Cyclic |
| 1 | Green | I | 100 | 80 | 60 | 70 |
| 5 | Green | IX | 30 | 85 | 35 | 55 |
| 17 | Green | I | 61 | 57 | 56 | 75 |
| 20 | Green | II | 116 | 77 | 37 | 69 |
| 26 | Yellow-green | IX | 33 | 20 | 42 | 20 |
| 33 | Green | V | 65 | 75 | 75 | 40 |
| 35 | Yellow-green | IV | 30 | 40 | — | — |
| 37 | Green | V | 30 | 80 | 30 | 60 |
| 41 | Green | III | 35 | 105 | 40 | 110 |
| 43 | Yellow-green | IV | 0 | 5 | 0 | 5 |
| 49 | Yellow-green | IX | 20 | 20 | — | — |
| 50 | Green | I | 75 | 48 | 20 | 40 |
| 325 | Yellow-green | I | 10 | 30 | — | — |

[a] Fluorescence classes are shown in Fig. 3.
[b] Electron transport was measured with potassium ferricyanide, DPIP, or dibromothymoquinone as electron acceptors for PS-II. PS-I was measured with ascorbate-reduced DPIP or diaminobenzidine as electron donors and methyl viologen as the electron acceptor. Data given are percent of sibling control plants and represent an average of several experiments. Dibromothymoquinone method was from C. D. Miles, *FEBS Lett.* **61**, 251 (1976).
[c] Photophosphorylation was determined essentially by the method of R. McC. Lilley and D. A. Walker [*Biochem. Biophys. Acta* **314**, 354 (1973)]. PMS or pyocyanin were cyclic cofactors and methyl viologen was the electron acceptor for noncyclic electron transport. Data are given as percent of sibling control plants and represent an average of several experiments.

$f$(hcf-6) and $b$-559 (hcf-9) are missing, are given in Fig. 6 and a listing of the cytochrome mutants is in Table VII.

The fourth classification is the carbon fixation mutant which exhibits good rates of electron transport and phosphorylation but limited $CO_2$ uptake (Table VIII).

At this point in mutant analysis, a large number of additional biochemical or analytical measurements have been or could be done. More complete details of selected maize mutants will be published elsewhere.[36] Additionally, since these mutants were selected in a genetically well-

[36] K. L. Leto and C. D. Miles, *Plant Physiol.* (in press, 1979).

TABLE V

PHOTOPHOSPHORYLATION MUTANTS[a]

| Mutant | Chlorophyll pigments | Fluorescence class | Electron transport | | ATP synthesis | |
|---|---|---|---|---|---|---|
| | | | PS-II | PS-I | Noncyclic | Cyclic |
| 7 | Yellow-green | IV | 80 | 90 | 40 | 10 |
| 8 | Yellow-green | VII | 34 | 38 | 0 | 0 |
| 15 | Yellow-green | IV | 85 | 100 | 0 | 0 |
| 18 | Yellow-green | I | 100 | 95 | 100 | 40 |
| 23 | Green | I | 75 | 80 | 23 | 22 |
| 30 | Green | IV | 105 | 112 | 38 | 60 |
| 34 | Yellow-green | IV | 50 | 70 | 15 | 40 |
| 48 | Yellow-green | IV | 90 | 85 | 0 | 0 |
| 317 | Yellow-green | IV | 75 | 100 | 0 | 0 |
| 323 | Green | I | 90 | 85 | 25 | 24 |
| 324 | Green | IV | 100 | 100 | 25 | 50 |

[a] See Table IV for procedures.

studied species, a variety of genetics tests can also be performed. Allelism tests, chromosome mapping, and the effect of two different mutant genes on photosynthesis are possible. More important is that the mode of inheritance is clear. When genetic and biochemical data are correlated additional information is available.

TABLE VI

CONTENT OF EXTRACTABLE PLASTOQUINONES IN NORMAL AND MUTANT MAIZE CHLOROPLASTS[a]

| Mutant | Plastoquinones | | | |
|---|---|---|---|---|
| | $\mu$moles/$\mu$g Chl | % | $\mu$moles/mg fresh wt | % |
| WT | 29.7 | 100 | 17.8 | 100 |
| 1 | 29.2 | 98 | 19.5 | 109 |
| 2 | 1.0 | 3 | 2.7 | 15 |
| 3 | 31.2 | 105 | 14.5 | 82 |
| 4 | 9.8 | 33 | 8.7 | 50 |
| 5 | 3.2 | 11 | 2.1 | 12 |
| 9 | 27.0 | 91 | — | — |
| 19 | 23.2 | 78 | — | — |

[a] Unpublished data of E. Lee and D. Miles.

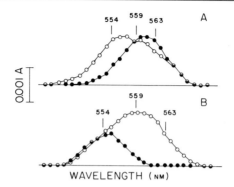

WAVELENGTH (nm)

FIG. 6. Oxidized-reduced difference spectra of maize chloroplasts. (A) Dithionite-ferricyanide, hcf-6 (solid) compared to wild type (open) showing loss of cytochrome $f$ at 554 nm. (B) Hydroquinone-ferricyanide, hcf-9 (solid) compared to wild type (open) showing loss of cytochrome b559. An aminco DW-2 spectrophotometer was used in the split-beam mode.

TABLE VII

ELECTRON-TRANSPORT CYTOCHROME MUTANTS[a]

| Mutant | Chlorophyll pigments | Fluorescence class | Electron transport | | ATP synthesis | | Cytochrome lost |
|---|---|---|---|---|---|---|---|
| | | | PS-II | PS-I | Noncyclic | Cyclic | |
| 3 | Green | VII | 0 | 120 | 0 | 50 | $b$-559 |
| 9 | Green | VII | 0 | 130 | 0 | 0 | $b$-559 |
| 16 | Green | VI | 40 | 75 | 50 | 75 | $b$-559 |
| 19 | Yellow-green | VI | 30 | 120 | 40 | 100 | $b$-559 |
| 2 | Yellow-green | IX | 5 | 100 | 0 | 0 | $f$ |
| 6 | Green | V | 30 | 100 | 0 | 30 | $f$ |
| 38 | Green | VI | 20 | 95 | 25 | 20 | $f$ |
| 44 | Green | VI | 20 | 30 | 30 | 40 | $f$ |
| 47 | Yellow-green | VI | 80 | 100 | — | — | $b_6$ |

[a] See Table IV for procedures.

Acknowledgment

Development of these methods was supported by research grants from the National Science Foundation, PCM 76-08831.

TABLE VIII
CARBON DIOXIDE FIXATION MUTANTS[a]

| Mutant | Chlorophyll pigments | Fluorescence class | Electron transport | | ATP synthesis | | $CO_2$ fixed |
| | | | PS-II | PS-I | Noncyclic | Cyclic | |
|---|---|---|---|---|---|---|---|
| 4 | Green | II | 100 | 100 | 110 | 100 | 46 |
| 11 | Green | II | 100 | 100 | 113 | 100 | 50 |
| 14 | Green | II | 100 | 100 | 100 | 248 | 62 |
| 21 | Green | II | 100 | 100 | 75 | 100 | 0 |
| 28 | Green | II | 100 | 100 | 106 | 110 | 24 |
| 29 | Yellow-green | II | 75 | 85 | 80 | 85 | 11 |
| 31 | Yellow-green | II | 100 | 105 | 216 | 215 | 30 |
| 42 | Green | II | 70 | 110 | 90 | 105 | 15 |

[a] See Table IV for procedures.

# [2] Isolation of Mutants of *Euglena gracilis:* An Addendum

*By* JEROME A. SCHIFF, HARVARD LYMAN and GEORGE K. RUSSELL

## Introduction

Our previous article[1] on the isolation of mutants of *Euglena gracilis* Klebs var. *bacillaris* Cori or strain Z Pringsheim presented methods for obtaining and selecting mutant strains as well as general conditions for growth and maintenance of the organisms. We also presented a brief listing of mutants reported in the literature. The present article adds a few more methods for obtaining mutants and presents a substantially increased list of published mutants to which has been appended a description of their unique properties.

Table IA and IB present a summary of published mutants. More complete descriptions of these strains are in the cited references.

## Methodology

Photosynthetic mutants of *Chlamydomonas* have been obtained using the drug metronidazole or Flagyl (2-methyl-5-nitroimidazole-1-ethanol).[2]

---

[1] J. A. Schiff, H. Lyman, and G. K. Russell, Vol. 23, p. 143.

[2] G. W. Schmidt, K. Matlin, and N. Chua, *Proc. Natl. Acad. Sci. U.S.A.* **74**, (1977).

METHODS IN ENZYMOLOGY, VOL. 69

TABLE IA
FULLY NAMED MUTANTS OF *Euglena gracilis* DESCRIBED IN RECENT PUBLICATIONS

| Mutant | Footnote | Reference | Comments[i] |
|---|---|---|---|
| GbZS | a, k | kk | High rate of formation of nongreen colonies |
| G₁BU | a, c | hh, m | Constitutive for p1 tRNA's.[b] Appreciable plastid structure and chlorophyll; substantial thylakoid nonstacking |
| M⁻₂BUL | d, j | p, o | Nonswimmer. Loses plastids irreversibly ("bleaches") during dark growth |
| Na1ʳ ₈₀ZNalL | d, k | q | Loses CPII in high light intensity. Pale green |
| O₁BS | a | s | Appreciable plastid structure and chlorophyll. Photosynthesis present. Substantial thylakoid nonstacking. |
| O₂BX | a | s, y | Appreciable plastid structure and |
| P₁BXL | c | z, y, g, h, i | chlorophyll. Photosynthesis present. Substantial thylakoid nonstacking |
| P₄ZUL | c | bb, jj | Photosynthetic mutant. Lacks "Q" (electron receptor for PS-II). Lacks plastoquinone A. Little pairing of lamellae |
| P₇ZN₈L | e | aa, bb | Ph⁻ mutant. Lacks "Q" |
| P₈ZN₈L | e | aa, bb | Ph⁻ mutant. Lacks "Q" |
| P₉ZN₈L | | j | Ph⁻ mutant. Lacks "Q" |
| P₁₀ZNalL | e | aa, bb | Ph⁻ mutant. Lacks "Q" |
| P₂₃ZUL | | r | Ph⁻ mutant |
| P₂₇ZNalL | | t | Loses irreversibly the blue light photomorphogenetic system during dark growth. Becomes Y₁₁ZD. (see below) |
| P₃₁ZNalD | | r | Ph⁻ |
| P₃₃ZNalL | | g | Cold sensitive. Makes substantially less chloroplast components at 19° |
| Smʳ ₁BN₈L | | u | Resistant to Sm bleaching. Lowered binding of Sm to 70 S ribosomes due to alteration of several ribosomal proteins |
| W₁BVL | f | cc | |
| W₁ZXL | | i | |
| W₂ZUL | | i | |
| W₃BUL | | y, dd, k, s, ee, g, k, ff | Plastid DNA, protochlorophyll(ide), and many other plastid constituents undetectable. Shows presence of blue light system including paramylum breakdown, CAP reduction, Cyt rRNA transcription, increased O₂ consumption. Plastid sulfolipid present |
| W₄BUL | | i | |

TABLE IA—Continued

| Mutant | Footnote | Reference | Comments[i] |
|--------|----------|-----------|-------------|
| $W_8BHL$ | | *y, s, g, k, hh* | Plastid DNA undetectable |
| $W_{10}BSmL$ | | *i, k, s, g* | Plastid DNA undetectable |
| $W_{14}ZNalL$ | | *v, ii, t* | Plastid DNA undetectable |
| $W_{30}BS$ | *a* | *s* | |
| $W_{33}ZUL$ | *g* | *l* | |
| $W_{34}ZUD$ | | *x* | Small amount of light-induced plastid development |
| $Y_1BXD$ | *h* | *y, s, h, k, hh* | Develops only very rudimentary plastid |
| $Y_2BUL$ | | *s, h, i, n* | |
| $Y_3BUD$ | | *y, g, h, k, hh* | See $Y_1BXD$ |
| $Y_6ZNalL$ | | *ll, r* | |
| $Y_9ZNalL$ | | *w, t, r* | Lacks red-blue (porphyrin-like) photocontrol for plastid synthesis |
| $Y_{11}ZD$ ($P_{27}ZNalL$) | | *w, t, r* | Lacks blue photocontrol for plastid synthesis. Derived by growing $P_{27}ZNalL$ (see above) in the dark for several generations, then returning to light |
| $Y_{12}ZUL$ ($P_{31}ZNalD$) | | *r* | |

*a* It is not known whether these were isolated in light or darkness.

*b* Abbreviations used: pl, plastid associated; cyt, cytoplasmically associated.

*c* Color designations: Y, yellow; O, olive; W, white; P, pale green; Gb, golden brown; (see also footnote *k*).

*d* Unless indicated all mutants are normally green when light grown.

*e* These four mutants were originally called $P_4$, $P_7$, $P_8$ and $P_9$, by Russell and Lyman,[aa] but the present designation is in agreement with the proposed nomenclature.

*f* This strain should probably be properly called $W_1BVL,L$ indicating that it was isolated in the light. It conflicts with a strain called $W_1BZX$ described earlier by Stern.

*g* This strain was originally called ZUV-1 by Scott and Smillie.[1]

*h* Inadvertently cited as $Y_1BXL$ previously. (See Schiff *et al.*,[1] Table I, footnote *m*)

*i* A series of mutants was described by J. Gross and T. Jahn [*J. Protozool.* **5**, 126 (1958)] before the present terminology was suggested. These mutants are listed in the Indiana Type Culture Collection [R. C. Starr, *Am. J. Bot.* **51**, 1013 (1964)]. Other mutant strains have been described by L. Moribor, B. Hershenov, S. Aaronson, and B. Bensky [*J. Protozool.* **10**, 80 (1963)], G. Braverman, C. Rebman, and I. Chargaff [*Nature (London)* **187**, 1037 (1960)], R. Neff [*J. Protzool.* **7**, 69 (1960)], A. Gibor and S. Granick [*J. Protozool.* **9**, 327 (1962)] and A. Gibor and H. Herron *in* "The Biology of *Euglena*" (D. Buetow, ed.), Vol. 2, p. 335. Academic Press, New York 1968.

*j* $M_2^-BUL$ was obtained in a collection of $M^-$ mutants kindly provided by R. Lewin and isolated by him using his highly effective selection technique (see footnote *o*).

*k* Meaning of the first letter of phenotypic designations: Gb, Golden-brown; G, Golden; $M^-$, Lacks motility; Nal[r], nalidixic acid resistant; O, olive-green; P, pale green; Ph[−], nonphotosynthetic; Sm[r], streptomycin resistant; W, white; Y, yellow. Second letter indicates strain: B, bacillaris; Z, Z strain. Third letter indicates mutagen: H, heat (growth at elevated temperatures); Nal, nalidixic acid; Ng, nitrosoguanidine; S, spontaneous; Sm, streptomycin; U, ultraviolet; X, X-ray. Fourth letter indicates whether the mutant was isolated from dark-grown (D) or light-grown (L) cells.

[l] N. S. Scott and R. M. Smillie, *Biochem. Biophys. Res. Commun.* **28**, 598 (1967).

[m] D. J. Goins, R. Reynolds, J. Schiffs, and W. Barnett, *Proc. Natl. Acad. Sci. U.S.A.* **70**, 1749 (1973).

[n] S. Bingham *et al., Plant Physiol.* **59**(6), Suppl. 9 (1977).

[o] R. Lewin, *Can. J. Microbiol.* **6**, 21 (1960).

[p] N. H. Goldstein, S. D. Schwartzbach, and J. A. Schiff, *J. Protozool.* **21**, 443 (1974).

[q] H. Lyman, R. Alberte, and P. Thornber, *Plant Physiol.* **57**(5), Suppl. 73 (1976).

[r] U. Srinivas and H. Lyman, *J. Cell Biol.* **75**, 304a, (1977).

[s] M. Edelman, J. Schiff, and H. Epstein, *J. Mol. Biol.* **11**, 769 (1965).

[t] G. Schmidt and H. Lyman, *Proc. Int. Congr. Photosynth. Res., 3rd, 1974* p. 1755 (1975).

[u] J. Diamond and J. A. Schiff, *Plant Sci. Lett.* **3**, 289 (1974).

[v] A. Uzzo and H. Lyman *Photosynth., Two Centuries Its Discovery Joseph Priestley, Proc. Int. Congr. Photosynth. Res., 2nd, 1971* p. 2601 (1972).

[w] G. Schmidt and H. Lyman, *J. Cell Biol.* **59**, 305a (1973).

[x] G. Salvador, V. Nison, F. Richard, and P. Nicolas, *Prototistologica* **8**(4), 533, (1972).

[y] J. A. Schiff and H. T. Epstein, *in* "The Biology of *Euglena*" (D. Buetow, ed.) Vol. 2, p. 285. Academic Press, New York, 1968.

[z] H. Hill, J. A. Schiff, and H. T. Epstein, *Biophys. J.* **6**, 125 (1966).

[aa] G. Russell and H. Lyman, *Plant Physiol.* **43**, 1284 (1968).

[bb] G. Russell, H. Lyman, and R. Heath, *Plant Physiol.* **44**, 929 (1969).

[cc] J. Leff and N. Krinsky, *Science* **158**, 1332 (1967).

[dd] J. A. Schiff and M. Zeldin, *J. Cell Physiol.* **72**, Suppl. 1, 103 (1968).

[ee] J. Leff, M. Mandel, H. Epstein, and J. Schiff, *Biochem. Biophys. Res. Commun.* **13**, 126 (1963).

[ff] M. Edelman *et al., Proc. Natl. Acad. Sci. U.S.A.* **52**, 1214 (1964).

[gg] M. Edelman, H. T. Epstein, and J. A. Schiff, *J. Mol. Biol.* **17**, 463 (1966).

[hh] G. C. Harris, M. Nasatir, and J. A. Schiff, *Plant Physiol.* **63**, 908 (1979).

[ii] H. Lyman, A. Jupp, and I. Larrinva, *Plant Physiol.* **55**, 390 (1975).

[jj] F. Schwellitz, R. Dilley, and F. Crane, *Plant Physiol.* **50**, 161 (1972).

[kk] C. L. Hershberger, D. Morgan, F. Weaver, and P. Pienkos, *J. Bacteriol.* **118**(2), 434 (1974).

[ll] G. Russell and H. Lyman, *J. Protozool.* **14**, Suppl. 13 (1967).

Cells with lesions in photosynthetic electron transport are viable after incubation with the inhibitor because reduction of ferredoxin by photosynthetic electron transport and reaction of reduced ferredoxin with the inhibitor are necessary to produce a product toxic to the cells. This inhibitor may prove useful in selecting nonphotosynthetic (Ph⁻) mutants in *Euglena*.

### Growth Media

Several additional growth media and trace metal mixes have been summarized recently.[3]

[3] H. Lyman and K. Traverse, *in* "Phycology Handbook" (E. Gantt, ed.), Vol. II. Cambridge Univ. Press, London and New York (in press).

TABLE IB
OTHER MUTANTS OF *Euglena gracilis* IN RECENT PUBLICATIONS[a,b]

| Mutant | Footnote | Reference | Comments |
|--------|----------|-----------|----------|
| PR-1,2,3 | Previously also cited by Gross and Jahn[a] | c | Light: contain: phytoene, phytofluene, $\eta$-carotene, $\beta$-zeacarotene, $\beta$-carotene |
| PR-4 | Previously also cited by Gross and Jahn[a] | | Light: lack the above, Dark: contain all above |
| SmL-1 | Previously also cited by Gross and Jahn[a] | d | Contains: unsaturated $C_{46}$ polyenes and phytoene |
| PBZ-G-3 | Previously also cited by Gross and Jahn[a] | | Contains the above |
| PBZ-G-4 | Previously also cited by Gross and Jahn[a] | | Lacks the above |
| HB-G | Previously also cited by Gross and Jahn[a] | | Contains the above |
| Y-1 | | e | No chlorophyll, but some membranes and carotenoids synthesized upon illumination |
| $E_sZHB$ | | f | Does not show wild-type light induced drop of malate synthase or aconitase |
| SM-V | | g | White. See also citation in Gross and Jahn[a] |
| SM-BS | | h | White. See also citation in Gross and Jahn[a] |
| m-28 | | i | Photosynthetic mutant; defect in photophosphorylation |
| m-50 | | i | Photosynthetic mutant; lesion unknown |
| m-52 | | i | Photosynthetic mutant; possibly missing or defective cytochrome 558 |
| m-102 | | i | Photosynthetic mutant; blocked on oxidizing side of PS-II |
| ZHB | | j | White strain derived by growth at 34°. No detectable plastid DNA |

[a] See footnote *i* in Table IA.
[b] These are mutants described in some recent publications which have come to our attention.
[c] J. A. Gross, R. S. Stroz, and G. Britton, *Plant Physiol.* **55**, 175 (1975).
[d] J. A. Gross and R. S. Stroz, *Plant Sci. Lett.* **3**, 67 (1974).
[e] E. Marčenko, *Protoplasma* **76**, 417 (1973).
[f] J. R. Cook and M. Carver, *Plant Cell Physiol.* **7**, 377 (1966).
[g] J. Vavra, *Arch. Mikrobiol.* **25**, 223 (1956).
[h] Y. Ben-Shaul and I. Ophir, *Planta* **91**, 195 (1970).
[i] A. Shneyour and M. Avron, *Plant Physiol.* **55**, 137 (1975).
[j] D. Gruol and R. Haselkorn, *Biochim. Biophys. Acta* **414**, 20 (1975).

## The Hereditary System of *Euglena*

Since the first review more data has appeared on plastid DNA in *Euglena*. These data are a source of several anomalies regarding the numbers of complete genomes in the plastid.[4-7] Older data based on target analysis[6,8] of uv inactivation of green colony forming ability implied that each cell had 30 to 60 copies of the plastid genome. Because the data appeared to fit curves derived from target theory, it was assumed that this type of analysis was appropriate. More recently, a direct isolation of circular plastid DNA molecules has yielded numbers as high as 720 per cell.[4] Other analyses have yielded values ranging from 30 to 90.[5-7] Since the number of plastid genomes may vary with growth conditions one must be cautious in assigning a definite number at this point.

Older target data had also indicated that the nucleus of *Euglena gracilis* var. *bacillaris* was polyploid, perhaps as high as octaploid.[9] On the basis of renaturation kinetics, Rawson has calculated that this organism may be diploid.[10] However, this still leaves open the question of the difficulty of obtaining nuclear mutants.[1] although diploidy might be sufficient to prevent completely the expression of recessive mutants should they arise in these asexually reproducing organisms. No mutants (such as auxotrophic mutants) have been described whose phenotypes indicate that they represent nuclear mutations. One can think of situations where some mutations to auxotrophy might be lethal but it remains to be shown whether any such conditions exist in *Euglena*.

Although mapping of the chloroplast genome by genetic recombination has not been possible yet, mapping by means of restriction endonucleases appears promising.[11-14]

Certain classes of mutants in *Euglena* may be difficult to obtain because their presence may be masked by other phenomena. For ex-

---

[4] J. Manning and O. Richards, *Biochim. Biophys. Acta* **259**, (1972).

[5] J. A. Schiff, *Adv. Morphog.* 10, 265 (1973).

[6] A. Uzzo and H. Lyman, *Photosynth., Two Centuries Its Discovery Joseph Priestley, Proc. Int. Cong. Photosynth. Res. 2nd, 1971* Vol. III, p. 2585 (1972).

[7] J. E. Manning, D. Wolstenholme, R. Ryan, J. Hunter, and O. Richards, *Proc. Natl. Acad. Sci. U.S.A.* **68**, 1169 (1971).

[8] H. Lyman, H. T. Epstein, and J. A. Schiff, *Biochim. Biophys. Acta* **50**, 301 (1961).

[9] H. Hill, J. A. Schiff, and H. T. Epstein, *Biophys. J.* **6**, 125 (1966).

[10] J. Rawson, *Biochim. Biophys. Acta* **402**, 171 (1975).

[11] E. Stutz, E. Crouse, L. Graft, B. Jenni, and H. Kopecka, *in* "Genetics and Biogenesis of Chloroplasts and Mitochondria" (T. Bücher *et al.*, eds.), p. 339. Elsevier, Amsterdam, 1976.

[12] P. Gray and R. B. Hallick, *Biochemistry* **16**, 1665 (1977).

[13] B. Atchison, P. R. Whitfield, and W. Bottomly, *Mol. Gen. Genet.* **148**, 263 (1976).

[14] J. Mielonz, J. Milner, and C. L. Hershberger, *J. Bacteriol.* **139**, 860 (1977).

ample, in *Euglena* and *Tetrahymena*[15-17] cycloheximide is an excellent inhibitor of translation on cytoplasmic ribosomes and incubation in this antibiotic leads to cessation of cell division. After a time, however, most of the cells begin to divide and appear to be resistant to the antibiotic. On transfer of the cells to cycloheximide-free medium, after a few divisions, sensitivity to cycloheximide is regained. These phenomena do not result from a heritable change in the cells, and alternative explanations must be sought. Thus selection of cycloheximide-resistant mutants would be very difficult. A similar situation with respect to DCMU resistance has been described.[18]

Ebringer[19] has reviewed a wide variety of drugs and their effects on *Euglena* chloroplasts. He has divided these agents into those which bleach cells permanently (i.e., causes irreversible loss of plastids) and those which do not. Among these drugs are several which act as mutagens. It would be of interest to determine whether other bleaching agents are also mutagenic.

Acknowledgment
    Supported by National Institutes of Health Grant GM-14595 and NSF B038262.

[15] J. Winick and J. A. Schiff, unpublished data.
[16] C. B. Heyer and J. Frankel, *J. Cell Physiol.* **78**, 411 (1971).
[17] J. V. Roberts, C. T. Orias, and E. Orias, *J. Cell Biol.* **62**, 707 (1974).
[18] D. Laval-Martin, G. Dubertret, and R. Calvayrac, *Plant Sci. Lett.* **10**, 185 (1977).
[19] L. Ebringer, *J. Gen. Microbiol.* **71**, 35 (1972).

# [3] Isolation of Mutants of Photosynthetic Bacteria

*By* BARRY MARRS, SAMUEL KAPLAN, and WILLIAM SHEPHERD

The discoveries of genetic exchange systems for two members of the genus *Rhodopseudomonas* have increased the potential usefulness of biochemical genetics approaches to research with photosynthetic bacteria.[1,2] Thus the isolation of mutants of photosynthetic bacteria is likewise becoming a more fruitful enterprise. Rhodopseudomonads present no special difficulties with respect to mutagenesis or selection techniques that are routinely applied to other microorganisms,[3] but the availability

[1] B. Marrs, *Proc. Natl. Acad. Sci. U.S.A.* **71**, 971 (1974).
[2] W. R. Sistrom, *J. Bacteriol.* **131**, 526 (1977).
[3] D. A. Hopwood, *in* "Methods in Microbiology" (J. R. Norris and D. W. Ribbons, eds.), Chapter VI, p. 363. Academic Press, New York, 1970.

of several alternative modes of energy metabolism affords some unusual opportunities for the isolation of mutants with altered energy transducing systems. Methods for isolating respiratory-deficient mutants of *R. capsulata* have been published,[4,5] as have methods for selection of mutants blocked in bacteriochlorophyll synthesis[6–8] and photosynthetic growth.[9,10] Respiratory-deficient mutants are propagated by photosynthetic growth, and nonphotosynthetic mutants by respiration. Mutants lacking both photosynthetic and respiratory activity have not been described, but anaerobic dark growth has been described for several species,[11,12] and mutants of this class should be sought.

Four techniques for the isolation of mutant rhodopseudomonads are presented below. These include two variations of the basic penicillin selection, one adapted for *R. sphaeroides* which is relatively resistant to penicillin,[13] and one that has been very useful with *R. capsulata* and involves solid media. The third technique described uses cycloserine as a selective agent, and although this might be expected a priori to function much as a penicillin selection, since cycloserine causes aberrant cell wall synthesis in growing cells,[14] an unexpectedly interesting collection of survivors are typically obtained. Cycloserine does not inhibit the growth of *R. capsulata,* thus the technique is not applicable to that species. The fourth technique selects for mutants defective in photosynthetically driven active transport, and may be termed tetracycline suicide. It is based on the observation of light-induced tetracycline accumulation by a phototrophic bacterium.[15]

### Isolation of *Rhodoseudomonas sphaeroides* Mutants Using Sodium Ampicillin

The penicillin enrichment technique is a powerful method for the isolation of many types of bacterial mutants.[16] Enrichment for a particular

[4] B. Marrs, C. L. Stahl, S. Lien, and H. Gest, *Proc. Natl. Acad. Sci. U.S.A.* **69,** 916 (1972).
[5] B. Marrs and H. Gest, *J. Bacteriol.* **114,** 1045 (1973).
[6] M. Griffiths and R. Y. Stanier, *J. Gen. Microbiol.* **14,** 698 (1956).
[7] J. Lascelles, *Biochem. J.* **100,** 175 (1966).
[8] G. Drews, I. Leutiger, and R. Ladwig, *Arch. Mikrobiol.* **76,** 349 (1971).
[9] W. R. Sistrom and R. K. Clayton, *Biochim. Biophys. Acta* **88,** 61 (1964).
[10] P. Weaver, *Proc. Natl. Acad. Sci. U.S.A.* **68,** 136 (1971).
[11] R. L. Uffen and R. S. Wolfe, *J. Bacteriol.* **104,** 462 (1970).
[12] H. C. Yen and B. Marrs, *Arch. Biochem. Biophys.* **181.** 141 (1977).
[13] P. F. Weaver, J. D. Wall, and H. Gest, *Arch. Microbiol.* **105,** 207 (1975).
[14] R. Curtiss, III, L. S. Charamella, C. M. Berg, and P. E. Harris, *J. Bacteriol.* **90,** 1238 (1965).
[15] J. Weckesser and J. A. Magnuson, *J. Supramol. Struct.* **4,** 515 (1976).
[16] J. Lederburg and N. Zinder, *J. Am. Chem. Soc.* **70,** 4267 (1948).

mutant phenotype is achieved by treating a population of bacteria with penicillin under conditions which favor the phenotypic expression of the desired mutation, and consequent nongrowth of the desired mutant. Under these conditions penicillin selectively kills the growing cells, hence increasing the proportion of mutants in the resulting population of bacteria.

*Rhodopseudomonas sphaeroides,* a gram-negative photoheterotroph, is remarkably resistant to penicillin as well as a number of synthetic penicillins. Concentrations as high as 20 mg/ml are required merely to achieve bacteriostasis. However, this organism is reasonably sensitive to sodium ampicillin with concentrations of 100 $\mu$g/ml being sufficient to kill growing populations of *R. sphaeroides.*

*Method.* The method given here is suitable for the selection of amino acid auxotrophs in *R. sphaeroides* and with appropriate modifications can be used for the selection of other mutant phenotypes.

*Materials Required*

1. Medium A of Sistrom[17] prepared with and without 0.1% casamino acids.

2. Sodium ampicillin solution (10 mg/ml) in Sistrom's medium A without casamino acids. Prepared fresh daily and filter sterilized.

3. Mutagenized culture of *R. sphaeroides.* Mutagenized cells of *R. sphaeroides* are freed of the mutagen by washing in medium A containing 0.2% casamino acids and resuspended to a density of $1 \times 10^7$ bacteria/ml as determined spectrophotometrically. The culture is allowed to grow to $5 \times 10^8$ cells/ml to permit expression of the mutant phenotype. Depending upon the particular needs of the investigator, either chemoheterotrophic or photoheterotrophic growth may be employed.

Following the incubation period, usually overnight, the bacteria are washed three times with 0.5 volume changes of ice cold medium A minus casamino acids. Following the third wash the culture is resuspended to a density of $1 \times 10^8$ bacteria/ml and allowed to incubate an additional 6 hr (approximately 1–2 doublings) to exhaust completely any remaining amino acid pools. At this time sodium ampicillin solution is added to the culture to a final concentration of 100 $\mu$g/ml and incubation continued for another 18 hr. After the ampicillin treatment, aliquots of the culture are plated on medium A containing 0.2% casamino acids.

Amino acid auxotrophs can be identified by either picking or replica plating colonies from the casamino acid supplemented plates onto plates containing nonsupplemented medium A and on nonsupplemented medium A + 20 $\mu$g/ml of the desired amino acid. Colonies which show

[17] W. R. Sistrom, *J. Gen. Microbiol.* **22,** 778 (1960).

an amino acid requirement should be restreaked, and a single colony isolate used for subsequent evaluation of the auxotrophs desired.

*Discussion.* It is essential to use freshly prepared sodium ampicillin for each selection. The lactam ring of penicillin is hydrolyzed in aqueous solutions, and the effective useful life of the stock penicillin solution is about 12 hr. When used in medium A the useful killing lifetime of sodium ampicillin is 18 hr, after this time it exerts mainly a bacteriostatic effect on the culture and eventually looses even this effect. Therefore, it is important to terminate the selection within 16–18 hr after sodium ampicillin addition to prevent subsequent overgrowth of undesired strains.

Wild-type cultures, with a generation time of 3 hr, are killed at a rate of approximately 1 log unit/generation time under the conditions described here. This rate of kill is also obtained using mutagenized populations of *R. sphaeroides*. It is important to begin the sodium ampicillin enrichments at cell densities at or below $5 \times 10^8$ bacteria/ml because lysis products from killed cells have a tendency to stabilize protoplasted cells, which under normal circumstances would be killed by the selection technique.

Finally, medium A of Sistrom contains both L-aspartic and L-glutamic acids, and it is unlikely that auxotrophs for these two amino acids will be isolated using normally prepared medium A. Removal of these amino acids will not appreciably affect growth of the wild type organism. Likewise casamino acids do not contain L-tryptophan, and the levels of L-methionine and L-tyrosine can also vary from batch to batch. Therefore, supplementation of casamino acids with these amino acids may be required. Alternatively, a proteolytic digest of casein can be used without further supplementation. Other precautions may be required for the isolation of vitamin-deficient mutants, purine or pyrimidine mutants, as well as cold-sensitive or temperature-sensitive mutants.

Penicillin Selection of Nonphotosynthetic Mutants of
  *Rhodopseudomonas capsulata* on Solid Medium

In 1953, Adelberg and Myers[18] published a modification of the penicillin technique which uses solid medium throughout. Their procedure has several advantages over the analogous liquid method, namely, that each mutation gives rise to only one colony, so a wide variety of independent mutations may be quickly gathered; the yield of mutants is high because penicillin is administered to microcolonies rather than single cells; and wild-type colonies which escape the initial killing can easily be

---

[18] E. A. Adelberg and J. W. Myers, *J. Bacteriol.* **65,** 348 (1953).

recognized and therefore avoided. The procedure involves many variables which must be adjusted for each organism, and a procedure which works well for *R. capsulata* is described below. The method is given for isolation of nonphotosynthetic mutants, but it can obviously be adapted to select for a wide variety of mutants. In contrast to the behavior of *R. sphaeroides* described above, *R. capsulata* is unusually sensitive to penicillin for a gram-negative organism,[13] thus ampicillin is not necessary, and complete removal of residual penicillin after killing is completed is essential.

*Method.* Five milliliters of molten RCV[13]–soft agar (0.6% agar) at 45° is allowed to solidify in the bottom of a sterile petri plate. This is then overlayered with 5.0 ml of the same soft agar containing $10^7$–$10^8$ mutagenized cells of *R. capsulata*. To take full advantage of this method, no period of outgrowth of the mutagenized culture should occur after mutagenesis, thus all mutant clones will arise from independent mutagenic events. Once solidified, this cell-containing layer is sealed under 5.0 ml of sterile RCV–soft agar. The sandwich thus formed is incubated photosynthetically until microcolonies of about 100 cells each are formed by wild-type cells. This requires about 14 to 22 hr at 35°, depending on how badly the cells were injured by the mutagenic treatment. During this incubation mutations blocking photosynthesis will express themselves, and preexisting components will be diluted out, eventually causing an arrest of the growth of the desired mutant clones. Since we have found it difficult to control this timing precisely, and since it is important for the success of the method, we routinely use a series of plates incubated for times varying by 2-hr increments. Microcolonies of 100 cells are just visible as pinpoints of light under a 20 × dissecting microscope. Photosynthetic incubation is achieved in Gas Pack Anaerobic Jars (Bioquest) at 30° between two banks of 60 W Lumiline incandescent bulbs, about 1 m apart. At the various times indicated, plates are removed from the anaerobic jars and immediately overlayered with 5.0 ml RCV–soft agar containing 200 U/ml penicillin (filter sterilized and freshly prepared). As soon as this agar layer has solidified, plates are returned to anaerobic jars, and the jars are returned to illumination at 30°. Because the anaerobic jars prevent drying out of the agar plates, the agar remains fragile and must be handled with care. The penicillin-containing plates are then incubated under photosynthetic growth conditions for about 18 hr, at which time they are removed from the anaerobic jars, overlayered with 5.0 ml of RCV–soft agar containing 200 U/ml penicillinase (Sigma Chemical; filter sterilized and freshly prepared), and then returned to anaerobic, lighted conditions. After 2 days plates are removed, and any visible colonies marked on the backs of the plates, since these colonies are clearly capable of photosynthetic growth and need not be examined further. The plates are now incubated aerobically at 35° for several days. Each day for up to 4 days, a new crop of colonies appears and may be picked for further screening or marked and picked at a later time.

*Discussion.* This method may obviously be used for selection of various types of auxotrophs, as discussed by Adelberg and Meyers.[18] As described above it has yielded a variety of mutants blocked in bacteriochlorophyll synthesis as well as mutants lacking active photophosphorylation systems, but we have also used the procedure with appropriate modifications to select amino acid and vitamin auxotrophs of *R. capsulata*.

Isolation of Ps⁻ Mutants of *Rhodopseudomonas sphaeroides* Using Cycloserine

Cycloserine exerts a bacteriocidal effect in both gram-positive and gram-negative bacteria by blocking the conversion of L-alanine to D-alanine.[14] As a result of the inhibition of cell wall biosynthesis, with continued macromolecular synthesis, cell death occurs upon subsequent lysis of treated cells.

As a result of attempts to perfect a mutant enrichment technique for *Rhodopseudomonas sphaeroides* using cycloserine as the enrichment agent, analogous to the use of penicillin,[16] we observed large numbers of pigmentation mutants for plates containing cycloserine spread with nonmutagenized wild-type *R. sphaeroides*. These cycloserine-treated survivors occurred at a frequency of approximately $1 \times 10^{-6}$ to $1 \times 10^{-7}$ and were subsequently found to be resistant to the action of cycloserine.

*Methods and Results. Rhodopseudomonas sphaeroides* strain 2.4.1 was grown and maintained on a medium consisting of 0.3% peptone, 0.3% yeast extract, pH 7.1, with KOH (PYE). This medium allows growth of *R. sphaeroides* as luxuriant as that obtained on medium A of Sistrom[17] supplemented with 0.2% casamino acids. When used in plates, PYE was solidified by the addition of 1.5% agar.

Cycloserine-resistant pigment mutants were isolated by plating 1 to $2 \times 10^{8}$ nonmutagenized chemoheterotrophically grown cells onto PYE plates containing 100 μg/ml cycloserine and incubating in air in dim light at 33°.

After a 4-day incubation period, approximately 100–200 colonies appeared on each plate, and 75% revealed abnormal colony pigmentation. In all other respects colony morphology was similar to that of wild type.

Eight abnormally pigmented colonies were picked at random from several plates containing colonies resistant to 100 μg/ml cycloserine. All of the isolates showed either poor growth or no growth when restreaked on plates with or without cycloserine and incubated under photohetero-trophic conditions. However, 7 of the 8 isolates will grow anaerobically in the light or dark, in the presence of 0.2% dimethyl sulfoxide.[12] Table I gives the phenotypes of the cycloserine resistant nonphotosynthetic (Ps⁻) mutants which have been characterized to date.

Figure 1 shows the absorption spectra of intact chromatophores from

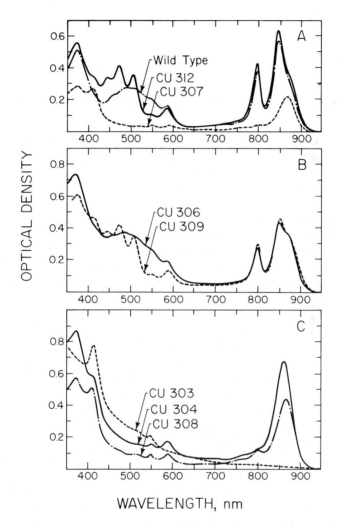

FIG. 1. Absorption spectra of *R. sphaeroides* Ps⁻ mutants and revertants selected for resistance to 100 μg/ml cycloserine. CU303, CU304, CU306, CU307, CU308, and CU309 are Ps⁻ mutants. CU312 is a Ps⁺ revertant of CU307. Chromatophores were prepared as described[19] except that the Sepharose 2B purification step was omitted.

TABLE I
PHENOTYPES OF CYCLOSERINE-RESISTANT NONPHOTOSYNTHETIC MUTANTS

| Isolate | Aerobic colony color | Reversion to Ps$^+$ | Anaerobic growth in presence of dimethyl sulfoxide |
|---------|---------------------|---------------------|---------------------------------------------------|
| CU302 | Green | $2.9 \times 10^{-6}$ | + |
| CU303 | Pink | $5.6 \times 10^{-6}$ | + |
| CU304 | Green | $5.8 \times 10^{-6}$ | + |
| CU305 | Pink | $1.8 \times 10^{-6}$ | + |
| CU306 | Deep red | $<1.0 \times 10^{-7}$ | − |
| CU307 | Green | $1.5 \times 10^{-5}$ | + |
| CU308 | Green | $3.8 \times 10^{-5}$ | + |
| CU309 | Pink | $<1.0 \times 10^{-7}$ | + |

several of the mutants listed in Table I. Photosynthetic membrane bio-genesis was induced by incubating mutants in PYE in the presence of 0.5 ppm dissolved oxygen in the dark. Under semiaerobic conditions growth of the mutants is not appreciable. Chromatophores from these mutants were isolated by previously described methods.[19]

Revertants (Ps$^+$) of these mutants can be isolated by plating Ps$^-$ strains under photoheterotrophic conditions such that only Ps$^+$ revertants can grow. Revertants selected under these conditions and in the presence of cycloserine retain their Cyc$^r$ phenotype. However, they do not always exhibit the normal pigmentation associated with wild-type cultures.

*Discussion.* A novel class of Ps$^-$ mutants of *R. sphaeroides* can be isolated by selecting for Cyc$^r$ mutants under conditions gratuitous for the Ps phenotype. Many of the Ps$^-$ mutants isolated in this manner will grow under anaerobic conditions when supplemented with 0.2% dimethyl sulfoxide (DMSO). These mutants were also characterized by alterations in the absorption spectra of their photosynthetic membrane system. Potential alterations in intracytoplasmic membrane-specific polypeptides are currently under investigation.

Tetracycline Suicide for Selection of *Rhodopseudomonas capsulata* Mutants

Many species of bacteria concentrate tetracycline from the medium by an energy-dependent process.[15] This observation has been adapted to

19 P. J. Fraker and S. Kaplan, *J. Bacteriol.* **108**, 465 (1971).

provide a method for selection of mutants defective in energy transduction. The method requires incubating cells anaerobically in the light in a growth medium containing a sufficiently low concentration of tetracycline such that only cells which actively accumulate it will be killed, and then plating survivors at a dilution such that residual tetracycline will not inhibit the respiratory growth of nonphotosynthetic survivors.

*Method.* A photosynthetic preculture of *R. capsulata,* either mutagenized or not, is grown in RCV[13] to stationary phase, and then diluted in fresh RCV to a density of about $7 \times 10^8$ cells/ml in a 17-ml screwcap tube fitted with an injection septum and filled to contain an air space of 0.1–0.2 ml. This tube is incubated at 35° between two banks of three 60-W Lumiline bulbs (General Electric) 35 cm apart until the turbidity doubles. This assures an anaerobic environment. Filter sterilized tetracycline (6.8 $\mu$l of 2.5 mg/ml stock in RCV medium) is injected through the septum. Incubation is continued for 2 days, during which time an initial turbidity rise is followed by a decrease. One-tenth milliliter of undiluted culture (containing a maximum 0.1 $\mu$g tetracycline) is spread on each of several 20-ml RCV agar plates (1.2% agar). The plates are incubated aerobically in the dark for 2 days at 35°. The concentration of tetracycline in each plate is thus 5 ng/ml, assuming none has been degraded. Control experiments show that 1 ng/ml has no effect on respiratory colony development, 10 ng/ml results in a noticeable partial inhibition, and 100 ng/ml gives no colony formation. About 100 survivors per plate are typically found. A high percentage of these are shown to be nonphotosynthetic by replica plating.

*Discussion.* This method has given a population of nonphotosynthetic mutants with phenotypes different from those previously described. Of particular interest are mutants lacking cytochrome $c_2$, which are currently under investigation. These were identified among the normally pigmented nonphotosynthetic survivors of tetracycline suicide by screening respiratory colonies for the Nadi reaction.[5]

Acknowledgments

The authors wish to thank Dr. H. C. Yen and Mr. T. Lucas for their contributions to the development of these techniques.

# [4] Mutants of Blue-Green Algae

By CHASE VAN BAALEN

The blue-green algae (the appellation Cyanobacteria is gaining popularity) have been much used in comparative studies of photosynthesis, while there is a growing interest in many facets of their physiology and biochemistry. Their prokaryotic nature, phycobiliprotein accessory pigments, and oxygen yielding photosynthesis set them apart from the diverse eukaryotic algae and from the photosynthetic bacteria. From an experimental standpoint, blue-green algae can be most recalcitrant, and selection and characterization of mutant forms has proceeded slowly. In the past few years, satisfactory growth conditions have been developed for plating coccoid and filamentous forms. This has allowed some standard chemical and physical agents to be tested for their effectiveness as mutagens. Herein is presented a overview of the existing methodology for induction of mutants, with comments on extant types of blue-green alga mutants. There are two recent summaries of this field.[1,2]

## General Growth Conditions

*Liquid Cultures.* It is not uncommon that the growth rate of a blue-green alga under a given set of growth conditions becomes erratic or depressed. The subtle toxic effects causing lowered growth rates have never been entirely pinned down. However, some of the following algal lore may help to alleviate problems. Cotton should be prewashed before use. It is autoclaved in distilled water, well-rinsed with distilled water, and then air dried on a clean surface. Culture tubes, bubbling tubes, and media flasks, after soaking in a common laboratory detergent such as Alconox (made up in distilled water), are then rinsed, and soaked overnight in 5% HC1, followed by tap water and distilled water rinses. Some plastic ware and rubber materials can be very toxic, and their use should be approached with caution. The use of stock solutions for media preparation lends uniformity, but troubles may develop here with the iron stock or low level fungal growth. Rapid temperature shifts of the inocu-

[1] C. Van Baalen, *in* "The Biology of the Blue-green Algae" (N. G. Carr and B. A. Whitton, eds.), p. 201. Blackwell, Oxford, 1973.
[2] S. F. Delaney, M. Herdman, and N. G. Carr, *in* "Genetics of Algae" (R. A. Lewin, ed.), p. 7. Blackwell, Oxford, 1976.

lum or of the culture medium after inoculation should be minimized, as some strains are cold sensitive, especially when rapidly shifted from 39° to 20° or below.

*Plate Growth.* An early problem with coccoid blue-green algae was the inability to grow single cells [colony forming units (CFU)] to visible colonies. The first experiments with *Anacystis nidulans,* strain Tx20, suggested that at least part of the problem was $H_2O_2$ sensitivity.[3,4] Subsequently conditions were found that did allow reasonably quantitative recovery of CFU with Tx20[4,5]; however, the basis for the $H_2O_2$ sensitivity of Tx20 is still not understood. Recent work with plating filamentous forms after break-up by mild sonication again confirms that plating efficiency (CFU plated to colonies recovered) can vary from species to species.[6] It, therefore, is important to verify plating efficiencies, especially of new isolates, by simply comparing a microscopic count with a visible colony count. There are also cultures which present no problems with plating,[6,7] and their use is recommended.

Commercial plastic petri dishes (100 × 15 mm) of several manufacturers have proved nontoxic; evaporation can be controlled by sealing with a clear plastic tape. In the absence of temperature–light-controlled incubators, satisfactory petri dish incubation space can be constructed using linear banks of seven to nine 60-W tungsten bulbs mounted on plywood strips 70 cm or so in length. It is useful to have five to seven such strips attached at each end to 4–6 ft, ½ inch diameter, aluminum rods positioned 50–60 cm above a plywood surface. The lamp banks can be moved from side to side or the number of lamps per bank can be varied, thereby controlling the plate incubation temperature and irradiance. Under comparable conditions, the early growth rates on agar plates are similar to those in liquid cultures. A need for protracted incubation periods indicates poor plate growth conditions, which should be avoided unless slow growing mutant forms are under consideration.

The usual agar preparations, such as Difco Bacto-Agar 0140, may be satisfactory for plating; however, washing by centrifugation with distilled water or 1 m$M$ Na$_4$EDTA followed by distilled water can improve recovery of or recognition of some mutant types. The commercially available agaroses are very satisfactory for plate growth and lessen metal and organic contamination of the medium.

[3] C. Van Baalen, *J. Phycol.* **1,** 19 (1965).
[4] J. E. Marler and C. Van Baalen, *J. Phycol.* **1,** 180 (1965).
[5] M. M. Allen, *J. Phycol.* **4,** 1 (1968).
[6] C. P. Wolk and E. Wojcivch, *Arch. Mikrobiol.* **91,** 91 (1973).
[7] C. Van Baalen, *J. Phycol.* **3,** 154 (1967).

Although growth in pour plates may be better for some strains, than when spread over the agar surface with a glass rod, the latter technique yields plates that are more amenable to muant recognition and recovery. The surviving colonies should be carefully examined under a dissecting scope (10–20×) by both transmitted and reflected illumination. Each plate should be examined several times during the incubation period.

Mutagenic Agents

Alkylating agents are effective mutagenic agents for blue-green algae, with N-methyl-N'-nitro-N-nitrosoguanidine (MNNG) being commonly used.[8-14] Nitrosomethylurea[15] and to a lesser extent ethane methane sulfonate[16] are also useful. Other chemical agents, such as nitrous acid or base analogs, do not induce mutation in blue-green algae. There is also some hesistancy in accepting X rays or UV radiation (primarily 254 nm) as generally mutagenic. In a comparative study of the mutagenic action of nitrosomethylurea, UV, and X rays, no pigment mutants were detected after various doses of UV or X rays, while under similar conditions nitrosomethylurea readily induced pigment mutants.[15] This view was subsequently reiterated.[17] In a study of mutant reversion frequency UV was judged 1000-fold less effective than MNNG.[18] However, after UV treatment of several organisms, a limited number of variants have been isolated and partially characterized.[9,19-22a] Part of the problem with UV induction of mutants in blue-green algae may lie with differences among

[8] C. Van Baalen, Science 149, 70 (1965).
[9] Y. Asato and C. E. Folsome, Mutat. Res. 8, 531 (1969).
[10] E. Padan and M. Shilo, J. Bacteriol. 97, 975 (1969).
[11] A. R. Kaney and M. P. Dolack, Genetics 71, 465 (1972).
[12] V. D. Zhevner and S. V. Shestakov, Arch. Mikrobiol. 86, 349 (1972).
[13] B. D. Sinha and H. D. Kumar, Ann. Bot. (London) [N. S.] 37, 673 (1973).
[14] W. F. Doolittle and R. A. Singer, J. Bacteriol. 119, 677 (1974).
[15] V. N. Stoletov, V. D. Zhevner, D. V. Garibyan, and S. V. Shestakov, Genetika 6, 77 (1965).
[16] S. F. Delaney and N. G. Carr, J. Gen. Microbiol. 88, 259 (1975).
[17] S. V. Shestakov, in "Taxonomy and Biology of Blue-green Algae" (T. V. Desikachary, ed.), p. 262. University of Madras, India, 1972.
[18] M. Herdman and N. G. Carr, J. Gen. Microbiol. 70, 213 (1972).
[19] R. N. Singh and H. N. Singh, Arch. Mikrobiol. 48, 109 (1964).
[20] R. N. Singh and D. N. Tiwari, Nature (London) 221, 62 (1969).
[21] B. S. Srivastava, Phycologia 9, 205 (1970).
[22] R. N. Singh, D. N. Tiwari, and V. P. Singh, in "Taxonomy and Biology of Blue-green Algae (T. V. Desikachary, ed.), p. 27. University of Madras, India, 1972.
[22a] C. Astier, F. Espardellier, and I. Meyer, Arch. Microbiol. 120, 93 (1979).

strains in enzymatic photoreactivation.[23] Deficiency in photoreactivation may allow greater UV induction of mutation.

## MNNG Treatment Conditions

$N$-Methyl-$N'$-nitro-$N$-nitrosoguanidine (MNNG) should be handled with great care. Besides being a very potent mutagen, it is a severe skin irritant and suspect as a carcinogen. It should be stored in the cold and solutions made and filter sterilized (0.4 $\mu$m Millipore or Nucleopore) just before use. There are several studies with MNNG on the conditions for maximum mutation frequency. Highest frequency of "yellow" (mostly $NO_3^-$ assimilation forms) mutants per 1000 survivors in the marine coccoid, *Agmenellum quaruplicatum* strain PR-6 was found after 5–15 min treatment with 300 $\mu$g MNNG/ml, at a surviving fraction of approximately 0.1%.[24] In *Anacystis nidulans* strain Tx20, maximum mutation frequency occurred at a dose where viability was 0.4%.[18] Highest frequency of "minute colony forming" types, in Tx20 was noted at 0.2% survival, 15 min after treatment with 50 $\mu$g/ml MNNG.[9] Studies such as these suggest that a high degree of kill is desirable for MNNG induction of mutation, but there are numerous instances in the literature wherein MNNG has been used at a low concentrations (15–50 $\mu$g/ml) allowing high survival, and a variety of mutant types have been recovered.[14,25–27] MNNG is particularly unstable under akaline conditions, and low concentrations should be used with dispatch.

The classic penicillin technique for enrichment of mutants has been applied to the blue-green algae, but there is little uniformity in methodology. Concentrations used have varied from 10 U/ml,[14] to 200 U/ml,[27] to as high as 1000 U/ml [25] with treatment times of 24, 48, or 120 hr, respectively.

There are several other physiological aspects of the mutagenic process which could take on more importance as the era of "we'll take whatever we can get" shifts more toward tailoring specific mutant types. Does prior cell history (growth rate, temperature, and medium composition) influence the frequency and recovery of certain mutant types? Is a dark period or even a prolonged starvation period desirable prior to MNNG treatment? Is an intermediate cultivation period helpful after MNNG treatment, and what multiple of the generation time should it be? Is

[23] C. Van Baalen, *Plant Physiol.* **43**, 1689 (1968).
[24] S. E. Stevens, Jr. and C. Van Baalen, *J. Phycol.* **5**, 136 (1969).
[25] M. Wilcox, G. J. Mitchison, and R. H. Smith, *Arch. Microbiol.* **103**, 219 (1975).
[26] C. L. R. Stevens, S. E. Stevens, Jr., and J. Myers, *J. Bacteriol.* **124**, 247 (1975).
[27] T. C. Currier, J. F. Haury, and C. P. Wolk, *J. Bacteriol.* **129**, 1556 (1977).

MNNG too selective in its action on blue-green algae, and are some types thereby missed? Do repair processes influence mutant recovery? A dark repair of MNNG damage has been noted but not further studied.[24]

## MNNG Treatment of *Agmenellum quadruplicatum*, Strain PR-6

Sterile technique is observed throughout. PR-6 is grown in 22.5 × 175 mm Pyrex test tubes in 20 ml of medium ASP-2 with 8 $\mu$g/liter vitamin $B_{12}$ [24] at 39° with four 20 W F20T12 Daylight fluorescent lamps two on each side of the water bath 7.5 cm from lamp front edge to test tube center). The tubes are continuously bubbled with 1 ± 0.1% $CO_2$ in air via a bubbling tube inserted through a hole in the cotton stopper. Under these conditions the generation time is about 3.5 hr. When cell density reaches 2 to 3 × $10^7$ cells/ml (36 to 25% T on a Lumetron Colorimeter Model 402E), the algal suspension is placed in the dark at room temperature for 18–24 hr. The cells are then collected by centrifugation (Corex tube No. 8446) and resuspended in fresh medium ASP-2. Two milligrams of freshly weighted MNNG is dissolved in 10 ml of distilled water; and the solution is rapidly sterilized (Swinney filter holder; 25 mm–0.4 $\mu$m Nucleopore filter). The MNNG solution (2.5 ml) is then mixed with 7.5 ml of cell suspension, followed by incubation in the dark at room temperature for 30 min. The cells are collected by centrifugation and resuspended to 10 ml in fresh ASP-2 medium; 0.1 and 0.2 ml of undiluted suspension and of suspension diluted 1:10 are placed on the agar surface (20 ml of medium ASP-2–$B_{12}$ per dish) and spread with glass spreaders. The petri dishes are sealed with transparent tape and incubated upside down. Beginning at 3 days the plates should be examined under a dissecting scope (10–20 ×), and desired colonies should be restreaked to a fresh plate and reincubated. After very careful scrutiny under a dissecting scope of colony color and morphology to establish colony uniformity, colonies are transferred to stock slants.

In an experiment such as this filamentous forms, pigment mutants, and nitrate assimilation impaired forms are common. Many of the colonies remain small and bleached and do not grow upon transfer. If auxotrophs are desired then some form of penicillin treatment[14,25,26] and replicate plating may be helpful.[6]

### Mutant Types

Representative mutant types are listed below by category. Mutants are referenced by parent culture, and unless otherwise stated all were

[28] R. Kunisawa and G. Cohen-Bazire, *Arch. Mikrobiol.* **71**, 49 (1970).

derived via MNNG mutagenesis. The taxonomy of the blue-green algae, especially of coccoid forms, remains controversial, and it behooves an investigator to obtain a desired strain from the original source. The American Type Culture Collection has recently established a collection of blue-green algal strains, and within time this should become a reliable source.

*Morphological Mutants*

    *1. Coccoid to Filamentous Transitions*
    *Anacystis nidulans*, strain Tx20: filaments 12–15 × normal, distorted and twisted, growth rate 80% of wild type.[8]
    *Anacystis nidulans*, strain Tx20: filamentous forms noted as being induced both by MNNG and UV, no characterization.[9]
    *Anacystis nidulans*, strain 6311: light microscope and EM characterization of several types of filamentous forms.[28]
    *Agmenellum quadruplicatum*, strain BG-1: two general classes of filaments noted, those with or those without cross walls. All filamentous forms had growth rates similar to the wild type under a variety of conditions. For the nonseptate mutant, 53SB2, no change in DNA content, base ratios, or growth rates were found as compared to wild type.[29] The cell walls (peptidoglycan) of two filamentous cell division mutants (FM10, 53S; from wild type BG-1) were examined.[30] A comparison of peptidoglycan composition in these two filamentous forms and drug induced phenocopies suggests that structural alterations in the peptidoglycan were not involved in the cell division impairment. Mutant SN12 (wild type BG-1) grew as multinucleoid filaments in dilute suspension but recovered normal phenotype with increased culture density.[31] Ethanol extracts of the supernatant after growth of SN12 stimulated cell division in a serpentine mutant SN29.[32]
    *Agmenellum quadruplicatum*, strain PR-6: EM study of a filamentous form, no notable structural variation in wall components (cell envelope).[33]
    *2. Filamentous to Short Trichome Form*
    *Plectonema boryanum*, Indiana Culture Collection No. 594: a short trichome form (1–8 cells/trichome) was enriched by repeated growth cycles and filtration. Pure lines of these short trichome

[29] L. O. Ingram and C. Van Baalen, *J. Bacteriol.* **102**, 784, (1970).
[30] L. O. Ingram, C. Van Baalen, and W. D. Fisher, *J. Bacteriol.* **111**, 614 (1972).
[31] L. O. Ingram and W. D. Fisher, *J. Bacteriol.* **113**, 999 (1972).
[32] L. O. Ingram and W. D. Fisher, *J. Bacteriol.* **113**, 1106 (1972).
[33] R. M. Brown, Jr. and C. Van Baalen, *Protoplasma* **70**, 87 (1970).

forms gave small colonies on agar, growth rate same as wild type.[34]

3. *Coccoid to Small-cell Form*

  *Agmenellum quadruplicatum,* strain BG-1: proposed blue-green algal equivalent of minicell mutants in bacterial systems.[35]

4. *Filamentous to Branching Form*

  *Nostoc linckia:* mutagenic agent, UV from germicidal lamp. Branching mutant found, also showed increased UV sensitivity and chains of heterocysts (see below).[20]

5. *Forms with Altered Heterocyst Spacing*

  *Anabaena cylindrica,* Cambridge Collection 1403/2a: forms with altered heterocyst spacing; terminal only, terminal and intercalary, or with groups of heterocysts, or those with no heterocysts.[25]

  *Anabaena variabilis,* Indiana Culture Collection No. 1444: mutants with abnormally high or low heterocyst frequencies; heterocyst abnormalities—thickened envelopes, incompletely developed pore regions, protoplasts separated from heterocyst envelope.[27]

*Mutants impaired in Nitrogen Metabolism*

1. *Nitrate Assimilation*

  *Anacystis nidulans,* strain Tx20: mutants blocked at different stages in reduction of nitrate.[8,18]

  *Agmenellum quadruplicatum,* strain PR-6: two mutants blocked at nitrate reductase, grew on nitrite or ammonium ion; four mutants blocked at nitrite reductase, grew on ammonium ion but had a functional nitrate reductase and readily reduced nitrate to nitrite.[36] One mutant of this series, AQ-6, has been studied in some detail. AQ-6 can be readily grown with $NH_4Cl$, then transferred to $NaNO_3$, whereupon the relationship of light to the induction of nitrite reductase and the production of nitrite from nitrate can be examined in essentially a resting cell preparation.[37] AQ-6 has also been used as a tool to examine the control of nitrate reductase.[38]

[34] E. Padan, and M. Shilo, *J. Bacteriol.* **97**, 975 (1969).
[35] L. O. Ingram, G. J. Olson, and M. M. Blackwell, *J. Bacteriol.* **123**, 743 (1975).
[36] S. E. Stevens, Jr. and C. Van Baalen, *Arch. Mikrobiol.* **72**, 1 (1970).
[37] S. E. Stevens, Jr. and C. Van Baalen, *Plant Physiol.* **51**, 350 (1973).
[38] S. E. Stevens, Jr. and C. Van Baalen, *Arch. Biochem. Biophys.* **161**, 146 (1974).

## 2. Nitrogen Fixation-less

*Anabaena doliolum:* treatment with 100 μg/ml MNNG, survivors were cultured for 18 hr then plated and exposed to UV. Frequency of nonnitrogen fixers (*nif⁻*) induced by the combination MNNG–UV treatment was approximately 4 × 10⁻⁴. The *nif⁻* mutant had a reduced heterocyst frequency, 1.3% as compared to wild type, 6%.[13]

*Nostoc muscorum,* Botany Department, Iowa State University: *nif⁻* mutant grows as short, nonheterocystous filaments, dies in nitrogen-free medium, and does not reduce acetylene.[39]

*Anabaena variabilis,* Indiana Culture Collection No. 1444: of six *nif⁻* mutants tested two, NF16 and NF76, were quite stable with no revertants being detected upon plating out. Four others, NF30, 14, 15, and 12 gave 2–4 revertants for 10⁸ cells plated on minimal agar. Coincident with reversion to the *nif⁺* condition, heterocyst frequency became more like the wild type.[27] Oxygen-sensitive nitrogenase mutants have also been isolated.[39a]

## Auxotrophs

*Agmenellum quadruplicatum,* strain BG-1: obligate requirement for L-tryptophan, approximately 2 μM required to saturate growth. Enzymological block was identified in the A activity of tryptophan synthetase.[40]

*Synechococcus cedrorum,* Indiana Culture Collection Nol 1911: isolation of three phenyalanine requiring mutants, two of which grew slowly without added phenyalanine, while the third one had an absolute requirement for L-phenylalanine, optimal doubling time 3.1 hr as compared to 2.6 hr (?) for the wild type.[41] Growth rate was maximum from 10 μM to 1.0 mM reversion frequency was estimated as 3 × 10⁻⁸. Prephenate dehydratase activity, while low in the wild type, was undectable in mutant.

*Anacystis nidulans,* strain Tx20: uracil requiring mutant, reversion frequency 2 × 10⁻⁸, at approximately 15 μM growth rate was 90% of wild type.[42] Strains requiring phenylalanine, methionine, biotin, and thiamin

[39] W. D. P. Stewart and H. N. Singh, *Biochem. Biophys. Res. Commum.* **62,** 62 (1975).

[39a] J. F. Haury and C. P. Wolk, *J. Bacteriol.* **136,** 688 (1978).

[40] L. O. Ingram, D. Pierson, J. F. Kane, C. Van Baalen, and R. A. Jensen, *J. Bacteriol.* **111,** 112 (1972).

[41] A. R. Kaney, *Arch. Mikrobiol.* **92,** 139 (1973).

[42] C. L. R. Stevens, S. E. Stevens, Jr., and J. Myers, *J. Bacteriol.* **124,** 247 (1975).

have been described[18] while subsequently phenotypes requiring thiamin p-aminobenzoic acid, serine or glycine, and adenine were found.[43]

## Pigment Mutants

Observations on pigment mutants in the coccoids, Tx20 and PR-6, have been often noted.[8,9,15,36,44] Recently, attention has been directed towards two classes of pigment mutants commonly found in Tx20, the "yellow-greens" and "blues" and the effects of pigment variation on the photosynthetic unit.[45] The "yellow-greens" have normal chlorophyll but only half the phycocyanin of the wild-type, while they are similar to the parent in number of reaction centers per cell, number of chlorophylls per reaction center, and maximum turnover of reaction center. The "blues" have somewhat higher phycocoyanin and only one-third or so the chlorophyll of the wild type. "Blues" have fewer reaction centers per cell, a smaller number of chlorophylls per reaction center, and a higher maximum turnover rate of reaction centers. These mutants have also been used in studies of action spectra for photosystem I and II.[46]

## Resistance Mutants

There are numerous examples of isolation of strains resistant to penicillin, streptomycin, polymixin B, chloramphenicol, erythromycin, and sulfanilamide.[1,2] Propionate resistant mutants were shown to lack acetate thiokinase and this was proposed as the biochemical basis for their resistance.[47] Mutants of Tx20, resistant to aromatic amino acid analogs, have been used to examine the regulation of DAHP synthetase (EC 4.1.2.15).[48] Ultraviolet-sensitive mutants in strain Tx20 have been induced with MNNG,[49] In *Synechocystis aquatilis,* Leningrad University, Collection No. 428, both MNNG and nitrosomethylurea were used to induce UV-sensitive mutants.[12] These UV mutants are useful for probing repair mechanisms in the blue-green algae. It has long been known that photosynthesis is sensitive to UV, most likely in photosystem II, and both UV-resistant and sensitive mutants could be useful in future studies.

[43] M. Herdman, S. F. Delaney, and N. G. Carr, *J. Gen. Microbiol.* **79**, 233 (1973).
[44] S. V. Shestakov and V. D. Zhevner, *Mikrobiologiya* **38**, 118 (1969).
[45] C. L. R. Stevens and J. Myers, *J. Phycol.* **12**, 99 (1976).
[46] R. T. Wang, C. L. R. Stevens, and J. Myers, *Photochem. Photobiol.* **25**, 103 (1977).
[47] A. J. Smith and C. Lucas, *Biochem. J.* **124**, 23p (1971).
[48] W. Phares and L. F. Chapman, *J. Bacteriol.* **122**, 943 (1975).
[49] Y. Asato, *J. Bacteriol.* **110**, 1058 (1972).

*Other Mutants*

Temperature-sensitive mutants of Tx20 have been isolated,[8,11] but no growth factor requirements were found at the nonpermissive temperature. An interesting mutant of Tx20 which behaves like the wild-type in the light but which rapidly loses viiality and respires little in darkness has been isolated and used to attempt to unravel the role(s) of endogenous metabolism in blue-green algae.[14]

# [5] Nitrogen Fixation (*nif*) Mutants of *Klebsiella pneumoniae*

By K. T. SHANMUGAM and R. C. VALENTINE

## Mutagenesis

Genes which code for nitrogenase in *Klebsiella pneumoniae* are clustered in the chromosome near the histidine biosynthetic operon.[1,2] A number of unlinked mutations (based on generalized transduction using phage $P_1$), which affect the production of nitrogenase have been described.[3] Nitrogen fixation mutants are readily isolated using any of the various procedures of mutagenesis described for isolation of auxotrophic mutants of *E. coli* (see Roth[4]). The commonly used chemical mutagens are EMS and NTG. Procedures involving the use of these mutagens for the isolation of nitrogenase mutants are described below. The use of virulent phages for selecting spontaneously occurring deletions of nitrogenase are also presented.

*Use of Ethyl Methane Sulfonate (EMS) as Mutagen.* To an actively growing culture of *K. pneumoniae*, in sucrose minimal medium with $NH_4^+$ as the nitrogen source[5] ($5 \times 10^8$ cells/ml; 10 ml in 125-ml flask at

[1] S. Streicher, E. Gurney, and R. C. Valentine, *Proc. Natl. Acad. Sci. U.S.A.* **68,**1174 (1971).

[2] S. L. Streicher, E. Gurney, and R. C. Valentine, *Nature (London)* **239**, 495 (1972).

[3] K. T. Shanmugam, C. Morandi, and R. C. Valentine, *in* "Iron–Sulfur Proteins" (W. Lovenberg, ed.), Vol. 3, p. 1. Academic Press, New York, 1977.

[4] J. R. Roth, Vol. 17, p. 3.

[5] The minimal medium contains per liter: $Na_2HPO_4$, 6.25 gm; $KH_2PO_4$, 0.75 gm; NaCl, 2.00 gm; sucrose, 15 gm; $FeSO_4 \cdot 7H_2O$, 0.01 gm; $Na_2MoO_4 \cdot 2H_2O$, 0.01 gm; $MgSO_4 \cdot 7H_2O$, 0.20 gm; and $(NH_4)_2SO_4$, 1.00; pH 7.0 [a modification of the medium by D. C. Yoch and R. M. Pengra, *J. Bacteriol.* **92**, 618 (1966)]. Broth contains per liter: Tryptone (Difco), 10 gm; yeast extract (Difco), 5 gm; NaCl, 10 gm and sucrose, 3 gm; pH 7.0 [a modification of L broth by S. E. Luria and J. W. Burrows, *J. Bacteriol.* **74**, 461 (1957)].

37° with shaking), EMS was added to obtain a final concentration of 1% (0.1 ml/10 ml). Incubation was continued for 1 hr with constant shaking. At the end of 1 hr, the cells were harvested by centrifugation (5000 $g$ for 5 min) to remove the EMS and the carbon source. The mutagenized cells were resuspended in 10 ml of the minimal medium containing no $NH_4^+$ or sucrose ($-C\&N$). The cell suspension in a 125-ml flask was heated in a water bath at 48° and held at that temperature for 40 min without shaking to help remove the alkylated guanine residues from the DNA.[6] Following the heat treatment, the cells were removed by centrifugation and resuspended in 10 ml of broth and incubated at 37° for 2 hr without shaking, to allow the mutagenized chromosomes to segregate. Mutants can be isolated from this culture using appropriate selection procedures described in a later section.

*Use of N-Methyl-N'-nitro-N-nitrosoguanidine (NTG) as Mutagen.* To obtain mutants, a culture of *K. pneumoniae* was treated with NTG (Aldrich Chemical Co.), according to the procedure of Adelberg *et al.*,[7] with modifications as follows[8]: cells grown overnight at 37° in 10 ml of sucrose minimal medium (with $NH_4^+$ as N source) in screwcap tubes (16 × 150 mm) were harvested by centrifugation at 12,000 $g$ for 5 min at 4° and then washed three times with 10 ml saline to remove any traces of the medium. For mutagenesis, washed cells were incubated in 10 ml of saline containing 300 $\mu$g/ml of NTG for 30 min at room temperature. Appropriate amount of NTG can be weighed and dissolved directly in saline. Alternatively, a stock solution of NTG can be prepared in acetone (due to poor solubility in $H_2O$) and diluted into saline to obtain the required final concentration. To remove NTG, the cells were washed by centrifugation (three times with 10-ml aliquots of saline) and finally resuspended in 1 ml of saline solution. To allow segregation of mutagenized genomes to occur, cells were grown in broth for 2 hr as described in the previous section.

Since NTG is known to induce multiple, nonrandom mutations which are closely linked,[9] EMS may be the mutagen of choice (especially regulatory mutants) for the isolation of mutants of nitrogenase.

## Selection Techniques for Mutants

*nif.* A mutagenized culture was appropriately diluted and plated on LB plates to obtain about 50–100 colonies per plate and incubated at 37°.

---

[6] B. S. Strauss, *J. Bacteriol.* **83**, 241 (1962).
[7] E. A. Adelberg, M. Mandel, and G. C. C. Chen, *Biochem. Biophys. Res. Commun.* **18**, 788 (1965).
[8] K. T. Shanmugam, I. Chan, and C. Morandi, *Biochim. Biophys. Acta* **408**, 101 (1975).
[9] N. Guerola, J. L. Ingraham, and E. Cerda-Olmedo, *Nature (London), New Biol.* **230**, 122 (1971).

TABLE I
PROPERTIES OF *nif⁻* MUTANTS OF *K. pneumoniae*[a]

| Strain | Nitrogenase activity[b] | Activity[c] I | Activity[c] II | CRM[d] I | CRM[d] II | MO–CO-factor[e] | Cotransduction with *hisD* (%) |
|---|---|---|---|---|---|---|---|
| | | | | | | | |
| UN (parent strain | 40.09 | + | + | + | + | + | − |
| UN 26 | ? | − | + | − | + | ? | 28 |
| UN 106 | 0.00 | − | + | + | + | − | 81 |
| UN 83 | ? | + | − | + | − | + | 21 |
| UN 179 | ? | − | − | − | − | − | 41 |

*Header: "Nitrogenase components" spanning Activity, CRM, MO–CO-factor columns.*

[a] Data from St. John *et al.*[15] See also Elmerich *et al.*,[10a] Merrick *et al.*,[10b] and Roberts *et al.*[10c]

[b] Nitrogenase activity in crude extracts (nmoles $C_2H_4$ produced per minute per milligram protein).

[c] Based on the ability of the crude extract to complement purified nitrogenase components "*in vitro*." + denotes the presence; and − denotes the absence (undetectable) of the component.

[d] Cross-reactive material (CRM) is based on immunodiffusion experiments using Ouchterlony plate technique with antiserum prepared against purified components.

[e] Mo–Co factor is the Mo-containing, acid-stable material isolated from nitrogenase component 1.

The colonies were transferred by replica plating techniques onto sucrose–minimal medium with and without $NH_4^+$. These plates were incubated under $N_2$ for 3 days at room temperature. Colonies which failed to grow in the N-free medium and grew normally in the presence of $NH_4^+$ were selected. These clones were further tested for their growth and nitrogenase activity in liquid medium. About 95% of these clones were found to be defective in the production of nitrogenase. These mutations are cotransducible with *his* by the phage $P_1$. Properties of some of these mutants are presented in Table I.

*Deletions of nif.* Strains carrying deletions of the *his, nif* region can be isolated from phage-resistant clones by using either one of the two methods described below.[10]

[10] K. T. Shanmugam, A. S. Loo, and R. C. Valentine, *Biochim. Biophys. Acta* **338**, 545 (1974).

[10a] C. Elmerich, J. Houmard, L. Sibold, L. Manheimer, and N. Charpin, *Molec. Gen. Genet.* **165**, 181 (1978).

[10b] M. Merrick, M. Filser, C. Kennedy, and R. Dixon, *Molec. Gen. Genet.* **165**, 103 (1978).

[10c] G. P. Roberts, T. MacNeil, D. MacNeil, and W. J. Brill, *J. Bacteriol.* **136**, 267 (1978).

A fresh culture of M5A1 was incubated into 10 ml of broth in a 16 × 150 mm screwcap tube and incubated at 37° without shaking until maximum cell density was reached. Incubation was continued at room temperature for about 3–5 days. For enriching deletion mutants resistant to *Klebsiella* phage 3 (K3), a small aliquot (0.1 ml) of this culture was inoculated into 9 ml of L broth in a 125-ml flask and grown at 37° with shaking to a cell density of about $10^8$ to $5 × 10^8$ cells/ml; this cell culture was infected with 0.1 ml of a phage stock of K3 ($1.8 × 10^9$ PFU/ml) and incubation was continued until cell lysis was complete. The lysed cell suspension was next centrifuged at 10,000 $g$ in a Sorvall centrifuge for 5 min, and the cell pellet was collected and washed once in 10 ml of sucrose–minimal medium and resuspended in the same medium. The cell suspension was incubated for 30 min at 37° with shaking. Penicillin G (5 mg/ml, 1635 U/mg) was added, and the incubation continued for another 4 hr to enrich for histidine auxotrophs. The cells were washed in sucrose–minimal medium three times and resuspended in 2 ml broth. Samples (0.1 ml) of cells escaping penicillin lysis were inoculated into 0.5 ml of L broth in a 16 × 100 mm tube and incubated at 37° until maximum density was reached. This culture was appropriately diluted for single colony formation in saline, spread onto broth plates, and incubated at 37°. These colonies were transferred by replica plating onto sucrose–minimal plates with and without histidine to score for their histidine requirement. Histidine-requiring clones were purified by restreaking and stored as stocks.

A second method of selection involves the use of phage alone, omitting the penicillin-enrichment step.[10] An actively growing bacterial culture (0.1 ml) was mixed with 0.1 ml of the phage stock (either K3 or K14) and 2 ml of LCTG–top agar and poured as a lawn over broth plates. After incubation overnight at 37°, about 100 phage-resistant colonies were picked and streaked onto broth plates. After 24 hr incubation, colonies were replica plated onto minimal medium plates supplemented with either histidinol or histidine, and also onto a nitrogen-free medium supplemented with histidine. The clones which fail to grow on minimal as well as on N-free medium, putative *nif* deletions, were picked for further analysis. About 70% of the deletions have been reported to extend through *his* and into (through) the *nif* segment of DNA.[10] Of the remaining 30%, the majority are deletions extending beyond the *nif* gene cluster.

Recently, phage Mu has also been utilized for generation of deletions of *nif*. Detailed procedures are available in the literature.[11,12]

*Regulatory Mutants.* Most of the reported NIF regulatory mutants

[11] R. N. Rao, *J. Bacteriol.* **128**, 356 (1976).
[12] M. Backhuber, W. J. Brill, and M. M. Howe, *J. Bacteriol.* **128**, 749 (1976).

carry defects in the assimilation of $NH_4^+$.[3,13] The $NH_4^+$ produced by nitrogenase in *K. pneumoniae* is assimilated by the following enzymatic reactions.[14, 15]

Glutamine synthetase:

$$\text{Glutamate} + NH_4^+ + \text{ATP} \rightarrow \text{Glutamine} + \text{ADP} + P_i$$

Glutamate synthase:

$$\text{Glutamine} + \text{2-Oxoglutarate} + \text{NADPH} \rightarrow 2\ \text{Glutamate} + \text{NADP}^+$$

$NH_4^+$ is also assimilated to the level of glutamate, when the concentration of $NH_4^+$ in the medium is greater than 1 m$M$, by the enzyme glutamate dehydrogenase which catalyses the following reaction.

$$\text{2-Oxoglutarate} + NH_4^+ + \text{NADPH} \rightarrow \text{Glutamate} + \text{NADP}^+$$

Mutations which affect the production or activity of any of these enzymes also leads to pleiotrophic defects on nitrogenase production. For example, strains lacking glutamine synthetase activity fail to produce nitrogenase activity under conditions in which the parental strain, M5A1, produces nitrogenase.[13] Some of the strains lacking glutamate synthase activity (Asm⁻), and thus are conditional glutamate auxotrophs (when the $NH_4^+$ concentration is much lower than 1 m$M$ in the medium), also produce decreased levels of nitrogenase activity. Mutant strains that induce nitrogenase even in the presence of $NH_4^+$ are found to be defective in the assimilation of $NH_4^+$ to the level of glutamate.[8] Isolation of these strains involve a two-step procedure, since there are two major pathways for the production of glutamate in *K. pneumoniae*. It is convenient to isolate Nif derepressed mutants starting with strains lacking glutamate synthase activity (Asm⁻). A variety of selective procedures have proved successful such as isolation of Asm⁻ mutants lacking glutamate synthase activity and isolation of a Glu⁻ or Gln⁻ mutant starting with Asm⁻ mutants as parents. One of these procedures is described below.

Isolation of Asm⁻ Mutants Lacking Glutamate Synthase Activity. Colonies which failed to grow under $N_2$ as sole N source following mutagenesis are picked and checked for their growth in a medium containing $NO_3^-$ as sole source of nitrogen under both aerobic and anaerobic conditions. Colonies which failed to grow under $N_2$ as well as under $NO_3^-$ (aerobic as well as anaerobic) are selected. Most of the Asm⁻ strains are also defective in their ability to use histidine, proline or xanthine as sole source of nitrogen. Most of the colonies which fail to

[13] S. L. Streicher, K. T. Shanmugam, F. Ausubel, C. Morandi, and R. Goldberg, *J. Bacteriol.* **120**, 815 (1974).

[14] H. Nagatani, M. Shimizu, and R. C. Valentine, *Arch. Mikrobiol.* **79**, 164 (1971).

[15] R. T. St. John, H. M. Johnston, C. Seidman, D. Garfinkel, J. K. Gordon, V. K. Shah, and W. J. Brill, *J. Bacteriol.* **121**, 759 (1975).

grow using $N_2$, $NO_3^-$, histidine, proline, or xanthine as sole N source but can grow in the presence of higher concentrations of $NH_4^+$ ($>1$ m$M$) are found to be defective in the production of glutamate synthase activity.

Glutamate or Glutamine-Requiring Strains Derepressed for Nitrogenase Biosynthesis. These strains lack activities of at least two of the three $NH_4^+$ assimilation enzymes and thus do not use $NH_4^+$ as sole N source. These strains require glutamate as N source. Gln⁻ strains require L-glutamine as an essential amino acid for growth. L-Glutamine can satisfy both the glutamate and glutamine requirement of the Gln⁻ strains. A major proportion of L-glutamine can be converted into L-glutamate by glutaminases in the cell. Auxotrophic mutants in the mutagenized population can be isolated using replica plating techniques. Colonies on the LB plates are transferred to minimal medium (with $NH_4^+$ as the N source) by replica plating techniques. Clones which failed to grow in the minimal medium but are capable of growing in broth are isolated and tested for ability to grow in a glutamine (1 mg/ml) or glutamate (1 mg/ml) supplemented minimal medium. The glutamate or glutamine auxotrophs are further tested for the production and derepression of nitrogenase activity. Properties of some of these mutants are presented in Table II.

TABLE II
PROPERTIES OF NITROGENASE-DEREPRESSED MUTANTS OF *K. pneumoniae*[a,b]

| Strain | Nitrogenase activity | | Glutamine synthetase | | Glutamate synthase | | Glutamate dehydrogenase | |
|---|---|---|---|---|---|---|---|---|
| | − | +$NH_4^+$ | − | +$NH_4^+$ | − | +$NH_4^+$ | − | +$NH_4^+$ |
| M5A1 | 4.04 | 0.00 | 878 | 229 | 37 | 53 | <5 | 126 |
| Asm-1 | 3.14 | 0.00 | 653 | 147 | <5 | <5 | 18 | 116 |
| SK-28 | 3.27 | 3.05 | 1006 | 1055 | <5 | <5 | <5 | <5 |
| SK-25 | 4.47 | 3.70 | 0 | 0 | <5 | <5 | 12 | 16 |
| SK-27 | 2.88 | 1.81 | 0 | 0 | <5 | <5 | 17 | 15 |

[a] Data from Shanmugam *et al.*[3]
[b] Nitrogenase activity is $\mu$moles $C_2H_4$ produced per hour per milligram cell protein. Other enzyme activities are expressed as nmoles per minute per milligram protein. Strain SK-28 is a Glu⁻ strain. Strains SK-25 and SK-27 are Gln⁻. The sucrose minimal medium contained L-glutamate (100 $\mu$g/ml) in all these experiments (L-glutamine replaced L-glutamate in the experiments involving Gln⁻ strains).

## Section II
# Cellular and Subcellular Preparations

# [6] Isolation of Leaf Mesophyll and Bundle Sheath Cells

By CLANTON C. BLACK, JR., and IRAJ ROUHANI

## I. Introduction

The use of unicellular organisms to study photosynthesis has a long and successful history. However, when photosynthesis was studied with intact multicellular tissues, such as leaves, many uncertainties arose because mature leaves contain various types of photosynthetic cells as well as nonphotosynthetic cells. The major types of photosynthetic cells present are mesophyll, palisade, bundle sheath, or guard cells, and each cell type has a different function presumably within the leaf which is reflected in leaf photosynthesis. Today the idea of a single type of photosynthetic $CO_2$ fixation occurring in all cells of all higher plant leaves or even in adjacent leaf cells of a single plant is not tenable.[1,2] For example, when "the chloroplasts" of a plant leaf (such as spinach or corn) are isolated, these chloroplast preparations likely represent a heterogenous population of chloroplasts from various leaf cell types. Thus separation and isolation of pure green leaf cell types is essential to characterize quantitatively the diverse photosynthetic cells in an intact leaf.

Higher plants also possess multiple pathways for photosynthetic $CO_2$ assimilation. Today at least three broad groups of higher plants can be distinguished by their $CO_2$ fixation pathway; namely, reductive pentose phosphate ($C_3$) plants, $C_4$-dicarboxylic acid ($C_4$) plants, and Crassulacean acid metabolism (CAM) plants. In addition, several distinct biochemical variations exist within these pathways in specific plant species.[2,3] To assist in characterizing such diversity in leaf photosynthesis methods were developed to isolate pure green leaf cell types in sufficient quantities for biochemical studies.

Isolated cells, homogeneously suspended in a liquid medium, also constitute a very useful experimental system in general for biochemical, physiological, and cytochemical studies of cellular processes in plants. The use of isolated cells to study leaf metabolism by-passes other problems such as stomatal function, free space between leaf cells, and the

---

[1] T. M. Chen, P. Dittrich, W. H. Campbell, and C. C. Black, *Arch. Biochem. Biophys.* **163**, 246 (1974).

[2] C. C. Black, *Annu. Rev. Plant Physiol.* **24**, 253 (1973).

[3] R. H. Burris and C. C. Black, eds., "$CO_2$ Metabolism and Plant Productivity," p. 431. Univ. Park Press, Baltimore, Maryland, 1976.

METHODS IN ENZYMOLOGY, VOL. 69

uncertainties about concentrations of $CO_2$, $O_2$, or water inside of intact leaves. Isolated cells offer at least three other advantages. First, the cells can be handled similarly to unicellular algae, thus facilitating the removal of uniform aliquots. Second, cells in a suspension can be uniformly exposed to equal concentrations of compounds in the suspending medium. Third, isolated cells contain all of the enzymes, organelles, and other constituents which are involved in photosynthesis as well as other aspects of cell metabolism so that an isolated cell type can be studied as a total entity and then integrated with whole leaf data to allow an understanding of photosynthesis and other features of cellular metabolism in the intact multicellular leaf.

The procedures which have proved useful in this laboratory are divided conveniently into $C_4$ plant, CAM plant, and $C_3$ plant cell isolation techniques. Each of these plant groups exhibit special characteristics which one should be aware of to isolate active cells successfully. In the isolation of leaf cell types we reasoned that plant cell walls generally are much stronger than internal cell membranes, such as those surrounding chloroplasts. Thus the release of cells with the cell wall intact is a means for protecting cells during isolation. We also have isolated leaf protoplasts (plant cells which lack a cell wall), and some of these isolation techniques are included.

## II. General Considerations about Isolating Leaf Cells

The identification of isolated cell types is essential and, therefore, some preliminary orientation work must be done with the leaves under study. Cell identification is done conveniently and routinely with a light microscope. Leaf cross sections and sections parallel to veins should be examined in the light microscope to allow one to visualize the cell types, cell sizes and shapes, and the general location of cells within the intact leaf. Thin leaf sections can be prepared by hand from fresh leaf material by holding a section of the leaf in a split carrot and slicing thin sections with a razor. The sections should be quickly examined microscopically. We routinely place the sections in a drop of water, add a cover slip, and examine them under 50 to 1000 × magnification depending upon the leaf.

An estimation of cell size is obtained by placing nylon nets[4] of various meshes alongside the leaf sections under the microscope. One can match

---

[4] Precisely manufactured nylon nets are used in these cell isolation procedures. Square weave monofilament nylon nets with mesh openings from 20 to 2000 $\mu$m can be purchased under trade names such as Nitex. One company address is Cistron Corporation, 27th and Cumberland Streets, Lebanon, Pennsylvania 17042.

the leaf cell size with the net mesh opening. This information is used to select the correct mesh for collecting the cells during isolation. The amount of cell breakage in isolated cell preparations can be analyzed visually with the light microscope. The viability and degree of intactness also can be assessed by using certain dyes. For example, Evans blue [0.05 to 0.25% (w/v) in water] can be added to the cell preparation, and dead cells will become pigmented in less than 5 min as the dye penetrates. Live cells retain their semipermeable properties and therefore exclude Evans blue.[5] Neutral red [0.01% (w/v) in water] is taken up by plant cells and sequestered in the vacuole. This takes about 5 to 10 min, and the intact cells will have red vacuoles when viewed in the light microscope.

### III. Isolation of $C_4$-Dicarboxylic Acid Cycle Mesophyll Cells and Bundle Sheath Strands and Bundle Sheath Cells

$C_4$ photosynthesis requires the presence of two distinct green cell types which have definite roles in photosynthesis. Green mesophyll cells carboxylate phosphoenol pyruvate (PEP) through the action of PEP carboxylase, while green bundle sheath cells carboxylate ribulose 1,5-bis-phosphate ($1,5-P_2$) via ribulose $1,5-P_2$ carboxylase.[2,3] In fact a test for purity after isolating $C_4$ mesophyll and bundle sheath cells is to assay PEP and ribulose $1,5-P_2$ carboxylases. Over 98% of the PEP carboxylase will be in the mesophyll cells, and over 98% of the ribulose $1,5-P_2$ carboxylase will be in bundle sheath strands or cells.[6,7]

Two methods have been employed for the separation of $C_4$ leaf cell types. One is gentle grinding in a mortar and pestle,[6,8] and the other is an enzymatic digestion.[7] Both are designed to release the cells with their cell walls sufficiently intact to then allow the isolation of the cell types on nylon nets by filtration.

### A. *Gentle Grinding for the Isolation of Digitaria Mesophyll Cells, Bundle Sheath Strands and Bundle Sheath Cells*

This cell-isolation procedure basically involves the application of mechanical force to release cells from fully differentiated *Digitaria* leaves

[5] D. F. Gaff and O. Okong'O-Ogola, *J. Exp. Bot.* **22**, 756 (1971).
[6] G. E. Edwards, S. S. Lee, T. M. Chen, and C. C. Black, *Biochem. Biophys. Res. Commun.* **39**, 389 (1970).
[7] T. M. Chen, W. H. Campbell, P. Dittrich, and C. C. Black, *Biochem. Biophys. Res. Commun.* **51**, 461 (1973).
[8] G. E. Edwards and C. C. Black, *Plant Physiol.* **47**, 149 (1971).

and tissues followed by a separation of cell types by filtration. *Digitaria sanguinalis* (crabgrass) is the plant of choice, but the method described in Fig. 1 works well with *D. decumbens, D. ischaemum, D. setivalva, D. pentzii, D. milanjiana,* and *D. eriantha.*[8,9]

Following the procedures outlined in Fig. 1, a gentle grinding of fully differentiated leaves in a mortar releases mesophyll cells which are isolated on nylon nets by filtration. More extensive grinding of the remaining tissue yields bundle sheath strands which are isolated by filtration with stainless steel sieves and nylon nets. Further grinding of bundle sheath strands (outlined in Fig. 2) in a tissue homogenizer releases bundle sheath cells which are collected on nylon nets. So in a sequential fashion these techniques (Figs. 1 and 2) result in the isolation first of mesophyll cells, then bundle sheath strands, and finally bundle sheath cells.[8] For many photosynthetic-type studies one can stop with bundle sheath strands rather than isolate bundle sheath cells, since these are the only green cells in strands.

Throughout the cell isolation, each fraction is examined with the light microscope at frequent intervals for such factors as efficiency of cell release, cell breakage, and contamination. The distinguishing features of crabgrass mesophyll cells in the light microscope are a tendency for the chloroplasts to be scattered in each cell, cells generally with rounded edges, and cells often having a rounded or egg-shaped appearance. The distinguishing features of bundle sheath cells in the light microscope are chloroplasts packed into the cells, often seeming to fill the cell completely; cells having angular edges; and the cells appearing to be shaped as rectangles, cylinders, or squares. Bundle sheath strands are easily recognized in the light microscope, since a strand consists of the vascular tissue in the center which is very tightly surrounded by bundle sheath cells. The bundle sheath cells fit very tightly, somewhat like bricks in a chimney, in an orderly arrangement that can be followed as spirals around the vascular tissue. Indeed two separate spirals are present, so that the bundle sheath cells form a double helix of two spirals throughout the length of the vascular tissue.[8,10]

*Medium.* The isolation medium used throughout the handling of tissues and cells is composed of sorbitol, 0.33 $M$; Tricine–NaOH buffer, pH 8.0, 0.05 $M$; $NaNO_3$, 2 m$M$; EDTA, 2 m$M$; $MnCl_2$, 1 m$M$; $MgCl_2$, 5 m$M$; and $K_2HPO_4$, 5m$M$.

[9] M. L. Salin, W. H. Campbell, and C. C. Black, *Proc. Natl. Acad. Sci. U.S.A.* **70,** 3730 (1973).
[10] C. C. Black, W. H. Campbell, T. M. Chen, and P. Dittrich, *Q. Rev. Biol.* **48,** 299 (1973).

B. *Partial Enzymatic Digestion of Cyperus rotundus Leaves to Isolate Mesophyll Cells and Bundle Sheath Strands.*

This cell isolation procedure[1,7] basically involves the incubation of leaf sections with cellulase and pectinase (crude preparations) to loosen cellular connections, such as between mesophyll and bundle sheath, or mesophyll and epidermis, or among mesophyll cells. Subsequently the partially digested leaves are gently ground to release cells and tissues which then are isolated as in the *Digitaria* method (Section III,A).

1. *Plants.* Use vigorous, fully differentiated leaves. Expose leaves to light at least 2 to 3 hours prior to tissue separation.

2. *Media.*
   a. Incubation medium.   The incubation medium (Fig. 3) contains 1% cellulysin and 1% macerase (Calbiochem Co.), 2% PVP 40 (Sigma Chemical Co.), 0.33 $M$ sorbitol, 2 m$M$ NaNO$_3$, 2 m$M$ EDTA, 1 m$M$ MgCl$_2$, and 5 m$M$ K$_2$HPO$_4$. Since cellulysin has a pH optimum of 4–5 and macerase a pH optimum of 5–6, the incubation medium is adjusted to pH 5.0.
   b. Grinding medium.   The grinding medium is composed of 0.33 $M$ sorbitol, 2 m$M$ NaNO$_3$, 2 m$M$ EDTA, 1 m$M$ MgCl$_2$, 5 m$M$ K$_2$HPO$_4$, 2% polyvinylpyrrolidone (PVP-40, Sigma Chemical Co.), and 10 m$M$ $\beta$-mercaptoethanol.

3. *Filter Assembly.* The filter assembly consists from top to bottom of a 20-mesh steel screen, 210, 80, 44, and then 20 $\mu$m pore nylon nets mounted on modified Falcon filter units. The bottom containers of Falcon units are cut off 1.0–1.5 cm from their extended rims, in order to stack the units so that a tissue suspension can be filtered through the entire assembly in one operation. Mesophyll cells still attached to bundle sheath strands are removed by using a Ten Broeck ground glass homogenizer (see Fig. 1).

4. *Incubation and Cell Isolation.* The flow diagram for the incubation of leaf segments and the isolation procedure is given in Fig. 3. The procedure involved the incubation of leaf sections with cellulase (cellulysin) from *Trichoderma viride* and pectinase (macerase) from *Rhizopus* sp. for 1 hr before mechanical grinding. As a result of the enzyme incubation, only one grinding and filtration step is needed as shown in the flow chart (Fig. 3).

Throughout the isolation procedure the results should be checked carefully with the light microscope as outlined earlier for *Digitaria* but checking for the cell sizes and shapes characteristic of *Cyperus rotundus*.[1]

Four grams of fresh, turgid, crabgrass leaves
↓
Wash with distilled water and cut in 0.5 to 1 mm segments with razor[1]
↓
Grind gently with mortar and pestle[2] using 6 ml medium/gram tissue
↓
Add equal volume of grinding medium to homogenate and filter through a 20-mesh stainless steel sieve in a Falcon filter unit[3] while stirring with glass rod

| Filtrate<br>PREPARATION OF MESOPHYLL<br>CELLS | Unmacerated tissue<br>PREPARATION OF BUNDLE SHEATH<br>STRANDS |
|---|---|
| ↓ | ↓ |
| Pass filtrate through a 30-$\mu$m nylon net in a Plexiglas filter unit (See Ref. 8) with magnetic stirring;[4] wash three times with 10 ml of medium with stirring | Return fragments on sieve to mortar<br>↓<br>Add 15 ml of medium and grind until all leaf material is macerated and suspension becomes somewhat homogeneous<br>↓<br>Add 30 ml of medium to homogenate and filter through 20 and 35 mesh stainless steel sieves in a Falcon filter unit while stirring with glass rod |

| Pass filtrate through a 20-$\mu$m nylon net[5] in a Plexiglas filter unit under slight vacuum without magnetic stirring; wash once with 10 ml of medium | Material on net (add to unmacerated tissue for bundle sheath preparation) | Filtrate<br>Pass filtrate through an 80-$\mu$m nylon net[4] in a Falcon filter unit with magnetic stirring | Debris on sieves containing mostly epidermis tissue (discard) |
|---|---|---|---|

| Resuspend cells on net in petri dish in 8 ml of medium | Filtrate contains cellular organelles from broken cells (discard) | Collect material on net in petri dish containing 8 ml of medium | Filtrate (discard) |
|---|---|---|---|
| ↓ | | ↓ | |
| Centrifuge at 1000 $g$ for 1 min and resuspend the pellet in grinding medium (mesophyll cells) | | Add to 15 ml capacity Ten Broeck[6] tissue grinder and grind by hand (two to three strokes) | |

Filter through an 80-$\mu$m nylon net with magnetic stirring, wash strands on net three times with 10 ml of medium; maintain magnetic stirring during wash

| Resuspend strands on net in 5 ml of medium in petri dish | Filtrate (discard) |
|---|---|
| ↓ | |
| Centrifuge at 1000 $g$ for 1 min and resuspend the pellet in grinding medium (bundle sheath strands) | |

Suspend bundle sheath strands in 5 ml of medium. Grind by hand in a 7-ml capacity Ten-Broeck tissue grinder making 10 to 15 strokes. Examine aliquots under the light microscope at intervals to determine when most of the bundle sheath cells have been separated from the vascular bundles

Filter through a 42-$\mu$m nylon net in the filter unit[1] with stirrer, and wash twice with 10 ml of medium.

Pass filtrate through a 20-$\mu$m net under          Material on net (discard)
slight vacuum without stirrer and wash
with 10 ml of medium.

Resuspend cells on net in 5 ml of grinding          Filtrate (discard)
medium

Centrifuge at 1000 $g$ for 1 min and resuspend pellet in grinding medium (bundle sheath cells)

FIG. 2. Flow chart of technique for isolating bundle sheath cells from bundle sheath strands of *Digitaria sanguinalis*. [1]The Plexiglas filter unit (Fig. 1) or Falcon filter units are used with the 42- and 20-$\mu$m nylon net.

FIG. 1. Flow chart of technique for isolating mesophyll cells and bundle sheath strands from *Digitaria sanguinalis* leaves. (Adapted from Edwards and Black[8]). [1]The isolation is conducted at room temperature. [2]The mortar used is porcelain, Coors 522, size No. 6 having an unglazed grinding surface. [3]Stainless steel wire cloth is purchased from W. S. Tyler Co. (Mentor, Ohio). Falcon filter units, Model 7102, are purchased from Scientific Glass Apparatus (Bloomfield, New Jersey) and adapted to hold the 20- and 30-mesh stainless screen and the 80 $\mu$m nylon net. Plexiglas filter units having inside diameters of 4 inches are built to hold the 20- and 30-$\mu$m nylon net. In the Plexiglas filter units, a 20 mesh sieve is placed beneath the net as a support (see Edwards and Black[8]). [4]The nylon net is Nitex Nylon Monofilament Bolting Cloth purchased from Tobler, Ernst, and Traber, Inc., (Saw Mill River Rd., Elmsford, New York) (see Ref. 4). The magnetic stirrer used is a Precision Senior Magmix by Precision Scientific Co., which has a 1⅛-inch diameter by 3-inch long magnet. [5]All nets are washed after each isolation, examined under the light microscope, and changed as they wear or clog. [6]Whereas most of the strands are without mesophyll cells, microscopic examinations show that some strands have mesophyll cells attached. By grinding in the tissue homogenizer most of the remaining mesophyll cells are released from the bundle sheath strands. Extensive grinding is to be avoided as the yield will be lowered due to breakage of the bundle sheath strands. Older plants which are flowering yield purer preparations of bundle sheath strands and in this case the above step can be omitted.

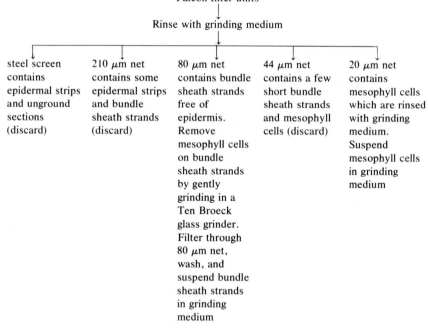

20 ml of incubation medium in a 250-ml Erlenmyer flask fitted for evacuation

2 gm of leaves cut with a razor blade into 1- to 2-mm sections

Sections infiltrated with 20 ml of incubation medium at pH 5.0 with a vacuum pump twice and flask either kept under vacuum or released

Incubate at 37°, water bath, for 1 hr

Decant incubation medium and gently grind leaf sections with a mortar and pestle in 15 ml of cold grinding medium until most leaf sections become pale green

Pass suspension through a filter assembly consisting of, from top to bottom, a 20-mesh steel screen, 210 $\mu$m, 80 $\mu$m, 44 $\mu$m, and then 20 $\mu$m nylon nets mounted on modified Falcon filter units

Rinse with grinding medium

| steel screen contains epidermal strips and unground sections (discard) | 210 $\mu$m net contains some epidermal strips and bundle sheath strands (discard) | 80 $\mu$m net contains bundle sheath strands free of epidermis. Remove mesophyll cells on bundle sheath strands by gently grinding in a Ten Broeck glass grinder. Filter through 80 $\mu$m net, wash, and suspend bundle sheath strands in grinding medium | 44 $\mu$m net contains a few short bundle sheath strands and mesophyll cells (discard) | 20 $\mu$m net contains mesophyll cells which are rinsed with grinding medium. Suspend mesophyll cells in grinding medium |

FIG. 3. Flow diagram of the procedures used to isolate mesophyll cells and bundle sheath strands from mature nutsedge leaves.[1]

## IV. Isolation of Crassulacean Acid Metabolism Plant Cells and Protoplasts

The green tissues of CAM plants generally are characterized by a marked diurnal fluctuation of malate, high in the dark and low in the light. The leaf stomata open principally at night, and starch accumulation and degradation occur principally in the day and night, respectively.[11-13] Environmental conditions, such as temperature,[14] day length,[14,17,18] or water regime,[19,20] strongly effect the magnitude of CAM, especially $CO_2$ metabolism.[21]

## A. Plant Growth Conditions

It is essential that plant growth be rapid and healthy and that a marked diurnal fluctuation of malate and starch occur in the green tissues. When leaves are selected the investigator should know where the cells are in their daily cycles of acid and starch metabolism, since this will drastically affect the activity of the cells as well as the isolation results.[22] The diurnal acid cycle can be evaluated by determining the titratable acid content of green tissues harvested through a 24-hr cycle.[14] Two methods will be described for the isolation of CAM cells.

## B. Mechanical Separation of Sedum telephium Cells

Cells are isolated from *Sedum telephium* leaves simply by gently mashing the leaves in a mortar and pestle and following the isolation techniques shown in Fig. 4.[22] In following the isolation technique the preparation is examined with the light microscope at frequent intervals during each step of the cell preparation for such factors as efficiency of cell release, cell breakage, and contamination.

The *S. telephium* plants used for cell isolation should be exposed to light for several (4 to 8) hours prior to harvesting the leaves to reduce the leaf acidity. Fully expanded leaves from plants of similar size are placed on crushed ice for about 30 min and then washed with distilled water. Healthy leaves (30 gm) are selected, and the margin of each leaf

[11] H. Beevers, M. L. Stiller, and V. S. Butt, *in* "Plant Physiology" (F. C. Steward, ed.), Vol. 4B, p. 119. Academic Press, New York.

[12] J. Bonner, *in* "Plant Biochemistry" (J. Bonner and J. E. Varner, eds.), 1st ed., p. 154. Academic Press, New York, 1950.

[13] K. Nishida, *Physiol. Plant.* **16**, 218 (1963).

[14] C. E. Crews, S. L. Williams, H. M. Vines, and C. C. Black, *in* "$CO_2$ Metabolism and Plant Productivity" (R. H. Burris and C. C. Black, eds.), p. 235. Univ. Park Press, Baltimore, Maryland 1976.

[15] T. F. Neales, *Aust. J. Biol. Sci.* **26**, 539 (1973).

[16] T. F. Neales, *Aust. J. Biol. Sci.* **26**, 705 (1973).

[17] O. Queiroz, *Physiol. Veg.* **6**, 117 (1968).

[18] W. W. Allaway, C. B. Osmond, and J. H. Troughton. *R. Soc. N. Z. Bull.* **12**, (1974).

[19] M. Kluge and K. Fischer, *Planta* **77**, 212 (1967).

[20] B. Bartholomew, *Photosynthetica* **7**, 114 (1973).

[21] M. M. Bender, I. Rouhani, H. M. Vines, and C. C. Black, *Plant Physiol.* **52**, 427 (1973).

[22] I. Rouhani, H. M. Vines, and C. C. Black, *Plant Physiol.* **51**, 97 (1973).

GRIND IN MOTAR AND PESTLE

10 g of peeled *Sedum telephium* leaves are gently ground with a motar and pestle in 10 ml of solution A[1]

↓

FILTER

10 ml of solution B[1] are added to the homogenate followed by filtration through a 20 mesh stainless steel sieve in a Falcon Filter Unit,[2] while stirring with a glass rod. The motar and pestle are washed with 10 ml of Solution B and poured onto the sieve

↓

REGRIND UNACERATED TISSUE

The unmacerated tissue is removed from the sieve and ground gently, then filtered as above. This step may be repeated as many as five or six times. Filtrates are saved

↓

SIEVE

About 20-ml lots of the filtrate are added to test tubes (24 × 120 mm) and shaken[3] for 30 sec and filtered through a 30 mesh stainless steel sieve, in a Falcon Filter Unit while being stirred with a glass rod. The entire filtrate is shaken before being passed through the sieve

↓

SEPARATION

The filtrate is again shaken as before for one minute, and passed through a 136 micron nylon net above an 80-μm micron nylon net in a Plexiglas filter unit with slow magnetic stirring, and then washed with 3 volumes of solution B

↓

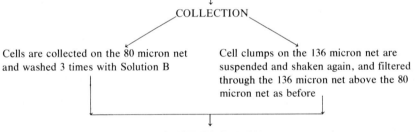

COLLECTION

Cells are collected on the 80 micron net and washed 3 times with Solution B

Cell clumps on the 136 micron net are suspended and shaken again, and filtered through the 136 micron net above the 80 micron net as before

↓

CENTRIFUGATION

The cell suspension is centrifuged at 750 *g* for one minute and then resuspended by shaking in solution C[1] (1 ml). These isolated spongy mesophyll cells are stored in an ice bucket until needed for assay.

Fig. 4. Flow diagram for the isolation of *Sedum telephium* spongy mesophyll cells. [1]For the isolation of cells, three solutions are used and the isolation is at room temperature. Solution A: (soaking medium) contained Trizma base (0.06 *M*) adjusted with MES Buffer to pH 8.5, 5 m*M* $MgCl_2$, 5 m*M* $K_2HPO_4$, 2 m*M* $NaNO_3$, 2 m*M* EDTA, and 1 m*M* $MnCl_2$. Solution B: (washing medium) contained 0.3 *M* sorbitol, 0.05 *M* Trizma base adjusted with MES to pH 8.5, and 5 m*M* $MgCl_2$. Solution C: (cell suspension medium) contained 0.05 *M* Trizma base adjusted with MES to pH 8.0, 2 m*M* EDTA, 1 m*M* $MnCl_2$, 2 m*M* $NaNO_3$, 5 m*M* $MgCl_2$, 5 m*M* $K_2HPO_4$, and 0.35 *M* sorbitol. [2]For details on filter units and nylon nets see Fig. 1. [3]The test tubes are shaken with a Deluxe Mixer at fast setting.

removed with a dissecting knife. The leaves are bent slightly; both the upper and lower epidermis peeled off by hand, and the midribs removed. From 30 gm of whole leaf tissue, about 10 gm of peeled leaf tissue can be separated. The peeled leaf portions are placed in 10 ml of medium A (Fig. 4) and kept at 10° to 15°. This portion of the preparation usually takes 1 to 2 hr. All solutions are kept at ice bath temperature, but the filtering and related operations in Fig. 1 are at room temperatures (21° to 23°).

The peeled leaf sections are ground "very gently." This grinding is critical, since excess force will break all cells. Since it is empirical, the preparation should be examined regularly with a light microscope to determine the degree of cell and tissue breakage. The yield of intact cells is determined with a light microscope. We obtain between 55 and 68% intact cells and between 32 and 45% broken or plasmolyzed cells.

The yield of cells can be determined by assaying for total leaf chlorophyll and for total chlorophyll in the isolated cell preparations. Our yield of intact mesophyll cell from *S. telephium* leaves is about 1%. The mesophyll cells of *Sedum* have a diameter between 80 and 136 $\mu$m.

The isolated cells are active in fixing $CO_2$ for at least 4 hr after isolation when stored in an ice bath. In the microscope they appear to be intact for periods up to 10 hr.[22]

## C. Enzymatic Cell Separation

Leaves low in acid content are washed with distilled water, and surfaces are sterilized by dipping it with 70% ethanol.[23] Remove the margin of each leaf with a dissecting knife. Leaves are bent slightly, and both the upper and lower epidermis peeled off by hand with the aid of a dissecting knife and the midribs removed. Peeled leaves are cut into small pieces approximately 2 mm$^2$ with a razor. Six grams are placed in a petri dish with 12.5 ml of medium containing 1.5% (weight to volume) cellulysin[10] (Calbiochem) and 0.6 $M$ mannitol. The pH of the solution is adjusted to pH 5.8 with NaOH. The preparation is incubated at 28° with gentle shaking (35 rpm) in the dark. Use the light microscope during the incubation period to determine when cells have been released. Figure 5 outlines the steps for separation and purification of intact *Sedum telephium* mesophyll cells.

This method is applicable for pineapple leaves and *Stapelia* which also are CAM plants. Each CAM plant or leaf system to be used to isolate cell presents its own particular and characteristic problems which should be studied as given in Section II.

[23] G. J. Wagner and H. W. Siegelman, *Science* **190**, 1298 (1975).

Six grams of peeled leaf segments are placed in 12.5 ml of a medium containing 1.5% w/v cellulysin and 0.6 $M$ mannitol, pH 5.8

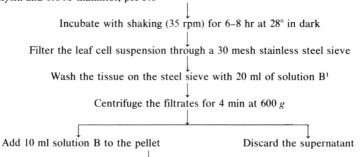

Incubate with shaking (35 rpm) for 6–8 hr at 28° in dark

Filter the leaf cell suspension through a 30 mesh stainless steel sieve

Wash the tissue on the steel sieve with 20 ml of solution B[1]

Centrifuge the filtrates for 4 min at 600 $g$

Add 10 ml solution B to the pellet        Discard the supernatant

Filter cells through a 136-micron nylon net, wash the tissue on the net with 10 ml solution B

Collect the cells on an 80-micron nylon net and wash the cells on the net with 10 ml solution B

Concentrate the cells on the net by washing into a petri dish with 1 ml of solution C

FIG. 5. Flow diagram for isolating *S. telephium* cells using an enzymatic separation procedure. [1]Solutions B and solution C are the same solutions used in the mechanical isolation of *Sedum* cells given in Fig. 4.

## D. *Preparation of CAM Protoplasts*

Protoplasts can be isolated from leaves of many CAM plants, such as *Sedum telephium, Cactacea grusonia, Stapelia vobilis, Stapelia gigantea*, and *Ananas cosmosus*, by enzymatic digestion of the leaf cell walls. By using digestive enzyme preparations such as cellulysin, protoplasts can be isolated from all of these plants, but we report here only the isolation technique for pineapple (*Ananas cosmosus*).

1. *Isolation of Pineapple Leaf Protoplasts.* Fully expanded leaves are removed from pineapple plants, washed with distilled water, blotted dry, and surface sterilized by treating with 70% ethanol. Leaves are cut to approx 7-cm lengths, the margin removed, and the upper epidermis removed with a razor. The leaves are cut into 1- to 2-mm sections and treated as outlined in Fig. 6.

These protoplasts are contaminated but they can be purified using a Ficoll step gradient made in solution B (Fig. 6). Suspend the protoplasts in 15% (w/v) Ficoll and layer with 10% and 2.5% Ficoll. A typical experiment would use a 1.5-ml aliquot of the final pellet (Fig. 6) with 15% Ficoll (w/v) in solution B in a 15 ml glass centrifuge tube, overlaid gently

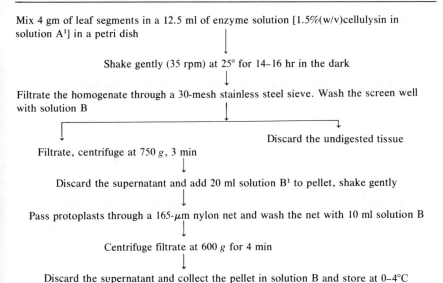

Mix 4 gm of leaf segments in a 12.5 ml of enzyme solution [1.5%(w/v)cellulysin in
solution A[1]] in a petri dish

Shake gently (35 rpm) at 25° for 14–16 hr in the dark

Filtrate the homogenate through a 30-mesh stainless steel sieve. Wash the screen well
with solution B

Discard the undigested tissue

Filtrate, centrifuge at 750 $g$, 3 min

Discard the supernatant and add 20 ml solution B[1] to pellet, shake gently

Pass protoplasts through a 165-$\mu$m nylon net and wash the net with 10 ml solution B

Centrifuge filtrate at 600 $g$ for 4 min

Discard the supernatant and collect the pellet in solution B and store at 0–4°C

FIG. 6. Flow scheme for the isolation of pineapple leaf protoplasts. [1]Solution A: 0.6 $M$
mannitol adjust to pH 5.6 with NaOH. Solution B: 0.6 $M$ mannitol, 20 m$M$ MES buffer,
pH 6.0.

with 2.5 ml of 10% (w/v) Ficoll in solution B, 1.25 ml of 2.5% (w/v) Ficoll
in solution B, and 1.25 ml of solution B. The tubes are centrifuged at 750
$g$ for 30 min.

Collect fractions from the test tube and locate protoplasts by using
neutral red and microscopic analysis of each fraction. Most protoplasts
will be in the 10% Ficoll; if they are not completely pure repeat the step
gradient. The protoplasts finally are centrifuged and collected in solution
B. The yield of protoplasts per gram of leaf tissue is about 0.9–1.2 × 10[6].

## V. Isolation of Spinach (a C₃ Plant) Mesophyll Cells

The techniques used with crabgrass leaves (Fig. 1) can be modified
to isolate mesophyll cells from spinach leaves.[8] Figure 7 is a flow chart
giving the procedure. These cells will fix $CO_2$ via the $C_3$ cycle without
added substrates. Yields of intact cells are near 1% on a chlorophyll
basis.[8] Cell purity, intactness, and other features should be evaluated
microscopically as given in previous sections.

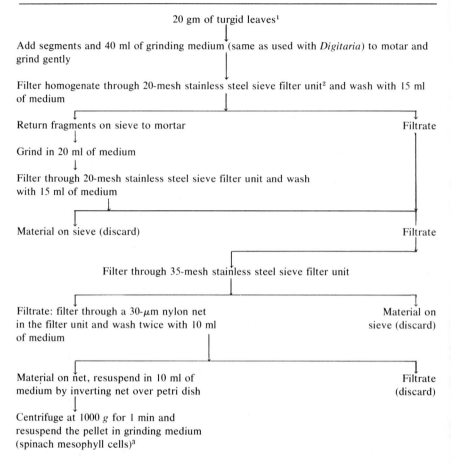

20 gm of turgid leaves[1]

Add segments and 40 ml of grinding medium (same as used with *Digitaria*) to motar and grind gently

Filter homogenate through 20-mesh stainless steel sieve filter unit[2] and wash with 15 ml of medium

Return fragments on sieve to mortar                                                        Filtrate

Grind in 20 ml of medium

Filter through 20-mesh stainless steel sieve filter unit and wash with 15 ml of medium

Material on sieve (discard)                                                        Filtrate

Filter through 35-mesh stainless steel sieve filter unit

Filtrate: filter through a 30-μm nylon net                          Material on
in the filter unit and wash twice with 10 ml                       sieve (discard)
of medium

Material on net, resuspend in 10 ml of                                        Filtrate
medium by inverting net over petri dish                                      (discard)

Centrifuge at 1000 *g* for 1 min and
resuspend the pellet in grinding medium
(spinach mesophyll cells)[3]

FIG. 7. Flow chart of technique for isolating mesophyll cells from spinach leaves. [1]Refer to flow chart for separation of cells of *Digitaria sanguinalis* in Fig. 1 for further specifications of materials used. [2]The Falcon filter unit is used with the 20 mesh and 35 mesh screen and the Plexiglas filter unit is used with the 30 μm nylon net (See Fig. 1). [3]The preparation contains single mesophyll cells as well as some cells which are still joined together.

# [7] Protoplasts of Plant Cells

## By ALBERT RUESINK

In 1960, enzymes were first used to remove the cell walls from living plant cells, releasing spheres of living cytoplasm known as protoplasts.[1] Thereafter, a number of reviews have dealt with the method of release, stabilization, and use of protoplasts in some detail.[2-7] This article will emphasize advances since 1975 and cite papers dealing with protoplasts from as many tissues and species as possible.

Though it is possible to find reports of protoplast release from almost every plant organ, the usual tissues of choice have been either leaf mesophyll or tissue culture. These tissues consist almost totally of cells bounded by only primary walls. Lignification seems markedly to inhibit the release of protoplasts.

The most important parameters for releasing protoplasts successfully include (1) choice of tissue, (2) preparation of tissue, (3) choice of enzymes, (4) conditions during incubation, (5) times of incubation, and (6) the osmoticum. Detailed descriptions of general methods have appeared. Cocking and Evans[5] described release from crown-gall callus and Paul's Scarlet rose callus, as well as release from leaves of cereal, legumes, tobacco, and petunia. Constabel[6] described release from soybean, Vicia, and carrot cultures and from leaves of pea, bean, and rapeseed. Deserving special mention are careful efforts to maximize protoplast release from two types of tissue—tobacco callus[8] and tomato leaves.[9]

Different tissues often require different approaches to release protoplasts effectively. Though general reviews exist in relative abundance,

[1] E. C. Cocking, *Nature (London)* **187,** 962 (1960).
[2] A. W. Ruesink, Vol. 23, p. 197.
[3] E. C. Cocking, *Annu. Rev. Plant Physiol.* **23,** 29 (1972).
[4] R. A. Miller, K. N. Kao, and O. L. Gamborg, in "Tissue Culture: Methods and Applications" (P. F. Kruse, Jr. and M. K. Patterson, eds.), p. 500. Academic Press, New York, 1973.
[5] E. C. Cocking and P. K. Evans, *Bot. Monogr. (Oxford)* **11,** 100 (1973).
[6] F. Constabel, in "Plant Tissue Culture Methods" (O. L. Gamborg and L. R. Wetter, eds.), p. 11. Nat. Res. Counc. Can., Ottawa, 1975.
[7] P. K. Evans and E. C. Cocking, in "New Techniques in Biophysics and Cell Biology" (R. H. Pain and B. J. Smith, eds.), Vol. 2, p. 127. Wiley, New York, 1975.
[8] H. Uchimiya and T. Murashige, *Plant Physiol.* **54,** 936 (1974).
[9] A. C. Cassells and M. Barlass, *Physiol. Plant.* **37,** 239 (1976).

nowhere is there a single source of information about the usual techniques found to be appropriate for particular situations, such as Table I.[1,6,8-65] A selection has been made, in some cases rather arbitrarily, of what seems to be either the most recent or the best method for releasing protoplasts from each of as many kinds of tissue as possible.

On the basis of the specific cases shown in Table I, it is possible to generalize about the typical protoplast release procedures.

[10] D. Bui-Dang-Ha and I. A. Mackenzie, *Protoplasma* **78**, 215 (1973).

[11] B. G. Hughes, F. G. White, and M. A. Smith, *Protoplasma* **90**, 399 (1976).

[12] R. Kanai and G. E. Edwards, *Plant Physiol.* **51**, 1133 (1973).

[13] S. C. Huber and G. E. Edwards, *Plant Physiol.* **55**, 835 (1975).

[14] R. Kaur-Sawhney, B. Staskawicz, W. R. Adams, and A. W. Galston, *Plant Sci. Lett.* **7**, 57 (1976).

[15] P. K. Evans, A. G. Keates, and E. C. Cocking, *Planta* **104**, 178 (1972).

[16] G. A. Strobel and W. M. Hess, *Proc. Natl. Acad. Sci. U.S.A.* **71**, 1413 (1974).

[17] G. J. Wagner and H. W. Siegelman, *Science* **190**, 1298 (1975).

[18] L. Taiz and R. L. Jones, *Planta* **101**, 95 (1971).

[19] H. Koblitz, *Biochem. Physiol. Pflanz.* **170**, 295 (1976).

[20] J. X. Hartmann, K. N. Kao, O. L. Gamborg, and R. A. Miller, *Planta* **112**, 45 (1973).

[21] F. Motoyoshi, *Exp. Cell Res.* **68**, 452 (1971).

[22] J. B. Power, S. E. Cummins, and E. C. Cocking, *Nature (London)* **225**, 1016 (1970).

[23] M. Ito, *Bot. Mag.* **86**, 133 (1973).

[24] A. W. Ruesink and K. V. Thimann, *Proc. Natl. Acad. Sci. U.S.A.* **54**, 56 (1965).

[25] Y. Zuily-Fodil and R. Esnault, *Physiol. Veg.* **14**, 109 (1976).

[26] E. Zeiger and P. K. Hepler, *Plant Physiol.* **58**, 492 (1976).

[27] K. Wakasa, *Jpn. J. Genet.* **4**, 279 (1973).

[28] Y. P. S. Bajaj, *Plant Sci. Lett.* **3**, 93 (1974).

[29] I. Potrykus and J. Durand, *Nature (London), New Biol.* **237**, 286 (1972).

[30] E. M. Frearson, J. B. Power, and E. C. Cocking, *Dev. Biol.* **33**, 130 (1973).

[31] J. F. Shepard and R. E. Totten, *Plant Physiol.* **60**, 313 (1977).

[32] I. Takebe, Y. Otsuki, and S. Aoki, *Plant Cell Physiol.* **9**, 115 (1968).

[33] Y. Chupeau, J.-P. Bourgin, C. Missonier, N. Dorion, and G. Morel, *C. R. Hebd. Seances Acad. Sci., Ser. D* **274**, 1565 (1974).

[34] F. J. Zapata, P. K. Evans, J. B. Power, and E. C. Cocking, *Plant Sci. Lett.* **8**, 119 (1977).

[35] E. W. Rajasekhar, *Nature (London)* **246**, 223 (1973).

[36] M. Lorenzini, *C. R. Hebd. Seances Acad. Sci., Ser. D* **276**, 1839 (1973).

[37] M. R. Davey, E. M. Frearson, L. A. Withers, and J. B. Power, *Plant Sci. Lett.* **2**, 23 (1974).

[38] D. W. Gregory and E. C. Cocking, *J. Cell Biol.* **24**, 143 (1965).

[39] L. E. Pelcher, O. L. Gamborg, and K. N. Kao, *Plant Sci. Lett.* **3**, 107 (1974).

[40] M. R. Davey, E. Bush, and J. B. Power, *Plant Sci. Lett.* **3**, 127 (1974).

[41] F. Constabel, J. W. Kirkpatrick, and O. Gamborg, *Can. J. Bot.* **51**, 2105 (1973).

[42] O. L. Gamborg, J. Shyluk, and K. K. Kartha, *Plant Sci. Lett.* **4**, 285 (1975).

[43] C. R. Landgren, *Am. J. Bot.* **63**, 473 (1976).

[44] M. R. Davey, E. C. Cocking, and E. Bush, *Nature (London)* **244**, 460 (1973).

[45] G. Shetter and D. Hess, *Biochem. Physiol. Pflanz.* **171**, 63 (1977).

Tissue Choice

As significant numbers of protoplasts are released from more and more kinds of tissues, investigators are able to choose a tissue for a particular series of experiments based on the need for particular physiological characteristics and to be moderately confident that some protoplasts can be obtained from the tissue. It is not always possible, however, to obtain enough protoplasts of a particular tissue to carry out biochemical types of experiments. Often one must try several potential tissues when seeking to develop a new system. Rapidly growing cells of nearly mature size but with thin walls provide the best chance of success.

Tissue Preparation

Though the prior growth conditions of the plant tissue obviously have an effect on subsequent protoplast release, little systematic attention has been devoted to the matter. When plants are grown in growth chambers, one may still have trouble releasing good protoplasts at certain times during the year. Even the old standby, tobacco, has sometimes given

[46] G. Weber, F. Constabel, F. Williamson, L. Fowke, and O. L. Gamborg, Z. Pflanzenphysiol. **79**, 459 (1976).
[47] H. Harada, Z. Pflanzenphysiol. **69**, 77 (1973).
[48] R. K. Horine and A. W. Ruesink, Plant Physiol. **50**, 438 (1972).
[49] M. Messerschmidt, Z. Pflanzenphysiol. **74**, 175 (1974).
[50] K. Wegmann and H.-P. Mühlbach, Biochim. Biophys. Acta **314**, 79 (1973).
[51] T. Eriksson and K. Jonasson, Planta **89**, 85 (1969).
[52] C. Ambid, M.-H. Delmestre, and J. Fallot, C. R. Hebd. Seances Acad. Sci., Ser. D. **279**, 1429 (1974).
[53] C. Buser and P. Matile, Z. Pflanzenphysiol. **82**, 462 (1977).
[54] K. K. Kartha, M. R. Michayluck, K. N. Kao, O. L. Gamborg, and F. Constabel, Plant Sci. Lett. **3**, 265 (1974).
[55] S. Poirier-Hamon, P. S. Rao, and H. Harada, J. Exp. Bot. **25**, 752 (1974).
[56] S. W. J. Bright and D. H. Northcote, J. Cell Sci. **16**, 445 (1974).
[57] L. C. Fowke, C. W. Bech-Hansen, O. L. Gamborg, and J. P. Shyluk, Am. J. Bot. **60**, 304 (1973).
[58] K. Glimelius, A. Wallin, and T. Eriksson, Physiol. Plant. **31**, 225 (1974).
[59] P. Vardi, P. Spiegel-Roy, and E. Galun, Plant Sci. Lett. **4**, 231 (1975).
[60] C. A. Beasley, I. P. Ting, A. E. Linkins, E. H. Birnbaum, and D. P. Delmer, in "Tissue Culture and Plant Science" (H. E. Street, ed.), p. 169. Academic Press, New York, 1974.
[61] W. R. Scowcroft, M. R. Davey, and J. B. Power, Plant Sci. Lett. **1**, 451 (1973).
[62] R. H. A. Coutts and K. R. Wood, Plant Sci. Lett. **9**, 45 (1977).
[63] O. L. Gamborg and J. P. Shyluk, Bot. Gaz. (Chicago) **137**, 301 (1976).
[64] D. Hess and R. Endress, Z. Pflanzenphysiol. **68**, 441 (1973).
[65] I. Potrykus, Nature (London), New Biol. **231**, 57 (1971).

TABLE I
Successful Protoplast Release

| Plant group | Enzyme(s)[a] | Osmoticum[b] | Temp[c] (°C) | Time (hr) | pH | Reference |
|---|---|---|---|---|---|---|
| **Monocots** | | | | | | |
| Leaf tissue | | | | | | |
| Asparagus | 1% MZ, 3% OC SS | 0.9 $M$ man | 32 | 3 | 5.2 | 10 |
| Barley | 1% MZ R-10, 1% OC R-10 | 0.45 $M$ man, 0.01 $M$ citrate | 30 | 3 | 5.6 | 11 |
| Corn | 2% OC R-10 | 0.6 $M$ sorb, 5 m$M$ Ca, 20 m$M$ MES | RT | 4 | 5.5 | 12 |
| Grasses (8) | 2% OC 4S, 0.1% MZ | 0.5 $M$ sorb, 1 m$M$ Mg, 1 m$M$ phosphate | 30 | 3 | 5.5 | 13 |
| Oat | 0.5% C | 0.6 $M$ man | 33 | 2 | 5.7 | 14 |
| Rye | 4% MC, 1% MA | 0.7 $M$ sorb or man, 1% K-DS | 25 | 17 | 5.8 | 15 |
| Sugar cane | 1% MA, 4% cellulase | 0.3 $M$ suc, 10 m$M$ citrate | 23 | 35 | 5.6 | 16 |
| Tulip | 2% C | 0.6 $M$ man | 25 | 20 | 5.8 | 17 |
| Other tissues | | | | | | |
| Barley aleurone | 7% OC | 0.7 $M$ sorb, NS | 35 | 3 | 5.0 | 18 |
| | then 50% G | 1.0 $M$ sorb, NS | 35 | 3 | — | |
| Barley callus | 5% OC P1500, 2% MZ | 0.3 $M$ man, 0.3 $M$ sorb | — | 3 | 5.7 | 19 |
| | then 0.25% D, 0.25% SE | 0.3 $M$ Man, 0.3 $M$ sorb | — | 16 | 5.7 | |
| Brome grass culture | 2% D, 1% P | 0.35 $M$ sorb 0.35 $M$ man, 7 m$M$ Ca | RT | 6 | 5.5 | 20 |
| Corn endosperm callus | 5% OC CLA696, 2% MZ | 0.6 $M$ man 0.01% DS | 37 | 3.5 | 5.4 | 21 |
| Corn and oat root tips | 10% OC P1500, 5% MZ | 0.56 $M$ suc | 22 | 24 | — | 22 |
| Lily micro-sporocytes | 4% OC CLA565, 2% MZ | 0.3 $M$ suc | 25 | 0.5 | — | 23 |
| Oat coleoptile | 2% SE | 0.5 $M$ man | RT | 1.5 | 6.0 | 24 |

TABLE I—Continued

| Plant group | Enzyme(s)[a] | Osmoticum[b] | Temp[c] (°C) | Time (hr) | pH | Reference |
|---|---|---|---|---|---|---|
| Oat coleoptile | 6% OC SS, 4% MZ, 0.4% helicase | 0.55 $M$ sorb | 25 | 4 | 5.7 | 25 |
| Onion guard cells | 4% C | 0.23 $M$ man | — | 6 | 5.0 | 26 |
| Rice callus | 0.2% CN, 5% OC 4S | 0.4 $M$ man 0.05% DS | 35 | — | 5.2 | 27 |
| Tulip petals | 2% C | 0.6 $M$ man | 25 | 20 | 5.8 | 17 |
| Wheat pollen tetrads and mother cells | 0.75% helicase | 0.27 $M$ suc | 24 | 1 | — | 28 |
| **Solanaceae** | | | | | | |
|   | | | | | | |
| **Leaf tissue** | | | | | | |
| Petunia | 2% SP then | 0.4 $M$ man | 26 | 0.5 | 5.4 | 29 |
| | 2% OC SS | 0.4 $M$ man | 26 | 1 | 5.8 | |
| Petunia | 1.2% OC P1500, 0.4% MZ | 0.7 $M$ man, NS | 26 | 18 | 5.8 | 30 |
| Potato | 1% MZ, 5% OC R-10 | 0.3 $M$ suc, NS | 28 | 4 | 5.6 | 31 |
| Tobacco | 0.5% MZ then | 0.8 $M$ man, 0.3% DS | 25 | 2 | 5.8 | 32 |
| | 2% OC P1500 | 0.8 $M$ man | 36 | 3 | 5:4 | |
| Tobacco (9 species) | 0.02% MZ R-10, 0.1% OC R-10, 0.05% D | 0.55 $M$ man, NS | 22 | 15 | — | 33 |
| Tomato | 0.5% D | 0.66 $M$ man | 25 | 15 | 5.8 | 9 |
| Tomato | 4% MC, 0.4% MZ, 0.25% D | 0.5 $M$ man, NS | 29 | 18 | 5.8 | 34 |
| **Other tissues** | | | | | | |
| *Datura* pollen tetrads | Snail juice and cellulase | 0.3 $M$ suc, 50 m$M$ acetate | 28 | 2 | 5.4 | 35 |
| Potato tuber | 2% MZ, 5% SL, 1.5% helicase then | 0.44 $M$ suc | 29 | 1 | — | 36 |
| | 5% Merck cellulase | 0.44 $M$ suc | 29 | 4 | — | |
| Tobacco epidermis | 1.5% OC R-10 0.5% MZ, 1.0% R HP150 | 0.65 $M$ man | 20 | 5 | 5.8 | 37 |
| Tobacco pollen tetrads and mother cells | 0.75% helicase | 0.27 $M$ suc | 24 | 1 | — | 28 |

(*continued*)

TABLE I—Continued

| Plant group | Enzyme(s)[a] | Osmoticum[b] | Temp[c] (°C) | Time (hr) | pH | Reference |
|---|---|---|---|---|---|---|
| Tobacco tissue culture | 1% OC, 0.2% MZ | 0.4 to 0.7 M man | 27 | 3 | 5.2 | 8 |
| Tomato fruit | 20% PL R-10 | 0.59 M suc | 27 | 2 | — | 38 |
| Tomato root | 5% SE | 0.6 M suc, 20 mM phosphate | 27 | 2 | 6.0 | 1 |
| **Leguminoseae** | | | | | | |
| Leaf tissue | | | | | | |
| Bean | 0.25% D, 0.25% P | 0.3 M man | 24 | 15 | 7.0 | 39 |
| Cowpea | 4% MC, 0.4% MZ | 0.71 M man, NS | 27 | 16 | 5.6 | 40 |
| Pea | 1% D, 1% R HP150, 0.5% P | 0.55 M sorb, NS | — | 3.5 | 5.5 | 6, 41 |
| Other tissues | | | | | | |
| Pea apices | 1% D, 1% OC P1500, 0.05% P, 0.5% HC | 0.25 M sorb, 0.25 M man, 6 mM Ca | 26 | 4.5 | 6.2 | 42 |
| Pea roots | 5% OC SS, 2% MZ | 0.6 M man, 0.02 M suc, NS | 25 | 17 | 7.0 | 43 |
| Soybean culture | 1% OC, 0.5% HC | 0.55 M sorb | RT | 4 | 5.5 | 6 |
| Soybean root nodule | 1.5% OC R-10, 0.5% MZ, 1% R HP150 | 0.65 M man | 20 | 5 | 5.8 | 44 |
| Soybean root nodule | 5% D | 0.6 M man, 3.5 mM Ca, 0.7 mM phosphate | RT | 3 | 5.8 | 45 |
| *Vicia* callus | 1% OC SS, 0.5% P, 0.125% R | 0.25 M man, 0.25 M sorb, 6 mM Ca | RT | 5 | 5.7 | 46 |
| **Convolvulaceae** | | | | | | |
| Leaf tissue | | | | | | |
| *Ipomea* and *Calystegia* | SE (snail gut) | 0.63 M man | 25 | 14 | 6.3 | 47 |
| Other tissue | | | | | | |
| *Convolvulus* culture | 4% SE | 0.14 M KCl, 0.10 M CaCl₂ | RT | 2 | — | 48 |
| *Pharbitis* cotyledon | 2% SP then | 0.3 M man | 28 | 0.3 | 5.4 | 49 |

TABLE I—Continued

| Plant group | Enzyme(s)[a] | Osmoticum[b] | Temp[c] (°C) | Time (hr) | pH | Reference |
|---|---|---|---|---|---|---|
| | 3% OC SS then | 0.3 $M$ man | 28 | 1.3 | 5.4 | |
| | 2% SE | 0.3 $M$ man | 28 | 1.6 | 5.4 | |
| Compositae | | | | | | |
| Sunflower leaf | 0.5% pectinase, 5% cellulase | 0.59 $M$ suc | — | 4 | 5.4 | 50 |
| Other tissues | | | | | | |
| *Haplopappus* culture | 5% OC P1500 | 0.5 $M$ sorb, 0.12 $M$ suc | 30 | 4 | 5.5 | 51 |
| Jerusalem artichoke | 8% OC, 4% MZ | 0.5 $M$ suc | 25 | 20 | 5.5 | 52 |
| Other dicot families | | | | | | |
| Leaf | | | | | | |
| Begonia | 0.5% MZ, 2% OC 4S | 0.5 $M$ man | 35 | — | 5.8 | 27 |
| *Bryophyllum* | 1.5% MZ then | 0.7 $M$ man, 25 m$M$ MES | 25 | 2.5 | 5.5 | 53 |
| | 1.5% OC SS | 0.7 $M$ man, 25 m$M$ MES | 36 | 2.5 | 5.5 | |
| Radish | 2% MZ then | 0.7 $M$ man, DS | 35 | — | 5.8 | 27 |
| | 2% OC 4S | 0.7 $M$ man, DS | 35 | — | 5.8 | |
| Rape | 0.5% OC, 0.5% R HP150 then | 0.25 $M$ man, 0.25 $M$ sorb, 7 m$M$ Ca | — | 2 | 6.2 | 54 |
| | 0.5% D, 0.5% hemicellulase | 0.25 $M$ man, 0.25 $M$ sorb, 7 m$M$ Ca | — | 3 | 6.2 | |
| Snapdragon | 0.5% MZ, 2% OC SS | 0.7 $M$ man | 32 | 2 | 5.5 | 55 |
| Other tissues | | | | | | |
| *Acer* cell culture | 2.5% OC, 2.5% MZ | 0.45 $M$ sorb | — | 6 | 5.1 | 56 |
| *Ammi visnaga* culture | 0.5% OC P5000, 0.25% R HP150 | 0.56 $M$ man, NS | — | 6 | — | 57 |
| Carrot culture | 2% D | 0.4 $M$ sorb | 28 | 4 | — | 58 |

*(continued)*

TABLE I—Continued

| Plant group | Enzyme(s)[a] | Osmoticum[b] | Temp[c] (°C) | Time (hr) | pH | Reference |
|---|---|---|---|---|---|---|
| Citrus ovular callus | 1% KP, 1% OC R-10 | 0.14 *M* suc, 0.28 *M* man, 0.28 *M* sorb | 27 | 3 | 5.7 | 59 |
| Cotton fibers | C, MA | — | RT | 3 | — | 60 |
| Crown gall (*Parthenociss us*) | 2% OC R-10 0.01% MZ R-10 | 0.71 *M* man | 22 | 16 | 5.6 | 61 |
| Cucumber cotyledons | 0.5% P, 1.5% D | 0.55 *M* man, 3 m*M* MES buffer | 25 | 3 | 5.8 | 62 |
| Flax hypocotyls | 0.5% D, 0.5% P, 1% OC | 0.25 *M* sorb, 0.25 *M* man, 6 m*M* Ca | — | 1 | 6.2 | 63 |
| | then 0.5% D, 0.5% P | 0.25 *M* sorb, 0.25 *M* man, 6 m*M* Ca | — | 4 | 6.2 | |
| *Nemesia* petals | 2% SP, 0.5% OC SS | 0.6 *M* man | 22 | 3 | 5.6 | 64 |
| Petals of various plants | 2% C | 0.6 *M* man | 25 | 20 | 5.8 | 17 |
| *Torenia* petals | 1% SC, 1% SP | 0.3 *M* sorb, 0.05 *M* suc | — | 0.5 | 5.8 | 65 |
| | then 2% Merck cellulase | 0.3 *M* sorb, 0.05 *M* suc | — | 0.5 | — | |

[a] Enzymes: C, cellulysin; CN, colonase; D, driselase; G, glusulase; HC, hemicellulase of Rohm and Haas; KP, Koch-Light pectinase; MA, macerase; MC, meicelase; MZ, macerozyme; OC, Onozuka cellulase; P, Sigma pectinase; PL, Rohm & Hass pectinol; R, rhozyme; SC, Serva cellulase; SE, special enzyme prepared in the lab using it; SL, Sigma cellulase; SP, Serva pectinase. Other letters or numbers refer to the particular type of enzyme preparation used, e.g., Onozuka cellulase has been produced as several different types; SS, 4S, etc.

[b] Osmotica: Ca, CaCl₂; DS, dextran sulfate; man, mannitol; NS, nutrient salts; sorb, sorbitol; suc, sucrose.

[c] RT, Room temperature.

trouble. The most complete study of the problem[66] found that to release good protoplasts tobacco plants had to be growing rapidly, in light of 10,000 to 30,000 lux, with regular fertilization. Plants from 40 to 60 days of age gave the best yield and any interruptions of growth decreased subsequent protoplast yields. The presence of starch from very high light intensities or from advanced leaf age made the protoplasts extremely fragile. In another case, light had to be totally excluded from a plant prior to protoplast release. Protoplasts from pea leaves were found to be stable only after the intact plants had been kept in the dark for at least 30 hr, with an optimum time of 100 hours.[4]

Sometimes leaf tissues yield better protoplasts if they are slightly wilted before being placed in releasing medium. Tissues with an epidermis release protoplasts more readily if it is first peeled off. An enzymatic method of removing the cuticle[67] has not been generally used. Sometimes protoplasts are more stable if the tissue is allowed to equilibrate for up to 1 hr in the osmoticum before enzymes are added. Presumably this prevents the uptake of exogenous enzymes into the cytoplasm during the plasmolysis process as well as eliminating osmotic shock at the time that enzymes are added. To facilitate subsequent protoplast culturing experiments, most plant tissues can be sterilized by 30 sec in 70% ethanol followed by 30 min in 2% sodium hypochlorite. After being rinsed in sterile distilled water, such tissues release protoplasts in the usual fashion.

The use of tissue culture cells has the important advantage that the cells are already axenic, and protoplasts can subsequently be cultured without contamination. Pretreatments of such tissue cultures prior to adding enzymes can alter protoplast release. With a *Haplopappus* culture, 5 mg/liter 2,4-D or NAA, 0.05 $M$ sucrose, and the presence of 0.5 m$M$ cysteine, methionine, or mercaptoethanol were found to optimize the release of viable protoplasts.[68]

Enzyme Choice

As indicated by Table I, a number of enzyme preparations have proven useful for releasing protoplasts. The sources of those most commonly used are shown in Table II.

In every case, almost certainly a mixture of enzyme activities rather than a single enzyme is important. Consistent product quality is not

---

[66] J. W. Watts, F. Motoyoshi, and J. M. King, *Ann. Bot. (London)* [N.S.] **38**, 667 (1974).
[67] L. Schilde-Rentschler, *Z. Naturforsch., Teil B* **27**, 208 (1972).
[68] A. Wallin, K. Glimelius, and T. Eriksson, *Physiol. Plant.* **40**, 307 (1977).

TABLE II

SOURCES OF WALL-DEGRADING ENZYMES

| Enzyme | Source |
|---|---|
| Pectinases | |
|   Colonase | Wakamoto Pharmaceutical Co., Ltd., Japan |
|   Pectinase | Sigma Chem. Co., St. Louis, Missouri |
|   Macerase | Calbiochem, San Diego, California |
|   Macerozyme | All Japan Biochemicals Co., Nishinomiya |
|   Rhozyme or pectinol | Rohm and Haas, Philadelphia,, Pennsylvania |
| Cellulases | |
|   Cellulysin | Calbiochem, San Diego, California |
|   Onozuka cellulase | All Japan Biochemicals Co., Nishinomiya |
|   Driselase | Kyowa Kakko USA, Lefcourt National Building, 521 Fifth Ave., New York, New York |
|   Meicelase | Meiji Seika Kaisha Ltd., Tokyo |
| Other | |
|   Hemicellulases: Rhozyme HP150 | Rohm and Haas, Philadelphia, Pennsylvania |
|   Glusulase | Endo Labs, Garden City, New York |

maintained when companies produce these enzymes. Therefore, whenever possible it is wise to test several lot numbers of an enzyme before buying a supply. If a particularly good lot is found, enough should be purchased for several years. When kept dry and refrigerated, the polysaccharide-degrading enzymes are quite stable with time.

For some tissues, particularly for tissue cultures, it is necessary to desalt enzyme preparations before using them. Even the interior of protoplasts may apparently be affected, since ATPases of the tonoplast are active only if the initial protoplast release was carried out with desalted enzyme.[69] The desalting is best carried out with a polyacrylamide gel column and not with dialysis tubing or a dextran gel, which are sugar derivatives and may bind or inactivate essential enzymes. The number of enzyme preparations that are sometimes mixed together to release healthy protoplasts indicates a surprising lack of toxicity in the enzyme preparations of Table II. Even the pectinases that are integral parts of the activity of pathogens on host plants kill the host cells not by poisoning the cytoplasm in some way but instead simply by digesting away the cell wall so that the plasma membrane is ruptured by the internal turgor pressure.[70]

[69] W. Lin, G. J. Wagner, H. W. Siegelman, and G. Hind, *Biochim. Biophys. Acta* **465,** 110 (1977).

[70] D. F. Bateman, *in* "Biochemical Aspects of Plant-Parasite Relationships" (J. Friend and D. R. Threlfall, eds.), p. 79. Academic Press, New York, 1976.

Most importantly, it is apparent that even though the walls of many higher plant cells are thought to have relatively similar structures,[71] the enzyme preparations that give optimum protoplast release are quite different. The enzyme preparation of choice for a particular situation can be determined only by monitoring the release of protoplasts from that system.

## Incubation Conditions

Many tissues have abundant air spaces that slow the penetration of digestive enzymes. An initial vacuum infiltration can help overcome this problem, though one must be careful not to damage cells by drawing too complete a vacuum or snapping it too abruptly when readmitting air. In general, the incubation temperature should be as high as possible consistent with avoiding damage to the tissue (thus about 30°), and the digestion time should be as short as possible. Gentle agitation often hastens release of the protoplasts, serving both to stir the enzyme solution so as to bring new enzymes in contact with the cell walls and to provide some physical force to help release protoplasts from within cell walls. Intermittent agitation will provide both of the above benefits while proving less damaging than continuous agitation to the protoplasts that are released.

## Incubation Time

The obvious goal is to minimize the digestion time, or essentially to strike the best compromise between the need for wall digestion and the problem of protoplast deterioration with increased time in the enzymes. In some cases, it has been possible to separate one type of cell from another by means of utilizing a digestion time that removes the wall from one and not the other. This technique has been most useful in separating mesophyll protoplasts from bundle sheath strands in the leaves of $C_4$ plants.[13] With the most commonly used tissue, tobacco leaves, it is possible to separate spongy mesophyll protoplasts from palisade parenchyma protoplasts on the basis of digestion time.[32]

Though polysaccharide-degrading enzymes are quite stable at room temperature, they are inactivated relatively quickly by exposure to native cell walls. This probably results in large part from their binding to sugar moieties in the cell wall that are very similar to the normal substrates yet not capable of being hydrolyzed so that the enzyme can be released. In many cases it has proved beneficial to renew the enzyme solution halfway

[71] D. Burke, P. Kaufman, M. McNeil, and P. Albersheim, *Plant Physiol.* **54,** 109 (1974).

through the protoplast release process, sometimes replacing the preparation with more of the same and sometimes with a different mixture.

## The Osmoticum

As can be seen in Table I, a variety of osmotic solutions have been used for releasing protoplasts. In many cases, it is possible to detect differences in protoplast release when comparing even such closely related osmotica as sorbitol and mannitol. Surprisingly, the uptake of leucine across the plasma membrane has been shown to be very similar in all the standard osmotica tested.[72] The osmotic potential of the release medium is usually adjusted so that the cells are just slightly plasmolyzed during wall digestion. With less solute present, many protoplasts will rupture during the release process. With more solute present, there is a marked reduction in both the wall regeneration process[73] and the uptake of leucine across the plasma membrane.[72] Temporary transfers to osmotica consisting of $NaNO_3$[22] or polyethylene glycol[74] strongly promote the fusion of protoplasts as is described in detail in another article in this series.[74a]

## Stabilizing Protoplasts

The presence of the appropriate osmotic concentration is vital to the release of stable protoplasts. Certain other factors can, however, provide a stabilizing influence. Magnesium, or especially calcium, at 0.1 to 10 m$M$ has often been used to enhance protoplast stability. Though neither appears often in Table I, a divalent cation is frequently added as soon as wall digestion is complete. In many cases, the pH is also modified upward as the protoplasts are washed free from enzyme, moving it from the 5.5 of the digestion process to the region from 6.5 to 7.0. Dextran sulfate (MW about 3000 and concentration about 0.3%) has often been beneficial, expecially for the first (pectinase) half of a two-step digestion process.[32] Larger dextran sulfates often have a deleterious rather than a beneficial impact. Two recent reports from Galston's laboratory suggest other additions that can enhance protoplast viability: (1) Cycloheximide pretreatment of excised leaves for 18 hr at 1 $\mu$g/ml promoted both protoplast viability and leucine incorporation by the isolated protoplasts.[14] (2) The

---

[72] A. Ruesink, *Plant. Physiol.* **44**, 48 (1978).
[73] R. S. Pearce and E. C. Cocking, *Protoplasma* **77**, 165 (1973).
[74] K. N. Kao and M. R. Michayluk, *Planta* **115**, 355 (1974).
[74a] A. Ruesink, Vol. 58, p. 359.

addition of 1 to 10 m$M$ arginine, lysine, or polyamine (cadaverine, pu-
trescine, spermidine, or protamine sulfate) markedly inhibited the rise in
RNase activity that usually accompanies protoplast release[75] Whether
these methods have general applicability remains to be seen.

### Cleaning up the Protoplasts

Once cell walls have been adequately digested, it is necessary to
separate healthy protoplasts from fragments of intact tissue, bits of undigested wall, cytoplasmic debris from broken cells, and the wall-degrading enzymes. As soon as a sufficient number of protoplasts have been
released, the digestion solution is typically diluted with more of the
osmoticum (without enzymes) and protoplasts are gently centrifuged
down, e.g., 3 min at 40 $g$. Swinging buckets should always be used for
centrifuging protoplasts to minimize damage at the tube walls. The supernatant is discarded, and the protoplasts are carefully resuspended in
fresh osmoticum and spun down again. This procedure can be repeated
a number of times, eliminating more and more enzymes. Since protoplasts
are quite fragile, however, some are lost upon each washing and a point
of diminishing returns is soon reached.

Other methods have been devised to clean up the protoplasts further.
The cytoplasmic debris and cell wall fragments around carrot protoplasts
stick to glass wool, while the protoplasts do not. Thus, a single passage
of a protoplast preparation through a 10-mm thickness of glass wool
packed into a funnel cleans them up considerably.[75a] Flotation can also
be used to purify protoplasts, causing less breakage than downward
centrifugation, since protoplasts are not thrown against the walls of the
tubes. One such method is to isolate protoplasts in sucrose alone (in
which many kinds of protoplasts will float) and then overlayer the sucrose
with successive layers containing decreasing sucrose and increasing sorbitol, keeping the total osmotic strength constant while the density decreases.[76] The synthetic polymer Ficoll (Pharmacia Fine Chemicals,
Inc.), formed by the copolyermization of sucrose and epichlorohydrin,
has also been used for preparing density gradients to collect good protoplasts.[77,77a] Since Ficoll alters density in much the same way that a
similar concentration of sucrose does and yet has little impact on
osmotic potential of the solution, discontinuous density gradients with

[75] A. Altman, R. Kaur-Sawhney, and A. W. Galston, *Plant Physiol.* **60**, 570 (1977).
[75a] W. Boss and A. Ruesink, *Plant. Physiol.* (in press, 1979).
[76] Y. Chupeau and G. Morel, *C. R. Hebd. Seances Acad. Sci., Ser. D* **270**, 2659 (1970).
[77] P. J. Larkin, *Planta* **128**, 213 (1976).
[77a] A. Ruesink, unpublished.

Ficoll as the only varying substituent can readily be prepared with the osmoticum held constant throughout the gradient. Protoplasts can be either layered on the bottom in the highest Ficoll concentration or layered on top in a Ficoll-free layer. Centrifugation for 5 min at forces up to 500 $g$ will usually move all protoplasts to a region of equilibrium density. With proper choice of Ficoll concentrations, which varies depending on the source of protoplasts, one can separate good protoplasts from all other components in the release medium. There are two cautions: (1) Following a centrifugation through Ficoll, the behavior of mitochondrial and plasma membranes on a standard sucrose density gradient was altered,[78] suggesting the Ficoll may have a long-term effect on membranes; and (2) Centrifugation at forces in excess of 500 $g$ may throw heavy organelles such as starch grains right out of the sides of protoplasts, leaving behind atypical cells.

An unusual two-phase system has been used to clean up protoplasts from several plants.[79] Protoplasts are prepared by a standard enzymatic means and washed once. They are then stirred quickly into a mixture containing the following final concentrations: 5.5% polyethylene glycol 6000, 10.0% dextran $T_{40}$ (Pharmacia), 10 m$M$ phosphate, 460 m$M$ sorbitol, 0.5 m$M$ MgCl$_2$, and 0.5 m$M$ Tricine, pH about 7.7. After centrifugation at 300 $g$ for 6 min, the protoplasts accumulate at the interface between the two phases that separate in the tube, while chloroplasts and broken protoplasts are suspended in the lower phase.

### Determining Protoplast Viability

If protoplasts will regenerate a wall and/or divide, they are obviously alive, but monitoring such processes quantitatively is time consuming and tedious. Careful inspection of some kinds of protoplasts for normal for normal cyclosis, also laborious, yields another estimate of viability,[24] but the most convenient and reliable estimate of viability can be obtained by using vital dyes. After screening 27 dyes, Widholm[80] concluded that only one could be used to stain living cells specifically, and that one (fluorescein diacetate) could be monitored only with a fluorescence microscope. The dye is concentrated from a 0.01% solution in 5 min. On the other hand, several dyes can be used to determine which cells in a protoplast preparation are nonviable. Those that have been used include

[78] R. T. Leonard and W. J. VanDerWoude, *Plant Physiol.* **57,** 105 (1976).
[79] R. Kanai and G. E. Edwards, *Plant Physiol.* **52,** 484 (1973).
[80] J. M. Widholm, *Stain Technol.* **47,** 189 (1972).

0.1% phenosafranine, 0.5% Evans blue, and 0.1% trypan blue. Within a few minutes after one of these dyes dissolved in the appropriate osmoticum has been added to a protoplast preparation on a microscope slide, one can see significant dye uptake by nonliving material under a brightfield microscope, while living cells remain unstained.

Isolation of vacuoles

The isolation off intact vacuoles from plant tissues has proved very difficult, but three recent methods have been described. In 1976, Wagner and Siegelman[17] reported the use of phosphate at pH 8 with gentle stirring as an appropriate method for releasing vacuoles from the protoplasts of various flower parts, with petal and certain leaf protoplasts giving the best results. The method currently being used by that laboratory[80a] involves starting with protoplasts released into 0.6 $M$ mannitol. The protoplasts are pelleted with a 70 $g$ spin and exposed quickly to 40 ml of 0.2 M $Na_2HPO_4$ or $K_2HPO_4$ in 0.5 m$M$ dithiothreitol, all adjusted to pH 8.0 with HCl. The phosphate solution is injected into the tube containing the pellet using a 50-ml syringe with a 5-mm opening of the catheter type tip. The suspension is stirred slowly and poured through plastic screen (1-mm mesh) onto two layers of glass wool in a funnel. Particulate materials are retained on the screen or glass wool, while the vacuoles pass through. The suspension is centrifuged for 2 min at 25 $g$ to remove heavy debris. Particulate matter (but not vacuoles) will stick to a wooden applicator stick used to stir some more. Pigmented petal vacuoles can be sedimented by 3 min at 100 $g$ while leaf vacuoles require 1100 $g$. The vacuole pellet can be washed by carefully transferring it with a Pasteur pipette, with stirring, into 40 ml of 0.7 $M$ mannitol, 5 m$M$ dithiothreitol (DTT), and 1 to 10 m$M$ HEPES at pH 8.0.

The second method of vacuole preparation uses strictly mechanical methods and was developed for beet roots.[81] Fresh beet roots are sliced by either a razor blade or a specially designed cutter into 2 $M$ sucrose (razor blade) or 1 $M$ sorbitol (cutter) containing 5 m$M$ EDTA, 0.1 mg/ml sodium 2-mercaptobenzothiazole, and 50 m$M$ Tris-HCl at pH 7.6. Subsequent operations at 4° include centrifuging down the vacuoles (2000 $g$ for 10 min) and resuspending them in 1.5 $M$ sorbitol, 15% metrizamide, 1 m$M$ EDTA, and 10 m$M$ Tris-HCl at pH 7.6. The resuspended vacuoles are overlaid with successive layers of resuspension medium containing 10, 2.5, and 0% metrizamide and centrifuged at 650 $g$ for 10 min. The

---

[80a] G. J. Wagner, personal communication.
[81] R. A. Leigh and D. Branton, *Plant Physiol.* **58,** 656 (1976).

majority of vacuoles are recovered at the 10–2.5% interface, but recovery amounts to only 2% of the initial pigment present.

Still a third method of vacuole preparation uses a polyanion to rupture the plasma membrane of *Bryophyllum* leaf cells.[53] Protoplasts are released in 0.7 *M* mannitol with 10 m*M* HEPES–NaOH at pH 7.3. DEAE-dextran is added to a concentration of 0.5 mg/ml for 5 × 10⁴ protoplasts/ml. After 15 to 30 sec, an equal amount of dextran sulfate is mixed in to bind free dextran and protect vacuolar membranes. The lysis of protoplasts and release of vacuoles takes place within 15 min at room temperature. The yield of free vacuoles is proportional to the number of those protoplasts that initially have most of their cytoplasm clumped in an aggregate on one side of the vacuole.

Of the three methods, only the one of Wagner and colleagues is claimed to have some general applicability. Even with that method, the isolation of good vacuoles in quantitative yield is apparently limited to tulip leaves and several kinds of petals.[80a] The isolation of vacuoles is by no means as routine as the isolation of protoplasts.

### Protoplast Culture and Fusion

The techniques needed for culturing and fusing protoplasts go beyond the scope of this article. For pertinent information on culturing protoplasts the reader should see "Plant Tissue Culture Methods,"[82] a paper on the optimization of protoplast growth,[83] or a paper on general protoplast culturing.[7] The methodology for the fusion of protoplasts of higher plant cells is considered in another article in this series[74a] and an earlier review exists.[84]

---

[82] O. L. Gamborg and L. R. Wetter, eds. "Plant Tissue Culture Methods." Nat. Res. Counc. Can., Ottawa, 1975.

[83] H. Uchimiya and T. Murashige, *Plant Physiol.* **57**, 424 (1976).

[84] K. N. Kao, *in* "Plant Tissue Culture Methods" (O. L. Gamborg and L. R. Wetter, eds.), p. 22. Nat. Res. Counc. Can., Ottawa, 1975.

# [8] Higher Plant Chloroplasts and Grana: General Preparative Procedures (Excluding High Carbon Dioxide Fixation Ability Chloroplasts)

*By* S. G. REEVES and D. O. HALL

In searching the literature for techniques of chloroplast isolation it soon becomes obvious that there are nearly as many techniques as there are research groups in the field. We will describe a basic technique of chloroplast isolation and modifications and adjustments that can be made. We believe that in all experiments on isolated chloroplasts the original preparation should be as biochemically active as possible, i.e., the chloroplast preparation should resemble as closely as possible the chloroplast *in vivo,* so that in the first instance, before subsequent experimentation, it should be capable of fixing $CO_2$ at high rates, and show high rates of coupled electron transport. The use of chloroplasts of this type might well help eradicate some of the discrepancies found in measurements of electron, proton, and ion flow, and phosphorylation in chloroplasts.

However, the isolation of chloroplasts of high activity is not easy. In the case of spinach, the most commonly used plant material, there is a large seasonal variation in rates of uncoupled electron transport. Using a standard technique for isolation of chloroplasts (see later) and assaying for $NH_4Cl$-uncoupled electron transport under saturating light in an oxygen electrode, we have found rates which varied from 30 $\mu$moles $O_2$/mg chlorophyll/hr to 520 $\mu$moles $O_2$/mg chlorophyll/hr. In our experiments on electron transport we now discard chloroplast preparations that initially show rates of electron transport below 150 $\mu$moles $O_2$/mg chlorophyll/hr, and we prefer preparations that give rates of double this value. We believe that chloroplasts showing low initial rates of electron flow are already partly "aged"[1] and that the state of their membranes and photosynthetic apparatus differs from chloroplasts which show high initial rates.

It is probable that some reports of beneficial additives to chloroplasts (e.g., on chloroplast stability) are merely improving "poor" chloroplasts, and have no effect on fresh, active chloroplasts. Thus the addition of bovine serum albumin (BSA) to fresh chloroplasts has little effect, but it

---

[1] S. G. Reeves and D. O. Hall, *Cell Biol. Int. Rep.* **1**, 353 (1977). See also S. G. Reeves and D. O. Hall, *Biochim. Biophys. Acta* **463**, 275 (1978).

increases rates of electron transport in chloroplasts that have been aged at 20° for 4 hr or frozen and stored in liquid nitrogen for 1 month.[2]

Before discussing actual techniques of chloroplast isolation a description of the types of chloroplasts that can be isolated is probably worthwhile. These are shown in Table I (adapted from Hall[3]). Most of the data compiled in this table were taken from papers relating to spinach, but they are applicable to most species. It is the authors' contention that high rates of coupled electron transport can only be obtained if the chloroplasts are first isolated as type A, i.e., capable of fixing $CO_2$ at high rates.

### Isolation of Spinach Chloroplasts

A general technique for isolation of active spinach chloroplasts is given below. Various points of the procedure and possible modifications are then discussed.

Although the authors do not endorse the idea, it is possible to prepare broken chloroplasts (types C and E) directly, instead of first preparing highly active (type A) chloroplasts and then breaking them as is described below. To prepare chloroplasts of types C and E it is usual to grind the leaves in isotonic salt or sucrose, wash the chloroplasts in hypotonic media, and resuspend in hypotonic media. References to these techniques, and the effects of the various wash steps on the activities of the chloroplasts, can be found in the literature.[4-7]

*Chloroplast Isolation Technique*

This is a standard technique for Type A chloroplasts.[8]
*Grinding medium*
   0.33 $M$ Sorbitol
   10 m$M$ $Na_4P_2O_7$
   4 m$M$ $MgCl_2$
   2 m$M$ Ascorbic acid
   Adjust to pH 6.5 with HCl

[2] P. Morris, G. Nash, and D.O. Hall, unpublished observations.
[3] D. O. Hall, *Nature (London)* **235**, 125 (1972) and J. Barber (ed.) "The Intact Chloroplast," p. 135, Elsevier, Amsterdam (1976).
[4] A. T. Jagendorf, *Plant Physiol.* **37**, 135 (1962).
[5] F. R. Whatley and D. I. Arnon, Vol. 6, p. 308.
[6] S. G. Reeves, D. O. Hall, J. West, *Photosynth., Two Centuries After Its Discovery By Joseph Priestley, Proc. Int. Congr. Photosynth. Res., 2nd, 1971* p. 1357. Junk, N. V., The Hague.
[7] D. O. Hall and S. Hawkins "Laboratory Manual of Cell Biology," p. 148. English Universities Press, London, 1975.
[8] R. McC. Lilley and D. A. Walker, *Biochim. Biophys. Acta* **368**, 269 (1974).

*Resuspending medium*
  0.33 *M* Sorbitol
  2 m*M* EDTA
  1 m*M* MgCl$_2$
  1 m*M* MnCl$_2$
  50 m*M* HEPES
  Adjust to pH 7.6 with NaOH

If broken (type C) chloroplasts are required from this preparation, these can be obtained by resuspending the chloroplast pellet in a "resuspending medium" which contains 0.05 *M* Sorbitol instead of 0.33 *M* Sorbitol.

*Technique*

The grinding medium (300 ml) is cooled (e.g., in a deep freeze) until it reaches the consistency of melted snow. Deribbed, washed spinach leaves (60 gm) are chopped briefly into strips about 5-mm wide with a knife and placed in a cooled, square Perspex grinding vessel (internal measurements 5 × 5 × 20 cm deep) and the grinding medium poured over them. They are macerated by two 1-sec bursts on a Polytron homogenizer at a speed setting of 6 (PT20 with a PT35 head), squeezed through two layers of muslin, and poured through eight more layers. The filtrate is then centrifuged in an MSE Super Minor bench top centrifuge with a precooled 6 × 100 ml angle head. The centrifuge is accelerated to approximately 8000 *g*, run at this speed for 5 sec, and then slowed down by hand. The total centrifugation time should be about 60–90 sec. The supernatant and the soft layer of chloroplasts on top of the pellet are discarded, and the pellet is resuspended in 2–3 ml of cooled resuspending medium with the aid of a *small* piece of cotton wool and a glass rod. All procedures should be carried out as quickly as possible, and all solutions and equipment kept at about 4°. The whole procedure should take less than 5 min and should yield approximately 9 mg chlorophyll.

**Modification and Notes**

The essence of this procedure lies in a fast and cold preparation. Once plant cells are ruptured many proteolytic and other degrading enzymes are released, and it is essential to remove the chloroplasts from this environment as fast as possible. The low temperature decreases any enzyme activities and helps prevent decay of chloroplast activity. The use of half-frozen grinding medium ensures that the temperature of the maceration mixture remains close to zero.

TABLE I

TYPES OF CHLOROPLAST PREPARATIONS AFTER THE CLASSIFICATION OF HALL[3] TOGETHER WITH AN OUTLINE OF THEIR MAIN PROPERTIES

| Chloroplast type | Description | Preparation method | Appearance under phase contrast microscopy | Envelope | Rate of $CO_2$ fixation ($\mu$moles $CO_2$/mg chlorophyll/hr) | Exogenous substrate penetration and requirements | Electron transport and photophosphorylation capacity |
|---|---|---|---|---|---|---|---|
| A | Complete chloroplasts | Rapid, in isotonic or hypertonic sugar, one centrifugation | Outer mobile jacket present. Bright and highly reflecting. Grana not obvious | Intact | 50–250 | NADP and ferricyanide do not penetrate. Slow uptake of ATP, ADP and $P_i$ | Presumed to be unimpaired. (ATP/$2e$ approaching 2.0 when assayed in hypotonic medium; good photosynthetic control) |
| B | Unbroken chloroplasts | In isotonic or hypertonic sugar or salt, with 2 or 3 centrifugations | Bright and highly reflecting. Smooth outline. Grana not obvious ("class I") | Morphologically but not functionally intact | <5 | NADP, ferricyanide and ADP penetrate. Ferredoxin may not be necessary to add for NADP reduction | Good ATP/$2e$ (greater than 1.0 and often approaching 2.0 when assayed in hypotonic medium). Good photosynthetic control |
| C | Broken chloroplasts | Vigorous in isotonic sugar or salt | Not bright in appearance. 2 to 3 times larger than types A and B Grana conspicuous (...) | Broken and usually lost in preparation (stroma also lost) | Little or none | NADP, ferricyanide and ADP penetrate. Ferredoxin needs to be added for NADP... | ATP/$2e$ >1.0 or <1.0 depending on isolation and assay conditions. Some photosynthetic... |

| | | | | | | | |
|---|---|---|---|---|---|---|---|
| D | Free-lamellar chloroplasts | Osmotic shock of type A chloroplasts immediately followed by return to isotonic medium | — | Lost from chloroplasts but retained in medium | High rates if carbon pathway intermediates and chloroplast extract added | — | Good photosynthetic control |
| E | Chloroplast fragments | Resuspend chloroplasts in hypotonic medium | — | Lost | None | Ferredoxin needs to be added for NADP reduction | Higher rates of electron transport and lower rates of photophosphorylation than Types B and C. $ATP/2e < 1.0$. No photosynthetic control |
| F | Sub-chloroplast particles | By sonication or detergent treatment or French press | — | Lost | None | Ferredoxin and plastocyanin (and some reductase) needs to be added for NADP reduction | Photophosphorylation (cyclic only) low or absent. Limited electron transport when electron donors added |

*Maceration*

The maceration can be carried out many different ways. Some years ago extracts were made by grinding leaves in a pestle and mortar. The more recent approach is to use a domestic blender, such as an Atomix, Braun, or Waring, or a more specialized one such as the Polytron described above. Success can be achieved with almost any technique as long as everything is kept at a low temperature, and as long as the process is carried out quickly. Longer grinding periods than that suggested can be used, but the activity of the chloroplast preparation will usually be lower.

*Filtration*

An older technique for separation of cell debris from the chloroplast involved a low-speed centrifugation.[5] In the interest of quickness this has been replaced by filtration, usually through a number of layers of muslin, with or without the addition of cotton wool in between the layers (which is believed to selectively adsorb chloroplast fragments). Muslin (cheesecloth) is generally preferred to nylon of different meshes or similar materials.[9]

*Centrifugation*

The ideal centrifuge for this has rapid acceleration and braking to allow a fast centrifugation. The authors use a bench centrifuge and slow it down by hand. It is obviously less dangerous to have a centrifuge that will do this automatically! In addition a swing-out head is preferable to an angle head, because it reduces degradation of the chloroplasts due to friction as they come into contact with the side wall of the centrifuge tube. Both MSE6L and Griffin Christ centrifuges have been used in this manner. The times of centrifugation given can be extended, but in the authors' experience the shorter times are preferable.

*Osmoticum*

It is now known that in order to extract chloroplasts that will fix $CO_2$ (type A) it is necessary that they retain their outer envelope. As the outer envelope is very fragile it is necessary to use a grinding and resuspending medium that is approximately isoosmotic with the chloroplast in the leaf. The most commonly used osmoticum is 0.33–0.4 $M$ sugar. Sucrose is often used but as this is a metabolizable sugar, sorbitol (a sugar alcohol) has been generally preferred, although mannitol, glucose, and fructose

[9] D. A. Walker, Vol. 23, p. 211.

have been used. Salts (such as 0.35 $M$ NaCl) have also been used, but generally they lead to poorer chloroplast preparations (see Table I and Reeves *et al.* [6]). In order to obtain protein synthesis by isolated chloroplasts it appears necessary to have 0.2 $M$ KCl in the resuspension medium instead of a sugar.[10]

### Buffer

The exact pH of the grinding medium is probably not vital—values between 6.5 and 8.5 have been used—but we think buffering at the lower pH values is preferable. Normally buffers at concentrations of 10 to 50 m$M$ are used. Similarly any buffer can be used that buffers in the correct range, except Tris, which is a mild uncoupling agent. We routinely use pyrophosphate because of its cheapness. Further details of buffers and their usefulness in photosynthetic research have been published elsewhere.[11]

### Ionic Content

The addition of ions to the grinding medium is merely a safeguard. It is unlikely that the ionic content will fall to a low enough level to cause damage to the chloroplast during a normal grinding process.

Similarly, if any of the ions used will interfere with subsequent work (e.g., $Mn^{2+}$ in the case of electron paramagnetic resonance) they can usually be safely omitted from the grinding medium or replaced. A technique has been described[12] that involves a wash step in sorbitol (0.33 $M$) and buffer only. This does not appear to harm the electron transport, and improves the percentage of intact chloroplasts.

### Ascorbic Acid

Often the addition of ascorbic acid to the grinding medium has no effect on the performance of the chloroplasts. However, at some times of the year it can stimulate the activity of a chloroplast preparation. It probably acts as an antioxidant, and thus as a protective agent. As it is unstable, it should be added to the grinding medium just prior to use.

### Resuspension

The resuspension medium described is one suitable for keeping chloroplasts with their outer membrane intact. When these chloroplasts are

[10] G. E. Blair and R. J. Ellis, *Biochim. Biophys. Acta* **319,** 223 (1973).
[11] N. E. Good and S. Izawa, Vol. 24, p. 53.
[12] H. Nakatani and J. Barber, *Biochim. Biophys. Acta* **461,** 510 (1977).

used in electron transport and phosphorylation studies, the outer membrane needs to be ruptured to allow the access of substrates. This can be achieved by either rupturing the membranes in a hypotonic reaction medium or by putting the chloroplasts through a hypotonic wash step. A typical wash would use 4 ml of the standard resuspending medium diluted 1:25 with $H_2O$ to resuspend the chloroplasts. This dilution is followed by a centrifugation at 8000 $g$ for 5 min, and resuspension of the chloroplasts in normal isotonic resuspending medium. The centrifugation times used have varied and can be adapted to the experimenter's purpose. This procedure has the advantage of removing all the stromal proteins, which can interfere with certain experiments. However, it does cause some decrease in the maximum rate of electron transport.

The technique described here has been successfully used to isolate highly active chloroplast preparations from *Spinacia oleracea,* (spinach), *Lactuca sativa* (lettuce), *Pisum sativum* (peas), and *Chenopodium quinoa* in this laboratory.

## Isolation of Chloroplasts from Other Species

Attempts have been made to isolate chloroplasts from many other species, but generally they are poor compared to those of spinach or peas. There appear to be two main problems, that of maceration of the leaves and the presence of polyphenols and polyphenol oxidases.

Tissue maceration can be very difficult in species of grasses. The long vascular strands that run parallel down the leaf cause problems in disrupting the leaves. Cutting the leaf blades into small (1 cm) squares can partially alleviate this problem, as can the use of an electric knife adapted so that razor blades can be fitted.[13]

Another technique of maceration has to be used to separate bundle sheath and mesophyll cells in $C_4$ plants (see [6], this volume). A differential grinding technique has been successfully used by Woo et al. [14] with *Zea mays* leaves. A short maceration was used to produce mesophyll chloroplasts, which were separated from the cell debris by filtration. A second grinding removed any remaining mesophyll chloroplasts, leaving bundle sheath cells attached to vascular strands. (The exact details of these macerations were changed from species to species.) A final preparation of bundle sheath chloroplasts was obtained by homogenizing in a mill with glass beads.

[13] C. A. Mitchell and C. R. Stocking, *Plant Physiol.* **55**, 59 (1975).
[14] K. C. Woo, J. M. Anderson, N. K. Boardman, W. J. S. Downton, C. B. Osmund, and S. W. Thorne, *Proc. Natl. Acad. Sci. U.S.A.* **67**, (1970).

The problems of the presence of polyphenols and polyphenol oxidases is more complicated. Baldry et al. [15,16] have studied the effects of polyphenols on photochemical and enzymic activities in isolated chloroplasts. These effects could be partially reversed by addition of thioglycolate, $\beta$-mercaptoethanol, polyethylene glycol, BSA, polyvinylpyrrolidone, or Polyklar AT, and the addition of these compounds to the grinding medium can mitigate some of the deleterious effects. Other useful techniques include low $O_2$ tension, reducing compounds (e.g., dithiothreitol), copper chelators (e.g., EDTA) and high pH, but the usefulness of all these procedures has to be balanced against any deleterious effects on the actual chloroplast isolation itself.

### Bulk Extraction of Chloroplasts

Obviously the techniques described above are unsuitable for extraction of the large quantities of chloroplasts necessary for certain procedures such as extracting large amounts of chloroplast coupling factor, or photosystem I and photosystem II particles. For reasons of expense, in large preparations the media can be simplified, e.g. 0.4 $M$ sorbitol, 10 m$M$ $Na_2P_4O_7$, 10 m$M$ NaCl, pH 6.5, for grinding and 0.4 $M$ sorbitol, 50 m$M$ tricine, pH 7.5, for resuspending. Tricine is much cheaper than HEPES; Tris is cheaper still but is not as desirable since it is a mild uncoupler. Sorbitol and sucrose are similar in price. NaCl as an osmoticum, as mentioned earlier, is not considered desirable. If precautions are taken as described earlier, keeping all solutions cold, and keeping the tissue–grinding medium ratio the same, it is still possible to retain much of the original activity. Again we would stress that however much the final product from the chloroplasts requires a destructive (degradative) procedure, it is still important to start with (a) the best leaf material possible, and (b) the most active chloroplasts possible.

### Isolation of Chloroplasts from Protoplasts

One underexploited technique for the isolation of functional chloroplasts, especially from species that pose problems using normal techniques, is by use of protoplasts as the starting material. There have been reports in the literature on this process,[17,18] and when used properly

[15] C. W. Baldry, C. Bucke, J. Coombs, and D. Gross, *Planta* **94**, 107 (1970).
[16] C. W. Baldry, C. Bucke, and J. Coombs, *Planta* **94**, 124 (1970).
[17] C. K. M. Rathnam and G. E. Edwards, *Plant Cell Physiol.* **17**, 177 (1976).
[18] M. Nishimura, D. Graham, and T. Akazawa, *Plant Physiol.* **88**, 309 (1976).

yields are comparable with other techniques; percentage intactness and membrane characteristics have been reported to be superior to those of mechanically isolated chloroplasts.[17] However, as yet, no thorough biochemical analysis has been carried out. The difficulty with this technique lies in the preparation of the protoplasts themselves, but as more work is carried out this may well present less of a problem (see [7], this volume).

# [9] Preparation of Higher Plant Chloroplasts

## By D. A. WALKER

The separation of intact chloroplasts from spinach (and some other $C_3$ species) was considered at some length in an earlier article in this series[1] but continues to present problems. The present account therefore concentrates on the possible sources of difficulty and makes mention of some relevant matters which have become clearer during the past six years. The isolation of protoplasts and chloroplasts from protoplasts[2,3] is too large a subject to be considered in this context, but it should be noted that fully functional chloroplasts may be readily separated from spinach protoplasts[2] and that in $C_4$ species there is no real alternative to the inclusion of this step if bundle sheath and mesophyll chloroplasts are to be separated without appreciable cross-contamination (see [6] and [7], this volume).

At the outset, attention should be drawn to two aspects which may appear self-evident but which are frequently overlooked. The first is that it is not possible to prepare "good" chloroplasts from "bad" material, and there are obvious advantages to be derived by starting with leaves capable of active photosynthesis. The second fact is even more obvious but for all that it has taken years to elucidate. It is simply that "good" chloroplasts will not perform well if they are badly treated. Thus, it is clear, in retrospect, that (apart from preliminary separation of protoplasts) there have been few real advances in the initial extraction of intact chloroplasts since the return in 1963 to Hill's custom of using a sugar as an osmoticum.[4] This together with brief grinding and rapid centrifugation[4]

---

[1] D. A. Walker, Vol. 23, p. 211.
[2] M. Nishimura, D. Graham, and T. Akazawa, *Plant Physiol.* **58**, 309 (1976).
[3] C. K. M. Rathnam and G. E. Edwards, *Plant Cell Physiol.* **17**, 117 (1976).
[4] D. A. Walker, *Biochem. J.* **92**, 22c (1964).

led to photosynthetic performances which were a great improvement on those which had previously been achieved but still fell short of those displayed by the parent tissue. Rates of more than 100 $\mu$moles $CO_2$ fixed/ mg Chl/hr (marginally greater than the parent leaf in air levels of $CO_2$ at 20°) were not recorded until 1966.[5,6] They did not then result from further improvements in isolation but simply from better treatment in assay. Improvement in assay was partly a consequence of the employment of Good's buffers[7] but, in the main, seems to have derived from inclusion of inorganic pyrophosphate in the assay medium.[6]

*Material.* Given a free choice, spinach still reigns supreme for a variety of reasons[1] but principally because of its lack of contaminating phenolics and the fact that there is already an enormous literature relating to chloroplasts from this species. For some purposes peas are a convenient alternative, although it is becoming increasingly clear[8-10] that chloroplasts isolated from young pea shoots have different permeability characteristics to those from mature spinach. Lettuce has been used extensively by Avron and others but mainly for work on electron transport and photophosphorylation.

*Culture.* Although true spinach (*Spinacia oleracea*) is an easy plant to use, it is a difficult plant to grow. In some parts of North America, Australia, India, and southern Europe it may be successfully cultivated in the field even in the height of summer, but in more northern latitudes (as in the United Kingdom) it produces virtually no leaf before flowering in long days and is grown, for experimental purposes, under conditions in which the photoperiod is limited to 8½ hr. In the absence of devices for shortening the day length, some extension of the effective growing season may follow the utilization of long-day varieties. It is doubtful whether any particular variety can be recommended for all environments. U.S. Hybrid 424 (Ferry-Morse Seed Co., P. O. Box 100, Mountain View, California 94042) which is also sold as Hybrid 102 Yates (244-254 Horsley Road, Milperra, N.S.W. 2214, Australia or The Seed Centre, Withyfold Drive, Macclesfield, Cheshire, United Kingdom) is evidently very useful in many circumstances. For example, the highest rates of carbon assimilation ever recorded (350 $\mu$moles/mg chlorophyll/hr at 20° and better) were obtained by Ulrich Heber using "Yates 102" grown in water culture

[5] C. Bucke, D. A. Walker, and C. W. Baldry, *Biochem. J.* **101**, 636 (1966).
[6] R. G. Jensen and J. A. Bassham, *Proc. Natl. Acad. Sci. U.S.A.* **56**, 1095 (1966).
[7] N. E. Good, G. D. Winget, W. Winter, T. N. Conolly, S. Izawa, and R. M. M. Singh, *Biochemistry* **5**, 467 (1966).
[8] S. P. Robinson and J. T. Wiskich, *Plant Physiol.* **58**, 156 (1976).
[9] S. P. Robinson and J. T. Wiskich, *Plant Physiol.* **59**, 422 (1977).
[10] Z. S. Stankovic and D. A. Walker, *Plant Physiol.* **59**, 428 (1977).

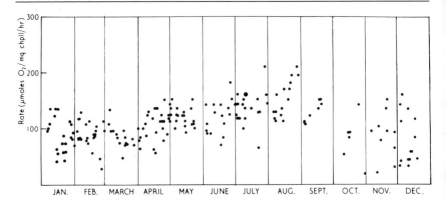

FIG. 1. Rates of $CO_2$-dependent $O_2$ evolution by intact spinach chloroplasts. Chloroplasts from spinach (U.S. hybrid 424 grown in water culture) were prepared and assayed (at 20°) in standard media (see text) during 1973–1976. It will be seen that rates above 100 μmoles/mg chlorophyll/hr were commonplace. Summer spinach was normally better than that grown in the winter although the worst of the December figures relate to an ill-advised and short-lived, change in the nature of the supplementary lighting (see text).

in Canberra.[11] The same variety grown in the same medium at Sheffield (United Kingdom) led to substantial improvements. Rates in excess of 100 became commonplace, and rates in excess of 200 were occasionally recorded (Fig. 1). These may of course may be equalled by field grown spinach at its best, but for many purposes there is no doubt that water culture is worth the extra effort involved. At Sheffield the standard nutrient solution (see Table I) now employed is based on that used in Canberra. This nutrient solution is drawn at 20 liters/min through a gravel-bed filter and delivered to the plant roots in d plastic gutters before return to a reservoir. Seeds are germinated on vermiculite (heat-expanded mica) moistened with nutrient solution and when the cotyle-dons are fully expanded are held in polyurethane foam plugs fitting into 3.5-cm apertures cut into the lid of aerated aquarium tanks. Each tank accommodates 12–20 plants, and when these have developed leaves of 2–3 cm they are transferred (in their plugs) to similar apertures in the lids of the gutters carrying the circulated nutrient. In the summer months, day-length is shortened to 8½ hr by an automatic blind which prevents illumination during the early morning and early evening. In the winter, daylight is augmented by 400-W mercury vapor and 150-W tungsten lamps at alternate 1-ft centers delivering approximately 50 W/m² of photosynthetically active light. (To date we have not been able to grow

[11] U. Heber, N. K. Boardman, and J. M. Anderson, *Biochim. Biophys. Acta* **432**, 275 (1976).

TABLE I
NUTRIENT SOLUTION FOR SPINACH WATER CULTURE

| No. | Solution | Stock concentration | Volume (ml) per 20 liters nutrient solution made up with distilled water |
|---|---|---|---|
| 1 | $KNO_3$ | 1 $M$ | 120 |
| 2 | $Ca(NO_3)_2$ | 1 $M$ | 80 |
| 3 | $MgSO_4$ | 1 $M$ | 40 |
| 4 | $KH_2PO_4$ | 1 $M$ | 20 |
| 5 | $MgCl_2$ | 1 $M$ | 80 |
| 6 | Trace elements | (B, Mn, Zn, Cu, Mo) | 20 |
| 7 | NaFe-EDTA | (3.86 gm/250 ml) | 20 |

| Trace elements | Quantity (mg) in 250 ml $H_2O$ |
|---|---|
| $H_3BO_3$ | 715 |
| $MnCl_2:4H_2O$ | 452 |
| $ZnSO_4:7H_2O$ | 55 |
| $CuSO_4:5H_2O$ | 20 |
| $NaMoO_4:2H_2O$ | 7.25 |

spinach which would yield really active class A chloroplasts under fluorescent light.)

In full daylight, high temperatures appear to be unimportant, and, indeed, some of the best material has been grown under glass at summer temperatures reaching 50° (in the sun). Conversely, in poor winter light, 25° should not be exceeded.

*Peas.* Seeds are soaked overnight in aerated water and germinated, in vermiculite, in flat trays at a density of 125 gm (dry weight) per tray (38 × 23 × 8 cm). Again the combination of low light and high temperature is to be avoided, but in relatively strong light (>50 W/m²) temperatures of 20–25° and photoperiods of 9–12 hr, good material can be grown in 9–11 days, by which time the shoots of a variety such as Progress No. 9 (Suttons Ltd., London Road, Earley, Reading, United Kingdom) are 3–4 cm high.

## Grinding Medium

*Spinach.* As previously indicated,[1] any one of a number of sugars or sugar alcohols may be used as osmotica, and any one of a number of buffers to maintain the pH. Some advantage is derived from the use of slightly acid pH and the inclusion of a small quantity of $P_i$. ("Clumping" sometimes occurs in $P_i$-free media and pea chloroplasts appear to do this more readily than those from spinach.) Small quantities of $MgCl_2$ and EDTA are often included (sometimes merely to maintain an established practice and to allow direct comparison with previously acquired results). (Originally there were apparently good reasons for including both Mn and Mg. EDTA was then added to avoid precipitation in the presence of $P_i$. The beneficial effect of these ions despite the presence of the chelate was attributed to partitioning between the EDTA and binding sites in/on the chloroplasts. It might also be noted that there is a frequent tendency to persist with well-tried media even though these may be difficult to justify on a strictly logical basis. The reason is immediately obvious if the variables are examined. Even the work involved in permutating the contents of a simple mixture containing osmoticum and buffer at several concentrations and pH values is very considerable and is not helped by the intrinsic variability imposed by the leaf material.)

Good chloroplasts can be prepared in the absence of Mg (for example, in 0.33 $M$ sorbitol and 10 m$M$ pyrophosphate at pH 6.5), but the presence of Mg together with EDTA appears to be beneficial even though the reasons remain obscure. In assay at pH 8.5, $MgCl_2$ has been reported to be inhibitory in the absence of EDTA,[12] but there is no detectable inhibition at the concentration and pH values normally used, and in the light of some recent reports it seems unlikely that $Mg^{2+}$ can pass easily into the chloroplast from the medium. This is not to say that it will be without effect but that observed responses are more likely to result from external events. For example, external hydrolysis of fructose diphosphate may have contributed to the inhibitions mentioned above,[12] and Mg will certainly inhibit in mixtures containing $PP_i$ and high proportions of ruptured chloroplasts (by facilitating the hydrolysis of $PP_i$ to $P_i$).

As previously indicated[1] the following solution continues to be of the most useful for many purposes.

> 0.33 $M$ sorbitol
> 10 m$M$ $Na_4P_2O_7$
> 5 m$M$ $MgCl_2$
> 2 m$M$ Na isoascorbate, all adjusted to pH 6.5 with HCl

Its principal disadvantage is that it comprises a potential source of $P_i$, and if [$P_i$] in assays is to be precisely controlled then an inert buffer such as MES[6] may be preferred. On the other hand, precise control of [$P_i$] is

[12] M. Avron and M. Gibbs, *Plant Physiol.* **53**, 140 (1974).

not needed in many applications and indeed at the levels of $P_i$ and chlorophyll concentrations normally used the [$P_i$] will change during the course of illumination as $P_i$ becomes incorporated into sugar phosphates. The main advantage of $PP_i$, apart from its efficacy, is its cheapness. A useful compromise is therefore to use $PP_i$ in the large volume of the initial grinding medium and to replace by a more inert buffer, following washing, in the much smaller volume of the resuspending medium.

*Peas.* The above medium is less useful for peas because of the recently established inhibition by $PP_i$ (both ADP and $PP_i$ inhibit when added singly but stimulate when added together[8-10]). The solution first used for the isolation of active pea chloroplasts[4] (below) is still as good as any for this purpose despite the high [$P_i$], although again a more inert buffer can be used at the outset or after washing.

> 0.33 $M$ glucose
> 50 m$M$ $Na_2HPO_4$
> 50 m$M$ $KH_2PO_4$
> 5 m$M$ $MgCl_2$
> 0.1% NaCl
> 0.2% Na isoascorbate
> 0.1% BSA, all adjusted to pH 6.5 with HCl

Media are best used as semi-frozen slush[1] (solutions may be stirred in chilled alcohol at about $-15°$ until a suitable consistency is achieved).

*Harvesting.* Most leaves yield better chloroplasts if freshly harvested but spinach has been known to give rates of 80 $\mu$moles/mg Chl/hr after 4 weeks cold storage. Conversely pea shoots deteriorate very rapidly and should always be used immediately.

*Preillumination.* If leaves are brightly illuminated for 20–30 min prior to grinding, chloroplast yield is increased and induction[13] shortened.

*Grinding.* This is a compromise between opening as many cells and as few chloroplasts as possible. The homogenization time obviously depends on the characterization of the apparatus employed, but 3–5 sec is appropriate for a conventional Waring blender (at full speed) and 3 sec or less for the Polytron. The latter gives a higher yield. For spinach, 80 gm of leaf tissue macerated for 3 sec in a Polytron yields chloroplasts containing 4–6 mg of chlorophyll.

*Filtration.* The brei is usually squeezed through two layers of muslin (to remove coarse debris) and filtered through eight layers and cotton wool. Muslin seems preferable to smooth fibers such as nylon, and there

---

[13] D. A. Walker, *in* "The Intact Chloroplast" (J. Barber, ed.), Chapter 7, p. 235. Elsevier, Amsterdam, 1976.

would appear to be some evidence that muslin and cotton wool retain ruptured chloroplasts more readily than whole chloroplasts.

*Centrifugation.* Many variants may be used, but it is important to separate the chloroplasts quickly from the supernatant. For example, if a brei of 200–300 ml is divided between four tubes of approximately 100-ml capacity in a swing-out head, adequate precipitation may be achieved by accelerating to approximately 6000 rpm and returning to rest in 90 sec. Advantage derives from the use of centrifuge tubes with surfaces which have not been scratched or cleaned with detergents.

*Resuspension.* The supernatant is decanted (no care is needed because the pellet adheres loosely to the tube). The pellet surface may be washed (e.g., with grinding or resuspending medium which is then discarded). Small quantities (0.5 ml/tube) of resuspending medium (see discussion of assay medium) are added and the chloroplasts resuspended (usually by using a small piece of cotton wool and a glass rod, or by gently rocking the tubes).

*Chlorophyll.* As previously discussed[1] this is often determined by Arnon's procedure. If 50 $\mu$l of chloroplast suspension is added to 20 ml of 80% (v/v) aqueous acetone and filtered (protect from light), then 9/ absorbance at 652 nm, gives the volume of the original suspension (in $\mu$l) containing 100 $\mu$g of chlorophyll. It is convenient to add chloroplasts to assay mixtures in 100 $\mu$g aliquots (or appropriate multiples) rather than fixed volumes of variable chlorophyll content.

Assay medium

This naturally depends on the nature of the investigation but the following, based on that of Jensen and Bassham[6] is useful for many purposes and (minus bicarbonate) is often used as a resuspending medium.

> 0.33 $M$ sorbitol
> 2 m$M$ EDTA
> 1 m$M$ MgCl$_2$
> 1 m$M$ MnCl$_2$
> 50 m$M$ HEPES
> 10 m$M$ NaHCO$_3$
> 5 m$M$ PP$_i$
> 0.5 m$M$ P$_i$, all adjusted to pH 7.6 with KOH

The role of pyrophosphate is too complex to discuss in detail here, but it seems clear that PP$_i$ does not readily cross the intact spinach chloroplast envelope, although it interferes with the action of Heldts P$_i$ translocator.[14] For these reasons PP$_i$ (in the presence of external Mg and

---

[14] D. A. Walker, *Curr. Top. Cell. Regul.* **11**, 203 (1976).

PP$_i$ase released from damaged chloroplasts) acts as an optimal P$_i$ source. Chloroplasts require P$_i$ for carbon assimilation because their major product is triose phosphate rather than free carbohydrate, but too much P$_i$ (in the absence of PP$_i$) is inhibitory, prolonging induction by promoting triose phosphate loss.[13-15] If PP$_i$ is not employed the optimal [P$_i$] is about 0.25 m$M$ (depending on the sugar phosphate levels within the chloroplast), but the optimum is so sharp it is very difficult to achieve and, of course, if the chlorophyll concentration is not kept very low the [P$_i$] will change rapidly during illumination. Recent work with chloroplasts from young pea leaves suggests that ADP and ATP within the chloroplasts may exchange with external PP$_i$.[8-10] For this reason PP$_i$ is inhibitory if used alone with pea chloroplasts but stimulates in the presence of small quantities (0.4 m$M$) of ADP and ATP.

In all experiments involving measurements of $CO_2$ fixation or $CO_2$-dependent oxygen evolution the initial lag[13] and the final rate depend on the balance between sugar phosphate within the chloroplast (and therefore the immediate history of the parent tissue) and the exogenous [P$_i$].[13,14] In a standard test it is therefore useful to examine PGA-dependent $O_2$ evolution (and/or PGA-dependent $CO_2$ fixation) because PGA will overcome the initial lag. (In the above assay 1 m$M$ PGA usually gives about 60% of the $CO_2$-dependent $O_2$ evolution but with only a very small initial lag. Contrary to some statements to the contrary, PGA is not inhibitory if added after induction. In any event PGA is a useful additive to chloroplasts of suspect activity. If PGA-dependent $O_2$ evolution can be achieved at good rates there is every reason to hope for good $CO_2$-dependent $O_2$ evolution following manipulation of [P$_i$], etc. If the preparation will not evolve $O_2$ with PGA, it is most unlikely to assimilate $CO_2$.)


Criteria of Intactness

Unless illuminated in mixtures containing additional ferredoxin, NADP, ADP, Mg, etc. (the constituents of the reconstituted system), envelope-free chloroplasts will not carry out $CO_2$-dependent $O_2$ evolution, $CO_2$ fixation, or PGA-dependent $O_2$ evolution. The best test of photosynthetic function is therefore simply to illuminate in near-saturating red light in an appropriate assay medium (above). At 20°, spinach chloroplasts which exhibit rates of $CO_2$ fixation or $CO_2$-dependent $O_2$ evolution of less than 50 $\mu$moles/mg chlorophyll/hr may be regarded as

[15] D. A. Walker and A. Herold, *Plant Cell Physiol., Spec. Issue* pp. 295–310 (1977).

poor. Rates of 50–80 are reasonable, 80–150 are good, and rates above 150 are exceptional.

Chloroplasts with intact envelopes (Class $A^{16}$) will not carry out many functions at fast rates because of the permeability barrier afforded by the inner envelope. They will not, for example, (1) reduce exogenous oxidants such as ferricyanide or NADP, (2) rapidly phosphorylate exogenous ADP, (3) support $O_2$ uptake with nonpermeating Mehler reagents such as ferredoxin, (4) fix $CO_2$ in the dark with ribulose diphosphate or ribose 5-phosphate and ATP as substrates, and (5) hydrolyse exogenous inorganic pyrophosphate. All of these could be used as the basis of "intactness" assays.

Possibly the simplest, and certainly that which has gained most favor is the use of ferricyanide as Hill oxidant. The reduction of this compound may be measured spectrophotometrically or followed in the $O_2$ electrode provided that $CO_2$-dependent $O_2$ evolution is inhibited by glyceraldehyde (Fig. 2). The assay probably overestimates intactness (there is evidence which suggests that rupture and loss of stromal protein may be followed by resealing[17]), but it provides a useful basis for comparison. The response to the uncoupler (Fig. 2) is also useful. Well-coupled chloroplasts may show as much as a 14-fold increase in rate following the addition of $NH_4Cl$. Chloroplasts which show a fourfold response, or less, are unlikely to support carbon assimilation at good rates.

### High-Percentage Intactness

The percentage intactness (as determined by the ferricyanide assay) for spinach chloroplasts prepared in PP$_i$ media from 1973–1976 at Sheffield is shown in Fig. 3. On average, pea chloroplasts, prepared in Pi media, are a little better. H. Y. Nakatani (personal communication) has recently achieved improved intactness with peas by avoiding high concentrations of small cations (e.g., by using media containing MES or HEPES adjusted to an appropriate pH by the addition of Tris). In our own hands, pea chloroplasts isolated in conventional $P_i$ media have sometimes shown significant improvement in intactness when washed in 0.33 $M$ sorbitol containing 2% HEPES (and Tris to pH 7.6), whereas washing in normal grinding or resuspending medium was ineffectual or deleterious.

[16] D. O. Hall, *Nature (London) New Biol.* **235**, 125 (1972).

[17] R. McC. Lilley, M. P. Fitzgerald, K. G. Rienits, and D. A. Walker, *New Phytol.* **75**, 1 (1975).

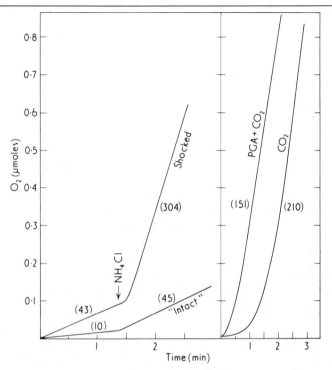

FIG. 2. Standard Assays. On the left, the standard assay for intactness[12] with ferricyanide as the Hill oxidant. The lower trace records $O_2$ evolution by "intact" chloroplasts. The reaction mixture giving the upper trace contained the same components as the lower, but the order of addition was varied. In this way the chloroplasts (100 μg chlorophyll in 2 ml) received an osmotic shock, which rendered them 100% envelope-free. At the arrow, $NH_4Cl$ was added as an uncoupling agent. Percentage intactness (85%) is calculated from the uncoupled rates, 100-[(45 × 100)/304], on the assumption that the intact envelope is completely impermeable to ferricyanide. On the right, $CO_2$-dependent $O_2$ evolution and PGA-dependent $O_2$ evolution (chloroplasts ≡ 200 μg chlorophyll in 2 ml). Under these conditions PGA gives a much shorter lag than $CO_2$ but a slower rate. Conversely it stimulates, if added after induction and not corrected for intactness, i.e., the rate of 210 is for 85% intactness and would rise to 247 if based on 100% Class A chloroplasts. Rates (given in parentheses) in μmoles $O_2$/mg chlorophyll/hr. All reactions at 20°.

## The Reconstituted Chloroplast System

This again is too large a subject to be dealt with adequately here, but it is perhaps worth stating that the reconstituted system[18] is much easier to prepare than first class intact chloroplasts and that it may also be of use in dealing with "difficult" species.[19] Essentially, it involves preparing

[18] R. McC. Lilley and D. A. Walker, in "Encyclopedia of Plant Physiology (new series). Photosynthesis, Vol. 2," (M. Gibbs and E. Latzko, eds.). Springer-Verlag, Berlin and New York, 1979.

[19] M. E. Delaney and D. A. Walker, Plant Sci. Lett. 7, 285 (1976).

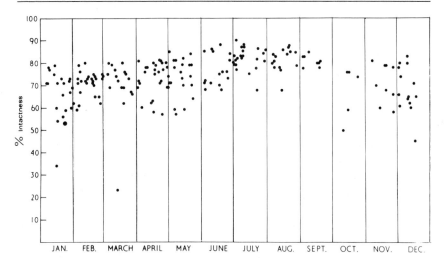

FIG. 3. Percentage Intactness. Intactness, based on the ferricyanide assay,[12] of spinach chloroplasts prepared in standard medium (text) during the period 1973–1976.

intact chloroplasts (but less care is needed and quality is less important), exposing these to osmotic shock, and adding back stromal protein plus additional ferredoxin, ADP, NADP, and Mg to the thylakoid fraction. With substrates such as R5P, rates of $CO_2$-dependent $O_2$ evolution in excess of 100 $\mu$moles/mg chlorophyll/hr (at 20°) are readily achieved (although care must be taken to avoid excess substrate).

### Gradient Centrifugation

Although this article has been primarily concerned with conventional isolation, it should be noted that the purity of chloroplast fractions can be improved by gradient centrifugation.[20–23] Further purification has sometimes been associated with loss of ability to assimilate $CO_2$ (not necessarily a disadvantage if, e.g., the prime objective is enzyme location) but rates of 130 $\mu$moles/mg chlorophyll/hr have been achieved with spinach chloroplasts which had been subjected to isopycnic centrifugation.[23] Whether these techniques will lead to routine improvement in intactness (cf. Fig. 3) remains to be established.

[20] R. M. Leech, *Biochim. Biophys. Acta* **79**, 637 (1964).
[21] B. J. Miflin and H. Beevers, *Plant Physiol.* **53**, 870 (1974).
[22] V. Rocha and I. P. Ting, *Arch. Biochem. Biophys.* **140**, 398 (1970).
[23] J. J. Morgenthaler, C. A. Price, J. M. Robinson, and M. Gibbs, *Plant Physiol.* **54**, 532 (1974).

# [10] Photosynthetically Active Algal Preparations[1]

*By* HARTMUT SPILLER and PETER BÖGER

Methodological details of algal cell-free systems are described for species that have been found suitable to yield thylakoid or chloroplast material exhibiting substantial rates of either light-induced $NADP^+$ reduction or noncyclic photophosphorylation, as well as activity in general Hill reactions or phosphorylation catalyzed by phenazine methosulfate (PMS). Consequently, cell-free assays exhibiting only the latter activities or subchloroplast or photosystem I and II preparations are not described. Further, the rates of the algal systems considered here are comparable with light-induced reactions of isolated spinach chloroplast material.

There are not many species which fulfil these requirements. This was stated previously,[2] and the situation has not essentially changed. Low rates of $NADP^+$ reduction and noncyclic photophosphorylation, weak coupling, and critical reproducibility of activity are the main problems. However, for algae, such as *Nostoc, Chlamydomonas, Scenedesmus,* or *Bumilleriopsis,* definite methodological improvements have been achieved.

Agal systems offer many advantages to study details of photosynthetic mechanisms. In contrast to higher plants, algae can be grown in the laboratory under sterile and controled environmental conditions (of, e.g., temperature, light, culture medium), and some species can be manipulated genetically. Furthermore, algal systems are of interest for comparative reasons. Differences in their complement of redox carriers (versus higher plants) have been reported recently.[3,4]

## I. Cell-Free Systems from Blue-Green Algae (Cyanophyceae)

### A. *Phormidium*

*Phormidium luridum var. olivaceae (Algal Culture Collection of Indiana University, Bloomington, Indiana).* The earlier work of Biggins[5]

---

[1] For methodology prior to 1970, refer to Vol. 23 of this series.[1a,2]

[1a] D. Graham and R. M. Smillie, Vol. 23, p. 228.

[2] P. Böger, Vol. 23, p. 242.

[3] P. M. Wood, *Eur. J. Biochem.* **72**, 605 (1977).

[4] P. Böger, *Proc. Int. Congr. Photosynth. Res., 4th, 1977* D. O. Hall *et al.*, eds., Biochem. Soc. London p. 755 (1978).

[5] J. Biggins, *Plant Physiol.* **42**, 1442 (1967).

METHODS IN ENZYMOLOGY, VOL. 69

for the preparation of photosynthetically active membranes was modified and expanded by first preparing spheroplasts.[6]

*Preparation.* Cells were grown in 250 ml medium of Kratz and Myers at pH 7.8, 25° under 2000 lux supplied by Sylvania Grolux lamps and gassed with air supplemented with 5% $CO_2$ (final pH 6.8). Cells were harvested during the late log phase (6–7 days) of growth by centrifugation at 3000 $g$ for 10 min. They were washed in medium A [mannitol 0.5 $M$; $N$-tris(hydroxymethyl)methylglycine (Tricine–NaOH) 10 m$M$; $MgCl_2$, 10 m$M$; and sodium/potassium phosphate buffer, pH 7.8, 5 m$M$]. The pellet from 250 ml culture was diluted in about 4 ml of medium A to give 0.3–0.5 mg chlorophyll/ml, and lysozyme was added to 0.1% (w/v). The resulting mixture was incubated at 35° for 90 min without stirring until 80–90% of the filaments were converted to spheroplast chains, then centrifuged for 1 min at 1000 $g$ and resuspended in medium A to give 1 mg chlorophyll (Chl)/ml.

*Reaction Conditions.* Active membranes with photosystem I + II and phosphorylation activity were obtained by diluting spheroplasts in a ratio 1:10 to 1:200 in reaction medium B (10 m$M$ Tricine–NaOH, 10 m$M$ $MgCl_2$, and 5 m$M$ sodium potassium phosphate, pH 7.8). Photosystem I membranes were obtained by diluting spheroplasts in a 1:10 to 1:200 ratio in 10 m$M$ Tricine–NaOH, pH 7.5. For measuring the activity of spheroplasts, samples were diluted in medium B containing 0.25 $M$ KCl. Light-induced dichlorophenolindophenol (DCIP) and $NADP^+$ reduction were measured with a Cary 17 spectrophotometer at 560 and 340 nm, respectively. Samples were illuminated with a 500-W slide projector, the light defined by Corning filter CS 2-62. The photomultiplier was protected by Corning filter CS 4-96 for DCIP reduction and by filters CS 4-97 and CS 7-59 for $NADP^+$ reduction. The latter reaction was carried out in a 0.6-ml cuvette of 2 mm light path placed at a 60° angle to the measuring beam (see Table I for rates).

Assays for cyclic and noncyclic phosphorylation were performed at 25° for 2 min with 140,000 lux white light. Labeled ATP was measured according to Avron.[7]

By using spheroplasts as starting material, high activities can also be achieved with membrane preparations (Table I). The presence of $Mg^{2+}$ and phosphate proved beneficial in stabilizing the activities. Photosytem I particles (inactive in photosytem II reactions) require the addition of ferredoxin, plastocyanin, reductase, $MgCl_2$, phosphate, and an uniden-

[6] A. Binder, E. Tel-Or, and M. Avron, *Eur. J. Biochem.* **67**, 187 (1976).
[7] M. Avron, *Anal. Biochem.* **2**, 535 (1961).

TABLE I

PHOTOSYNTHETIC ACTIVITIES OF SPHEROPLASTS, ACTIVE MEMBRANES, AND
PHOTOSYSTEM I PARTICLES FROM *Phormidium luridum*[a]

| | Rates in microequivalents electrons or micromoles ATP/mg Chl $\times$ hr | | |
| Assay system | Spheroplasts | Thylakoids | Photosystem I particles |
| --- | --- | --- | --- |
| Photosystem I + II | | | |
| $H_2O \rightarrow$ DQ | 300–500 | 400–800 | 0 |
| $H_2O \rightarrow$ NADP | 100–350 | 150–400 | 0 |
| Photosystem II | | | |
| $H_2O \rightarrow$ FeCy + DCMU + silicomolybdate | — | 610 | 0 |
| Photosystem I | | | |
| DCIP/asc $\rightarrow$ DQ + DCMU | 1200–2900 | 1500–3200 | 800–1800 |
| DCIP/asc $\rightarrow$ NADP$^+$ + DCMU | 80–200 | 150–350 | 100–300 |
| Photophosphorylation | | | |
| $H_2O \rightarrow$ FeCy | — | 100–200 | 0 |
| $H_2O \rightarrow$ NADP$^+$ | — | 100–250 | 0 |
| PMS/ascorbate $\rightarrow$ NADP$^+$ + DCMU | 150–250 | 900–1800 | 0–50 |

[a] Activities of spheroplasts, thylakoids and photosystem-I particles were tested in medium B as follows: FeCy and silicomolybdate reduction by oxygen evolution and by DQ-mediated $O_2$ uptake in the presence of 0.5 m$M$ sodium azide: FeCy 1 m$M$, silicomolybdate 0.2 mg/2 ml; diquat (DQ) 50 $\mu M$; DCMU 10 $\mu M$; DCIP 50 $\mu M$; ascorbate 10 m$M$. Where indicated, NADP$^+$-reduction was carried out in 0.6 ml mixture with 30–10 $\mu$g Chl/ml in the presence of 0.5 m$M$ NADP$^+$, 10 $\mu$g ferredoxin plus ferredoxin–NADP$^+$ reductase from Swiss chard (amount not specified); with 20 $\mu M$ DCIP, 20 m$M$ ascorbate, and 10 $\mu M$ DCMU where indicated. Photophosphorylation was assayed in the respective medium used for electron transport or with PMS (50 $\mu M$ and 10 m$M$ ascorbate plus 5 m$M$ ADP plus P$_i$ including 6 $\times$ 10$^6$ cpm $^{32}$P). Data from Binder *et al.*[6]

tified protein factor for maximum NADP$^+$ reduction. This factor seems to be involved in both $O_2$ and NADP$^+$ reduction.

## B. Nostoc

*Preparation. Nostoc muscorum* strain 7119 (from the culture collection of the University of Washington) was grown autotrophically under nitrogen-fixing conditions in 10-liter batches under continuous illumination (3000 lux) at 20°. Composition of the culture medium is given in

Arnon *et al.*[8] Cultures were bubbled with a continuous flow (20 ml/min) of $N_2/CO_2$ (98/2,v/v). The yield of cells grown for 4–5 days was 1–2 g/ liter culture medium. The cells were washed twice in suspending solution S (sucrose, 0.5 $M$; $MgCl_2$, 10 m$M$; Tricine–KOH, 50 m$M$, pH 7.7). Disruption was achieved by a Ribi cell press between 15,000–18,000 psi at 0°–15°. One milligram of DNase and 1 mg of RNase were added per 60 ml of slurry. Cells and cell debris were removed by centrifugation for 5 min at 2500 $g$, and the supernatant recentrifuged at 35,000 $g$ for 15 min. The pellet was resuspended in medium S to give fraction A: 0.5 mg Chl/ ml, exhibiting high activity.[8] Spheroplasts could be prepared by lysozyme treatment (Binder *et al.*,[6] H. Spiller and P. Böger, unpublished).

*Reaction Conditions.* Details are given in Table II. Photoreduction of $NADP^+$ or ferricyanide was monitored by absorption changes at 340 and 420 nm and measured under the same assay conditions as photophosphorylation.[9] $NADP^+$ reduction was enhanced by adding ferredoxin plus the reductase from spinach. Particles retain high activity of $NADP^+$ reduction from water in spite of considerable loss of phycocyanin (maximum of 300 $\mu$moles $NADP^+$ reduced/mg Chl × hr). Noncyclic photophosphorylation was one-half of the corresponding rate of electron transport from water to $NADP^+$; cyclic phosphorylation is fairly high (see Table II, lines b and d). Spheroplast fragments show remarkable rates of photosystems I + II and photosystem II electron transport (Table II, part B); addition of ferredoxin–$NADP^+$ reductase decreases the rate due to oxygen reduction (see line i in Table II). Cytochrome *b*-559 exists in several forms; its high-potential form is photooxidized by photosystem II and appears to be located in the main pathway of electron transport from water to $NADP^+$.[10] Two types of P700 seem to exist, one which functions only in the DCIP/ascorbate → $NADP^+$ pathway and seemingly does not participate in general linear electron transport.[11] The roles of cytochromes *f* and *b*$_6$ have recently been assessed.[12]

## C. Anabaena

*Thylakoid preparation. Anabaena variabilis* (from the Institute of Applied Microbiology, Tokyo) was autotrophically grown under 5000 lux incandescent light at 25° and was gassed with air supplemented by 0.5%

[8] D. I. Arnon, B. D. McSwain, H. Y. Tsujimoto, and K. Wada, *Biochim. Biophys. Acta* **357**, 231 (1974).

[9] B. D. McSwain and D. I. Arnon, *Proc. Natl. Acad. Sci. U.S.A.* **61**, 989 (1968).

[10] H. Y. Tsujimoto, B. D. McSwain, T. Hiyama, and D. I. Arnon, *Biochim. Biophys. Acta* **423**, 303 (1976).

[11] T. Hiyama, B. D. McSwain, and D. I. Arnon, *Biochim. Biophys. Acta* **460**, 76 (1977).

[12] D. Knaff, *Arch. Biochem. Biophys.* **182**, 540 (1977).

TABLE II
CELL-FREE PHOTOSYNTHETIC ACTIVITIES OF *Nostoc muscorum*

| Assay system | Acceptor reduced ($\mu$moles/mg Chl/hr) | ATP formed ($\mu$moles/mg Chl × hr) |
|---|---|---|
| (A) Thylakoids (fraction A)[a] | | |
| (a) $H_2O \rightarrow$ FeCy | 392 | 72 |
| (b) $H_2O \rightarrow NADP^+$ | 177 | 85 |
| (c) DCIP/ascorbate $\rightarrow NADP^+$ | 84 | 36 |
| (d) Phosphorylation with PMS | | 164 |

| | Micromoles oxygen evolved, taken up, or $NADP^+$ reduced per mg Chl per hr | |
|---|---|---|
| | (−) DCMU | (+) DCMU |
| (B) Spheroplasts[b] | | |
| (e) $H_2O \rightarrow$ FeCy | 174 | 0 |
| (f) $H_2O \rightarrow$ silicomolybdate | 90 | 60 |
| (g) DAD/ascorbate $\rightarrow$ MV | — | 700 |
| (h) $H_2O \rightarrow NADP^+$ | 150 | (j) 130 |
| (i) $H_2O \rightarrow NADP^+$ (plus reductase) | 137 | (j) 104 |

[a] (A): Data from Arnon et al.[8] The reaction mixture contained in a 1 ml final volume 50 mM Tricine, pH 7.7, 10 mM $MgCl_2$, 2 mM ADP, and 2 mM $K_2H^{32}PO_4$. Additions in (a) 4 mM FeCy; (b) 0.1 mM spinach ferredoxin, 2mM $NADP^+$, plus saturating amount of spinach ferredoxin–$NADP^+$ reductase; (c) 10 mM ascorbate, 0.1 mM DCIP, and 0.001 mM DCMU; (d) 0.05 mM PMS. Conditions: gas phase air, 20°, 3 min of illumination with red light of $2.6 \times 10^5$ ergs/cm² × sec defined by Corning filters No. 2-64 and 1-69.

[b] Data inpart from H. Spiller and P. Böger, *Proc. Int. Congr. Photosynth Res., 4th, 1977*, Abstract 362 (unpublished). Reaction conditions and illumination according to Spiller and Böger.[20] (e) and (f): Reaction mixture (1.8 ml) contained 20 mM Tricine–NaOH, pH 7.8; 50 mM sucrose; 2 mM FeCy; 2.5 mM potassium phosphate, 10 mM $MgCl_2$; and 0.4 mg silicomolybdate. (g)–(j): Reaction mixture (1 ml) contained 20 mM Tricine–NaOH, pH 7.8; 50 mM sucrose; and 2.5 mM potassium phosphate, and 10 mM $MgCl_2$. Additions as indicated: 1.5 mM ascorbate, 0.12 mM DAD, 0.1 mM MV, 0.001 mM DCMU, 0.2 mM $NADP^+$, 4 nmoles ferredoxin, and ferredoxin–$NADP^+$ reductaase in saturating amounts, both proteins from the alga *Bumilleriopsis*. Purification following Böger.[17]

$CO_2$.[13] Membrane fragments were used to study the specificity of plastocyanin and cytochrome c-553 in photosystem I-mediated ascorbate photooxidation. The assay medium contained in a 2.5-ml final reaction volume 125 $\mu$moles tris(hydroxymethyl)aminomethane (Tris–HCl), pH 7.5, 25 $\mu$moles Na ascorbate, 0.25 $\mu$moles methylviologen (MV), 3 $\mu$g

[13] Y. Fujita and R. Suzuki, *Plant Cell Physiol.* **12**, 641 (1971).

of Chl, plus nonsaturating amounts of plastocyanin ($2 \times 10^{-5}$ $M$) or cytochrome c-553 ($10^{-5}$ $M$). They produced rates of 4000 and 1800 $\mu$moles $O_2$ uptake per mg chlorophyll per hour, respectively. Cytochrome $b$-559 was shown to be oxidized by both photosystems and reducible by photosystem II.[14].

*Heterocysts.* *Anabaena cylindrica* was obtained from the Cambridge Culture Collection No. 1403/2a and grown in axenic culture under nitrogen-fixing conditions in the medium of Allen and Arnon[15] at 26° and 3000 lux.[16] Preparation of isolated heterocysts was achieved by lysozyme treatment and subsequent rupture in the Yeda press. Using ascorbate plus dichlorophenolindophenol, ruptured heterocysts reduced oxygen via MV and $NADP^+$ with ferredoxin plus ferredoxin–$NADP^+$ reductase yielding 252 and 95 $\mu$moles substrate reduced per mg Chl per hr, respectively. Photophosphorylation proceeded at 60 $\mu$moles ATP/mg Chl $\times$ hr.

## II. Chloroplast Material from Xanthophyceae

*Bumilleriopsis filiformis* Vischer (from the stock culture of P.B.) is a coccoid xanthophycean species which has been used in this laboratory for years. It can be grown under sterile conditions as described, either as mass[2,17] or synchronized culture.[18,19]

Chloroplasts with high electron-transport and photophosphorylation rates can be isolated,[2] and by recently developed procedures we were able to obtain class A chloroplasts capable of $CO_2$ fixation and class C preparations with high coupling.[20,21]

*Intact Chloroplasts (Class A).* Two grams of algal cells were washed once and resuspended in a mixture of 8 ml of 0.1 $M$ potassium phosphate, pH 5.8, and 0.2 $M$ sorbitol. Hemicellulase, 300 mg, and 100 mg of pectinase were added and the suspension incubated at 35° for 90 min. After dilution to 20 ml, the algae were centrifuged at 1000 $g$ for 3 min and washed with homogenization medium A, the digested cell material (capable of fixing $CO_2$ at 60–100 $\mu$moles/mg Chl/hr) was mixed with 17 ml of medium A, brought to 50 ml volume with glass beads (1-mm diameter) and cooled to $-5°$ [medium A: sorbitol 0.4 $M$; 2-($N$-morpholino)ethanesulfonic acid (Mes–NaOH) 50 m$M$, pH 6.2; sodium

[14] Y. Fujita, *Plant Cell Physiol.* **16**, 1037 (1975).
[15] M. B. Allen and D. I. Arnon, *Plant Physiol.* **30**, 366 (1955).
[16] E. Tel-Or, and W. D. P. Stewart, *Biochim. Biophys. Acta* **423**, 189 (1976).
[17] P. Böger, *Z. Pflanzenphysiol.* **61**, 85 (1969).
[18] M. Hesse, *Planta* **120**, 135 (1974).
[19] M. Hesse, P. Bley, and P. Böger, *Planta* **132**, 53 (1976).
[20] H. Spiller and P. Böger, *Photochem. Photobiol.* **26**, 397 (1977).
[21] H. Böhme, K.-J. Kunert, and P. Böger, *Biochim. Biophys. Acta* **501**, 275 (1978).

pyrophosphate ($Na_4P_2O_7$) 5 m$M$; isoascorbate 2 m$M$; NaCl 10 m$M$; bovine serum albumin (BSA) 0.5%; Ficoll 1.5%]. After a short burst of 4–5 sec at 2000 rpm in a Merkenschlager homogenizer (Braun, Melsungen, Germany) the homogenate was quickly separated from the glass beads by dilution in 40 ml medium B (Tricine–NaOH, 25 m$M$, pH 7.0; soribitol, 0.4 $M$; $Na_4P_2O_7$, 1 m$M$; $MgCl_2$, 2 m$M$; BSA, 0.5%; Ficoll, 1.5%; potassium phosphate, 0.5 m$M$). Unbroken cells were discarded by centrifugation at 500 $g$ for 90 sec; chloroplasts were pelleted at 1000 $g$ for 3 min and resuspended in 2 ml of medium B.

The degree of intactness was checked by phase contrast microscopy and ferricyanide-dependent $O_2$ evolution before and after osmotic shock in Tricine buffer. Fifty to seventy-five percent of the chloroplasts retained the outer membrane (electron microscopy) and exhibited $CO_2$ fixation rates of 12–20 $\mu$moles/mg Chl/hr. Based on only the intact fraction of chloroplasts in the sample, this translates to rates of 20–30 $\mu$moles $CO_2$ assimilated. The $^{14}CO_2$ fixation medium contained in 1 ml: 25 $\mu$moles Tricine–NaOH, pH 7.0, 400 $\mu$moles sorbitol, 1 $\mu$mole sodium pyrophosphate, 0.5 $\mu$mole potassium phosphate; 2 $\mu$moles $MgCl_2$, 10 $\mu$moles $NaH^{14}CO_3$, 0.5% BSA, 1.5% Ficoll, and 20–50 $\mu$g chlorophyll. No other cofactors were added (H. Spiller and P. Böger, unpublished).

*Coupled Chloroplasts.*[20,21] Cells were washed once with homogenization medium, then 10–12 gm of wet algal paste added to 40 ml of glass beads (1-mm diameter) and 25 ml of cold homogenization medium. The cells were broken in a 50-ml flask using ten bursts at 4000 rpm (1-sec burst followed by 3-sec interval at half-speed in a Merkenschlager homogenizer). Since appropriate cooling of beads, flask, etc., had been achieved beforehand, cooling during homogenization was unnecessary. The homogenization medium consisted of 0.6 $M$ sorbitol; 50 m$M$ Mes–NaOH, pH 6.2, 1 m$M$ Na isoascorbate, 3.5% polyvinylpyrrolidone M 2500 (PVP), 5 m$M$ $MgCl_2$, 1% BSA. After removal of the glass beads the homogenate (about 25 ml) was diluted to 70 ml with homogenization medium and centrifuged for 1 min at 1000 $g$. The resulting supernatant was recentrifuged for 3 min at 3000 $g$. The chloroplast pellet was evenly suspended in 0.6 $M$ sorbitol containing 50 m$M$ N-2-hydroxyethylpiperazine-$N'$-2-ethanesulfonic acid (HEPES–NaOH), pH 7.8, 5 m$M$ $MgCl_2$, and 1% BSA. Remaining cell material was removed by two successive centrifugations at 1000 $g$ for 30 sec each. Chloroplasts were then obtained from the supernatant by a 2000 $g$ centrifugation for 2 min. They were again suspended in the medium just mentioned and the chlorophyll concentration adjusted to 2 mg/ml.

The reaction mixture contained in a 3-ml final volume 0.6 $M$ sorbitol, 5 m$M$ $MgCl_2$, 50 m$M$ HEPES–NaOH, pH 7.8, 10 m$M$ $K_2HPO_4$, and either 0.5 m$M$ NADP$^+$ + 10 $\mu$ *Bumilleriopsis* $M$ ferredoxin or 0.1 m$M$ methylviologen (MV), respectively. For photophosphorylation the reaction mixture contained 1 m$M$ inorganic phosphate (P$_i$) + $^{32}$P and 1 m$M$ ADP (or limiting amounts when necessary). Chlorophyll was about 10 $\mu$g/ml; saturating red light was provided by a tungsten iodine lamp (with a Filtraflex heat filter and a 610-nm Schott cut-off filter).

The chloroplasts exhibit a photosynthetic control of 2–3 (i.e., ratio of the rate in the presence of ADP divided by the rate in its absence) and P/2$e$ value of 1. Uncouplers such as 2.5 m$M$ NH$_4$$^+$ or gramicidin (5 $\mu$g/ml) inhibit phosphorylation completely[21] (see Table III).[21] The chloroplasts so isolated retained almost all soluble plastidic cytochrome $c$-553. On average they had 4 molecules of soluble cytochrome $c$-553 and 1 molecule of bound $c$-type cytochrome $f$-553 per P700. Chloroplasts from iron-deficient cultures had a decreased amount of cytochrome $c$-553, while the content of bound cytochrome $f$-553 remained unaltered (approximately 1–1.5 molecule per P700). As shown in Table III the rate of linear photosynthetic electron transport (and phosphorylation) depends on the amount of soluble cytochrome $c$-553 retained in the thylakoids.

## III. Chloroplast Material from Volvocales

### A. *Chlamydomonas*

This alga was introduced into the biochemistry of photosynthesis by the elegant studies of Levine and co-workers.[22–24] However, varying degrees of success have been achieved with cell-free systems from this alga. In our hands, mechanical cell disruption has not been successful. Recently, two methods of breakage by sonication have been described which yield reasonable rates of both electron transport and noncyclic phosphorylation.

*Method of Togasaki and Collaborators.*[25,26] *Chlamydomonas reinhardii* (wild type, strain 137-C,[+]) was grown according to Levine.[23] Cells were washed in 0.05 m$M$ Tricine–NaOH, pH 7.5, and a suspension equivalent to 60 $\mu$g Chl/ml was preilluminated with fluorescent light (10$^4$

[21a] D. O. Hall, *in* "The Intact Chloroplast" (J. Barber, ed.), p. 135. Elsevier, Amsterdam, 1976.

[22] R. P. Levine, *Science* **162**, 768 (1968).

[23] R. P. Levine, *Plant Physiol.* **41**, 1648 (1966), and references therein.

[24] A. L. Givan and R. P. Levine, *Plant Physiol.* **42**, 1264 (1967).

[25] V. A. Curtis, J. J. Brand, and R. K. Togasaki, *Plant Physiol.* **55**, 183 (1975).

[26] J. J. Brand, V. A. Curtis, R. K. Togasaki, and A. San Pietro, *Plant Physiol.* **55**, 187 (1975).

TABLE III

PHOTOSYNTHETIC ELECTRON TRANSPORT AND COUPLING IN ISOLATED CHLOROPLASTS
FROM *Bumilleriopsis filiformis*[a]

| Chloroplasts | Total cytochrome[c] per Chl | per P700 | Assay system | Rates ($\mu$moles $O_2$/mg Chl $\times$ hr) −ADP | +ADP | +NH$_3$Cl | ADP/O[d] |
|---|---|---|---|---|---|---|---|
| Normal | 5 | 4 | H$_2$O → MV[c] | 51 | 104 | 199 | 0.93 |
| culture | | | H$_2$O → NADP$^+$ | 37 | 95 | 130 | 1.02 |
| | | | H$_2$O → DAD | 230 | — | — | — |
| (−) Iron | 2.3 | 1.9 | H$_2$O → MV[c] | 38 | 82 | 95 | 1.2 |
| culture | | | H$_2$O → NADP$^+$ | 22 | 54 | 46 | 1.1 |
| (−) Iron | 1.9 | 1.5 | H$_2$O → MV[c] | 27 | 54 | 78 | — |
| culture | | | H$_2$O → DAD | 200 | — | — | — |

[a] Adapted from Böhme *et al.*[21]

[b] Mole/mole. All three chloroplast samples had the same content of (bound) cytochrome *f*-553 (approximately 1–1.5 mole per mole of P700).

[c] Rates denote oxygen uptake; the other oxygen evolution.

[d] For calculation of the ADP/O ratios see Hall.[21a] PMS-catalyzed ATP formation (with 15 $\mu$M PMS and 1 m$M$ ascorbate under air) was 211 $\mu$moles/mg Chl per hr.

ergs/cm$^2$ × sec) for at least 15 min. Then 5-ml aliquots were sonified in 10-ml glass beakers without cooling. Sonication time had to be 3–10 sec depending on the assay system used, with 80% of the maximum power setting of a Quigley-Rochester dismembrator.

Sonication is the crucial step. Thereafter, electron transport activities such as H$_2$O → benzoquinone (BQ); dichlorophenolindophenol (DCIP)/ascorbate → methylviologen (MV) remained stable for over 2 hr at 0°[25] (for rates check Table IV). This was different for photophosphorylation. The half-life of both cyclic and noncyclic phosphorylation was about 20 min. The reaction mixture for photophosphorylation contained in a final volume of 1.5 ml, besides sonicated cells equivalent to 18 $\mu$g of chlorophyll, 50 $\mu$moles Mes–Tricine buffer, pH 7.8, 1 $\mu$mol ADP, 3.8 $\mu$moles P$_i$ with appropriate $^{32}$P, 0.5 $\mu$mole MgCl$_2$, and either 0.1 $\mu$mole PMS [together with 0.01 $\mu$mole 3-(3′,4′-dichlorophenyl)-1,1-dimethylurea (DCMU)], 1.5 $\mu$mole ferricyanide (FeCy), 0.5 $\mu$mol or NADP$^+$ plus 0.03 $\mu$mole spinach ferredoxin. Magnesium ions are necessary but should not exceed 0.3 m$M$. The gas phase is air.

Coupling was reported with a P/2*e* value of 0.72 (system H$_2$O → FeCy). In contrast to the method described below, BSA neither increased the rates nor did it stabilize activity. Methylamine (even at 40 m$M$) uncoupled by less than 50%, the effect of ammonia was not reported.

TABLE IV
PHOTOSYNTHETIC ELECTRON TRANSPORT AND PHOTOPHOSPHORYLATION WITH ISOLATED
CHLOROPLAST MATERIAL FROM VOLVOCALES[a]

| Method | Micromoles reduced/mg Chl × hr (assay system) | Micromoles ATP/ mg Chl × hr | P/2e | Micromoles ATP/mg Chl × hr | |
|---|---|---|---|---|---|
| | | | | PMS | DCIP/ ascorbate → MV (DQ) |
| *Chlamydomonas reinhardi* | | | | | |
| (1) Togasaki[b,c] | 360 (H$_2$O → BQ) | 118 | 0.33 | 390–450 | |
| | 372 (H$_2$O → FeCy) | 133 | 0.72 | | |
| | 250 (H$_2$O → NADP$^+$) | | | | |
| (2) Yannai[d] | 656 (H$_2$O → FeCy) | 177 | 0.54 | 484 | 43 |
| | 104 (H$_2$O → MV) | 172 | 0.83 | | |
| *Dunaliella parva* | | | | | |
| (3) Ben-Amotz[e] | 60 (H$_2$O → FeCy) | 65 | | 200 | 35–50 |
| | 80 (H$_2$O → DQ) | 95 | | | |
| | 14 (H$_2$O → NADP$^+$) | 30 | | | |
| (4) Böger[f] | 490 (H$_2$O → FeCy) | | | | |
| | 130 (H$_2$O → NADP$^+$) | — | | | |
| (5) Gimmler[g] | 300 (H$_2$O → FeCy) | 100 | — | 200 | 120 |

[a] Chl, chlorophyll; BQ, benzoquinone; FeCy, potassium ferricyanide; MV, methyl viologen (1,1′-dimethyl-4,4′-dipyridylium dichloride); DQ, 1,1′-ethylene-2,2′-dipryidylium
dibromide; PMS, 5-methylphenazoniummethyl sulfate (phenazine methosulfate);
DCIP, dichlorophenolindophenol.
[b] Data from Curtis *et al.*[25]
[c] Adopted from Brand *et al.*[26]
[d] From Yannai *et al.*[27]
[e] See Ben-Amotz.[30]
[f] Unpublished, see text.
[g] Unpublished. (pers. commun.)

*Method of Neumann and Collaborators.*[27] A medium of high osmolarity containing serum albumin was applied. About 2 ml of cell suspension (8 × 10$^8$ cells of *Chlamydomonas reinhardii,* wild type, strain 137-C
from the early log phase of growth) with about 3 mg of chlorophyll was
added to 3 ml solution of human serum albumin (45 mg/ml) and 3 ml of
medium containing 400 m$M$ sucrose, 10 m$M$ Tricine–NaOH buffer (pH
7.8), and 10 m$M$ NaCl. Cells were disrupted for 20 sec at 0° and 60 W
with the microtip of a Branson sonifier. Then, the homogenate was

[27] Y. Yannai, B. L. Epel, and J. Neumann, *Plant Sci. Lett.* **7,** 295 (1976).

centrifuged for 7 min at 2000 $g$, the pellet discarded, and the supernatant centrifuged again for 7 min at 18,800 $g$. The pellet was resuspended in a small volume of homogenization medium containing albumin (4.5 mg albumin added per 1.1 ml of suspended chloroplast material). The extent of cell breakage was not reported.

ATP formation and electron-transport rates (see Table IV) were determined under aerobic conditions in 3-ml reaction volume containing, besides 60 $\mu$g of chlorophyll, 50 $\mu$moles Tricine buffer, pH 8.0, 100 $\mu$moles NaCl, 2.5 $\mu$moles MgCl$_2$, 10 $\mu$moles P$_i$, 3 $\mu$moles ADP, Hill reagents, and further additions in concentration ranges as generally applicable. Excellent stability was found. Noncyclic photophosphorylation activity remained with a halflife of about 24 hr and appeared to be due to the albumin. The coupling was quite high (P/2$e$ = 0.83 at maximum); no uncoupling was observed even with 5 m$M$ NH$_4$Cl in both the H$_2$O $\rightarrow$ MV and DCIP/ascorbate $\rightarrow$ MV systems.

## B. Dunaliella

*Dunaliella parva* was grown according to Ben-Amotz and Ginzberg[28] with minor modifications made by some authors. The mineral medium included 1.5 $M$ NaCl for optimum growth. The alga can be synchronized according to Gimmler *et al.*[29] Cells were osmotically broken by dilution, i.e., introducing them into a solution containing about 25 m$M$ NaCl and 0.01 $M$ Tris–NaOH.[30,31] The reaction media used were more or less standard. ATP formation is satisfactory although less than achieved with *Chlamydomonas*. Noteworthy is the low activity of linear electron transport from water to NADP$^+$ which was but poorly stimulated by as much as 40 $\mu M$ *Dunaliella* ferredoxin.[30] However, the hypothesis proposing ferredoxin as unnecessary for photosynthetic NADP$^+$ reduction[32] could not be confirmed with cell-free systems obtained by cell breakage with glass beads (P. Böger, unpublished). Chloroplast material prepared and assayed according to the procedures applied to *Bumilleriopsis*[2] exhibited a strict dependence on ferredoxin, saturating at about 3 $\mu M$ with rates approaching those of other systems[33] (Table IV).

Photophosphorylation with *Dunaliella* chloroplast material was generally found to be rather poor, as was the P/2$e$ ratio. Chloroplasts pre-

[28] A. Ben-Amotz and B. Z. Ginzburg, *Biochim. Biophys. Acta* **183**, 144 (1969).
[29] H. Gimmler, R. Schirling, and U. Tobler, *Z. Pflanzenphysiol.* **83**, 145 (1977).
[30] A. Ben-Amotz, *Plant Physiol.* **49**, 240 (1972).
[31] H. Gimmler, *Z. Pflanzenphysiol.* **68**, 289 (1973); also personal communication.
[32] A. Ben-Amotz and M. Avron, *Plant Physiol.* **49**, 244 (1972).
[33] P. Böger, *Z. Naturforsch., Teil B* **26**, 807 (1971).

pared by homogenization with glass beads as done with *Bumilleriopsis*[2] exhibited almost no photophosphorylation activity, although electron transport rates were generally higher than those of the assay system using chloroplast material obtained by osmotic shock. Activity of light-induced $NADP^+$ reduction remained constant for 3–5 hr when kept at $0°$. This could not be altered by adding diphenylcarbazide and/or ferredoxin–$NADP^+$ reductase.

## IV. Chloroplast Material from Chlorococcales

### A. *Scenedesmus*

An improved method was reported[34] to isolate active chloroplast material from heterotrophically and autotrophically grown *Scenedesmus obliquus*, strain $D_3$ (see Bishop and Senger[35] and Kessler *et al.*[36] for growth conditions). Cells from 250 ml suspension (equivalent to 1–2.5 ml packed cell volume) were washed and transferred to a freshly prepared homogenization medium of 20 m$M$ Tricine–KOH (pH 7.5), 10 m$M$ KCl, 0.05 m$M$ EDTA, and 0.5 m$M$ dithiothreitol [19.5 ml in a 50-ml stainless steel cup of a Vibrogen Zellmühle, (Bühler, Tübingen, Germany)]. The remainder of the cup was filled with glass beads of 0.35 mm diameter. The vessel was closed and after cooling to $0°$ was shaken for 5 min at full speed. The homogenate was separated from the beads using a washing solution (20 m$M$ Tricine–KOH, pH 7.5, and 10 m$M$ KCl) and was made up with sucrose to 0.4 $M$. After appropriate centrifugation the top layer of the pellet was used and suspended in about 1 ml of storage solution [0.4 $M$ sucrose, 30 m$M$ KCl, and 1%, (w/v) BSA]. After centrifugation (5 min, 620 $g$) the supernatant was diluted to 250 $\mu$g Chl/ml with a solution containing 0.4 $M$ sucrose, 30 m$M$ KCl, 20 m$M$ Tricine–KOH, pH 7.8, and 1% albumin.

The assay for light-induced $NADP^+$ reduction included in a 3 ml volume 400 $\mu$moles sucrose, 50 $\mu$moles Tricine–KOH, pH 7.8, 60 $\mu$moles $Cl^-$, 3 $\mu$moles $NH_4^+$, 1 $\mu$mole $NADP^+$, and 87–143 $\mu$g *Scenedesmus* ferredoxin. A dramatic increase (threefold) of the rate was obtained by adding ferredoxin–$NADP^+$ reductase (12 units).[37] Rates were 240–260 $\mu$moles NADPH formed/mg Chl per hr with material from heterotrophically grown cells. Rates were about half with autotrophic cells. Apparently, this is due to higher chlorophyll content of the latter material

[34] R. J. Berzborn and N. I. Bishop, *Biochim. Biophys. Acta* **292**, 700 (1973).
[35] N. I. Bishop and H. Senger, Vol. 23, p. 53.
[36] E. Kessler, W. Arthur, and J. E. Brugger, *Arch. Biochem. Biophys.* **71**, 326 (1957).
[37] A. T. Jagendorf, Vol. 6, p. 430.

(which is about twice that of the heterotrophic cultures). The presence of $Cl^-$ was obligatory. Activity was preserved for about 6 hr at 0°. Neither photophosphorylation, nor Hill activity, nor the degree enhancement by ammonia were reported. Applying this method, high rates of $NADP^+$ reduction were achieved also with *Chlamydomonas*.

Substantial and reproducible Hill activity was measured with chloroplast material from *Scenedesmus acutus* (strain No. 276-3a) for the systems $H_2O \rightarrow MV$ or diaminodurene (DAD)/ascorbate $\rightarrow MV$.[38] Cells were harvested from mass cultures after about 20 hr of growth in the logarithmic phase. A homogenization medium as described for the preparation of *Bumilleriopsis* chloroplast material was used except for the omission of PVP.[2] The ratio of glass beads to algae was found to be critical. Reproducible and optimum rates were obtained with 13–15 gm of wet algal paste, 13–15 ml homogenization medium, and 42 gm glass beads of 0.5 mm diameter. Good cell breakage was achieved in a Merkenschlager homogenizer by 1-min homogenization time at 4000 rpm. Rates were about 80–90 $\mu$mol $O_2$ uptake/mg Chl per hr (system $H_2O \rightarrow MV$). Standard assay procedures and mixtures were used. Photosystem I activity was as high as 250 $\mu$moles provided exogenous plastocyanin or cytochrome $c$-553 from *Scenedesmus* was added.[38] Photophosphorylation activity was virtually absent.

Senger and co-workers were also successful in isolating an active preparation from synchronized *Scenedesmus obliquus* strain D.[39] A Vibrogen Zellmühle (see above) was used for shaking cell samples mixed with glass beads in stainless steel cups. The volume of the cup, the size of the glass beads, and the ratio of glass beads to cup volume are important for optimum breakage. The procedure was similar to Berzborn and Bishop.[34] The activity was tested after centrifugation and resuspending the pelleted chloroplast material in phosphate buffer (0.05 $M$, pH 7.5). The chloroplast material yielded rates of 485 micromoles of reduced DCIP/mg Chl per hr. Photosystem I-mediated reduction of methylviologen (0.2 m$M$) with ascorbate (5 m$M$) and DCIP (0.1 m$M$) as electron donor yielded values of 1130 $\mu$moles $O_2$/mg Chl per hr in the presence of 2.5 $\mu M$ DCMU. The activities related to photosystem I and II varied within the cell cycle. Values for chloroplast material from cells of minimum activity were lower (30% for system I and 50% for system II) than the corresponding values from synchronized cells with highest photosynthetic rates. No data on light-induced $NADP^+$ reduction or photophosphorylation were given.

[38] K.-J. Kunert, H. Böhme, and P. Böger, *Biochim. Biophys. Acta* **449**, 541 (1976).
[39] H. Senger and V. Mell, *Methods Cell Biol.* **15**, 201 (1977).

Further disintegration of the chloroplast material with 1% digitonin and subsequent separation on a sucrose gradient yielded three distinct bands.[40] The upper band contained photosystem I particles without Hill activity. The photosystem I-mediated reduction of methylviologen, measured as $O_2$ uptake under the same conditions as above, yielded rates of 3000 $\mu$moles $O_2$/mg Chl per hr. Particles from cells of the maximum and minimum stages of photosynthetic capacity attained the same values. The constant activity of photosystem I during the cell cycle with subchloroplast particles and the variability of this activity with untreated chloroplast material as obtained above indicates that the chloroplast material still contains most of the plastoquinone, which is claimed to limit electron flow according to the developmental stage. Difference in photosystem I and II activity with isolated chloroplasts was also observed with synchronized *Bumilleriopsis*.[19]

## B. Chlorella[41,42]

*Chlorella fusca* Shihira et Kraus (strain 211-15 from the Agal Culture Collection at Göttingen) was grown autotrophically.[41] Cells (1.5–2.5 gm fresh weight) were washed and resuspended in 4.5 ml medium of 50 m$M$ $N$-tris(hydroxymethyl)methyl-2-aminoethanesulfonic acid (TES), pH 7.5, 0.25 $M$ sucrose, 20 m$M$ $MgCl_2$, 1 m$M$ $CaCl_2$, 5 m$M$ isoascorbate, 5% Ficoll, 2.5% dextran, and 0.5% BSA. Glass beads (12 ml, 0.1 mm diameter) and 4 ml of glass beads (0.35 mm) were added and the viscous mixture was shaken three times for 5 sec in a Nossal Shaker in the cold. The resulting homogenate was diluted with 20 ml of the medium mentioned above and filtered through two layers of Miracloth. Whole cells were discarded by low-speed centrifugation at 300 $g$ for 10 min. After centrifugation at 1000 $g$ for 20 min, the pellet was resuspended in 0.3 $M$ sucrose containing 0.01 $M$ KCl, phosphate 0.05 $M$ buffer, pH 7.2, and 0.5% BSA. It was centrifuged at 10,000 $g$ for 10 min, and the pellet resuspended in the same medium. This fraction could be further treated with digitonin and Triton X-100[41] to obtain subchloroplast particles. Reaction conditions are given in Grimme and Boardman.[43]

One fraction ($P_1$) resembles whole cells in absorption and fluorescence properties and exhibits good rates of $NADP^+$ photoreduction with $H_2O$ as donor (Table V), provided saturating amounts of ferredoxin and plas-

[40] H. Senger and V. Mell, in preparation.
[41] L. H. Grimme and N. K. Boardman, *Biochem. Biophys. Res. Commun.* **49**, 1617 (1972).
[42] L. H. Grimme and N. K. Boardman, *Proc. Int. Congr. Photosynth. Res., 3rd, 1974* (M. Arron, ed.), Elsevier, North Holland, Amsterdam, New York, Oxford. Vol III, p. 2115 (1975).
[43] L. H. Grimme and N. K. Boardman, *Hoppe-Seyler's Z. Physiol. Chem.* **354**, 1499 (1973).

TABLE V

PHOTOCHEMICAL ACTIVITIES OF *Chlorella fusca* CHLOROPLAST MATERIAL[a]

| | Rates in $\mu$moles reduced/mg Chl $\times$ hr | |
| --- | --- | --- |
| Assay system[b] | ($-$) DCMU | ($+$) DCMU |
| $H_2O \rightarrow NADP^+$ | 122 | 1.7 |
| $H_2O \rightarrow TCIP^{+c}$ | 69 | 0.9 |
| $H_2O \rightarrow FeCy$ | 176 | 0 |
| DCIP/ascorbate $\rightarrow NADP^+$ | 96 | 87 |
| DPC $\rightarrow$ DCIP | 117 | 31 |

[a] From Grimme and Boardman.[41]
[b] With fraction $P_1$ of Grimme and Boardman.[41]
[c] TCIP, trichlorophenolindophenol.

tocyanin were added. Of the detergent-treated material, one fraction can be associated with photosystem II (128 $\mu$moles DCIP reduced/mg Chl per hr with diphenylcarbazide as electron donor); another with photosystem I activity, yielding a high rate of light-induced $NADP^+$ reduction (DCIP/ascorbate $\rightarrow NADP^+$ : 430 $\mu$moles/mg Chl per hr).

Neither photophosphorylation nor the influence of ammonia were reported.

## V. Chloroplast Material from *Euglena* (Euglenophyceae)

*Crude Chloroplast Preparations from Euglena gracilis.* The earlier method of Katoh and San Pietro[44] was recently modified.[45] Cells were grown under autotrophic conditions with 5% $CO_2$ (v/v) in air according to Böger and San Pietro[46] and illuminated with incandescent light of 150 lux. After harvesting, the cells were washed once and resuspended in 40 m$M$ Tricine–NaOH, pH 7.8, containing 0.33 M mannitol and 5 m$M$ $MgCl_2$ at a ratio of 10 gm of wet weight of cells per 100 ml medium. The material was broken either by passage through a Yeda press at 650 psi or in a Sorvall Ribi RF-1 (French press) at 6000 psi. The homogenate was spun down at 500 $g$ for 1 min and the supernatant centrifuged at 1000 $g$ for 5 min to obtain the chloroplast fraction.[45]

[44] S. Katoh and A. San Pietro, *Arch. Biochem. Biophys.* **118**, 488 (1967).
[45] G. F. Wildner and G. Hauska, *Arch. Biochem. Biophys.* **164**, 127 (1974).
[46] P. Böger and A. San Pietro, *Z. Pflanzenphysiol.* **58**, 70 (1967).

*Reaction Conditions.* The assay for $NADP^+$ reduction contained, in 1 ml, 50 m$M$ HEPES–NaOH, pH 6.5, 50 m$M$ NaCl, 5 m$M$ $MgCl_2$, 0.25 m$M$ $NADP^+$, 10 $\mu M$ ferredoxin (either from spinach or *Euglena*), and 10 $\mu$g chlorophyll. The assay was performed in an open cuvette in a Zeiss PMQ II photometer equipped with appropriate cross-illumination; light intensity was $4 \times 10^4$ erg/cm$^2$ × sec. Cyclic phosphorylation was measured according to McCarty and Racker[47] by incorporation of $^{32}P_i$ into ADP. Rates of electron transport ($H_2O \rightarrow NADP^+$) and PMS-catalyzed cyclic photophosphorylation were 29 and 83 $\mu$moles/mg Chl × hr and correlated with the content of endogenous cytochrome *c*-552 retained in the chloroplasts (see findings with *Bumilleriopsis*[21]).

*Isolation of Intact Chloroplasts from Euglena gracilis.* This isolation was recently reported.[48] Cells were suspended at a ratio of 1 gm wet weight per 2 ml solution made of 0.15 $M$ sucrose, 0.15 $M$ sorbitol, 1% Ficoll (w/v), 2 $\mu$g/ml polyvinyl sulfate, 15 m$M$ NaCl, 5 m$M$ mercaptoethanol, 5 m$M$ HEPES–NaOH, pH 6.8, and broken in a French press at 105 kg/cm$^2$. The resulting homogenate was fractioned on a 20–70% linear silica sol gradient (Ludox AM) containing 10% polyethylene glycol (w/v), glutathione in place of 5 m$M$ mercaptoethanol, and 1% BSA in the above-mentioned solution. Although light-driven $CO_2$ incorporation could not be demonstrated, light-dependent protein synthesis was shown.

## VI. Isolation of Chloroplast Material from Other Algae

### A. *Acetabularia mediterranea (Siphonales)*

The preparation of intact chloroplasts capable of fixing $CO_2$ at physiological rates was reported earlier.[49] Recently, the isolation of photosystem II particles from this alga was reported.[50] The chloroplast material was treated with 1% Triton X-100 and separated by sucrose gradient centrifugation. A photosystem II-active fraction photoreduces DCIP at 25 $\mu$moles/mg Chl × hr but shows no activity in light-induced $NADP^+$ reduction.[50]

### B. *Codium vermilara (Siphonales)*

Thalli of this alga were grown and harvested as reported.[51] Class I chloroplasts were obtained from *Codium* by hand-grinding and centrifu-

[47] R. E. McCarty and E. Racker, *J. Biol. Chem.* **242**, 3435 (1967).
[48] J. L. Salisbury, A. Vasconcelos, and G. L. Floyd, *Plant Physiol.* **56**, 399 (1975).
[49] R. G. S. Bidwell, W. B. Levin, and D. C. Shepard, *Plant Physiol.* **49**, 946 (1969).
[50] K. Apel, L. Bogorad, and C. L. F. Woodcock, *Biochim. Biophys. Acta* **387**, 568 (1975).
[51] M. Schönfeld, M. Rahat, and J. Neumann, *Plant Physiol.* **52**, 283 (1973).

gation.[51] Maximum $CO_2$ fixation rate (62 $\mu$moles/mg Chl $\times$ hr) was obtained in high salt concentration (0.5 $M$ NaCl). Ferricyanide was reduced at 240 $\mu$moles/mg Chl $\times$ hr but only with osmotically shocked chloroplasts. Apart from the quite unique functional stability of $CO_2$ fixation which is retained for several hours and the physical integrity as demonstrated by electron micrographs, no further properties were reported. Several recent articles should be consulted for further details on plastids from siphonous algae.[52–54]

## C. Fucus serratus L. (Phaeophyceae)

Fronds of *Fucus* were collected in the littoral zone of Helgoland and chloroplasts were obtained after short homogenization in a complex medium.[55] Chloroplasts, 80% intact, fix $CO_2$ at 2 $\mu$moles/mg Chl $\times$ hr; $O_2$ evolution is 20 $\mu$moles/mg Chl $\times$ hr and is increased to 105 $\mu$moles after osmotic shock upon addition of ferricyanide. No further photosynthetic properties were given.

[52] A. H. Cobb, *Protoplasma* **92**, 137 (1976).
[53] S. W. Wright and B. R. Grant, *Plant Physiol.* **61**, 768 (1978).
[54] S. Katoh, A. Yamagishi, and T. Yamaoka, *Plant Cell Physiol.* **16**, 1093 (1975).
[55] G. Nordhorn, M. Weidner, and J. Willenbrink, *Z. Pflanzenphysiol.* **80**, 153 (1976).

# [11] Isolation of Chloroplast Envelope Membranes

*By* RAYMOND P. POINCELOT

Since $CO_2$ and other metabolites required for photosynthesis must be transported across the chloroplast envelope membrane prior to $CO_2$ reduction, and photosynthetic products must also cross this barrier, the regulatory functions of the envelope membrane are obviously of great importance. However, chloroplast envelope membranes were not isolated until the early 1970's, when procedures were developed independently in three laboratories.[1–3] Because of the need for isolated chloroplasts with intact envelope membranes, these procedures were confined to mesophyll chloroplasts derived from spinach and broad bean. These envelope membrane preparations suffered from one drawback: the ma-

[1] R. L. Mackender and R. M. Leed, *Nature (London)* **228**, 1347 (1970).
[2] R. Douce, R. B. Holz, and A. A. Benson, *J. Biol. Chem.* **248**, 7215 (1973).
[3] R. P. Poincelot, *Arch. Biochem. Biophys.* **159**, 134 (1973).

jority of the isolated membranes were incomplete, i.e., they had lost one of the two membranes which collectively comprise the envelope membrane. Subsequently, my earlier method was refined and resulted in preparations which consisted mainly of complete envelope membranes isolated from mesophyll chloroplasts of spinach, sunflower, and maize.[4,5]

Envelope membranes from bundle sheath chloroplasts are more difficult to isolate, primarily because of the difficulties involved in isolating undamaged bundle sheath chloroplasts. However, the usefulness of enzymic digestion for the isolation of protoplasts led to the use of enzymic techniques for the isolation of envelope membranes from bundle sheath chloroplasts.[6,7] Together with envelope membranes from mesophyll chloroplasts, these provide a useful system for making comparative biochemical and permeability studies among plants that vary in their anatomy, biochemical pathways, and photosynthetic efficiency.

### Isolation of Envelope Membranes from Mesophyll Chloroplasts

*Plant Material and Growing Conditions.* Spinach (*Spinacia oleracea* L., var. Viroflay, Asgrow Seed Co.), maize (*Zea mays* L., hybrid 595S, Agway, Inc.), and sunflower (*Helianthus annuus* L., hybrid 896 [cmsHA 89 × RHA 266], a gift from G. Fick, United States Department of Agriculture at North Dakota State University) were grown in trays of vermiculite. Nutrient solution was added twice weekly. During late fall through early spring, plants were grown in a greenhouse maintained at day/night temperatures of 22°/16°. During the rest of the season plants were grown in trays outdoors.

*Chloroplasts.* The best preparations of envelope membranes are derived from chloroplasts that are predominately intact. Fresh, unwilted leaves give the best results. Starch was generally not present in amounts large enough to cause disruption of the envelope membranes during the isolation of spinach, sunflower, or maize chloroplasts. If large amounts of starch are encountered with other plant species, or with certain growing conditions, the percentages of intact chloroplasts are higher if starch levels are reduced by darkening of the plants several hours prior to harvesting.

Intact mesophyll chloroplasts were isolated from 10-gm batches of freshly harvested, rinsed leaves from 4- to 6-week-old plants of spinach,

[4] R. P. Poincelot and P. R. Day, *Plant Physiol.* **54**, 780 (1974).
[5] R. P. Poincelot and P. R. Day, *Plant Physiol.* **57**, 334 (1976).
[6] A. W. Ruesink, Vol. 23, p. 197.
[7] R. P. Poincelot, *Plant Physiol.* **60**, 767 (1977).

sunflower, and maize by procedures established by others.[8-10] Mutiples of 40 gm of leaves were used, since four chloroplast pellets were required to produce a suspension of chloroplast envelope membranes of adequate volume for the 34-ml ultracentrifuge tubes subsequently used (Beckman Model L ultracentrifuge).

Bundle sheath chloroplasts were isolated from bundle sheath strands prepared by the following method.[7] Maize leaves were freshly harvested from plants 4 to 6 weeks old. The leaves were halved by removing the midvein, and the halves were cut into sections 2 to 4 cm in length. About 56 gm of leaf sections were divided into 7-gm batches. These were ground at full speed for 2 min in 30 ml of a modified Jensen and Bassham's solution A in a semi-micro Waring blender at 4°.[9] Solution A contained the following: 0.33 $M$ sorbitol, 50 m$M$ 2-($N$-morpholino)ethanesulfonic acid adjusted to pH 6.1 with NaOH, 20 m$M$ NaCl, 2 m$M$ NaNO$_3$, 2 m$M$ EDTA (dipotassium salt), 1 m$M$ MgCl$_2$, 1 m$M$ MnCl$_2$, and 0.5 m$M$ K$_2$HPO$_4$. Each homogeneate was filtered through six layers of cheesecloth. The residual material, which consists of bundle sheath strands and some adhering mesophyll cells, was collected from the eight separate cheese-cloths and combined. The combined material was added to 150 ml of solution A containing 1.0% (w/v) each of combined preparations of Cellulysin™ (Calbiochem), hemicellulase (Sigma), and pectinase (Sigma). The resulting suspension was stirred gently for 45 min at room temperature to allow enzymic digestion of the remaining mesophyll cells. Afterward the suspension was filtered through six layers of cheesecloth, and the residue washed with 50 ml of solution A to remove any remaining enzyme. The residue, which consists of bundle sheath fibers, was divided into four parts. Each part was homogenized separately in a Waring blender for 45 sec at full speed in 25 ml of solution A at 4°. The resulting homogenate was filtered through six layers of cheesecloth and the filtrate was centrifuged at 2000 $g$ for 10 min at 4°. The pellet consisted of bundle sheath chloroplasts.

*Mesophyll Chloroplast Envelope Membranes.* Each pellet of intact mesophyll chloroplasts from spinach, sunflower, or maize was suspended in 3 ml of hypotonic medium, which consisted of 50 m$M$ Tricine buffer at pH 7.6 and 4°. A 5 ml hypodermic syringe with a 10-cm 14-gauge cannula facilitated the suspension. These suspensions of about 0.50 to 0.80 mg of chlorophyll/ml were maintained at 4° for 20 min with occa-

[8] B. R. Grant, C. A. Atkins, and D. T. Canvin, *Planta* **94,** 60 (1970).
[9] R. G. Jensen and J. A. Bassham, *Proc. Natl. Acad. Sci. U.S.A.* **56,** 1095 (1966).
[10] K. C. Woo, J. M. Anderson, N. K. Boardman, W. J. S. Downton, C. B. Osmond, and S. W. Thorne, *Proc. Natl. Acad. Sci. U.S.A.* **67,** 18 (1970).

sional swirling. If desired, the rupture of the chloroplasts, the distension of the envelope membrane, and its complete separation can be monitored under a light microscope at 1200 × with oil immersion. A 1% solution of osmium textroxide used as a stain (handle with extreme caution) aids greatly in the visualization of the envelope membrane. If plant chloroplasts other than spinach, sunflower, or maize are used, it is advisable to monitor the separation of the envelope membrane as a function of time.

After removal of the envelope membrane by osmotic shock, suspensions from four chloroplasts pellets were combined and homogenized by making three passes in a Ten Broeck glass homogenizer. This was found to improve the complete separation of the envelope membranes from the damaged chloroplasts. Twelve milliliters of this homogenized suspension was diluted at 4° to 16 ml with 48% (w/v) sucrose in 50 m$M$ Tricine buffer at pH 7.6 to produce a final suspension of 12% (w/v) sucrose.

The suspension was fractionated on a three phase discontinuous sucrose gradient in a Beckman Model L preparative ultracentrifuge. The sucrose gradient, buffered throughout with 50 m$M$ Tricine buffer at pH 7.6, consisted from top to bottom of 16 ml of the osmotically shocked suspension of chloroplasts in 12% sucrose (w/v), 6 ml of 23% sucrose (w/v), and 9 ml of 30% sucrose. Tubes of the above sucrose gradient were centrifuged in a swinging bucket rotor (type SW 25.1) at 4° for 90 min at 23,000 rpm (75,000 $g$ at $R_{max}$). After centrifugation two pale bands, visible as milky, white bands with back lighting, were present at the two interfaces. The upper band consists primarily of single membrane vesicles. The lower band consists of 75, 70, and 40% complete (double) envelope membranes for spinach, sunflower, and maize, respectively. The green pellet at the bottom of the centrifuge tube consists mostly of lamellae.

*Bundle Sheath Chloroplast Envelope Membranes.* Envelope membranes were not removed from bundle sheath chloroplasts by osmotic shock, since the envelope membranes were so fragile that they were fragmented. Each of the four chloroplast pellets was suspended in 3.0 ml of 12% (w/v) sucrose in 50 m$M$ Tricine at pH 7.6. A 5-ml hypodermic syringe fitted with a 10-cm 14-gauge cannula was used to suspend the chloroplasts; the hydrodynamic shear forces developed in the cannula released the envelope membranes. The further purification on the discontinuous sucrose gradient differed in the following manner. The sucrose gradient, buffered throughout with 50 m$M$ Tricine buffer at pH 7.6, consisted from top to bottom of 12 ml of the osmotically shocked chloroplasts in 12% sucrose (w/v), 9 ml of 18% sucrose, and 6 ml of 25% sucrose (w/v). Ultra-centrifugation conditions were the same as for mesophyll chloroplast envelope membranes. After centrifugation two pale bands, visible more readily with back lighting, were present at the two interfaces. The upper and lower bands contained incomplete and complete envelope membranes, respectively. Lamellae were present in the bottom of the tube.

*Measures of Purity.* A nonlatent $Mg^{2+}$-dependent ATPase activity, insensitive to $N,N'$-dicyclohexylcarbodiimide is present in the chloroplast envelope membrane.[2–5] As such, it serves as a marker enzyme for monitoring the increasing purity of the envelope membrane during the isolation procedure. Activity levels of the envelope membrane ATPase decrease as the envelope membrane is damaged, are highest with outdoor plant material grown under optimal conditions, and vary with plant species.[2–5] Optimal values reported for purified envelope membranes from mesophyll chloroplasts of spinach, sunflower, and maize are 80, 163, 126 $\mu$moles $P_i$ released/hr/mg protein, respectively.[5] For envelope membranes from bundle sheath chloroplasts of maize, the author found a value of 40.[7]

Stromal contamination, brought about by the trapping of stromal material in the resealed envelope membrane, does not occur. This is verifiable by assaying for the activity of stromal marker enzymes such as ribulosediphosphate carboxylase, P-glycolate phosphatase, fructose 1,6-diphosphatase, and carbonic anhydrase.[2–5]

Membrane contamination from other sources does not occur. Chlorophyll and ferredoxin–NADP$^+$ oxidoreductase, markers for lamellae, are not present, nor is NAD(P)H:cytochrome $c$ oxidoreductase, a marker for mitochondrial or microsomal membranes.[11]

Chloroplast envelope membranes were monitored for any bacterial or fungal contamination by plating on nutrient agar. If any were present, Millipore filtration (0.22 $\mu$m) of the preparative solutions and other aseptic techniques eliminated contamination.[4] A potential source of contamination was caused by aphids on leaves, since their sticky carbohydrate secretions were found to harbor yeasts that would appear in preparations of chloroplast envelope membranes. This was not a problem if aphids were kept under control.

Purity can also be assessed by electron microscopy. A typical electron micrograph of a field of chloroplast envelope membranes derived from mesophyll chloroplasts of spinach is shown in Fig. 1. Under higher magnification (Fig. 2) many of the envelope membranes were observed to have the double membrane structure typical of the chloroplast envelope. Regardless of plant species, the chloroplast envelope membranes all had a similar appearance and size.

[11] R. Douce, C. A. Mannella, and W. D. Bonner, *Biochim. Biophys. Acta* **292,** 105 (1973).

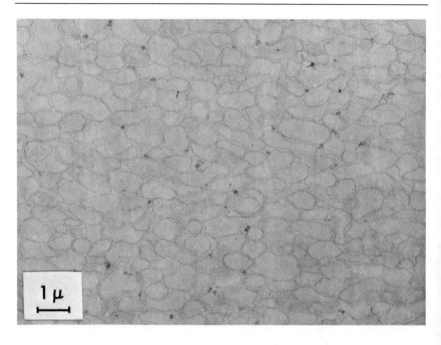

FIG. 1. A representative field of purified envelope membranes from spinach chloroplasts.

*Characterization.* Envelope membranes from mesophyll chloroplasts of different plant species show similarities in their lipid compositions, which have been extensively characterized by myself and others.[2,3,12–14] The glycolipids are the major lipids, and they consist mainly of monogalactosyldiglyceride and digalactosyldiglyceride. Phospholipids are present in lesser amounts, and the major lipids in this group are phosphatidylcholine and phosphatidylglycerol. These four lipids account for 75 to 90% of the lipids in the chloroplast envelope membrane. A number of lesser constituents, sterylglycoside, acylated sterylglycoside, sterol, and steryl ester are found only in the envelope membrane and not the lamellae.

The fatty acid compositions of the envelope membranes from mesophyll chloroplasts were predominately palmitic ($C_{16:0}$) and linolenic ($C_{18:3}$).[12,13] While the types of fatty acid were similar among the exam-

[12] R. P. Poincelot, *Plant Physiol.* **58**, 595 (1976).
[13] R. O. Mackender and R. M. Leech, *Plant Physiol.* **53**, 496 (1974).
[14] H. Hashimoto and S. Murakami, *Plant Cell Physiol.* **16**, 895 (1975).

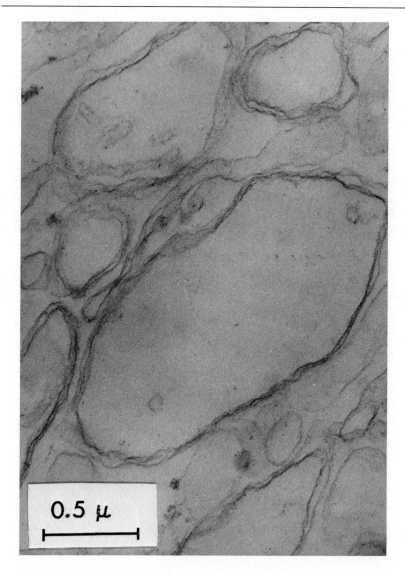

FIG. 2. An enlargement of a chloroplast envelope membrane from spinach showing the typical spacing between the two membranes comprising the envelope.

ined plant species, variation was noted in the overall unsaturated/saturated ratio for sunflower (1.4), spinach (0.9), and maize (0.3).[12]

The envelope membranes from mesophyll chloroplasts are also rich in hexosamine, and hence glycoprotein.[15] The envelope membrane would appear to be a glycolipid–glycoprotein membrane. About 50 to 60% of the membrane consists of lipid.[12]

*Permeability.* Permeability studies can be readily conducted with envelope membranes.[5,16,17] Experiments on permeability of envelope membranes were completed within an hour or two after their isolation, since their ability to transport metabolites declines with time.[16] Envelope membranes were maintained at about 4° in an ice bath prior to and during the experiment. Experiments were done in a $10 \times 75$ mm test tube maintained at 30° in a water bath. The reaction mixture was allowed to equilibrate for 30 sec, and the reaction was initiated (zero time) by the addition of the desired radioactive permeant. The final volume was 0.5 ml and contained 50 m$M$ Tricine (pH 7.6), 0.33 $M$ sorbitol, 3 m$M$ $Mn^{2+}$, 3 m$M$ $Mg^{2+}$ (chloride salts), 2 to 4$\mu$g of envelope membrane protein, and 1 m$M$ radioactive permeant (0.1 to $2.0 \times 10^6$ cpm/100 $\mu$l). Each addition was made with automatic pipettes and was followed by vigorous shaking on a mechanical mixer. Experiments were terminated by rapid removal of the sample with a 1-ml syringe, followed by vacuum filtration with a Millipore filter (25-mm diameter, GS, 0.22 $\mu$m). Completion of filtration constituted the end time. The filter was washed by syringe injection of 10 ml of Tricine buffer and 0.33 $M$ sorbitol at 4° and pH 7.6. This removed extra- and intramembranous radioactive permeant, but not permeant inside the membrane vesicle. The filter was removed after a 15 sec wait to ensure completion of filtration. Radio-activity of the filter was determined with a scintillation counter. Loss in counting efficiency resulting from the Millipore filter was corrected by using a known [14]C standard. Results were expressed on a milligram envelope membrane protein basis.

[15] D. Racusen and R. P. Poincelot, *Plant Physiol.* **57**, 53 (1976).
[16] R. P. Poincelot, *Plant Physiol.* **54**, 520 (1974).
[17] R. P. Poincelot, *Plant Physiol.* **55**, 849 (1975).

## [12] Subchloroplast Particle Enriched in P700 and Iron–Sulfur Protein

*By* JOHN H. GOLBECK

Chloroplast membrane-bound iron-sulfur proteins are currently the focus of considerable attention due to their participation in the primary photochemistry of photosystem I.[1] Detailed characterization of the structure and function of these proteins in photosynthesis requires large quantities of highly purified reaction center particles enriched in iron-sulfur protein and P700. The procedure described[2] employs the nonionic detergent Triton X-100 to liberate an iron-sulfur and P700-containing reaction center particle from the chloroplast lamellae and to free a significant amount of light-harvesting chlorophyll from the P700-containing fragment. The use of gel filtration and ion-exchange chromatography leads to the isolation of a photoactive particle having a chlorophyll/P700 ratio of 25 and containing 10-12 moles of nonheme iron (NHI) and acid-labile sulfide (ALS) per mole of P700. The particle is nearly devoid of chlorophyll $b$, cytochromes $f$, $b_6$, and $b$-559, and has a greatly diminished content of $\beta$-carotene.

### I. Preparation of the Particle

*A. Chloroplast Preparation*

A flowchart is provided in Fig. 1 which depicts the scheme used to prepare spinach chloroplasts for detergent fractionation. Depetiolated spinach leaves (1000 gm) are homogenized for 20 sec in 1000 ml of ice-cold buffer containing 0.05 $M$ Tris-Cl [tris(hydroxymethyl)aminomethane] (pH 7.8), 0.4 $M$ sucrose, and 0.01 $M$ KCl in a high-speed Waring blender. The homogenate is filtered through three layers of cheesecloth and one layer of Miracloth to remove debris, and the filtrate is centrifuged at 2500 g for 10 min. The chloroplasts are osmotically shocked by resuspending the pellet (Precipitate I) in 0.025 $M$ Tris-Cl buffer, pH 7.8, and 0.01 $M$ KCl for 5 min. The shocked chloroplasts are pelleted by centrifugation at 5000 g for 30 min. Coupling factor ($CF_1$) and other loosely bound proteins are removed from the chloroplast membrane by resuspending

---

[1] R. Malkin and A. R. Bearden, *Proc. Natl. Acad. Sci. U.S.A.* **68**, 16 (1971).

[2] J. H. Golbeck, S. Lien, and A. San Pietro, *Arch. Biochem. Biophys.* **178**, 140 (1977).

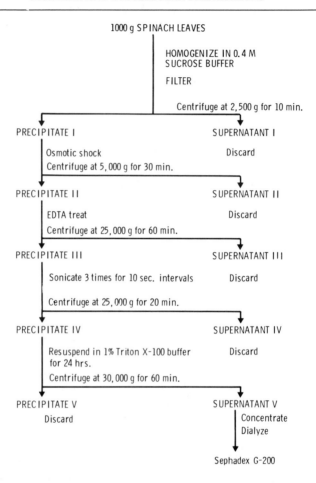

FIG. 1. Fractionation scheme for spinach chloroplasts. Adapted from Golbeck *et al.*[2]

the shocked chloroplasts (Precipitate II) in 0.7 m$M$ EDTA to a chloro-phyll concentration of 300 $\mu$g/ml, incubating 10 min at room temperature, and centrifuging at 25,000 $g$ for 60 min. The pellet (Precipitate III) is suspended in 300 ml of ice-cold buffer containing 0.05 $M$ Tris-Cl (pH 7.8) and 0.01 $M$ KCl and sonicated three times at high power for 10-sec intervals, insuring that the temperature in the vessel does not rise above 4°. The resulting thylakoid fragments are pelleted by centrifugation at 25,000 $g$ for 20 min. All steps are performed at 0° except for the EDTA wash. Thylakoid fragments (Precipitate IV) have been stored at −15° for periods up to 1 year without detectable loss of nonheme iron, labile sulfide, or P700 photoactivity.

## B. Detergent Fractionation

Thylakoid fragments (Precipitate IV) are resuspended in buffer containing 0.05 $M$ Tris-Cl (pH 9.0), 1% Triton X-100, and 0.2 $M$ KCl (Triton-KCl buffer) to a chlorophyll concentration of 300 $\mu$g/ml and stirred for 24 hr at 4°. Since the light photosystem I fragment and solubilized light-harvesting chlorophyll are released into the supernatant, the remaining heavier membrane fragments are removed by centrifugation at 30,000 $g$ for 60 min. A tenfold concentration of the supernatant is achieved by loading it (Supernatant V) into a well-washed dialysis sack and surrounding by crystalline sucrose for 8-12 hr. After concentration, the syrupy solution is squeezed to one end, the sack is retied and dialyzed against Triton-KCl buffer to remove the sucrose. Even though the dialysis sack may become very taut, breakages seldom occur. The concentrated supernatant may be stored at $-15°$ for at least 6 months.

## C. Chromatography

Approximately 50 ml of the concentrated supernatant is loaded onto a 2000-ml (5 × 100 cm) Sephadex® G-200 or Sephacryl® S-200 (Superfine) column and eluted with Triton-KCl buffer. Two chlorophyll-containing peaks emerge, as shown in Fig. 2. The first peak, labeled the G-I particle (G-200, fraction I) is considerably enriched in iron-sulfur protein and P700 relative to chlorophyll; the second peak, labeled the G-II particle (G-200, fraction II) contains cytochromes $f$ and $b_6^2$ and the majority of the light-harvesting chlorophyll.

The G-I particle is collected and dialyzed against 0.05 $M$ Tris-Cl (pH 8.3) and 1% Triton X-100 for 12 hr and applied to a 250 ml (4.5 × 18 cm) column containing DEAE-BioGel® A. After the entire preparation is loaded, the column is washed with 0.05 $M$ Tris-Cl (pH 8.3) and 1% Triton X-100 until the eluate appears colorless. The chlorophyll-containing flow-through, labeled the A-I fraction (agarose, fraction I), is photochemically inactive and devoid of protein (Table I). The DEAE column retains two populations of chlorophyll-containing protein. The intensely green band occupying the first several centimeters of the column is labeled the A-III particle (agarose, fraction III); a green band of lighter intensity, labeled the A-II particle (agarose, fraction II), occupies the remainder of the column. Frequently, the boundary between the two is indistinct. Both bands are eluted with a pulse of Triton-KCl buffer. The particles may be stored at 4° for up to 3 weeks without loss of photochemical activity.

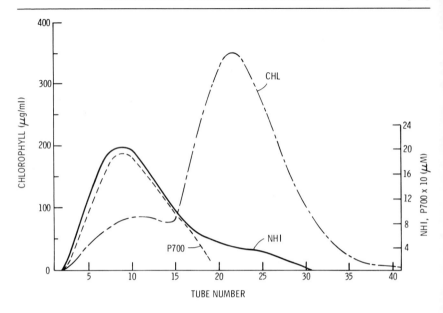

FIG. 2. Sephadex G-200 chromatography of supernatant V; distribution of P700 and nonheme iron (NHI). Fractions were collected before the first appearance of chlorophyll and analyzed for nonheme iron and P700. The flow rate was maintained at 80 ml/hr and 8-ml samples were collected. The contents of tubes 2-14 were pooled and constitute the G-I particle; the contents of tubes 16-32 were collected and constitute the G-II fraction.

## II. Properties of the Subchloroplast Fragments

### A. Fractionation of Spinach Chloroplasts

A detailed composition of spinach thylakoid fragments is shown in Table I. Treating the chloroplast fragments with Triton X-100 under the specified conditions results in the release of a photoactive particle (supernatant V) that contains 75-80% of the membrane-bound nonheme iron and 100% of the P700 and membrane-bound labile sulfide.[3] The remaining nonheme iron is most likely associated with PS-II since it pellets with this photosystem during the 30,000 $g$ spin. Assuming $\epsilon_{700\ nm} = 64\ \text{m}M^{-1}\ \text{cm}^{-1}$ for P700,[4] supernatant V is shown to contain 13-15 moles of nonheme iron and acid-labile sulfide per mole of P700 (Table II).

One caution regarding the chemical difference assay for P700 should

---

[3] J. H. Golbeck and A. San Pietro, *Anal. Biochem.* **73**, 539 (1976).
[4] T. Hiyama and B. Ke, *Biochim. Biophys. Acta* **267**, 160 (1972).

TABLE I

DISTRIBUTION OF COMPONENTS DURING THE FRACTIONATION OF SPINACH CHLOROPLASTS[a]

| Sample | Chl (mg) | a/b | NHI (μmole) | ALS (μmole) | P700[b] (μmole) | Cytochrome f (μmole) | Protein (mg) |
|---|---|---|---|---|---|---|---|
| Thylakoid fragments | 170 | 2.6 | 8.3 | 6.9 | 0.49 | — | 1030 |
| Supernatant V | 95 | 5.3 | 6.4 | 7.1 | 0.50 | — | 780 |
| Precipitate V | 71 | 1.3 | 1.9 | ND[d] | ND | — | 260 |
| Sephadex G-I | 14.8 | >6 | 3.29 | 3.15 | 0.36 | 0.02 | 410 |
| Sephadex G-II | 93.7 | — | — | 1.77 | 0.06 | 0.49 | 293 |
| DEAE flowthrough (A-I)[c] | 4.65 | — | ND | ND | ND | ND | ND |
| DEAE A-II[c] | 2.09 | >6 | 0.82 | 0.80 | 0.08 | ND | 62 |
| DEAE A-III[c] | 2.26 | >6 | 1.07 | 1.05 | 0.10 | ND | 76 |

[a] From Golbeck et al.[2]
[b] P700 is assayed by chemical difference spectroscopy.
[c] Only 60% of the G-I particle is used for ion-exchange chromatography.
[d] ND, not detectable.

be mentioned. In spinach chloroplasts or in chloroplast fragments, the P700 content is easily underestimated when determined by chemical-difference spectroscopy because ferricyanide is not totally effective in oxidizing 100% of the endogenous P700. For a more accurate estimation, Triton X-100 should be added to the cuvettes to a concentration of 1%; the chloroplast suspension will clarify immediately, and the P700 will become more accessible to the oxidant. After the chemical difference spectrum has been run, the baseline is determined by adding excess

TABLE II

CHLOROPHYLL/P700, NONHEME IRON/P700, AND LABILE SULFIDE/P700 IN ISOLATED FRACTIONS[a]

| Sample | Chl/P700[b] | NHI/P700[b] | ALS/P700[b] |
|---|---|---|---|
| Thylakoid fragments | 345 | 17.0 | 14.2 |
| Supernatant V | 190 | 12.8 | 14.3 |
| Sephadex G-I | 41.5 | 9.3 | 8.9 |
| Sephadex G-II | 1500 | — | — |
| DEAE A-II | 27.1 | 10.6 | 10.3 |
| DEAE A-III | 22.6 | 10.7 | 10.5 |

[a] From Golbeck et al.[2]
[b] P700 is determined by chemical difference spectroscopy.

ascorbate to the ferricyanide-containing cuvette and rescanning the pertinent wavelength range. This method assures that only the reversible component at 700 nm is determined. The P700 content, nevertheless, should be verified by photochemical estimation; since ascorbate or ascorbate-DPIP are effective in reducing P700 in the dark, the photochemical content should match the chemically determined P700 content for the determination to be considered reliable (see Fig. 4).

The data shown in Table I indicate that approximately 50% of the chlorophyll contained in the thylakoid fragments is released by detergent treatment into supernatant V, with a significant enrichment in chlorophyll *a*. Cytochromes *f* and $b_6$ are also solubilized in supernatant V, but the high chlorophyll content precludes an accurate estimation. $\beta$-Carotene is also liberated and rises to the surface during the centrifugation where it is carefully removed by aspiration.

## B. G-I Particle

Gel-permeation chromatography of supernatant V using Sephadex G-200 or Sephacryl S-200 (Superfine) results in two chlorophyll-containing peaks (Fig. 2). As shown in Tables I and II, the G-I particles is considerably enriched in iron-sulfur protein and P700 relative to chlorophyll. The NHI/P700 ratio has declined slightly from about 12-14 in supernatant V to 10 in the G-I particle due to the partition of a small population of iron–sulfur protein into the cytochrome-containing G-II fraction.

The visible spectrum of the G-I particle shows a red shift of the main chlorophyll peak from 669 nm in the thylakoid fragments to 673 nm in the G-I particle, indicating an enrichment in the far-red forms of chlorophyll *a*. $\beta$-Carotene is diminished, as indicated by the decline in absorbance between 450 and 500 nm. The particle is also characterized by a very low level of cytochromes *f* and $b_6$. The chlorophyll *a/b* ratio is greater than 6.

The G-I particle is photochemically active in the photooxidation of P700 and in the anaerobic photoreduction of ferredoxin and methyl viologen. The light-induced oxidation of P700 requires ascorbate to produce maximum photobleaching; the addition of submicromolar amounts of DPIP or $N,N,N',N'$-tetramethyl-*p*-phenylenediamine (TMPD) may be added to increase the dark rate of P700$^+$ re-reduction.

The particle also catalyzes a large, viologen-independent oxygen uptake in the light. Rates greater than 8000 $\mu$moles $O_2$/mg Chl/hr have been measured using ascorbate as the electron donor. There is no measurable uptake in the dark. The uptake is not mediated by the primary photochemical event, since destruction of P700 by heating is not accompanied by a decline in the rate of oxygen uptake.

## C. A-III Particle

Ion-exchange chromatography of the G-I particle on DEAE Bio-Gel A results in the isolation of two nearly identical photosystem I particles. Properties of the spinach A-II and A-III reaction center particles are given in Tables I and II. Ion-exchange chromatography provides no further fractionation of iron-sulfur protein from from the photoactive center, but an enrichment in nonheme iron, labile sulfide and P700 has occurred with respect to chlorophyll and total (Lowry) protein. The Chl/P700 ratio has declined to 25, and the NHI/P700 and ALS/P700 ratios remain between 10 and 12. Since both A-II and A-III fractions elute at the same salt concentration and have similar amounts of chlorophyll, protein, P700 and iron-sulfur centers, they may separate primarily on the basis of size. We will focus our attention on the spinach A-III particle at this point, although both particles are identical in the chemical characterizations that have been performed.

The visible spectrum of the A-III particle, shown in Fig. 3, shows an even further decline in absorbance between 450 and 500 nm, indicating the additional removal of $\beta$-carotene. Chemical analysis, moreover, in-

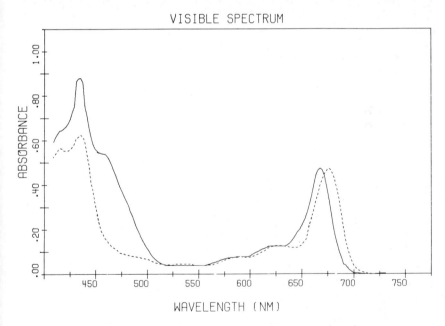

FIG. 3. Visible spectrum of spinach thylakoid fragments (Precipitate IV) and the A-III particle. The spectra were recorded with a Cary 14 spectrophotometer. The curves were normalized to 0.45 absorbance unit in the red.

dicates the presence of only 0.30 to 0.45 mole of $\beta$-carotene per mole of P700, depending somewhat on the preparation. The chlorophyll maximum in the red occurs at 675 nm.

The molecular weight of the A-III particle, as determined by gel filtration with Sephadex G-200, is between 150 and 200 kilodaltons. The polypeptide composition of the A-III particle, determined by SDS gel electrophoresis, indicates the presence of one large polypeptide chain (apparent molecular weight $\cong$ 80,000 daltons) and at least three low-molecular-weight polypeptide chains ($\sim$15,000 to 20,000 daltons). It should be noted that these molecular weights are with reference to "soluble" marker enzymes; the molecular weights of membrane-bound proteins may not be accurately reflected in this method since the polypeptide chains may not completely unfold in 1% SDS.

The P700 content determined during continuous illumination of the A-III particle is nearly identical to the amount determined by chemical difference spectroscopy (Fig. 4). The peak maximum in the red occurs at 697.5 nm and the shape is asymmetrical, showing a sharper drop in the blue than in the red. Flash-induced changes in P700 are shown in Fig. 5. In the presence of excess potassium ferricyanide, no significant absorption changes occur during a 5-$\mu$sec flash. In the presence of sodium ascorbate and DPIP, the full extent of photobleaching is produced but is followed by an immediate dark decay having a half-life of about 10 msec (Fig. 5A). Because methylviologen suppresses the decay completely (Fig. 5B), the dark decay represents the back reaction between P430$^-$ and P700$^+$ that occurs in the absence of a secondary electron acceptor.[5] Due to the greater net quantum efficiency of electron flow through the reaction center in the presence of viologen, weak light produces a larger amount of P700 photooxidation in its presence than in its absence. Spinach ferredoxin is able to suppress most of the dark decay (data not shown).

The intermediate electron acceptor, $A_2$, which functions immediately following the primary photoevent but prior to P430,[6] becomes visible in the A-III particle. Figure 6 shows the photooxidation of P700 after consecutive flashes in a spinach A-III particle incubated with dithionite at pH 10.0. The first flash induces an absorbance decrease at 700 nm which decays gradually in the presence of dithionite. (The absence of the 10-msec back reaction seen with ascorbate alone indicates that dithionite is most likely reducing P700$^+$, thereby complementing the P430$^-$ $\rightarrow$ P700$^+$ back reaction.) After a few flashes, a large acceleration in the P700$^+$

---

[5] T. Hiyama and B. Ke, *Proc. Natl. Acad. Sci. U.S.A.* **68**, 1010 (1971).

[6] K. Sauer, P. Mathis, S. Acker, and J. A. Van Best, *Biochim. Biophys. Acta* **503**, 120 (1978).

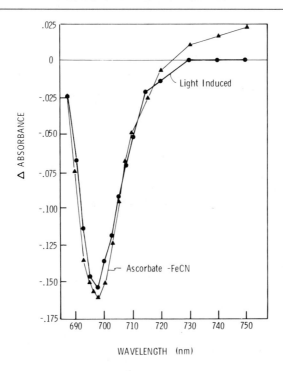

WAVELENGTH (nm)

FIG. 4. Spectral characteristics of P700 in the spinach A-III particle. The chemical difference spectrum of the A-III particle was obtained at a chlorophyll concentration of 58 $\mu$g/ml in buffer containing 0.05 M Tris-Cl (pH 8.8) and 1% Triton X-100 using a Cary 14 spectrophotometer. Light-induced changes were measured in a double-beam spectrophotometer during continuous illumination with saturating blue light. The reaction mixture contained spinach A-III particles at a chlorophyll concentration of 58 $\mu$g/ml, 0.003 $M$ sodium ascorbate, and 0.11 m$M$ DPIP in buffer containing 0.05 $M$ Tris-Cl (pH 8.8) and 1% Triton X-100.

decay is obtained, stabilizing in the rapid decay following the fourth or fifth flash. The half-life of the rapid decay is estimated to be 200-300 $\mu$sec; accordingly, the decay is ascribed to the reduced intermediate acceptor $A_2^-$, backreacting with $P700^+$.[7]

The ESR spectrum of the A-III particle indicates the presence of ESR Center A ($g$ = 1.86, 1.94, and 2.05) and Center B ($g$ = 1.89, 1.92, and 2.05) following chemical reduction with dithionite and methyl viologen (Fig. 7a). Both centers are partially photoreduced at 15° K when incubated with ascorbate in the dark prior to freezing (Fig. 7b). Recent studies

---

[7] J. H. Golbeck, B. Velthuys, and B. Kok, *Biochim. Biophys. Acta* **504**, 226 (1978).

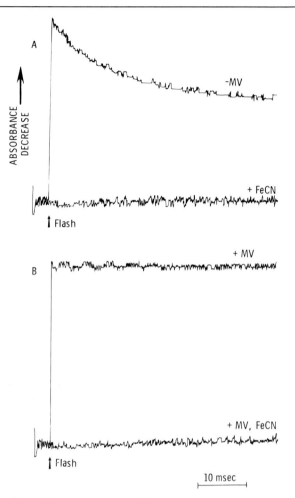

FIG. 5. Absorbance changes at 700 nm during a 5 $\mu$sec saturating blue flash. The reaction mixture contained spinach A-III particles at a chlorophyll concentration of 45 $\mu$g/ml in buffer containing 0.05 M Tris-Cl (pH 8.8) and 1% Triton X-100. The arrow indicates the onset of the flash. (A) Absorbance changes with 1.0 m$M$ sodium ascorbate and 0.076 m$M$ DPIP (top trace) and with 0.01 $M$ potassium ferricyanide (bottom trace). (B) Same as (A) except in the presence of 0.1 m$M$ methyl viologen.

have indicated that ESR Centers A and B may be composed of only [4Fe, 4S] cores.[8] Assuming there are 12 moles of nonheme iron and labile sulfide per mole of P700 in the photosystem I reaction center (Table II), ESR centers A and B alone may account for nearly two-thirds of the nonheme iron and labile sulfide in a spinach A-III particle. The data

[8] R. Cammack and M. C. W. Evans, *Biochem. Biophys. Res. Commun.* **67**, 544 (1975).

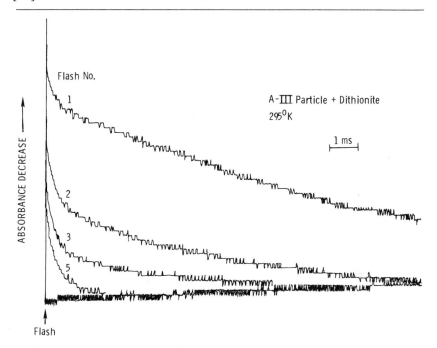

FIG. 6. Absorbance changes at 700 nm in a spinach A-III particle after the addition of a few grains of dithionite. The particles were suspended in glycine buffer (0.05 *M*, pH 10.0) containing 1% Triton X-100 and 0.3 *M* KCl and illuminated with a flash the number of times indicated. The baseline was determined using the measuring beam *only* prior to the first flash. The signal was stored in a Fabri-Tek 1052 prior to readout on a Hewlett-Packard 7004B X-Y Recorder.

indicate that a third iron-sulfur center remains to be found in photosystem I (ESR component "X"? See Evans *et al.*[9] and Golbeck and Kok[10] for discussion.)

The A-III particles may be stored at 4° for several weeks before degradation in labile sulfide or P700 photooxidation is detected. Long-term storage at −15°, however, is not advised since a progressive loss of activity has been noticed after several freeze-thaw procedures. Since the G-I particle is stable when frozen, it may be readily converted to an A-III particle when needed.

Under certain circumstances, it may become necessary to concentrate the enriched PS I particle in order to study changes in the chemical

[9] M. C. W. Evans, C. K. Sihra, J. R. Bolton, and R. Cammack, *Nature (London)* **256,** 668 (1975).
[10] J. H. Golbeck and B. Kok, *Arch. Biochem. Biophys.* **188,** 233 (1978).

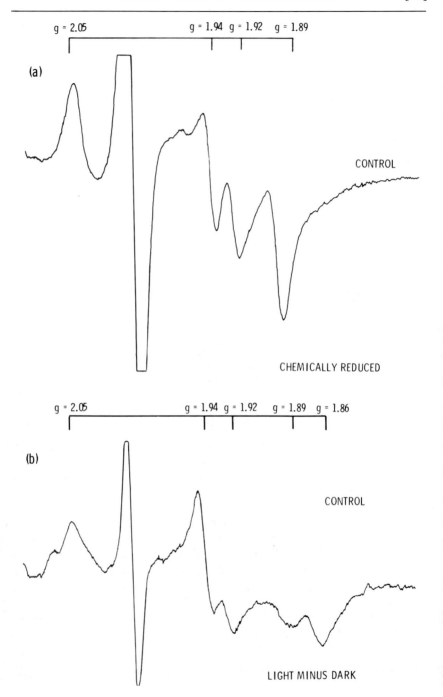

composition. Three methods have been used that allow concentration of the A-III particle without loss of photochemical activity.

1. The particles may be concentrated in a dialysis sack by surrounding with dry, crystalline sucrose. The sucrose is subsequently removed from the particles by dialysis against buffer containing 0.05 $M$ Tris-Cl (pH 8.3) and 1% Triton X-100 for 24 hr.

2. Ammonium sulfate may be added to a final concentration of 25% and the resulting precipitate spun at 20,000 $g$ for 20 min. The pellet is resuspended to the desired volume in buffer containing 0.05 $M$ Tris-Cl (pH 8.3) and 1% Triton X-100.

3. The particle may be brought to 40% acetone by the slow addition of acetone that has been chilled to dry ice temperature to the rapidly stirring mixture. The precipitate is spun at 20,000 $g$ for 20 min and resuspended to the desired volume. The concentration of acetone is critical; less than 40% will not precipitate the entire particle and more than 45% will remove chlorophyll from the photosystem I protein.

## III. Closing Remarks

Ultracentrifugation studies of the A-III particle using a 55% sucrose gradient shows the migration of only one chlorophyll and P700-containing band. Additional ion-exchange chromatography using DEAE-BioGel A and a linear 0-0.4 $M$ KCl gradient produces little or no additional fractionation of iron-sulfur protein or chlorophyll from the P700-containing fragment. The purity, ease in isolation, and extensive characterization of the spinach A-III reaction center particle therefore renders it ideally suited for investigations involving electron transfer reactions involving photosystem I.

Acknowledgments

This research was supported in part by Grant No. PCM74-20736 from the National Science Foundation.

FIG. 7. ESR spectra of the bound iron-sulfur proteins at 15°K in a spinach A-III particle. The spectra were recorded at a chlorophyll concentration of 250 $\mu$g/ml in buffer containing 0.025 $M$ Tris-Cl (pH 10.0) and 1% Triton X-100. The sample in (a) was chemically reduced with dithionite and methyl viologen prior to freezing; amplifier gain 56. The light-minus-dark difference spectrum (b) was determined after illumination at 15°K; amplifier gain 116. Instrument settings: frequency 9.21 GHz, microwave power 10 mW, modulation amplitude 10 G. The author is indebted to Dr. Richard Malkin for kindly performing the ESR analysis.

## [13] The P700-Chlorophyll a-Protein of Higher Plants

### By JUDITH ANN SHIOZAWA

It is now known that the chlorophylls in higher plant chloroplasts are organized into at least two, but probably more, chlorophyll–proteins.[1] One, the light-harvesting chlorophyll a/b–protein, has been isolated in a homogeneous preparation and partially characterized.[2] A second pigment–protein, the P700–chlorophyll a–protein, had been previously isolated from sodium dodecyl sulfate (SDS)-treated photosynthetic membranes of a cyanobacterium.[3] A homologous chlorophyll–protein, complex I, was known to exist in higher plants[4] but, until recently, could not be isolated with the expected characteristics. Chromatography of Triton X-100-solubilized chloroplast lamellae on hydroxylapatite has yielded a fraction containing the chlorophyll–protein with the desired characteristics.[5] This chromatographic procedure has been successfully used to isolate P700–chlorophyll a–protein fractions from several species of higher plants, algae, and cyanobacteria.[4]

### Isolation Procedure

*Material.* Chloroplasts are isolated from young leaves by differential centrifugation. The chloroplasts are suspended in a hypotonic buffer consisting of 0.1 $M$ NaCl, 50 m$M$ Tris-Cl, 20 m$M$ sodium ascorbate, and 1 m$M$ EDTA (pH 8.0) with the aid of a glass homogenizer. The lamellae are pelleted by centrifugation at 40,000 $g$ for 10 min, resuspended in the same buffer, and again pelleted. These washed lamellae can be used immediately or suspended in a minimal volume of 50 m$M$ Tris-Cl buffer containing 20 m$M$ sodium ascorbate and 5 m$M$ $MgCl_2$ (pH 8.0) and stored below 0°.

*Hydroxylapatite Chromatography.* A summary of the chromatographic procedure is shown in Fig. 1. Chloroplast lamellae (containing about 28 mg Chl) are solubilized in 1% Triton X-100, 50 m$M$ Tris-Cl, 20

---

[1] J. P. Thornber, R. S. Alberte, F. A. Hunter, J. A. Shiozawa, and K. S. Kan, *Brookhaven Symp. Biol.* **28,** 132 (1977).

[2] J. P. Thornber, this volume.

[3] J. P. Thornber, Vol. 23, p. 682.

[4] J. P. Thornber, *Annu. Rev. Plant Physiol.* **26,** 127 (1975).

[5] J. A. Shiozawa, R. S. Alberte, and J. P. Thornber, *Arch. Biochem. Biophys.* **165,** 388 (1974).

METHODS IN ENZYMOLOGY, VOL. 69

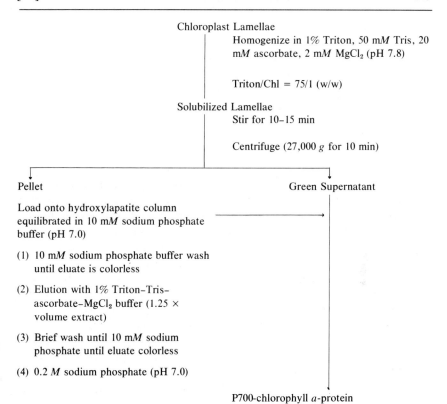

Chloroplast Lamellae
  Homogenize in 1% Triton, 50 m$M$ Tris, 20
  m$M$ ascorbate, 2 m$M$ MgCl$_2$ (pH 7.8)

  Triton/Chl = 75/1 (w/w)

Solubilized Lamellae
  Stir for 10–15 min

  Centrifuge (27,000 $g$ for 10 min)

Pellet                                          Green Supernatant

Load onto hydroxylapatite column
equilibrated in 10 m$M$ sodium phosphate
buffer (pH 7.0)

(1) 10 m$M$ sodium phosphate buffer wash
    until eluate is colorless

(2) Elution with 1% Triton–Tris–
    ascorbate–MgCl$_2$ buffer (1.25 ×
    volume extract)

(3) Brief wash until 10 m$M$ sodium
    phosphate until eluate colorless

(4) 0.2 $M$ sodium phosphate (pH 7.0)

                                          P700-chlorophyll *a*-protein

FIG. 1. Flow diagram of the isolation of the P700–chlorophyll *a*-protein from Triton X-100-solubilized chloroplast lamellae.

m$M$ sodium ascorbate, 2 m$M$ MgCl$_2$ (pH 7.8) using a glass homogenizer. The detergent to chlorophyll ratio is 75:1 (w/w). The pH of the solubilization buffer was changed from 7.4 to 7.8 to minimize the chance of generating pheophytin during the subsequent chromatography.[6] The Triton-solubilized lamellae are stirred for 10 to 15 min at room temperature and then centrifuged at 27,000 $g$ for 10 min. The pellet which contains mostly starch is discarded, and the supernatant is applied to the top of a hydroxylapatite[7] column (3.0 × 4.0 cm wide) which has been previously equilibrated with 10 m$M$ sodium phosphate buffer (pH 7.0). The column is then washed with 10 m$M$ sodium phosphate buffer until a colorless eluate is obtained. This is followed by a wash with the solubilizing buffer. The volume of the wash is equal to about 1.25-times that of the extract.

[6] J. A. Shiozawa, Ph.D. Dissertation, University of California, Los Angeles (1976).
[7] H. S. Siegelman, G. A. Wieczorek, and B. C. Turner, *Anal. Biochem.* **13,** 402 (1965).

The column is then washed with 10 m$M$ sodium phosphate buffer until the eluate is again colorless. The P700–chlorophyll $a$–protein is eluted with 0.2 $M$ sodium phosphate (pH 7.0). The isolated fraction is centrifuged at 27,000 $g$ for 10 min to remove particles of hydroxylapatite that may be present.

Methods of analysis have been described elsewhere.[5,6]

## Function

The P700–chlorophyll $a$–protein comprises the heart of photosystem I. It is believed to be ubiquitous in all chlorophyll $a$-containing plants with the possible exception of some photosystem I-deficient mutants.[4] Depending on the plant source, 10–18% of the chlorophyll in higher plant chloroplasts is found in this pigment–protein. The isolated complex contains one photoactive P700 entity per 40 chlorophyll molecules; this represents a seven- to tenfold enrichment in P700 over the starting detergent-solubilized chloroplast lamellae.

## Properties

Table I shows the recovery of P700–chlorophyll $a$–protein from a typical experiment. The yield of P700 and chlorophyll in the 0.2 $M$ sodium phosphate eluate varies between 30–55% and 3–5%, respectively. The chlorophyll/P700 ratio (40–45:1) remains constant in every preparation.

### Spectral

ABSORPTION. The room temperature absorption spectrum of the isolated P700–chlorophyll $a$–protein fraction is shown in Fig. 2; the general characteristics of the spectrum are identical to that of the cyanobacterium P700–chlorophyll $a$–protein (shown in Fig. 2, inset). However, the rela-

TABLE I
RECOVERY OF P700–CHLOROPHYLL $a$–PROTEIN[a]

|  | Chl (%) | Chl[b] $a/b$ | Chl/P700[b] | P700 (%) |
|---|---|---|---|---|
| Triton extract | 100 | 2.9 | 360/1 | 100 |
| 10 mM Sodium phosphate eluate | 66 | 2.9 | 900/1 | 27 |
| 1% Triton eluate | 17 | 1.6 | $\infty$ | 0 |
| 0.2 M Sodium phosphate eluate | 5 | >7 | 40/1 | 45 |

[a] Reprinted from Shiozawa et al.[5]
[b] Molar ratio.

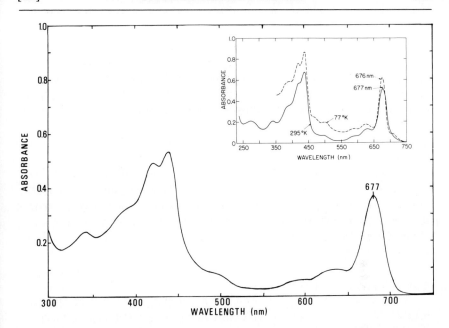

FIG. 2. Room temperature visible light absorption spectrum of the 0.2 M sodium phosphate buffer fraction isolated from Triton X-100-treated N. tabacum chloroplast lamellae. Inset shows the room temperature and 77°K spectra of the P700–chlorophyll a–protein isolated from SDS-solubilized cyanobacterium photosynthetic membranes.[3] Reprinted from Shiozawa et al.[5]

TABLE II

MILLIMOLAR ABSORPTIVITY VALUES[a] OF P700-CHLOROPHYLL a-PROTEIN BASED ON CHLOROPHYLL a

|  | Wavelength (nm) | | | | |
|---|---|---|---|---|---|
|  | 342 | 420 | 437 | 629 | 677 |
| Tobacco[b] | 43 | 75 | 80 | 15 | 59 |
| Cyanobacterium[b] | 35 | 75 | 78 | 15 | 60 |

[a] Limit of error estimated as ±3%.
[b] Absorptivity $\epsilon$ (m$M^{-1}$ cm$^{-1}$).

tive absorbance intensity at 342 nm in the higher plant preparation is more variable[5] and is reflected in the millimolar absorptivity values shown in Table II. This variability has been attributed to an unidentified yellow pigment having an absorption maximum at 342 nm and which is observed eluting immediately behind the green, P700-containing band.[6] Most preparations do not have an absorbance at about 540 nm. This is the wavelength at which the $Q_x$ band of pheophytin $a$ absorbs. It is tentatively concluded that pheophytin $a$ is not a component of the P700–chlorophyll $a$–protein.[1]

The spectral species of chlorophyll occurring in the isolated P700–chlorophyll $a$–protein were determined by two procedures: computer-assisted curve fitting[8] and fourth derivative analysis[9] of the liquid nitrogen temperature red wavelength absorption band. The chlorophyll $a$ spectral forms found by curve fitting were Chl $a$-662, 669, 677, 686, and a minor component at 690 nm. Those determined by fourth derivative analysis were Chl $a$-664, 677, 688, and 705. The values obtained by the two methods agree fairly well.

The addition of SDS to the isolated fraction results in the concomitant loss of P700 activity and a blue shift in the red wavelength absorption maximum.[5] Changes in the spectral species of chlorophyll $a$ present are also observed.[6]

FLUORESCENCE EMISSION. The isolated complex, as is characteristic of photosystem I-enriched preparations, is not strongly fluorescent when compared to whole lamellae or photosystem II-enriched fractions. A liquid nitrogen temperature emission spectrum[6,10] has a major peak at 697 nm with minor emission components in the region of 725 nm. Since the far-red emission band is very prominent in preparations less enriched in P700,[11] it is possible that the long wavelength fluorescence arises from a light-harvesting chlorophyll entity which feeds light energy preferentially to P700[1] or from the interaction of this entity with the P700–chlorophyll $a$–protein. Brown[10] has postulated that the fluorescence band at 697 nm arises from oxidized P700.

LIGHT-INDUCED SPECTRAL CHANGES. Figure 3 shows the 77°K light-induced absorption difference spectrum of the P700–chlorophyll $a$–protein. The wavelength of maximum bleaching is 699 nm. The absorption increase at 688 nm has been observed by several investigators and has been used to explain the asymmetry of the red wavelength trough.[6] The

[8] C. S. French, J. S. Brown, and M. C. Lawrence, *Plant Physiol.* **49**, 421 (1972).
[9] W. L. Butler and D. W. Hopkins, *Photochem. Photobiol.* **12**, 439 (1970).
[10] J. S. Brown, *Carnegie Inst. Washington, Yearb.* **75**, 460 (1976).
[11] N. K. Boardman, *Annu. Rev. Plant Physiol.* **21**, 115 (1970).

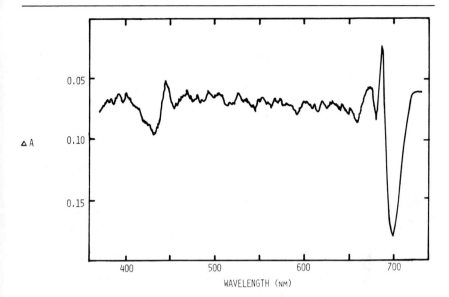

ΔA

WAVELENGTH (NM)

FIG. 3. Liquid nitrogen temperature light-induced difference spectrum of the P700–chlorophyll *a*-protein isolated from *N. tabacum*. Recorded in Dr. W. L. Butler's laboratory, University of California, San Diego.

Triton complex also shows light-inducible electron paramagnetic resonance signals attributable to oxidized P700 and to two photoreduced iron–sulfur centers. One iron–sulfur center ($g$ = 2.05, 1.94, and 1.86) has been attributed to be the primary electron acceptor of photosystem I.[12] Little is known about the second iron-sulfur center ($g$ = 2.05, 1.92, and 1.88); it has been proposed that both iron–sulfur centers occur in the same molecule of ferredoxin.[13,14] Under the experimental conditions used,[15] the signals were irreversible.

*Composition.*

CHEMICAL. The composition of the Triton–P700–chlorophyll *a*-protein complex, as known thus far, is shown in Table III. Chlorophyll *a* and β-carotene are probably the only photosynthetic pigments occurring in the Triton complex. The chlorophyll *a/b* ratio of the isolated material

[12] R. Malkin and A. J. Bearden, *Proc. Natl. Acad. Sci. U.S.A.* **68,** 16 (1971).
[13] R. Cammack and M. C. W. Evans, *Biochem. Biophys. Res. Commun.* **67,** 544 (1975).
[14] E. H. Evans, R. Cammack, and M. C. W. Evans, *Biochem. Biophys. Res. Commun.* **68,** 1212 (1976).
[15] R. Malkin, A. Bearden, F. A. Hunter, R. S. Alberte, and J. P. Thornber, *Biochim. Biophys. Acta* **430,** 389 (1976).

TABLE III
COMPOSITION OF TRITON—P700—CHLOROPHYLL $a$—PROTEIN COMPLEX

| Component | Quantity (nmoles) |
|---|---|
| P700 | 1 |
| Chlorophyll $a$ | 40 |
| $\beta$-Carotene | 1.1 |
| Phylloquinone | 1.1 |
| Cytochrome $f$ | 0.9 |
| Cytochrome $b_6$ | 1.8 |
| Protein[a] | 1.6[b] |

[a] Assay method, C. A. Lang, *Anal. Chem.* **30**, 1692 (1958).
[b] Value in mg.

is greater than 7:1 (Table I), a value at which the reliability of the equations used for determination of chlorophyll $b$ content is poor.[5] In addition, thin-layer chromatography of lipid extracts of the isolated material showed that most preparations were devoid of chlorophyll $b$.[5] In all species studied (spinach, *Hordeum vulgare*, *Nicotiana tabacum*, and *N. excelsior*), phylloquinone was the only quinone detected in the isolated complex.[1,6] Cytochromes $f$ and $b_6$ also occur in the isolated fraction[5]; no low potential cytochrome $b$-559 was detected.[6] Brown[10] has suggested that cytochrome $f$ is not an integral component of the Triton complex. Other investigators[16–19] have been able to isolate fractions greatly enriched in P700 and from which cytochromes $f$ and $b_6$ are absent.

POLYPEPTIDES. The native complex contains one pigment–protein zone (molecular weight about 100 kilodaltons) when subjected to SDS-polyacrylamide gel electrophoresis. Other lower molecular weight protein zones are also observed.[6] The fully dissociated Triton complex contains several polypeptides; the molecular weights are about 51, 49, 45, 35, and 25 kilodaltons, and two zones with molecular weights of less than 15 kilodaltons.[6] For comparison, complex I, the SDS-altered form of the P700–chlorophyll $a$–protein,[4] was isolated and analyzed for polypeptides.[6] Complex I contained only the 51 and 49 kilodalton polypeptides. In both chlorophyll–protein preparations, the 51 kilodalton polypeptide is the major zone. It is, therefore, concluded that the two polypeptides,

[16] C. Bengis and N. Nelson, *J. Biol. Chem.* **250**, 2783 (1975).
[17] B. Ke, K. Sugahara and E. Shaw, *Biochim. Biophys. Acta* **408**, 12 (1975).
[18] R. Malkin, *Arch. Biochem. Biophys.* **169**, 77 (1975).
[19] J. H. Goldbeck, S. Lien, and A. San Pietro, *Arch. Biochem. Biophys.* **178**, 140 (1977).

51 and 49 kilodaltons, comprise the P700–chlorophyll $a$-protein. The remaining polypeptides on the Triton complex have been tentatively attributed to cytochrome $b_6$ (43 kilodaltons), cytochrome $f$ (35 kilodaltons), apoprotein of the light-harvesting chlorophyll $a/b$-protein, and iron–sulfur protein (<15 kilodaltons).[6]

The association of two polypeptides, one major (51 kilodaltons) and one minor (49 kilodaltons), with the higher plant P700–chlorophyll $a$-protein complement data recently obtained on the P700–chlorophyll $a$-protein isolated from cyanobacteria.[1] Hunter (cf. Thornber $et$ $al.$[1]) has found that the fully dissociated pigment–protein isolated from SDS-treated photosynthetic membranes of $Phormidium$ $luridum$ contain two polypeptides having molecular weights of 48 and 46 kilodaltons. As is the case in the higher plant preparation, the higher molecular weight polypeptide is the dominant component. The occurrence of two polypeptides in the $P.$ $luridum$ chlorophyll–protein is further substantiated by the detection of two N-terminal amino acids.[1] These data are in conflict with those obtained from polypeptide analyses of chloroplast lamellae and other P700-enriched preparations (see refs. 1, 16, 20) from which it was concluded that a single polypeptide of 50 to 70 kilodaltons was associated with the photosystem I chlorophyll–protein.

## Other Related Preparations

Recently, several new procedures for the isolation of photosystem I preparations having approximately the same chlorophyll/P700 ratio (25–40:1) as the Triton complex have become available.[16–19] Most of these fractions are obtained after further purification of digitonin or Triton X-100 photosystem I subchloroplast particles.[16–18,20a] These purification steps involve treatment with a second detergent,[16,18] chromatography,[16,18,19] and high-speed centrifugation.[16,17] Solubilization of photosynthetic lamellae from some plants in a mixture of detergents (i.e., Triton X-100, SDS, and lauryldimethylamine oxide) followed by hydroxylapatite chromatography yields a fraction containing one mole P700 per 16–20 moles chlorophyll.[20a] Particles enriched in P700 (Chl/P700 = 5–30:1) have also been obtained by organic solvent extraction of photosystem I subchloroplast particles[21–23] but are not practical for further

[20] S. Bar-Nun, R. Schantz, and I. Ohad, $Biochim.$ $Biophys.$ $Acta$ **459**, 451 (1977).

[20a] R. S. Alberte and J. P. Thornber $FEBS$ $Lett.$ **91**, 126 (1978).

[21] P. V. Sane and R. B. Park, $Biochem.$ $Biophys.$ $Res.$ $Commun.$ **41**, 206 (1970).

[22] L. P. Vernon and E. R. Shaw, Vol. 23, p. 277.

[23] I. Ikegami and S. Katoh, $Biochim.$ $Biophys.$ $Acta$ **376**, 588 (1976).

studies on the nature of the P700-containing entity and its relationship to the photosynthetic membrane.

## Summary

Hydroxylapatite chromatography of Triton X-100-solubilized photosynthetic membranes is a relatively quick method of isolating the P700–chlorophyll $a$–protein in a photoactive state, in good yields, and from a diverse group of photosynthetic plants.[4] It is probable that other P700-enriched preparations[16–19,20a] are comprised, in part, of this chlorophyll-protein. However, this idea and the subunit structure of the P700–chlorophyll $a$–protein need clarification and substantiation.

## [14] The Light-Harvesting Chlorophyll $a/b$–Protein

By J. PHILIP THORNBER[1] and JANET M. THORNBER

This pigment–protein complex accounts for some 50% of the total chlorophyll in chloroplasts of angiosperms, gymnosperms, and green algae; it may also occur in lesser amounts in *Euglena*.[1a] Thus this chlorophyll–protein is probably found in all organisms that contain chlorophyll $b$. The light-harvesting chlorophyll $a/b$–protein is the pigment–protein complex that has variously been referred to as complex II, component II, pigment–protein complex I, photosystem II chlorophyll-protein, CP2, or LHPP. The chlorophyll–protein is bound to the chloroplast thylakoid membrane and requires the use of an anionic detergent to dissociate it from those membranes; however, addition of large excesses of the detergent destroys its native structure by removing the chlorophyll from its conjugation with protein. Its presence during the isolation procedure and its purity are monitored spectrally and by electrophoresis on sodium dodecyl sulfate (SDS)–polyacrylamide gels; the complex has a unique absorption spectrum (Fig. 1). The chlorophyll—protein was initially isolated by electrophoresis of anionic detergent extracts of chloroplasts on polyacrylamide gels; however, since the composition of the isolated material varied slightly between different runs, it has been found preferable to use the chromatographic procedure described below.

[1] Most of the research described in this article was supported by a grant from the National Science Foundation to the author.

[1a] J. P. Thornber, *Annu. Rev. Plant Physiol.* **26**, 127 (1975).

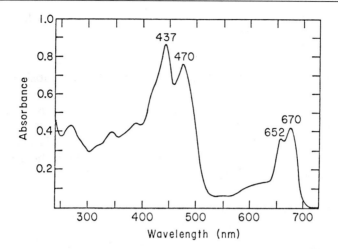

FIG. 1. Room temperature absorption spectrum of the light-harvesting chlorophyll *a*/*b*—protein in 50 m*M* Tris-HCl, pH 8.0.

## Purification Procedure[2,3]

The isolation is carried out at room temperature.

*Step 1.* Exhaustively washed, broken chloroplasts are treated with 50 m*M* Tris-1% (w/v) SDS, pH 8.0 (detergent : chlorophyll = 5 : 1, w/w), and centrifugated at 100,000 *g* for 20 min. The resulting green supernatant (SDS extract) is used as the starting material for the isolation.

*Step 2.* A column (1.5 × 9 cm) is packed with hydroxylapatite[4] to a height of 6—7 cm, and then equilibrated with 10 m*M* sodium phosphate, pH 7.0. A flow rate of 1.5 ml/min is maintained throughout the packing of the column and the subsequent chromatography by use of a peristaltic pump. About 3 ml of extract (containing about 6 mg chlorophyll) are run into the column, and washed in with 15 ml of the equilibration buffer. All the color in the extract is absorbed to the hydroxylapatite.

*Step 3.* Increasing concentrations of sodium phosphate, pH 7.0 (15 ml each of 0.1, 0.2, and 0.3 *M*) are added to the column. A small portion of the absorbed chlorophyll is eluted by 0.2 and 0.3 *M* sodium phosphate. This eluate contains, among other components, the SDS-altered form of the P700–chlorophyll *a*–protein.[1a]

[2] S. D. Kung and J. P. Thornber, *Biochim. Biophys. Acta* **253,** 285 (1971).
[3] K.-S. Kan and J. P. Thornber, *Plant Physiol.* **57,** 47 (1976).
[4] H. W. Siegelman and J. H. Kycia, *in* "Handbook of Phycological Methods" (J. Cragle and J. A. Helleburst, eds.). Cambridge Univ. Press, London and New York (in press).

*Step 4.* The column is washed with 0.4 *M* sodium phosphate–1 m*M* MgCl₂ containing 0.05% SDS, pH 7.0, and the fractions having an absorption spectrum with a pronounced 650 nm shoulder on a 670 nm peak are pooled.

*Step 5.* The pooled fractions are diluted with an equal volume of 10 m*M* sodium phosphate—1 m*M* MgCl₂, pH 7.0, and run into a second, identical hydroxylapatite column, previously equilibrated with 10 m*M* sodium phosphate—1 m*M* MgCl₂, pH 7.0. This column is eluted with 0.01, 0.1, 0.2, and 0.3 *M* sodium phosphate—1 m*M* MgCl₂, pH 7.0.

*Step 6.* The green material eluted by the latter two concentrations is further purified by repeating step 5, followed by ammonium sulfate fractionation. The required chlorophyll–protein is precipitated at an ammonium sulfate concentration between 10 and 15% (w/v), and the precipitate is dissolved and stored, in 50 m*M* Tris–1 m*M* MgCl₂, pH 7.0.

## Alternative Methods of Preparation

Hydroxylapatite can be used in the form of a slurry rather than as a column.[5] The pigment–protein can also be isolated using polyacrylamide gel electrophoresis: A sample of an anionic detergent extract is electrophoresed into a column of gel to give three distinct green bands.[6,7] The middle band is cut out of the column and the disc is chopped up. The chlorophyll-containing protein is eluted from the gel particles by use of a Sephadex column[6] or electrophoresed out of the particles into free solution.[7]

## Function

This chlorophyll–protein contains some 50% of the total chlorophyll present in chloroplasts of higher plants and green algae. It is the location of all of the chlorophyll *b* present in these plants. Since a photosynthetically competent barley mutant has been found which lacks this chlorophyll–protein,[8] then the chlorophyll *a/b*-protein is not essential for photosynthetic electron transfer reactions, and its main function is therefore as an antenna component for the two photochemical systems (particularly for photosystem II). The transfer of the energy in photons absorbed by chlorophyll *b* to chlorophyll *a*, in a plant, almost certainly takes place

[5] D. J. Davis and E. L. Gross, *Biochim. Biophys. Acta* **387**, 557 (1975).
[6] J. P. Thornber, J. C. Stewart, M. W. C. Hatton, and J. L. Bailey, *Biochemistry* **6**, 2006 (1967).
[7] T. Ogawa, F. Obata, and K. Shibata, *Biochim. Biophys. Acta* **112**, 223 (1966).
[8] J. P. Thornber and H. R. Highkin, *Eur. J. Biochem.* **41**, 109 (1974).

entirely within the chlorophyll *a/b*-protein and with a 100% efficiency.[9]

It has also been postulated that the chlorophyll–protein acts as a store of protein nitrogen and carbon skeletons for plant's metabolic processes.[10]

## Properties

### Purity

The chromatographically isolated complex electrophoreses as a single pigmented zone of approximately 30,000 daltons on SDS–polyacrylamide gels.[1a] Analytical ultracentrifugation of the isolated chlorophyll–protein exhibits a boundary of 3.1 ± 0.1 S; all the color in the preparation sediments with this boundary. No impurity is detected by ultracentrifugation.

### Spectral Characteristics

*Absorption* (Fig. 1). The room temperature spectrum of the chlorophyll *a/b*-protein shows a double peak in the red spectral region ($\lambda_{max}$ 670–672 and 652–653 nm for chlorophylls *a* and *b*, respectively). Two Soret peaks at 437–438 and 470–471 nm which correspond to those of chlorophyll *a* and *b*, and a shoulder on the longer wavelength side of the Soret peaks, probably contributed largely by carotenoids present in the complex[9] are observed. The extinction coefficients of the red peaks are 34 m$M^{-1}$ cm$^{-1}$ at 672 nm and 31 m$M^{-1}$ cm$^{-1}$ at 653 nm.[3] At 77°K several spectral forms of chlorophyll *a* [chlorophyll *a* 662, 670, 677 and 684 (minor) nm] have been resolved.[1a,9] Circular dichroism studies indicate strong coupling between the chlorophyll *b* molecules in the complex while this is not indicated for the chlorophyll *a* molecules.[9]

*Emission.* The complex exhibis a single fluorescence emission peak at ~685 nm.

### Chemical Composition

There are equimolar quantities of chlorophyll *a* and *b* present.[1a,9] For the *Chlamydomonas* component the chlorophyll–protein ratio is 6 moles per 29,000 gm of protein; a similar ratio is likely to occur in the higher plant complex.[1] Every carotenoid present in the intact chloroplast is contained in this chlorophyll–protein, although not in the same stoichiometric ratio. The majority of the carotenoid is represented by lutein and β-carotene; a large proportion of a chloroplast's neoxanthin is associated

[9] R. L. Van Metter, *Biochim. Biophys. Acta* **462**, 642 (1977).

with this component. The carotenoid–chlorophyll molar ratio varies between 1:3 to 7. Trace amounts of galacto- and phospholipids occur as well as traces of carbohydrates.

### Amino Acid Composition

For the *Chlamydomonas* complex[3] this is (moles per 29,000 gm protein): Asp (24), Thr (14), Ser (10), Pro (20), Glu (25), Gly (34), Ala (31), Val (12), Met (5), Ile (12), Leu (32), Tyr (9), Phe (19), Lys (15), His (4), Arg (9), and Cys (1.5). The amino acid composition of the higher plant component is very similar.[1a,6] Both show a very high content of nonpolar amino acid residues as expected for a membrane-bound protein, and both have an unusually high proline content. One or more threonine residues in this protein become phosphorylated when isolated chloroplasts are exposed to light[10]; dephosphorylation occurs on returning the chloroplasts to darkness.

### Molecular Weight

Values ranging from 27,000–35,000 daltons have been reported for the native complex of higher plants or green algae.[1a] When the chlorophyll is removed from the pigment–protein, the resulting apoprotein electrophoreses on poly-acrylamide gel as a band(s) of slightly lower apparent molecular weight. The number and size of the polypeptide(s) present is equivocal[1a]; some investigators support the view that just one polypeptide is present, while others believe that two (sometines three) polypeptides of similar amino acid composition, but differing in size by 2000–4000 daltons, are contained in the native pigment–protein. The dimeric form of the native complex has been described.[11] It is anticipated that higher oligomeric forms of the chlorophyll *a/b*-protein occur in each intact photosynthetic unit.[1a,3]

### Summary

The light harvesting chlorophyll *a/b*-protein is isolated in its monomeric form (MW.~30,000). Each monomer most probably contains 3 molecules of chlorophyll *a*, 3 of chlorophyll *b*, and 1 of carotenoid, and a polypeptide (or polypeptides) of approximately 28,000 daltons. The reader is referred to Thornber[1a] for finer details and other references on this chlorophyll–protein.

[10] J. Bennett, *Nature (London)* **269**, 344 (1977).
[11] R. G. Hiller, S. Genge, and D. Pilger, *Plant Sci. Lett.* **2**, 239 (1974).

## [15] Bacterial Reaction Center (RC) and Photoreceptor Complex (PRC) Preparations

### By Paul A. Loach

As progress in the isolation of photosynthetic membrane components continues, new terms are used which attempt to describe a particular preparation in terms of the function it served while part of the intact membrane. Three kinds of preparations will be discussed here. The reaction center (RC)[1] is the term used to describe those minimal components (bacteriochlorophyll, bacteriopheophytin, ubiquinone, protein, etc.) that are required for carrying out the primary photochemical event. The photoreceptor complex (PRC), is the term used for a preparation which contains the reaction center and also a specific quantity of antenna pigments (bacteriochlorophyll and carotenoids), and, according to our present understanding, it also contains specific protein components that bind the antenna pigments. The third term is that which describes a light-harvesting pigment–protein complex, abbreviated LH, and represents an antenna bacteriochlorophyll and/or carotenoid complex which contains a specific protein component but no reaction center.

Because of the very high efficiency with which light energy absorbed in LH is transferred to the reaction center, and because some PRC preparations can be obtained which are small (MW approximately 125,000) and of high integrity, the PRC may be the fundamental unit of the photosynthetic membrane and perhaps the biochemical equivalent of the photosynthetic unit.[1a] There is evidence that some photosynthetic bacteria may have more than one LH associated with one RC.[2]

---

[1] Abbreviations used: RC, reaction center or phototrap; PRC, photoreceptor complex which contains a reaction center as well as antenna pigments; LH, light-harvesting complex; LDAO, lauryldimethylamine oxide; Triton X-100, octylphenoxypolyethoxyethanol; Brij, a polyethoxy ether of a fatty alcohol; CTAB, cetyltrimethylammonium bromide; AUT, alkaline-urea-Triton X-100 treatment; SDS-PAGE, sodium dodecyl sulfate-polyacrylamide gel electrophoresis; Asc, ascorbic acid.

[1a] R. Emerson and W. Arnold, *J. Gen. Physiol.* **15,** 391 (1932).

[2] R. L. Hall, M. Chu Kung, M. Fu, B. J. Hales, and P. A. Loach, *Photochem. Photobiol.* **18,** 505 (1973).

## Reaction Center Preparations

Since the last description of RC preparations in "Methods in Enzymology,"[3] the methods have been applied widely to a variety of bacteria. Such preparations are summarized in Table I. For more detail, the reader is directed to the original literature and several recent review articles which summarize the work of selected laboratories.[4–6]

*Assay.* The standard tests for RC activity in preparations from photosynthetic bacteria have been the demonstration at low temperature ($\leq 77°K$) of appropriate light-induced absorbance changes and an electron spin resonance (ESR) signal characteristic of formation of the primary electron donor bacteriochlorophyll cation radical with a quantum yield near 1.0. Usually, demonstration of normal decay kinetics at low temperature ($\leq 77°K$) is also required.

Physical properties that could be added to the assay requirements, but usually are not, are (a) demonstration of appropriate $E_0'$ values for the primary electron donor and acceptor species as well as the pH dependency of these values, (b) unchanged circular dichroism properties, and (c) efficient secondary electron transport activity when the normal secondary electron trnasport components are added to the system.

Although it is still a little premature to specify the minimal components which should be presented in such a preparation, most laboratories would agree that 4 bacteriochlorophyll molecules, 2 bacteriopheophytin molecules, and 1 or 2 ubiquinone molecules (or possibly other quinones[7,8]) should be present. Of somewhat less certainty is the requirement for iron or the exact number and size of required polypeptide components.

*Methods of Preparation.* As indicated by the data listed in Table I,[2,4,5,9–41] all RC isolation procedures require the use of a detergent in

[3] This series, Vol. 23, see R. K. Clayton and R. T. Wang, p. 695 and J. P. Thornber, p. 688.

[4] J. M. Olson and J. P. Thornber, *in* "Energy Transduction" (R. A. Capaldi, ed.). Dekker, New York (in press).

[5] G. Gingras, *in* "The Photosynthetic Bacteria" (R. K. Clayton and W. R. Sistrom, eds.), p. 119, Plenum, New York, 1978.

[6] G. Feher and M. Okamura, *in* "The Photosynthetic Bacteria" (R. K. Clayton and W. R. Sistrom, eds.), p. 349, Plenum, New York, 1978.

[7] M. Y. Okamura, L. C. Ackerson, R. A. Isaacson, W. W. Parson, and G. Feher, *Biophys. J.* **16,** 223, Abstr. F-PM-D8 (1976).

[8] G. Feher and M. Y. Okamura, *Brookhaven Symp. Biol.* **28,** 183 (1976).

[9] D. W. Reed and R. K. Clayton, *Biochem. Biophys. Res. Commun.* **30,** 471 (1968).

[10] D. W. Reed, *J. Biol. Chem.* **244,** 4936 (1969).

[11] R. K. Clayton and R. T. Wang, Vol. 23, Part A, p. 696 (1971).

[12] R. K. Clayton and R. Haselkorn, *J. Mol. Biol.* **68,** 97 (1972).

order to separate these integral protein systems from the membrane. Nonionic detergents (LDAO, Triton X-100, Brij) have been used most often, although the anionic detergent SDS has also been used successfully. Cationic detergents (e.g., CTAB) have rarely been used. In each case, the form of the biological sample has been membrane vesicles (chromatophores) prepared by centrifugation after sonic oscillation of the intact bacteria or after breakage with a high pressure cell. The suspension buffer has usually been Tris or phosphate at a pH near 7. For some preparations, the choice of buffer is important for success with the procedure used.

[13] G. Feher, *Photochem. Photobiol.* **14**, 373 (1971).

[14] M. Y. Okamura, L. A. Steiner, and G. Feher, *Biochemistry* **13**, 1394 (1974).

[15] G. Jolchine, F. Reiss-Husson, and M. D. Kamen, *Proc. Natl. Acad. Sci. U.S.A.* **64**, 650 (1969).

[16] B. J. Segen and K. D. Gibson, *J. Bacteriol.* **105**, 701 (1971).

[17] F. Reiss-Husson and G. Jolchine, *Biochim. Biophys. Acta* **256**, 440 (1972).

[18] G. Jolchine and F. Reiss-Husson, *FEBS Lett.* **40**, 5 (1974).

[19] G. Jolchine and F. Reiss-Husson, *FEBS Lett.* **52**, 33 (1975).

[20] L. Slooten, *Biochim. Biophys. Acta* **256**, 452 (1972).

[20a] L. Slooten, *Biochim. Biophys. Acta* **275**, 208 (1972).

[21] G. Gingras and G. Jolchine, *Prog. Photosynth. Res.* **1**, 209 (1969).

[22] R. T. Wang and R. K. Clayton, *Photochem. Photobiol.* **17**, 57 (1973).

[23] J. Oelze and J. R. Golecki, *Arch. Microbiol.* **102**, 59 (1975).

[24] R. Bachofen, K. W. Hanselman, M. Snozzi, H. Zürrer, P. Cuendet, and H. Zuber, *Brookhaven Symp. Biol.* **28**, 365, Abstr G2 (1976).

[25] M. Snozzi, *Ber. Dtsch. Bot. Ges.* (in press).

[26] W. R. Smith, Jr., C. Sybesma, and K. Dus, *Biochim. Biophys. Acta* **267**, 609 (1972).

[27] H. Noel, M. Van der Rest, and G. Gingras, *Biochim. Biophys. Acta* **275**, 219 (1972).

[28] M. Van der Rest, H. Noel, and G. Gingras, *Arch. Biochem. Biophys.* **164**, 285 (1974).

[29] R. C. Prince and A. R. Crofts, *FEBS Lett.* **35**, 213 (1973).

[30] K. F. Nieth and G. Drews, *Arch. Microbiol.* **96**, 161 (1974).

[31] K. F. Nieth, G. Drews, and R. Feick, *Arch. Microbiol.* **105**, 43 (1975).

[32] R. K. Clayton and B. J. Clayton, *5th Annu. Meet. Am. Soc. Photobiol.* Abstr. FAM D2 (1977).

[32a] B. J. Clayton and R. K. Clayton, *Biochim. Biophys. Acta* **501**, 470 (1978).

[33] L. Lin and J. P. Thornber, *Photochem. Photobiol.* **22**, 37 (1975).

[34] Y. D. Halsey and B. Byers, *Biochim. Biophys. Acta* **387**, 349 (1975).

[35] D. M. Tiede, R. C. Prince, and P. L. Dutton, *Biochim. Biophys. Acta* **449**, 447 (1976).

[36] J. P. Thornber, J. M. Olson, D. M. Williams, and M. L. Clayton, *Biochim. Biophys. Acta* **172**, 351 (1969).

[37] J. P. Thornber, Vol. 23, Part A, p. 688 (1971).

[38] N. L. Pucheu, N. L. Kerber, and A. F. Garcia, *Arch. Microbiol.* **101**, 259 (1974).

[39] N. L. Pucheu, N. L. Kerber, and A. F. Garcia, *Arch. Microbiol.* **109**, 301 (1974).

[40] T. L. Trosper, D. L. Benson, and J. P. Thornber, *Biochim. Biophys. Acta* **460**, 318 (1977).

[41] J. P. Thornber, this volume.

TABLE I
REACTION CENTER PREPARATIONS

| Bacteria and Reference | Detergent | Purification Steps | | | | | | | Polypeptide components (kilodaltons) |
|---|---|---|---|---|---|---|---|---|---|
| | | Centrifugation | $(NH_4)_2SO_4$ | DEAE | Gel filtration | Celite | Hydroxylapatite | AUT-e | |
| **1. *Rhodopseudomonas sphaeroides* (R-26)** | | | | | | | | | |
| a. Reed and Clayton[9,10a] | (3%) TX-100 | + | | | + | | | | |
| b. Clayton and Wang[11]; Clayton and Haselkorn[12] | (1%) LDAO | + | + | | + | | | | 19, 22, 27 |
| c. Feher[13] | (1%) LDAO | + | + | + | | + | | | 21, 24, 28 |
| d. Okamura et al.[14] | (1%) LDAO | + | + | + | | + | | | 21, 24 |
| e. Hall et al.[2] | (1.5%) TX-100 | + | | | | | | + | 10, 21, 24 |
| **2. *Rhodopseudomonas sphaeroides* (wild type)** | | | | | | | | | |
| a. Jolchine et al.[15a] | (0.3%) CTAB | + | | | | | | | |
| b. Segen and Gibson[16a] | | | | | | | | | |
| c. Reiss-Husson and Jolchine[17] | (0.3%) CTAB | + | | | | | | | |
| d. Jolchine and Reiss-Husson[18,19] | (0.5%) CTAB | + | + | | + | + | | | 22, 24, 27 |
| | (0.25%) LDAO | + | + | | + | + | | | |
| e. Slooten[20] | (0.3%) SDS | + | | | | | | + | |
| **3. *Rhodospirillum rubrum* (G-9)** | | | | | | | | | |
| a. Gingras and Jolchine[21,a] | (7%) TX-100 | + | | | | | | | 21, 24, 28 |
| b. Wang and Clayton[22] | (0.3%) LDAO | + | | | + | | | | 21, 24, 28 |
| c. Okamura et al.[14] | (0.4%) LDAO | + | + | | | + | | | 24, 28, 31, 32 |
| d. Oelze and Golecki[23] | (0.4%) LDAO | + | + | | | | | | |
| e. Bachofen et al.[24] | (0.3%) LDAO | + | | + | + | | | | 22, 26, 29 |
| f. Snozzi[25] | (0.6%) LDAO | + | | + | + | | | | 21, 24, 29 |
| **4. *Rhodospirillum rubrum* (wild type)** | | | | | | | | | |

| Source | Detergent | | | | | | | | References |
|---|---|---|---|---|---|---|---|---|---|
| b. Noel et al.[27]; Van der Rest et al.[28]; Gingras[5] | (0.25%) LDAO | + | | | + | | | | | 21, 24, 32 |
| | (0.3%) LDAO | + | | | + | | | | | 21, 26, 28 |
| | (0.3%) LDAO | + | + | | + | | | | | |
| c. Slooten[20,20a] | (0.3%) SDS | | | | + | | | | ± | |
| 5. Rhodopseudomonas capsulata (carotenoidless) | | | | | | | | | | |
| a. Prince and Crofts[29] | (1%) LDAO | + | | | + | | | | | 23, 28, 32 |
| b. Nieth and Drews[30] | TX-100 | + | | + | + | | | | | 21, 24, 28 |
| c. Nieth et al.[31] | (1%) LDAO | | | + | | | | | | 21, 24 |
| | (0.5%) SDS | | | | | | | | | |
| 6. Rhodopseudomonas gelatinosa (carotenoidless) | | | | | | | | | | |
| a. Clayton and Clayton[32,32a] | (1%) LDAO | | | + | + | | | | | 24, 34, 43 |
| | (4–8%) LDAO | | | | + | | | | | 24, 34 |
| 7. Chromatium vinosum[b] | | | | | | | | | | |
| a. Lin and Thornber[33]; Olson and Thornber[4] | LDAO | + | | + | + | | | | | 23, 27, 30 |
| b. Halsey and Byers[34] | SDS-Brij | | | | | | | | | 12, 22, 27, 30, 45–50 |
| c. Tiede et al.[35] | TX-100, cholate | + | | + | + | | | | | |
| 8. Rhodopseudomonas viridis (wild type)[b] | | | | | | | | | | |
| a. Thornber et al.[36] | (1%) SDS | + | + | + | | | | | | |
| b. Thornber[37] | (1%) SDS | + | + | + | | + | | | | |
| c. Pucheu et al.[38,39] | LDAO | + | + | + | | | | | | 23, 29, 37, 45 |
| d. Trosper et al.[40] | SDS + $Na_2S_2O_4$ | + | + | + | | | | | + | |
| e. Thornber[41] | (1%) LDAO | + | + | + | + | | | | + | 23, 29, 37, 45 |

[a] Although these preparations contained components other than those of the RC and were large in size, they stimulated much of the subsequent work.

[b] The RC's prepared from these bacteria are not as well defined as those for the other bacteria listed in this table, but substantial progress toward isolation of pure RC's is indicated.

One of the most purified and best characterized preparations is that of Feher,[13] as modified by Okamura *et al.*[14] A summary of their procedure follows[6]: Chromatophores prepared from the R-26 mutant of *Rhodopseudomonas sphaeroides* are suspended in 10 m$M$ Tris buffer at 4° at a concentration of 0.3 gm (wet weight)/ml and stirred overnight. After adding LDAO to give a final concentration of 1.2%, the solution is immediately centrifuged at 250,000 $g$ for 1 hr. The supernatant fraction which contains the RC's is brought to room temperature and concentrated ammonium sulfate added to a final concentration of 24% (w/v). The solution is centrifuged (50,000 $g$, 10 min), and the floating solid material is resuspended in 10 m$M$ Tris, pH 8. To the remaining solution additional LDAO is added (0.4%), and the solution is centrifuged again to obtain an additional floating pellet which is combined with the first. After standing overnight at room temperature, undissolved material in the Tris buffer is removed by centrifugation, and the supernatant is added to Celite (approximately 1 gm Celite/$A$802 = 150). To the Celite slurry, concentrated $(NH_4)_2SO_4$ containing 0.1% LDAO in 10 m$M$ Tris at pH 8 is then added with stirring to give a final concentration of 25%. This mixture is then placed with vacuum on a 10-cm diameter filter containing a bed of 3 gm Celite which was prewashed with 25% ammonium sulfate in 10 m$M$ Tris buffer containing 0.1% LDAO. The material is further washed with the latter solvent until the eluate is clear or RCs begin to elute. The RC–Celite mixture is then transferred to a chromatographic column, and the RC is eluted with a decreasing ammonium sulfate gradient (25–15%) in 10 m$M$ Tris at pH 8 containing 0.1% LDAO. The RC fraction is dialyzed and then applied to a DEAE column (Whatman DE 52) which was prewashed with 1 $M$ NaCl and equilibrated with 10 m$M$ Tris buffer, pH 8, containing 0.1% LDAO. After washing the RC, which binds to the column with the equilibrating buffer, the same solvent but containing 0.06 $M$ NaCl is applied until the RC's begin to be eluted. At that point, the solvent is made 0.12 $M$ in NaCl to speed RC removal from the column. The RC's are dialyzed at 4° and stored in 10 m$M$ Tris, pH 8, containing 0.025% LDAO. Sometimes the last DEAE step is repeated if there are still some impurities. The overall yield is of the order of 40%. This is one of the best yields reported for the more extensive RC preparations. In most other procedures, the yield was reported to be closer to 10–20%.[5]

Although a very similar procedure to that described above can be used to isolate RC's from other photosynthetic bacteria, the exact conditions for each preparation seem to require some fine tuning. The trick is to use enough detergent, etc., to obtain the RC free from antenna pigments and other unnecessary components, but not so much that there is undue loss of phototrap activity because of degradation or loss of key components (e.g., ubiquinone). Therefore, slightly different conditions may be required with each bacteria in order to isolate the RC successfully.

*Properties of Reaction Center Preparations.* This is perhaps the most interesting aspect because it begins to deal with structure–function relationships and comparative biochemistry of this most fundamental system. For all the RC preparations listed in Table I, the primary property that they have in common is their absorbance spectra in the near infrared and the light-induced absorbance change observed in that region of the spectra. In each case, from this evidence, the RC is presumed to contain 4 bacteriochlorophyll molecules and 2 bacteriopheophytin molecules.

Another aspect that the RC preparations have in common, although in this case the data do not yet exist for all the bacteria listed, is the presence of a quinone molecule which is important as the first stable electron acceptor molecule.

When the first RC's were isolated, there seemed to be a parallel polypeptide pattern in each. Thus, the idea that all RCs would have a triad of polypeptides of molecular weights of about 21,24, and 29 kilodaltons was prominent. However, since Feher[13] and Okamura *et al.*[14] (also see Hall *et al.*[2]) demonstrated that only two polypeptides (those of apparent MW 21 and 24 kilodaltons) may, in fact, be required, the triad requirement was considerably weakened. It might then be concluded that the 21 and 24 kilodalton polypeptides will be found in all RC preparations, but a perusal of Table I will show that there are several RC's which do not seem to have a 21 kilodalton polypeptide (3d, 5b, 6a, 7a, 8c, and 8e in Table I). Although these variations might be ascribed to the variation of the SDS–PAGE method as used in different laboratories, it could also indicate that only one polypeptide may eventually be found to be required and that it may have variant forms as a result of evolutionary changes.

One additional comparative result that again points up the substantial differences in the polypeptides isolated from different reaction centers is the fact that no antibody cross-reactions could be demonstrated between antibodies to the RC's prepared from the R-26 mutant of *Rhodopseudomonas sphaeroides* and chromatophores from *Rhodospirillum rubrum, Rhodopseudomonas capsulata, R. palustris, R. gelatinosa,* or *R. viridis.*[12]

The only way to answer with certainty questions pertaining to the comparative role of the polypeptides in these RC's is to establish the amino acid sequence of the polypeptides isolated in each case and then to probe structure–function questions at the fundamental level of protein structure. First efforts toward obtaining the sequence of one of these polypeptides were recently reported by Rosen *et al.*[42] for the R-26 RC. The amino terminal sequence of the 21 kd polypeptide was

Ala-Leu-Leu-X-Phe-Glu-Arg-Lys-Tyr-Arg-Val-Pro-Gly-Gly-Thr-Leu-Val-Gly-Gly-Asn-

Leu-Phe-Asp-Phe

Although sequencing these integral protein components is exceedingly difficult, this information will be extremely important and interesting.

The RC's isolated from wild-type bacteria have a carotenoid molecule still bound which, of course, those isolated from the carotenoidless mutants do not. Thus, for *R. sphaeroides* (wild type) RC's, spheroidene was present,[19] and for *R. rubrum* (wild type), spirilloxanthin was found in a 1:1 ratio with the primary electron donor.[42a] Even though these RC's from the wild-type bacteria seem to have this additional component relative to RC's from the carotenoidless mutants (i.e., *R. sphaeroides* R-26 and *R. rubrum* G-9) of the same bacteria, there does not appear to be an additional polypeptide in the wild-type RC's (Table I).

Although RC's obtained from *C. vinosum* and *R. viridis* are listed in Table I, it should be noted that these preparations usually contain cytochrome components (although see Feher and Okamura[6]) and are larger in size than the comparable preparations from *R. sphaeroides* and *R. rubrum*. Nevertheless, substantial progress toward obtaining pure RC's from these species has been made.

### Photoreceptor Complex Preparations

*Assay.* Phototrap (reaction center) activity is measured by determining the light-induced difference spectra, usually in the near infrared, and the light-induced ESR signal at low temperature. Additional tests for native activity include the decay (charge recombination) kinetics at low temperature, quantum yield for charge separation, and room temperature reactivity with secondary electron transport components, such as ubiquinone and cytochrome *c*. The unique part of the photoreceptor complex that differentiates it from the reaction center preparation is the presence of light-harvesting antenna pigments. Thus, the preparation should be relatively small in size (e.g., molecular weight less than 200,000) with perhaps only one polypeptide beyond those found in a reaction center preparation from the same species, but with an absorbance spectrum that

---

[42] D. Rosen, M. Y. Okamura, G. Feher, L. A. Steiner, and J. E. Walker, *Biophys. J.* **17**, 67a, Abstr. W-PM-F15 (1977).
[42a] M. Van der Rest and G. Gingras, *J. Biol. Chem.* **249**, 6446 (1974).

shows the presence of antenna pigments which appear to be in the same molecular environment as they were *in vivo*. In addition, the quantum yield for transfer of excitation energy from the antenna pigments to the phototrap should be as high as *in vivo*.

*Method of Preparation.* Our laboratory seems to be the only one which has pursued the isolation of this kind of complex. The first well-defined preparations of PRC were reported about the same time as the first well-defined RC preparations.[43-45] Table II summarizes our results with several different bacteria.[46-48] The basic approach makes use of alkaline solution, addition of urea, and addition of a few percent of a nonionic detergent such as Triton X-100 (referred to as the AUT method). Each condition was chosen because we anticipated a need to disaggregate protein complexes in order to release integral units of the membrane. We felt that the more extreme the pH, the highest urea concentration, and the highest detergent concentration that could be used, without loss of the activity referred to above, would result in the smallest membrane fragments. In practice, one must compromise with the detergent concentration and the pH, as a combination of high pH and high detergent can cause rapid loss of activity. In hindsight, the urea is probably stabilizing membrane proteins rather than causing their denaturation or dispersion.

An outline of the steps of a typical preparation of PRC from *R. rubrum* is given in Fig. 1. Details of a typical preparation are as follows[2]: The bacteria are propagated in liquid media which contains malic acid as a carbon source (modified Hutner's medium[49]) at 30° and harvested by centrifugation in their final stages of logarithmic growth, usually after 3 to 4 days. Our inoculum volume from the older wet culture has usually been between 0.5 to 1% the volume of the new wet culture. Illumination during growth is provided by two banks of fluorescent lights (40 W) which result in a light intensity at the sample of about $10^4$ ergs/cm²/sec. One-liter culture flasks with ground glass stoppers are used for growing the bacteria. After adding the inoculum volume to some media, the flasks are filled with media to overflowing and sealed with Parafilm. A typical yield of cells is 1 to 1.2 gm of lyophilized cells per liter of growth medium.

[43] P. Loach, W. R. Heftel, R. M. Hadsell, and F. J. Ryan, *Prog. Photobiol., Proc. Int. Congr., 5th, 1968* Abstr. Bf-1 (1969).

[44] P. Loach, W. R. Heftel, R. M. Hadsell, and A. Stemer, *Proc. Int. Bot. Congr., 11th, 1969* Abstr. 54 (1969).

[45] P. Loach and R. Hall, *Biophys. J.* **10,** 201a, Abstr. FAM-E2 (1970).

[46] P. Loach, R. M. Hadsell, D. L. Sekura, and A. Stemer, *Biochemistry* **9,** 3127 (1970).

[47] P. Loach, D. Sekura, R. M. Hadsell, and A. Stemer, *Biochemistry* **9,** 724 (1970).

[48] J. A. Runquist and P. A. Loach, unpublished work.

[49] G. Cohen-Bazire, W. R. Sistrom, and R. Y. Stanier, *J. Cell. Comp. Physiol.* **49,** 25 (1957).

TABLE II
PHOTORECEPTOR COMPLEX PREPARATIONS

| Bacteria | Polypeptide composition (kilodaltons) | | |
| | Phototrap–Antenna– protein complex | Antenna–protein complex only | Reference |
| --- | --- | --- | --- |
| R. rubrum | 10, 21, 24, 29 | — | 2, 46 |
| R. sphaeroides (wild type) | 10, 12, 21, 24[a] | 12, (32), (45) | 2, 47 |
| R. sphaeroides (R-26) | 10, 21, 24[a] | 10 | 2, 47 |
| R. capsulata | 9, 12, 21, 24, 30 | — | 48 |

[a] Because of evidence for a slight separation upon electrophoresis of a phototrap (RC) component from the antenna pigments, these may not be tightly associated PRC preparations. The apparent 10 and 12 kilodalton polypeptides tended to move with the center of the antenna pigments while the 21 and 24 kilodalton polypeptides moved with the phototrap activity.

Cytoplasmic membrane vesicles (chromatophores) are prepared either from freshly grown whole cells or from frozen whole cells stored in a pellet at $-20°$. As outlined in Fig. 1, the most optimal conditions for dissolution of these membrane vesicles by the AUT procedure were found to be pH 12 (0.05 $M$ phosphate buffer; the pH is adjusted before adding urea and Triton), 6 $M$ urea, 3% Triton X-100, and 0.001 $M$ $MgCl_2$. The membrane material is centrifuged into this media at $2°$ by layering 1 ml of chromatophores (A 880 nm = 100) suspended in water on top of 24 ml of the AUT mixture and centrifuging at 80,000 $g$ for 45 min. Because of the presence of 6 $M$ urea, the chromatophore sample will layer neatly on top of the AUT ingredients with a little care. As a result of centrifugation, the membrane material interacts with the AUT ingredients and a sharp, densely colored band is formed about 5 ml from the top of the 25-ml centrifuge tube. This band is carefully removed in a 1- to 2-ml volume with a drawn out disposable pipette, the pH is brought to near 7 by adding the appropriate amount of $KH_2PO_4$, and the sample dialyzed at $3°$ for 1–2 hr with at least three changes of the dialysate buffer (0.001 $M$ phosphate buffer, pH 7.5, 0.001 $M$ $MgCl_2$). The yield of phototrap activity and antenna pigments at this stage was typically better than 80%.

For the final stage of purification, about 16 ml of AUT particles prepared as described above (the pooled product from about eight centrifuge tubes) are applied to a 440-ml column for electrophoresis against a sucrose gradient. If necessary, the AUT particles may be stored frozen

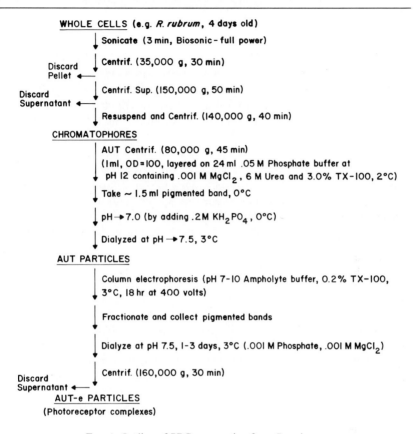

WHOLE CELLS (e.g. *R. rubrum*, 4 days old)
  Sonicate (3 min, Biosonic – full power)

Discard     Centrif. (35,000 g, 30 min)
Pellet ◄—

Discard     Centrif. Sup. (150,000 g, 50 min)
Supernatant ◄—

  Resuspend and Centrif. (140,000 g, 40 min)

CHROMATOPHORES

  AUT Centrif. (80,000 g, 45 min)
  (1ml, OD = 100, layered on 24 ml .05 M Phosphate buffer at
  pH 12 containing .001 M MgCl$_2$, 6 M Urea and 3.0% TX-100, 2°C)

  Take ~ 1.5 ml pigmented band, 0°C

  pH → 7.0 (by adding .2 M KH$_2$PO$_4$, 0°C)

  Dialyzed at pH → 7.5, 3°C

AUT PARTICLES

  Column electrophoresis (pH 7-10 Ampholyte buffer, 0.2% TX-100,
  3°C, 18 hr at 400 volts)

  Fractionate and collect pigmented bands

  Dialyze at pH 7.5, 1-3 days, 3°C (.001 M Phosphate, .001 M MgCl$_2$)

Discard     Centrif. (160,000 g, 30 min)
Supernatant ◄—

AUT-e PARTICLES
(Photoreceptor complexes)

FIG. 1. Outline of PRC preparation from *R. rubrum*.

overnight, but some degradation occurs. For column electrophoresis, an LKB Instruments Electrofocusing apparatus (Rockville, Maryland) is employed using pH 7-10 mixed ampholytes as the buffer system. 0.2% Triton X-100 is present throughout the column to retard aggregation. The columns were prepared as though for an electrofocusing experiment (LKB Instruction Manual I-8100-EO1) with the AUT sample applied about one-fifth the distance from the top of the column. The bacteriochlorophyll content of the sample is near 0.1 mg/ml. The electrodes are connected so that the negatively charged components migrate toward the bottom of the column. A period of 18 hr at 400 V is used for development of the column with the temperature maintained at 3°. A single pigmented band is usually maintained as the sample moves down the column upon electrophoresis. The band is collected as the column is fractionated and the sample is dialyzed against 0.001 $M$ phosphate, pH 7.5, 0.001 $M$ $MgCl_2$ at 3° for 1–3 days. The PRC's are then easily recovered by centrifugation. Overall, greater than 90% of the phototrap activity and more than 70% of the original pigment content can be recovered in PRC.

*Properties of Photoreceptor Complex Preparations.* Composition of the PRC prepared from *R. rubrum* as described above is summarized in Table III and compared with the chromatophore fraction from which it was prepared. The light-induced absorbance changes and quantum yield for photooxidation of the primary electron donor unit were the same as for intact bacteria (0.95).[46,50,51] The $E_0'$ value for the primary electron donor unit was also 0.44 V as for the *in vivo* system. Especially interesting is the fact that the absorbance spectrum, which largely reflects antenna bacteriochlorophyll and carotenoids, is little changed from the *in vivo* state.[46] Because of the sensitivity of both the bacteriochlorophyll antenna aggregate and the carotenoid pigments to changes in their environment (bacteriochlorophyll *in vivo* in *R. rubrum* has its far red band shifted substantially relative to its location in organic solvents or in aqueous detergent systems), the *in vivo*-like absorbance spectrum is taken as strong evidence that the preparation represents a unique structural and functional entity and not an artifactual pigment–protein complex arising because the treatment displaced pigments from their normal habitat.

The low iron content is especially noteworthy (Table III) as these were the first preparations in which it was demonstrated that iron is not required in the primary photochemical event. These were also the first preparations in which the first stable electron acceptor (often called the primary electron acceptor or X) was shown to be ubiquinone by both room temperature and low temperature ESR and by quantum yield measurements.[51]

From Table III, it may be seen that the major polypeptide components found are those of 21, 24, and 29 kilodaltons as are most often found in RC preparations from *R. rubrum* (Table I), but also there is a major component of apparent molecular weight of 12 kilodaltons. Because the PRC has antenna pigments still bound, it is tentatively assumed that this additional polypeptide is a light-harvesting (LH) protein component. It should be noted that the 21- and 24-kilodalton polypeptides have appeared only in very small quantities in many PRC preparations, but the 29–32 and 12-kilodalton components are always major polypeptides.

Of special interest is the fact that the 12-kilodalton component may be quantitatively separated from the other polypeptides because it alone

[50] P. A. Loach and D. L. Sekura, *Biochemistry* **7**, 2642 (1968).
[51] P. A. Loach and R. L. Hall, *Proc. Natl. Acad. Sci. U.S.A.* **69**, 786 (1972).

TABLE III
COMPOSITION OF PHOTORECEPTOR COMPLEXES FROM *R. rubrum*

|  | PRC | Chromatophores |
| --- | --- | --- |
| Molecular weight | 125,000 ± 25,000 | >30,000,000 |
| Bacteriochlorophyll $a^a$ | 24 | 28 |
| Bacteriopheophytin $a^a$ | 2 | 2 |
| Carotenoid$^a$ | 8 | 9 |
| P-lipid (%) | < 0.3 | 25 |
| Iron$^a$ | 0.1 to 0.3 | 5 |
| Copper$^a$ | < 0.1 | <0.2 |
| Manganese$^a$ | < 0.2 | — |
| Ubiquinone + Rhodoquinone$^a$ | 3.5 | 8 |
| Polypeptide molecular weight | 12,000 | — |
| (by SDS-PAGE) | (21,000) | |
|  | (24,000) | |
|  | 30,000 | |

$^a$ Number per phototrap (per primary electron donor unit).

dissolves in organic solvents either from lyophilized PRC or from lyophilized chromatophores.[52] This protein component is further described in the next section on light-harvesting (LH) preparations.

### Light-Harvesting Pigment–Protein Complexes

*Assay.* About the only known native property of LH complexes that is readily measured is the absorbance spectrum. The ability to reconstitute PRC with RC and LH preparations could also be considered; however, this may not be a very specific test because I suspect most integral protein complexes, regardless of their source, will associate with other integral protein components when added together. In addition, a relatively long distance of approach (30 to 60 Å) would suffice for efficient transfer of excitation energy from LH to RC.

There is also the problem of artificial species arising during treatment with detergent because the pigments are small molecules, highly hydrophobic, and present at high concentration. The overall yield of pigment obtained in LH preparations rarely exceeds 50% of the original pigment content, the rest presumably being lost due to degradation or solubilization. Nevertheless, several other factors seem to have made it possible to isolate antenna pigment–protein complexes that may have been little

[52] S. J. Tonn, G. E. Gogel, and P. A. Loach, *Biochemistry* **16**, 877 (1977).

modified from their native state. Two factors which are important are the reasonably tight and specific binding of the pigments to protein in photosynthetic bacteria and the very high concentration of the antenna pigment–protein complex in the specialized membranes called chromatophores. In fact, as much as 50 to 70% of some chromatophore preparations may consist of LH complexes. Perhaps it is more constructive to think of the preparation of LH in terms that suggest removal of minor contaminants from the chromatophore starting material rather than in terms that suggest isolation of a minor component as is standard practice in enzymology.

*Methods of Preparation and Properties.* In Table IV is a summary of LH preparations that have been reported.[2,9,24,53–68] In each case where the protein component has been sufficiently studied, with the exception of the water-soluble bacteriochlorophyll–protein complex from *C. limicola* which does not contain membrane integral protein, one or more polypeptides of apparent molecular weight between 7 and 12 kilodaltons have been found (by SDS–PAGE) to be the predominant, or the only polypeptide component(s) present. Within the limitation that the apparent molecular weight determined by SDS–PAGE can be very misleading for such highly hydrophobic and integral protein components, the small molecular weight polypeptide seems to be characteristic of membrane-bound LH complexes.

The methods used for obtaining LH from chromatophores all utilize detergents with generally fewer steps of purification than is true for RC preparations. For example, a typical LH preparation from *R. sphaeroides*

[53] P. J. Fraker and S. Kaplan, *J. Bacteriol.* **108**, 465 (1971).

[54] P. J. Fraker and S. Kaplan, *J. Biol. Chem.* **247**, 2732 (1972).

[55] J. W. Huang and S. Kaplan, *Biochim. Biophys. Acta* **307**, 317 (1973).

[56] J. W. Huang and S. Kaplan, *Biochim. Biophys. Acta* **307**, 332 (1973).

[57] R. K. Clayton and B. J. Clayton, *Biochim. Biophys. Acta* **283**, 492 (1972).

[58] P. Heathcote and R. K. Clayton, *Biochim. Biophys. Acta* (in press).

[59] K. Sauer and L. A. Austin, *Biochemistry* **17**, 2011 (1978).

[60] L. P. Vernon and A. F. Garcia, *Biochim. Biophys. Acta* **143**, 144 (1967).

[61] B. Ke, M. Green, L. P. Vernon, and A. F. Garcia, *Biochim. Biophys. Acta* **162**, 467 (1968).

[62] U. Schwenker and G. Gingras, *Biochem. Biophys. Res. Commun.* **51**, 94 (1973).

[63] U. Schwenker, M. St.-Onge, and G. Gingras, *Biochim. Biophys. Acta* **351**, 246 (1974).

[64] P. A. Cuendet and H. Zuber, *FEBS Lett.* **79**, 96 (1977).

[65] J. M. Olson and C. A. Romano, *Biochim. Biophys. Acta* **59**, 728 (1962).

[66] J. P. Thornber and J. M. Olson, *Biochemistry* **7**, 2242 (1968).

[67] R. E. Fenna, B. W. Matthews, J. M. Olson, and E. K. Shaw, *J. Mol. Biol.* **84**, 231 (1974).

[68] R. E. Fenna and B. W. Matthews, *Brookhaven Symp. Biol.* **28**, 170 (1976).

may involve only addition of detergent (LDAO) and centrifugation of the mixture.[57]

Because it has been possible to prepare stable crystals of the water-soluble antenna bacteriochlorophyll–protein complex from *C. limicola*,[67] X-ray diffraction studies have been successfully applied to this system.[67,68] They show that 7 bacteriochlorophyll molecules are bound in each of three identical 50-kilodalton polypeptides which make up a trimeric unit. The pigment molecules are not very close to one another (average center–center nearest neighbor distance between porphyrin rings is 12 Å), and, to a first approximation, their planes lie parallel with one another. The major near infrared absorbance band of this complex is at 809 nm. The question arises as to what extent one might expect the number and distribution of bacteriochlorophyll molecules in LH complexes isolated from membrane systems to be reminiscent of this water-soluble complex from *C. limicola*. At this point, the three major properties of LH complexes isolated from membrane systems are very different from those of the water-soluble complex. They are (1) the much greater preponderance of hydrophobic amino acid residues in the protein from membrane LH complexes, (2) the apparently much smaller size of the membrane LH polypeptides, and (3) the much greater red shift of the *in vivo* near infrared absorbance band of bacteriochlorophyll (often between 850 and 890 nm) in the membrane LH complexes. These substantial differences would lead one to be surprised indeed if a close structural relationship exists between LH complexes in membranes and the water soluble complex from *C. limicola*.

The LH preparation derived from *R. sphaeroides* (wild type) by the method of Clayton and Clayton[57] has been used for a "cross" reconstitution experiment with RC's prepared from *R. sphaeroides* (R-26).[58] Heathcote and Clayton[58] demonstrated reconstitution of excitation energy transfer from the antenna pigments to the phototrap with a high quantum yield as well as restoration of the fluorescent yield and the transient behavior of fluorescence as exhibited by chromatophores of *R. sphaeroides*.

Of particular interest is the fact that the polypeptides of both LH components reported by Hall *et al.*[2] as prepared from *R. sphaeroides* (wild type) are soluble in organic solvents.[52,69] In addition, as indicated earlier, the small polypeptide of the PRC complex prepared from *R. rubrum* also totally dissolves in organic solvents.[52] This feature has allowed our laboratory to prepare large quantities of this polypeptide in

[69] S. J. Tonn, PhD Thesis, Northwestern University, Evanston, Illinois (1975).

TABLE IV

LIGHT-HARVESTING PIGMENT-PROTEIN COMPLEXES

| Bacteria and Reference | Detergent | Purification steps | | | | | Components present | | | |
|---|---|---|---|---|---|---|---|---|---|---|
| | | Centrifugation | (NH$_4$)$_2$SO$_4$ | Gel filtration | Prep. SDS-PAGE | AUT-e | Pigments[b] | | P-Lip. | Polypeptide (kilodaltons) |
| | | | | | | | BChl (nm) | Car | | |
| 1. *Rhodopseudomonas sphaeroides* (wild type) | | | | | | | | | | |
| a. Fraker and Kaplan[53-56] | 2-Chloroethanol + ~0.1 N HCl | + | | + | + | | 760[a] | +(?) | + | 7-11 |
| b. Clayton and Clayton[57] | (1%) LDAO | + | + | | | | 800, 850 | + | | 9 |
| c. Hall *et al.*[2] | (1.5%) TX-100 | + | | | | + | 800, 850 | + | | 10 |
| d. Heathcote and Clayton[58] | (1%) LDAO | + | | + | | | 800, 850 | + | | 9,12 9 |
| e. Sauer and Austin[59] | TX-100 | + | | | | | 800, 850 | + | | 10 |

| | Detergent | | | | | | |
|---|---|---|---|---|---|---|---|
| 2. *Rhodopseudomonas sphaeroides* (R-26) | | | | | | | |
|   a. Reed and Clayton[9] | (3%) TX-100 | + | | | | 870 | 10 |
|   b. Hall *et al.*[2] | (3%) TX-100 | + | | | | 770 | 10 |
|   c. Sauer and Austin[59] | TX-100 | + | | | | | |
| 3. *Rhodospirillum rubrum* (wild type) | | | | | | | |
|   a. Vernon and Garcia[60]; Ke *et al.*[61] | TX-100 | + | | | + | | |
|   b. Schwenker and Gingras[62,63] | (1%) SDS + β-mercapto-ethanol | + | + | + | + | | 11 |
|   c. Hall *et al.*[2] | TX-100 (3%) | + | | + | + | – | |
| 4. *Rhodospirillum rubrum* (G-9) | | | | | | | |
|   a. Cuendet and Zuber[64]; Bachofen *et al.*[24] | LDAO (1%; .1%) | + | + | + | + | 770, 820 | 12–14 |
| 5. *Chlorobium limicola* | | | | | | | |
|   a. Olson and Romano[65]; Thornber and Olson[66]; Fenna *et al.*[67]; Fenna and Matthews[68] | None | + | + | + | | | 50 |

[a] The BChl had undergone pheophytinization due to acidic conditions of the preparation.

[b] Abbreviations: BChl, bacteriochlorophyll; Car, cartenoid; P-Lip, phospholipid.

pure form, and we are in the process of determining its amino acid sequence. Cuendet and Zuber[64] using the organic solvent extraction procedure developed in our laboratory have also prepared large quantities of a similar polypeptide from *R. rubrum* G-9 (carotenoidless mutant) and are also initiating amino acid sequence determination studies. The amino acid contents of these two polypeptides, as isolated and studied in two different laboratories, are quite similar.[52,64] An eventual comparison between species of the sequence and membrane distribution of this kind of LH polypeptide should be exceedingly interesting. Indeed, it will probably not be possible to draw intelligent conclusions about the relationship of these LH polypeptides to the phototrap until substantial amino acid sequence information is available on each of the polypeptides apparently involved in the PRC.

Acknowledgment

Supported by research grants from the National Institutes of Health (GM 11741) and the National Science Foundation (PCM 74-12588 AO2).

## [16] The Photochemical Reaction Center of the Bacteriochlorophyll *b*-Containing Organism, *Rhodopseudomonas viridis*

By J. Philip Thornber[1] and Janet M. Thornber

The photochemical reaction center of this bacteriochlorophyll *b*-containing organism consists of a photobleachable pigment, P985 (P960 in isolated reaction centers and dried films of lamellae) and a pigment, P830, which shifts to the blue upon illumination.[1a] Bacteriochlorophyll *b*-containing reaction centers have one particular advantage over all other equivalent preparations in that larger differences in energy levels occur between the different spectral forms of bacteriochlorophyll *b* *in situ* which enables the forms to be more easily distinguished in spectra of this reaction center. Isolation and some characteristics of this component have been described in a previous volume in this series,[2] but since then, improved isolation procedures have been discovered which have yielded preparations that lack the 685 nm-absorbing contaminant present in ma-

[1] The author's research reported here was supported by National Science Foundation grants and a Guggenheim Memorial Fellowship.

[1a] A. S. Holt and R. K. Clayton, *Photochem. Photobiol.* **4**, 829 (1965).

[2] J. P. Thornber, Vol. 23, p. 688.

terial obtained using the original method. This enabled the reaction center of this organism to be better characterized.

## Isolation Procedures

### Procedure A of Pucheu et al.[3]

*Step 1.* Photosynthetic membranes of *Rhodopseudomonas viridis* are suspended in 10 m$M$ Tris-HCl, pH 8.0, and the volume adjusted to give an absorbance at 1015 nm of 50 cm$^{-1}$. Lauryldimethylamine oxide (LDAO) is added as a 30% solution to give a final detergent concentration of 3 mg/mg of protein.

*Step 2.* Fifty percent (w/v) ammonium sulfate is added to give a final concentration of 23% (w/v), and the suspension is centrifuged at 9000 $g$ for 20 min. The floating pellet is discarded. Further 50% ammonium sulfate is added to make the solution 28% (w/v), and this time the floating precipitate obtained after centrifugation is resuspended in 50 m$M$ Tris-HCl, pH 8.0.

*Step 3.* One gram of Hyflo Supercel per 50 absorbance units cm$^{-1}$ at 830 nm and sufficient ammonium sulfate to give 28% (w/v) are added to this solution. The slurry is packed into a 2.5 cm diameter column and washed with a 50 m$M$ Tris–0.1% LDAO–28% ammonium sulfate solution. Reaction center-containing material is eluted by a decreasing gradient of ammonium sulfate (28–10%, w/v) in Tris and LDAO. The required material is solubilized at approximately 20% (w/v) ammonium sulfate.

*Step 4.* P960-containing fractions are dialyzed against 10 m$M$ Tris–1 m$M$ EDTA–0.1% LDAO, pH 8.0, for at least 6 hr (preferably overnight), and then applied to a DEAE–cellulose column (2 × 25 cm) from which the reaction center is eluted as a single band by the dialysis buffer.

*Step 5.* Further purification is achieved by centrifugation of the DEAE–cellulose eluate on a linear sucrose gradient (0.4 → 1.0 $M$) in 50 m$M$ Tris—1 m$M$ EDTA—0.6% LDAO, pH 8.0, at 45,000 rpm for 15 hr. The reaction center sediments as a brown band. Repetition of this step improves the purification.

### Procedure B of Pucheu et al.[3]

*Step 1.* Same as procedure A (above).

*Step 2.* The LDAO-treated membranes are dialyzed against 10 m$M$ EDTA–0.1% LDAO (preferably overnight). Thereafter the solution is

---

[3] N. L. Pucheu, N. L. Kerber, and A. Garcia, *Arch. Microbiol.* **109,** 301 (1976).

centrifuged (40,000 $g$; 30 min), and the supernatant applied to a DEAE–cellulose column (2 × 25 cm).

*Step 3.* The column is washed with dialysis buffer until no further protein is eluted; green and brown material is removed during the wash. The reaction center is eluted by 10 m$M$ Tris—1 m$M$ EDTA—0.1% LDAO—150 m$M$ NaCl.

*Step 4.* Steps 2 and 3 are repeated, and then the resulting solution is subjected to sucrose gradient centrifugation as described under step 5 [procedure A (above)].

*Method of Thornber*[4,5]

A much shorter procedure based on the findings of Pucheu *et al.*[3] has been devised.

*Step 1.* Packed cells (60 ml) are suspended in 50 m$M$ Tris-HCl, pH 8.0, to give a final volume of 500 ml, and the mixture is homogenized in a Waring blender for 30 sec. Cells in this suspension are broken by sonication or by passage through a needle valve disintegrator, and the treated material is centrifuged at 20,000 $g$ for 5 min. The lightly colored supernatant is discarded, and the pellet is resuspended in 50 m$M$ Tris-HCl (250 ml) by homogenization in a Waring blender. LDAO (Onyx Chemical Co., New Jersey) is added to the suspension to give a final concentration of 1%, and the mixture incubated at room temperature for 1 hr.

*Step 2.* The incubated solution is centrifuged at 35,000 $g$ for 10 min and the supernatant stored. The pellet is resuspended as described above to give 100 ml of solution with LDAO concentration of 1%, incubated for 1 hr, and then centrifuged. The green supernatant is bulked with the first supernatant, and the pellet is discarded. A spectrophotometric check is made to ensure that all the 1015-nm antenna chlorophyll has been converted to the 810- and/or 685-nm spectral form. If it has not, the solution can be warmed to 30/–37° until the conversion is complete.

*Step 3.* The supernatants are dialyzed against 50 m$M$ Tris–0.1% LDAO (5 liters) overnight, or diluted with water to give a LDAO concentration of 0.1%. A DEAE–cellulose column (packed volume 80 ml) is poured in a 100-ml glass syringe barrel, and equilibrated with 50 m$M$ Tris-HCl, pH 8.0. The dialyzed extract is applied to the column until the

[4] J. P. Thornber, P. L. Dutton, J. Fajer, A. Forman, D. Holten, J. M. Olson, W. W. Parson, R. C. Prince, D. M. Tiede, and M. W. Windsor, *Proc. Int. Congr. Photosynth. Res., 4th, 1977* pp. 55–70 (1977).

[5] R. C. Prince, D. M. Tiede, J. P. Thornber, and P. L. Dutton, *Biochim. Biophys. Acta* **462**, 467 (1977).

top one-third of the DEAE–cellulose is colored. After eluting with 50 m$M$ Tris–0.1% LDAO, pH 8.0, until the eluate is colorless (it is green at first), the gray-brown P960-containing material is washed from the column with 10 m$M$ sodium phosphate–135 mM NaCl, pH 7.0. The remaining dialyzed extract can be chromatographed on the same column after reequilibration of the DEAE–cellulose in 50 m$M$ Tris–0.1% LDAO.

*Step 4.* A hydroxylapatite[6] column (50 ml) is equilibrated with 10 m$M$ sodium phosphate—200 mM NaCl, pH 7.0, and the DEAE–cellulose eluate adsorbed to the hydroxylapatite. After eluting the column with 50–100 ml of equilibration buffer reaction center-containing material is eluted with 150 m$M$ sodium phosphate–200 m$M$ NaCl, pH 7.0. A peristaltic pump is used to maintain a reasonable flow rate for the hydroxylapatite column.

*Step 5.* A saturated solution of ammonium sulfate is added to the P960-containing hydroxylapatite column eluate until a precipitate forms [usually at about 22% (w/v) ammonium sulfate]. Centrifugation at 20,000 g for 10 min gives a floating precipitate which can be collected on a coarse sintered glass filter. Finally, the precipitate is dissolved in 50 m$M$ Tris-HCl, pH 8.0; several milliliters of ~200 $\mu M$ P960 can be obtained in this manner.

## Method of Trosper et al.[7]

All the above procedures use LDAO as the membrane-solubilizing agent, while this one uses sodium dodecyl sulfate and a fractionation method analogous to that in the original procedure.[2] The Trosper *et al.*[7] procedure uses sodium dithionite in all buffers so that formation of the 685 nm-contaminant is prevented. Sodium dodecyl sulfate, unlike LDAO, does not remove this contaminant from the reaction center preparation. However, the sodium dithionite addition is not always successful in its purpose, but nevertheless the procedure[7] remains the best available for preparing the sodium dodecyl sulfate–reaction center.

The preparation is carried out at room temperature in dim light. All solutions contain 0.01 $M$ Na$_2$S$_2$O$_4$ which is added immediately prior to using the solution.

*Step 1.* Photosynthetic lamellae are treated with sodium dodecyl sulfate[2] (70 mg detergent/$\mu$mole chlorophyll); bacteriochlorophyll *b* content is determined from its absorbance at the near infrared maximum (~1015

[6] H. W. Siegelman and J. H. Kycia, *in* "Handbook of Phycological Methods" (J. Cragle and J. A. Helleburst, Cambridge Univ. Press, London and New York, 1978.
[7] T. L. Trosper, D. L. Benson, and J. P. Thornber, *Biochim. Biophys. Acta* **460**, 318 (1977).

nm) using $\epsilon = 10^5 M^{-1}$ cm$^{-1}$. The detergent-treated material is diluted so that the detergent concentration is less than 1.5%, and the suspension stirred for 15 min. A spectrophotometric check is made to ensure that the 1015 nm form has been converted to the 810-nm form (cf. Thornber method above).

*Step 2.* The suspension (containing less than 10 $\mu$mole of chlorophyll) is centrifuged (20,000 $g$ for 20 min), and the supernatant applied to a hydroxylapatite column (70 ml in a 100 ml glass syringe barrel) equilibrated with 10 m$M$ sodium phosphate–200 m$M$ NaCl–0.01 $M$ Na$_2$S$_2$O$_4$, pH 7.0. The column is washed with 50 ml of equilibration buffer and then with 50 ml of 100 m$M$ sodium phosphate–200 m$M$ NaCl–0.01 $M$ Na$_2$S O$_4$. At this stage, the column appears green throughout most of its length with a brownish band at the front. This brownish band (P960-containing) is eluted by increasing the sodium phosphate to 150 m$M$, and to 200 m$M$ if necessary. A peristaltic pump maintains the flow rate at 2–3 ml/min.

*Step 3.* Ammonium sulfate is added to the P960-containing eluate to 8% (w/v) and the precipitate (810 nm-absorbing material) obtained by centrifugation is discarded. The supernatant is made to 25% (w/v) ammonium sulfate, and this time the precipitate containing P960 is recovered and solubilized in 50 m$M$ Tris, pH 8.0.

*Step 4.* If the desired spectral purity has not been obtained the product can be taken through steps 2 and 3 again, but this time a 10-ml hydroxylapatite column in a 30-ml syringe barrel is used.

The product can be stored in closed containers at 4° in the dark for several weeks. Alternatively it can be frozen to −17° and kept indefinitely.

### Properties

The characteristics described below are those of the LDAO-reaction center preparation except where otherwise stated.

*Spectral.*[4] The room temperature absorption spectra of the reduced and light-oxidized reaction center preparation are shown in Fig. 1. The reduced spectrum exhibits a band at 958 nm due to P960, a peak at 832 nm due to some other bacteriochlorophyll $b$ molecules closely associated with P960, and a shoulder at ~790 nm due to bacteriopheophytin $b$. On chemical or photooxidation of P960 most of the 958-nm absorbance is lost, some of 605-nm absorbance (Q$_x$ band of bacteriochlorophyll $b$) disappears, a peak appears at 1310 nm (due to P$^+$960), and the 832-nm peak shifts to 828 nm.

At 77° the 832-nm band is better resolved.[7] Its 790 nm shoulder is now seen as a peak, and the 832 nm peak is resolved into a major component

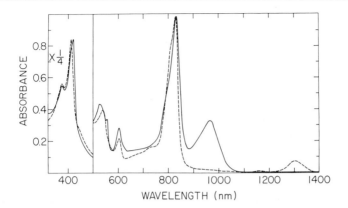

FIG. 1. Room temperature absorption spectra of the reaction center preparation in 50 mM Tris-HCl, pH 8.0, recorded in the IR1 (solid line) and IR2 (dashed line) modes of the Cary 14R spectrophotometer. In the IR1 mode the sample is exposed to a weak beam of monochromatic light, while in the IR2 the sample is illuminated by a strong beam of white light, and P960 is photooxidized.

at 836 nm, a lesser peak at 817 nm, and a shoulder at 850 nm. When P960 is oxidized, the 850 nm shoulder is lost, the 836 nm band shifts to 830 nm and a band appears at 808 nm. Thornber *et al.*[4] describe a possible interpretation of the spectral changes that occur on oxidation of P960.

*Composition.*[4] For each P960 entity ($\Delta\epsilon = 100$ m$M^{-1}$ cm$^{-1}$ at 958 nm) there are four bacteriochlorophyll *b* molecules (two of which constitute P960), two bacteriopheophytin *b* molecules, two hemes of cytochrome *c*-558 ($E_m = \pm330$ mV) and 2–3 hemes (probably 2) of cytochrome *c*-553 ($E_m = -12$ mV). A carotenoid, dihydrolycopene, is present in the sodium dodecyl sulfate preparation, whereas two as yet unidentified carotenoids occur in the LDAO preparation. A menaquinone- like substance occurs in the proportion of 0.3 mole/P960; no ubiquinone is present.[3] Other proteins in addition to those of the cytochromes are contained in the isolated complex.

*Size and Subunit Composition.*[3] Gel filtration and calibrated gel electrophoresis give values of 230–240 kilodaltons (cf. 110 kilodaltons daltons for the sodium dodecyl sulfate preparation.[2]) The polypeptides present in the LDAO preparation are of 45, 37, 29, and 23 kilodaltons).

*Characteristics of the Primary Photochemical Reactants in the Preparation.* The primary electron donor (P960) is a special pair of bacteriochlorophyll *b* molecules (BChl-*b*)$_2$. The *g* value of P$^+$960 is 2.0026, and the linewidth of this signal is 11.8 $\pm$ 0.2 G at 130° and 230°K; ENDOR transitions occur at 1.6, 1.9, 3.3 and 4.0 G. A light-induced spin-polarized triplet signal with zero field splitting characteristics of $D = 157 \times 10^4$ cm$^{-1}$ and $E = 39 \times 10^{-4}$ cm$^{-1}$ is seen when reaction centers are illuminated in the presence of sodium dithionite at 5°K.[5]

The traditionally defined "primary" electron acceptor, X, is a quinone–iron complex (presumably the menaquinone-like substance forms the quinone). In LDAO reaction centers X$^-$ produces a well-characterized $g_y = 1.82$ signal, whereas in sodium dodecyl sulfate–reaction centers, in which the iron has been displaced so that it cannot interact with the quinone, a $g = 2.0045$ signal is obtained.[5]

The intermediate electron carrier I is a bacteriopheophytin $b$ molecule that functions between P960 and X. Its $E_m$ is $-410$ mV, and it can be reduced by illuminating the preparation in the presence of Na$_2$S$_2$O$_4$ for 30 sec at room temperature or for 3 min at 200°K. The ESR spectrum of I$^-$ at 130°K exhibits a signal at $g = 2.0035 \pm 0.0002$[4] with a linewidth of 13 G and ENDOR transitions at ~3.2 and ~2.7 G. However, at <15°K, I$^-$ in LDAO reaction centers shows a mixture of a broad split signal centered close to $g = 2.003$ and a signal at $g = 2.003$; I$^-$ in sodium dodecyl sulfate–reaction centers lack the broad split signal and has only the 13-G wide free radical centered at $g = 2.003$.

P-960 is oxidized and I reduced within 6 psec of excitation.[8] The $t_{1/2}$ for the I$^-$ → X reaction is 240 psec[9] and for cytochrome $c$-558 → P$^+$960 at room temperature is 200 nsec; at lower temperatures only cytochrome $c$-553 can feed electrons to P$^+$960 ($t_{1/2} = 1$ msec at 77°K).

[8] T. Netzel, P. M. Rentzepis, D. M. Tiede, R. C. Prince, and P. L. Dutton, *Biochim. Biophys. Acta* **460**, 467, 1977.
[9] D. Holten, M. W. Windsor, W. W. Parson, and J. P. Thornber, *Biochim. Biophys. Acta* **501**, 112 (1978).

# Section III
# Components

# [17] Chloroplast Cytochromes $f$, $b$-559, and $b_6$

## By Aaron R. Wasserman

Higher-plant chloroplast particles contain three tightly bound cytochromes which in reduced form each exhibit a characteristic $\alpha$-absorption maximum: cytochrome $f$ (553.5 nm), cytochrome $b$-559 (559 nm), cytochrome $b_6$ (563 nm).[1] Like many other tightly bound membrane components throughout biology (e.g., the succinoxidase components of animal mitochondria), they are not extracted by sonication in salt solutions, with or without prior delipidation of their parent membrane. The present procedures achieve a disc-electrophoretically homogeneous preparation of each of the cytochromes with good recovery and retention of known biochemical properties. (No alternate procedures exist for $b$-559 and $b_6$.) Of possible value for purifying other membrane components are step 1 of the $b$-559 and $b_6$ procedures (extraction of all three cytochromes by Triton–4 $M$ urea, pH 8) and step 3 of all three procedures (preparative disc gel electrophoresis in nonionic detergent). Cytochromes $f$, $b$-559, and $b_6$ are shown to be intrinsically aggregatable species. Thus, a rationale for the procedures is that aqueous nonionic detergent solution maintains both the desired components and their impurities as separate monomolecular species, thus facilitating their purification. Since both cytochromes $b$-559 and $b_6$ are isolated as lipoproteins and the lipid content of cytochrome $f$ has never been examined, we prefer the term "component" at present rather than "protein."

## Assay

Cytochrome $b_6$ is reduced by dithionite ($Na_2S_2O_4$) only, $b$-559 and $f$ are reduced by either ascorbate or dithionite. Assay of each component at its $\alpha$ maximum is performed by difference spectrophotometry (reduced versus oxidized) with sample in both cuvettes. Spectra are recorded from 500 to 600 nm in a Phoenix-Chance spectrophotometer. A baseline is drawn (by extrapolation) below the observed peak using points on both sides of it.

*Cytochrome f (in Cytochrome f Procedure).* Neither $b_6$ nor $b$-559 are present in the extracts; $f$ is usually present in reduced form. Assay is therefore either as untreated versus oxidized (a few crystals of potassium

---

[1] D. S. Bendall, H. E. Davenport, and R. Hill, Vol. 23, p. 327.

METHODS IN ENZYMOLOGY, VOL. 69

ferricyanide) or ascorbate-reduced (a few drops of 0.1 $M$ sodium ascorbate, pH 7.2) versus oxidized (ferricyanide). [For pure samples, crystals of ammonium persulfate (colorless) replace ferricyanide (yellow) to allow more complete spectra below 500 nm.] One unit of cytochrome $f$ per milliliter, by definition, produces 1 absorbance unit at 553.5 nm (1 cm cell) in an absolute spectrum. Difference spectra values are converted to absorbance units (as in Table I) by using the factor of 1.4.

*Cytochrome b-559.* In fresh Triton–4 $M$ urea extracts (pH 8) the difference spectrum, reduced (ascorbate) versus untreated, gives a reliable measure of $b$-559 since cytochrome $f$ is present in reduced form and its spectrum is cancelled. After the extracts are supplemented with dithiothreitol (DTT), both $f$ and $b$-559 are fully reduced, and a difference spectrum of untreated (or ascorbate-reduced) versus oxidized (ferricyanide) shows both cytochromes, thus allowing only an approximation of $b$-559 content. After electrophoretic separation of cytochrome $f$, $b$-559 is again assayed reliably by a difference spectrum (untreated or ascorbate versus ferricyanide) or by an absolute spectrum. One unit of $b$-559 per milliliter, by definition, gives 1 absorbance unit (559 nm, 1-cm cell). Units in Table II are from difference spectra, except where indicated otherwise.

*Cytochrome $b_6$.* The difference spectrum, reduced (dithionite) versus ascorbate, is used. One unit of $b_6$ per milliliter, by definition, gives one absorbance unit (563 nm, 1-cm cell). Units in Table III are from difference spectra, except where indicated otherwise.

## Methods

*Starting Material.* Preparation from spinach leaves and storage of total chloroplast particle fraction (TCPF)[2] and of washed $P_1S$[3] (washed chloroplast grana) has been described. The pure component isolated ($f$, $b$-559, $b_6$) is the same from either membrane suspension. TCPF, easier to prepare and in better yield, is suitable for larger preparations.

*Preparation of Ethanol-Extracted Particles at 0° (Used in b-559, $b_6$ Procedures.)*[4] Suspensions of TCPF (or $P_1S$) were extracted by adding 9 volumes of absolute, 0° ethanol to achieve a concentration of 90% ethanol (Waring blender, 30 sec), centrifuged (30,000 $g$, 5 min), and the dark-green supernatant was discarded. The pellet was reextracted as before but in absolute ethanol. (More complete delipidation should be avoided.) The particles were then washed twice with 7–10 volumes of 0.05 $M$ Tris-

[2] J. Singh and A. R. Wasserman, *J. Biol. Chem.* **246**, 3532 (1971).
[3] A. R. Wasserman and S. Fleischer, *Biochim. Biophys. Acta* **153**, 154 (1968).
[4] H. S. Garewal, J. Singh, and A. R. Wasserman, *Biochem. Biophys. Res. Commun.* **44**, 1300 (1971).

Cl, pH 8, dispersing gently to prevent excessive foaming (Waring blender, 1 min, 20 V) and recentrifuging (30,000 $g$, 5 min). The washed "ethanol-extracted particles" could either be further processed or stored as a frozen suspension at $-10°$ to $-20°$ in 0.05 $M$ Tris-Cl, pH 8, containing either 20% glycerol or 15% dimethyl sulfoxide.

*Sonic Disruption.* The instrument (Blackstone Ultrasonics, Inc., Sheffield, Pennsylvania) consisted of probe model BP2 with 5/8-inch probe tip and generator model SS-2 capable of maximum power output of 200 W. It was set to position 40, out of a possible range of 100, and tuned to maximum pitch.

Protein was assayed by the biuret method of Gornall et al.[5] (throughout the $b$-559 and $b_6$ procedures and in assaying pure cytochrome $f$). Less sample (50 to 100 $\mu g$) can be analyzed when a sensitive spectrophotometer is used. Crystalline bovine serum albumin serves as standard. The biuret color of pure cytochrome samples (aliquots dialyzed versus phosphate) is corrected for heme absorption by that of an alkaline control. During the purification of cytochrome $f$, the procedure of Lowry et al.[6] is employed, using aliquots dialyzed-free of DTT.

*Concentration of Cytochromes by Sephadex G-25.* A threefold concentration per operation was found to be optimal (24 gm of prechilled dry Sephadex G-25 added to each 100 ml of solution to be concentrated). The beads, after swelling for 30 min, are transferred to coarse, 60 ml sintered glass funnels placed in 250 ml centrifuge cups. Centrifugation at 1000 rpm for 15 min at 0° recovers the concentrated solution at the bottom of the cups. (Fifteen milliliter funnels are used in 50 ml tubes for smaller volumes.)

*Preparative Disc Gel Electrophoresis.* An inexpensive apparatus is available from Kontes Glass Co., Vineland, New Jersey.

## Purification of Cytochrome $f^2$

Cytochrome $f$ is completely stable in reduced form (50 m$M$ Tris-HCl, pH 8, 1 m$M$ DTT, $0°$–$5°$). Exposure of cytochrome $f$ to aqueous butanol should be limited to 2 days, the temperature should be kept to 5° or less, and media should be supplemented with 1 m$M$ DTT and preferably gassed with $N_2$. All operations are performed in a cold room (2°–5°) or in an ice bath, unless indicated otherwise. A typical preparation is summarized in Table I.

[5] A. G. Gornall, C. J. Bardawill, and M. M. David, *J. Biol. Chem.* **177**, 751 (1949).
[6] O. H. Lowry, N. J. Rosebrough, A. L. Farr, and R. J. Randall, *J. Biol. Chem.* **193**, 265 (1951).

TABLE I
PURIFICATION OF CYTOCHROME $f^a$

| Fraction | Total cyt $f$ (units) | Total protein (mg) | Specific content (units/mg protein) |
|---|---|---|---|
| 1. Particles | 64 | 16,900 | 0.0038 |
| Aqueous butanol–Triton suspension | 64 | 16,900 | 0.0038 |
| Aqueous supernatant | 41.5 | 2,700 | 0.015 |
| 2. Eluate from DEAE–cellulose column | 35 | | |
| Concentration (dialysis versus butanol) | 35 | 93 | 0.38 |
| 3. Preparative polyacrylamide electrophoresis, elution, and concentration by Sephadex G-25 | 27 | 57 | $0.47^b$ |

[a] From Singh and Wasserman.[2]
[b] The final sample had a value, based on biuret protein, of 0.475.

## Procedure

*Step 1: Extraction of Cytochrome f.* A frozen suspension of TCPF or "washed $P_1S$" (500 ml, 1500 mg of total chlorphyll) is thawed to 0°, and DTT is added to 1 m$M$ final concentration. Five volumes of 0° $n$-butanol is added with continuous stirring, the suspension is blended (5 sec, maximum speed, 1-liter Waring blender), and stirred 1 hr at 0°–4° (magnetic stirrer). The suspension is centrifuged (15 min, 30,000 $g$), and the dark green upper phase is carefully removed and discarded. The aqueous–butanol lower phase (12% butanol) and interfacial particles are hand homogenized, and Triton X-100 is added to 5% (v/v) final concentration. Eighty milliliter portions of the suspension are each sonically disrupted in three 1-min intervals under $N_2$ with alternate 1-min cooling intervals. The portions are combined, rechilled to 0°, centrifuged (30,000 $g$, 15 min, 50-ml tubes), and the lower aqueous layers (greenish brown) combined. The interfacial particles, containing 23% of the cytochrome $f$ content, are discarded since reextraction is relatively ineffective.

*Step 2: Binding of Protein Impurities to DEAE–Cellulose.* The column medium is prepared by mixing 1 volume of $n$-butanol (0°) with 5 volumes of 50 m$M$ Tris-HCl, pH 8, 1 m$M$ DTT, 10% (v/v) glycerol in a large separatory funnel. After overnight equilibration at 4°, the aqueous–butanol lower phase is removed, made to 1% Triton X-100 (v/v), and gassed with $N_2$. A 2.5 × 50 cm column of DEAE–cellulose (50 mM Tris-HCl, pH 8) is equilibrated with this medium. The freshly prepared extract of cytochrome $f$ is applied to the column, and 1–2 holdup volumes of equilibration medium is passed through the column at maximum rate (about 60 ml/hr) to elute cytochrome $f$ quantitatively in a dark green, clear solution (400 ml). Considerable amounts of brown, yellow, and red pigments, as well as colorless proteins, are bound to the top quarter of the column; cytochrome $f$ does not bind.

Concentration and Further Delipidation by Dialysis of the Aqueous Eluate against $n$-Butanol. The dark green eluate (400–600 ml) is put in 1.6 cm diameter bags and dialyzed twice (magnetic stirrer) under $N_2$, each time versus 10 volumes of $0°$ $n$-butanol containing 1 m$M$ DTT for a total time of about 24 hr. By this time, the aqueous, lower, pink-brown phase that has gradually separated should be one-tenth of the original volume. The bag contents are centrifuged in narrow tubes (2 min, maximum speed, clinical centrifuge), and the top, green phase is quantitatively pipetted off and discarded. The lower aqueous phase is then dialyzed extensively under $N_2$ against 50 m$M$ Tris-HCl, pH 8, 1 m$M$DTT to remove the remaining butanol and Triton X-100. The dialyzed cytochrome (75 ml) can be stored at $-20°$ or purified immediately.

*Step 3: Preparative Disc Gel Electrophoresis.* Cooling is accomplished by circulating ethylene glycol at $-1°$ to keep the temperature below $4°$. A 2 cm long, 4% stacking gel and a 3 cm long, 5% resolving gel provide adequate resolution. Electrophoretic media are a modification of those of Davis[7] [4% gel: 0.056 $M$ Tris-HCl, pH 6.7, 1 m$M$ DTT, 0.1% Tween 80; 5% gel: 0.38 $M$ Tris-HCl, pH 8.9, 1 m$M$ DTT, 0.1% Tween 80; electrode buffer: 0.005 $M$Tris base, 0.039 $M$ glycine, pH 8.3, 1 m$M$ DTT, 0.1% Tween 80; sample: 0.05 $M$ Tris-HCl, pH 8, 1 m$M$ DTT, 0.1% Tween 80, 20% sucrose, 30–100 mg of total protein (about 1.25 mg protein/ml)]. After 9 hr of electrophoresis at 680 V, 45 mA (for Büchler apparatus; about 300–400 V, 25 mA for Kontes apparatus), the pink cytochrome $f$ band (now about 1 cm long) is about halfway down the 5% resolving gel, and its closest impurity, a thinner brown band, is completely resolved about 0.5 cm ahead.

Elution of Cytochrome $f$ from gel. The single, cytochrome $f$-containing band is sliced off, not frozen, and fragmented in a hand homogenizer together with about 7 volumes of electrode buffer supplemented with 0.1% Tween 80, 1 m$M$ DTT. The suspension is stirred ($0°$, 1 hr) and then centrifuged (30,000 $g$, 20 min). After the supernatant is collected, the pellet is reextracted as before, but with 2–3 volumes of the previous medium. The polyacrylamide pellet, once pink, is now colorless. The two supernatants are combined and then concentrated using dry Sephadex G-25. The cytochrome $f$ solution is dialyzed extensively against 0.05

---

[7] B. J. Davis, *Ann. N.Y. Acad. Sci.* **121**, 404 (1964).

M Tris-HCl, pH 8, 1 m$M$ DTT (under N$_2$) to remove Tween 80 and then stored at $-20°$. Overall recovery is 42%.

### Properties[2]

*Purity.* Analytical disc gel electrophoresis of cytochrome $f$ (5% gel, 0.1% Tween 80 or Triton X-100) in undenatured, nonaggregated form shows a single cytochrome band with no protein contaminants. The undenatured molecule (in 1% Triton) behaves as a single molecular species in sedimentation equilibrium in the analytical ultracentrifuge. The denatured cytochrome in sodium dodecyl sulfate (SDS)-polyacrylamide electrophoresis shows only a single polypeptide band.

*Spectrophotometric Properties.* The reduced cytochrome (22°, absolute spectrum) shows absorption maxima at 553.5 nm ($\alpha$), 522 nm ($\beta$), and 421 nm (Soret); the oxidized cytochrome has its Soret at 410 nm (Fig. 1). The millimolar extinction coefficient at 553.5 nm ($\alpha$) is 29. The difference spectrum (reduced minus oxidized) at 22° shows maxima at 553.5, 522, and 421 nm (Fig. 2). At $-196°$ the reduced molecule (absolute or difference spectrum) shows maxima at 551.5, 548, and 542 nm ($\alpha$ peaks); 522.5 and 531 nm ($\beta$ peaks); and 420 nm (Soret) (Fig. 3). At $-196°$, the oxidized Soret (not shown) is at 409 nm.

*Chemical Properties and Composition.* Amino acid analysis[8] shows a high content of nonpolar (55%) and of neutral amino acids (at least 22%). Once extracted, cytochrome $f$ is completely soluble (pH 8) but aggregated (without detergent) and reversibly, nondestructively disaggregated by 0.1 to 1% nonionic detergent. It is irreversibly inactivated (0°, pH 8) by 6 $M$ urea, 5.4 $M$ guanidine, or 0.1% sodium dodecyl sulfate. A $c$-type heme is present but no nonheme iron. The cytochrome is not autoxidizable (in absence of detergent) and does not combine with carbon monoxide. The equivalent weight is 61,000 gm of protein per mole of heme, but only one size of polypeptide chain is found (32,500 after denaturation in SDS and mercaptoethanol). This would suggest one heme-containing chain and one nonheme chain per 61,000 equivalent weight. The hydrodynamic molecular weight is unknown: although a $\bar{v}$ value (partial specific volume) of 0.745 ml/gm was calculated from amino acid analysis, an experimentally determined $\bar{v}$ value for the cytochrome-detergent complex is crucial for a correct molecular weight via sedimentation equilibrium.

[A subsequent spinach cytochrome $f$ preparation[9] has a denatured

[8] J. Singh and A. R. Wasserman, unpublished results (1971).
[9] N. Nelson and E. Racker, *J. Biol. Chem.* **247**, 3848 (1972).

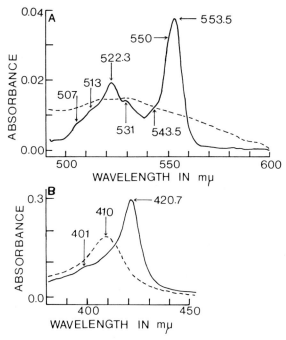

FIG. 1. Absolute absorbance spectra of cytochrome *f* at 22°. ———, reduced cytochrome; - - -, cytochrome oxidized by a few grains of ammonium persulfate. (The reference cell for the oxidized spectrum also contained the oxidant.) From Singh and Wasserman.[2]

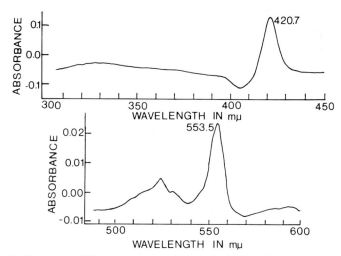

FIG. 2. Absorbance difference spectrum (reduced minus oxidized) of cytochrome *f* at 22°. From Singh and Wasserman.[2]

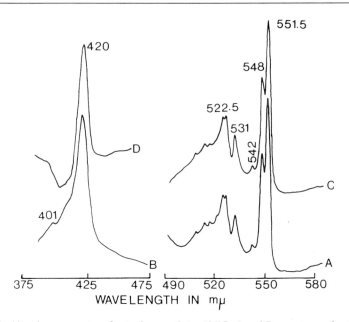

FIG. 3. Absorbance spectra of cytochrome $f$ at $-196°C$. A and B, spectrum of reduced cytochrome; C and D, difference spectrum (reduced minus oxidized). Samples were in medium containing 0.025 $M$ Tris-HCl, pH 8, and 50% (v/v) glycerol. [The spectrum of the oxidized cytochrome (not shown) showed a Soret peak at 409 mm.] From Singh and Wasserman.[2]

polypeptide chain size of 34,000, in close agreement with that of the above preparation. The reported equivalent weight of 36,000 gm protein per mole of heme (versus 61,000 above) is *unsupported by a direct determination for protein* but is calculated from amino acid analysis. The other criterion presented for greater purity than the present preparation, that of the 421/278 nm absorbance ratio, is irrelevant since the present preparation contains detergent absorbing at 278 nm. The absorbance spectra of the two preparations indicate no differences, except at 278 nm.]

*Autoreduction.*[10] The pure molecule at pH 8 has the curious property of spontaneous reversion to the reduced state on the removal of oxidant, without exogenously added reducing agent. A pure cytochrome $f$ preparation prepared differently as a by-product of the $b$-559 procedure in the following section also exhibits this behavior.[11] The autoreduction is light-

[10] H. S. Garewal and A. R. Wasserman, *Biochim. Biophys. Acta* **275**, 437 (1972).
[11] H. S. Garewal, A. L. Stuart, and A. R. Wasserman, *Can. J. Biochem.* **52**, 67 (1974).

TABLE II
PURIFICATION OF CYTOCHROME $b$-559[a]

| Fraction | Total cytochrome $b$-559 (units) | Total protein (mg) | Specific content (units/mg protein) |
|---|---|---|---|
| 1. a. Ethanol-extracted particles (in Triton-4 $M$ urea) | 16.9 | 4180 | 0.0041 |
| b. Supernatant after 30,000 $g$ centrifugation of Triton-4 $M$ urea extract | 16.9 | 1615 | 0.011 |
| 2. DEAE-cellulose eluate | — | 53.6 | — |
| 3. Preparative polyacrylamide electrophoresis, elution, and concentration by Sephadex G-25 | 5.6 (7.6)[b] | 14.5 | 0.39 (0.52)[b] |

[a] Reprinted from Garewal and Wasserman,[12] *Biochemistry* **13**, 4063 (1974). Copyright 1974 by the American Chemical Society. Reproduced by permission of the copyright owner.
[b] Based on absolute spectra.

dependent and blocked by 10 $\mu M$ concentration of the photosynthetic inhibitor DCMU [3-(3,4-dichlorophenyl)-1,1-dimethylurea].[11]

## Purification of Cytochrome $b559$[12]

All operations are performed at 0° or in a 4° cold room. The procedure, as described, is for a preparation using chloroplast particles originally containing about 550 mg of chlorophyll and about 5800 mg of protein. A typical preparation is summarized in Table II.

*Step 1: Extraction of Cytochromes.* A suspension of ethanol-extracted particles is centrifuged (30,000 $g$, 5 min), and the pellet is resuspended by hand homogenization in 10 pellet volumes (about 100 ml) of 2% Triton X-100, 4 $M$ urea, 0.05 $M$ Tris-HCl, pH 8. The suspension is sonicated (30–40 ml batches) for 2 min total time (four 30-sec sonication periods in ice interspersed with 1-min cooling intervals). After sonication, the combined dispersion is immediately cooled in ice to 0° to 2° and centrifuged (30,000 $g$, 30 min, the pellet is discarded). A second, optional centrifugation (90,000 $g$, 30 min) allows the use of a smaller amount of DEAE-cellulose in the next step, and the procedure is thus shortened. After an

[12] H. S. Garewal and A. R. Wasserman, *Biochemistry* **13**, 4063 (1974).

aliquot is withdrawn for analyses of cytochromes, the clarified solution is made to 5 m$M$ in DTT and processed immediately.

*Step 2: Binding of Impurities to DEAE-Cellulose to Permit Subsequent Electrophoresis.* A column of DEAE-cellulose is previously prepared with a bed volume about twice that of the sample to be applied. It is then equilibrated with 2 $M$ urea, 2% Triton, 2 m$M$ DTT, 50 m$M$ Tris-HCl, pH 8. The sample solution from step 1 (95 ml, brownish-green) is then applied. The column equilibration medium is used to wash in the sample and to elute cytochromes $f$ and $b$-559, which do not bind to the column and emerge in a green eluate fraction. Collection of $f$ and $b$-559 is begun when the green solution first emerges and is continued until the solution collected, including a later yellow fraction, is about twice the original sample volume. This treatment of the extract on DEAE–cellulose in Triton–urea removes most of the protein impurities (as indicated by analytical disc gel electrophoresis in 1% Triton). Secondary treatment on a smaller column is obligatory whenever bound, brown impurities extend more than three-fourths of the length of the column. (Spectrophotometric assays of eluted fractions are performed using the DTT-containing sample versus ferricyanide-oxidized sample. Both $f$ and $b$-559 are reduced by DTT in this medium and their absorption peaks overlap, but a semiquantitative assay of each cytochrome can be obtained.)

*Step 3: Preparative Disc Gel Electrophoresis in Triton X-100.* The previous light-green eluate (215 ml) containing $f$ and $b$-559 is concentrated twice at 0° with dry Sephadex G-25 to achieve a volume of 25 to 35 ml. The sample is then made to 10% glycerol and 4 m$M$ Tris-thioglycolate. Polyacrylamide gel electrophoresis is performed using a 2-cm long, 4% stacking gel and an 8-cm long, 10% resolving gel. Electrophoretic media are 4% gel: 0.1% Triton X-100, 1 m$M$ DTT, 10% glycerol, 0.056 $M$Tris-HCl, pH 6.7; 10% gel: 0.1% Triton X-100, 4 m$M$ DTT, 10% glycerol, 0.38 $M$ Tris-HCl, pH 8; electrode buffer: 0.1% Triton X-100, 4 m$M$ Tris-thioglycolate (no glycerol), 1 m$M$ DTT, 0.005 $M$ Tris base-0.039 $M$ glycine, pH 8.3. The sample is applied beneath the electrode buffer using a 10 ml syringe fitted with long, thin rubber tubing. Electrophoresis is performed in a Kontes Glass Co. preparative electrophoresis apparatus at 25 mA, 400 V (250-ml chamber). Cooling is achieved by circulating ethylene glycol at −1°. After the cytochromes have stacked and are about to enter the 10% gel, electrophoresis is stopped briefly and chlorophyll above the 4% gel is removed. Fresh electrode buffer containing thioglycolate is now placed above the gel and replaced again later at 16-hr intervals. Within 2 to 3 hr of electrophoresis in the 10% gel, the red bands of cytochrome $f$ and $b$-559 are visibly resolved, but 25–30 hr more are required for separation sufficient for easy cutting of the two bands without risking cross-contamination. After cytochrome $b$-559 has moved about 0.5 cm ahead of cytochrome $f$, the gel is sliced without freezing. The $b$-559 is eluted by hand homogenization to a fine slurry in 7 volumes of 5 m$M$ DTT, 15% glycerol, 50 m$M$ Tris-HCl, pH 8 (no detergent). The suspension is stirred for 1 to 2 hr under nitrogen and the gel removed by centrifugation (30,000 $g$, 1 hr). (Foaming during the elution process or later should be avoided as this inactivates the cytochrome.) The pink-orange solution is concentrated with dry Sephadex to about 20 ml and is best stored frozen ($-20°$) under nitrogen in this medium.

Overall recovery of pure $b$-559 is 30–50% of that present in ethanol-extracted particles. No single step resulted in any substantial loss, and, as protected by thiols and glycerol, the cytochrome behaves as a single species during purification. About 20% loss is encountered after electrophoresis and elution, approximately 12% loss results from step 2, and the remaining loss is due to cumulative losses from the concentrations with Sephadex G-25 (10 to 15% per concentration operation).

### Properties[4,12,13]

*Purity.* Analytical disc gel electrophoresis of reduced $b$-559 (10% gel, 0.1% Triton X-100, 10% glycerol, pH 8) in undenatured, nonaggregated form showed a single protein band (cytochrome band) representing over 99% of the applied protein. Spectrophotometric analyses showed no cytochrome $f$ or $b_6$ present. The denatured preparation in SDS–polyacrylamide electrophoresis showed only one polypeptide band. Analyses of the undenatured $b$-559 molecule by sedimentation equilibrium in the analytical ultracentrifuge (a less sensitive criterion) also showed only a single molecular species.

*Stability.* Solutions of reduced $b$-559 (5 m$M$ DTT, 15% glycerol, 50 m$M$ Tris-HCl, pH 8, under nitrogen) are stable for several months when stored frozen at $-15°$ to $-20°$; in the liquid state at $0°$, such solutions are stable for several weeks provided the DTT level is maintained. Ascorbate (5 m$M$) as reducing agent can replace DTT at $0°$ but for shorter storage periods. The reduced cytochrome is also stable at pH 7 in phosphate buffer but is unstable at pH 6 and pH 8.9. The oxidized cytochrome is completely stable only for a few hours; additional (artifactual) electrophoretic bands of the cytochrome are gradually produced, followed later by irreversible inactivation of the cytochrome. Of importance to future work in photosynthesis is that, as a consequence of its instability in

[13] H. S. Garewal and A. R. Wasserman, *Biochemistry* **13**, 4072 (1974).

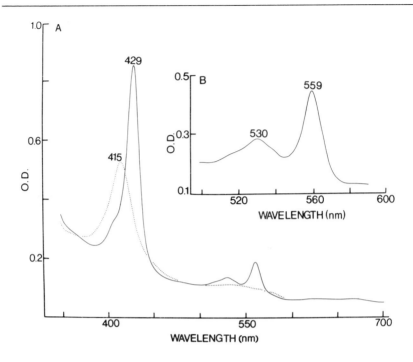

FIG. 4. Absolute spectra of cytochrome $b$-559 at 22° (A) Oxidized spectrum (····) with ammonium persulfate as oxidant. Reduced spectrum (——) in the presence of dithiothreitol. (B) Reduced spectrum of $\alpha$ and $\beta$ peaks using a more concentrated sample to show greater detail. From Garewal *et al.*[4]

oxidized form, cytochrome $b$-559 may well be present in inactivated form in smaller chloroplast particles, allegedly free of $b$-559, unless such particles are prepared with reducing agents and glycerol as supplements at pH 7–8.

*Spectrophotometric Properties.* Absorption maxima of the reduced cytochrome at 22° (absolute spectrum) are at 559 nm ($\alpha$), 530 nm ($\beta$), and 429 nm (Soret); the oxidized molecule has a Soret peak at 415 nm (Fig. 4). The millimolar extinction coefficient of the reduced molecule at its $\alpha$ maximum is 21 ($A_{559\ nm} - A_{600nm}$). Absorption maxima at 22° of the difference spectrum (reduced minus oxidized) are at 559, 530, and 429 nm (Fig. 5). At −196°, there is an expected shift in the difference spectrum of the $\alpha$ band to shorter wavelength but no split of this alpha peak: maxima are at 556 nm ($\alpha$); 529, 513, 536 nm ($\beta$ bands); and 429 nm (Soret) (Fig. 6). [For qualitative analysis of cytochrome content in chloroplast particles at −196° in 50% glycerol, it is worth noting that cytochrome $b_6$

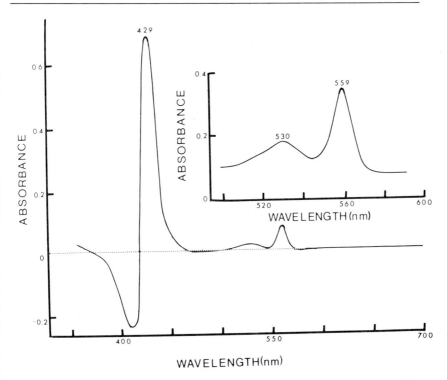

Fig. 5. Difference spectrum (reduced versus oxidized) of pure cytochrome $b$-559 at 22°C. The cytochrome was present in 0.05 $M$ Tris-HCl (pH 8) containing 15% glycerol. The sample cuvet contained the cytochrome reduced in the presence of DTT, and the reference cuvet contained an identical sample oxidized by a few colorless crystals of ammonium persulfate. The inset shows the $\alpha$ and $\beta$ peaks in greater detail using a more concentrated sample. Reprinted from Garewal and Wasserman,[12] *Biochemistry* **13**, 4063 (1974). Copyright 1974 by the American Chemical Society. Reproduced by permission of the copyright owner.

(*vide infra* cytochrome $b_6$, Fig. 9) has two $\alpha$ peaks of practically equal intensity at 561 and 557 nm. The latter peak practically coincides with the single $b$-559 $\alpha$ band at 556 nm. When no active $b_6$ is present, the 561 nm band should be absent. When equimolar amounts of $b_6$ and $b$-559 are present, the 556–557 band should be much larger than that at 561 nm.] Carbon monoxide produced no change in the spectrum of reduced $b$-559 (in the absence or presence of 0.1% Triton X-100).

*Redox Properties.* The pure molecule is autoxidizable (pH 8, 0°) but is subsequently completely reducible by ascorbate (or DTT or dithionite) provided reduction is performed within 1–2 hr. At no stage in the puri-

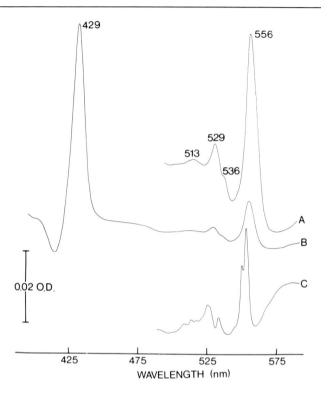

FIG. 6. Low temperature ($-196°$) difference spectra (reduced versus oxidized) of pure cytochrome $b$-559 in 50% glycerol, 0.05 $M$ Tris-HCl, pH 8. A and B, cytochrome $b$-559 (spectrum A was recorded at about 4× the magnification of spectrum B). C, cytochrome $f$ recorded for comparison purposes. From Garewal et al.[4]

fication procedure was a "dithionite-reducible" form of $b$-559 observed (i.e., a low potential form reducible by dithionite but not by ascorbate). In ethanol-extracted particles, complete reduction by ascorbate is slow and variable (minutes), but dispersion in Triton-2 or 4 $M$ urea, pH 8, produces complete ascorbate reduction of $b$-559 without time lag or loss of cytochrome, suggesting that the particles are only slowly permeable to ascorbate before dispersion. Aging of the pure preparation results in some $b$-559 which is reducible by dithionite but not by ascorbate. Pure $b$-559 showed no detectable reduction by hydroquinone at pH 8. Since photosynthetically active chloroplast particles (of uncertain internal redox conditions or solution parameters) exhibit both hydroquinone-reducible (as well as ascorbate-reducible) $b$-559 and also ascorbate-reducible (only) $b$-559, it is uncertain whether a distinct, separate high-potential $b$-559 molecule is destroyed at the very start of the purification (via partial extraction of lipid with 90% ethanol) or whether hydroquinone reducibility *in vitro* is potentially possible for the preparation given parametric changes in its assay (e.g., a different pH or the addition of micellar chloroplast lipids).

*Chemical Composition and Molecular Weight.* Cytochrome $b$-559 exhibits 0.52 absorbance units ($A_{559-600 \, nm}$) per milligram of protein. The heme prosthetic group is indistinguishable from protohemin (that of hemoglobin) on the basis of its reduced pyridine haemochromogen. The heme is probably not covalently linked to protein in that all of it is extractable by acid–acetone at 0°. No nonheme iron is present in $b$-559. The equivalent weight is 45,900 gm protein per mole of heme. Only a single size (5600 ± 1000) of polypeptide chain is found; about 8 small chains of 5600 are calculated to be present per mole of heme. Analysis shows Glu, Asp, and Thr as N-terminal amino acids, indicating that at least three different polypeptide chains are present. The equivalent dry weight (of Triton-depleted samples) is 111,000 gm/mole of heme of which (41% (45,900 gm) is contributed by protein and 59% by nonprotein substances. Noncovalently linked lipid content is 56(±6)% of the molecule (using Triton-depleted samples). The lipid composition is relatively simple (relative to that of the original chloroplast membranes) and is comprised of four lipid components. Two nonpolar lipids are present—chlorophyll $a$ (3%, 4 moles/heme) and $\beta$-carotene (1.6%, 3 moles/heme). (Tocopherols, plastoquinones, and chlorophyll $b$ are not present.) Two unknown polar lipids, one major (about 40%) and one minor (5–8%) comprise most of the total lipid present. (Triton X-100 content is less than 5–6% and migrates as a single spot different from those of the two unknown polar lipids.) Neither polar lipid is a glycolipid, and no detectable hexose is found in the $b$-559 preparation. The major lipid is not a phospholipid. The hydrodynamic molecular weight of $b$-559, likely as a complex with Triton X-100, is 117,000 with a partial specific volume of 0.91 ml/gm. The molecular weight of $b$-559 is estimated as about 110,000 (that of the equivalent dry weight per mole of heme). The partial specific volume as calculated from protein and lipid composition is 0.89 ml/gm, in approximate agreement with the experimental value of 0.91. Amino acid analysis indicates that nonpolar amino acids comprise at least 60% of the total (Trp being unknown). A relatively high content is present of the hydroxylated amino acids Ser and Thr (25 residues per total of 119 residues). From the above composition data, cytochrome $b$-559 appears to be a composite or mosaic lipohaemoprotein molecule of moderate size comprised of multiple small protein and lipid component pieces. Like cytochromes $f$ and $b_6$, cytochrome $b$-559 is an intrinsically (reversibly)

TABLE III

PURIFICATION OF CYTOCHROME $b_6$[a]

| Fraction | Total units[b] of cytochrome $b_6$ | Total protein (mg) | Specific content (units/mg protein) | Recovery of cytochrome $b_6$ (%) |
|---|---|---|---|---|
| 1. Unextracted green particles[c] | 15.3[d] | 6285 | 0.0024 | 100 |
| 2. Ethanol-extracted particles in Triton X-100–4 $M$ urea | 15.3 | 4230 | 0.0036 | 100 |
| 3 Concentrated 100,000 $g$ supernatant | 14.0 | 1555 | 0.0090 | 93 |
| 4. Eluent from BioGel A-1.5 m column | 9.8 | 61.3 | 0.16 | 65[e] |
| 5. Polyacrylamide gel electrophoresis eluent after concentration | 4.6 (4.9)[g] | 9.4 | 0.49 (0.52)[g] | 30[f] |

[a] From Stuart and Wasserman.[14]

[b] Determined by difference spectra.

[c] These particles contained 750 mg of chlorophyll.

[d] Determined from ethanol-extracted particles assuming no loss during extraction. Cytochrome content in green particles could not be reliably measured because of optical limitations.

[e] Seventy to 80% of the units applied to the column were routinely recovered.

[f] Recovery after electrophoresis was routinely at least 80% of the applied units. Concentrating the eluents after steps 4 and 5 by dry Sephadex G-25 contributed the bulk of the loss in this case, although each concentration can incur, with greater care, less than 10% loss.

[g] Numbers in parentheses employ units of cytochrome $b_6$ from absolute spectra (563–600 nm).

aggregatable molecule: when Triton is depleted, a soluble undenatured aggregate is formed which cannot penetrate a 10% polyacrylamide gel. The reduced molecule is denatured but not dissociated by either 8 $M$ urea or 6 $M$ guanidine at 37°; SDS denatures and completely dissociates the molecule to its polypeptide chains.

## Purification of Cytochrome $b_6$[14]

All steps are performed at 0°–4°. The procedure, as described, is for a preparation using chloroplast particles originally containing about 750

[14] A. L. Stuart and A. R. Wasserman, *Biochim. Biophys. Acta* **314**, 284 (1973).

mg of chlorophyll and about 6300 mg of protein. A typical preparation is summarized in Table III.

*Step 1: Extraction of Cytochromes.* The brown pellet of washed "ethanol-extracted particles" containing 0.05 $M$ Tris-HCl buffer, pH 8, is resuspended by hand homogenization in about 10 pellet volumes (100 ml) of 2% Triton X-100, 4 $M$ urea, 0.05 $M$ Tris-HCl buffer, pH 8. The Triton to protein ratio (mg/mg) is at least 0.4. Final volume is usually 110–115 ml. The suspension is sonicated for 2 min total time (four 30-sec sonication periods in ice interspersed with 1-min cooling intervals) and then centrifuged twice, each time at 27,000 $g$ for 15 min. The supernatant is further centrifuged, this time at 100,000 $g$ for 60 min, to remove noncytochrome components. After spectrophotometric assay for content of cytochromes $b_6$, $b$-559, and $f$, the extract is concentrated twice with dry Sephadex G-25 to reduce the volume to about 25 ml. Recovery of $b_6$ after extraction and concentration with Sephadex is essentially complete.

*Step 2: Chromatography on a BioGel A-1.5m Column Containing Triton–4 M Urea.* A 5.0 × 84 cm column of BioGel A-1.5 m (200–400 mesh) is prepared and equilibrated with the extraction medium (2% Triton X-100, 4 $M$ urea, 0.05 $M$ Tris-HCl, pH 8). The 25-ml extract from step 1 is applied and chromatographed using the same extraction medium in a descending direction at a flow rate of 60 ml/hr. Large size, noncyto-chrome components, which otherwise severely interfered with purifica-tion in step 3 (disc electrophoresis), are effectively removed as an ex-cluded front fraction. Cytochrome $b_6$ moves as a single retarded fraction (peak at 924 ml versus $V_0$ for the column of 528 ml—$K_{av} = 0.36$) which is virtually completely separated from cytochrome $f$ ($K_{av} = 0.59$). (Due to the lack of thiols, which protect $b$-559 but inactivate $b_6$, no $b$-559 is observed in the eluted fractions.) The tubes containing cytochrome $b_6$ are pooled, supplemented to 10% glycerol, and then concentrated once or twice with dry Sephadex G-25 to achieve a volume of about 15 ml.

*Step 3: Preparative Disc Gel Electrophoresis (0.5% Triton X-100).* A Kontes Glass Co. Preparative Electrophoresis Apparatus (250-ml cham-ber) is used. Electrode buffer is Tris–glycine, pH 8.3 (0.005 $M$ Tris base, 0.039 $M$ glycine) plus 0.5% Triton X-100. The running gel (9%, 100 ml volume) contains 10% glycerol, 0.5% Triton X-100, and 0.38 $M$ Tris-HCl buffer, pH 8.0. Prewashing of the gel with electrode buffer is performed for 2 hr at 20 mA, 300 V prior to applying the sample; this prewashing is necessary to protect the cytochrome as a single form. The 15 ml sample, greenish brown in color, is then applied, and electrophoresis at 20 mA, 300 V is performed for about 40 hr. After the cytochrome has penetrated the gel (about 12–16 hr) the electrophoresis is briefly stopped, and a chlorophyll layer (light green) remaining above the gel is carefully pipetted off; electrophoresis is then resumed until the sharp, orange $b_6$

band has penetrated 1 cm into the gel. After cutting out the orange $b_6$ band, the gel slice is fragmented in a hand homogenizer together with about 40 ml of 0.05 $M$ Tris-HCl, pH 8, 10% glycerol (no Triton X-100). The suspension is diluted with additional amounts of this medium so that the total amount of fluid added is about 10 times (about 60 ml) that of the original volume of the gel slice. The gel suspension is stirred about 3 hr, then centrifuged (27,000 $g$, 15 min), and the supernatant is concentrated with dry Sephadex G-25. The concentrated $b_6$ solution contains about 0.3% Triton X-100, estimated from the volume of the gel slice and its Triton concentration of 0.5%, from the known dilution factor during extraction from the slice, and from the known concentration after treatment with dry Sephadex G-25. (Triton X-100 micelles are quantitatively concentrated together with protein by dry Sephadex G-25.) Overall yield of pure $b_6$ was 30% of that present in ethanol-extracted particles.

### Properties[14,15]

*Purity.* Analytical disc gel electrophoresis of $b_6$ (9% gel, 1% Triton, 10% glycerol at pH 8, 7, or 8.9) in undenatured, nonaggregated form showed a single cytochrome band with no protein contaminants. Sedimentation equilibrium in the analytical ultracentrifuge (a less sensitive criterion) also showed only a single protein species. Several properties of the cytochrome (*vide infra*), such as absorption spectra, equivalent weight, polypeptide composition, and lipid content also serve as additional indices of purity.

*Stability.* Solutions of $b_6$ (0.05 $M$ Tris-HCl, pH 8, 10% glycerol) are stable for at least 3 months when stored frozen at $-15°$ and for at least 7 days in the liquid state at 0°. Although otherwise unusually stable, $b_6$ is slowly inactivated by thiols, such as 1–5 m$M$ dithiothreitol, in contrast to cytochromes $f$ and $b$-559 which are protected by thiols.

*Spectrophotometric Properties.* The dithionite-reduced molecule at 22° shows three absorption maxima at 563 ($\alpha$), 536 ($\beta$), and 434 nm (Soret) in its absolute spectrum (Fig. 7); the air-oxidized molecule has a Soret peak at 413 nm. The 563 nm ($\alpha$) band of the reduced molecule has a millimolar extinction coefficient, $\Delta\epsilon$ (m$M$) = 21 (563 nm − 600 nm). Absorption maxima for the difference spectrum (reduced minus oxidized) are at 563, 536, and 434 nm (Fig. 8). At the temperature of liquid nitrogen ($-196°$) the difference spectrum (reduced minus oxidized) shows $\alpha$ maxima at both 561 and 557 nm and a shoulder at about 550 nm; the $\beta$ region shows peaks at 539 and 532 nm; the Soret peak remains at 434 nm (Fig.

[15] A. L. Stuart and A. R. Wasserman, *Biochim. Biophys. Acta* **376,** 561 (1975).

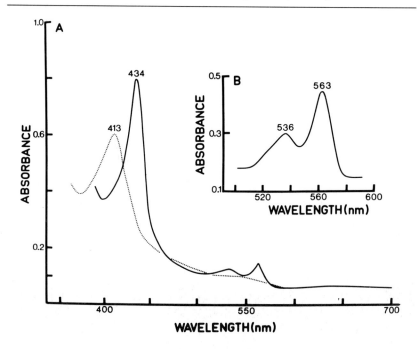

FIG. 7. (A) Absolute reduced (dithionite) (solid line) and absolute oxidized (untreated) absorbance spectra of pure cytochrome $b_6$ (22°). For insert (B), the sample aliquot was four times the concentration of that in (A), so as to reveal more detail. Solution medium was 0.05 $M$ Tris-HCl, pH 8. From Stuart and Wasserman.[14]

9). (Fifty percent glycerol is used to facilitate the splitting of the bands at −196°.) Of precautionary importance for cytochrome analysis in particles is that $b$-559 and $b_6$ both exhibit a peak at about 557 nm at −196°. Like cytochrome $b_6$ *in situ*, the spectrum of the reduced molecule is unaffected by carbon monoxide, provided the sample is depleted of its Triton X-100 content. In 0.5% Triton, $b_6$ is CO-reactive with the disappearance of its typical spectrum and the appearance of a Soret band at 428 nm.

*Redox Properties.* Like particulate $b_6$, the pure molecule is autoxidizable and reduced by dithionite but not by ascorbate. Oxidation–reduction proceeds *via* a transfer of one electron, and the redox potential is independent of pH in the region pH 7–8. The redox potential, $E_0'$, is only slightly affected by the presence of 0.5% Triton, suggesting that both the aggregated and monomolecular forms of $b_6$ have the same redox properties. $E_0'$ values are −0.084 V (pH 8, 0.5% Triton); −0.080 V (pH 7, 0.5% Triton); −0.120 V (pH 7, Triton absent).

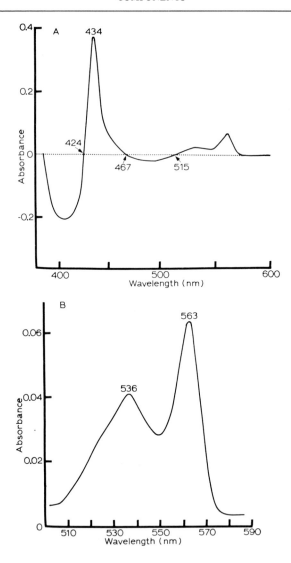

FIG. 8. Difference spectrum at 22° of pure cytochrome $b_6$ [reduced (with dithionite) minus oxidized (untreated)]. The spectrum in (B) was recorded at ten times the magnification of that in (A) to show the α and β peaks in greater detail. Solution medium was 0.05 M Tris-HCl buffer, pH 8. From Stuart and Wasserman.[14]

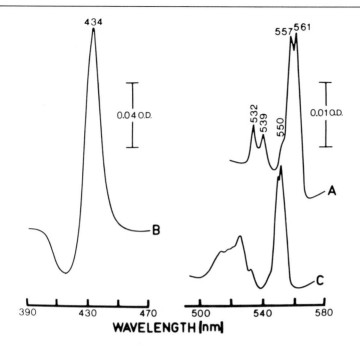

Fig. 9. Absorbance difference spectrum [reduced (dithionite) versus oxidized (untreated)] of pure cytochrome $b_6$ at the temperature of liquid nitrogen ($-196°C$). A and B cytochrome $b_6$. Magnification in A is four times that in B. The concentration of the sample was such that it gave an absorbance of 0.040 (563–600 nm) in a 1-cm cell at 22°. Dialysis for 5 days removed sufficient Triton X-100 to allow optical measurements at $-196°$; $b_6$ was stable during this dialysis. C, horse heart ferrocytochrome $c$ spectrum ($-196°$) as reference to demonstrate validity of the technique as performed. Medium in A, B, and C was 0.05 $M$ Tris-HCl, pH 8, 50% glycerol. The slit width was made as narrow as possible for good optical resolution. From Stuart and Wasserman.[14]

*Chemical Composition and Molecular Weight.* Cytochrome $b_6$ has 0.52 absorbance units ($A_{563-600\ nm}$) per milligram of protein. The equivalent weight is 40,000 gm protein per mole of heme. The heme prosthetic group is indistinguishable from protohemin on the basis of its reduced pyridine hemochromogen. No nonheme iron is present. The protein content (40,000 gm) is comprised of four polypeptide chains: 20,000 ± 1500 (one), 9600 ± 1000 (one), 6600 ± 1000 (two). The equivalent dry weight is 60,000 gm (corrected for detergent) per mole of heme, of which 20,000 gm is extractable as lipid (corrected for detergent). A simple noncovalently bound lipid composition (relative to the original membranes) is present in the $b_6$ preparation: about 7 moles of chlorophyll $a$ and 6 moles of cardiolipin per mole of $b_6$. These two lipids thus comprise about 75–80% of the lipid content. An unidentified minor neutral lipid and minor polar lipid were also detected. The hydrodynamic molecular weight of the cytochrome $b_6$–Triton complex is 67,000 with a partial specific volume of 0.84. The hydrodynamic molecular weight of the detergent-free molecule is estimated as 60,000 (that of the equivalent dry weight). From the above data, cytochrome $b_6$ appears to be a composite or mosaic lipohemoprotein molecule of moderate size comprised of multiple protein and lipid component pieces. Like cytochromes $f$ and $b$-559, cytochrome $b_6$ is an intrinsically (reversibly) aggregatable molecule: when Triton is depleted, a soluble undenatured aggregate is formed which cannot penetrate a 9% polyacrylamide gel.

Acknowledgment

This work was supported by funds from the Medical Research Council (Canada).

## [18] Theoretical Time Dependence of Oxidation– Reduction Reactions in Photosynthetic Electron Transport: Reduction of a Linear Chain by the Plastoquinone Pool

By J. WHITMARSH and W. A. CRAMER

## I. Introduction

In recent years, primarily due to the widespread use of signal averaging and new developments in fast pulsed light sources, the time resolution in measurements of redox kinetics in photosynthetic systems has improved greatly.[1-4] Currently, discussions of kinetic data involving photosynthetic electron-transport components include not only considerations based on measurements of half-times, but also inferences based on the shape of the trace describing the detailed time course of the absorb-

[1] B. Ke, Vol. 24, p. 25.
[2] P. Mathis, "Primary Processes of Photosynthesis" (J. Barber, ed.), p. 270. Elsevier, Amsterdam, 1977.
[3] P. M. Rentzepis, Vol. 54, p. 3.
[4] D. DeVault, Vol. 54, p. 32.

ance changes.[5-10] Interpretations of kinetic data concerning electron-transport pathways often involve comparisons of different reaction rates and require knowledge of the detailed time course expected for the electron-transport pathway(s) under consideration. Since a precise explanation of the expected kinetic behavior is seldom available, results are often interpreted in terms of qualitative predictions. One example is provided by high potential cytochrome $b$-559, which had been suggested to be an electron transport component in the chloroplast chain between photosystem II and plastoquinone (PQ).[11] In this linear sequence, the predicted rate of reduction of cytochrome $b$-559 by photosystem II should be comparable to the rate of reduction of the plastoquinone pool. In fact, kinetic measurements indicate that cytochrome $b$-559 is photoreduced at least ten times more slowly than plastoquinone.[12] In this case, a detailed description of the electron-transfer reactions is not needed to argue that the relatively slow reduction of the cytochrome is inconsistent with its functioning in the linear electron-transport chain. Another example, extensively discussed here, is provided by cytochrome $f$ which, on the basis of a large quantity of biochemical, genetic, and biophysical data (summarized in Whitmarsh and Cramer[10,13]), has been inferred as an obligatory component in the transfer of electrons between plastoquinone and P700. Recently, however, it has been argued that the kinetic data may not be consistent with an obligatory function in the main chain.[5,6,14] It was found that the amplitude of the dark reduction of cytochrome $f$ was smaller than expected,[5,6] and that the time course of its reduction did not exhibit a pronounced sigmoidicity.[5,6,14] A sigmoidal time course for the dark reduction of cytochrome $f$ by the plastoquinone pool is expected if cytochrome $f$, plastocyanin and P700 are initially oxidized and are in a linear chain PQ $\rightarrow$ Cyt $f \rightarrow$ PC $\rightarrow$ P700 in which the electron transfer is largely irreversible. Following illumination, electrons from the reduced plastoquinone pool will be transferred first to P700 and plastocyanin, resulting in a delay in cytochrome $f$ reduction.

[5] W. Haehnel, *Biochim. Biophys. Acta* **305**, 618 (1973).

[6] W. Haehnel, *Biochim. Biophys. Acta* **459**, 418 (1977).

[7] B. Bouges-Bocquet, *Biochim. Biophys. Acta* **462**, 362 (1977).

[8] B. Bouges-Bocquet, *Biochim. Biophys. Acta* **462**, 371 (1977).

[9] K. L. Zankel and B. Kok, Vol. 24, p. 218.

[10] J. Whitmarsh and W. A. Cramer, *Biophys. J.* **26**, 223 (1979).

[11] P. Horton, H. Böhme, and W. A. Cramer, *Proc. Int. Congr. Photosynth. Res., 3rd, 1974*, p. 535, (1975).

[12] J. Whitmarsh and W. A. Cramer, *Biochim. Biophys. Acta* **460**, 280 (1977).

[13] W. A. Cramer and J. Whitmarsh, *Annu. Rev. Plant Physiol.* **28**, 133 (1977).

[14] R. P. Cox, *Eur. J. Biochem.* **55**, 625 (1975).

In the work presented here, a theoretical treatment is provided that is capable of predicting the kinetics of oxidation–reduction reactions for a set of linear, irreversible electron transfer reactions involving separated chains. The assumption of separate chains[15] is reasonable considering the membrane-bound nature of the components, although there are suggestions in the literature of electron exchange at the level of photosystem I.[16,17] Consideration of the membrane-bound nature of the components is an important aspect in the development of the kinetic treatment presented here. The particular application of this treatment is the determination of the time course of the electron-transfer reactions, immediately following illumination, between plastoquinone and the initially oxidized components of the photosystem I complex. One question addressed in this context is whether the delay in cytochrome $f$-reduction is expected, on theoretical grounds, to be large[5] or small[10] relative to its half-time for reduction. The assumption of irreversibility in the electron-transfer steps ignores the possibility of electron sharing between the approximately isopotential cytochrome $f$ and plastocyanin. Nevertheless, the model can be used to determine the maximum delay expected in the dark reduction of cytochrome $f$, since consideration of reversible electron transfer between plastocyanin and cytochrome $f$ would be expected to shorten the delay involved in reducing cytochrome $f$. One result of the theoretical treatment presented here is that the delay predicted for the reduction of cytochrome $f$ is much shorter than predicted by Haehnel,[5] and consistent with the data of Whitmarsh and Cramer.[10] In addition, the calculation predicts the rate of cytochrome $f$ dark reduction relative to the rate-limiting step. Comparison of the theoretically predicted values to those determined experimentally indicates that the measured half-time for dark reduction of cytochrome $f$ relative to that of the rate-limiting step is longer than indicated by the simplest application of the model. This leads to the suggestion that the rate of electron transfer from plastoquinone to cytochrome $f$ may depend upon the charge state of the entire system I donor complex.

Before describing the details of the theoretical model, it is necessary to summarize relevant general properties of electron transport components functioning between plastoquinone and P700, and their qualitative behavior in the dark following a long flash. In the dark, electron-transport components with midpoint potentials lower than approximately $+100$ mV tend to be oxidized, while those with midpoint potentials higher than $+250$ mV are reduced, implying that the ambient redox potential in

[15] S. Malkin, *Biophys. J.* **9,** 489 (1969).
[16] B. Bouges-Bocquet, *Biochim. Biophys. Acta* **396,** 382 (1975).
[17] W. Haehnel, *Proc. Int. Congr. Photosynth. Res. 4th, 1977,* p. 777, (1978).

thylakoid membranes is between $+100$ and $+250$ mV. The situation is reversed in the light, as electrons from photosystem II reduce the plastoquinone pool, while photosystem I oxidizes those components subsequent to plastoquinone, indicating that the rate-limiting step in electron transport is the oxidation of plastoquinone. When the light is switched off, the electron-transport chains return to the dark state in at least two different time domains. Although the capacity of the plastoquinone pool under steady state light is 8 to 12 electrons,[18-20] when the light is switched off only 3 electrons are observed to rapidly ($t_{1/2} < 100$ msec) leave the pool.[5,12,18,19,21] The first electron released by the plastoquinone pool reduces P700 in a time comparable to the rate-limiting step. The electrons subsequently transferred reduce the remaining oxidized components between P700 and plastoquinone. Once the components between plastoquinone and P700 are reduced, the transfer of the electrons remaining in the plastoquinone pool is slow (5-10 sec)[20], at least in the absence of a high potential oxidant.[22] We are interested in deriving a quantitative expression to describe the transient behavior of the electrons leaving the plastoquinone pool immediately following a long light flash, that is, the reduction kinetics of P700 and the intermediates between P700 and plastoquinone, as well as the rapid oxidation kinetics of the plastoquinone pool. We will not consider the slow equilibration of the pool with endogenous oxidant.

As previously stated, measurements of the number of electrons leaving the plastoquinone pool subsequent to a light flash indicate that there are two electron-transport components between plastoquinone and P700. Although this result is consistent with interpretations based upon a large amount of data supporting the involvement of cytochrome $f$ and plastocyanin in a linear electron-transport chain, the uncertainty in determining the electron capacity of the plastoquinone pool is large enough that the participation of additional components remains a possibility. Firstly, there may be two active plastocyanin molecules per chain.[6] In addition two new components have been proposed. One, an iron–sulfur center observed by low temperature EPR, has been suggested as an intermediate between plastoquinone and cytochrome $f$.[23-26] The other component, PD

[18] H. H. Stiehl, and H. T. Witt, *Z. Naturforsch., Teil B* **24**, 1588 (1969).

[19] T. V. Marsho, and B. Kok, *Biochim. Biophys. Acta* **223**, 240 (1970).

[20] J. Whitmarsh, unpublished data.

[21] R. W. Keck, R. A. Dilley, and B. Ke, *Plant Physiol.* **46**, 699 (1970).

[22] J. Whitmarsh, and W. A. Cramer, *Biochim. Biophys. Acta* **501**, 83 (1978).

[23] R. Malkin, and P. J. Aparicio, *Biochem. Biophys. Res. Commun.* **63**, 1157 (1975).

[24] R. Malkin, and A. J. Bearden, *Biochim. Biophys. Acta* **505**, 147–181 (1978).

[25] H. Koike, K. Satoh, and S. Katoh, *Plant Cell Physiol.* **19**, 1371 (1978).

[26] J. Whitmarsh, and W. A. Cramer, *Proc. Natl. Acad. Sci. U.S.A.*, **76**, 4417 (1979).

(primary donor), has been inferred from light-induced kinetic measurements indicating electron transfer to P700 faster than oxidation of plastocyanin or cytochrome $f$.[27-29] In view of these considerations we have chosen to treat three linear chains:

$$PQ \rightarrow Cyt f \rightarrow PC \rightarrow P700$$

$$PQ \rightarrow Cyt f \rightarrow PC \rightarrow PD \rightarrow P700$$

$$PQ \rightarrow FeS \rightarrow Cyt f \rightarrow PC \rightarrow P700$$

These models appear to be the simplest electron-transport pathways that are capable of accommodating the majority of biochemical and biophysical data. We have chosen not to consider the linear chain including all of the components mentioned above because of space limitations, although the analysis presented in the next section can readily be extended to more complex chains.

## II. Rate Equations and Solutions

The problem dealt with below is the derivation of rate equations describing the time course of the oxidation of the plastoquinone pool and the reduction of P700, plastocyanin, and cytochrome $f$ immediately following illumination. Once we have derived the rate equations, we choose rate constants for the individual electron-transfer reactions in order to plot the solutions. This leads to a qualitative discussion of the dependence of the oxidation rate of plastoquinone on the number of electrons in the plastoquinone pool.

The key to our approach is to treat a single electron-transport chain and to consider that the electrons exit from the plastoquinone pool in sequence. In order to represent this sequential flow mathematically we insert into the rate equations for the second and third electrons a linear constraint to insure that only one electron at a time can be transferred to cytochrome $f$. The analysis is based upon the electron transport model shown in Fig. 1 and the following assumptions:

1. The rate-limiting step is the oxidation of plastoquinone, which is at least an order of magnitude slower than the electron transfer from cytochrome $f$ or plastocyanin.

2. The three electrons leaving the plastoquinone pool are treated in sequence. $k_{11}$ is the rate constant for the first electron leaving the pool,

[27] W. Haehnel, G. Döring, and H. T. Witt, Z. Naturforsch., Teil B **26**, 1171 (1971).
[28] R. Delosme, A. Zickler, and P. Joliot, Biochim. Biophys. Acta **504**, 165 (1978).
[29] B. Bouges-Bocquet, FEBS Lett. **94**, 100 (1978).

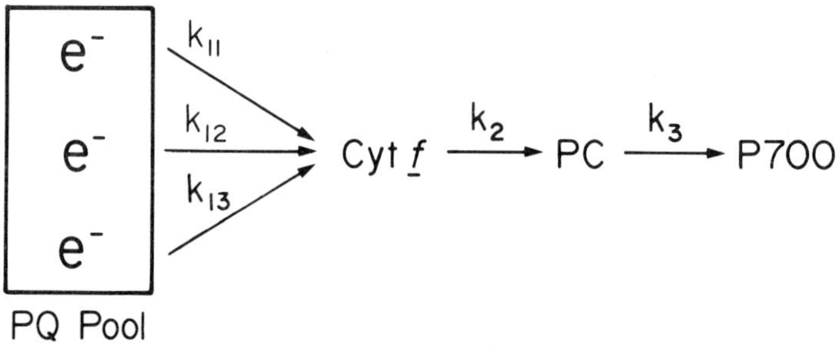

FIG. 1. Electron-transport chain used for deriving the rate equations in Section I. PQ, plastoquinone; cyt $f$, cytochrome $f$; PC, plastocyanin; P700, photosystem I reaction center. The electrons shown in the plastoquinone pool correspond to the number that exit rapidly to reduce oxidized P700, PC, and cytochrome $f$.

and $k_{12}$ and $k_{13}$ the rate constants for the second and third electrons, respectively.

3. Each electron-transfer reaction is irreversible.

4. Electron-transport chains are considered to be identical and electrons may not be transferred between different chains subsequent to the rate-limiting step.

Although we deal with a single electron-transport chain and individual electrons, considering for example the partial reduction of a cytochrome $f$ molecule, the functions derived below for the redox state of a single component can be considered as representative of the entire population of the component. Under experimental conditions a large number of chains are active concurrently, so that the partial reduction of a component may be thought of as the probability that the molecule is reduced at a particular time. This conceptual simplication is acceptable if we assume the electron-transport chains are synchronized by the light flash. That is, for a sufficiently long light flash, at the moment the light is switched off ($t = 0$) the electron-transport chains are poised in the steady-state level such that the plastoquinone pool is reduced and all components of the system I complex are oxidized.

*Example I: PQ → Cyt f → PC → P700*

Let $A_n(t)$ be the fraction of the $n$th electron remaining in the plasto-quinone pool, $B_n(t)$ be the fraction of cytochrome $f$ reduced by the $n$th electron, $C_n(t)$ be the fraction of plastocyanin reduced by the $n$th electron, and $P(t)$ be the fraction of P700 reduced, where $n = 1$, 2 or 3, corresponding to the first, second, or third electron exiting from the plasto-quinone pool. The initial conditions for a single chain can be written:

$$A_n(t = 0) = 1 \qquad B_n(t = 0) = 0$$

$$C_n(t = 0) = 0 \qquad P(t = 0) = 0$$

As indicated above, our approach in determining the rate equations is to consider the electrons exiting from the plastoquinone pool in sequence. The first electron is transferred to cytochrome $f$, then to plastocyanin, and finally to P700. If we assume first order processes, then the rate equations describing the initial electron are those for three consecutive reactions.[30]

$$\frac{dA_1}{dt} = -k_{11}A_1$$

$$\frac{dB_1}{dt} = k_{11}A_1 - k_2B_1$$

$$\frac{dC_1}{dt} = k_2B_1 - k_3C_1$$

and

$$P = 1 - A_1 - B_1 - C_1$$

In our treatment the exit of a second electron from the plastoquinone pool is also described by a first order constant $(k_{12})$. However, since cytochrome $f$ can accept only one electron at a time, the rate equation is subject to the constraint that transfer of the second electron must be subsequent to that of the first electron. The sequential order is introduced into the rate equations by multiplying by the factor $(1 - A_1)$, which is the probability that the first electron has left the plastoquinone pool. Including this constraint we obtain the following rate equations for the second electron.

$$\frac{dA_2}{dt} = -k_{12}A_2(1 - A_1)$$

$$\frac{dB_2}{dt} = k_{12}A_2(1 - A_1) - k_2B_2$$

$$C_2 = 1 - A_2 - B_2$$

The possibility that the transfer of the second electron from plastoquinone may be influenced by the presence of the first electron on cytochrome $f$

[30] N. M. Rodiguin, and E. N. Rodiguina, "Consecutive Chemical Reactions," D. van Nostrand Co., Princeton, New Jersey, 1964.

has been ignored, since the rate of electron transfer from cytochrome $f$ is several times faster than the rate of electron transfer to the cytochrome. P700 does not appear in the equations for the second electron because it is assumed to be fully reduced by the first electron. This is a consequence of assuming irreversible transfer.

Introducing a similar constraint for the third electron, i.e., $(1 - A_2)$, the rate equation becomes:

$$\frac{dA_3}{dt} = - k_{13}A_3 (1 - A_2)$$

and

$$B_3 = 1 - A_3$$

Finally the redox state of the components in the chain in terms of the quantities defined above is given by:

[P700] $= 1 - C_1 - B_1 - A_1 =$ fraction of P700 reduced

[PC] $= C_1 + C_2 =$ fraction of plastocyanin reduced

[Cyt $f$] $= B_1 + B_2 + B_3 =$ fraction of cytochrome $f$ reduced

[PQ] $= A_1 + A_2 + A_3 =$ fraction of rapidly turning over plastoquinone that is reduced

In order to plot solutions for the equations it is necessary to select values for the various rate constants. Ideally, the rate constants would be determined by the experimental data. However, measurements of the half-times for the oxidation of plastoquinone, cytochrome $f$, plastocyanin, and P700 appearing in the literature exhibit a wide range of values and, in general, have not been carried out under identical conditions of temperature, pH, light intensity, etc. Since the rate of oxidation of plastoquinone is much slower than the oxidation of cytochrome $f$ and plastocyanin, the shape of the kinetic traces is determined primarily by the rate constants $k_{11}$, $k_{12}$, and $k_{13}$. The plots are relatively insensitive to the rate constants $k_2$ and $k_3$, which can be changed by more than a factor of two without significantly altering the traces. For the oxidation rate constants of cytochrome $f$ and plastocyanin we have chosen the experimental values $k_2 = 6.9 \times 10^3$ sec$^{-1}$ and $k_3 = 9.9 \times 10^3$ sec$^{-1}$ corresponding to oxidative half-times of 100 and 70 $\mu$sec for cytochrome $f$ and plastocyanin, respectively.[7] The choice of rate constants for the electrons leaving the plastoquinone pool is more complex.

We first consider that the plastoquinone pool contains 12 electrons and that each electron has an equal probability of reducing cytochrome $f$. If we assume that the rate of electron transfer from plastoquinone is proportional to the electron pressure of the pool, then the rate constants for transfer of the second and third electrons, $k_{12}$ and $k_{13}$, are slightly smaller than $k_{11}$, i.e., $k_{12} = (11/12)k_{11}$ and $k_{13} = (5/6)k_{11}$. We will label this choice of rate constants $k_{1i}$ as case A; comparison with experimental data will lead us later to consider another choice of the $k_{1i}$'s (case B). A computer (HP9830A)-generated solution for case A, using a value of $k_{11}$ corresponding to a rate-limiting step of 20 msec ($k_{11} = 34.6$ sec$^{-1}$) for transfer of the first electron from the pool under basal conditions of electron transfer, is shown in Fig. 2A. The half-time for oxidation of plastoquinone is 44 msec. The half-times for reduction of P700, plastocyanin, and cytochrome $f$ are 20, 44, and 67 msec, respectively. Under uncoupled conditions, the rate-limiting step would be approximately 5

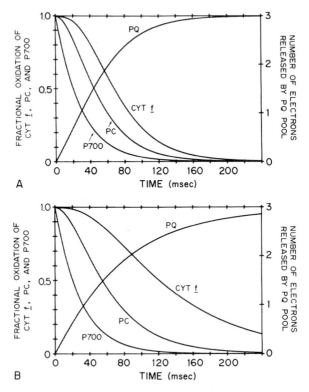

Fig. 2. Time course of the oxidation of plastoquinone and the reduction of cytochrome $f$, plastocyanin, and P700 following illumination predicted by the theoretical treatment given in example I for basal conditions (rate-limiting step = 20 msec). (A) The rate constants used were (in sec$^{-1}$): $k_{11} = 35$, $k_{12} = 32$, $k_{13} = 29$, $k_2 = 6940$, and $k_3 = 9900$. (B) The rate constants used were (in sec$^{-1}$): $k_{11} = 35$, $k_{12} = 23$, $k_{13} = 11$, $k_2 = 6940$, and $k_3 = 9900$.

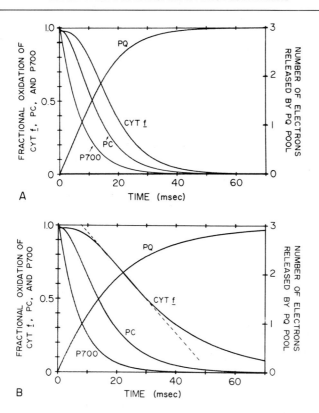

Fig. 3. Time course of the predicted redox changes in example I for uncoupled conditions (rate-limiting step = 5 msec). As in Fig. 2 except the rate constants used in (A) are (sec⁻¹): $k_{11} = 139$, $k_{12} = 127$, $k_{13} = 116$, and in (B) are: $k_{11} = 139$, $k_{12} = 93$, and $k_{13} = 46$.

msec and, utilizing the same relationship between $k_{11}$, $k_{12}$, and $k_{13}$, the half-times for reduction of P700, plastocyanin, and cytochrome $f$ are 5, 11, and 17 msec, respectively (Fig. 3A). From the half-times calculated in Figs. 2A and 3A, the ratio of the reductive half-time of cytochrome $f$ to that of P700, or the rate limiting step, is seen to be 3.4.[30a] This ratio is lower than that determined experimentally under both uncoupled and

[30a] Experimentally, the half-time for the rate-limiting step is determined by measuring on a chlorophyll basis (i) the rate of electron transport from $H_2O$ to a photosystem I acceptor under saturating light, (ii) the number of electron transport chains, and then calculating the time for half an electron to traverse a single chain. This measured half-time for the transfer of a single electron from a fully reduced plastoquinone pool through the rate-limiting step to cytochrome $f$ should be very nearly equal to the half-time for P700 reduction, since the oxidation of cytochrome $f$ and plastocyanin is much faster than the rate-limiting step.

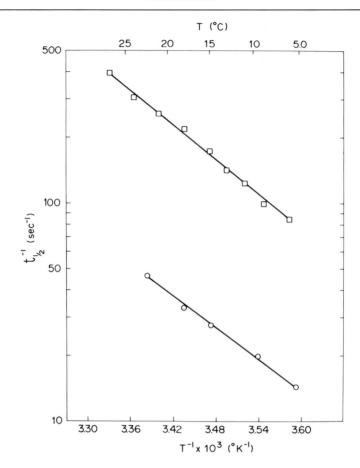

FIG. 4. Arrhenius plots of cytochrome $f$ dark reduction rate (○) after a 0.3 sec flash and the rate of electron transport (□) under steady-state conditions, in the presence of 2 $\mu M$ gramicidin.[10]

basal conditions. As shown in Fig. 4, the rate of cytochrome $f$ reduction in the presence of uncoupler is 6–7 times slower than that of the rate-limiting step over a wide range of temperatures.[10] Under basal conditions in the presence of valinomycin this ratio is approximately 5, when the half-time for cytochrome $f$ reduction is 135 msec (Fig. 5). In other words, comparison with experimental results indicates that the rate constant, $k_{13}$, for reduction of cytochrome $f$ by plastoquinone is smaller than indicated by the considerations invoked thus far. An *ad hoc* explanation for choosing a small value of $k_{13}$ might be that the number of electrons effectively

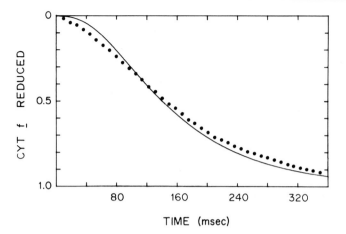

FIG. 5. Comparison of theoretical (solid curve) and experimentally measured (dotted curve) time course for the dark reduction of cytochrome $f$. The rate constants used were (sec$^{-1}$): $k_{11} = 30$, $k_{12} = 20$, $k_{13} = 10$, $k_2 = 6940$, and $k_3 = 9900$ using the $f$-PC-P700 model of example I. The data was obtained in the presence of 0.8 $\mu M$ valinomycin as described in Whitmarsh and Cramer.[10]

able to reduce the group of oxidized donors is small. The experimental data for the time course of cytochrome $f$ reduction in the presence of valinomycin can be fit reasonably well (Fig. 5) if $k_{12} = (2/3)k_{11}$ and $k_{13} = (1/3)k_{11}$. (We will use this relationship between the $k_{1i}$'s to generate solutions of the equations, and define this as case B.) If these relationships between the rate constants were determined by plastoquinone pool size, it would imply that the effective size of the fully reduced pool would be only 3 electrons. Since there is no precedent for an effective plastoquinone pool so much smaller than the measured pool capacity, we are led to consider other factors that can account for the relatively slow reduction of cytochrome $f$ (Section III). The observation that the reduction of cytochrome $f$ after illumination is substantially slower than that of P700 has been made previously in experiments done at subzero temperatures in a medium containing 50% ethylene glycol, where it was found that the two half-times differed by a factor of 3–5.[14,30b] It was argued that the relatively slow time course of the dark reduction of cytochrome $f$ implied a complexity in its interaction with P700 (14). The model presented here indicates that a factor of 3.4 would be expected for sequential electron transfer using the most straightforward choice of rate constants (case A).

[30b] L. F. Olsen, and R. P. Cox, *FEBS Lett.* **103**, 250 (1979).

The comparison of the computer-generated solutions for cases A and B for basal electron transport ($t_{1/2}$ = 20 msec) is shown in Figs. 2A, B. The solutions for coupled or uncoupled electron transport are essentially the same as for basal electron transport except that the time base must be divided by a factor of 2–4. The solutions for cases A and B applied to uncoupled electron transport ($t_{1/2}$ = 5 msec) are shown in Figs. 3A, B. The half-times for plastoquinone oxidation, and reduction of P700, plastocyanin, and cytochrome $f$ for the cases A and B under basal and uncoupled conditions are summarized in Table I (I, A, B—Basal and Uncoupled). The reduction of cytochrome $f$ under these four conditions exhibits a delay that is pronounced, but much shorter than the half-time for its reduction. For case B applied to the uncoupled system, the lag or delay in cytochrome $f$ reduction, extrapolated to zero absorbance change, as shown in Fig. 3B, is 7–10 msec. The ratio of this lag to the half-time of 25–30 msec for cytochrome $f$ reduction is smaller than predicted in ref. 5. As noted above, the theoretically predicted time course under basal conditions for case B fits fairly well the experimental data for the kinetics of cytochrome $f$ reduction measured in the presence of valinomycin (Fig. 5). The $k_{11}$ for this experiment was chosen to be 30 sec$^{-1}$, somewhat smaller than used in Fig. 2, since valinomycin causes a small inhibition of the basal electron-transport rate. The predicted delay in cytochrome $f$ reduction (~40 msec) is somewhat greater than obtained from the experimental trace (~20 msec) by extrapolating the data of Fig. 5 to zero absorbance change. The experimental problems of measuring

TABLE I

THEORETICAL HALF-TIMES FOR REDOX REACTIONS IMMEDIATELY FOLLOWING ILLUMINATION FOR THE LINEAR CHAINS I, II AND III[a]

| Chain | PQ (oxid)[b] | FeS (red)[b] | Cyt $f$ (red) | PC (red) | PD (red) | P700 (red) |
|---|---|---|---|---|---|---|
| IA (Fig. 2A, Basal) | 44 | — | 67 | 44 | — | 20 |
| IB (Fig. 2B, Basal) | 55 | — | 121 | 55 | — | 20 |
| IA (Fig. 3A, Uncoupled) | 11 | — | 17 | 11 | — | 5 |
| IB (Fig. 3B, Uncoupled) | 14 | — | 30 | 14 | — | 5 |
| IIA (Fig. 6A, Basal) | 56 | — | 93 | 69 | 45 | 20 |
| IIB (Fig. 6B, Basal) | 71 | — | 181 | 94 | 51 | 20 |
| IIIA (Fig. 7A, Basal) | 56 | 93 | 69 | 45 | — | 20 |
| IIIB (Fig. 7B, Basal) | 71 | 181 | 94 | 51 | — | 20 |

[a] Half-times in msec.
[b] oxid., oxidation; red., reduction.

small absorbance changes at early times to a level of precision sufficient to accurately determine a delay have been discussed elsewhere.[10,14]

The model predicts that plastocyanin should also exhibit a small delay in its reduction, the half-times increasing from 44 msec in case A to 55 msec in case B, the same half-times as calculated for oxidation of plastoquinone. The reduction of P700 is pseudo first-order, with the same half-time as the rate-limiting step (Figs. 2A, B).

*Example II: PQ → Cyt f → PC → PD → P700*

The existence of an additional electron carrier, PD (primary donor), in the system I donor–complex has been inferred from observation of a very fast component in the time course of the dark reduction of P700 after a flash.[27–29] It is of interest to include this unidentified component in the calculations of the time course of the redox reactions, both for the sake of completeness and because the additional component in the chain will increase the delay in the dark reduction of cytochrome $f$, since one more electron must be transferred in the model of the irreversible chain before cytochrome $f$ can be reduced.

In this case, four electrons are sequentially released by the plastoquinone pool. The derivation follows from the same arguments presented above, where $A_n(t)$, $B_n(t)$, $C_n(t)$, and $P(t)$ are as defined previously, $D_n(t)$ = fraction of PD reduced by the $n$th electron, and $n$ = 1, 2, 3, or 4, corresponding to the first, . . . , or fourth electron exiting from the plastoquinone pool. The initial conditions may be written:

$$A_n(t = 0) = 1 \qquad B_n(t = 0) = 0$$
$$C_n(t = 0) = 0 \qquad D_n(t = 0) = 0$$
$$P(t = 0) = 0$$

For the first electron, the rate equations are:

$$\frac{dA_1}{dt} = -k_{11}A_1$$

$$\frac{dB_1}{dt} = k_{11}A_1 - k_2B_1$$

$$\frac{dC_1}{dt} = k_2B_1 - k_3C_1$$

$$\frac{dD_1}{dt} = k_3C_1 - k_4D_1$$

$$P = 1 - A_1 - B_1 - C_1 - D_1$$

For the second electron, the rate equations are:

$$\frac{dA_2}{dt} = -k_{12}A_2 (1 - A_1) \qquad\qquad \frac{dC_2}{dt} = k_2B_2 - k_3C_2$$

$$\frac{dB_2}{dt} = k_{12}A_2 (1 - A_1) - k_2B_2 \qquad D_2 = 1 - A_2 - B_2 - C_2$$

For the third electron, the rate equations are:

$$\frac{dA_3}{dt} = -k_{13}A_3 (1 - A_2)$$

$$\frac{dB_3}{dt} = k_{13}A_3 (1 - A_2) - k_2B_3$$

$$C_3 = 1 - A_3 - B_3$$

Finally, for the fourth electron the rate equations are:

$$\frac{dA_4}{dt} = -k_{14}A_4 (1 - A_3)$$

$$B_4 = 1 - A_4$$

The redox state of the components in the chain is given by:

$$[P700] = 1 - A_1 - B_1 - C_1 - D_1$$

$$[PD] = D_1 + D_2$$

$$[PC] = C_1 + C_2 + C_3$$

$$[Cyt\ f] = B_1 + B_2 + B_3 + B_4$$

$$[PQ] = A_1 + A_2 + A_3 + A_4$$

For the 4-electron problem, the choice of rate constants $k_{1i}$ for cases A and B as discussed in example I are: $k_{12} = (11/12)k_{11}$, $k_{13} = (5/6)k_{11}$, $k_{14} = (3/4)k_{11}$ (case A): $k_{12} = (3/4)k_{11}$, $k_{13} = (1/2)k_{11}$, $k_{14} = (1/4)k_{11}$ (case B). The computer-generated solutions for these two cases applied to the chain including PD are shown in Figs. 6A, B, with the rate constant for PD oxidation based on an approximate half-time for oxidation of 10 $\mu$sec.[27-29] A summary of the half-times of the redox changes for this four component chain under basal conditions of electron transport is shown in Table I (IIA, B). The presence of the additional electron transfer component results in an increase in the half-time of plastoquinone oxidation, and as well longer times for reduction of plastocyanin and cytochrome $f$. Whereas the ratio of the half-times for cytochrome $f$ reduction to those of P700 are 3.4 (Fig. 2A) and 6.1 (Fig. 2B) for the three component PS–I donor complex, these ratios are 4.7 (Fig. 6A) and 9.1 (Fig. 6B) when the component PD is included in the chain. The choice of rate

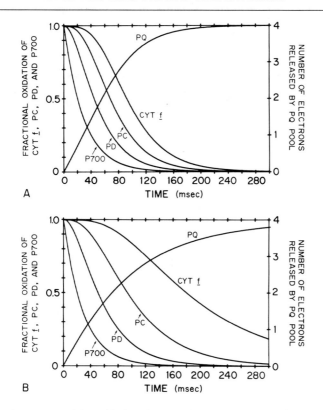

FIG. 6. Time course of the oxidation of plastoquinone and the reduction of cytochrome $f$, plastocyanin, PD, and P700 following illumination predicted by the theoretical treatment given in example II. (A) The rate constants used were (in sec$^{-1}$): $k_{11} = 35$, $k_{12} = 32$, $k_{13} = 29$, $k_{14} = 26$, $k_2 = 6940$, $k_3 = 9900$, and $k_4 = 69,000$. (B) The rate constants used were (in sec$^{-1}$); $k_{11} = 35$, $k_{12} = 26$, $k_{13} = 17$, $k_{14} = 8.7$, $k_2 = 6940$, $k_3 = 9900$, and $k_4 = 69,000$.

constants, $k_{1i}$, for case B does not generate a good fit to the data for a four component chain including PD.

As noted earlier, there are data in the literature to support the function of two plastocyanin molecules in the chain.[6] Because of the insensitivity of the solution for cytochrome $f$ to the values of the rate constants $k_2 - k_4$, the time course shown for cytochrome $f$ reduction in Fig. 6 is the same as that for the linear sequence cyt $f \rightarrow PC \rightarrow PC \rightarrow$ P700.

*Example III: $PQ \rightarrow FeS \rightarrow Cyt f \rightarrow PC \rightarrow P700$*

The above chain is included since there is now an appreciable amount of evidence for participation of an Fe–S center in the PS I–donor complex.[23-26] Since there are no kinetic data for the Fe–S center we have chosen the rate constant for its oxidation to be the same as that of cytochrome $f$, and the rate constants for the oxidation of plastoquinone to be the same as those used in example II, cases A and B. The formalism is otherwise identical to that presented for model II. The computer-generated solutions are shown in Figs. 7A, B, and the half-times for the redox changes of the chain are given in Table I. The half-time for oxidation of plastoquinone is the same as that for the chain of example II. The ratio of the half-times for reduction of cytochrome $f$ to P700 are 3.5 and 4.7 for cases A and B (Figs. 7A,B).

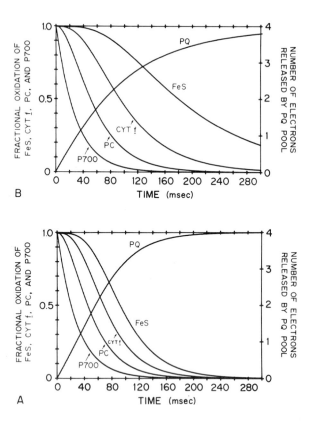

FIG. 7. Time course of the oxidation of plastoquinone and the reduction of the iron-sulfur center (FeS), cytochrome $f$, plastocyanin, and P700 following illumination predicted by the theoretical treatment given in example III. (A) The rate constants used were the same as in Fig. 6A except $k_3 = 6940$, and $k_4 = 9900$. (B) The rate constants used were the same as in Fig. 6B, except that $k_3 = 6940$ and $k_4 = 9900$.

### III. An Explanation for the Choice of Rate Constants: Application of Semi-Classical Electron-Transfer Theory

To obtain the best fit of the theoretical curves with those determined experimentally, it is necessary to make the rate constants $(k_{1i})$ for the transfer of electrons from plastoquinone decrease much more sharply than would be expected from a large homogeneous pool of reduced plastoquinone (e.g., example I, Section II). In an attempt to provide a physical basis for the sharp decrease of rate constants in the presence of a fairly constant electron pressure from the plastoquinone pool, we put forward the following hypothesis: The rate of electron transfer from plastoquinone to cytochrome $f$ depends in part on the net charge of the system I complex. The basis for this statement is that in semi-classical models of electron transfer the rate of transfer in solution depends on the change in free energy of the reaction.[32-34] Simple considerations of electrochemistry indicate that the effective oxidation-reduction potential of a redox couple will depend upon the local charge environment. In the present case, we assume that the effective midpoint oxidation-reduction potential $(E_m)$ of cytochrome $f$ will depend upon the net charge on plastocyanin, P700, and any other neighboring electron carriers. The net charge of the photosystem I complex is more positive immediately after illumination than it is after dark electron transfer from plastoquinone, which results in partial reduction of the complex. These considerations lead to the expectation that the affinity of the photosystem I complex for electrons donated by plastoquinone increases with increasing positive charge of the complex. Thus, the rate of electron flow to cytochrome $f$ from the plastoquinone pool may depend upon the charge state of the subsequent electron acceptors, regardless of the redox state of the pool.

The linear chain considered in example I can be described as being in the state $f^+$-PC$^+$-P700$^+$ immediately after illumination, in the state $f^+$-PC$^+$-P700 after transfer of one electron, and in the state $f^+$-PC-P700 after transfer of two electrons. Because of the proximity of the components in the membrane, the increase in positive charge on plastocyanin and P700 would cause a positively directed increase in the effective midpoint potential of cytochrome $f$, the magnitude of the increase depending upon the proximity of the redox centers. That is, the value of the midpoint potentials $(E_m)$ for cytochrome $f$ corresponding to the three different charge states of the system I would decrease in the order, $E_m$ $(3+) > E_m (2+) > E_m (1+)$. $E_m (1+)$ is assumed to describe the midpoint potential of cytochrome $f$ in solution. This decrease in effective midpoint potential of cytochrome $f$ with increasing reduction of the system I complex can account for the slow reduction of cytochrome $f$ relative to P700 as follows.

From absolute reaction rate theory, the rate constant, $k$, for electron transfer can be written as $k = A e^{-\Delta G^{\ddagger}/k_0 T}$, where A(T) is the pre-exponential coefficient having a value of $10^4$–$10^{13}$ for different biochemical electron-transfer reactions,[34] $k_0 T$ is the Boltzmann energy $\approx 1/40$ eV at room temperature, and $\Delta G^{\ddagger}$ is the free energy of activation of the complex. In the electron-transfer theory of Marcus,[32-34] the activation energy can be related to the free energy change between reactants and products ($\Delta G$) by the following equation:

$$\Delta G^{\ddagger} = (\lambda - \Delta G)^2/4\lambda$$

for the approximation in which the reactant and product potential energy curves are drawn as equivalent parabolae in one dimension.[32] $\lambda$ is the energy that would be required for the nuclear coordinate changes of the reaction in the absence of electron transfer.[34] Since $\Delta G^{\ddagger} = (\lambda - \Delta G^2/4\lambda$, the value of $\lambda$ can be obtained for a reaction in which the free energy of activation and the $\Delta G$ are known. We will continue to assume that the rate-limiting step is the electron transfer from plastoquinone to cytochrome $f$. The activation energy for the rate-limiting step and dark reduction of cytochrome $f$, measured in the presence of uncoupler, is 11 ± 1 kcal/mole from 5 to 25° in spinach chloroplasts (Fig. 4). The midpoint potentials of the PQ–PQH$_2$ couple and of the cytochrome $f$ couple in solution at pH 7.8 are +30[35] and +365 mV,[36] respectively. The $\Delta G$ for a one-electron reaction operating between the midpoints of the PQ and cytochrome $f$ couples would be +0.335 eV. Solving for $\lambda$ in the equality $\Delta G^{\ddagger} = (\lambda - \Delta G)^2/4\lambda$, yields a value for $\lambda$ of approximately 2.5 eV. This is similar to the value of +2.2 eV calculated for the photooxidation of *Chromatium* cytochrome $c$.[34]

Considering the consecutive transfer of the three electrons to the system I complex, changes in the rate constants will depend only on changes in the effective redox potential change of the reaction if $\lambda$ and $A$ remain constant. Transfer of the three electrons with rate constants $k_{11}$, $k_{12}$, and $k_{13}$ in the ratio 3:2:1 describe the data reasonably well (Fig. 5). These three rate constants would correspond to free energy changes $\Delta G(3+)$, $\Delta G(2+)$, and $\Delta G(1+)$ for the sequential transfer of the three electrons from plastoquinone to cytochrome $f$. We assume that the value of $\Delta G = 0.335$ eV corresponds to the effective free energy change, $\Delta G(1+)$, operating for the transfer of the third electron with rate constant

[31] P. Horton, and W. A. Cramer, *Biochim. Biophys. Acta* **368,** 348 (1974).
[32] R. A. Marcus, *J. Phys. Chem.* **67,** 853 (1963).
[33] R. A. Marcus, in "Tunneling in Biological Systems" (B. Chance *et al.*, eds.) p. 109. Academic Press, New York, 1979.
[34] D. DeVault, *Q. Rev. Biophys.*, in press.

$k_{13}$. Using the value of $\lambda$ calculated above, one can then compute the values of $\Delta G(3+)$ and $\Delta G(2+)$ that would apply to the transfer of the first and second electrons with the choice of rate constants $k_{11} = 3k_{13}$ and $k_{12} = 2k_{13}$. The computed values of $\Delta G(3+)$ and $\Delta G(2+)$ are $+0.40$ and $+0.375$ eV, respectively, corresponding to cytochrome $f$ midpoint potentials of $+0.43$ and $+0.405$ V, respectively. In summary, assuming that the difference in rate constants for sequential transfer of three electrons from plastoquinone can be explained entirely by changes in the effective potential of cytochrome $f$, the effective midpoint potentials of cytochrome $f$ are $E_m(1+) = 0.365$ V (assumed), $E_m(2+) = +0.405$ V, and $E_m(3+) = +0.430$ V.

The changes in effective potential of cytochrome $f$ needed to explain the otherwise anomalously large differences in $k_{1i}$'s for plastoquinone oxidation are not very large. By this criterion the mechanism by which the cytochrome $f$ potential and the rate constants are altered by the charge of the entire system I donor complex may be considered reasonable. On the other hand, there are other factors in the membrane environment of cytochrome $f$ that may change (e.g., local hydrophobicity, membrane potential) and perhaps cause large effects in the effective redox potential. In relation to this question, we note that the anomalously slow reduction of cytochrome $f$, and inferred large difference in the $k_{1i}$'s, is observed in the presence of gramicidin so that changes in transmembrane potential do not appear to be important in this problem. The above discussion using the electron-transfer theory of Marcus to explain the changes in rate constants for plastoquinone oxidation has been directed toward cytochrome $f$. We have also considered above a model for the electron transport chain in which Fe–S center is an intermediate between plastoquinone and cytochrome $f$. The ideas invoked here can equally well be applied to the Fe–S center as an acceptor of electrons from plastoquinone.

## IV. Concluding Remarks

The theoretically predicted half-times for the oxidation of the plastoquinone pool and the reduction of the subsequent components following illumination are shown in Table I for the various models considered above. In order to compare the predicted half-times with experimental values, it is necessary to consider the ratio of a particular half-time to either the rate-limiting step or the reduction half-time of P700 (which in our treatment are equivalent) because the rate-limiting step in photosynthetic electron transport ranges from a few milliseconds to several times longer depending upon the experimental conditions (e.g., coupling, temperature, etc.). Another problem in attempting to interpret experimental half-times for different components is the limited number of measurements obtained under the same conditions. This is even a problem for cytochrome $f$, probably the most thoroughly investigated component in the intermediate chain. Since in our models cytochrome $f$ is reduced by the third or fourth electron released by the plastoquinone pool, its dark reduction is three to nine times slower than the rate-limiting step (Table I). This is in the range of experimentally measured values. The ratio of the half-time for reduction of cytochrome $f$ to that of P700, or that of the rate-limiting step, has been found in spinach chloroplasts to be 4.3,[5,6] 3.3–5,[14,30b] and 5–7.[10] The initial conditions of the model presented here have been chosen to match the experimental conditions of Whitmarsh and Cramer.[10] The theoretical treatment emphasizes the initial *ad hoc* adjustments in the rate constants of the sequential electron transfer from plastoquinone that must be made to fit the data reasonably well. In seeking an explanation for this choice of rate constants, one is led to consider the effect of the net charge of the system I complex on the rate constants for transfer to cytochrome $f$, and to an application of electron-transfer rate theory.

The delay or lag in the dark reduction of cytochrome $f$ is an area of disagreement in the literature. It has been apparent in some measurements[10,37] and absent in others.[5,6,14] The extrapolated delay predicted by the analysis presented here, 7–10 msec and ~40 msec in the presence of gramicidin or valinomycin, respectively, is shorter relative to the half-time for reduction of cytochrome $f$ than had been proposed previously.[5] The delay predicted by a model for an irreversible chain is a maximum estimate. The effect of considering reversible electron transfer between cytochrome $f$ and plastocyanin in the model would be to increase the rate of reduction of cytochrome $f$ while decreasing the rate at which plastocyanin is reduced, thus further decreasing the value of the predicted delay.

Given the degree of uncertainty that presently exists in the data and in the number of components in the system I donor complex, the kinetic analysis presented here cannot yet be used to fit the data unambiguously. For example, the required degree of variation in the $k_{1i}$ is smaller if component PD is present in the chain (example II) than if it is absent (examples I and III). More data are needed in order to determine the position and stoichiometry of the Fe–S and PD components in the system

[35] S. Okayama, *Biochim. Biophys. Acta* **440**, 331 (1976).

[36] H. E. Davenport, and R. Hill, *Proc. R. Soc. London, Ser. B.* **139**, 327 (1952).

[37] M. W. Bugg, J. Whitmarsh, C. E. Riecke, and W. S. Cohen, *Plant Physiol.*, in press.

I complex. It is hoped that the framework and ideas presented here will allow better understanding of the more refined kinetic data available from the newer pulse and averaging techniques.

Acknowledgments

We thank Dr. D. C. DeVault for discussions from which the ideas presented here on the possible application of electron transfer theory of Marcus were drawn, Dr. A. T. Winfree for his help in developing the programs needed to solve the kinetic equations and the use of his Hewlett Packard 9830A computer, Dr. R. P. Cox for sending us a preprint of his work, and Mrs. Mona Imler for her devoted help with the manuscript. This work has been supported by National Science Foundation grant # PCM 77-25196. J. W. is a recipient of a National Needs Postdoctoral Fellowship from the National Science Foundation.

Note Added in Proof:

A more general mathematical approach to problems of kinetics of multicomponent respiratory chains, which has not yet been applied to a specific experimental problem, has appeared recently (PNAS, **76**, 3203, 1979).

# [19] Plastocyanin

By WILLIAM L. ELLEFSON, ELDON A. ULRICH, and DAVID W. KROGMANN

Plastocyanin is a small copper-containing protein which donates electrons to photosystem I. Plastocyanin has been found in many higher plants, green, red and some, but not all blue-green algae. The protein appears to be absent from photosynthetic bacteria and has not been detected in *Euglena gracilis*. Amino acid sequences have been published for a number of plastocyanins,[1,2] and the crystal structure has been derived from X-ray analysis[3] and the structure of plastocyanin in solution will be derived from nuclear magnetic resonance studies.[4]

Katoh had described the assay of plastocyanin by direct light absorbance spectroscopy (sensitive to 10 $\mu$moles) and the isolation of plastocyanin from spinach leaves.[5] Here we will describe several catalytic assay procedures for the detection of nanomolar amounts of plastocyanin in plant extracts. This is followed by procedures which can be used for

[1] A. Aitkin, *Biochem. J.* **149**, 675 (1975).

[2] J. A. M. Ramshaw, M. D. Scawen, E. A. Jones, R. H. Brown, and D. Boulter, *Phytochemistry* **15**, 1199 (1976).

[3] P. M. Coleman, H. C. Freeman, J. M. Guss, M. Murata, V. A. Norris, J. A. Ramshaw, and M. P. VenKatappa, *Nature (London)* **272**, 314 (1978).

[4] J. L. Markley, E. L. Ulrich, S. P. Berg, and D. W. Krogmann, *Biochemistry* **4**, 4428 (1975).

[5] S. Katoh, Vol. 12, p. 408.

the purification of plastocyanin from spinach and from the blue-green alga *Anabaena variabilis*.

## Assay

*Principle*. Chloroplast electron transport activity is eliminated by removal of plastocyanin and this activity can be restored by the addition of catalytic amounts of plastocyanin.[6] The amount of plastocyanin required for restoration of chloroplast activity is usually higher than the amount present *in vivo* and is somewhat variable depending on chloroplast composition and residual detergent concentration.

*Preparation of Plastocyanin Depleted Chloroplasts*. Spinach chloroplasts, prepared according to any one of the numerous procedures which give chloroplasts of high specific activity, are used.[7,8] Pelleted chloroplasts are resuspended in a solution of cold 1% Tween 20 to a concentration of 1 mg chlorophyll per 50 ml.[9] This suspension is centrifuged in the cold at 48,000 $g$ for 10 min, and the resulting pellet is resuspended in 0.4 $M$ sucrose containing 0.05 $M$ NaCl to give a concentration of 1 mg chlorophyll per ml.

*Spectrophotometric Assay*. The photosystem I-dependent oxidation of large and easily measured amounts of exogenous horse heart cytochrome $c$ is dependent on the addition of catalytic amounts of plastocyanin. Reduced cytochrome $c$ is prepared by making a solution of 20 mg/ml of commercially available cytochrome $c$ in water, adding 4 mg of sodium ascorbate per milliliter, and dialyzing overnight to remove the ascorbate. The resulting reduced cytochrome $c$ is frozen as 5-ml samples until needed. Reduced cytochrome $c$ photooxidation is conveniently measured in a 3-ml reaction volume containing the following components: 150 $\mu$moles Tricine buffer, pH 8.0, p 0.03 $\mu$moles 3-(3,4-dichlorophenyl)-1,1-dimethylurea, 10 $\mu$moles sodium azide, 0.33 $\mu$moles reduced cytochrome $c$ (0.02 ml of stock solution described above), and 0.4 $\mu$moles methyl viologen. Plastocyanin-depleted chloroplasts containing approximately 10 $\mu$g of chlorophyll are added. Following illumination of the above reaction mixture the oxidation of reduced cytochrome $c$ in the presence of additions of plastocyanin is measured at 550 nm ($\epsilon$ m$M$ of reduced cytochrome $c$ = 19.5). Various methods for illuminating the reaction mixture and simultaneously recording the absorbance change or simply recording the absorbance before and after illumination outside the

[6] M. Plesnicar and D. S. Bendall, *Biochim. Biophys. Acta* **216**, 192 (1970).
[7] M. Avron, A. T. Jagendorf, and M. Evans, *Biochim. Biophys. Acta* **26**, 262 (1957).
[8] R. Ouitrakul and S. Izawa, *Biochim. Biophys. Acta* **305**, 105 (1973).
[9] R. Barr, D. Rosen, and F. L. Crane, *Proc. Indiana Acad. Sci.* **84**, 147 (1975).

measuring instrument are equally good. At saturating light intensity and a saturating amount of plastocyanin, a rate of 250 $\mu$moles of reduced cytochrome $c$ oxidized per milligram of chlorophyll per hour is routinely observed. To calibrate the assay system, a sample of spinach plastocyanin is prepared according to the method described below and its concentration determined by measuring the absorption difference between ferricyanide-oxidized and ascorbate-reduced samples at 597 nm ($\epsilon$ m$M$ of plastocyanin = 4.5). The assay gives a linear response in the range of 1 to 10 nomoles of added plastocyanin. This assay could be ambiguous since some redox dyes will substitute for plastocyanin. However, we have regularly checked the plastocyanin-active fractions of higher plant extracts by absorption on, and elution from, ion exchange cellulose. In all cases the material which supported catalytic activity in the above assay was found in the column fraction in which plastocyanin is eluted.

*Oxygen Electrode Assays.* An oxygen electrode can be used to measure plastocyanin with a sensitivity and convenience similar to that of above spectrophotometric assay. The assay is based on the measurement of oxygen consumption in a methyl viologen Hill reaction. The same treatment for removal of endogenous plastocyanin by washing with Tween 20 is used. The assay mixture contains the following: 250 $\mu$moles Tris-HCl, pH 7.6, 0.4 $\mu$mole methyl viologen; 0.002 $\mu$mole and gramicidin in a final volume of 1.5 ml. Plastocyanin-depleted chloroplasts containing approximately 50 $\mu$g of chlorophyll are added. Light-dependent oxygen consumption is the result of the autooxidation of methyl viologen which is reduced by chloroplasts in a plastocyanin-dependent reaction. The response is linear between 0.5 and 5 nmoles of added plastocyanin and gives a rate of 1000 $\mu$moles $O_2$ consumed/mg chlorophyll/hr at saturating light intensity and saturating plastocyanin.

### Purification of Spinach Plastocyanin

On the day prior to preparation of the crude extract, 10 liters of 0.4 $M$ sucrose containing 0.05 $M$ NaCl is prepared and stored at 2°, 10 liters of acetone are stored at −15°, and the leaves from one bushel of spinach are removed from the stems, washed, drained, and stored at 2°. This gives approximately 5 kg of washed leaves. One kilogram batches of leaves are ground for 30 sec in 1.5 liters of cold sucrose–NaCl solution in a 1 gallon capacity Waring blender. The homogenate is filtered through a heavy cloth sack to remove debris. A "Geer Press Floor King" mop wringer has proved very convenient for squeezing the aqueous extract from the leaf pulp. This device, usually used to squeeze the water from a floor mop, is mounted on the side of a bucket and will hold the sack containing the homogenate in the mop frame. A lever arm brings a perforated steel plate against the sack and squeezes out most of the aqueous extract. The pulp remaining in the sack is discarded. The aqueous extract is filtered through a fine mesh cloth then centrifuged at $10,000 \times g$ at a slow flow rate through a continuous flow centrifuge to bring down the chloroplasts. The precipitated chloroplasts are resuspended in 200 to 400 ml of 0.5 $M$ Tris-HCl buffer, pH 8. It is important to note that the chloroplasts must not be frozen at this stage, since the freeze-thaw process activates a polyphenolase which will destroy the plastocyanin. The chloroplast suspension is transferred to a beaker surrounded with ice and stirred vigorously; cold acetone ($-15°C$) is added slowly to give a final concentration of 35% (v/v). The mixture is stirred an additional 10 min and then centrifuged at $10,000 \, g$ for 10 min. The pellet is discarded, and the supernatant liquid is brought to a concentration 80% (v/v) acetone by the slow addition, with stirring, of more cold acetone. A flocculent precipitate settles rapidly, and after 30 min most of the liquid can be decanted. The remaining suspension is centrifuged at $10,000 \, g$ for 5 min. The precipitated protein is resuspended in 200 ml of cold 50 m$M$ potassium phosphate buffer, pH 7.6, and 1 gm of Bio-Rad AG1-X8 resin (100–200 mesh) is added to the suspension. The resin increases the yield of plastocyanin by absorbing phenolic substances. The suspension is dialyzed against 10 liters of 20 m$M$ potassium phosphate buffer, pH 7.6 in a 15,000 mw cut-off dialysis tubing.

The dialyzed extract is centrifuged at $10,000 \times g$ for 10 min to remove insoluble material. The solution is loaded onto a $3 \times 10$ cm DEAE-cellulose column (coarse grade) which had been equilibrated with 20 m$M$ potassium phosphate buffer, pH 7.6. The column is first washed with 500 ml of the equilibrating buffer and is then eluted with 200 ml of the equilibrating buffer containing 0.2 $M$ NaCl which removes both the plastocyanin and the ferredoxin–NADP oxidoreductase.

Solid sodium pyrophosphate to give a concentration of 0.05 $M$ is added to the 0.2 $M$ NaCl eluant from the DEAE column. Solid (NH$_4$)$_2$SO$_4$ is then added to 75% of saturation and the solution is allowed to stand for 2 hr. This is followed by centrifugation at $10,000 \, g$ for 10 min to remove the ferredoxin–NADP oxidoreductase, and supernatant liquid containing plastocyanin (which is easily detected by the appearance of blue color following addition of a trace of potassium ferricyanide) is dialyzed against 10 liters of 20 m$M$ potassium phosphate buffer, pH 7.6.

The dialyzed plastocyanin is loaded on a $1 \times 1.5$ cm DEAE–cellulose column equilibrated with 20 m$M$ potassium phosphate buffer, pH 7.6, and eluted with a minimum volume of 2 $M$ NaCl in the same buffer. The plastocyanin (4 ml) is loaded onto a $3 \times 60$ cm Sephadex G-50 column equilibrated with 50 m$M$ potassium phosphate buffer, pH 7. Fractions containing plastocyanin are recognized by the appearance of blue color after addition of 1 drop of 0.01 $M$ potassium ferricyanide. The plastocyanin-containing fractions are pooled and dialyzed against 50 m$M$ potassium phosphate buffer, pH 7.

The oxidized, dialyzed plastocyanin is loaded onto a DEAE–Sephadex A-50 Column (1.5 × 25 cm) equilibrated with 50 m$M$ potassium phosphate buffer, pH 7. DEAE–Sephadex A-50 is better than the A-25 since the latter resin causes more extensive reduction of the plastocyanin and lowers the yield of pure protein. The column is eluted with a linear gradient between 500 ml of 50 m$M$ and 500 ml of 200 m$M$ potassium phosphate, pH 7. Fractions of 6 ml are collected, and the oxidized plastocyanin emerges at about 130 m$M$ phosphate. Fractions having a ratio of $A_{597nm}$ to $A_{278 nm}$ of 1.5 or more are at least 97% pure. The reduced plastocyanin which emerges at 150 mM phosphate gives much higher UV to visible peak absorbance ratios but can be purified by recycling through this column after oxidation and dialysis. The yield of plastocyanin is usually 30 to 50 mg of pure protein from 5 kg of spinach leaves.

### Purification of *Anabaena variabilis* Plastocyanin

*Anabaena variabilis* is grown in the laboratory using Kratz and Meyers media C.[10] The cells are harvested by continuous flow centrifugation and stored as a thick paste at −15°. After thawing, the cells are suspended (200 gm wet weight per liter) in potassium phosphate buffer (50 m$M$, pH 7) and the solution is run through a Manton-Gaulin homogenizer at 10,000 psi. The homogenizer is kept at 10°, and the cells are recooled to 4° before a second pass through the homogenizer. Most of the phycobiliproteins and membrane fragments are removed by bringing the suspension slowly with stirring to 45% $(NH_4)_2SO_4$ saturation at 4°. The solution is stirred for 2 hr and then the precipitate removed by centrifugation at 10,000 $g$ for 10 min. The supernatant liquid is taken to 100% $(NH_4)_2SO_4$ saturation at 4°, stirred for 2 hr, and stored for 48 hr to allow complete precipitation of the plastocyanin. Centrifugation at 10,000 $g$ for 10 min gives a precipitate which is taken up in a minimum amount of distilled water. This protein solution is dialyzed against 1 m$M$ potassium phosphate buffer, pH 7, before loading on a 4.5 × 14 cm CM-52 carboxymethyl cellulose column equilibrated with the same buffer used in dialysis. The remaining phycobiliproteins are washed off the column with the equilibrating buffer after loading. The plastocyanin is eluted with a linear potassium phosphate gradient from 1 to 25 m$M$, pH 7, total volume 1 liter. A yellow protein with diaphorase activity coelutes with the plas-

[10] W. A. Kratz and J. Myers, *Am. J. Bot.* **42**, 282 (1954).

tocyanin. All of the fractions containing plastocyanin are concentrated and loaded on a 2.5 × 90 cm G-50 medium Sephadex gel column equilibrated with 1 m$M$ potassium phosphate buffer, pH 7. At this step the yellow protein elutes ahead of the plastocyanin and is completely removed. Potassium ferricyanide is added to the plastocyanin fractions from the Sephadex column, and the protein is loaded on a 1.7 × 32 cm C-25 CM-Sephadex gel column equilibrated with 1 m$M$ phosphate buffer, pH 7 containing 0.1 m$M$ potassium ferricyanide. A linear phosphate gradient from 1 to 50 m$M$ pH 7, total volume 1000 ml, is used to elute the plastocyanin. Potassium ferricyanide (0.1 m$M$) is also included in the linear gradient to maintain the plastocyanin in an oxidized state which causes the plastocyanin to elute as a single band. Reduced plastocyanin elutes from cation exchange resins at a different phosphate concentration than the oxidized form. Passing the plastocyanin a second time through the C-25 CM-Sephadex column gives a single symmetrical band of plastocyanin.

## [20] The Isolation and Primary Structures of Various Types of Ferredoxin

*By* Kerry T. Yasunobu and Masaru Tanaka

### Introduction

Ferredoxin is an iron–sulfur protein[1–11] that is found in certain anaerobic bacteria, algae, and plants. There are various types of ferredoxin, but they are all characteristically acidic and low-molecular-weight redox

[1] See addendum on p. 825.
[1a] W. Lovenberg, ed., "Iron-Sulfur Proteins, Vol. 1. Academic Press, New York, 1973.
[2] W. Lovenberg, ed., "Iron-Sulfur Proteins, Vol. 2. Academic Press, New York, 1973.
[3] W. Lovenberg, ed., "Iron-Sulfur Proteins, Vol. 3. Academic Press, New York, 1977.
[4] W. Lovenberg, *in* "Microbial Iron Metabolism" (J. B. Neilands, ed.), p. 161. Academic Press, New York, 1974.
[5] W. H. Orme-Johnson, *Annu. Rev. Biochem.* **42**, 159 (1973).
[6] D. O. Hall, R. Cammack and K. K. Rao, *in* "Iron in Biochemistry and Medicine" (A. Jacobs and M. Worwood, eds.), p. 279. Academic Press, New York, 1974.
[7] B. B. Buchanan and D. I. Arnon, *Adv. Enzymol.* **33**, 119 (1970).
[8] K. Yasunobu, H. Mower, and O. Hayaishi, eds., "Iron and Copper Proteins." Plenum, New York, 1976.
[9] R. Malkin and J. C. Rabinowitz, *Annu. Rev. Biochem.* **36**, 113 (1967).
[10] A. San Pietro, ed., "Non-Heme Iron Proteins: Role in Energy Conversion," Antioch Press, Yellow Springs, Ohio, 1965.
[11] R. C. Valentine, *Bacteriol. Rev.* **28**, 497 (1964).

METHODS IN ENZYMOLOGY, VOL. 69

proteins with very negative redox potentials (about $-0.4$ V).[1-11] They play a central role in the anaerobic metabolism of many anaerobic bacteria.[4,7] Ferredoxins act as redox proteins and in their oxidized forms catalyze the phosphoroclastic cleavage of certain $\alpha$-keto acids, and the resulting reduced forms then are electron sources for a wide variety of reductive reactions. The strongly reducing ferredoxin electrons can also be shunted toward the reduction of pyridine nucleotides for biosynthetic reactions or to protons in bacteria which contain hydrogenase, with the final product hydrogen being liberated. In photosynthetic bacteria, algae, and plants, ferredoxins play mutiple key roles[4,7] such as the photo-reduction of $NADP^+$, $CO_2$ fixation in photosynthetic bacteria, and the regulation of fructose-1, 6-diphosphatase. In nitrogen-fixing microbes, ferredoxin shunts electrons to nitrogenase during the conversion of nitrogen to ammonia.[12] Other reductive processes, such as sulfite and nitrite reduction, appear to use ferredoxin as a source of reducing power. For more detailed discussions on the function of ferredoxin, the article by Buchanan and Arnon[7] and by Lovenberg[4] are excellent.

The term ferredoxin was used by Tagawa and Arnon[13] to include iron–sulfur proteins required for the photoreduction of NADP by chloroplasts, and this definition will be strictly adhered to in this article.

Physicochemical studies[1-3,14] have established that ferredoxin types found thus far can be classified as 2Fe–2S, 4Fe–4S or 2(4Fe–4S) containing types. Physicochemical studies[1-3] performed on ferredoxin include: (a) absorption spectroscopy; (b) EPR, ENDOR, and Mössbauer spectroscopy; (c) NMR spectroscopy; and (d) redox potential measurements. The three-dimensional structures of the *Peptococcus aerogenes* (8Fe–8S type)[15,16] and the *Chromatium* high potential iron–sulfur protein (Hipip,[17] which is not a ferredoxin) have been resolved by crystal X-ray diffraction studies. Recently, crystals suitable for X-ray diffraction studies have been isolated for the *Spirulina platensis* ferredoxin (a 2Fe–2S type) and in the preliminary studies,[18] the two iron atoms were shown to

[12] R. W. Hardy and R. C. Burns, *in* "Iron-Sulfur Proteins" (W. Lovenberg, ed.), Vol. 1, p 65. Academic Press, New York, 1973.

[13] K. Tagawa and D. I. Arnon, *Nature (London)* **195**, 537 (1962).

[14] J. C. W. Tsibris and R. W. Woody, *Coord. Chem. Rev.* **5**, 417 (1970).

[15] E. T. Adman, L. C. Sieker, and L. H. Jensen, *Proc. Natl. Acad. Sci. U.S.A.* **248**, 3987 (1973).

[16] L. H. Jensen, *Annu. Rev. Biochem.* **43**, 461 (1974).

[17] C. W. Carter, S. T. Freer, Ng. H. Xuong, R. A. Alden, and J. Kraut, *Cold Spring Harbor Symp. Quant. Biol.* **36**, 381 (1971).

[18] K. Ogawa, T. Tsukihara, H. Tahara, Y. Tatsube, Y. Matsuura, N. Tanaka, and H. Matsubara, *J. Biochem. (Tokyo)* **81**, 529 (1977).

be closely situated. Model Fe–S compounds have been synthesized, crystallized, and their physicochemical properties shown to correspond closely to the Fe–S chelate structures of the ferredoxins, namely, of the 2Fe–2S and the 4Fe–4S types. It should be emphasized that the protein has a dramatic effect on the redox potential valance state of the iron as well as on the biological properties of the Fe–S proteins.[19] The *Azotobacter* ferredoxin I contains both a ferredoxin type of active center ($E_0$ = −0.4 V) and an Hipip type of active center ($E_0$ = +0.27 V).[20] In addition, there are other 2Fe–2S type of iron–sulfur proteins, such as the iron–sulfur proteins from *Pseudomonas putida* (putidaredoxin),[21] the bovine adrenal glands (adrenodoxin),[22] *Escherichia coli*,[23] *Halobacterium halobium*,[24] and animal mitochondria,[25] but they are not involved in the photo-reduction of $NADP^+$ by chloroplasts and therefore are not a ferredoxin.

In the present article, a brief summary of the various types of ferredoxins isolated, the procedure for the isolation of algal and plant type of ferredoxin and the primary structures of some ferredoxins are presented.

## Isolation of Various Types of Ferredoxin

The procedure for the isolation of clostridial ferredoxin has been aptly covered by Rabinowitz[26] in a previous volume of this series. In the section to follow, the more important papers which describe the isolation of the various types of ferredoxins will be briefly reviewed.

*8Fe–8S Type of Ferredoxin Isolation.* As mentioned above, the procedure for the isolation of the clostridial ferredoxin has been described by Rabinowitz.[26] Other 8Fe–8S type of ferredoxin which have been iso-

[19] T. Kerskovitz, B. A. Averill, R. H. Holm, J. A. Ibers, W. D. Phillips, and J. R. Weiher, *Proc. Natl. Acad. Sci. U.S.A.* **69**, 2437 (1972).

[20] D. C. Yoch and D. I. Arnon, *J. Biol. Chem.* **247**, 4515 (1972).

[21] I. C. Gunsalus and J. D. Lipscomb, *in* "Iron–Sulfur Proteins" (W. Lovenberg, ed.), Vol. 1, p. 151. Academic Press, New York, 1973.

[22] R. W. Eastbrook, K. Suzuki, J. I. Mason, J. Baron, W. E. Taylor, E. R. Simpson, J. Purvis, and J. McCarthy, *in* "Iron-Sulfur Proteins" (W. Lovenberg, ed.), Vol. 1, p. 193. Academic Press, New York, 1973.

[23] H.-E. Knoell and J. Knappe, *Eur. J. Biochem.* **50**, 245 (1974).

[24] L. Kerscher, D. Oesterhelt, R. Cammack, and D. O. Hall, *Eur. J. Biochem.* **71**, 101 (1976).

[25] H. Beinert, *in* "Iron and Copper Proteins" (K. T. Yasunobu, H. Mower, and O. Hayaishi, ed.), p. 137. Plenum, New York, 1976.

[26] J. C. Rabinowitz, Vol. 24, p. 431.

lated from photosynthetic bacteria include the proteins from *Chromatium vinosum*,[27] *Chlorobium limicola* (ferredoxin I and II[27-29]), *Rhodo-spirillum rubrum* (ferredoxin I[30]), and *Rhodospirillum spheroides* (ferredoxin I[30]). Basically, the procedures used are adaptations of the method used to isolate the clostridial ferredoxin,[26] but an additional Sephadex G-100 chromatography step, a modified DEAE-cellulose chromatography, or a hydroxylapatite chromatography step may be necessary to separate multiple forms when present. A unique type of 8Fe-8S iron–sulfur protein has been isolated from *Azotobacter vinelandii*[19] (ferredoxin I) which contains both a ferredoxin type and Hipip type of 4Fe-4S active site.

*4Fe–4S Ferredoxin Isolation.* The isolation of this type of iron–sulfur proteins was first reported from *Desulfovibrio gigas*[31] but has now also been isolated from other strains of *Desulfovibrio*, e.g., *Desulfovibrio desulfuricans*.[32] Moreover, this type of iron–sulfur protein has been isolated from *Spirochaeta aurantia*,[33] *Bacillus polymyxa* (ferredoxin I and II),[34] and *Rhodospirillum rubrum* (ferredoxin II).[28] The 4Fe-4S type of ferredoxin can also be isolated by slight modifications of the procedure used for the isolation of clostridial ferredoxin. (For *Bacillus polymyxa*, ferredoxin II is isolated first by 0.02 $M$ phosphate buffer, pH 7.4, containing 0.28 $M$ NaCl during the DEAE-cellulose chromatography step and ferredoxin I is eluted from the same column by buffer containing 0.35 $M$ NaCl.[34] Another 4Fe-4S iron sulfur protein has been isolated from the thermophile, *B. stearothermophilus*.[35]

*2Fe–2S Ferredoxin Isolation.* Plant and algal ferredoxins contain this type of active center. Algal ferredoxins which have been isolated include the proteins from *Scenedesmus quadricauda*,[36] *Spirulina maxima*,[37]

---

[27] R. Bachofen and D. I. Arnon, *Biochim. Biophys. Acta* **120**, 259 (1966).

[28] K. K. Rao, H. Matsubara, B. B. Buchanan, and M. C. W. Evans, *J. Bacteriol.* **100**, 1411 (1969).

[29] B. H. Gray, C. W. Fowler, N. A. Nugent, N. Rigapoulas, and R. C. Fuller, *Abstr. Annu. Meet. Am. Soc. Microbiol.* p. 156 (1972).

[30] D. C. Yoch, D. I. Arnon, and W. V. Sweeney, *J. Biol. Chem.* **250**, 8330 (1975).

[31] J. LeGall and N. Dragoni, *Biochem. Biophys. Res. Commun.* **23**, 145 (1966).

[32] J. A. Zubieta, R. Mason, and J. R. Postgate, *Biochem. J.* **133**, 851 (1973).

[33] P. W. Johnson and E. Canale-Parola, *Arch. Mikrobiol.* **89**, 341 (1973).

[34] D. C. Yoch, *Arch. Biochem. Biophys.* **158**, 633 (1973).

[35] R. N. Mullinger, R. Cammack, K. K. Rao, D. O. Hall, D. P. E. Dickson, C. E. Johnson, J. D. Rush, and A. Smopoulos, *Biochem. J.* **151**, 75 (1975).

[36] H. Matsubara, *J. Biol. Chem.* **243**, 370 (1968).

[37] D. O. Hall, K. K. Rao, and R. Cammack, *Biochem. Biophys. Res. Commun.* **47**, 798 (1972).

*Spirulina platensis*,[37] *Aphanothece sacrum* (ferredoxin I[38] and II[39]), *Anacystis nidulans*,[40] *Nostoc mucorum*,[41] *Phormidium persicinum*,[42] *Bumilleriopsis filiformis*,[43] and *Anabaena variabilis*.[43] The procedure for the isolation of the *Azotobacter* ferredoxin II, a 2Fe–2S ferredoxin, has been reported by Yoch and Arnon,[20] and it is separated from ferredoxin I on a Sephadex G-100 column. Since all plants and algae contain the 2Fe–2S type of ferredoxin, a list of all plant ferredoxins isolated would be too extensive for this article. Some representative ferredoxin isolation procedures are reported for the isolation of the protein from spinach,[13] *Lucaena glauca* (koa),[44] alfalfa,[45] *Colocasia esculenta* (taro),[46] cotton,[47] *Equisetum* (horsetail),[48] barley,[42] swiss chard,[42] parsley,[42] and maize.[42]

## Detailed Procedure for the Isolation of Algal and Plant Ferredoxin

The procedure used for the isolation of these ferredoxin is a modification of the one developed for the isolation of clostridial ferredoxin. It is reported in detail here since some of the assays used are different than for the one used for clostridial ferredoxin and because new modifications which simplify the ferredoxin isolation have been incorporated. The major difficulty in the isolation of plant ferredoxin is the contamination of the plant extracts by colored pigments which obscure the presence of the pinkish-brown colored ferredoxin.

*Assays.* At least four different assays can be utilized for monitoring the purification of ferredoxin. These are (1) phosphoroclastic cleavage of pyruvate which is described by Rabinowitz[26] in Volume 24B of this

[38] T. Hase, K. Wada, and H. Matsubara, *J. Biochem. (Tokyo)* **78**, 605 (1975).

[39] T. Hase, K. Wada, and H. Matsubara, *J. Biochem. (Tokyo)* **79**, 329 (1976).

[40] T. Yamanaka, S. Taenami, K. Wada, and K. Okunuki, *Biochim. Biophys. Acta* **180**, 196 (1969).

[41] T. Hase, K. Wada, M. Ohmiya, and H. Matsubara, *J. Biochem. (Tokyo)* **80**, 993 (1976).

[42] E. Tel-Or, S. Fuchs, and M. Avron, *FEBS Lett.* **29**, 156 (1973).

[43] E. Tel-Or, R. Cammack, and D. O. Hall, *FEBS Lett.* **53**, 135 (1975).

[44] A. M. Benson and K. T. Yasunobu, *J. Biol. Chem.* **244**, 955 (1969).

[45] S. Keresztes-Nagy and E. Margoliash, *J. Biol. Chem.* **241**, 5955 (1966).

[46] K. K. Rao, *Phytochemistry* **8**, 1379 (1966).

[47] D. J. Newman, J. N. Ihle, and L. Dure, III, *Biochem. Biophys. Res. Commun.* **6**, 947 (1969).

[48] S. J. Aggarwal, K. K. Rao, and H. Matsubara, *J. Biochem. (Tokyo)* **69**, 601 (1971).

series. (2) Cytochrome $c$ reduction[49] is described by Shin in Volume 23 of this series. (3) NADP photoreduction in which A Perkin-Elmer Model 356 spectro-photometer with actinic illumination at 670 nm is used. The reaction mixture contains: 0.05 $M$ Tris-HCl buffer, pH 8, 1 $\mu M$ of bovine serum albumin, 0.25 $M$ ascorbate, 2.5 m$M$ NADP$^+$, and a chloroplast suspension which contained 20 $\mu M$ chlorophyll. Chloroplasts are prepared by the procedure of Whatley and Arnon.[50] The grinding medium consisted of 0.04 $M$ Tris-HCl–0.35 $M$ NaCl, pH 8, and the lysing medium was 0.01 $M$ Tris-HCl–0.35 $M$ NaCl. Chlorophyll content is determined by the method of Arnon.[51] (4) The ferredoxin content can be monitored in the ESR apparatus.[52] The reduced (sodium dithionite) ferredoxin concentration should be about $10^{-4}$ $M$; the temperature below 25°K; microwave frequency, 9.17 GHz; microwave power, 0.9 mW; modulation frequency, 100 kHz; modulation width, 6 G; a field sweep rate, 100 G/min; time constant, 0.25 sec. The $g_y$ signal (1.96) can be used to quantitate the ferredoxin content.

*Extraction.* With most plants, ferredoxin can be readily extracted from the leaves by use of a Waring blender. However, there are certain types of plants from which ferredoxin cannot be solubilized by this technique, e.g., ferns and some strains of *Equistem* (horsetail). The algal ferredoxin can be extracted by hypertonic salt solution[38] or a Waring blender–glass bead procedure.[36]

*Purification.* The method which works well for the purification of *Lucaena glauca* and taro ferredoxin is a modification of the procedure described previously by San Pietro and Lang[53] and Tagawa and Arnon.[13] About 6 kg of leaves are collected and stored in a freezer at $-20°$ to break up the integrity of the leaves. Each batch is homogenized in a Waring blender at 4° for 2–10 min at high speed with 750 ml of 12.5 m$M$ Tris-HCl buffer, pH 7.2, which was also 44 m$M$ in NaCl. Twelve batches of leaves yielded 15 liter of homogenate. To each 2.5 liters of homogenate were added slowly and with vigorous stirring, 810 ml of acetone cooled to $-10°$. Each batch was immediately filtered through a layer of Solka-floc in a Büchner funnel at 4°. All particulate matter was retained by the filter and approximately 2.3 liters of brown filtrate were obtained per batch of homogenate. To each 2 liters of filtrate were added, slowly with

[49] M. Shin. Vol. 23, p. 440.
[50] F. R. Whatley and D. I. Arnon, Vol. 6, p. 308.
[51] D. I. Arnon, *Plant Physiol.* **24**, 1 (1949).
[52] W. H. Orme-Johnson and R. H. Sands, *in* "Iron-Sulfur Proteins" (W. Lovenberg, ed.), Vol. 2, p. 195. Academic Press, New York, 1973.
[53] A. San Pietro and H. M. Lang, *J. Biol. Chem.* **231**, 211 (1958).

stirring, 3.2 liters of acetone at $-10°$. The flask was packed in Dry Ice, and the precipitate was permitted to settle for a few minutes. Most of the supernatant was siphoned off, and the remaining suspension was centrifuged at $-10°$ for 5 min at 5000 rpm in the Sorval RC-2B centrifuge. All steps after this point were carried out at $4°$.

When twelve batches of leaves had been processed to this point, the precipitates thus obtained were combined and extracted with 1.2 liters of 0.01 $M$ Tris buffer, pH 7.3. After removal of the residual precipitate by centrifugation at 8500 rpm for 10 min at $0°$, the supernatant was dialyzed overnight against 4 liters of 0.01 $M$ Tris-HCl (pH 7.3) buffer at $4°$, centrifuged again to remove any precipitate which formed during dialysis, and then applied to a column (5 × 50 cm) of DEAE–cellulose which had been equilibrated with 0.1 $M$ Tris buffer. All Tris buffers were adjusted to pH 7.3 with HCl. One liter of this same buffer was then passed through the column to remove residual acetone. A dark brown band could be observed at the top of the column, with a red band immediately below it. The material was left at this stage for 24 hr, while another 6 kg of leaves were processed to this point. The material thus obtained is applied to the same DEAE–cellulose column. The column is then washed with 2 liters of 0.1 $M$ Tris buffer followed by 1 liter of 0.2 $M$ Tris buffer. Ferredoxin was eluted from the column in 1650 ml of 0.5 $M$ Tris buffer.This solution was diluted $2\frac{1}{2}$-fold to yield a Tris concentration of 0.2 $M$ and applied to a 3 × 20 cm column of DEAE–cellulose. Development of this and subsequent columns followed the procedure used for the initial column. After chromatography on another 3 × 20 cm column followed by a 3 × 6 cm column, the ferredoxin was contained in 85 ml of buffer. (Some of the rechromatography steps can be avoided by chromatographing the ferredoxin on a Sephadex G-100 column which separates the low molecular ferredoxin from larger proteins and plant pigments and can be recognized by the brownish-red color.) Ammonium sulfate (0.5 gm/ml) is then added. After removal of the precipitate by centrifugation, the supernatant solution is applied directly without dilution to a 1 × 15 cm column of DEAE–cellulose. Passage of about 200 ml of 0.1 M Tris buffer containing 0.5 gm of ammonium sulfate per milliliter through the column yields a yellow effluent which is discarded. Ferredoxin is eluted in 35 ml of 1 $M$ Tris and is then applied to a 4 × 70 cm column of Sephadex G-75 which had been equilibrated and developed with 0.1 $M$ Tris buffer; fractions of 16 ml are collected. The fractions with 420 to 277 nm ratios of 0.49 are collected and crystallized by the addition of 0.5 gm of ammonium sulfate per milliliter. The yield of crystalline ferredoxin amounts to 20 mg/kg of leaves. The crystalline preparations are quite stable in an inert gas atmosphere in the presence of ascorbic acid.

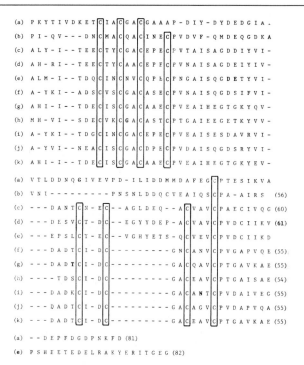

Fig. 1. Amino acid sequences of the 4Fe–4S and 8Fe–8S types of ferredoxins. In the figure, (a)–(k) stands for iron–sulfur proteins isolated from (a) *Bacillus stearothermophilus,* a thermophilic facultative bacterium; (b) *Desulfovibrio gigas,* a sulfate reducer; (c) *Chlorobium limicola* (ferredoxin I), a green photosynthetic bacterium; (d) *C. limicola* ferredoxin II; (e) *Chromatium vinosum,* a purple photosynthetic bacterium; (f) *Clostridium pasteurianum,* an obligate anaerobe; (g) *C. tartarivorum,* a heat-stable anaerobe; (h) *Peptostreptococcus elsdenii,* an aerobe; (i) *Clostridium ME,* an obligate anaerobe; (j) *C. acidi-urici,* an obligate anaerobe; and (k) *C. thermosaccharolyticum,* a thermophilic anaerobe. The first two are 4 Fe proteins while the remainder are 8Fe proteins. In all sequences shown, the IUB one letter code for amino acids is shown. The NH$_2$-terminal region starts from the upper left-hand portion and gaps have been inserted to indicate sequence homologies. The cysteine residues probably involved in iron chelation are blocked off. The numbers at the end refer to the total number of amino acids present in the proteins.

## Amino Acid Sequences of Various Ferredoxins

The amino acid sequences of the 2(4Fe–4S), the 4Fe–4S and the 2Fe–2S types of ferredoxins are discussed in this section.

*8Fe–8S Ferredoxin Sequences.* These sequences are summarized in Fig. 1 and include the proteins isolated from *C. pasteurianum,*[54] *C.*

[54] M. Tanaka, T. Nakashima, A. M. Benson, H. R. Mower, and K. T. Yasunobu, *Biochemistry* 5, 1666 (1966).

*butyricum,*[55] *C. acidi-urici,*[56] *P. aerogenes,*[57] *P. elsdenii,*[58] *C. thermosaccharolyticum,*[59] *C. tartarivorum,*[60] and *Clostridium ME.*[61] The 8Fe–8S type of ferredoxin from photosynthetic bacteria which had been sequenced are also shown in Fig. 1. Included are the *Chromatium*[62] and the *Chlorobium limicola* ferredoxin.[63,64] The *Chromatium vinosum* sequence[62] has been revised[65] and in the corrected version, residues 50–58 are Val-Glu-Val-Cys-Pro-Val-Asp-Cys-Ile rather than Val-Asp-Cys-Val-Glu-Val-Cys-Pro-Ile, and the total amino acids is 82 rather than 81 due to the presence of an additional Ile at position 58. An interesting feature of the clostridial ferredoxin is that the sequence repeats itself and residues 1–28 and 29–54 show sequence homology possible due to gene duplication. The cysteine residues involved in iron chelation in one Fe–S cluster involve the first three cysteine residues from the NH$_2$-terminal portion of the protein and the last cysteine residue, while in the other iron cluster the remainder of the four cysteine residues are chelated to the iron. Chemical modification of the NH$_2$-terminal region shows that this region is required for stability of the 8Fe–8S type of ferredoxin.

*4Fe–4S Ferredoxin Sequences.* Only the *Desulfovibrio gigas*[66] and the *Bacillus stearothermophilus*[67] proteins have been sequenced and the sequences are also shown in Fig. 1. The amino acid sequence of the *Chromatium* Hipip iron–sulfur protein has been determined by Dus *et*

[55] A. M. Benson, H. F. Mower, and K. T. Yasunobu, *Arch. Biochem. Biophys.* **121,** 563 (1967).
[56] S. C. Rall, R. E. Bolinger, and R. D. Cole, *Biochemistry* **8,** 2486 (1969).
[57] J. N. Tsunoda, K. T. Yasunobu, and H. R. Whiteley, *J. Biol. Chem.* **243,** 6262 (1968).
[58] P. Azari, M. Glantz, J. N. Tsunoda, S. Mayhew, and K. T. Yasunobu, unpublished results.
[59] M. Tanaka, M. Haniu, K. T. Yasunobu, R. Himes, and J. Akagi, *J. Biol. Chem.* **248,** 5215 (1973).
[60] M. Tanaka, M. Haniu, G. Matsueda, K. T. Yasunobu, R. Himes, J. Akagi, E. M. Barnes, and T. Devanathan, *J. Biol. Chem.* **246,** 2953 (1971).
[61] M. Tanaka, M. Haniu, K. T. Yasunobu, J. B. Jones, and T. C. Stadtman, *Biochemistry* **13,** 528 (1974).
[62] H. Matsubara, R. M. Sasaki, D. K. Tsuchiya, and M. C. W. Evans, *J. Biol. Chem.* **245,** 2121 (1970).
[63] M. Tanaka, M. Haniu, K. T. Yasunobu, M. C. W. Evans, and K. K. Rao, *Biochemistry* **13,** 2853 (1974).
[64] M. Tanaka, M. Haniu, M. C. W. Evans, and K. K. Rao, *Biochemistry* **14,** 1938 (1975).
[65] T. Hase, H. Matsubara, and M. C. W. Evans, *J. Biochem. (Tokyo)* **81,** 1745 (1977).
[66] J. Travis, D. J. Newman, J. LeGall, and H. D. Peck, Jr., *Biochem. Biophys. Res. Commun.* **45,** 452 (1966).
[67] T. Hase, N. Ohmiya, M. Matsubara, R. N. Mullinger, K. K. Rao, and D. O. Hall, *Biochem. J.* **159,** 55 (1976).

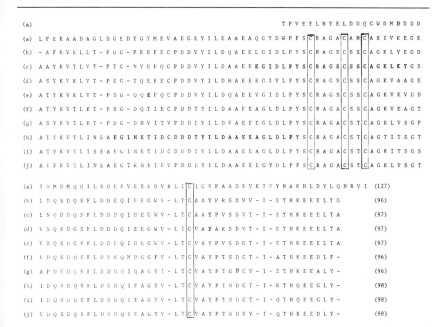

FIG. 2. Amino acid sequences of the 2Fe–2S type of ferredoxin. In the figure, (a)–(j) stand for the ferredoxin or iron–sulfur protein from (a) *Halobacterium halobium;* (b) *Lucaena glauca* (Koa); (c) spinach; (d) alfalfa; (e) taro; (f) *Scenedesmus,* green alga; (g) *Aphanothece sacrum,* blue-green alga; (h) *Spirulina platensis,* blue-green alga; (i) *Spirulina maxima,* blue-green alga; (j) *Nostoc mucorum,* blue-green alga. Cysteine residues chelated to iron are blocked off. See legend to Fig. 1 for other details.

*al.,*[68] but this 4Fe–4S iron–sulfur protein has neither the proper redox potential nor the biological activity to be classified as a ferredoxin.

*2Fe–2S Ferredoxin Sequences.* Ferredoxins from blue-green alga sequenced to date include the 2Fe–2S ferredoxins from *Spirulina maxima,*[69] *Spirulina platensis,*[70] *Nostoc mucorum,*[71] and *Aphanothece sacrum.*[72] The sequence of ferredoxin from the green alga, *Scenedesmus*[73] has also been determined and is shown in Fig. 2. Plant ferredoxins sequenced to

[68] K. Dus, S. Terdro, R. G. Bartsch, and M. D. Kamen, *Biochem. Biophys. Res. Commun.* **43**, 1239 (1971).

[69] M. Tanaka, M. Haniu, S. Zeitlin, K. T. Yasunobu, M. C. W. Evans, K. K. Rao, and D. O. Hall, *Biochem. Biophys. Res. Commun.* **64**, 399 (1975).

[70] M. Tanaka, M. Haniu, K. T. Yasunobu, K. K. Rao, and D. O. Hall, *Biochem. Biophys. Res. Commun.* **69**, 759 (1975).

[71] T. Hase, K. Wada, M. Ohmiya, and H. Matsubara, *J. Biochem. (Tokyo)* **80**, 993 (1976).

[72] T. Hase, K. Wada, and H. Matsubara, *J. Biochem. (Tokyo)* **79**, 329 (1976).

[73] K. Sugeno and H. Matsubara, *J. Biol. Chem.* **244**, 2979 (1969).

date include the proteins from spinach,[74] *Lucaena glauca*,[44] alfalfa,[75] and taro,[76] and the sequences are summarized in Fig. 2. The partial amino acid sequence of the *Equisetum* ferredoxin has been determined,[77] and this study established that cysteine residue 18 is not chelated to iron. Other 2Fe–2S non-ferredoxin iron–sulfur proteins whose sequences have been determined include adrenodoxin[78] from bovine adrenal glands and putidaredoxin[79] from *Pseudomonas putida*.

An excellent review of the three-dimensional structures and the active sites of the iron–sulfur proteins can be found in the article by Carter.[80] The type of evolutionary information contained in the amino acid sequences of the Fe–S proteins can be found in articles by Dayhoff[81] and Fitch and Yasunobu.[82]

Acknowledgments

These studies were supported in part by United States Public Health Service Grant No. GM 22556 and Grant GB 43448 from the National Science Foundation.

[74] H. Matsubara, R. M. Sasaki, and R. K. Chain, *Proc. Natl. Acad. Sci. U.S.A.* **57,** 439 (1967).

[75] S. Keresztes-Nagy, F. Perini, and E. Margoliash, *J. Biol. Chem.* **244,** 981–995 (1969).

[76] K. K. Rao and H. Matsubara, *Biochem. Biophys. Res. Commun.* **38,** 500 (1970).

[77] H. Kagamiyama, K. K. Rao, D. O. Hall, R. Cammack, and H. Matsubara, *Biochem. J.* **145,** 121 (1975).

[78] M. Tanaka, M. Haniu, K. T. Yasunobu, and T. Kimura, *J. Biol. Chem.* **248,** 1141 (1973).

[79] M. Tanaka, M. Haniu, K. T. Yasunobu, K. Dus, and I. C. Gunsalus, *J. Biol. Chem.* **249,** 3689 (1974).

[80] C. W. Carter, Jr., *in* "Iron-Sulfur Proteins" (W. Lovenberg, ed.), Vol. 3, p. 157. Academic Press, New York, 1977.

[81] M. O. Dayhoff, ed., "Atlas of Protein Sequence and Structure," Vol. 5, p. D-35. Natl. Biomed. Res. Fund., Washington, D.C., 1972.

[82] W. M. Fitch and K. T. Yasunobu, *J. Mol. Evol.* **5,** 1 (1975).

## [21] Bound Iron–Sulfur Centers in Photosynthetic Membranes (Higher Plants and Bacteria): Physical Detection and Occurrence

*By* RICHARD MALKIN and ALAN J. BEARDEN

Iron–sulfur proteins of the ferredoxin type are constituents of all photosynthetic cells. In addition to these soluble proteins, bound iron–sulfur groups have been identified in photosynthetic membranes. At this stage of characterization, the chemical nature of these groups is not known, and it therefore seems appropriate to refer to them in more general terms as iron–sulfur clusters or centers, since their association with specific proteins is not yet established.

Iron–sulfur centers in soluble proteins have oxidized minus reduced absorption bands in the visible spectral region (350 to 450 nm) that have been useful in their characterization, but such absorbance changes have been difficult to detect in membrane fragments from photosynthetic organisms. The presence of large amounts of chlorophyll with strong absorption in the 400-nm region makes observation of iron–sulfur bands almost impossible. Analytical procedures, such as those for the determination of nonheme iron and acid-labile sulfide, lack resolution in that one cannot distinguish on the basis of such measurements between functionally different centers. The physical technique that has been applied most successfully to the study of these centers is low-temperature electron paramagnetic resonance (EPR) spectroscopy. The basic principles of this technique have been described in a monograph[1] and in other chapters of these volumes.[2] In addition, a volume dealing specifically with experimental aspects of the EPR technique is available.[3]

### EPR Spectroscopy of Iron–Sulfur Centers

Reduced iron–sulfur centers display prominent EPR signals at temperatures below 80°K; these signals are caused by molecular antiferromagnetism between the high-spin iron(III) and high-spin iron(II) that gives rise to an $S = \frac{1}{2}$ molecular paramagnetism with characteristic EPR $g$ values centered below $g = 2.0$, the so-called "$g = 1.94$" type of EPR signal. This signal has a principal $g$-value in the range from 1.96 to 1.89, with $g = 1.94$ being the most commonly found. Reduced iron–sulfur centers in photosynthetic systems display either "rhombic" ($g_x \neq g_y \neq g_z$) or "axial" ($g_\perp \neq g_\parallel$) EPR spectra. Characterization of such centers generally has been based on one of the two or three "first-derivative" resonance lines, usually the central $g = 1.94$ line because of less interference from other paramagnetic species in this $g$-value region. The advantage of the EPR technique is that it affords a rapid, sensitive method for the unequivocal identification of an iron–sulfur center. It also can be used in quantitative experiments to determine the concentration of a center in any material. The EPR method also makes possible the detection of different types of centers in the same material because each

[1] H. M. Swartz, J. R. Bolton, and D. C. Borg, "Biological Applications of Electron Spin Resonance." Wiley (Interscience), New York, 1972.
[2] C. Poole, Jr., "Electron Spin Resonance-A Comprehensive Treatise on Experimental Techniques." Wiley (Interscience), New York, 1967.
[3] See Vol. 10 [93] and Vol. 24, [5].

usually will have different EPR properties. For a more complete discussion of theoretical aspects of the EPR properties of iron–sulfur centers, see Orme-Johnson and Sands.[4]

EPR spectroscopic analysis of reduced iron–sulfur centers is commonly performed on an X-band spectrometer equipped with a liquid-helium cryogenic system capable of achieving temperatures down to 4°K. With unfractionated chloroplasts from higher plants, chlorophyll concentrations of 1 to 3 mg/ml give signals sufficiently intense to be readily observed. With more enriched subchloroplast fragments, lower chlorophyll concentrations can be used with little loss of signal intensity. In chromatophores from photosynthetic bacteria, chlorophyll concentrations of 1 to 5 mg/ml are required, the concentration depending on the size of the photosynthetic unit in the respective organism. Usual experimental conditions are a field setting of 3400 ± 250 G (frequency approximately 9.20 GHz), a microwave power of 5 to 10 mW, a modulation amplitude of 5 to 10 G, and a temperature of approximately 15°K. Different centers have different saturation properties and different temperature sensitivities, and these should be kept in mind for the characterization of any centers.

To obtain EPR signals from bound iron–sulfur centers in photosynthetic membranes it is necessary to reduce these centers. Three techniques are available for reduction: (1) photochemical reduction at either physiological or cryogenic temperature, (2) complete chemical reduction by strong reducing agents, and (3) potentiometric reduction at a defined oxidation–reduction potential. The basic procedures of each of these techniques will be described.

*Photoreduction of Iron–Sulfur Centers.* The photoreduction of iron–sulfur centers utilizes the photosynthetic light-driven electron-transport chain. The procedure has been used extensively in the study of low-potential iron–sulfur centers of chloroplast photosystem I. Two types of reduction are possible: (a) reduction at cryogenic temperature and (b) reduction at physiological temperature. Photoreduction at cryogenic temperature can be accomplished only for components closely linked to primary photochemical processes. It is necessary to have a dark-adapted sample in a photochemically competent state prior to illumination, i.e., a state where the primary electron donor is reduced and the primary electron acceptor is oxidized. In chloroplast samples this is accomplished by dark incubation with a mild reductant (ascorbate, about 2 to 5 m$M$) and the inclusion of a mediator dye, such as 2,6-dichlorophenolindo-

[4] W. H. Orme-Johnson and R. H. Sands, *in* "Iron–Sulfur Proteins" (W. Lovenberg, ed.), Vol. 2, p. 195. Academic Press, New York, 1973.

phenol (DCIP). Samples are illuminated most conveniently directly in the EPR cavity at the optimum temperature of measurement (about 15° to 20°K for most iron–sulfur centers), and illumination in the cavity eliminates any errors that might arise because of placement, removal, and replacement of the sample. Either white or monochromatic light of high intensity ($\geq 10^5$ ergs/cm²/sec) is suitable for illumination. By direct-cavity illumination, it is possible to record the EPR spectrum with the light turned on as well as after cessation of illumination to check for reversibility of photoinduced signals.

The second photochemical procedure involves the addition of an electron-donor system, such as ascorbate plus DCIP, illuminating the sample at 300°K instead of 15°K for about 20 sec, and then freezing the sample to 77°K in the light. It is important to illuminate the sample during freezing to trap the centers in the reduced state because electron flow occurs under these conditions. Because reactions other than those related to primary processes can occur in these experiments, it is likely that components involved in secondary electron-transfer processes will also be photoreduced. Therefore, to differentiate between centers involved in the various electron-transfer processes, it is best to study photoreduction at cryogenic as well as physiological temperatures.

*Dark Chemical Reduction with Reducing Agents.* The most widely used reducing agents in photosynthesis research are sodium dithionite and sodium ascorbate. Sodium dithionite is a strong reducing agent that is used for the reduction of low-potential centers ($E_m \leq -200$ mV); ascorbate is a relatively mild reductant that can be used conveniently for higher potential centers ($E_m \geq +100$ mV).

Dithionite requires special precautions because of its reaction with oxygen in aqueous solution. Concentrated solutions (0.1 $M$) usually are made in alkaline solution (0.1 $M$ Tris-HCl buffer, pH 9.0, or 0.02 $M$ KOH) that have been deoxygenated by flushing with $N_2$ or Ar gas. Small amounts of the dithionite solution are removed and transferred to the EPR tubes containing the reaction mixture; it is most convenient to flush the sample with an inert gas while making dithionite additions. Mixing is accomplished with a long stainless steel needle that extends to the bottom of the EPR tube.

It also is possible to use hydrogen gas and a partially purified hydrogenase to obtain strongly reducing conditions. In this case, the EPR sample is equilibrated with hydrogen gas at a specific pH, and a small amount of crude clostridial hydrogenase[5] is added. Equilibration time is allowed (about 10 min is sufficient), and the effective oxidation–reduction

[5] K. Tagawa and D. I. Arnon, *Biochim. Biophys. Acta* **153,** 602 (1968).

potential of the system can be controlled by making changes in the pH of the medium.

In all of the above reactions, it is necessary to include low-molecular-weight dye mediators that make possible equilibration between the reductant and the membrane-bound carriers. With ascorbate, DCIP or phenazine methosulfate (PMS) are commonly used; with dithionite or hydrogen gas, methyl viologen is the mediator of choice.

*Oxidation–Reduction Potentiometry in Conjunction with EPR Spectroscopy.* An important advance in the characterization of iron–sulfur centers in photosynthetic systems has come from the application of oxidation–reduction potentiometry to these membrane systems. By this technique, it has been possible to distinguish multiple iron–sulfur centers in chloroplasts and bacterial chromatophores on the basis of their midpoint oxidation–reduction potentials ($E_m$). Theoretical considerations of this technique in relation to membrane-bound electron carriers have been described.[6]

For oxidation–reduction titrations, the titration assembly must permit introduction of a suitable oxidant or reductant, while the oxidation–reduction potential of the sample is continuously monitored. An inert gas atmosphere is desirable for titrations with oxygen-sensitive mediators and is essential for achievement of electronegative potentials. In addition, samples must be removed at the desired oxidation–reduction potential. A titration vessel supplied by Metrohm (Model EA 876-5) is ideal in that it is water-jacketed for temperature control, has a relatively small sample volume (5 to 10 ml), and can be fitted with a tight upper portion which has five ground-joint openings. The center opening is fitted with a ground-joint combination platinum microelectrode with a Ag/AgCl reference system (Metrohm Model EA-234). This electrode requires about 5 ml of sample to cover the sensitive portion. The remaining four ports in the electrode assembly are closed with puncture-type rubber closures through which (1) inert gas enters, (2) inert gas exits, (3) titrant is added, and (4) samples are removed under anaerobic conditions. A magnetic stirring bar is placed at the bottom of the vessel to stir the reaction mixture during the titration.

To ensure equilibration of the platinum electrode with the bound electron carriers in the chromatophore or chloroplast membrane, it is essential to include oxidation–reduction mediators in the potential region being investigated. Commonly used mediators and their midpoint potentials are the following: potassium ferricyanide (+430 mV), diaminodurene (+240 mV), $N,N,N',N'$-tetramethylphenylenediamine (+260 mV), 2,6-

[6] P. L. Dutton and D. F. Wilson, *Biochim. Biophys. Acta* **346**, 165 (1974).

dichlorophenolindophenol (+217 mV), 2,5-dimethylbenzoquinone (+180 mV), 1,2-naphthoquinoe (+145 mV), phenazine methosulfate (+80 mV), phenazine ethosulfate (+55 mV), 5-hydroxy-1,4-naphthoquinone (+30 mV), 1,4-naphthoquinone (+60 mV), duroquinone (+5 mV), pyocyanine (−35 mV), 2-hydroxy-1,4-naphthoquinone (−145 mV), potassium indigotetrasulfonate (−46 mV), potassium indigosulfonate (−125 mV), sodium anthraquinone-2-sulfonate (−225′mV), neutral red (−325 mV), benzyl viologen (−360 mV), methyl viologen (−440 mV), and Triquat (−520 mV).

When adding mediators, it is important that the concentration is sufficiently high to permit rapid equilibration of the bound carriers with the electrode, but the mediator must not interfere with subsequent EPR analysis, i.e., should not contribute significant signals in the spectral region under investigation. This is particularly important if one attempts to monitor electron carriers in both the $g = 2.00$ region and the $g = 1.94$ region. The $g = 2.00$ EPR signal of reaction center chlorophyll usually is obscured by signals from the mediator unless the mediator is very low in concentration (5 to 10 $\mu M$). Mediator interference in the $g = 1.94$ region of iron–sulfur centers is, however, not common; and concentrations of mediators in the range of 50 to 100 $\mu M$ usually do not cause any interference problems, yet they give good equilibration with the electrode.

The reaction mixture, containing an appropriate buffer and mediators, is flushed thoroughly with argon to remove oxygen before the addition of the photosynthetic sample. Traces of oxygen introduced with the sample are removed by the first addition of reductant and, therefore, are not a serious problem. Titrants commonly used are 0.1 $M$ sodium dithionite (in 0.02 $M$ KOH) for reductive titrations or 0.1 $M$ potassium ferricyanide for oxidative titrations. These are added in microliter amounts from a suitable microsyringe (syringe microburet Model 5B2; Micrometric Industrial Company, Cleveland, Ohio). The titrants are freshly prepared and made anaerobic by cycles of evacuation and flushing with argon prior to introduction into the titrant syringe.

The reaction mixture is adjusted to the desired potential by addition of an appropriate titrant. At each point, sufficient time is allowed for equilibration; during this period, little or no potential drift should occur and equilibration usually is complete in about 5 min. At this time, a 0.3-ml sample is removed with a 0.5-ml Hamilton gas-tight syringe with gas-flush modification which makes possible flushing of the syringe with argon prior to withdrawal of the sample. The syringe has a 10 inch needle to inject samples into calibrated 3-mm i.d. quartz EPR tubes which are continuously flushed with argon. The tubes are sealed with rubber clo-

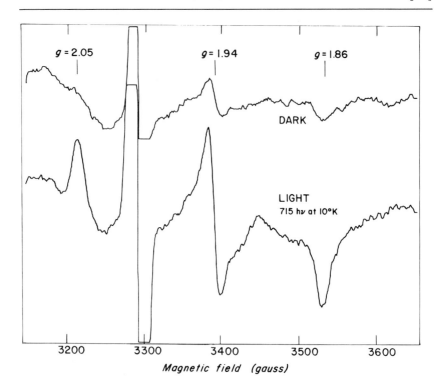

$g = 2.05$      $g = 1.94$      $g = 1.86$

DARK

LIGHT
715 h$\nu$ at 10°K

3200      3300      3400      3500      3600

*Magnetic field (gauss)*

FIG. 1. Photoreduction of 10°K of a bound iron–sulfur center in unfractionated spinach chloroplasts.

sures and frozen rapidly by immersion in a mixture of isopentane and methylcyclohexane (5 : 1, v/v) which has been precooled in liquid nitrogen to about 80°K.

It is possible to test the apparatus and the EPR response in several ways. The platinum electrode should be standardized against a saturated quinhydrone solution at pH 7.0 ($E_m$ = +296 mV). Components with known midpoint potential can then be titrated: benzyl viologen ($E_m$ = −369 mV, $g$ = 2.00 free-radical signal); soluble chloroplast ferredoxin ($E_m$ = −420 mV, $g$ = 1.96 iron–sulfur EPR signal). Titrations should be reversible and independent of mediator concentration.

The amplitude of the signal under investigation is then plotted against the measured oxidation–reduction potential. Because most iron–sulfur centers are involved in a one-electron transition, a Nernst plot derived from the data should give the midpoint potential of the center and an $n$ value of about 1.0. Values greater than unity usually indicate that the

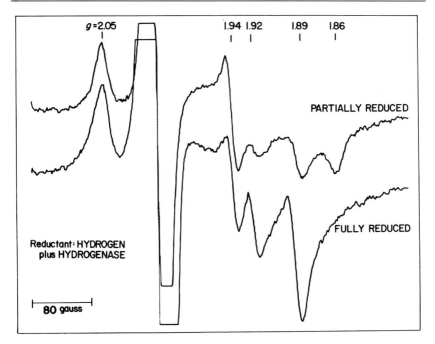

Fig. 2. Dark chemical reduction of bound iron–sulfur centers in unfractionated spinach chloroplasts. Reductant: Hydrogen gas in the presence of clostridial hydrogenase.

system is poorly equilibrated, i.e., the mediator concentration is too low. Such systems should be titrated with higher mediator concentrations.

## Occurrence of Membrane-Bound Iron–Sulfur Centers in Photosynthetic Systems

*Oxygen-Evolving Organisms.* Three different iron–sulfur centers have been detected by EPR spectroscopy at cryogenic temperatures in membrane fragments from oxygen-evolving organisms. Two of these are associated with the photosystem I reaction center.[7,8] As shown in Fig. 1, one center (center A; g values of 2.05, 1.94, and 1.86) is photoreducible at cryogenic temperatures after illumination with monochromatic light which activates photosystem I. The photoreduction of center A has been shown to be irreversibly linked to the photooxidation of the photosystem

[7] R. Malkin and A. J. Bearden, *Proc. Natl. Acad. Sci. U.S.A.* **68,** 16 (1971).
[8] A. J. Bearden and R. Malkin, *Brookhaven Symp. Biol.* **28,** 247 (1976).

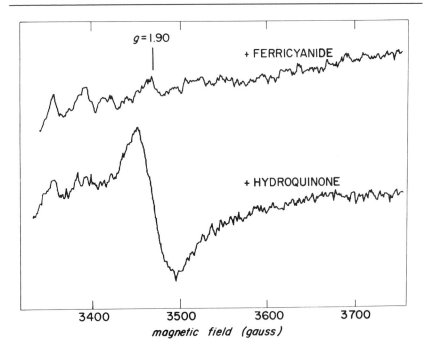

Fig. 3. Dark chemical reduction of the "Rieske" iron–sulfur center in unfractionated spinach chloroplasts. Reductant: hydroquinone. Taken from Malkin and Aparicio.[10]

I reaction center chlorophyll, P700.[9] Quantitative EPR studies of P700 and center A have demonstrated a stoichiometric relationship between P700$^+$ and the reduced bound iron–sulfur center.[9] Center A has been proposed to function as the stable primary electron acceptor of photosystem I.[8] As shown in Fig. 2, dark reduction with a strong reductant (hydrogen plus hydrogenase in the presence of methyl viologen, dithionite yields similar results) reduces an additional low-potential iron–sulfur center (center B; $g$ values of 2.05, 1.92, and 1.89) as well as center A. The function of center B in electron-transfer processes is not known.

The third iron–sulfur center in oxygen-evolving organisms has different EPR and oxidation–reduction properties.[10] As shown in Fig. 3, this center is reducible by hydroquinone and is characterized by a $g$ value of about 1.90. A similar center was first discovered in mitochondria by Rieske.[11] In chloroplasts, evidence has been obtained that this "Rieske-

[9] A. J. Bearden and R. Malkin, *Biochim. Biophys. Acta* **283,** 456 (1972).
[10] R. Malkin and P. J. Aparicio, *Biochem. Biophys. Res. Commun.* **63,** 1157 (1975).
[11] See Vol. 10 [64].

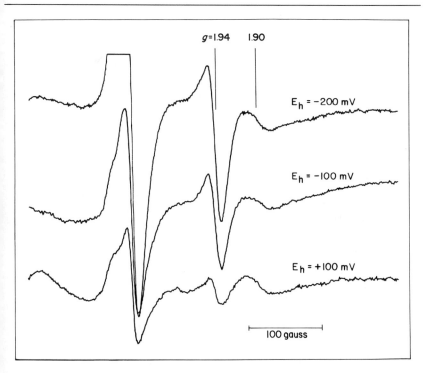

FIG. 4. Oxidation–reduction titration of bound iron–sulfur centers in *Chlorobium* chromatophores. Taken from Knaff and Malkin.[15]

type" iron–sulfur center functions as an electron carrier in the dark electron-transport chain between the two light reactions.[12]

The EPR and oxidation–reduction properties of these centers as well as those of soluble chloroplast ferredoxin are shown in Table I.[5,10,13,14]

*Chromatophore Fragments from Photosynthetic Bacteria.* Multiple iron–sulfur centers have been observed in chromatophore preparations from all photosynthetic bacteria studied. These centers differ in EPR properties as well as in oxidation–reduction properties. A typical potentiometric titration of the centers in *Chlorobium* chromatophores is shown in Fig. 4. Centers with $g$ values at 1.94 and 1.90 are found in these fragments.[15] Several centers characterized by $g = 1.94$ EPR signals are found in the following organisms: *Chlorobium* ($-25$, $-175$, and $\sim -550$

[12] R. Malkin and H. B. Posner, *Biochim. Biophys. Acta* **501,** 522 (1978).
[13] B. Ke, R. E. Hansen, and H. Beinert, *Proc. Natl. Acad. Sci. U.S.A.* **70,** 2941 (1973).
[14] M. C. W. Evans, S. G. Reeves, and R. Cammack, *FEBS Lett.* **49,** 111 (1974).
[15] D. B. Knaff and R. Malkin, *Biochim. Biophys. Acta* **430,** 244 (1976).

TABLE I

ELECTRON PARAMAGNETIC RESONANCE PARAMETERS AND OXIDATION–REDUCTION PROPERTIES OF CHLOROPLAST BOUND IRON–SULFUR CENTERS AND SOLUBLE CHLOROPLAST FERREDOXIN

| Iron-sulfur center or protein | $E_m$ (mV) | Principal g values | Linewidth (gauss)[a] | Optimum temperature for observation[b] (°K) | Optimum microwave power (mW) | Quantity detectable[c] (pmoles) |
|---|---|---|---|---|---|---|
| Bound Fe–S center A[d] | ~ −530[g] | 1.86,1.94,2.05 | 15–20 | 15–20 | 10–20 | 75 |
| Bound Fe–S center B | ~ −600[g] | 1.89,1.92,2.05 | 15–20 | 15–20 | 10–20 | 75 |
| "Rieske" Fe–S center | +290[h] | 1.78,1.89,2.02 | 40 | 15–30 | 10–20 | 250 |
| Soluble chloroplast Ferredoxin | −420[i] | 1.89,1.96,2.05 | 40[e] | 10–40 | 20–50 | 250 |

[a] Linewidth dependent on preparation; centers in chloroplasts display narrowest linewidth and increased in linewidth observed in purified photosystem I preparations.

[b] Assuming X-band (~9.2 GHz), cavity-Q (loaded) = 5000.

[c] X-band spectrometer, 15°K; power, 10 mW; 100 KHz modulation amplitude, 10 G (1 mtesla); cavity-Q (loaded), 5000; chlorophyll concentration, 1 mg/ml; reaction center content, 1/400 chlorophyll molecules.

[d] Photoreducible at cryogenic temperature.

[e] In contrast to bound centers, signal from soluble ferredoxin shows inhomogeneous broadening and linewidth depends on microwave power.

[f] Ke et al.[13]

[g] Evans et al.[14]

[h] Malkin and Aparicio.[10]

[i] See Tagawa and Arnon.[5]

TABLE II
OXIDATION–REDUCTION PROPERTIES OF THE $g$ = 1.89 "RIESKE" CENTER IN
PHOTOSYNTHETIC BACTERIA

| Organism | $E_m$ (mV) |
|---|---|
| *Chromatium* | +285 (pH 8.0)[a] |
| *Rhodopseudomonas sphaeroides* | +285 (pH 5.8–8.2)[b] |
| *Rhodopseudomonas capsulata* | +310 (pH 5.8–8.2)[c] |
| *Rhodospirillum rubrum* | +265 (pH 7.7)[d] |
| *Chlorobium* | +160 (pH 7.0)[e] |

[a] From Evans *et al.*[16]
[b] From Prince *et al.*[18]
[c] From Prince *et al.*[17]
[d] D. C. Yoch, unpublished observations (1977).
[e] From Knaff and Malkin.[15]

mV),[15] *Chromatium* (−50 and −290 mV),[16] *Rhodopseudomonas capsulata* (+30, −235, and −335 mV),[17] *Rhodopseudomonas sphaeroides* (+40, −200, and −350 mV),[18] and *Rhodospirillum rubrum* (+20, −175, and −390 mV).[19] Solubilization experiments and effect of various substrates has led to the identification of some centers with bacterial succinic dehydrogenase.[19,20]

In addition, chromatophores contain a $g$ = 1.89 "Rieske-type" center. The oxidation–reduction midpoint potential of the latter centers are shown in Table II. Although the midpoint potential of the centers was first reported to be independent of pH,[18] further experiments with *Chlorobium*,[15] *Rhodopseudomonas sphaeroides*,[21] and *Chromatium*[22] indicate that the center does have a pH-dependent potential and that it follows a −60 mV/pH unit dependence. The site of function of the "Rieske" center is not known, although it has been shown to undergo photooxidation in chromatophores that have been illuminated at physiological temperatures.[16,18]

[16] M. C. W. Evans, A. V. Lord, and S. G. Reeves, *Biochem. J.* **138**, 177 (1974).
[17] R. C. Prince, J. S. Leigh, Jr., and P. L. Dutton, *Biochem. Soc. Trans.* **2**, 950 (1974).
[18] R. C. Prince, J. G. Lindsay, and P. L. Dutton, *FEBS Lett.* **51**, 108 (1974).
[19] R. P. Carithers, D. C. Yoch, and D. I. Arnon, *J. Biol. Chem.* **252**, 7461 (1977).
[20] W. J. Ingledew and R. C. Prince, *Arch. Biochem. Biophys.* **178**, 303 (1977).
[21] R. C. Prince and P. L. Dutton, *FEBS Lett.* **65**, 117 (1976).
[22] R. C. Prince, D. B. Knaff, and R. Malkin, unpublished observations (1977).

# [22] Ferredoxin–NADP$^+$ Oxidoreductase

By GIULIANA ZANETTI and BRUNO CURTI

2 Reduced ferredoxin + NADP$^+$ ⇌ 2 oxidized ferredoxin + NADPH

The flavoprotein ferredoxin–NADP$^+$ reductase (EC 1.6.7.1) is a membrane-bound component of the photosynthetic electron-transport system. Extensive purification, crystallization, and kinetic and structural studies have been carried out on the spinach enzyme. More recently the enzyme has been isolated from other sources: *Pinus pinea*,[1] the alga *Bumilleriopsis filiformis*,[2] *Tsuga canadensis*,[3] and *Phaseolus vulgaris L.*[4] Immunochemical experiments suggest a localization for the spinach enzyme on the outer surface of the thylakoid membrane of the chloroplast.[5-7] A stoichiometric ratio of 5:3:4 for ferredoxin to reductase to plastocyanin respectively, has been determined per mole of P700 in spinach chloroplast.[7a] Several catalytic roles have been shown for this enzyme. They include (a) the photoreduction of NADP$^+$ by ferredoxin as electron donor[8]; (b) the reduction of NAD$^+$ by NADPH or its analogs (transhydrogenase)[9]; (c) the oxidation of NADPH by K$_3$Fe(CN)$_6$ or dichlorophenolindophenol (DCPIP)[10] or 2-(p-iodophenyl)-3-nitrophenyl-5-phenyltetrazolium chloride (INT)[11] (diaphorase); and (d) the reduction of cytochrome *f* by NADPH.[12]

## Assay Method

The more reliable assay methods involve the reduction of ferredoxin[13] or K$_3$Fe(CN)$_6$ by NADPH in the presence of a regenerating system;

[1] A. M. Fiorenzuoli, G. Ramponi, P. Vanni, and A. Zanobini, *Life Sci.* **7**, 905 (1968).

[2] P. Böger, *Z. Pflanzenphysiol.* **61**, 447 (1969).

[3] J. Riov and G. Brown, *Physiol. Plant.* **38**, 147 (1976).

[4] C. M. T. Sluiters-Scholten, W. A. W. Moll, and D. Stegwee, *Planta* **133**, 289 (1977).

[5] R. Berzborn, *Z. Naturforsch., Teil B* **23**, 1096 (1968).

[6] D. Hiedemann-van Wyk and C. Gamini Kannangara, *Z. Naturforsch., Teil B* **26**, 46 (1971).

[7] H. Böhme, *Eur. J. Biochem.* **72**, 283 (1977).

[7a] H. Böhme, *Eur. J. Biochem.* **83**, 137 (1978).

[8] M. Shin, K. Tagawa, and D. I. Arnon, *Biochem. Z.* **338**, 84 (1963).

[9] D. L. Keister, A. San Pietro, and F. E. Stolzenbach, *J. Biol. Chem.* **235**, 2898 (1960).

[10] M. Avron and A. T. Jagendorf, *Arch. Biochem. Biophys.* **65**, 475 (1956).

[11] W. W. Fredricks and J. M. Gehl, *Arch. Biochem. Biophys.* **174**, 666 (1976).

[12] G. Forti and E. Sturani, *Eur. J. Biochem.* **3**, 461 (1968).

[13] This series, Vol. 23 [40].

alternative procedures have been described according to the different catalytic roles of the enzyme.[13]

*Cytochrome c* Reduction Method

   *Reagents*
      Cytochrome $c$, 1.0 m$M$
      Ferredoxin, 0.1 m$M$ in Tris-HCl buffer 0.15 $M$, pH 7.3 (20°)
      NADPH, 10 m$M$ in Tris-HCl buffer, 0.1 $M$, pH 8.5 (20°)
      Tris-HCl buffer, 0.5 $M$, pH 7.8
   *Procedure.* In a cuvette, in a final volume of 1 ml, 0.1 ml of Tris-HCl buffer, 0.05 ml of cytochrome $c$, 0.1 ml of ferredoxin, and 20 $\mu$l of NADPH are mixed at 25°. The reaction is started by the addition of the enzyme and the increase in absorbance at 550 nm is measured.

*$K_3Fe(CN)_6$ Reduction Method*

   *Reagents*
      $K_3Fe(CN)_6$, 30 m$M$
      NADP⁺, 10 m$M$
      Glucose-6-phosphate, 50 m$M$
      Glucose-6-phosphate dehydrogenase, 1 mg/ml
      Tris-HCl, 0.5 $M$, pH 8.2 (20°)
   *Procedure.* In a 3 ml final volume at 25°, 0.5 ml of Tris-HCl buffer, 50 $\mu$l of NADP⁺, 0.2 ml of glucose-6-phosphate and 5 $\mu$l of glucose-6-phosphate dehydrogenase are preincubated for 1–2 min. The reaction is started by the addition of 0.1 ml of ferricyanide and the enzyme. The decrease in absorbance at 420 nm is measured.

Purification Procedure

   The present method of purification of ferredoxin–NADP⁺ reductase has the advantage of shortening the working time and avoiding the use of large volumes of acetone, without decreasing the yield and the specific activity. Alternative purification methods for the spinach enzyme have been published.[12–14] The present procedure is essentially the same as described by Gozzer *et al.*[15] All the manipulations are carried out at 4° unless otherwise stated.

   *Step 1.* Ten kilograms of spinach leaves are cleaned and rinsed in distilled water. Lots of 250 gm of leaves are ground for 20–30 sec at low

[14] M. T. Borchert and J. S. C. Wessels, *Biochim. Biophys. Acta* **197**, 78 (1970).
[15] C. Gozzer, G. Zanetti, M. Galliano, G. A. Sacchi, L. Minchiotti, and B. Curti, *Biochim. Biophys. Acta* **485**, 278 (1977).

speed and 75 sec at high speed in a Waring blender with 300 ml of 10 m$M$ potassium phosphate buffer, pH 7.4, containing 1 m$M$ EDTA, 1 m$M$ $\beta$-mercaptoethanol, and 10 $\mu M$ phenylmethylsulfonyl fluoride (PMSF). The homogenate is filtered through a double layer of cheese cloth to remove debris and the filtrate is centrifuged at 11,000 $g$ for 45 min.

*Step 2.* To the supernatant solid ammonium sulfate is slowly added to 35% saturation (209 gm/liter), being aware of maintaining the pH around 7.3–7.4 by addition of 100 m$M$ NaOH. After standing for 30 min, the suspension is centrifuged and the supernatant is brought to 70% saturation (238 gm/liter), the pH always being maintained at 7.3–7.4. After standing for 30 min, the suspension is centrifuged, and the precipitate is redissolved in a minimum volume of 5 m$M$ potassium phosphate, pH 7.4. The suspension is then exaustively dialyzed against several changes (about ten times with 5 liters) of the same buffer containing 10 $\mu M$ PMSF, to get rid of the yellow-green material, followed by two changes with 5 liters of 50 m$M$ potassium phosphate, pH 7.4, containing 10 $\mu M$ PMSF.

*Step 3.* After dialysis, the solution is clarified by centrifugation and an acetone fractionation is carried out on the supernatant. During this step the maintenance of a low temperature is critical for a good recovery of the enzyme activity. Precooled acetone at $-20°$ is added under continuous stirring to a final concentration of 45% by volume, while maintaining the temperature of the solution at $0°–2°$. The precipitate is removed by centrifugation, and cold acetone is again added to the supernatant to reach a final concentration of 75% by volume, the temperature being maintained at $-5°$ through the aid of an ice-salt bath. After standing for 60 min at the same temperature, the solution is centrifuged at 9500 $g$ for 50 min. The precipitate is collected, dissolved in a minimum volume of 35 m$M$ Tris-HCl buffer, pH 7.4 (20°), and dialyzed against the same medium.

*Step 4.* After dialysis, the solution is centrifuged, and the small precipitate is washed once with a minimum volume of the Tris-HCl buffer. The solution and the washing are applied to a DEAE–cellulose column (3.4 × 130 cm) previously equilibrated with 35 m$M$ Tris-HCl buffer pH 7.4 (20°). The effluent rate is set at about 50 ml/hr. The column is washed with 1.5–2 volumes of the same buffer, after which the molarity of the Tris-HCl is increased to 100 mM. The yellow band of the enzyme moves down the column and is eluted in a narrow sometime biphasic peak. The yellow fractions are collected, pooled, and precipitated with ammonium sulfate at 90% saturation. The precipitate, after centrifugation is dissolved in 10 m$M$ Tris-HCl buffer, pH 7.4 (20°), and dialyzed against the same medium.

*Step 5.* The dialyzed solution, after centrifugation, is applied to a Whatman P-11 phosphocellulose column (1.9 × 40 cm), previously equilibrated with 10 m$M$ Tris-HCl buffer pH 7.4 (20°). The column is then washed with 1.5–2 volumes of the same buffer to eliminate protein contaminants. The enzyme is eluted stepwise, first with 2.5 volumes of 100 m$M$ potassium phosphate, pH 7.4, then with 200 m$M$ of the same buffer. The yellow band comes out of the column generally in more than one peak the elution pattern corresponding to the enzyme multiple forms. The active fractions are collected and concentrated by precipitation with ammonium sulfate at 90% saturation. The precipitate is dissolved and dialyzed against 10 m$M$ Tris-HCl, pH 7.4 (20°); the concentrated enzyme can be stored at −20° for several months without substantial loss of activity. Ferredoxin could be purified as a side fraction from the DEAE-cellulose column, as described by Buchanan and Arnon.[16]

Properties[13]

Multiple molecular forms are present in the ferredoxin–NADP⁺ reductase from spinach.[11,15,17,18,18a] Five forms *a, b, c, d,* and *e* are resolved by isoelectric focusing with p$I$ values of 6.0, 5.5, 5.2, 5.0, and 4.8, respectively. Form *c* seems largely predominant over the other components. Two classes of molecular weights of 34,000 and 37,000 have been found for the enzyme forms *a, b,* and *c, d,* respectively. By different methods,[15] form *c* gives a value in the range of 37,000–42,000, although values between 33,000 and more than 120,000 have been reported for ferredoxin–NADP⁺ reductase.[11,19,19a] The existence of dimeric forms of the enzyme has also been recently reported.[19b] Amino acid analysis of the enzyme have been performed[12,15,19a]; the enzyme contains one disulfide and four SH groups per mole of FAD.[20] The flavoprotein purified from spinach has very little fluorescence in comparison to that shown by the free FAD (0.6%[21]). A method for the preparation of the partially active apoprotein upon reconstitution with FAD has been reported.[21a]

[16] This series, Vol. 23 [39].
[17] J. J. Keirns and J. H. Wang, *J. Biol. Chem.* **247**, 7374 (1972).
[18] G. Zanetti, *Biochim. Biophys. Acta* **445**, 14 (1976).
[18a] W. L. Ellfson and D. W. Krogmann, *Arch. Biochem. Biophys.* **194**, 593 (1979).
[19] R. Schneeman and D. W. Krogmann, *J. Biol. Chem.* **250**, 4965 (1975).
[19a] H. Hasumi and S. Nakamura, *J. Biochem. (Tokyo)* **84**, 707 (1978).
[19b] M. Shin and R. Oshino, *J. Biochem. (Tokyo)* **83**, 357 (1978).
[20] G. Zanetti and G. Forti, *J. Biol. Chem.* **244**, 4757 (1969).
[21] M. Shin, *Biochim. Biophys. Acta* **292**, 13 (1973).
[21a] G. Bookjans, A. San Pietro, and P. Böger, *Biochem. Biophys. Res. Commun.* **80**, 759 (1978).

The enzyme forms a neutral semiquinone upon anaerobic reduction with EDTA–light or with NADPH.[22] Two different values of the $\epsilon_{600}$ (1810 [23] and 3600 $M^{-1}$ cm$^{-1}$)[17] are reported for the semiquinone. Full reduction of the enzyme is difficult to achieve owing to its low redox potential, which has been measured at pH 7 to be $-0.360$ V[17]; two electron equivalents per flavin are required for the complete reduction of the flavoprotein.[22,23] Rapid reaction kinetics of the enzyme in turnover experiments at different wavelengths, with $K_3Fe(CN)_6$ as electron acceptor, showed the formation of an intermediate whose spectrum clearly indicates that the neutral semiquinone is a participant in the catalytic mechanism of the diaphorase reaction.[22] Studies carried out *in vivo* using Chlorella cells yielded spectral evidence for semiquinone formation during the light reaction.[22a,b] Steady state kinetics of the enzyme depend upon the type of electron acceptor used.[12,23,24] Cytochrome $f$ reduction gives, on the Lineweaver–Burk plot, a series of converging lines, whereas the kinetics of all the other electron acceptors show a series of parallel lines. It is still unclear whether or not this different behavior could be ascribed to different catalytic mechanisms. The $K_m$ values for NADPH in the ferredoxin reductase,[12] ferricyanide reductase,[17] DCPIP reductase,[25] and cytochrome $f$ reductase[12] reactions are 10, 10, 2.2, and 1.6 $\mu M$, respectively. $K_m$ and $V_{max}$ in the diaphorase reactions are strongly influenced by the ionic strength.[17,25] The $K_m$ for NADPH in the ferricyanide reductase reaction is dependent upon the pH, with a p$K_1 = 6.3$ and a p$K_2 = 8.9$.[17] Pyrophosphate is an inhibitor of ferredoxin-dependent photoreduction of the enzyme.[23] Brief preincubation of the flavoprotein with NADPH inhibits the diaphorase and the ferredoxin reductase activities, whereas under the same conditions the cytochrome $f$ reductase reaction is not influenced.[12,24] Ferredoxin shows a partially competitive inhibition with NADPH in the DCPIP reductase reaction.[25] Mercurials inactivate the enzyme and cause FAD dissociation.[12,17,20] NADPH enhances the enzyme inactivation; in contrast, NADP$^+$ decreases the rate of inactivation in the case of polar mercurials, whereas it increases the rate if nonpolar reagents are used.[20] The activation of diaphorase and

[22] V. Massey, R. G. Matthews, G. P. Foust, L. G. Howell, C. H. Williams, Jr., G. Zanetti and S. Ronchi, *in* "Pyridine Nucleotide Dependent Dehydrogenases" (H. Sund, ed.), p. 393. Springer-Verlag, Berlin and New York, 1970.

[22a] B. Bouges-Bocquet, *FEBS Lett.* **85**, 340 (1978).

[22b] B. Bouges-Bocquet, *FEBS Lett.* **94**, 95 (1978).

[23] G. Forti, B. A. Melandri, A. San Pietro, and B. Ke, *Arch. Biochem. Biophys.* **140**, 10 (1970).

[24] G. Zanetti and G. Forti, *J. Biol. Chem.* **241**, 279 (1966).

[25] S. Nakamura and T. Kimura, *J. Biol. Chem.* **246**, 6235 (1971).

transhydrogenase activities by stoichiometric amounts of ferredoxin has been reported[25–27]; polylysine and ammonium ions are also activators of the ferredoxin–NADP$^+$ reductase.[19,28] The complex formation between enzyme and ferredoxin[29] has been confirmed by CD spectra[21,30] as well as by gel filtration[21] and by isoelectric focusing[18] (p$I$ of the complex = 4.4 [28]). One sulfhydryl group seems to be essential for the catalytic activity of the enzyme.[20] A lysyl residue with a p$K_a$ of 8.7 has been identified at the pyridine nucleotide binding site of the enzyme by the dansyl chloride reaction.[18] A role for arginyl residues in the active site region has been demonstrated for the spinach enzyme[31] as well as for the algal ferredoxin–NADP$^+$ reductase.[32,33]

[26] W. W. Fredricks and J. M. Gehl, *J. Biol. Chem.* **246**, 1201 (1971).
[27] P. Böger, *Planta* **99**, 319 (1971).
[28] G. Zanetti, unpublished observations.
[29] G. P. Foust, S. G. Mayhew, and V. Massey, *J. Biol. Chem.* **244**, 964 (1969).
[30] R. Cammack, J. Neumann, N. Nelson, and D. O. Hall, *Biochem. Biophys. Res. Commun.* **42**, 292 (1971).
[31] G. Zanetti, C. Gozzer, G. Sacchi, and B. Curti, *Biochim. Biophys. Acta* **569**, 127 (1979).
[32] G. Bookjans and P. Böger, *Arch. Biochem. Biophys.* **190**, 459 (1978).
[33] G. Bookjans and P. Böger, *Arch. Biochem. Biophys.* **194**, 387 (1979).

# [23] Ferredoxin–Nitrite Reductase

*By* JOSÉ M. VEGA, JACOBO CÁRDENAS, and MANUEL LOSADA

$$NO_2^- + 6\ e + 8\ H^+ \rightarrow NH_4^+ + 2\ H_2O$$

Ferredoxin–nitrite reductase (EC 1.7.7.1) is the second enzyme component of the photosynthetic nitrate-reducing system. It was first identified by Losada *et al.*[1,2] as a ferredoxin-dependent chloroplast enzyme which catalyzes the 6-electron reduction of nitrite to ammonia. During the last years, several comprehensive reviews dealing with the assimilatory reduction of nitrate have been published.[3–5]

[1] M. Losada, A. Paneque, J. M. Ramirez, and F. F. del Campo, *Biochem. Biophys. Res. Commun.* **10**, 298 (1963).
[2] J. M. Ramirez, F. F. del Campo, A. Paneque, and M. Losada, *Biochim. Biophys. Acta* **118**, 58 (1966).
[3] M. Losada and A. Paneque, Vol. 23 [44].
[4] E. J. Hewitt, *Annu. Rev. Plant Physiol.* **26**, 73 (1975).
[5] M. Losada, *J. Mol. Catal.* **1**, 245 (1975–1976).

## Assay Methods

*Dithionite Assay*

*Principle.* The usual assay involves sodium dithionite as reductant and either ferredoxin or its artificial substitute, methyl viologen, as electron carrier.[2] Enzymatic activity can be best followed by measuring colorimetrically the rate of disappearance of nitrite.[6]

*Reagents*

Tris-HCl buffer, 0.5 $M$, pH 8.0
Sodium nitrite, 20 m$M$
Ferredoxin,[7] 5 m$M$ (for routine assay of the enzyme, 10 m$M$ methyl viologen can be used)
Sodium dithionite, 25 mg/ml in 0.29 $M$ NaHCO$_3$
Diazo-coupling reagents, 1% sulfanilamide in 3 $M$ HCl, and 0.02% $N$-(1-naphthyl)ethylenediamine hydrochloride[6]

*Procedure.* The reaction is run in open test tubes. Mix 0.3 ml of Tris buffer, 0.2 ml of sodium nitrite, either 0.2 ml of ferredoxin or 0.15 ml of methyl viologen, 0.1–0.2 unit of enzyme (see definition below), 0.3 ml of fresh dithionite solution, and water to make up 2 ml. After incubation for 10 min at 30°, the reaction is stopped by vigorous shaking in a Cyclomixer until the dithionite is completely oxidized and the dye becomes colorless. One milliliter of each of the diazo coupling reagents is then added to a 3-ml aliquot of a 100-fold dilution of the reaction mixture. After 10 min, the absorbance of the solution is determined at 540 nm, and the nitrite content is calculated from a standard curve. Minor variants of this procedure have been described.[8–11]

*Definition of Unit and Specific Activity.* One unit of activity is defined as the amount of enzyme which catalyzes the reduction of 1 $\mu$mole of nitrite per minute, under the standard assay conditions. Specific activity is expressed as units per milligram of protein. The protein content of the enzyme preparation is determined by the method of Lowry *et al.*[12]

---

[6] F. D. Snell and C. T. Snell, "Colorimetric Methods of Analysis," 3rd ed., Vol. 2. Van Nostrand Reinhold, Princeton, New Jersey, 1949.

[7] For the preparation of ferredoxin, see this series, Vol. 23 [39].

[8] K. W. Joy and R. H. Hageman, *Biochem. J.* **100**, 263 (1966).

[9] A. Hattori and I. Uesugi, *Plant Cell Physiol.* **9**, 689 (1968).

[10] W. G. Zumft, *Biochim. Biophys. Acta* **276**, 363 (1972).

[11] S. Ida and Y. Morita, *Plant Cell Physiol.* **14**, 661 (1973).

[12] O. H. Lowry, N. J. Rosebrough, A. L. Farr, and R. J. Randall, *J. Biol. Chem.* **193**, 265 (1951).

*Reduced Methyl Viologen Assay*

*Principle*. This assay follows the nitrite reductase activity of a purified preparation by measuring colorimetrically at 604 nm the rate of nitrite-dependent oxidation of reduced methyl viologen (MVH),[13] which can be prepared with $H_2$ and platinum asbestos as described by Siegel *et al.*[14]

*Reagents*
Potassium phosphate buffer, 1 *M*, pH 7.7
Potassium nitrite, 10 m*M*
MVH, 10 m*M*, in 50 m*M* potassium phosphate buffer, pH 7.7

*Procedure*. The reaction is run under anaerobic conditions in Thunberg cuvettes fitted with serum caps; 0.25 ml of phosphate buffer, 0.25 ml of potassium nitrite, and 1.8 ml of water are placed in the main compartment of a Thunberg cuvette, and 0.1 ml of enzyme solution is added to the side arm. The system is bubbled with $O_2$-free $N_2$ for 15 min. The enzyme is then tipped in and 0.1 ml of MVH is added with a gas-tight Hamilton syringe to start the reaction. Control mixture contains buffer instead of electron acceptor. Nonenzymatic oxidation of MVH in the absence or presence of nitrite is usually small. Molar extinction coefficient for MVH[15] at 604 nm is $1.14 \times 10^4 \ M^{-1} \ cm^{-1}$.

*NADPH Assay*

*Principle*. This assay is based on the use of NADPH and ferredoxin–NADP reductase[16] as electron-donor system to reduce ferredoxin, and follows nitrite reductase activity by measuring colorimetrically the rate of disappearance of nitrite.[6]

*Reagents*
Tris-HCl buffer, 0.5 *M*, pH 8.0
Sodium nitrite, 20 m*M*
Ferredoxin,[7] 5 m*M* in 50 m*M* Tris-HCl, pH 8.0
Ferredoxin–NADP reductase,[16] 1 m*M*, in 50 m*M* Tris-HCl, pH 8.0
NADPH, 20 m*M*
Diazo coupling reagents (see above)

*Procedure*. The reaction is carried out under anaerobic conditions in test tubes fitted with rubber serum caps. Mix 0.3 ml of Tris buffer, 0.2

---

[13] M. J. Murphy, L. M. Siegel, S. R. Tove, and H. Kamin, *Proc. Natl. Acad. Sci. U.S.A.* **71**, 612 (1974).

[14] L. M. Siegel, P. S. Davies, and H. Kamin, *J. Biol. Chem.* **249**, 1572 (1974).

[15] K. K. Eisenstein and J. H. Wang, *J. Biol. Chem.* **244**, 1720 (1969).

[16] For the preparation of ferredoxin-NADP reductase, see this series, Vol. 23 [40].

ml of nitrite, 0.2 ml of ferredoxin, 0.2 ml of ferredoxin–NADP reductase, about 0.1 unit of nitrite reductase, and water up to 1.8 ml. The system is made up anaerobically, and then 0.2 ml of the anaerobic solution of NADPH is added with a gas-tight Hamilton syringe to start the reaction. Control mixture contains buffer in place of nitrite reductase. After incubation for 10 min at 30°, nitrite is determined as indicated above.

## Purification Procedure

Nitrite reductases from spinach,[17,18] calabash,[19] *Curcubita pepo*,[20] and *Chlorella*[10] have been purified to homogeneity.

The purification procedure presented below has been described for the nitrite reductase of spinach leaves.[18] Unless otherwise indicated, all steps were performed at 0°–4°. The standard buffer was 10 m$M$ Tris-HCl, pH 8.0. Saturation of ammonium sulfate solutions were calculated as described in this series.[21]

*Step 1: Extraction.* Thirty kilograms of fresh spinach leaves were treated with 30 liters of standard buffer in a Waring blender, using 1 kg batches and blending for 3 min at high speed. The homogenate was forced through eight layers of cheesecloth and allowed to drain overnight.

*Step 2: Acetone Fractionation.* Acetone at −10° was added to cold crude filtrate in a −10° room until 35% (v/v) final concentration, i.e., 538 ml acetone per liter of filtrate. After 15 min stirring, the suspension was centrifuged at 10,000 rpm for 10 min, and the resulting precipitate was discarded. Acetone was then added to the supernatant to give a final concentration of 70% (v/v), i.e., 1.167 liters acetone per liter supernatant, and the suspension was stirred for 15 min and allowed to stand at −10° for 1–2 hr for the precipitate to settle. Most of the liquid phase was decanted and the precipitate was removed by centrifugation at 10,000 rpm for 10 min. This pellet was carefully resuspended in 3 liters (1 liter per 10 kg of fresh leaves) of standard buffer, 100 m$M$ in NaCl. After stirring for 2 hr at 0°, the pellet was collected and discarded by centrifugation at 10,000 rpm for 10 min. The supernatant was dialyzed against

[17] J. Cárdenas, J. L. Barea, J. Rivas, and C. G. Moreno, *FEBS Lett.* **23**, 131 (1972).

[18] J. M. Vega and H. Kamin, *J. Biol. Chem.* **252**, 896 (1977).

[19] J. Cárdenas, J. Rivas, and J. L. Barea, *Rev. R. Acad. Cienc. Exactas, Fis. Nat. Madrid* **66**, 565 (1972).

[20] D. P. Hucklesby, D. M. James, M. J. Banwell, and E. J. Hewitt, *Phytochemistry* **15**, 599 (1976).

[21] A. A. Green and W. L. Hughers, Vol. 1 [10].

20-liter batches of standard buffer, 200 m$M$ in NaCl, first for 6 and then for 14 hr.

*Step 3: First DEAE–Cellulose Treatment and First Ammonium Sulfate Fractionation.* The above solution was passed through a DEAE–cellulose column (4 × 25 cm) equilibrated with standard buffer, 200 m$M$ in NaCl, to remove ferredoxin and other acidic proteins. Solid ammonium sulfate, 242 gm/liter of eluate, was added slowly with stirring over 30 min. The precipitate was removed by centrifugation at 10,000 rpm for 10 min, and 241 gm of solid ammonium sulfate per liter of supernatant were added slowly, maintaining the pH of 8.0 by addition of 1 $M$ Tris solution. After 1 hr of stirring at 0°, the precipitate was collected by centrifugation for 10 min at 10,000 rpm and dissolved in a minimum amount of standard buffer. The enzyme solution was dialyzed twice against 20 liters of standard buffer containing 100 m$M$ NaCl, for 20 hr; dialysis buffer was changed after the first 6 hr.

*Step 4: DEAE–Cellulose Chromatography.* Nitrite reductase from step 3 was adsorbed on a DEAE–cellulose column (3.5 × 30 cm) previously equilibrated with standard buffer, 100 m$M$ in NaCl. After washing the bed column with 500 ml of standard buffer, 135 m$M$ in NaCl, nitrite reductase was eluted in 20 ml fractions with standard buffer, 170 m$M$ in NaCl, at a rate of 60 ml/hr.

*Step 5: DEAE–Sephadex A-50 Chromatography.* Pooled step 4 fractions with nitrite reductase activity were concentrated to half-volume by ultrafiltration with an Amicon concentrator with a Diaflo PM-10 membrane and diluted with standard buffer to a final NaCl concentration of 100 m$M$. The enzyme solution was then applied to a DEAE-Sephadex A-50 column (3 × 25 cm), previously equilibrated with standard buffer, 135 m$M$ in NaCl. After washing the bed column overnight with the same buffer, nitrite reductase was eluted in 20 ml fractions with standard buffer, 170 m$M$ in NaCl, at a rate of 40 ml/hr.

*Step 6: Second Ammonium Sulfate Fractionation.* For each 100 ml of the combined active fractions, 56.1 gm of solid ammonium sulfate were added, maintaining the pH at 7.5–8.0 by addition of 1 $M$ Tris. After 1 hr stirring at 0°, the precipitate was collected by centrifugation for 20 min at 10,000 rpm. The supernatant was discarded and the sediment resuspended in 15 ml of a 67% saturated ammonium sulfate solution in 0.1 $M$ potassium phosphate buffer, pH 7.7. After centrifugation, the sediment was extracted for 15 min with 10 ml of a 55% saturated ammonium sulfate solution in 0.1 $M$ potassium phosphate buffer, pH 7.7. The bulk of nitrite reductase was recovered after three extractions. Solid ammonium sulfate was added to the extracted enzyme solution up to 70% saturation, final concentration, maintaining the pH at 7.5–8.0 by addition of 1 $M$ Tris solution. After stirring 30 min at 0°, the precipitated protein was collected by centrifugation at 15,000 rpm for 15 min and dissolved in a minimum volume of 0.1 $M$ potassium phosphate, pH 7.7.

*Step 7: Chromatography on Sephadex G-200.* Three to five milliliters of the nitrite reductase solution from step 6 were filtered (downward) through a Sephadex G-200 column (2.6 × 105 cm), equilibrated with 0.1 $M$ potassium phosphate buffer, pH 7.7, at a rate of 13 ml/hr, and fractions were collected at 10-min intervals. Nitrite reductase activity eluted with the major peak of protein. A nearly constant specific activity was observed throughout most of the peak. Fractions of 90 $U/A_{280}$ specific activity or greater were pooled. A constant $A_{280}/A_{386}$ ratio of about 2.0 was observed in the fractions of the major protein peak. The best preparation had an $A_{280}/A_{386}$ ratio of 1.8.

The purification procedure, summarized in Table I, yielded 31 mg of nitrite reductase from the original 30 kg of fresh spinach leaves, with a total purification of about 2000-fold, a yield of 13% and a specific activity of 108 U/mg, which corresponds to a turnover number of 6588 min$^{-1}$.

*Ferredoxin–Sepharose Affinity Chromatography for the Purification of Higher Plants Nitrite Reductase.* Ida et al.[22] have described a method for purification of spinach nitrite reductase which includes 40–75% ammonium sulfate fractionation, DEAE–cellulose chromatography, second ammonium sulfate fractionation, DEAE–Sephadex A-50 and ferredoxin-Sepharose 4B chromatography. The final enzyme preparation had 70 U/mg specific activity, $A_{276}/A_{386}$ 5.0, with a total purification of 1045-fold and a yield of 24%.

### Properties

*Stability.* The purified enzyme can be stored in the deep-freeze for several months with no loss in activity. Purified spinach nitrite reductase stored in 50% glycerol at −20° was stable for at least 6 months.[11] The enzyme solutions in 0.1 $M$ potassium phosphate buffer, pH 7.7 are stable at 4° for 1 week; concentrated solutions are more stable than diluted ones. The half-time of nitrite reductase activity in phosphate buffer solution (5 mg protein/ml) at 25° is 30–35 hr. Heating of this solution at 60° for 5 min causes a total loss of the nitrite reductase activity.[2,8,9,23] Repetitive freezing-thawing of a nitrite reductase solution, as well as prolonged dialysis against low ionic strength buffer, may produce a loss of

[22] S. Ida, K. Kobayakawa, and Y. Morita, *FEBS Lett.* **65,** 305 (1976).
[23] C. H. Ho and G. Tamura, *Agric. Biol. Chem.* **37,** 37 (1973).

TABLE I

PURIFICATION OF NITRITE REDUCTASE FROM SPINACH LEAVES[a]

| Step and fraction | Volume (ml) | Protein (gm) | Activity (units) | Specific activity (U/mg) | Yield (%) | $\dfrac{A_{280}}{A_{386}}$ | $\dfrac{A_{386}}{A_{573}}$ |
|---|---|---|---|---|---|---|---|
| 1. Homogenate | 40,000 | 440 | 25,320 | 0.057 | 100 | — | — |
| 2. (35–70%) acetone ppt | 3,250 | 35 | 19,496 | 0.557 | 77 | — | — |
| 3. (40–70%) $(NH_4)_2SO_4$ ppt | 600 | 13.74 | 12,407 | 0.903 | 49 | — | — |
| 4. DEAE–cellulose eluate | 820 | 1.746 | 10,887 | 7.37 | 43 | 11.43 | 4.34 |
| 5. DEAE–Sephadex A-50 eluate | 1,600 | 0.246 | 7,089 | 28.82 | 28 | 3.77 | 3.67 |
| 6. (55–67%) $(NH_4)_2SO_4$ ppt | 66 | 0.046 | 4,304 | 93.56 | 17 | 2.76 | 3.69 |
| 7. Sephadex G-200 eluate | 35 | 0.031 | 3,342 | 107.80 | 13 | 2.00 | 3.66 |

[a] From Vega and Kamin,[18] and Lancaster et al. (see addendum).

activity and of absorption at 386 nm.[23] The nitrite reductase from *Chlorella* is unstable in crude extracts, and it is rapidly inactivated in solutions below pH 7.0.[10]

*Substrate Affinities.* A $K_m$ for nitrite, when ferredoxin or methyl viologen reduced with dithionite served as electron donor, of the order of 0.1 to 0.6 m$M$ has been reported for the spinach,[2,8,11,24] maize[8] and *Dunaliella tertiolecta*[25] enzymes. Lower values, 1 and 50 $\mu M$, have been reported for the *Cucurbita pepo*[26] and *Anabaena cylindrica*[9] enzymes, respectively. Exceptionally higher values, 2 m$M$ for the enzyme of barley root and leaf[27] and 40 m$M$ for the enzyme of *Citrus* leaves,[28] can be found in the literature. Ten times higher $K_m$ values are reported for hydroxyalmine.[4] $K_m$ values of the spinach enzyme for ferredoxin are usually between 10 and 70 $\mu M$,[2,23] and for methyl viologen, 70 $\mu M$.[2,11] A value of 60 $\mu M$ was determined by Cresswell *et al.*[26] for benzyl viologen with the *Curcurbita pepo* nitrite reductase.

*Specificity.* Purified ferredoxin–nitrite reductase catalyzes the stoichiometric reduction of nitrite to ammonia. There is a marked specificity for reduced ferredoxin as the natural electron donor, and the purified enzyme is inactive with either reduced pyridine nucleotides or reduced flavin nucleotides as electron donors.[1,2,29] Among the artificial substitutes for ferredoxin, methyl viologen is the most effective one examined.[2,8,9] Physiologically, the following electron donor systems are effective as the source of reducing power for the reaction catalyzed by different ferredoxin–nitrite reductases: Illuminated chlorophyll-containing particles, NADPH plus ferredoxin–NADP reductase, and $H_2$-hydrogenase.[1,2,8,9,29–32] In the presence of fresh grana, ferredoxin, and nitrite reductase, the light-dependent reduction of nitrite to ammonia is coupled to the evolution of oxygen and to the formation of ATP. The molar ratio between nitrite reduced, ammonia formed, orthophosphate incorporated, and oxygen evolved is 1:1:3:3.[29]

The *Chlorella* enzyme also functions with illuminated chloroplasts

[24] J. Shimizu and G. Tamura, *J. Biochem.* (*Tokyo*) **75**, 999 (1974).

[25] B. R. Grant, *Plant Cell Physiol.* **11**, 55 (1970).

[26] C. F. Cresswell, R. H. Hageman, E. J. Hewit, and D. P. Hucklesby, *Biochem. J.* **94**, 40 (1965).

[27] W. F. Bourne and B. J. Miflin, *Planta* **111**, 47 (1973).

[28] A. Shaked, A. Bar-Akiva, and K. Mendel, *Plant Cell Physiol.* **14**, 1151 (1973).

[29] A. Paneque, J. M. Ramirez, F. F. del Campo, and M. Losada, *J. Biol. Chem.* **239**, 1737 (1964).

[30] M. G. Guerrero, C. Manzano, and M. Losada, *Plant Sci. Lett.* **3**, 273 (1974).

[31] C. Manzano, P. Candau, C. Gómez-Moreno, A. M. Relimpio, and M. Losada, *Mol. Cell. Biochem.* **10**, 161 (1976).

[32] P. Candau, C. Manzano, and M. Losada, *Nature* (*London*) **262**, 715 (1976).

and flavodoxin,[10] a low-molecular-weight flavoprotein produced in response to iron deficiency.[33]

Nitrite reductase also catalyzes the reduction of hydroxylamine to ammonia. However, no sulfite reductase activity has been found in pure enzyme preparations.[18] Highly purified enzyme preparations vary in their ability to reduce hydroxylamine. The ratio of nitrite to hydroxylamine reductase activity using ferredoxin or methyl viologen as electron donor ranges between 6:1 in *Chlorella* and 100:1 in marrow.[4]

*Molecular Weight.* Spinach nitrite reductase has a molecular weight between 60,000 and 63,000 determined by gel filtration,[17,18,23,34] sedimentation studies, or amino acid composition.[18] The enzyme appears to be a single polypeptide chain[18]; alternatively, it has been proposed to consist of two indentical subunits of 37,000 daltons each.[11] The enzymes from different sources have molecular weights similar to the spinach enzyme as determined in marrow,[19,34,35] *A. cylindrica,*[9] *Chlorella,*[10] or *Dunaliella tertiolecta.*[25] An homogeneous preparation from *Chlorella* [10] consists of two electrophoretically distinguishable proteins, both independently capable of nitrite reduction and also of hydroxylamine reduction but at a lower rate. Sedimentation coefficients ($s_{20,w}^{\circ}$) of 4.26,[18] 4.90,[17] 4.57 S[11] and a Stokes radius ($R_s$) of 33.5 Å[18] have been reported for the spinach enzyme.

*Inhibitors.* The inhibition by potassium cyanide appears to be of the competitive type, $K_i = 0.37 \mu M$, with respect to nitrite for the *Cucurbita pepo* nitrite reductase. [19] Cyanide inhibition of nitrite reductase activity has been also reported for the spinach enzyme.[2,18,23]

Carbon monoxide inhibits nitrite reductase of higher plants and forms a complex with the reduced enzyme.[18,20] Formation or dissociation of the spectrophotometrically detectable CO complex with the spinach enzyme correlates with inhibition or inhibition reversal of nitrite reduction.[18] Nitrite, hydroxylamine, sulfite, or cyanide prevent the reaction of nitrite reductase with CO and, therefore, impede CO inhibition of enzyme activity.[18] Carbon monoxide also inhibits *Chlorella fusca* nitrite reductase after prolonged treatment.[10]

*p*-Chloromercuribenzoate (pCMB) has been reported to inhibit nitrite reductase from higher plants[18,26] and *Chlorella.*[10] Titration of pure spinach enzyme with pCMB has shown that maximal mercaptide formation and enzyme inactivation are achieved at 6 moles pCMB/mole nitrite reductase in the preincubation mixture. In addition, the absorption spectrum of the

[33] W. G. Zumft and H. Spiller, *Biochem. Biophys. Res. Commun.* **45,** 112 (1971).
[34] E. J. Hewitt and D. P. Hucklesby, *Biochem. Biophys. Res. Commun.* **25,** 689 (1966).
[35] D. P. Hucklesby and E. J. Hewitt, *Biochem. J.* **119,** 615 (1970).

pCMB-treated enzyme is markedly altered in the absorption band corresponding to the siroheme prosthetic group. Enzyme activity or native spectrum cannot be recovered by dialyzing the pCMB-treated enzyme against 0.1 $M$ potassium phosphate buffer, pH 7.0, in the presence of 0.2 m$M$ reduced glutathione. Titration of the enzyme with 5,5-dithiobis(2-nitrobenzoic acid) (DTNB) under denaturing conditions reveals that 8 SH groups/enzyme molecule may react with DTNB indicating that spinach nitrite reductase has two classes of cysteines: 6 available to pCMB and 2 accessible only under reducing denaturing conditions.[18] *Cucurbita pepo* nitrite reductase is inactivated by treatment with mersalyl, and the absorption spectrum of the enzyme is also altered in the bands corresponding to siroheme. After 30–60 min treatment, the enzyme inhibition cannot be reversed.[20]

*Intracellular Location.* Isolated chloroplasts contain nitrite reductase.[1,2,36,37] After breaking the chloroplast, most of the enzyme is recovered in the chloroplast extract, and only a small part remains bound to the grana.[2] In the blue-green alga, *Anacystis nidulans,* nitrite reductase is associated with the chlorophyll-containing particles.[30–32]

*pH Optimum.* Nitrite reductase from spinach,[2,8,11] *Cucurbita pepo,*[35] *Chlorella fusca,*[10] *Dunaliella tertiolecta,*[25] and *Anabaena cylindrica*[9] shows a broad range of optimum pH for activity between 7.0 and 8.0.

*Absorption Spectrum.* Plant and algal nitrite reductases are reddish-brown and show similar absorption spectrum, characteristic of a heme protein.[10,13,17,22,23,27] Spinach nitrite reductase exhibits an absorption spectrum with wavelength maxima at 276, 386 (Soret), 573 ($\alpha$), and 690 nm with shoulders at 290 and 640 nm (Fig. 1), and $A_{276}/A_{386}$ absorptivity ratio of 1.8. The extinction coefficients of the enzyme, measuring the protein concentration by dry weight and using 61,000 as molecular weight, are $7.15 \times 10^4 \, M^{-1} \, cm^{-1}$ at 276 nm, $3.97 \times 10^4 \, M^{-1} \, cm^{-1}$ at 386 nm, and $1.00 \times 10^4 \, M^{-1} \, cm^{-1}$ at 573 nm.[18]

Anaerobic addition of dithionite, which alone is a relatively inefficient donor, to spinach nitrite reductase results in a bleaching of the 573 and 690 nm bands and the appearance of two new peaks in this region at 545 and 585 nm. After oxidation of dithionite by gently shaking the enzyme preparation with air, the absorption band at 573 nm, but not at 690 nm, is recovered. In addition, the Soret band of native enzyme is shifted from 386 to 400 nm.[18] These data and those corresponding to the absorption spectrum of nitrite reductase from *Cucurbita pepo*[20] and *Chlorella fusca*[10] are presented in Table II.

[36] G. F. Betts and E. J. Hewitt, *Nature (London)* **210,** 1327 (1966).
[37] G. L. Ritenour, K. W. Joy, J. Bunnin, and R. H. Hageman, *Plant Physiol.* **42,** 233 (1967).

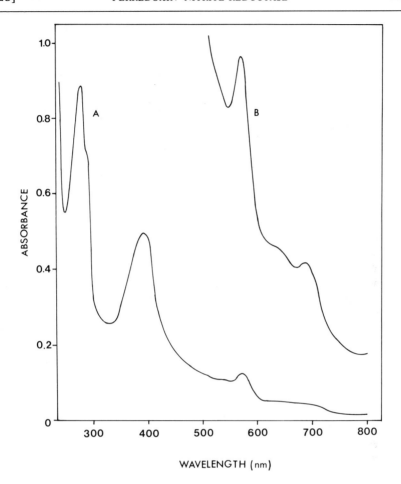

FIG. 1. Absorption spectra of spinach nitrite reductase, (A, 12.5 $\mu M$; B, 96.1 $\mu M$) in 0.1 $M$ potassium phosphate buffer, pH 7.7.[18]

*EPR Spectra of Nitrite Reductase.* Oxidized spinach nitrite reductase shows an EPR spectrum with resonance absorption at $g$ values 6.72, 5.21, and 2.03 characteristic of high-spin $Fe^{3+}$ heme with rhombically distorted tetragonal symmetry.[18,38,39] Under strongly reducing conditions (reduced ferredoxin plus CO), EPR signals of $g = 2.05$ and 1.94 (reduced

[38] P. J. Aparicio, D. B. Knaff, and R. Malkin, *Arch. Biochem. Biophys.* **169**, 102 (1975).
[39] J. M. Vega, H. Kamin, and W. H. Orme-Johnson, *Fed. Proc. Fed. Am. Soc. Exp. Biol.,* **35**, 1597 (1976).

TABLE II
SPECTROPHOTOMETRIC PROPERTIES OF FERREDOXIN–NITRITE REDUCTASE OF DIFFERENT ORIGINS

| Property | Spinacea[a] | Cucurbita pepo[b] | Chlorella fusca[c] |
|---|---|---|---|
| Absorption maxima[d] of native enzyme | 276; 290[e]; 386; 573, 640, 690 | 280; 384; 532; 572; 635; 697 | 278; 384, 530; 560; 573; 635; 692 |
| $A_{276}/A_{386}$ ratio | 1.8 | 2.0 | 3.0 |
| Molar extinction coefficient at 386 nm | $3.97–7.60 \times 10^4$ | $3.79 \times 10^4$ | $2.20 \times 10^4$ |
| Absorption maxima[d] of dithionite-reduced enzyme | 400; 545; 585 | 400; 590 | 555; 585 |
| Absorption maxima[d] of reoxidized enzyme | 400; 573 | 400; 572 | 525; 573 |

[a] Vega and Kamin,[18] and Lancaster et al. (See addendum).
[b] Hucklesby et al.[20]
[c] Zumft.[10]
[d] Wavelengths are in nanometers.
[e] Values in italics indicate shoulders or small peaks.

iron–sulfur) appear.[18,39] Nitrite reductase incubated with dithionite and cyanide shows at 20°K a set of resonance lines with $g$ values of 2.03, 1.94, and 1.91 indicative also of the presence of a reduced iron–sulfur center.[38]

The EPR spectrum of marrow nitrite reductase exhibits resonance absorption at $g$ values 6.86, 4.98, and 1.95. Upon addition of dithionite plus methyl viologen signals at $g = 2.038$, 1.944, and, 1.922 are observed[40].

*Composition.* It is well established that nitrite reductases from higher plants[17,19,20] and *Chlorella*[10] are iron-containing enzymes. Murphy *et al.*[13] have identified the heme prosthetic group in spinach nitrite reductase as "siroheme," an iron tetrahydrophorphyrin of the isobacteriochlorin type with eight carboxylic acid-containing side chains. Although it has been reported that nitrite reductase from green tissues contains 2 atoms of Fe/enzyme molecule,[10,17,19,20] and Vega and Kamin[18] showed 3 Fe atoms/molecule, further EPR studies (see addendum) showed the enzyme contains 1 tetranuclear center ($Fe_4$–$S_4^*$) and 1 siroheme, suggesting a minimum of 5 Fe atoms/active enzyme molecule. The amino acid compositions of the *Chlorella*,[10] *Curcubita pepo*,[20] and spinach[18] nitrite reductases are known.

*Oxidation–Reduction Properties.* The midpoint redox potential of the siroheme prosthetic group of plant nitrite reductase has been determined following the disappearance of the high spin heme EPR signal at $g$ values of 6.7 and 5.2 by reductive titration with mehyl viologen reduced with metallic zinc. The data give a good fit for an $n = 1$ titration with a midpoint potential of $-50$ mV, at pH 7.8 for the spinach enzyme[41] and $-120$ mV at pH 8.5 for that of marrow.[40]

The midpoint potential of the iron–sulfur center of plant nitrite reductase has also been determined by reductive titration with dithionite but following the increase of the EPR signal at $g = 1.94$. By this method, a midpoint potential of $-550$ mV (assuming $n = 1$), at pH 9.0, has been reported for spinach nitrite reductase.[41] Similarly, values of $-570$ mV (pH 8.1), $-615$ mV (pH 8.7) and $-660$ mV (pH 9.8), have been found for the marrow enzyme.[40]

The maximum size of the EPR signal of reduced iron–sulfur center, i.e., higher potential, is observed in the presence of cyanide or CO, typical heme binding agents.[38,39,40] These results are interpreted of considerable significance for the enzyme mechanism, since, presumably, a similar increase in potential, i.e., higher reducibility, also occurs when the physiological substrate binds at the enzyme heme.[40] In the spinach

[40] R. Cammack, D. P. Hucklesby, and E. J. Hewitt, *Biochem. J.* **171**, 519 (1978).

[41] M. L. Stoller, R. Malkin, and D. B. Knaff, *FEBS Lett.* **81**, 271 (1977).

enzyme, however, the presence of cyanide (5 m$M$) seems not to alter the midpoint potential of the iron–sulfur center as determined by titration with dithionite.[41]

*Enzyme Complexes.* Nitrite (the physiological substrate) or hydroxylamine (an alternate substrate) alter the spectrum of the oxydized enzyme in the absorption bands corresponding to siroheme.[18,38] The addition of nitrite to the enzyme produces a shift in the absorbance at 573 to 567 nm, a small peak at 528 nm, and bleaching of the absorbance beyond 640 nm. The nitrite reductase–hydroxylamine complex spectrum is characterized by an $\alpha$ band at 580 nm, a small $\beta$ band at 542 nm, and bleaching of the absorption at 690 nm present in the native enzyme.[18] Titration experiments indicate that 1 mole of nitrite or hydroxylamine binds per mole of enzyme with a $K_{diss}$ of $3.2 \times 10^{-6}$ $M$ for the enzyme–nitrite complex and $4.2 \times 10^{-3}$ $M$ for the enzyme–hydroxylamine complex.[18] Addition of sulfite to spinach nitrite reductase shifts the Soret absorption band to 400 nm. Alterations in the $\alpha$ band and bleaching of the absorbance at 690 nm are also observed.[18] Cyanide–nitrite reductase complex is characterized by absorption bands at 405 nm (Soret), 573 ($\alpha$), a small peak at 535 nm, and bleaching of the absorbance at 690 nm.

Nitrite, hydroxylamine, cyanide, and sulfite cause disappearance of the EPR high spin heme signal ($g = 6.72$ and 5.21) of spinach nitrite reductase.[18]

Dithionite-reduced nitrite reductase may also form complexes with nitrite, hydroxylamine, cyanide, or carbon monoxide (Table III).

Recently, a 1:1 complex between spinach ferredoxin and nitrite reductase has been proposed by spectrophotometric titration studies.[42] The possible role of this complex in the enzymic reduction of nitrite to ammonia remains to be demonstrated.

*Mechanism of Enzyme Catalysis.* Present evidence[18] supports an interaction between the siroheme prosthetic group and the substrates or inhibitors. During steady state turnover experiments with dithionite and nitrite, the enzyme forms a complex with an $\alpha$-band at 580 nm and a peak at 543 nm. This appears to be the active species of the enzyme during turnover. The predominant form of the enzyme during steady state turnover with hydroxylamine and dithionite shows two broad peaks at 494 and 562 nm. These differences and the above indicated kinetic data suggest that free hydroxylamine is not formed during the enzyme-catalyzed reduction of nitrite to ammonia.[18] However, steady state turnover experiments indicate that nitrite reductase may form complexes with nitrogen compounds of more than one oxidation state.[18] Electron para-

[42] D. B. Knaff, J. M. Smith, and R. Malkin, *FEBS Lett.* **90**, 195 (1978).

TABLE III

SPECTROPHOTOMETRIC PROPERTIES OF REDUCED NITRITE REDUCTASE COMPLEXES WITH
SUBSTRATES OR INHIBITORS

| Complex | Absorption maxima (nm) | |
|---|---|---|
| | Spinacea[a] | Cucurbita pepo[b] |
| Enzyme–nitrite | 543; 580 | 544; 582 |
| Enzyme–hydroxylamine | 494; 562 | – |
| Enzyme–cyanide | 400; 410; 540 | 398; 407 |
| Enzyme–CO | 396; 543; 589 | 398; 543; 589 |

[a] Vega and Kamin.[18]
[b] Hucklesby et al.[20]

magnetic resonance analysis of the enzyme in the presence of dithionite plus nitrite may indicate the formation of a NO-heme complex.[38]

Steady state kinetic studies have suggested a role for both the iron-sulfur center and the ferroheme·NO complex in the reaction mechanism of marrow nitrite reductase.[40]

Anaerobic titration of spinach nitrite reductase with dithionite in the presence of cyanide indicates that three electrons (1.5 molecules of dithionite) are required to reduce completely one molecule of enzyme.[18] The most probable distribution of these electrons in the enzyme molecule is one to two in the iron–sulfur cluster ($Fe_4$–$S_4^*$) and one in the siroheme, which binds nitrite and reduces it directly to ammonia. The enzyme appears to supply six electrons to one nitrite molecule in rapid steps of one electron each. The pathway of electrons from ferredoxin to nitrite seems, therefore, to proceed according to the scheme shown in Fig. 2.

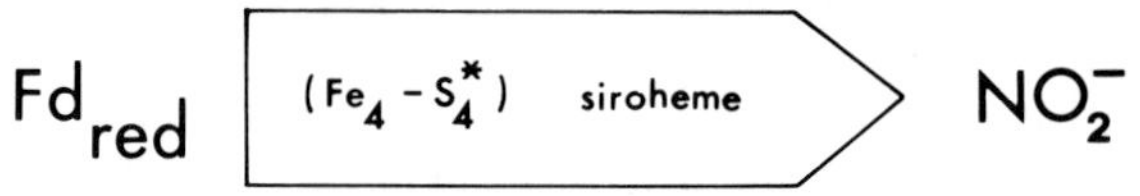

$$Fd_{red} \rightarrow \boxed{(Fe_4-S_4^*) \quad siroheme} \Rightarrow NO_2^-$$

FIG. 2. Schematic representation of the electron pathway catalyzed by ferredoxin–nitrite reductase.

Addendum

In a recent paper [J. R. Lancaster, J. M. Vega, H. Kamin, N. R. Orme-Johnson, W. H. Orme-Johnson, R. J. Krueger, and L. M. Siegel. J. Biol. Chem. 254, 1268 (1979)] the identification of the iron–sulfur center of spinach nitrite reductase is reported and preliminary EPR studies of the enzyme mechanism are presented. The EPR spectrum of reduced nitrite reductase in the presence of 80% dimethyl sulfoxide, taken at 20°K and 10 mW of power, is nearly axial, with $g$ values of 2.04 and 1.93 typical of a tetranuclear $(Fe_4S_4^*)$ iron-sulfur center-containing protein. Chemical analysis for total iron and acid-labile $S^{2-}$ confirms the presence of 3 moles of iron and 2 moles of $S^{2-}/61,000$ gm of protein. However, a reinterpretation of these data leads them to suggest that the enzyme contains 6 moles of iron and 4 moles of acid-labile $S^{2-}/mol$ of siroheme. The EPR results show that there is one iron–sulfur center/siroheme.

Rapid kinetic studies show that the high spin ferriheme and the iron–sulfur center of spinach nitrite reductase are reduced by dithionite on a comparable time scale ($k$ = 3 to 4 sec$^{-1}$). The iron–sulfur center of the enzyme prereduced with dithionite is rapidly reoxidized upon addition of nitrite ($k$ = 100 sec$^{-1}$), what strongly supports a role for the iron–sulfur center in the mechanism of nitrite reduction.

Finally, they proposed an extinction coefficient of $\epsilon_{386}$ = 7.6 × 10$^4$ cm$^{-1}$ ($M$ active center)$^{-1}$.

# [24] Nitrate Reductase from Higher Plants[1]

*By* R. H. HAGEMAN and A. J. REED

$$NO_3^- + NAD(P)H + H^+ \rightarrow NO_2^- + NAD + H_2O$$

This article is an update of the same topic published in 1971.[2] Consequently limited details of standard extraction and assay procedures will be given and major emphasis placed on new developments.

## Preparation

*Material.* Chlorophyllous lamina tissue from illuminated plants well supplied with nitrate is normally used as source material because of its high activity. However, nonchlorophyllous organs, such as corn scutella[3] or roots,[4] have been used. Soybean leaves,[5] corn scutella,[3] and cultured rice seedlings[6] are the best known sources of NAD(P)H nitrate reductase.

*Extraction.* A standard procedure for the extraction of the enzyme is as follows: The material is homogenized for 30 to 90 sec in a medium of 1 m$M$ EDTA, 1 to 25 m$M$ cysteine, and 25 m$M$ potassium phosphate,

[1] EC 1.6.6.1 nitrate oxidoreductase is the most prevalent enzyme in plants; however, EC 1.6.6.2 NAD(P)H nitrate oxidoreductase is present in some tissues.
[2] R. H. Hageman and D. P. Hucklesby, Vol. 23, p. 491.
[3] J. E. Elsner, D. P. Hucklesby, and R. H. Hageman, *Agron. Abstr.* p. 20 (1971).
[4] W. Wallace, *Plant Physiol.* **55**, 774 (1975).
[5] S. A. Jolly, W. H. Campbell, and N. E. Tolbert, *Arch. Biochem. Biophys.* **174**, 431 (1976).
[6] T. C. Shen, *Plant Physiol.* **49**, 546 (1972).

adjusted to a final pH of 8.8 with KOH. Six milliliters of grinding medium are added for each gram of fresh weight of tissue. The homogenate is pressed through four layers of cheesecloth or a single layer of Miracloth (Chicopee Mills, New York, New York) and the filtrate centrifuged for 15 min at 30,000 g. The supernatant fluid is then decanted through glass wool and used for assays. The homogenates and extracts are kept cold (2° to 3°) throughout. The optimum concentration of cysteine must be verified for each type of tissue and homogenization technique. Glutathione and dithiothreitol are usually slightly more effective (10%) than cysteine.

A marked improvement in the extraction of nitrate reductase from soybean leaf tissue was obtained with the following procedure.[7] One gram fresh tissue is frozen with liquid $N_2$ in a precooled mortar. After evaporation of the $N_2$, the frozen tissue is rapidly (15 sec) ground to a powder. The extraction medium (20 volumes of 25 m$M$ potassium phosphate, pH 7.8, 1 m$M$ cysteine, 5 m$M$ $KNO_3$, 5 m$M$ EDTA and 25 $\mu M$ FAD) is added and allowed to freeze in the mortar. The tissue is suspended by grinding as the mixture thaws. The homogenate is then treated as previously described.

*Special Protectants.* With tobacco leaves, the addition of 0.1% (w/v) bovine serum albumin to sucrose (plus other additives) extraction medium prevented the indiscriminate and variable binding of nitrate reductase to various organelles during subsequent sucrose density separation[8] and enhanced (twofold) the total recovery of activity. The enhanced activity observed with the use of exogenous protein was reported to be due to protection from a small-molecular-weight, heat-stable inhibitor. The addition of casein or bovine serum albumin (1 to 3%) to the standard extraction medium results in increased (up to 15-fold) recovery and stability of nitrate reductase from leaves of corn, oat, and tobacco plants[9] and corn roots.[4] It was suggested that the added protein protected the enzyme from proteolytic enzymes. With seedling tissue from corn or soybeans, extraction with added protein often slightly decreases recoverable activity (unpublished data). With cotton cotyledons, the addition of bovine serum albumin (3%, w/v) or Dowex 1-Cl, anion exchange resin (10%, w/v) to the extraction medium significantly improved the activity of the extracted enzyme.[10] Neither protein nor resin enhanced stability

[7] R. L. Scholl, J. E. Harper, and R. H. Hageman, *Plant Physiol.* **53**, 825 (1974).
[8] M. J. Dalling, N. E. Tolbert, and R. H. Hageman, *Biochim. Biophys. Acta* **283**, 505 (1972).
[9] L. E. Schrader, D. E. Cataldo, and D. M. Peterson, *Plant Physiol.* **53**, 688 (1974).
[10] A. C. Purvis, L. R. Tischler, and R. C. Fites, *Plant Physiol.* **58**, 95 (1976).

of the enzyme. The protective action was attributed to phenolic binding rather than decreased degradation by proteases. The inclusion of NADH (1 mg/ml) in the crude extract of beans[11] or rice[12] stabilized nitrate reductase activity.

The variation in interpretation given for the mode of action of the various protectants is attributed to the variation in inhibitors, proteases, and inherent stability (dissociation into inactive components) of the enzyme among the different plant species and tissues utilized. Polyvinylpyrrolidone may be the best protectant against the inhibitory effects of phenolics.[13] No reports were found where two or more protectants were added concurrently.

## Assay Method (*in Vitro*)

$$NO_3^- + AH_2 \rightarrow NO_2^- (+ A + H_2O$$

*Principle.* Nitrate reductase is capable of utilizing the reduced form of pyridine nucleotides, flavins, or benzyl viologen as electron donors for reduction of nitrate to nitrite. Because NADH-dependent nitrate reductase is most prevalent in plants, NADH is the most frequently used donor. Activity is usually measured by colorimetric determination[14] of nitrite produced; however, it can also be measured by following the oxidation of pyridine nucleotides at 340 nm. The stoichiometric relationship between nitrite produced and pyridine nucleotide or flavin oxidized has been established.[15,16]

*Reagents for Reaction*
Potassium phosphate buffer, 0.1 $M$ (pH 7.5)
Potassium nitrate, 0.1 $M$
NADH, 2 m$M$
*Reagents for Nitrite Assay*
Sulfanilamide, 1% (w/v) in 1.5 to 3.0 $N$ HCl
$N$-(1-Naphthyl)ethylenediamine dihydrochloride, 0.02% (w/v)

*Procedure (NADH).* The assay mixture contains in micromoles, potassium phosphate, 50; KNO$_3$, 20; NADH, 0.8; and enzyme, 0.1 to 0.2

[11] C. M. Sluiters-Scholten, *Planta* **123**, 175 (1975).
[12] A. P. Gandhi, S. K. Sawhney, and M. S. Naik, *Biochem. Biophys. Res. Commun.* **55**, 291 (1973).
[13] W. D. Loomis and J. Battaile, *Phytochemistry* **5**, 423 (1966).
[14] F. D. Snell and C. T. Snell, "Colorimetric Methods of Analysis." Van Nostrand Reinhold, Princeton, New Jersey (1949).
[15] H. J. Evans and A. Nason, *Plant Physiol.* **28**, 233 (1953).
[16] L. E. Schrader, G. L. Ritenour, G. L. Eilrich, and R. H. Hageman, *Plant Physiol.* **43**, 930 (1968).

ml of crude extract (equivalent to 0.2 to 1.0 mg of trichloroacetic acid precipitable protein) in a final volume of 2.0 ml. The optimum pH is usually 7.5; however, with crude extracts from soybean leaves the optimum pH is 6.5. The reaction can be initiated either by addition of NADH or enzyme. A zero time and minus NADH reaction mixture are used for controls. After incubation at 30° for 15 min the reaction is terminated by rapid addition of 1.0 ml of sulfanilamide reagent followed by 1.0 ml of the $N$-(1-naphthyl)ethylenediamine dihydrochloride reagent. The color is allowed to develop for 30 min prior to reading at 540 nm. The enzyme reaction rate is linear over a 30-min period.

The same procedure is used when NADH oxidation is followed spectrophotometrically at 340 nm. The linear initial reaction rate is used to compute the activity.

Activity is expressed as $\mu$moles of $NO_2^-$ produced per minute per gram fresh or dry tissue weight or per milligram of protein. The protein is precipitated by 5% trichloroacetic acid from the extract which is determined by any of several standard protein methods.

Details for preparation of other reductants and other details are as previously published.[2]

*Coupled Assays.* In addition to the coupled enzyme reactions previously reported,[2] nitrate reduction can be coupled via NADH generated by the addition of malate (20 m$M$) and NAD (2 m$M$) as a substitute for NADH (0.4 m$M$) in the standard assay.[17] With crude extracts of corn leaves containing high levels of malate dehydrogenase, the activity of the coupled reaction was 80% of the standard (NADH) assay.

*Precautions.* Residual NADH at the end of the assay, and unknown factors present in the crude extract, interfere with $NO_2^-$ color development. While NADH can be oxidized enzymatically,[18] this does not eliminate the extract factors. Postassay treatments have been found that minimize these problems.[7] The reaction is stopped by addition of zinc acetate (50 $\mu$moles/ml reaction mixture) or by placing the reaction tubes in boiling water. After clarification by centrifugation (1000 $g$ for 10 min), aliquots of the cooled (25°) supernatant liquid are treated with phenazine methosulfate (15 nmoles/ml reaction mixture) to oxidize the residual NADH. After 20 min, color is developed in the aliquot as previously described.

The enzyme from most plant species is very unstable even at 0°. Therefore, minimum time should elapse between extraction and assay even when stabilizers of enzyme activity have been added.

[17] C. A. Neyra and R. H. Hageman, *Plant Physiol.* **58**, 726 (1976).
[18] D. Spencer, *Aust. J. Biol. Sci.* **12**, 181 (1959).

## Assay Method (*in Vivo*)

*Principle.* Sections of plant tissue, containing adequate amounts of endogenous nitrate and substrates for NADH generation, produce and accumulate $NO_2^-$ when placed under dark anaerobic conditions. The assay as originally developed[19] has been adapted to provide a more accurate estimate of nitrate reduction *in situ*[20] or to study nitrate assimilation in tissue from species having high inhibitor content that prevents extraction of active enzyme.[2,21] Rates of activity obtained with the *in vivo* assay are lower (50 to 80%) than *in vitro* rates, however, they are correlated.[20,22]

*Procedure.*[20,23] Sections or discs of fresh tissue (0.2 to 0.5 gm) are placed in a test tube or small flask containing the incubation medium. The medium is composed of 0.1 $M$ potassium phosphate, pH 7.5, 1% (v/v) 1-propanol or 0.04% (v/v) Neutronyx 600 (Onyx Chemical Co., Jersey City, New Jersey) and 0.03 to 0.05 $M$ $KNO_3$. Stainless steel screens may be placed in each tube to hold the tissue below the solution surface. The samples are then evacuated (5 mm Hg) in a vacuum desiccator. Nitrogen gas is bled into the desiccator and the process repeated. The tubes may be stoppered before transferring to a shaking water bath and incubated (0.25 to 4 hr) at 30° in the dark. Aliquots can be removed from the same reaction tube at timed intervals to establish linearity of $NO_2^-$ production. Usually there is a 5- to 10-min lag period prior to establishment of a linear rate. The reaction is usually linear for 60 to 90 min. Normally the reaction is stopped by transferring the tubes to a boiling water bath for 2 min to release $NO_2^-$ remaining in the tissue. After cooling, aliquots are removed for colorimetric determination of $NO_2^-$. The difference between the amount of $NO_2^-$ produced at 30 min minus the production at 5 or 10 min, with a comparable set of materials, is used to calculate the rate of $NO_2^-$ production (equated with $NO_3^-$ reduction).

*Comments.* The amount of nitrite produced slightly underestimates nitrate reduction, as some nitrite disappears under dark anaerobic conditions.[21] However, very little nitrite appears to be assimilated to amino-N under these conditions.[24] Anaerobiosis is essential,[24] and other systems (continuous $N_2$ bubbling) can be devised to establish the anaerobic condition.

[19] E. G. Mulder, R. Borma, and W. L. VanVeen, *Plant Soil Sci.* **10**, 335 (1959).
[20] N. Brunetti and R. H. Hageman, *Plant Physiol.* **58**, 583 (1976).
[21] J. W. Radin, *Plant Physiol.* **51**, 332 (1973).
[22] J. G. Streeter and M. E. Bosler, *Plant Physiol.* **49**, 448 (1972).
[23] J. C. Nicholas, J. E. Harper, and R. H. Hageman, *Plant Physiol.* **58**, 731 (1976).
[24] D. T. Canvin and C. A. Atkins, *Planta* **116**, 207 (1974).

With tissue deficient in sugars or organic acids (prolonged exposure of plants to dark prior to sampling), marked stimulation of nitrate reduction can be obtained by addition of sugar phosphates or 3-phosphoglyceraldehyde (6 m$M$) to the incubation medium.[25] With dark pretreated soybean leaf tissue, the addition of 10 and 100 m$M$ glucose to the incubation medium increased activity 4.5- to 5.2-fold, respectively, over the control.[26] Addition of 10 m$M$ pyruvate, citrate, succinate, or malate was without effect, while activity was increased (two- to threefold) when added at 100 m$M$.

### Purification

Nitrate reductase extracted from higher plants is extremely unstable and difficult to purify to homogenity with reasonable recoveries. These difficulties are illustrated by a comparison of specific activities of 0.02, 0.2, 0.7, 0.7 and 2.5 $\mu$moles $NO_3^-$ reduced/min/mg protein for enzymes from soybeans,[5] maize,[27] marrow,[16] maize,[16] and spinach,[28] respectively with the specific activity of 83.1 for *Chlorella*,[29] when conventional purification techniques were used. By combining conventional purification procedures with the recently developed blue-dextran agarose affinity chromatography,[30] a homogenous nitrate reductase (specific activity 24.1 and 7% recovery) was prepared from spinach leaves.[31] This procedure is detailed as follows.

*Extraction.* Freshly harvested and chilled spinach leaves (1 kg) were homogenized in 5 m$M$ Tris-HCl, pH 7.5, and 1 m$M$ EDTA (2 ml buffer/gm leaf). The macerate was filtered through two layers of cheesecloth and centrifuged at 60,000 $g$ for 20 min. The supernatant was used as the crude extract.

*Purification.* Nucleic acids were removed by adding streptomycin (5 mg/gm leaf) to the crude extract. After 1 hr, the precipitate was removed (60,000 $g$ for 20 min) and discarded. Solid $(NH_4)_2SO_4$ (ARISTAR grade, British Drug House, Poole, England) was added to the supernatant to attain 50% saturation. After 20 min the precipitated protein was collected (60,000 $g$ for 20 min) and saved. The protein was redissolved in 0.1 $M$

[25] L. Klepper, D. Flesher, and R. H. Hageman, *Plant Physiol.* **48**, 580 (1971).
[26] J. C. Nicholas, J. E. Harper, and R. H. Hageman, *Plant Physiol.* **58**, 736 (1976).
[27] J. Roustan, M. Neuberger, and A. Fourey, *Physiol. Veg.* **12**, 527 (1974).
[28] B. A. Notton and E. J. Hewitt, *Plant Cell Physiol.* **12**, 465 (1971).
[29] L. P. Solomonson, G. H. Lorimer, R. L. Hall, R. Borchers, and J. L. Bailey, *J. Biol. Chem.* **250**, 4120 (1975).
[30] L. P. Solomonson, *Plant Physiol.* **56**, 853 (1975).
[31] B. A. Notton, R. J. Fido, and E. J. Hewitt, *Plant Sci. Lett.* **8**, 165 (1977).

phosphate, pH 7.5, and desalted on a Sephadex G-25 (Pharmacia, Uppsala, Sweden) column equilibrated with the same buffer. Fractions containing the enzyme were transferred to a column of hydroxylapatite (Bio-Rad Laboratories, Richmond, California) equilibrated with 0.05 $M$ phosphate pH 7.5. The column was then washed with buffer until absorbance (280 nm) of the eluate was less than 0.01. The enzyme was eluted from the column using the washing buffer fortified with 5% $(NH_4)_2SO_4$. The appropriate fractions were combined and the enzyme precipitated by adding solid $(NH_4)_2SO_4$ to attain 50% saturation. After 20 min the precipitate was collected (60,000 $g$ for 20 min) and redissolved in 0.1 $M$ phosphate, (pH 7.5) containing 0.1 $M$ KCl. This solution was transferred to a BioGel A (0.5 m, 100 to 200 mesh, Bio-Rad Laboratories) column. Column equilibration and elution were with the dissolving solution. Fractions containing the enzyme were bulked and reprecipitated with $(NH_4)_2SO_4$ at 50% saturation. The precipitate was collected by centrifugation and dissolved in 0.08 $M$ phosphate, pH 7.5, and the solution transferred to a blue-dextran Sepharose column equilibrated with the same buffer. (The blue-dextran Sepharose was made as previously described.[32]). The column was washed with the same buffer until the eluate gave negligible absorbance at 280 nm. Nitrate reductase was eluted by applying a 0 to 3 $M$ KCl gradient (in 0.08 $M$ phosphate, pH 7.5). Fractions containing the enzyme were bulked and concentrated with $(NH_4)_2SO_4$ (50% saturation) and dissolved in a small amount of 0.08 $M$ phosphate, pH 7.5. The BioGel A and blue-dextran Sepharose steps increased specific activity 4- and 22-fold, but lost 42 and 30% of the total activity, respectively.

Unlike the enzyme from *Chlorella*,[30] nitrate reductase from higher plants cannot be eluted, effectively, from the blue-dextran Sepharose with NADH. However, this problem was circumvented by the use of an alternative affinity medium, blue-Sepharose (Cibacron blue F3GA, Polysciences, Inc., Warrington, Pennsylvania, coupled to Sepharose CL4B, Pharmica, Uppsala, Sweden, as described[33]). The blue-Sepharose was added to the clarified homogenate and stirred for 1 hr at 4° to bind the nitrate reductase.[34] (The homogenate was prepared by grinding 100 gm squash cotyledons in 100 ml of 100 m$M$ phosphate, pH 7.5, 1 m$M$ EDTA, 1 m$M$ cysteine, and 20 gm polyvinylpyrrolidone.) The Sepharose was then collected by filtration and washed with the grinding medium prior

[32] L. D. Ryan and C. S. Vestling, *Arch. Biochem. Biophys.* **160,** 179 (1974).
[33] H. J. Böhme, G. Kopperschlager, J. Schulz, and E. Hofmann, *J. Chromatogr.* **69,** 209 (1972).
[34] W. H. Campbell and J. Smarrelli, Jr., *Plant Physiol.* **57,** 637 (1976).

to transfer to a column (2.5 cm diameter). The enzyme was eluted with the extraction buffer containing 0.1 m$M$ NADH, and the appropriate fractions combined. The enzyme was concentrated by precipitation with 45% saturated $(NH_4)_2SO_4$ and desalted on a Sephadex G-25 column with 100 m$M$ phosphate, pH 7.5, 5 m$M$ EDTA and 1 m$M$ cysteine. Activity was 2 and 12 $\mu$moles $NO_3^-$ reduced/min/mg protein for the bulked sample and maximum fraction, respectively. Recovery in the pooled sample was 30% of the initial activity. This procedure was not as effective with the enzyme extracted from corn leaves and the bulk of the activity was not eluted with NADH, but was washed off with 300 m$M$ $KNO_3$.

## Characteristics

The properties of nitrate reductases from higher plants have been recently reviewed.[35,36]

*Molecular Weight.* Values reported are 160,000 (maize[2]), 197,000 (spinach[36]), 230,000 (barley[37]), 220,000 (NADPH-enzyme, soybean[5]), 330,000 (NADH-enzyme, soybean[5]) and 500,000 (spinach[38]). For the spinach enzyme[36] the following properties were found: Stokes radius, 60 Å; S value, 8.1; diffusion constant, $d_{20,W}$ of $3.4 \times 10^7$; frictional ratio, $f/f_0$ of 1.55; asymmetry, $r_1/r_2$ of 10; isoelectric point, pH 5.0; and an $E_{280}^{0.1\%}$ of 1.62.

*Substrate Affinity.* A Michaelis constant $(K_m)$ for nitrate of 0.2 m$M$ has been reported for several plant species.[2] $K_m$'s for nitrate of 0.11 and 4.5 m$M$ were reported for the NADH and NAD(P)H enzymes from soybeans.[5,39]

*Pyridine Nucleotide Specificity.* The enzyme extracted from most higher plants preferentially utilizes NADH as the electron donor; however, the specificity is not absolute. The $K_m$ for NADH is approximately 2.5 $\mu M$.[40] A phosphatase, that converts NADPH to NADH and is not completely removed by conventional purification procedures, may account for some apparent NADPH activity. In maize, leaves, but not in maize scutella, the NADPH activity in crude or partially purified extracts

[35] E. J. Hewitt, D. P. Hucklesby, and B. A. Notton, *in* "Plant Biochemistry" (J. Bonner and J. E. Varner, eds.), 3rd ed., p. 633. Academic Press, New York, 1976.

[36] B. A. Notton and E. J. Hewitt, *in Nitrogen Assimilation of Plants* (E. J. Hewitt and C. V. Cutting, eds.) p. 708. Academic Press, New York, 1979.

[37] J. L. Wray and P. Filner, *Biochem. J.* **119**, 715 (1970).

[38] A. M. Relimpio, P. J. Aparicio, A., Panaque, and M. Losada, *FEBS Lett.* **7**, 226 (1971).

[39] W. H. Campbell, *Plant Sci. Lett.* **7**, 239 (1976).

[40] L. Beevers, D. Flesher, and R. H. Hageman, *Biochim. Biophys. Acta* **89**, 453 (1964).

can be attributed to phosphatase activity.[3,41] Two enzymes, NAD(P)H (EC 1.6.6.2) and NADH (EC 1.6.6.1) have been separated from soybean leaves. The $K_m$'s for NADPH and NADH for the NAD(P)H enzyme were 1.5 and 39 $\mu M$ and for the NADH enzyme 200 and 8 $\mu M$, respectively.[5] In the crude extracts the total NADPH activity accounted for approximately 10% of the total.[39]

*Flavin Requirement.* FAD is a constituent of the enzyme.[15,36] Addition of FAD often but not always stimulates the *in vitro* reduction of nitrate. It has been proposed that this may reflect the strength of attachment of the indigenous FAD and be species dependent. Reduced FAD or FMN can also serve as electron donors for nitrate reduction, however, there is disagreement over the apparent $K_m$ value (0.02 to 1.0 m$M$) and whether the flavins are physiological or incidental electron donors.[16,42] Supplemental FAD protects nitrate reductase from heat inactivation,[43] and stabilizes the enzyme during sucrose density centrifugation or dilution.[36]

*Involvement of Metals.* Molybdenum has been shown[44] to be a constituent of nitrate reductase of spinach by the use of [99]Mo. A *b*-type cytochrome with absorption bands at 560, 528, and 425 nm has been shown to be a functional constituent of the spinach enzyme.[31] Tungsten can be incorporated into the enzyme as a replacement for molybdenum; however, the analog will not reduce nitrate.[45]

*Sulfhydryl.* The protective effect of sulfhydryl compounds on nitrate reductase during extraction and purification is well established.[2] The sulfhydryl group on the enzyme from higher plants is considered to be involved in binding of the pyridine nucleotide.[16]

*Phosphate Requirement.* Nitrate reductase extracted from higher plant tissue with a media devoid of phosphate is stimulated by the addition of phosphate to the assay medium.[2] The reason for the stimulation is not known; however, it has been proposed that it may complex with the molybdenum of the enzyme thereby facilitating its reduction.

*Sequence of Electron Transfer.* The sequence of electron transfer of higher plant nitrate reductase is thought to be similar if not identical with that of *Neurospora*.[2,35,36]

[41] G. N. Wells and R. H. Hageman, *Plant Physiol.* **54**, 136 (1974).
[42] A. Panaque, F. F. Del Campo, and M. L. Losada, *Biochim. Biophys. Acta* **109**, 79 (1965).
[43] W. G. Zumft, P. J. Aparicio, A. Panaque, and M. L. Losada, *FEBS Lett.* **9**, 157 (1970).
[44] B. A. Notton and E. J. Hewitt, *Biochem. Biophys. Res. Commun.* **44**, 702 (1971).
[45] A. R. J. Eaglesham and E. J. Hewitt, *Biochem. J.* **122**, 18 (1971).

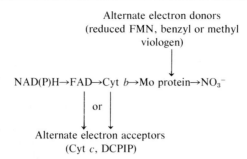

Alternate electron donors
(reduced FMN, benzyl or methyl
viologen)

NAD(P)H→FAD→Cyt $b$→Mo protein→$NO_3^-$

or

Alternate electron acceptors
(Cyt $c$, DCPIP)

The enzyme exhibits two functions: as a dehydrogenase in the reduction of cytochrome $c$ and as a reductase for nitrate.[36] It is not known whether the cytochrome $b$ is associated with the dehydrogenase reaction or not.

*Subunits and Proposed Structure.* The holoenzyme (8.1 S fraction from sucrose density centrifugation) is composed of a pyridine nucleotide–cytochrome $c$ reductase (3.7 S) bonded noncovalently to a molybdenum-containing subunit (minimum effective molecular weight of 30,000).[36] These functional units can be physically separated and recombined. It has been proposed[36] that the holoenzyme has four subunits having cytochrome $c$ reductase activity and one molybdenum-containing complex.

*Kinetics.* The reaction is essentially irreversible to the right. With purified spinach enzyme, an ordered sequential ping-pong mechanism was reported.[45] Work with nitrate reductase purified from squash and corn suggested standard ping-pong kinetics (initial velocity plots) or a two-sided ping-pong mechanism (product inhibition plots).[34]

*Inhibitors.* Nitrate reductase is sensitive to inhibition by PCMB (1 $\mu M$ to 1 m$M$) especially when pyridine nucleotides rather than $FMNH_2$ is the electron donor. Sulfhydryl compounds protect the enzyme from PCMB inhibition.[2,15] The enzyme is sensitive to reagents that react with metals (8-hydroxyquinoline and $o$-phenanthroaline) and cyanide and azide are especially effective. Atabrin inhibits marrow nitrate reductase at 5 m$M$.[2] Neither carbon monoxide nor fluoride (1 m$M$) inhibit soybean nitrate reductase.[15] Nitrate reductase from *Chlorella* inactivated by NADH and HCN can be reactivated by oxidation with ferricyanide.[46] This system has been proposed as a mechanism for regulation of nitrate reductase *in situ*.[47] Purified nitrate reductase from corn leaves can be inactivated by addition of NADH and HCN *in vitro;* however, the inactivation can be measured only when the assays are done under suboptimal

[46] L. P. Solomonson, *Biochim. Biophys. Acta* **334,** 287 (1974).
[47] L. P. Solomonson and A. M. Spehar, *Nature (London)* **265,** 373 (1977).

(0.2 m$M$) levels of nitrate. The presence of the normal level (10 m$M$) nitrate in the assay system apparently causes a rapid activation of the enzyme.[48,49]

*Cellular Location.* Although there is disagreement over the cellular location of nitrate reductase, the bulk of the evidence strongly suggests that it is located in the cytoplasm.[35] In plants with $C_4$-type photosynthesis, nitrate reductase is confined primarily to the mesophyll cells.[35]

*Induction.* Nitrate reductase is inducible by its substrate, nitrate,[2] and the increased activity is due to *de novo* synthesis.[50] Nitrite, the product, will induce nitrate reductase in rice seedlings[6] and beans[51]; however, it is not an effective inducer with many plant species.[52,53] Chloramphenicol and certain other organic nitro compounds will preferentially induce NAD(P)H nitrate reductase in rice seedlings.[6] In contrast, nitrate or nitrite will preferentially induce the NADH enzyme.[6] The two types of enzymes were separated by means of blue dextran Sepharose chromatography.[54]

*Stability.* Nitrate reductase is unstable both *in vivo* and *in vitro*. In excised whole maize seedlings at 30°, the half-life of the enzyme was estimated to be 4 hr. With intact maize plants placed in the dark at 25° the half-life was 12 to 14 hr.[2] This decrease in activity follows first-order kinetics.[16] Using triple labeling techniques, it was shown that synthesis and degradation of the enzyme were concurrent.[50] The stability *in vitro* and *in vivo* varies greatly with plant age, species, and tissue.

[48] D. Loussaert and R. H. Hageman, *Plant Physiol.* **57S**, 539 (1976).
[49] R. H. Hageman, *in Nitrogen Assimilation of Plants,* (E. J. Hewitt and C. V. Cutting, eds.) p. 708. Academic Press, New York, 1979.
[50] H. R. Zielke and P. Filner, *J. Biol. Chem.* **246**, 1772 (1971).
[51] S. H. Lips, D. Kaplan, and N. Roth-Bejerano, *Eur. J. Biochem.* **37**, 589 (1973).
[52] M. M. R. K. Afridi, and E. J. Hewitt, *J. Exp. Bot.* **16**, 628 (1965).
[53] L. Beevers, L. E. Schrader, D. Flesher, and R. H. Hageman, *Plant Physiol.* **40**, 691 (1965).
[54] T. C. Shen, E. A. Funkhouser, and M. G. Guerrero, *Plant Physiol.* **58**, 292 (1976).

# [25] P700 Detection

*By* T. V. MARSHO and B. KOK

The pigment designated P700[1] is the photoreactive center of photosystem I and as such is an obligate component in the overall photochemical transport of electrons from an appropriate donor ($H_2O$ or artificial

[1] B. Kok, *Biochim. Biophys. Acta* **22**, 399 (1956).

reductants) to NADP.[2-4] The pigment has been detected in all examined algae and green plants that are capable of growing autotropically.[5]

P700 occurs in low concentrations relative to the bulk chlorophyll(s) and is typically characterized by its main absorption bands around 430 and 700 nm. The location of these bands and the solubility properties of the pigment suggest that P700 is a form of chlorophyll $a$ segregated as a result of its unique environment within the photosynthetic apparatus. P700 behaves as a one-electron redox component ($E_0' \approx 430$ mV). Quanta absorbed by the harvesting pigment of photosystem I are preferentially transferred to P700, due to the lower energy level of its main absorption band in the red. The "trapping" of quanta by P700 results in the photochemical transfer of an electron to an unidentified acceptor (X) of low potential [Eq. (1)].

$$P700 + X \underset{\text{dark cycle}}{\overset{k_L \text{ system I}}{\rightleftharpoons}} P700^+ + X^- \tag{1}$$

This photoevent, which has a quantum efficiency approaching 1, leaves P700 in the oxidized state and is accompanied by a (partial?) loss of the pigment's main absorption bands at 430 and 700 nm. Correspondingly, P700 can also be chemically oxidized (e.g., with ferricyanide) in the absence of light. Restoration (reduction) of P700$^+$ occurs *via* one of several dark reactions [Eq. (2)].

$$P700^+ + \text{donor} \xrightarrow[k_D]{\text{dark}} P700 + \text{donor}^+ \tag{2}$$

Most kinetic evidence obtained with higher plant material suggests that the primary donor to P700$^+$, *in vivo*, is plastocyanin.[6,7] Although still unresolved, cytochrome $f$, in these systems, appears to operate via plastocyanin or perhaps in parallel to plastocyanin. In addition, reduction can also occur via a cyclic back flow of electrons from (X$^-$) as shown in Eq. (1). It has been suggested that cyclic rates of P700 reduction are minimal during normal steady state photosynthesis *in vivo*.[8] In isolated chloroplasts, exogenous reducing agents, e.g., ascorbate and phenazine methosulfate (PMS) or ascorbate + 2,6-dichlorophenolindophenol

[2] B. Kok, *in* "Plant Biochemistry" (J. Bonner and J. Varner, eds.), 3rd ed., p. 846. Academic Press, New York, 1976.

[3] G. E. Hoch, *in* "Encyclopedia of Plant Physiology" (A. Trebst and M. Avron, eds.), New Ser., Vol. 5, p. 136. Springer-Verlag, Berlin and New York, 1977.

[4] B. Ke, *Biochim. Biophys. Acta* **301**, 1 (1973).

[5] B. Kok, *Acta Bot. Neerl.* **6**, 316 (1957).

[6] W. Haehnel, *Biochim. Biophys. Acta* **459**, 418 (1977).

[7] W. A. Cramer and J. Whitmarsh, *Annu. Rev. Plant Physiol.* **28**, 133 (1977).

[8] P. C. Maxwell and J. Biggins, *Biochemistry* **15**, 3975 (1976).

(DCIP), can also act (directly or indirectly) as electron donors for P700. It should be noted, however, that the accessibility of P700$^+$ to exogenous reductants, particularily in freshly prepared chloroplasts, is limited and variable. With the possible exception of reduced PMS, reductants appear to react via plastocyanin. Similarly, chemical oxidation of P700 by ferricyanide may be quite slow (minutes). The accessibility of P700 to chemical titration is considerably improved in disrupted (e.g., detergent-treated) preparations.

## Assay Methods

Since P700 has never been obtained free of all nonphotoreactive chlorophylls, its detection has generally involved measurement of absorption changes at ~700 nm (or ~430 nm) induced by light or chemical oxidation–reduction. Observation at 700 nm is usually preferred due to the occurrence of other light-induced absorption changes in the blue end of the spectrum. Under some circumstances, e.g., to avoid interference by fluorescence changes, it might be advantageous to carry out measurements at ~820 nm. This wavelength is the maximum of a wide band of light-induced absorption attributable to P700$^+$. However, 820-nm absorption increases due to P700 photooxidation display a rather low differential molar extinction ($<\frac{1}{3}$ of the absorption decrease at 700 nm) and are potentially complicated by similar shifts due to the trapping center of photosystem II.[9-11] Alternate assay methods, such as the measurement of a specific electron spin resonance (ESR) signal or NADP or methyl viologen reduction rates, are discussed elsewhere in this series. Although these techniques avoid the inherent difficulties in optically detecting P700, they are indirect and require additional assumptions.

*Principle.* To measure P700 by means of light–dark spectroscopy, a monochromatic (~700 nm) "detecting light" ($I_{det}$) is passed through a sample and monitored by a photocell. The change of transmission of $I_{det}$ is observed upon illumination of the sample by a second "actinic light" ($I_{act}$) which is blocked from the photocell. Since this change is small ($\Delta I / I_0 < 1\%$), the detecting light must be stable and the noise in the photocurrent minimal. The signal-to-noise ratio is proportional to the square root of the number of quanta ($I_{det} \times t$) which are "seen" by the photocathode and used for the determination. It is desirable to use a strong

[9] D. C. Borg, J. Fajer, R. H. Felton, and D. Dolphin, *Proc. Natl. Acad. Sci. U.S.A.* **67**, 813 (1970).
[10] Y. Inoue, T. Ogawa, and K. Shibata, *Biochim. Biophys. Acta* **305**, 483 (1973).
[11] J. Haveman and P. Mathis, *Biochim. Biophys. Acta* **440**, 346 (1976).

detecting beam and to collect efficiently the light transmitted and scattered by the sample on a sensitive photocathode (S20 response for 700 nm). However, a strong 700 nm light perturbs the system since it sensitizes the photooxidation of P700 [Eq. (1)]. If $I_{det}$ provokes a rate of oxidation $k_{L(det)}$, and the dark reduction rate is $k_D$, the system will attain a steady state in which $[P700^+]/[P700] = k_{L(det)}/k_D$. To avoid photooxidation by the detecting beam and thus an underestimation of $P700_{total}$ we desire $k_{L(det)} \ll k_D$.

In chloroplast preparations, the rate of P700 reduction ($k_D$) can be increased by adding exogenous electron donors such as ascorbate plus catalytic amounts of PMS or DCIP to override the effects of the detecting beam. On the other hand, a high value of $k_D$ (such as in the presence of high concentrations of reduced DCIP or PMS) requires a strong $I_{act}$ to convert P700 into the oxidized form

$$[P700^+]/[P700] = k_{L(act + det)}/k_D$$

Although one should be aware of and check this aspect, it poses no serious problem, at least in chloroplasts, since it is not difficult to select $I_{act} \gg I_{det}$ and a suitable $K_D$ value. As outlined below, proper conditions can be readily chosen using detergent-treated material in which photosystem II and plastocyanin do not interfere. Measurements *in vivo*, however, may require considerable caution.

*Instrumentation.* A variety of different types (custom-built and commercial instruments) of sensitive spectrophotometers[12-15] have been used to measure P700. In general, the appropriate instrument is dictated by the type of observation one wishes to make. For example, under conditions where the P700 changes occur rapidly (<1 sec), one can use single-beam instruments where absorption changes induced by a brief saturating flash are amplified and recorded directly. The (photo)oxidized state can be induced rapidly by a light flash of sufficient quantum content. Such can be obtained using a laser (~ nanoseconds), a xenon discharge lamp (microseconds), or a rotating disk cutting through the image of a bright continuous source such as a xenon arc or incandescent lamp (~ milliseconds). In reaction systems where the rate of P700 reduction ($k_D$) is rapid, the redox state of P700 can be alternated repetitively by giving a sequence of ($n$) properly spaced flashes. As long as the reaction system is stable so that the recurrent event remains constant, one can average

[12] R. K. Clayton, "Molecular Physics in Photosynthesis." Ginn (Blaisdell), Boston, Massachusetts, 1965.
[13] B. Ke, Vol. 23, p. 25.
[14] G. E. Hoch, Vol. 23, p. 297.
[15] B. Chance, Vol. 23, p. 322.

the observations and so greatly improve the signal to noise ratio ($S/N$ $\simeq n^{1/2}$).[13,15]

In the measurement of slower events (>1 sec), interferences such as settling of the particles or light-induced swelling or shrinking generally prohibit single-beam measurements. Such interfering variations of $I_{det}$ can be balanced out by using a time-sharing, second light beam ($I_{ref}$) having either (1) a different wavelength (e.g., 720 nm) which passes through the same sample and varies as $I_{det}$ with the interfering effects, but is not affected by the actinic light (dual-wavelength apparatus) or (2) the same wavelength which passes through a parallel sample which is not illuminated with the actinic light (split-beam apparatus).

In measuring P700 at 700 nm, several potential problems should be considered. The actinic beam and the chlorophyll fluorescence which it excites (some 5% of $I_{act}$ at wavelengths between 680 and 750 nm) should not be seen by the photocathode. This problem can be solved with either time separation or color separation. In time separation, a phosphoroscope arrangement is used. The photocell is darkened during the time the actinic beam hits the sample. Single-beam, split-beam, or dual-wavelength spectrophotometers have been built along this principle.[5,16] Time separation avoids interference of $I_{act}$, and its fluorescence and allows actinic illumination with any wavelength. There is, however, an inherent loss of time during the switching from $I_{act}$ to $I_{det}$. Delayed light emission induced by $I_{act}$ (~0.1% of it) can interfere with the subsequent sampling of $I_{det}$. This interference may be checked for by observing the Δ signal in the absence of $I_{det}$.

The alternate solution is color separation. The actinic light contains only wavelengths longer (> ≃720 nm) or shorter (< ≃690 nm) than the detection wavelength (700 nm), and an appropriate color filter transmits only the latter to the photocathode. Fluorescence and delayed light emission excited by $I_{act}$ still interfere and pose major problems in single-beam instruments. Again these artifacts can be checked by decreasing or removing $I_{det}$ and recording the apparent transmission changes using the normal settings of $I_{act}$ and instrument sensitivity. Since the intensity of fluorescence decreases with the square of the distance, its interference can be selectively minimized by placing the photomultiplier some distance back from the sample. Fluorescence and luminescence occur in a wide spectral range, and their interference can be diminished further by placing a narrow band 700-nm filter in front of the photocell. In a split-beam or dual-wavelength apparatus, where the measuring and reference beams are alternated at a certain frequency and their signals subtracted,

[16] B. Kok, *Plant Physiol.* **34,** 184 (1959).

the residual transmission and the fluorescence from a continuous exciting beam should not, in principle, affect the measurements. In practice, one cannot take this for granted and checks as mentioned above must be made.

A potential fluorescence problem inherent in all measurements at 700 nm with photosystem II operative is a change of the yield of the fluorescence excited by $I_{det}$ caused by $I_{act}$. Although this effect is generally assumed to be small it can again be checked for and minimized by moving the photomultiplier back from the sample and/or inserting a 700 nm interference filter between the sample and photocell.

## Procedure

*Intact Systems.* In complete photosynthetic systems (whole cells or chloroplasts), the oxidized state of P700 can be induced rapidly by strong, rate-saturating light or by weak light which excites predominantly photosystem I ($\geq$710 nm). The rate of P700 reduction ($k_D$) generally is rapid in whole cells or fresh chloroplasts due to the simultaneous excitation of photosystem II by $I_{act}$ and/or cyclic back flow of electrons from (X$^-$). The latter effect can be minimized in isolated chloroplasts by the addition of a low potential, autoxidizable electron acceptor, such as methyl viologen, which removes electrons from (X$^-$). The uncertain rate, $k_D$, due to partial photosystem II excitation, however, can present a problem and, depending upon conditions such as instrumental response time, choice of wavelength, or intensity of $I_{act}$, a complete photooxidation of P700 may not be observed.

In most cases, chloroplast preparations may be assayed for P700 in a medium containing a suitable buffer such as Tris-HCl, Tricine-KOH, or phosphate ($\sim$50 m$M$), pH 7–8°, methyl viologen (10–100$\mu M$); and MgCl$_2$ (5 m$M$). Cell preparations may be assayed in the appropriate buffered growth medium. Chlorophyll concentrations should be kept at a minimum ($<$25$\mu$g Chl/ml) to avoid excessive attenuation of $I_{act}$ and $I_{det}$. In slow measurements, continuous stirring of the sample will not only minimize this problem but also prevent significant settling of the sample. To determine the amount of P700 present, samples should be preilluminated with continuous ($\geq$ 1 sec) strong (white or red) light to ensure that all P700 returns to the reduced state following the strong light. The photooxidation of P700 is subsequently elicited upon addition of the appropriate intensity of continuous far-red ($\geq$710 nm) light to the sample. Complete oxidation of P700 should be checked by adding DCMU to the sample (see below) and/or increasing $I_{act}$.

Kinetic studies of P700 in complete photosynthetic systems are more difficult and require considerable caution. In particular, "flash titration" experiments indicate polyphasic P700 reduction kinetics with half-times ranging from ~20 msec to 20 $\mu$sec.[17,18] The rapid decay times presumably reflect fast electron transport between P700 and its primary donor(s). The net result is that flash-induced measurements of P700 concentration or turnover, *in vivo*, necessitate rapid instrumental response times ($\leq$a few microseconds). However, samples can be preilluminated with far-red light in order to fully oxidize donors on the reducing end of P700. Now, a reducing equivalent generated by photosystem II, in a subsequently given flash, will reach P700 more slowly (msecs), due to the rate-limiting step between the photosystems.[17-19] Again, the end points, i.e., all P700 or all P700$^+$, can be checked following a long flash or continuous far-red illumination ($\pm$DCMU), respectively.

*Disrupted Systems.* The detection of P700 can be greatly facilitated after photosystem II is inactivated or uncoupled from photosystem I and its action replaced by that of an exogenous electron donor. For example any of the following treatments can be used successfully.

a. Suspension of chloroplasts in buffer containing ~1% Triton X-100 or digitonin [digitonin/chlorophyll ~12–80/1 (w/w)].

b. Addition of 1–10 $\mu M$ DCMU to isolated chloroplasts or whole cells.

c. Sonication or other disruption of chloroplasts or whole cells. Time and method required for breakage varies with the plant material. Fragments sedimenting between 10,000 and 144,000 $g$ are collected and suspended in the reaction mixture.[20]

An additional benefit of the first treatment is that samples are clarified, avoiding excessive scattering problems. The rate $k_D$ can now be controlled by varying the concentration of donor and catalyst. The proper donor concentration will again depend upon the type and response time of the instrument used. It is important in selecting a particular donor concentration to verify that P700 is completely reduced in the absence of $I_{act}$, (i.e., $k_D \gg I_{det}$) and that $I_{act}$ (flash or continuous) is sufficiently strong to completely oxidize P700. In general, photosystem II inactivated chloroplast or cell preparations may be assayed in reaction media (as above) containing 1–2 m$M$ ascorbate and ~10–50 $\mu M$ DCIP or ~1–5 $\mu M$

[17] W. Haehnel, G. Doring, and H. T. Witt, *Z. Naturforsch. Teil B* **26**, 1171 (1971).
[18] W. Haehnel and H. T. Witt, *Photosynth., Two Centuries Its Discovery Joseph Priestley, Proc. Res., Int. Cong. Photosynth. 2nd, 1971* p. 469 (1972).
[19] W. Haehnel, *Biochim. Biophys. Acta* **305**, 618 (1973).
[20] B. Ke, S. Katoh, and A. San Pietro, *Biochim. Biophys. Acta* **131**, 538 (1967).

PMS provided that instruments with response times $\leq 1$ msec (single-beam instruments) and saturating flashes or very strong $I_{act}$ are used. Using weaker actinic light or in measurements where $I_{act}$ is separated from $I_{det}$ (phosphoroscope arrangement), the reaction mixture should contain ascorbate alone ($\sim 1$ m$M$) or ascorbate plus a lesser amount of PMS (0.05–0.2 $\mu M$) or DCIP (1 $\mu M$).

*Chemically Induced P700 Measurements.* Chemically induced P700 spectra are most conveniently measured with a double-beam instrument and in preparations enriched in P700 (see below). However, similar spectra can also be obtained, albeit less simply, using detergent-treated or sonicated chloroplast or cell preparations as described above. Particles are suspended in Tris-HCl or phosphate buffer ($\sim 50$ m$M$) at pH 7–8. Equal samples containing about 25 $\mu$g of chlorophyll/ml (giving an absorbance of about 2 at 680 nm) are placed in identical cuvettes and used to record a baseline. One of the two samples is then treated with ferricyanide ($\sim 1$ m$M$), the other with ascorbate ($\sim 2$ m$M$), and allowed to equilibrate prior to recording a difference spectrum.

*Enrichment of P700.* Original procedures for P700 purification, involving acetone extraction or detergent treatment of chloroplast materials, typically yielded a two to sevenfold enrichment of P700 concentration per unit chlorophyll.[21] More recently, preparations containing chlorophyll/P700 ratios of 20 to 40 (in part due to the assumption of a lower $\Delta$ molar extinction for P700) have been obtained following disruption of chloroplast membranes with Triton X-100. The reader is referred to other chapters in this series[22,23] for an outline of these procedures (see also Shiozawa *et al.*[24]). It should be noted that light-induced measurements of P700 in these preparations requires the addition of a suitable exogenous reductant and electron acceptor as above.

## Properties

*Absorption Characteristics.* P700 has several minor (difference) absorption bands in addition to the main (light–dark) absorption minima observed between 698 and 709 nm in the red and between 430 and 435 in the blue. The exact location of these bands appears to vary slightly depending on the species and type of preparation used. Small light–dark

[21] T. V. Marsho and B. Kok, Vol. 23, p. 515.
[22] J. Goldbeck, this volume [12].
[23] L. P. Vernon and E. R. Shaw, Vol. 23, p. 277.
[24] J. A. Shiozawa, R. S. Alberte, and J. P. Thornber, *Arch. Biochem. Biophys.* **165**, 388 (1974).

difference bands have been reported at about 682 nm, 460 to 570 nm, and 740 to 850 nm using P700 enriched preparations.[10,25] Contrary to the 430, 700, and 682 nm peaks, the green and far-red bands are typically represented by very broad regions of absorption increase. The appearance of these light-induced absorption increases is consistent with the occurrence of a chlorophyll cation radical.[3,9]

*Extinction Coefficient and Concentration of P700.* It was originally assumed that the main 700 nm band of P700 had a difference millimolar extinction coefficient similar to chlorophyll *a* ($\approx$80 m$M^{-1}$ cm$^{-1}$).[21] More recently, stoichiometric measurements of the coupling between $N,N,N',N'$-tetramethylphenylenediamine (TMPD) oxidation and P700 reduction, using enriched P700 particles from spinach and *Anabaena*, suggest a differential millimolar extinction of 64 and 70 m$M^{-1}$ cm$^{-1}$, respectively, at ~700 nm. Correspondingly, a differential millimolar extinction of 44 and 60 m$M^{-1}$ cm$^{-1}$ were reported for the blue difference band (~430 nm) of P700 in Spinach and *Anabaena* preparations, respectively[4,25] (see also Hachnel[26]). It should be noted that acceptance of the lower molar extinction value for P700 increases earlier estimates of the P700 concentration relative to bulk chlorophylls and is more compatible with a quantitative correlation between light or chemically induced optical changes of P700 nm and ESR determined spin concentrations.

*Redox Potential.* Based on earlier measurements with acetone-extracted or aged spinach chloroplast preparations, it is still generally thought that P700 oxidation involves a single, pH-independent, electron transfer step displaying a midpoint potential of +430 to +460 mV.[27] It should be noted that some midpoint potential estimates for P700, using detergent preparations, have indicated a higher potential of +480 to +520 mV[28,29] (see also Ruuge and Izawa[30]). This, plus some of the kinetic anomalies observed in certain P700 measurements, is disturbing and implies that P700 may display a variable midpoint potential responsive to changes, either natural or unnatural, in the local environment surrounding the pigment.

*ESR Signal.* One of the light-induced ESR signals observed in algae and chloroplasts (signal *I*) peaks at $g$ = 2.0025 $\pm$ 0.005, is 7.2 G wide, and displays no hyperfine structure. The occurrence and kinetic behavior

[25] T. Hiyama and B. Ke, *Biochim. Biophys. Acta* **267**, 160 (1972).
[26] W. Haehnel, *Biochim. Biophys. Acta* **423**, 499 (1976).
[27] B. Kok, *Biochim. Biophys. Acta* **48**, 527 (1961).
[28] H. Y. Yamamoto and L. P. Vernon, *Biochemistry* **8**, 4131 (1969).
[29] D. B. Knaff and R. Malkin, *Arch. Biochem. Biophys.* **159**, 555 (1973).
[30] E. K. Ruuge and S. Izawa, *Fed. Proc., Fed. Am. Soc. Exp. Biol.* **31**, 901 (1972).

of this signal correlate with those of the oxidized form of P700 suggesting that P700[+] is the molecular species responsible for signal *I*.[31,32] More recently, another ESR signal with properties indistinguishable from that of P700 has been reported.[33,34] Although the origin of this signal is unresolved,[32,35,36] adequate recognition should be given to the possibility that ESR studies may actually reflect mixed signals which could cause problems in P700 measurements.

*Concentration.* In nature, the amount of P700 relative to total chlorophyll varies (~1/200 to 1/1000) dependent upon growth conditions, photosystem I enrichment in certain cell types, e.g., bundle sheath cells or blue-green algal heterocysts, and the presence of phycobilin accessory pigments, e.g., vegetative blue-green algal cells. In general, these cell types display higher P700/chlorophyll ratios than typically founded in other higher plant tissues.

*Stability.* The light-induced oxidation of P700 is unaffected by pH between 5 and 11 and remains observable in chloroplasts after heating at 55° for 5 min.[37] At 0°–5° it is stable for over 1 week, and at lower temperatures, indefinitely.

Acknowledgments

Preparation of this article was supported, in part, by National Science Foundation Grants GB-38237, PCM 74-20736, AER 73-03291, and Energy Research and Development Agency Grant E (11-1)–3326.

[31] R. A. Baker and E. C. Weaver, *Photochem. Photobiol.* **18**, 237 (1973).
[32] E. C. Weaver and G. A. Corker, *in* "Encyclopedia of Plant Physiology" (A. Trebst and M. Avron, eds.), New Ser., Vol. 5, p. 169. Springer-Verlag, Berlin and New York, 1977.
[33] A. J. Bearden and R. Malkin, *Biochim. Biophys. Acta* **325**, 266 (1973).
[34] H. J. van Gorkom, J. J. Tamminga, J. Haveman, and I. K. van der Linden, *Biochim. Biophys. Acta* **347**, 417 (1974).
[35] R. H. Lozier and W. L. Butler, *Biochim. Biophys. Acta* **333**, 465 (1974).
[36] D. B. Knaff, *Photochem. Photobiol.* **26**, 327 (1977).
[37] B. Rumberg and H. T. Witt, *Z. Naturforsch., Teil B* **19**, 693 (1964).

## [26] Chloroplast Envelope Lipids: Detection and Biosynthesis

By ROLAND DOUCE and JACQUES JOYARD

The two membranes, or envelopes, which surround the chloroplast stroma are characterized by a significantly different lipid composition than that found in thylakoid lamellae.[1-4] Recently, it has been found that the envelope is the site of synthesis of phosphatidic acid,[5] diacylglycerol,[5] and galactolipids[6,7] [monogalactosyldiacylglycerol (MGDG), digalacto-syldiacylglycerol (DGDG)] in spinach chloroplast.

This article describes the preparative procedures, lipid content and enzymatic activities of the envelope of spinach chloroplast.

### Chloroplast Envelope Preparation

The procedure is fully described in Fig. 1 and can be divided in two steps.

*Isolation of Chloroplasts.* Chloroplasts are prepared from washed, 8-week-old spinach (*Spinacia oleracea* L.) leaves. From 2 kg of leaves, the yield of intact and purified chloroplasts (P4 pellet) is equivalent to about 30–50 mg of chlorophyll (500–900 mg of proteins). This low recovery represents 2–3.5% of the total chlorophyll contained in the leaves.

*Isolation of Chloroplast Envelope Membranes.* The intact and purified chloroplasts are suspended in the swelling medium. Under these conditions, immediate swelling of intact chloroplasts causes rupture and detachment of the envelope membranes with the liberation of the stroma

[1] R. Douce, R. B. Holtz, and A. A. Benson, *J. Biol. Chem.* **248**, 7215 (1973).
[1a] Abbreviations used: MGDG, monogalactosyldiacylglycerol; DGDG, digalactosyldiacyl-glycerol; TGDG, trigalactosyldiacylglycerol; TTGDG, tetragalactosyldiacyl- glycerol; SL, sulfoquinovosyldiacylglycerol; PC, phosphatidylcholine; PE, phos- phatidylethan-olamine; PG, phosphatidylglycerol; DPG, diphosphatidylglycerol; PI, phosphatidylinos-itol; PS, phosphatidylserine; di, diacylglycerol.
[2] R. P. Poincelot, *Arch. Biochem. Biophys.* **159**, 143 (1973).
[3] R. O. Mackender and R. M. Leech, *Plant Physiol.* **53**, 496 (1974).
[4] H. Hashimoto and S. Murakami, *Plant Cell Physiol.* **16**, 895 (1975).
[5] J. Joyard and R. Douce, *Biochim. Biophys. Acta* **486**, 273 (1977).
[6] R. Douce, *Science* **183**, 852 (1974).
[7] H. C. Van Hummel, T. J. M. Hulsebos, and J. F. G. M. Wintermans, *Biochim. Biophys. Acta* **380**, 219 (1975).

Chop 1 kg (750 mg chlorophyll) of deribbed spinach leaves. Homogenize in a 1 gallon Waring blender at low speed for 5 sec with 2 liters of grinding medium (sucrose: 0.3 $M$, Tricine-NaOH: 30 mM, pH 7.6, defatted bovine serum albumin: 1 gm per liter). Pass through 6 layers of muslin and one of nylon Blutex (50-$\mu$m width). Centrifugations as follows:

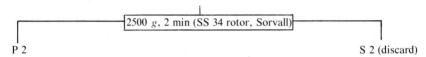

P 1                                                                    S 1 (discard)

Resuspended in 200 ml of suspension medium (sucrose: 0.3 $M$,Tricine-NaOH:10 mM, pH 7.6)
Mixed with a second batch (from another 1 kg of leaves).

P 2                                                                    S 2 (discard)

Resuspended in 120 ml of suspension medium. 30 ml of the mixture layered on top of a sucrose gradient (4 tubes, 150 ml each) made of 3 layers (30 ml each):1.5,1, and 0.75 $M$ sucrose containing 10 m$M$ Tricine-NaOH, pH 7.6.

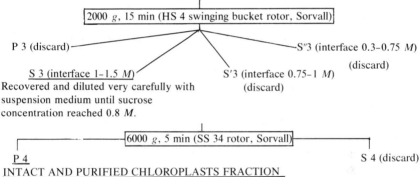

P 3 (discard)                                              S"3 (interface 0.3–0.75 $M$)
                                                                    (discard)
   S 3 (interface 1–1.5 $M$)          S'3 (interface 0.75–1 $M$)
Recovered and diluted very carefully with       (discard)
suspension medium until sucrose
concentration reached 0.8 $M$.

P 4                                                    S 4 (discard)
INTACT AND PURIFIED CHLOROPLASTS FRACTION
(chlorophyll concentration: 8 to 12 mg/ml). Diluted in 90 ml of swelling medium (Tricine-NaOH: 10 m$M$, pH 7.6, MgCl$_2$: 4 m$M$) 15 ml of the mixture layered on top of a sucrose gradient (6 tubes, 40 ml each) made of 2 layers (12 ml each): 0.93 and 0.6 $M$ sucrose containing 10 m$M$, Tricine-NaOH, pH 7.6 and 4 m$M$ MgCl$_2$.

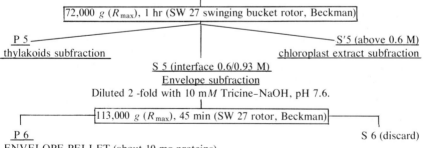

P 5                                                    S'5 (above 0.6 M)
thylakoids subfraction                      chloroplast extract subfraction
                    S 5 (interface 0.6/0.93 M)
                    Envelope subfraction
           Diluted 2 -fold with 10 m$M$ Tricine-NaOH, pH 7.6.

P 6                                                    S 6 (discard)
ENVELOPE PELLET (about 10 mg proteins)
Resuspended in 0.3 $M$ sucrose containing 10 m$M$ Tricine-NaOH, pH 7.6.

FIG. 1. Preparation of the envelope of spinach chloroplasts. All operations are performed at 2°.

material. The presence of magnesium in the medium serves to reduce breakage of thylakoids during swelling.

Chloroplast components are then separated by a sucrose density gradient procedure. After centrifugation, three subfractions can be distinguished[1,8]: a tightly packed, dark-green pellet (P5) at the bottom of the tube (thylakoid subfraction); a yellow band (S5) at the interface of the two sucrose layers (envelope membranes subfraction); and a clear brown supernatant (S'5) (soluble subfraction or chloroplast extract). There is no band at the top interface (except sometimes a whitish band rich in nucleic acid material). The envelope fraction is concentrated by a further centrifugation (P6 pellet).

The envelope preparation ($d = 1.12$) can be satisfactorily checked for cross-contamination by several methods. First, the color of the envelope membranes is examined by transmitted light. If free of thylakoid membranes they are clear and yellow in color. In contrast, in the case of contamined envelope preparations a tiny green "pinhead" spot occurs at the base of the centrifuge tube embedded at the center of the yellow pellet. Second, electron micrographs of purified envelope fraction show relatively large vesicles (average diameter 0.5 $\mu$m) or elongated profiles bordered by a single or a double membrane. In many instances, the vesicles contain smaller round vesicles bordered by a single membrane of the same thickness as the surrounding vesicle. At present, it is not clear whether both the inner and the outer envelope membranes are retained in the preparation. No plastoglobuli are trapped in the network of the envelope membranes and no thylakoids are observed. The use of purified intact chloroplasts minimized the dangers of cross-contamination.[1,8] The envelope preparations obtained from unpurified chloroplasts are heavily contaminated by other chloroplast constituents, particularly by small pieces of thylakoids rich in plastoglobuli.[1,9] Third, NADH-cytochrome $c$ oxidoreductase activity is negligible in the envelope fraction.[1,8] Hence, this fraction is essentially free of microsomal and mitochondrial membrane contaminations.

The yield of envelope membranes obtained is small (10 mg of protein) but sufficient for most lipid determinations and enzymatic activities.

## Determination of the Envelope Lipids

### Equipment and Reagents

Silica gel 60 precoated thin-layer chromatography (TLC) plates (Merck) without fluorescent indicator

---

[8] J. Joyard and R. Douce, *Physiol. Veg.* **14**, 31 (1976).
[9] B. Sprey and W. H. Laetsch, *Z. Pflanzenphysiol.* **75**, 38 (1975).

Methanol–chloroform (2:1, v/v)

Chloroform–methanol–water (65:25:4, v/v)

Chloroform–acetone–methanol–acetic acid–water (100:40:20:20: 10, v/v)

Transesterification mixture: methanol–sulfuric acid–benzene (100: 5:5, v/v)

Internal standard: behenic acid ($C_{22:0}$) 1 mg/ml in absolute methyl alcohol

*Extraction of Envelope Lipids.* In order to form one liquid phase and simultaneously to extract the lipids, 4.5 ml of methanol–chloroform (2:1) is added to the envelope suspension (1.2 ml, 10 mg protein). After 5 min, 1.5 ml of chloroform and 1.5 ml of water are added, and the mixture is agitated by bubbling with nitrogen and centrifuged. The chloroform layer is withdrawn with a Pasteur pipette and evaporated to dryness in a stream of nitrogen. The residue is dissolved in 1 ml of chloroform.

*Separation and Quantification of Envelope Lipids.* Thin-layer chromatography (TLC) provides a convenient and rapid means for the separation and precise quantitative analysis of lipids of envelope preparations. Silica gel 60 precoated TLC plates (Merck) are used and are the most successful. In order to remove any traces of lipid, the plates are carefully washed before use by ascending chromatography with a chloroform–methanol–water solvent system. The solvent front which contained the bulk of lipid contamination is discarded. The washed plates are then activated in an oven at 110° for at least 1 hr.

For two-dimensional chromatography, envelope lipids dissolved in chloroform are applied with a 50-$\mu$l Hamilton syringue to the lower right portion of the plate about 3 cm from each edge (200–600 $\mu$g of lipids are spotted). This amount allows the detection of lipids which make up even less than 1% of the lipid mixture. Solvent evaporation is hastened by a gentle jet of nitrogen. Immediately after spotting, plates are placed in rectangular chromatography chambers lined on all sides with solvent-saturated Whatman filter paper No. 1. About 150 ml of the desired solvent mixture is placed in each chamber. For the first direction a solvent system of chloroform–methanol–water is allowed to run to the front. The plate is removed and carefully dried at 35° under a slow stream of nitrogen. The sample is then chromatographed in the chloroform–acetone–methanol–acetic acid–water system in the second direction.

To bring out the lipid spots, the plates dried in nitrogen for several minutes are placed in a glass jar saturated with iodine vapor. After a few seconds, light yellow spots appear on the white background. The spots fade rapidly with time. Spots can also be located by spraying with anilinonaphthalene sulfonate (0.2% in MeOH) and viewing under UV light (360 nm). Correlation of $R_f$ values on thin layer chromatograms with standards obtained by preparative TLC and reactivity to specific spray reagents[10] provide the basis for characterization of envelope lipids. Chromatographic identification can be confirmed by mild alkaline deacylation followed by paper chromatography for separation of hydrolysis products.[11]

For quantitative fatty acid analysis, lipid areas are outlined with a scribe, scraped from the plate and transferred to a 20 ml carefully washed screw-capped vial containing 4 ml of the transesterification mixture and an accurately measured quantity of the internal standard. For major lipid spots (PC, DGDG) 100 $\mu$g of behenic acid in methanolic solution will be convenient. However, for minor lipid spots (TTGDG, PI) a fifth of this amount will be sufficient. The vials are bubbled with nitrogen, securely tightened with a Teflon-lined cap and placed in an oven at 68° for 2 hr. The separation of fatty acid methyl esters by gas chromatography and quantitation of their parent lipids are done according to the classical and very useful method of Allen and Good.[10]

### Results

The lipid and fatty acid composition of the envelope membranes and, for comparison, of the thylakoids[1] are shown in Tables I and II.

Table I clearly indicates that qualitatively the polar lipids of both types of chloroplast membranes, thylakoid, and envelope are identical, but the proportion in which they are present is different.[1,3] In contrast to the situation in mitochondria[12] and microsomal membranes[6] it is interesting to note that the envelope, when carefully prepared, is devoid of PE.[8]

Table II shows that the fatty acids are more saturated in the envelope membranes than in the thylakoids.[1,3] However, the difference found is not as dramatic as the one reported recently by Poincelot.[13]

### Lipid Biosynthesis

#### Galactolipids

##### Assay Procedure
*Principle.* The assay is based on the extent of conversion of galactose from UDPgalactose (water soluble) to several envelope lipids (gal-

[10] C. F. Allen and P. Good, Vol. 23, p. 523.

[11] M. Kates, *in* "Techniques of Lipidology. Isolation, Analysis and Identification of Lipids" (T. S. Work and E. Work, eds.), p. 558. North-Holland Publ., Amsterdam, 1972.

[12] R. Douce, *C. R. Hebd. Seances Acad. Sci.* **260**, 4067 (1965).

[13] R. P. Poincelot, *Plant Physiol.* **58**, 595 (1976).

TABLE I

Lipid Composition (Percent by Weight) of Thylakoids and Envelope Obtained after Swelling of Intact Spinach Chloroplasts by a Slight Osmotic Shock

| Lipids | Thylakoids[a] | Envelope |
|---|---|---|
| Monogalactosyldiacylglycerol (MGDG) | 52 | 16 |
| Digalactosyldiacylglycerol (DGDG) | 26 | 27 |
| Trigalactosyldiacylglycerol (TGDG) | nd | 7 |
| Tetragalactosyldiacylglycerol (TTGDG) | nd | 1 |
| Sulfoquinovosyldiacylglycerol (SL) | 6,5 | 6,5 |
| Phosphatidylcholine (PC) | 4,5 | 20 |
| Phosphatidylethanolamine (PE) | 0 | 0 |
| Phosphatidylglycerol (PG) | 9,5 | 11 |
| Diphosphatidylglycerol (DPG) | 0 | 0 |
| Phosphatidylinositol (PI) | 1,5 | 1,5 |
| Phosphatidylsérine (PS) | 0 | 0 |
| Diacylglycerol (di) | nd | 10 |
| Total lipids (mg/mg proteins) | 0,45 | 1,7 |

[a] nd, not determined.

actolipids). The enzymes involved (galactosyltransferases[14] and galactosidase[15]) catalyze the following reactions

$$\text{Diacylglycerol} + \text{UDPgal} \xrightarrow{1} \text{MGDG} + \text{UDP}$$

$$\text{MGDG} + \text{UDPgal} \xrightarrow{2} \text{DGDG} + \text{UDP}$$

$$\text{MGDG} + \text{MGDG} \xrightarrow{2'} \text{DGDG} + \text{diacylglycerol}$$

where 1 is UDPgal:diacylglycerol galactosyltransferase (first galactosylation enzyme), 2 is UDPgal:MGDG galactosyltransferase (second galactosylation enzyme), and 2' is galactolipid: galactolipid galactosyltransferase (galactosidase).

*Reagents*

MgCl$_2$ 0.1 $M$

Tricine–NaOH buffer, 10 m$M$, pH 7.2

UDP-[$^{14}$C]galactose, 2 $\mu$Ci/$\mu$mole, 10 m$M$

Silica gel 60 precoated TLC plates, Merck

Methanol–chloroform (2:1, v/v)

[14] A. Ongun and J. B. Mudd, *J. Biol. Chem.* **243**, 1558 (1968).
[15] A. Van Besouw and J. F. G. M. Wintermans, *Biochim. Biophys. Acta* **529**, 44 (1978).

TABLE II
Fatty Acid Composition of the Different Lipids from Thylakoids and
Envelope Obtained after Swelling of Intact Chloroplasts by a Slight
Osmotic Shock[a]

| Fatty Acid[a] | $C_{14:0}$ | $C_{16:0}$ | $C_{16:1}$ | $C_{16:2}$ | $C_{16:3}$ | $C_{18:0}$ | $C_{18:1}$ | $C_{18:2}$ | $C_{18:3}$ |
|---|---|---|---|---|---|---|---|---|---|
| Total lipids | | | | | | | | | |
| T | tr | 7.5 | 4.7 | tr | 12.9 | tr | 2.2 | 1.8 | 70.2 |
| E | tr | 13.4 | 3.2 | tr | 8.9 | tr | 5.2 | 8.8 | 58.6 |
| MGDG | | | | | | | | | |
| T | tr | tr | tr | tr | 26.0 | tr | 0.6 | 1.0 | 72.0 |
| E | tr | 3.9 | tr | tr | 19.7 | tr | 0.8 | 2.0 | 70.7 |
| DGDG | | | | | | | | | |
| T | 1.0 | 4.6 | tr | tr | 4.0 | tr | 2.5 | 0.8 | 87.0 |
| E | tr | 15.3 | tr | tr | 4.1 | 1.0 | 5.3 | 3.5 | 69.5 |
| TGDG | | | | | | | | | |
| E | tr | 4.0 | tr | tr | 19.1 | tr | 0.8 | 1.5 | 74.1 |
| PC | | | | | | | | | |
| T | tr | 13.2 | tr | tr | 1.1 | tr | 7.7 | 19.8 | 57.2 |
| E | 1.0 | 22.6 | 1.0 | tr | tr | tr | 15.8 | 21.3 | 37.3 |
| PG | | | | | | | | | |
| T | 1.0 | 18.3 | 39.6 | tr | tr | tr | 0.5 | 1.6 | 38.5 |
| E | 1.4 | 21.2 | 38.0 | tr | tr | tr | 1.6 | 3.6 | 32.1 |
| SL | | | | | | | | | |
| T | tr | 47.5 | tr | tr | tr | tr | 0.8 | 4.4 | 46.9 |
| E | tr | 47.1 | tr | tr | tr | 1.0 | 2.3 | 5.9 | 43.8 |
| PI | | | | | | | | | |
| T | 2.0 | 42.1 | 0.5 | 4.1 | 2.0 | 1.0 | 7.0 | 14.3 | 26.3 |
| E | 2.1 | 42.4 | tr | 3.0 | 1.0 | 1.0 | 8.0 | 20.1 | 22.2 |
| Diacylglycerol | | | | | | | | | |
| E | 4.0 | 3.3 | tr | tr | 20.1 | 1.3 | 8.7 | 3.3 | 58.8 |

[a] T, thylakoids; E, envelope; tr, trace.

Chloroform–methanol–water (65:25:4, v/v)

Envelope membranes suspended in 0.3 $M$ sucrose containing 10 m$M$ Tricine–NaOH buffer, pH 7.2 (10 mg protein/ml)

Kodirex Film, Kodak

Aquasol, New England Nuclear

*Procedure.* The incubation is performed in small test tubes for various time at 25°. The incubation mixture contains 2$\mu$moles of Tricine–NaOH buffer, 0.1 $\mu$mole of UDP-[$^{14}$C]galactose, 0.2 $\mu$mole of MgCl$_2$, and enzyme (50 $\mu$gm envelope protein) in a total volume of 0.2 ml. The incubation is stopped by the addition of 750 $\mu$l of methanol–chloroform mixture. The tubes are left at room temperature for 20 min. After addition of 250 $\mu$l of chloroform followed by 250 $\mu$l of water, the mixture is agitated by bubbling with nitrogen and centrifuged. The lower phase is withdrawn with a Pasteur pipette and evaporated to dryness under a stream of nitrogen. The residue containing the labeled galactolipids is dissolved in 1 ml of chloroform. Aliquots (200 $\mu$l) are taken for determination of radioactivity. Chloroform is carefully evaporated before counting. The activity is reported as nanomoles of galactose converted to lipid per milligram of envelope protein. Labeled galactolipids are chromatographed in one dimension on silica gel 60 precoated TLC plates; a solvent system of chloroform–methanol–water is used. The radioactivity is located by autoradiography (Kodirex film). The radioactive areas (MGDG, DGDG, . . .) are scraped into scintillation vials containing methanol (5 ml) and, after adding Aquasol (10 ml), counted by liquid scintillation.

*Results.* Figure 2 shows that the synthesis of MGDG starts very rapidly and stops after 20 min of incubation because of the envelope diacylglycerol (see Table I) consumption.[16] In marked contrast, the synthesis of DGDG starts smoothly after a lag phase but continues during a longer period of time.[16] The galactosylation enzyme system is specifically localized in the envelope membranes and has not been obtained in soluble form.

The envelope membranes contain saturating concentrations of diacylglycerol and MGDG (see Table I). For these reasons, additions of diacylglycerol and MGDG to the incubation medium are without effect on the initial rates of MGDG and DGDG formation.

The first galactosylation enzyme has its pH optimum above pH 8. In contrast, the second galactosylation enzyme has its maximum activity around pH 6.5.

The apparent Michaelis constant ($k_m$) for UDPgalactose is approximately 90 $\mu M$ for the first galactosylation enzyme. Others have not been measured.

Triton X-100 (0.9% by volume) is a strong inhibitor of the second galactosylation enzyme, but is practically without effect on the first galactosylation enzyme.

## Phosphatidic Acid and Diacylglycerol

### Assay Procedure

*Principle.* The assay is based on the extent of conversion of *sn*-glycerol-3-phosphate (water soluble) to several envelope lipids (lysophosphatidic acid, phosphatidic acid, and diacylglycerol). The enzymes

---

[16] J. Joyard and R. Douce, *Biochim. Biophys. Acta* **424**, 125 (1976).

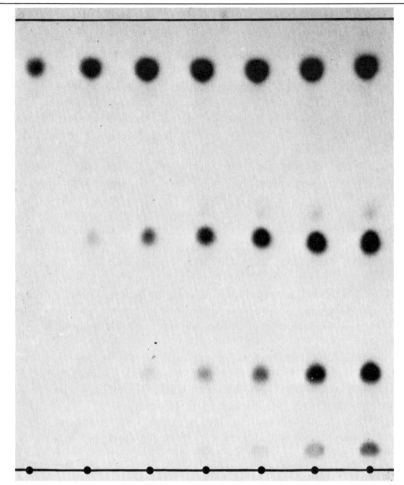

Fig. 2. Incorporation of UDP-[$^{14}$C]galactose into the lipids of the envelope of spinach chloroplasts. MGDG, Monogalactosyldiacylglycerol (1st row); DGDG, diagalactosyldiacylglycerol (2nd row); TGDG, Trigalactosyldiacylglycerol (3rd row); TTGDG: tetragalactosyldiacylglycerol (4th row).

involved, operating in a multienzyme sequence, catalyze the following reactions.[5]

$$ATP + fatty\ acid + CoASH \xrightarrow{1} acyl\text{-}CoA + AMP + PP_i$$

$$sn\text{-}Glycerol\text{-}3\text{-}phosphate + acyl\text{-}CoA \xrightarrow{2} lysophosphatidic\ acid + CoASH$$

$$Lysophosphatidic\ acid + acyl\text{-}CoA \xrightarrow{3} phosphatidic\ acid + CoASH$$

$$Phosphatidic\ acid \xrightarrow{4} diacylglycerol + P_i$$

where 1 is acyl-CoA synthetase, 2 is acyl-CoA:*sn*-glycerol-3-phosphate acyltransferase (first acylase), 3 is acyl-CoA:Acyl-*sn*-glycerol-3-phosphate acyltransferase (second acylase), and 4 is phosphatidic acid phosphatase.

*Reagents*

Tricine–NaOH buffer, 10 m$M$, pH 7.4
*sn*-[$^{14}$C]Glycerol-3-phosphate, 5 $\mu$Ci/$\mu$mole, 10 m$M$
CoASH, 10 m$M$
ATP, 0.1 $M$
Silica gel 60 precoated TLC plates, Merck
Methanol–chloroform (2:1, v/v)
Chloroform–methanol–water (65:25:4, v/v)
Envelope membranes suspended in 0.3 M sucrose containing 10 m$M$ Tricine–NaOH buffer, pH 7.4 (10 mg protein/ml); Chloroplast extract[5] (40 mg protein/ml$^{-1}$)
Kodirex film, Kodak
Aquasol, New England Nuclear

*Procedure*. The incubation is performed in small test tubes for various time at 25°. The incubation mixture contains 2 $\mu$moles of Tricine–NaOH buffer, 1 $\mu$mole of ATP, 0.2 $\mu$mole of CoASH, 0.1 $\mu$mole of *sn*-[$^{14}$C]glycerol-3-phosphate, and enzyme (100$\mu$g of envelope protein + 2 mg of chloroplast extract protein) in a total volume of 0.2 ml. Addition of fatty acids to the incubation mixture is not necessary. The incubation is stopped by addition of 750$\mu$l of the methanol–chloroform mixture and the labeled lipids are separated and analyzed as previously described. The activity is reported as nanomoles of sn-glycerol-3-phosphate converted to lipid per milligram of envelope protein.

*Results*. Figure 3 shows that the main product formed is phosphatidic acid which is then slowly dephosphorylated to yield the corresponding diacylglycerol. Under these assay conditions, little lysophosphatidic acid accumulates, suggesting that the second acylase very rapidly utilizes the first acylation product (lysophosphatidic acid).

The acyl-CoA synthetase,[5,17] the acyl-CoA:acyl-*sn*-glycerol-3-phosphate acyltransferase,[5] and the phosphatidic acid phosphatase[5] are firmly and specifically bound to the envelope membranes. In marked contrast, the acyl-CoA:*sn*-glycerol-3-phosphate acyltransferase is soluble[5,18] and present in the chloroplast extract (stroma). Without chloroplast extract in the incubation medium the envelope membranes are unable to synthesize phosphatidic acid. However, it is very likely that the added soluble

[17] P. G. Roughan and C. R. Slack, *Biochem. J.* **162**, 457 (1977).
[18] M. Bertrams and E. Heinz, *Planta* **132**, 161 (1976).

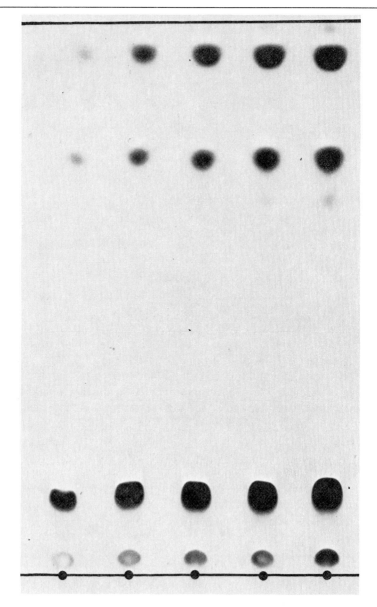

Fig. 3. Incorporation of sn-[<sup>14</sup>C]glycerol 3-phosphate into the lipids of the envelope of spinach chloroplasts. Lyso PA, Lysophosphatidic acid (4th row); PA, phosphatidic acid (3rd row); mono, monoacylglycerol (2nd row); di, diacylglycerol (1st row).

acylase recognizes some specific sites on the surface of the envelope membranes.

An assay of 60 min shows that the conversion of $sn$-glycerol-3-phosphate to envelope lipids is proportional to the amount of chloroplast extract added over a limited range.[5]

The different enzymes have different optimum pH. The soluble acylase has its maximum activity at pH 7.0; the bound acylase has its pH optimum around 7.8, and the phosphatidic acid phosphatase is an alkaline phosphatase; the maximum of its activity is at pH 9.0.

The apparent $K_m$ for $sn$-glycerol-3-phosphate is approximately 0.5 m$M$.


## [27] Coupling Factors from Higher Plants

*By* NATHAN NELSON[1]


The coupling device of energy transduction is essentially a proton translocating ATPase that is composed of two sectors, a membrane-bound ATPase that can readily be released to a soluble form and a membrane sector that conducts protons across the membrane. Proton ATPases from various organisms resemble each other in many respects. It is assumed the mitochondrial coupling device from plants should be similar to those from yeast and animals; therefore, most of the work on plant material was carried out on the chloroplast coupling device which possesses some unique properties. Chloroplast coupling factor 1 (CF$_1$) is the ATPase enzyme from chloroplasts. A comprehensive description of the purification and assay of this enzyme has been described in a previous volume.[2] Alternative or modified procedures and new developments concerning CF$_1$ and the membrane sector of the chloroplast coupling device are described here.


### Assay of ATPase Activity of CF$_1$

The latent ATPase activity of CF$_1$ can be activated by heat treatment or by trypsin digestion or by prolonged incubation at high concentrations of dithiothreitol as previously described.[2] An alternative procedure for

[1] This research was supported by a grant from the United States–Israel Binational Science Foundation (BSF), Jerusalem, Israel.
[2] S. Lien, and E. Racker, Vol. 23 [49].

trypsin activation is to include 25 $\mu$g of trypsin in the ATPase assay medium.

*Reagents*

Reaction mixture: 60 m$M$ Tricine (pH 8), 8 m$M$ ATP, 16 m$M$ CaCl$_2$, and 2 × 10$^5$ cpm of [$\gamma$-$^{32}$P]ATP

Trypsin [treated with L-1-tosylamido-2-phenylethylchloromethyl ketone (TPCK) a specific chymotrypsin inhibitor], 5 mg/ml in 3 m$M$ H$_2$SO$_4$

Trichloroacetic acid (TCA) 30%

Ammonium molybdate, 1.2% in 1.2 $N$ HCl

Isobutanol–benzene–acetone 5 : 5 : 1 (v/v) saturated with water

*Procedure*. The ATPase reaction is performed in a medium containing 0.5 ml reaction mixture, 5 $\mu$l trypsin, and 1 to 2 $\mu$g of CF$_1$ in a final volume of 1 ml. The reaction is started by the addition of CF$_1$ and allowed to proceed for 10 min at 37°. The reaction is then terminated by the addition of 0.1 ml TCA. After centrifugation at 1500 $g$ for 10 min, 0.5 ml of the deproteinized supernatant fluid is transferred to 16 × 150 mm tubes for extraction and measurement of the $^{32}$P$_i$ liberated from the [$\gamma$-$^{32}$P]ATP. To each tube. 0.5 ml water, 4 ml of the ammonium molybdate solution, and 7.0 ml of the isobutanol–benzene–acetone solution are added. The tubes are agitated vigorously an a Vortex mixer for 20 sec, and the phases are allowed to separate. An aliquot of 1.0 ml is then removed from the organic phase for determination of radioactivity. One of the reaction tubes, in which CF$_1$ is omitted, serves as a control. In this tube, 1 ml of the organic phase is counted for free $^{32}$P$_i$ that was present prior to the ATPase reaction, and the radioactivity of 1 ml from the aqueous phase is measured to determine the total counts of [$\gamma$-$^{32}$P$_i$]ATP in each tube. The amount of $^{32}$P$_i$ formed in each reaction tube is calculated as follows

$$\mu\text{moles }{}^{32}\text{P}_i\text{ formed} = \frac{(\text{cpm in reaction tube} - \text{cpm in control tube}) \times 7 \times 4}{(\text{cpm in aqueous phase of control tube}) \times 5}$$

*Remarks*. A similar procedure is used with CF$_1$ that was activated by heat, trypsin, or dithiothreitol except that trypsin is omitted from the reaction medium.

Since the amount of protein in the reaction medium is low the reaction can be performed in 16 × 150 mm tubes and terminated by addition of 4 ml of ammonium molybdate; the extraction can take place in the same tubes.

The procedure described can be used for the assay of Mg$^{2+}$-ATPase activity of heat or trypsin activated CF$_1$. The CaCl$_2$ in the reaction mixture is replaced by 4 m$M$ MgCl$_2$, and 60 $\mu$moles of sodium maleate (pH 8) or sodium bicarbonate are added to the reaction tubes.[3]

### Preparation of [γ-$^{32}$P]ATP

A convenient method for the preparation of [γ-$^{32}$P]ATP is photophosphorylation of ADP in the presence of $^{32}$P$_i$, followed by purification of [γ-$^{32}$P]ATP on a Dowex 1-Cl column.[4,5]

*Procedure.* The reaction mixture contained in a volume of 3 ml; 100 $\mu$moles of Tricine, 30 $\mu$moles of MgCl$_2$, 20 $\mu$moles of ADP, 20 $\mu$moles of sodium P$_i$ and 2 mCi of carrier-free $^{32}$P$_i$ (The $^{32}$P$_i$ was incubated at 100° for 10 min in 0.5 ml of 1 $N$ HCl). The reaction mixture is adjusted to pH 8 by 1 $N$ NaOH then 0.15 $\mu$mole of $N$-methylphenazonium methosulfate, chloroplasts equivalent to 0.5 mg of chlorophyll and water are added to give a final volume of 5 ml. After illumination by white light (3 × 10$^5$ ergs/cm$^2$/sec) for 5 min the suspension is centrifuged at 20,000 $g$ for 10 min. The clear supernatant is applied to a Dowex 1-Cl column (6 × 80 mm). The column is washed with 30 ml water, then 30 ml of 3 m$M$ HCl, then 50 ml of 10 m$M$ HCl and 20 m$M$ NaCl. The ATP is eluted with 50 ml of a solution containing 10 m$M$ HCl and 200 m$M$ NaCl. Fractions of 4 ml are collected in tubes containing 40 $\mu$mole of Tricine (pH 8). The pH is readjusted to 8 with 1 $N$ NaOH and the fractions are stored frozen.

*Remarks.* The Dowex column can be regenerated by washing with 30 ml of 1 $N$ HCl and distilled water and used for many preparations. Usually less than 2 nmole of ATP thus prepared are required to give the 2 × 10$^5$ cpm for each ATPase reaction tube. Therefore, the [γ-$^{32}$P]ATP can be considered as carrier-free.

The amount of free $^{32}$P$_i$ in the [γ-$^{32}$P]ATP tubes is usually 0.2 to 1.0%

### Assay of Coupling Activity of CF$_1$

Soluble form of CF$_1$ retains the ability to hydrolyse ATP, but loses the ability to catalyse ATP synthesis. In the presence of Mg$^{2+}$, the soluble preparation readily recombines with suitable CF$_1$-depleted chloroplasts, and the ability to synthesize ATP is regained. Stable EDTA-treated chloroplasts are used for the assay of coupling activity of CF$_1$.

*Preparation of Stable EDTA-Treated Chloroplasts.* The conditions for the removal of CF$_1$ from the chloroplast membranes vary from one

---

[3] N. Nelson, H. Nelson, and E. Racker, *J. Biol. Chem.* **247**, 6506 (1972).

[4] M. Avron, *Anal. Biochem.* **2**, 535 (1961).

[5] W. E. Cohn, and C. E. Carter, *J. Am. Chem. Soc.* **72**, 4273 (1950).

plant species to another and are dependent on the season of the year. Therefore, the concentration of EDTA during the treatment should be adjusted, prior to the large-scale preparation, so that the residual photophosphorylation activity of the EDTA particles is 5 to 10 $\mu$moles ATP formed/mg Chl/hr.

Chloroplasts suspended in a medium containing 0.4 $M$ sucrose, 10 m$M$ NaCl, 10 m$M$ Tricine (pH 8), and 10 mg/ml bovine albumin (Sigma, fraction V) at a chlorophyll concentration of about 1.5 mg/ml are diluted ten times in 2 m$M$ (or other desirable concentration) EDTA solution of 0°. Immediately after dilution the suspension is centrifuged at 20,000 $g$ for 10 min, and the pellet is suspended with a glass–Teflon homogenizer in a medium containing 0.4 $M$ sucrose, 10 m$M$ NaCl, 10 m$M$ Tricine (pH 8), and 10 mg/ml bovine albumin at a chlorophyll concentration of about 1.5 mg/ml. The $CF_1$-depleted particles are divided into 0.5 ml portions, frozen immediately at −70°, and thawed just before use. The inclusion of 1% bovine serum albumin stabilizes the particles, and they can be stored at −70° for several months.

The reconstitution of the particles with $CF_1$ and measurement of photophosphorylation are carried out as previously described.[6]

*Preparation of Chloroplasts Highly Depleted of $CF_1$ with a High Degree of Reconstitution.* A solution of 5 $M$ NaBr is added to chloroplasts suspended in a medium containing 0.4 $M$ sucrose, 10 m$M$ NaCl, 5 m$M$ dithiothreitol (DTT), and 10 m$M$ Tricine (pH 8), to give a final concentration of 2 $M$ NaBr and a chlorophyll concentration of 1 mg/ml. After incubation at 0° for 30 min, an equal volume of water is added and the suspension is centrifuged at 20,000 $g$ for 10 min. The pellet is suspended in 10 m$M$ Tricine (pH 8) at ten times the original volume with a glass–Teflon homogenizer and centrifuged at 20,000 $g$ for 10 min. The resulting pellet is homogenized in a medium containing 0.4 $M$ sucrose, 10 m$M$ NaCl, 5 m$M$ DTT, and 10 m$M$ Tricine (pH 8) at a chlorophyll concentration of about 2 mg/ml.

The depleted chloroplasts are essentially free of $CF_1$ as judged by lack of ATPase activity and the absence of the $\alpha$- and $\beta$-subunits in the SDS gels of the particles. Upon reconstitution with $CF_1$, the photophosphorylation activity of the particles increased from 0 up to 300 $\mu$moles $P_i$ mg Chl/hr, and over 60% of the extent of proton uptake was restored. After addition of 10 mg/ml bovine serum albumin, the $CF_1$-depleted chloroplasts could be stored at 4° for a week showing only a small decrease in their reconstitution activity. Upon addition of 25% glycerol the particles can be stored at −70° for several weeks with negligible decrease in the reconstitution activity.

[6] R. E. McCarty, Vol. 23 [23].

Preparation of $CF_1$

An elaborate procedure for large-scale purification of $CF_1$ was developed by Lien and Racker.[2] Using a slightly modified procedure, $CF_1$ has been purified from various plant sources with good yields and high degrees of purity. Quite frequently, large quantities of ribulose diphosphate carboxylase follow the $CF_1$ through the purification procedure. In order to eliminate this, the washing procedure during the preparation of chloroplasts was altered. The first chloroplast pellet, containing 200 to 500 mg of chlorophyll, is homogenized with a glass–Teflon homogenizer in 1.5 liters of medium containing 0.15 $M$ NaCl and 10 m$M$ Tricine (pH 8). After centrifugation for 10 min at 25,000 $g$, the supernatant is discarded and the pellet is homogenized with a glass–Teflon homogenizer in 1.5 liters of medium containing 10 m$M$ Tricine (pH 8). The suspension is centrifuged for 20 min at 25,000 $g$, and the pellet is homogenized in 10 m$M$ Tricine (pH 8) to give a chlorophyll concentration of 2 to 4 mg/ml. The EDTA treatment and the purification procedure are performed essentially as previously described.[2] The only exception is the swelling of DEAE–Sephadex A-50 that is performed as follows: 10 gm of dry gel are added to 1.9 liters distilled water while being stirred by a magnetic stirrer. The gel is allowed to swell for 10 min and then 4.8 gm of solid Tris, 20 ml of 0.1 $M$ ATP (Sigma grade II), 20 ml of 0.2 $M$ EDTA (pH 7.3), and 60 ml of 2 $M$ $(NH_4)_2SO_4$ are added. After an additional 10 min of stirring, the gel is allowed to sediment, the supernatant is decantated, and the column is packed without delay. This procedure is simple and brings about faster flow rates through the column.

Using this procedure, $CF_1$ was purified from spinach, Swiss chard, lettuce, and pea chloroplast. Upon addition of sodium dodecyl sulfate (SDS) and electrophoresis on SDS gels, $CF_1$ is dissociated into five different polypeptides.[7] They were designated as $\alpha$, $\beta$, $\gamma$, $\delta$, and $\epsilon$-subunits, according to the order of their decreasing molecular weights of 59,000, 56,000, 37,000, 17,500, and 13,000, respectively. A subunit composition of $2\alpha$, $2\beta$, $1\gamma$, $1\delta$, and $2\epsilon$ was reported.[7]

Preparation of $CF_1$ That Is Composed of $\alpha$- and $\beta$-Subunits[8]

An ammonium sulfate precipitate containing 30 mg of purified $CF_1$ is suspended in 10 ml solution containing 10 m$M$ Tricine (pH 8), 2 m$M$ EDTA, and 1 m$M$ ATP. Two hundred microliters of a 0.5% solution of TPCK–trypsin in 3 m$M$ $H_2SO_4$ are added, and the mixture is incubated

[7] N. Nelson, *Biochim. Biophys. Acta* **456**, 314 (1976).
[8] D. W. Deters, E. Racker, N. Nelson, and H. Nelson, *J. Biol. Chem.* **250**, 1041 (1975).

for 6 hr at 20°. Then an equal volume of saturated ammonium sulfate is added, and the suspension is centrifuged at 10,000 $g$ for 10 min. The pellet is dissolved in 2 ml of 10 m$M$ Tricine (pH 8), 2 m$M$ EDTA, 1 m$M$ ATP, and 100 m$M$ NaCl and chromatographed on a column (1.25 × 95 cm) of BioGel A-0.5 m that had been equilibrated with the same buffer. The flow rate is adjusted to give about 4 ml of effluent per hour. Fractions of 2 ml are collected and the tubes with peak ATPase activity are precipitated by addition of an equal volume of saturated ammonium sulfate. The preparation is kept at 4° and, when desired, is precipitated, dissolved, and assayed as $CF_1$ except that activation is not required. The purification procedure of trypsin-treated $CF_1$ is summarized in Table I.

*Properties.* Sodium dodecyl sulfate–acrylamide gel electrophoresis usually revealed one band in the position of the $\alpha$- and $\beta$-subunits of $CF_1$. Antiserum against $CF_1$ inhibited the ATPase activity of trypsin-treated $CF_1$ and antiserum against trypsin-treated $CF_1$ resembled the antiserum against $CF_1$. both antisera against $\alpha$- and $\beta$-subunits gave precipitation lines on immonudiffusion plates with trypsin-treated $CF_1$, whereas the antisera against $\gamma$-, $\delta$-, and $\epsilon$-subunits did not. Therefore, it seems that the trypsin treatment completely removed the $\gamma$-, $\delta$-, and $\epsilon$-subunits and also cleaved out the small polypeptide of the $\alpha$-subunit.

Like $CF_1$, trypsin-treated $CF_1$ is also cold labile, and loses its ATPase activity upon incubation at 0°. The presence of ATP stabilizes the enzyme against cold inactivation. In contrast to $CF_1$ the trypsin-treated enzyme is insensitive to the $CF_1$ inhibitor ($\epsilon$-subunit). Quercetin, 7-chloro-4-nitrobenzo-2-oxa-1,3-diazole and ADP inhibit the ATPase activity of the tripsin-treated enzyme in a fashion resembling their effects on $CF_1$. Trypsin-treated $CF_1$ does not bind to EDTA-treated chloroplasts and, therefore, does not reconstitute their photophosphorylation activity.

## Preparation of $CF_1$ Deficient of $\delta$-Subunit

The firmness of the $\delta$-subunit binding to $CF_1$ from different plant species is variable. Purified $CF_1$ from lettuce leaves is partially depleted of its $\delta$-subunit.[9] Using chloroform extraction, $CF_1$ lacking its $\delta$-subunit can be obtained.[10]

Chloroplasts are prepared as described for the preparation of $CF_1$ and finally suspended in a solution containing 0.25 $M$ sucrose, 10 m$M$ Tris-$SO_4$ (pH 7.6), 1 m$M$ EDTA, 2 m$M$ ATP, and 5 m$M$ DTT to give a chlorophyll concentration of about 3 mg/ml. To the chloroplast suspen-

[9] N. Nelson, and O. Karny, *FEBS Lett.* **70**, 249 (1976).
[10] H. M. Younis, G. D. Winget, and E. Racker, *J. Biol. Chem.* **252**, 1814 (1977).

TABLE I
SUMMARY OF THE PURIFICATION PROCEDURE OF TRYPSIN-TREATED $CF_1$

| Purification steps | Total protein (mg) | Specific activity ($\mu$mole $P_i$/mg protein/min) | Total units | Recovery (%) |
|---|---|---|---|---|
| Activated $CF_1$ | 30 | 22 | 660 | 100 |
| Ammonium sulfate pellet after 6 hr digestion | 21 | 24 | 504 | 76 |
| After BioGel column | 15 | 30 | 450 | 68 |

sion which contains up to 300 mg of chlorophyll, one-half volume of chloroform is added at room temperature, and the mixture is stirred with a magnetic stirrer for 15 sec. The emulsion is separated into two phases by centrifugation at about 300 $g$ for 3 min. The aqueous layer is removed and centrifuged at 20,000 $g$ for 10 min at room temperature. The yellow supernatant, which contains the $CF_1$, is applied on a DEAE–Sephadex A-50 column (2.5 × 30 cm) that was prepared as described for the preparation of $CF_1$. The enzyme is eluted by a linear gradient of ammonium sulfate from 0.1 to 0.4 $M$ (200 ml in each chamber) in a solution containing 20 m$M$ Tris-$SO_4$ (pH 7.6), 2 m$M$ EDTA, and 1 m$M$ ATP. The $CF_1$ appears in the fractions containing 0.2 to 0.25 $M$ ammonium sulfate. To these fractions an equal volume of saturated ammonium sulfate is added, and the resulting suspension is stored at 4°.

Only four polypeptides ($\alpha$-, $\beta$-, $\gamma$-, and $\epsilon$-subunits) are apparent on SDS gels. The preparation possesses a latent ATPase that can be activated by heat or trypsin treatments. The four-subunit $CF_1$ lost its capacity to bind to chloroplast membranes. Consequently this preparation fails to restore photophosphorylation activity of $CF_1$-depleted chloroplast particles.

### Purification and Properties of $\delta$- and $\epsilon$-Subunits of $CF_1$

Purification procedures for the $\delta$- and $\epsilon$-subunits of $CF_1$ have been described.[9,11] The procedures are based on differential precipitation of $CF_1$ subunits by pyridine and DEAE–cellulose chromatography in the presence of urea. A general purification procedure for the $\delta$- and $\epsilon$-subunits has been developed.

A suspension of $CF_1$ in 50% saturated ammonium sulfate, containing

[11] N. Nelson, H. Nelson, and E. Racker, *J. Biol. Chem.* **247**, 7657 (1972).

about 50 mg of protein, is centrifuged at 10,000 $g$ for 10 min. The precipitate is dissolved in 20 ml of a solution containing 10 m$M$ Tricine (pH 8) and 2 m$M$ EDTA. Twenty milliliters of pyridine are added with vigorous stirring, and the clear solution is allowed to stand for 10 min at room temerature. Sixty milliliters of distilled water and 1 ml of saturated $(NH_4)_2SO_4$ are added with stirring. The resulting turbid solution is kept at 4° for 6 hr and centrifuged at 10,000 $g$ for 10 min. The pellet contains mainly the $\alpha$-, $\beta$-, and $\gamma$-subunits of CF$_1$. To the supernatant, which contains the $\delta$- and $\epsilon$-subunits, 100 ml of ethanol are added and the solution is incubated overnight at 4°. The slightly turbid solution is centrifuged at 10,000 $g$ for 10 min. The pellet is freed of excess pyridine by a stream of nitrogen gas and is then dissolved in 5 ml of a solution containing 10 m$M$ Tris-Cl (pH 8) and 9 $M$ urea. The mixture is applied on a DEAE–cellulose column (1 × 12 cm)) that was equilibrated with the same buffer. The elution is carried out by a linear gradient of 0 to 300 m$M$ NaCl in a solution containing 10 m$M$ Tris-Cl (pH 8) and 9 $M$ urea (20 ml in each chamber). Fractions of 1.2 ml are collected, and the purity of the fractions is tested by SDS gel electrophoresis.

Figure 1 shows the elution profile of the DEAE–cellulose column. The early fractions contain the purified $\epsilon$-subunit and the late fractions contain the purified $\delta$-subunit with recovery of 10 to 20%. Fractions which are not pure $\delta$- or $\epsilon$-subunits are pooled separately, and can be farther purified by similar DEAE–cellulose column after threefold dilution with 9 $M$ urea.

*Properties of the Purified $\epsilon$-Subunit.* A molecular weight of 13,000 for the purified $\epsilon$-subunit was estimated by SDS gel electrophoresis.[11] Amino

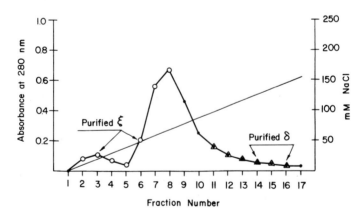

FIG. 1. Elution pattern of $\delta$- and $\epsilon$-subunits from a DEAE–cellulose column. (○), Fractions that contain the $\epsilon$-subunit; Δ, fractions that contain the $\delta$-subunit.

acids analysis revealed single residues of histidine, methionine, cysteine, and phenylalanine per ε polypeptide. Purified ε-subunit inhibits ATPase activity of heat activated $CF_1$. Therefore, it was termed as the "$CF_1$ inhibitor." The $CF_1$ inhibitor is heat stable but trypsin sensitive. It is specific for $CF_1$ and has no effect on the mitochondrial ATPase. It has been concluded that the ε-subunit takes part in the latency of $CF_1$. Activation of latent ATPase of $CF_1$ probably occurs via the removal of the ε-subunit from the active site, either by trypsin digestion or by conformational changes caused by heat or DTT treatments.

Antibody against γ-subunit prevented the inhibition of heat-activated $CF_1$ by the isolated ε-subunit. Furthermore, the $CF_1$ inhibitor does not inhibit the ATPase activity of trypsin-treated $CF_1$ in which the γ-subunit is missing.[8] This suggests that the binding site for $CF_1$ inhibitor is on γ-subunit, while its inhibitory effect is probably on the β-subunit.

*Properties of the Purified δ-Subunit.*[9] A molecular weight of 17,500 was estimated for the δ-subunit by SDS gel electrophoresis. The purified δ-subunit, like the other subunits of $CF_1$, is not soluble in water. It can be kept in 7 to 9 $M$ urea solution. Purified δ-subunit has no effect on ATPase activity of $CF_1$. The presence of δ-subunit within $CF_1$ is obligatory for its binding properties to the membrane and therefore for its coupling activity. Preparations of $CF_1$ which are depleted of the δ-subunit fail to bind to $CF_1$-depleted chloroplast membranes. Upon addition of purified δ-subunit, the binding properties and coupling activity are restored.

## Purification and Properties of a Chloroplast Proteolipid

Chloroplast membranes depleted of $CF_1$ become permeable to protons. This specific proton leak can be blocked by $N,N'$-dicyclohexylcarbodiimide (DCCD). It is assumed that the DCCD closes a specific proton channel which is part of the chloroplast coupling device. The isolation of chloroplast proteolipid can be achieved by butanol solubilization.[12] Upon reconstitution of this proteolipid into liposomes composed of chloroplast lipids, a DCCD-sensitive proton channel is formed.

*Isolation of the Chloroplast Proteolipid.* Two milliliter of a chloroplast suspension in 10 m$M$ Tricine (pH 8) containing 3 to 4 mg Chl/ml are injected into 100 ml of $n$-butanol at 0° under vigorous stirring. Following agitation for 30 min, the suspension is centrifuged twice at 20,000 $g$ for 10 min. To the butanol supernatant, 500 ml of diethyl ether are added,

---

[12] N. Nelson, E. Eytan, B. Notsani, H. Sigrist, K. Sigrist-Nelson, and C. Gitler, *Proc. Natl. Acad. Sci. U.S.A.* (in press).

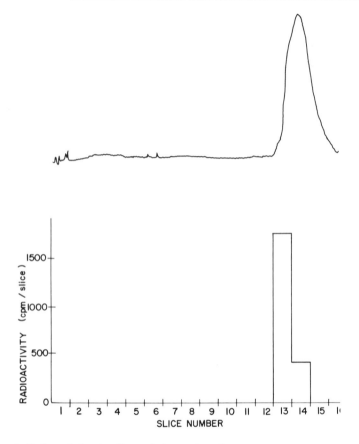

FIG. 2. Sodium dodecyl sulfate gel electrophoresis pattern of the purified proteolipid and [$^{14}$C]DCCD incorporation into it. Chloroplast membranes (3.5 mg chlorophyll in 1 ml) were incubated for 1 hr at room temperature with 100 nmole of [$^{14}$C]DCCD (45 mCi/m$M$ in ethanol). The proteolipid was isolated as described in the text. A sample containing 10 $\mu$g protein was incubated for 2 hr at room temperature with 2% sodium dodecyl sulfate and 2% mercaptoethanol and electrophoresed in the presence of sodium dodecyl sulfate. Following fixation the Coomassie blue-stained gels were scanned at 600 nm and subsequently cut into 0.5 cm slices. The slices were incubated at 70° for 2 hr in 0.5 ml Soluene 350 (Packard). Upon bleaching of the Coomassie blue stain the radioactivity was measured in 10 ml of toluene scintillation fluid.

and the mixture is incubated for 30 min at 0°. The proteolipid is collected by centrifugation in four 30-ml glass tubes at 10,000 $g$ for 10 min. The pellets are suspended in 1 ml of distilled water or other desired aqueous solutions.

Figure 2 depicts SDS gel pattern of the proteolipid that was isolated

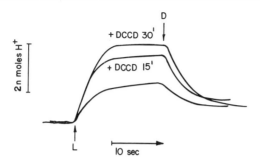

FIG. 3. Enhancement of light-induced proton uptake by DCCD in proteolipid bacteriorhodopsin vesicles. The experimental conditions are as described in the text. The concentration of DCCD, when present, was 20 $\mu M$.

from chloroplast membranes which were incubated for 1 hr with [$^{14}$C]DCCD. When the purification procedure is monitored by SDS gel electrophoretic patterns of protein and $^{14}$C incorporation, the radioactivity is found to be incorporated predominantly into the proteolipid.

*Isolation of a Proteolipid Active in DCCD-Sensitive Proton Translocation.* Six milliliters of a chloroplast suspension in 10 m$M$ Tricine (pH 8) containing 3 to 4 mg Chl/ml are injected into 300 ml of $n$-butanol at 0° under vigorous stirring. Following agitation for 30 min the suspension is centrifuged twice at 20,000 $g$ for 10 min. The supernatant that contains all of the chloroplast pigments is then evaporated to dryness in a rotary evaporator in which the reduced pressure is provided by a vacuum pump, the condenser is cooled to about −20°, and the flask with the butanol supernatant is maintained at room temperature. The dry material is dissolved in 3 ml of $n$-butanol and kept at 4° for up to 10 days.

*Assay.* Aliquots of 0.5 ml in 12 × 140 mm glass tubes are dried by a stream of nitrogen gas to evaporate the $n$-butanol. To the dry chloroplast, lipids containing the proteolipid 0.4 ml of 0.15 $M$ NaCl solution are added followed by sonication in a bath-type sonicator[13] for approximately 15 min. Then 0.1 ml of bacteriorhodopsin (3 mg protein/ml, prepared as described previously[14]) is added and sonication is continued for an additional 10 min.

Fifty microliters of the liposomes are assayed for DCCD-sensitive light-induced proton uptake in final volume of 1 ml of 0.15 $M$ NaCl at an initial pH of 6.7. The light-induced proton uptake is monitored by meas-

---

[13] C. Miller, and E. Racker, *J. Membr. Biol.* **26**, 319 (1976).
[14] D. Oesterhelt, and W. Stoeckenius, Vol. 31 [69].

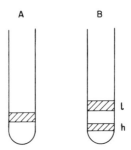

FIG. 4. The position of liposomes prepared from chloroplast lipids on ficoll gradient. Tube A: liposomes prepared by sonication of chloroplast lipids containing proteolipid. Tube B: liposomes prepared by sonication of chloroplast lipids containing proteolipid in the presence of 1% sodium cholate.

uring the light-induced increase in the external pH in the absence and presence of DCCD.

Figure 3 shows that the liposomes are active in light-induced proton uptake. After addition of 20 nmoles of DCCD and incubation for 30 min, the extent of the light-induced proton uptake was more than doubled. Since the dark decay kinetics with and without DCCD are similar, the most plausible explanation is that a mixture of liposomes is present. Some without proteolipid show the proton uptake in the absence of DCCD. The rest, containing the proteolipid which show no net proton uptake unless DCCD is present to block the proteolipid-induced leak. If,

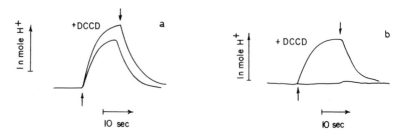

FIG. 5. Light-induced proton uptake in liposomes that were separated on ficoll gradient. (a) Lipsomes (0.1 ml) from the light band (l) from tube B-in Fig. 4 were sonicated with 50 $\mu$l of bacteriorhodopsin (3 mg protein/ml) for 10 min. The resulting vesicles (50 $\mu$l) were assayed for light-induced proton uptake and DCCD-sensitivity as described in the text. The concentration of DCCD, when present, was 100 $\mu M$ and the light-induced proton uptake was measured after a 10 min incubation. (b) The experiment was performed as described in (a) except that the liposomes were from the heavy band (h) from tube B in Fig. 4.

indeed, this is the case, one should be able to select between the two populations of liposomes.

#### SEPARATION OF LIPOSOMES ENRICHED WITH PROTEOLIPID BY FICOLL GRADIENT.

Aliquots of 0.2 ml of the *n*-butanol solution containing the chloroplast lipids and proteolipid are dried under stream of nitrogen gas. To the dry material 0.3 ml of a solution containing 50 m$M$ NaCl and 1% sodium cholate is added. After sonication for 10 min in a bath-type sonicator, the samples are placed on Ficoll gradients from 1 to 10%. The gradients are centrifuged at 5° in a SW50 rotor at 45,000 rpm for 15 hr. Two distinct green bands are formed and collected separately. A sample that was sonicated in a solution containing only 50 m$M$ NaCl appeared in the Ficoll gradient as a single green band. Figure 4 illustrates the position of the bands in the Ficoll gradients.

Upon sonication of the green band of tube A with bacteriorhodopsin the liposomes that were formed showed DCCD-sensitive light-induced proton uptake similar to that described in Fig. 3.

The bacteriorhodopsin liposomes that were formed from the light band (1) of tube B in Fig. 4 show a light-induced proton uptake which is almost insensitive to DCCD. The light-induced proton uptake into liposomes that are formed from the heavy band (h) of tube B in Fig. 4 was absolutely dependent on the addition of DCCD (Fig. 5). The amount of proteolipid and the incorporation of [$^{14}$C]DCCD are in good correlation with the DCCD sensitivity of the various liposomal preparations.

## [28] Coupling Factors from Photosynthetic Bacteria

*By* ASSUNTA BACCARINI-MELANDRI and BRUNO ANDREA MELANDRI

Coupling factors (ATPase) have been isolated and purified from three genera of photosynthetic bacteria, namely, *Rhodopseudomonas capsulata, Rhodospirillum rubrum* (Rhodospirillaceae), and *Chromatium vinosum* (strain D) (Chromatiaceae). This article will be divided into three sections, one for each species of photosynthetic microorganism. A review on the physical and chemical properties of these bacterial enzymes and on the modulation of membrane-bound ATPase has been recently published.[1]

---

[1] B. A. Melandri, and A. Baccarini-Melandri, *J. Bioenerg.* **8**, 109 (1976).

*Rhodopseudomonas capsulata,* Strain St. Louis (American Type Culture Collection No. 23782)

The first successful isolation and characterization of a coupling factor from a facultative photosynthetic bacterium was achieved for *R. capsulata.*[2] The methods for detachment and partial purification of this enzyme from membranes of photosynthetically grown cells were described in a previous volume of this series.[3] More recently, this coupling factor was shown to be able to reconstitute oxidative phosphorylation in membranes prepared from aerobically grown cells and depleted of ATPase.[4] A specific antibody prepared against the photosynthetic coupling factor can inhibit oxidative phosphorylation,[4] a finding also confirmed in *R. rubrum* chromatophores by immunological techniques (B. C. Johansson, personal communication).

Using the same procedure, described for the photosynthetic coupling factor,[3] a protein can be prepared from aerobic membranes, which can reconstitute both respiratory and photosynthetic phosphorylation. This protein is indistinguishable structurally and functionally from that prepared from photosynthetic membranes.[5,6] This enzyme, therefore, constitutes the first example of a coupling factor operative in respiratory and photosynthetic energy conversion.

The first hint of the role of the protein isolated from *R. capsulata* came from comparison of light-induced proton uptake in coupled and uncoupled chromatophores. In contrast to chloroplasts, no decrease in proton uptake was found in particles largely deprived of coupling factor,[7] indicating that the factor was not involved in the light-dependent cyclic electron transport. A more direct demonstration of the specific involvement of this protein in the last step of ATP synthesis came from evidence of an absolute requirement of the coupling factor for the formation of a proton gradient, only if induced by ATP hydrolysis and not by electron flow.[8] Similar data have also been reported for *R. rubrum.*[9,10]

[2] A. Baccarini-Melandri, H. Gest, and A. San Pietro, *J. Biol. Chem.* **245**, 1224 (1970).

[3] A. Baccarini-Melandri, and B. A. Melandri, Vol. 23, p. 556.

[4] B. A. Melandri, A. Baccarini-Melandri, A. San Pietro, and H. Gest, *Science* **174**, 514 (1971).

[5] A. Baccarini-Melandri, and B. A. Melandri, *FEBS Lett.* **21**, 131 (1972).

[6] S. Lien, and H. Gest, *Arch. Biochem. Biophys.* **159**, 830 (1973).

[7] B. A. Melandri, A. Baccarini-Melandri, A. San Pietro, and H. Gest, *Proc. Natl. Acad. Sci. U.S.A.* **67**, 477 (1970).

[8] B. A. Melandri, A. Baccarini-Melandri, A. R. Crofts, and R. Codgell, *FEBS Lett.* **24**, 141 (1972).

[9] B. C. Johansson, M. Baltscheffsky, and H. Baltscheffsky, *Photosynth., Two Centuries Its Discovery Joseph Priestley, Proc. Int. Congr. Photosynth. Res., 2nd, 1971* p. 1203 (1972).

[10] Z. Gromet Elhanan, *J. Biol. Chem.* **249**, 2522 (1974).

## Rhodospirillum rubrum S1

Three methods have been reported for purification to homogeneity of the coupling factor from *R. rubrum*.

### Method A[11,12]

The procedure is largely based on the method first used for the partial purification of coupling factor from *R. capsulata*.[3] The purification protocol, includes three steps: extraction of an acetone powder of chromatophores with an ATP containing buffer (step 1); precipitation with ammonium sulfate between 30 and 55% saturation (step 2); and column chromatography on Sepharose 6B (step 3). Two additional steps have been introduced.

*Step 4:* Active fractions from Sepharose 6B gel chromatography are pooled and loaded onto a DEAE–Sephadex column (8 × 2.4 cm), equilibrated with Tris-SO$_4$, 20 m$M$; (NH$_4$)$_2$SO$_4$, 80 m$M$, pH 7.5. The column is washed with 100 ml of 20 m$M$ Tris-SO$_4$, (NH$_4$)$_2$SO$_4$, 160 m$M$, pH 7.5, and then the enzyme is eluted with a linear gradient from 160 to 350 m$M$ (NH$_4$)$_2$SO$_4$, 20 m$M$ Tris -SO$_4$, pH 7.5 (both solutions are 150 ml each).

Analogous to what was found for chloroplast CF$_1$,[13] the ATPase-enriched fractions can be detected in the eluate monitoring the fluorescence emission ratio at 300 versus 350 nm, due to the low tryptophan content of the protein. The fractions showing the highest fluorescence emission ratio at 300/350 nm are pooled, concentrated by ultrafiltration, and precipitated with 55% (NH$_4$)$_2$SO$_4$.

*Step 5:* The ammonium sulfate precipitate is centrifuged and the pellet is dissolved in 0.5 ml of 50 m$M$ Tris-HCl, pH 7.5, containing 1 m$M$ ATP. The sample is layered on a sucrose density gradient (8–25%) and centrifuged in a Spinco SW Rotor 40 at 35,000 rpm for 20 hr. All the purification steps are carried out at 0°–4°.

*Properties of the Purified Enzyme.* In contrast to the preparation from *R. capsulata,* the coupling factor from *R. rubrum* when purified by this procedure is endowed with a very high Ca$^{2+}$-dependent ATPase activity. The specific activity of ATPase increases from 0.5 up to 900 $\mu$mole hr mg protein and the fluorescence emission ratio at 300/350 nm from 0.2 to 5.3–6. The increase in specific activity, however, does not only reflect the enrichment of the enzyme, since throughout the purification the total

[11] B. C. Johansson, M. Baltscheffsky, H. Baltscheffsky, A. Baccarini-Melandri, and B. A. Melandri, *Eur. J. Biochem.* **40,** 109 (1973).

[12] R. J. Berzborn, B. C. Johansson, and M. Baltscheffsky, *Biochim. Biophys. Acta* **396,** 360 (1975).

[13] S. Lien, and E. Racker, Vol. 23, p. 547.

units of ATPase increases by about tenfold,[12,14] indicating also the possibility that in this bacterial system an endogenous inhibitor of ATPase is removed during purification.[15,16]

The purified $Ca^{2+}$-ATPase has a $K_m$ value for Ca–ATP of 1.2 m$M$ and is competitively inhibited by Mg–ATP and by Ca–ADP with a $K_i$ of 170 and 42 $\mu M$, respectively. The $Ca^{2+}$-ATPase of the soluble enzyme is insensitive to oligomycin and inhibited by azide and $p$-hydroxymercuribenzoate.

When reassociated with depleted membranes the purified factor restores light-dependent phosphorylation; the coupling factor activity is coincident with the $Ca^{2+}$-dependent ATPase activity throughout all steps of purification. The interaction of the enzyme with the membrane restores also the $Mg^{2+}$-dependent ATPase activity sensitive to oligomycin, which is present in coupled chromatophores. The appearance of this activity is paralleled by the partial masking of the $Ca^{2+}$-dependent ATPase of the soluble enzyme, which becomes sensitive to oligomycin.

The sedimentation constant at 1 mg protein/ml was reported to be $s_{20,w} = 13.1 \times 10^{-13}$s and the approximate molecular weight 350,000 daltons. The enzyme appears homogenous, as judged by polyacrylamide gel electrophoresis according to Davis[17] and by immunoelectrophoresis. Five classes of subunits were detected[14] after treatment at 37° with 1% SDS plus 1% mercaptoethanol and polyacrylamide gel electrophoresis in SDS. On the basis of mobility, the subunits correspond to molecular weights of 54,000 ($\alpha$), 50,000 ($\beta$), 32,000 ($\gamma$), 13,000 ($\delta$), and 7500 ($\epsilon$), in agreement with data reported for coupling factors from other sources.[15,16]

The main disadvantage of this purification procedure is the high instability and the rather low yield of the purified enzyme. No real stabilizing conditions have been found so far: the enzyme can be stored as an ammonium sulfate precipitate at 0°C but loses half the activity in about 1 week. An additional difficulty is encountered in the preparation of membranes deprived of the coupling factor, due to the remarkable resistance of $R.$ $rubrum$ chromatophores to the dislocation of the coupling factor by sonication in EDTA-containing buffers.

*Method B*[18]

The main difference between this procedure and that described above consists in the technique used for solubilizing the coupling factor from the membranes.

[14] B. C. Johansson, and M. Baltscheffsky, *FEBS Lett.* **53**, 221 (1975).
[15] A. E. Senior, *Biochim. Biophys. Acta* **301**, 269 (1973).
[16] P. L. Pedersen, *Bioenergetics* **6**, 243 (1975).
[17] B. J. Davis, *Ann. N. Y. Acad. Sci.* **121**, 404 (1964).
[18] F. K. Lücke, and J. H. Klemme, *Z. Naturforsch.* **31**, 272 (1976).

*Step 1: Dialysis of Chromatophores.* Washed chromatophores are suspended in 1 m$M$ glycylglycine, pH 7.5, 3 m$M$ MgCl$_2$ at about 1.3 mg bacteriochlorophyll (Bchl)/ml, and dialyzed against 200 volumes of 1 m$M$ Tris-HCl, pH 8.0 (four to five changes). The dialyzed suspension is then centrifuged at 140,000 $g$ for 2 hr, and the supernatant discarded.

*Step 2: French Pressure Cell Passage.* The pellet is suspended in 1 m$M$ Tris-HCl, pH 8.0, at 0.3 mg BChl/ml and passed through an Aminco French pressure cell at 16,000 psi. This step releases the ATPase from chromatophores, which can be used as depleted chromatophores following high speed centrifugation.

*Step 3: Ammonium Sulfate Fractionation.* The supernatant, containing the solubilized ATPase, is made 40 m$M$ in Tris-HCl, pH 7.5, and the fraction precipitating between 30 and 60% (NH$_4$)$_2$SO$_4$ saturation is recovered and dissolved in about 6 ml of 50 m$M$ Tris-HCl, pH 8.0, containing 1 m$M$ ATP and 0.5 m$M$EDTA.

*Step 4.* After high-speed centrifugation (140,000 $g$ for 60 min), the solution containing the enzyme is chromatographed on a Sepharose 6B column (2.5 × 65 cm) equilibrated with 50 m$M$ Tris-HCl, pH 8.0, containing 1 m$M$ ATP and 0.5 m$M$ EDTA. The fractions showing high Ca$^{2+}$-dependent ATPase activity (700–750 $\mu$moles/hr/mg protein) are pooled and precipitated with 60% ammonium sulfate. The protein is then dissolved in 50 m$M$ glycylglycine, pH 7.5, containing 1 m$M$ ATP and 0.5 m$M$ EDTA.

The main advantages of this simple procedure, which gives an homogeneous preparation, are high specific activity (about 750 $\mu$moles/hr/mg protein) and remarkable stability. The pure enzyme can in fact be stored as an ammonium sulfate precipitate (50% saturation) at 4° for several months with only a 10% decrease in the Ca$^{2+}$-ATPase and recoupling activities.

The properties of the enzyme purified by this method are quite similar to those of the enzyme purified by method A ($K_m$ ATP = 1.4 m$M$ and inhibition of Ca$^{2+}$-ATPase activity by Mg$^{2+}$ ions), except that this enzyme showed also ATPase activity in the presence of Mg$^{2+}$ ions (about 6–7% of the rate obtained with Ca$^{2+}$).

*Method C*

A third procedure set up recently by Müller and collaborators[19] utilizes an affinity chromatography step for the final rapid stage of purification. The enzyme solubilized from the membranes by sonication in the presence of 1 m$M$ EDTA[11] is partially purified by (NH$_4$)$_2$SO$_4$ precipitation

---

[19] H. W. Müller, M. Schmitt, E. Schneider, and K. Dose, *Biochim. Biophys. Acta.* **565**, 77 (1979).

(between 30 and 55% saturation) and gel filtration on Sepharose CL-6B (2.5 × 100 cm column) essentially as described in steps 2 and 3 of method A.

*Step 4: Affinity Chromatography* The fractions from the column chromatography of step 3 containing ATPase activity are concentrated by ultrafiltration followed by dialysis and purified to homogeneity by affinity chromatography through a column of a Sepharose 4B-iminobisisopropylaminyl-N-acetylhomocysteinyl-6-thio-9-β-ribofuranosylpurine 5'-triphosphate.[20] Two milliliters of the Sepharose CL-6B filtered enzyme (1 mg/ml) are applied to a column packed with 4 ml of affinity matrix, previously equilibrated with 50 mM Tris-HCl buffer (pH 7.5) at a flow rate of 2 ml/hr. The column is then washed with 6–7 volumes of 50 mM Tris-HCl, pH 7.5, and eluted with a linear gradient from 0 to 5 mM ATP in 50 mM Tris-HCl buffer, pH 7.5. These last two operations are performed at a flow rate of 16–20 ml/hr.

The purified enzyme has a specific activity comparable to that of the two preparations described above. The main advantage of this procedure rests on its rapidity (3–4 hr after the first column chromatography) and in the remarkable yield, which was reported to be above 35%.

Information on the structure and properties of this preparation have not appeared.

Another procedure for the isolation of a factor required for ATP synthesis and for preparation of depleted chromatophores from *R. rubrum* has been reported by Binder and Gromet-Elhanan.[21] A 4 M LiCl solution containing 8 mM ATP is added to an equal volume of chromatophores suspended in Tricine–NaOH pH 7.6, 0.25 M sucrose and 4 mM MgCl$_2$ at a concentration of 0.4 mg/ml. The suspension is stirred for 30 min at 4° and then centrifuged. The supernatant contains a factor, which can restore photophosphorylation, whereas the sedimented chromatophores, washed only once, can be used as decoupled particles for reconstitution experiments.

The factor has been subsequently purified by extensive dialysis to remove LiCl, by (NH$_4$)$_2$SO$_4$ fractionation (30–60% saturation), and by DEAE–cellulose chromatography.

More recently,[22] it has been reported that LiCl extraction releases only one subunit of the coupling factor, which shows a molecular weight similar to that obtained for the β-subunit of *R. rubrum* coupling factor

[20] F. W. Hulla, M. Hockel, S. Risiand, and K. Dose, *Eur. J. Biochem.* **67,** 469 (1976).
[21] A. Binder, and Z. Gromet-Elhanan, *Proc. Int. Congr. Photosynth. Res., 3rd, 1974* p. 1163 (1975).
[22] S. Philosoph, A. Binder, and Z. Gromet-Elhanan, *J. Biol. Chem.* **252,** 8767 (1977).

(prepared by method A) as judged from SDS gel electrophoresis and sedimentation pattern in the analytical ultracentrifuge. This subunit has no ATPase activity by itself, but can reconstitute photophosphorylation and ATPase activity in LiCl-depleted chromatophores.

The technical approach proposed by Beechey et al.[23] for the solubilization of the mitochondrial ATPase has also been utilized for detachment of the coupling factor from R. rubrum[24]. A suspension of chromatophores at a BChl concentration of 0.25 m$M$ is shaken with half a volume of chloroform and then centrifuged in order to break the emulsion.

The aqueous layer, centrifuged at room temperature, at 70,000 g for 150 min, is dialyzed overnight, and then 10% glycerol is added. The enzyme isolated by this technique appears 70–80% pure. It shows a $Ca^{2+}$-dependent ATPase activity with properties similar to those reported for coupling factor purified by methods A, B, and C.

### Chromatium vinosum, Strain D

The isolation of a crude coupling factor preparation from Chromatium vinosum, strain D, was achieved by incubation, for 10 min, of chromatophores in a low ionic strength medium (Tricine, 3 m$M$, pH 7.8) and subsequent centrifugation at 144,000 g for 1 hr at 0°C.[25] The supernatant obtained could restore partially photophosphorylation and ATP-$P_i$ exchange activities in the chromatophores extracted with the low ionic strength buffer. Recently Gepshtein et al.[26] have succeeded in purifying from the supernatant an enzyme endowed with a rather active but latent ATPase activity using the procedure outlined below.

Step 1: DEAE–Cellulose Chromatography. To the supernatant, obtained after centrifugation of chromatophores (see above), Tris-Cl, pH 8.0, and Triton X-100, to a final concentration of 10 m$M$ and 0.5% (v/v), respectively, are added. This solution is applied to a DEAE-cellulose DE-11 column (2 × 15 cm) equilibrated at room temperature with 10 m$M$ Tris-Cl, pH 8.0. After washing with 50 ml of the same buffer, ATPase is eluted with a linear gradient from 0 to 0.5 $M$ $(NH_4)_2SO_4$ in 10 m$M$ Tris-Cl, pH 8.0 (flow rate 5 ml/min).

Step 2: $(NH_4)_2SO_4$ Precipitation. The active fractions from DEAE chromatography are pooled together, and the enzyme is precipitated with

[23] R. B. Beechey, S. A. Hubbard, P. E. Linnett, A. D. Mitchell, and E. A. Munn, Biochem. J. **148**, 533 (1975).

[24] G. D. Webster, P. A. Edwards, and J. B. Jackson, FEBS Lett. **76**, 29 (1977).

[25] A. Hochman, and C. Carmeli, FEBS Lett. **13**, 36 (1971).

[26] A. Gepshtein, C. Carmeli, and N. Nelson, FEBS Lett. **85**, 219 (1978).

$(NH_4)_2SO_4$ at 66% saturation. After centrifugation the pellet is suspended in 2.5 ml of 10 m$M$ Tris-Cl, pH 8.0.

*Step 3: Sucrose Density Gradient.* A linear gradient of sucrose from 5 to 30% (w/v) in 10 m$M$ Tris-Cl, pH 8.0, to which the enzyme is applied, is centrifuged at 28,000 rpm in a Spinco SW 40 rotor at 15° for 18 hr. Fractions of 0.4 ml are collected.

*Step 4: DEAE–Cellulose Chromatography.* The fractions from the sucrose density gradient showing ATPase activity are combined, applied to a DEAE–cellulose column, and eluted as described in step 1. The active fractions are stored as a suspension in $(NH_4)_2SO_4$ (66% saturation).

*Properties of the Purified Enzyme.* The purified preparation shows a very high ATPase activity in the presence of $Ca^{2+}$, but only after activation by digestion with trypsin (10–25 $\mu$g/ml for 1 min); a maximal specific activity of 535 $\mu$moles of ATP hydrolyzed/hr/mg protein, obtained at the final stage of purification, has been reported. This activity corresponds to a 24-fold increase in comparison to that of the initial crude extract; the final yield is, however, very low and corresponds only to 6% of the total initial activity. In spite of the high latent ATPase activity, and in contrast with the crude extract, the purified enzyme preparation is not capable of restoring photophosphorylation in extracted chromatophores.

The purified ATPase presents the usual five subunit composition typical of most coupling ATPases studied so far, when analyzed on standard SDS gel electrophoresis. In comparison with the starting crude extract, however, only a small amount of the δ-subunit is retained in the purified preparation. This depletion in δ-subunit could be the explanation for the lack of effectiveness in reconstituting photophosphorylation in depleted chromatophores. The role of coupling factor of this ATPase has been, however, clearly demonstrated, since a monospecific antibody against purified ATPase can effectively inhibit photosynthetic ATP synthesis.

No data have been reported so far for the cation specificity and the kinetic constants of the purified ATPase from *Chromatium;* some of these data are, however, available for the ATPase contained in the crude preparation extracted from chromatophores. After trypsin activation, both a $Mg^{2+}$- and $Ca^{2+}$-dependent ATPase activity are observed.

The optimal ion concentration and the maximal activity measured are, however, dependent upon the nature of the cation: in the presence of 3.3 m$M$ ATP the maximal activity is observed with 10 m$M$ $Ca^{2+}$ and is twofold higher than the maximal activity measured in the presence of $Mg^{2+}$, for which the optimal concentration is only 2 m$M$.[27] The activity

[27] A. Gepshtein, and C. Carmeli, *Eur. J. Biochem.* **44,** 593 (1974).

appears to depend on the concentration ratio between free cation(s) and cation–ATP complex, the free cation being a competitive inhibitor against the ion–ATP complex, the real substrate for the enzyme.[28]

A specific and unusual property of this ATPase, as well as of the ATPase contained in the crude extract, is its sensitivity to the inhibitor DCCD. Since this compound is thought to bind specifically and covalently to a hydrophobic subunit of the ATP synthetase complex, and not to any subunit of the hydrophilic portion of the ATPase ($F_1$ complex), this result could indicate that a DCCD binding component, not detectable in SDS gel electrophoresis with the usual staining techniques, is still present in this preparation.

[28] A. Gepshtein, and C. Carmeli, *Eur. J. Biochem.* **74**, 463 (1977).

# [29] Nucleotides Tightly Bound to Chloroplast Membranes

*By* N. Shavit and H. Strotmann

The presence of tightly bound nucleotides on isolated and membrane-bound chloroplast coupling factor ($CF_1$ was reported by several workers.[1–6] These nucleotides remain noncovalently bound to the membrane or to the isolated enzyme after repeated washings and are only partially removed by gel filtration. The wide distribution of such nucleotides in a variety of energy-transducing membrane and the fact that these nucleotides undergo exchange or can be phosphorylated, upon energization of the membrane, support the idea of their involvement in energy transduction.

The total amount of nucleotides present in washed chloroplast membranes and isolated coupling factor varies and depends upon the isolation conditions. Harris and Slater[1] found 2 moles of ADP and 1 mole of ATP per mole $CF_1$. Lower numbers were reported by Strotmann *et al.*[3] and Magnusson and McCarty.[4] Direct determinations of the nucleotide con-

[1] D. A. Harris and E. C. Slater, *Biochim. Biophys Acta* **387**, 335 (1975).
[2] C. Aflalo and N. Shavit, *Biochim. Biophys. Acta* **440**, 522 (1975).
[3] H. Strotmann, S. Bickel, and B. Huchzermeyer, *FEBS Lett.* **61**, 194 (1976).
[4] R. P. Magnusson, and R. E. McCarty, *J. Biol. Chem.* **251**, 7417 (1976).
[5] D. J. Smith, and P. D. Boyer, *Proc. Natl. Acad. Sci. U.S.A.* **73**, 4314 (1976).
[6] N. Shavit, *in* "Encyclopedia of Plant Physiology" (A. Trebst and M. Avron, eds.), New Ser., Vol. 5, Photosynthesis I p. 350. Springer-Verlag, Berlin and New York. 1977.

METHODS IN ENZYMOLOGY, VOL. 69

tent of purified $CF_1$ preparations[7] gave a molar ratio of 1 for ADP, < 0.1 ATP, and even less for AMP. Adenine nucleotides firmly bound to the membrane-bound $CF_1$ can undergo a rapid exchange with free ADP or ATP upon energization of the membrane.[1,3,8] The ADP present can also be phosphorylated with free $P_i$ to form ATP.[2,5] Both the exchange and phosphorylation reactions depend upon energization of the membrane and are sensitive to uncouplers. The exchange reaction is insensitive to arsenate and the energy transfer inhibitor phlorizin,[9] while phosphorylation is.[2] The exchange of adenylates is also highly specific for the adenine base moiety; while GDP and IDP are effective substrates in photophosphorylation, they do not serve as substrates in the exchange reaction.

The phosphorylation of membrane-bound ADP by short saturating flashes and the energy-induced adenine nucleotide exchange are described. The exchange can easily be measured either by the incorporation of $^3H$- or $^{14}C$-labeled ADP or ATP into membrane-bound $CF_1$ (forward exchange) or by the release of previously incorporated labeled nucleotides in the presence of unlabeled adenylates (back exchange).

### Exchange of Membrane-Bound Nucleotides

Chloroplasts isolated by usual procedures are washed at least three times in a medium containing 2 m$M$ Tricine, pH 8.0, 50 m$M$ NaCl, and 1 m$M$ $MgCl_2$ and resuspended in the same medium at a chlorophyll concentration of about 2 mg/ml. This preparation is used without further treatment in the "forward" exchange reaction or can be preloaded with $^3H$- or $^{14}C$-labeled nucleotides for measurement of the "back" exchange reaction.

### Forward Exchange

Thrice-washed chloroplasts are energized either by illumination or by an acid–base transition. The reaction mixture for illumination contains in a final volume of 0.5 ml at pH 8.0: 25 m$M$ Tricine, 50 m$M$ NaCl, 5 m$M$ $MgCl_2$, 0.5 m$M$ methyl viologen, 25–50 $\mu M$ [$^{14}C$]ADP, and chloroplasts containing 0.4–0.6 mg chlorophyll. The samples are illuminated in small glass vessels or test tubes with white light ($9 \times 10^5$ erg/cm²/sec) for 5 sec at 20°. The procedure for energization by an acid–base transition is the following. Thrice-washed chloroplasts are incubated for 15 sec in a 1-ml

[7] V. Shoshan, and N. Shavit, and D. M. Chipman, *Biochim. Biophys. Acta* **504,** 108 (1978).
[8] H. Strotmann and S. Bickel-Sandkötter, *Biochim. Biophys. Acta* **460,** 126 (1977).
[9] S. Bickel-Sandkötter and H. Strotmann, *FEBS. Lett.* **65,** 102 (1976).

reaction mixture at pH 4.0 containing: 10 m$M$ sodium succinate, 25 $\mu M$ 3-(3, 4-dichlorophenyl)-1, 1-dimethylurea) (DCMU), and 0.5 mg chlorophyll of chloroplasts. This suspension is then mixed with a 1-ml basic reaction mixture containing: 110 m$M$ Tricine, pH 8.3, 50 m$M$ NaCl, 5 m$M$ MgCl$_2$, and 50 $\mu M$ [$^{14}$C]ADP or [$^{14}$C]ATP. After 1 min at 20° the samples are processed as described below.

*Washing Procedure.* The incubation mixtures after illumination or acid–base treatment are centrifuged at 15,000 $g$ for 1 min at 4°. Supernatants are removed and the pellets are washed three times as follows: pellets from illumination, with 1 ml of an ice-cold medium containing 25 m$M$ Tricine, pH 8.0, and 50 m$M$ NaCl; pellets from acid–base, with 1 ml of an ice-cold medium containing 10 m$M$ Tricine, pH 8.3, 50 m$M$ NaCl, and 5 m$M$ MgCl$_2$. The resuspension of the pellets is aided by a small plastic pestle adapted to the form of the tube. The washed pellets are finally resuspended in 0.5 ml of the corresponding washing medium, and 0.1 ml is taken to determine the chlorophyll concentration. The remainder is deproteinized by addition of perchloric acid to a final concentration of 4%. Radioactivity is measured by liquid scintillation counting using Unisolve 1 scintillator (Koch-Light Laboratories). Samples can also be counted directly without deproteinization but in this case a correction should be made for quenching by the chlorophyll present.

## Back Exchange

The procedure for measurement of the release of labeled nucleotide (back exchange) involves prelabeling of thylakoid membrane with radioactive labeled nucleotides and reincubation in a medium containing unlabeled nucleotides. As for the forward exchange reaction, energization of the membranes is achieved by illumination or by an acid–base treatment.

*Prelabeling of Chloroplast Membranes.* Thrice-washed chloroplasts are illuminated (10$^6$ erg/cm$^2$/sec) for 1 min in a 2-ml reaction mixture containing 25 m$M$ Tricine, pH 8.0, 50 m$M$ NaCl, 1 m$M$ MgCl$_2$, 0.5 m$M$ methyl viologen, 2.5 $\mu M$ [$^{14}$C]ADP, and chloroplasts containing 1–2 mg chlorophyll. The suspension is then centrifuged at 15,000 $g$ for 1 min, and the pellets are washed three times with 2 ml of an ice-cold medium containing 25 m$M$ Tricine, pH 8.0, and 50 m$M$ NaCl. The pellets are either resuspended in the washing medium and the chlorophyll concentration adjusted to 1 mg/ml for measurement of the light-induced exchange or directly resuspended in the acid reaction mixture for the acid–base energization.

Reaction mixtures for light-induced exchange contain in 0.3 ml final volume: 25 m$M$ Tricine, pH 8.0, 50 m$M$ NaCl, 5 m$M$ MgCl$_2$, 0.5 m$M$

methyl viologen, 0.1 m$M$ ADP, and prelabeled chloroplasts containing 60–100 $\mu$g chlorophyll. The samples are illuminated in semitransparent plastic centrifuge tubes (capacity 0.4 ml, Beckman Instruments) inserted in a Microfuge 152 Beckman centrifuge for the desired time. After the light is turned off, the suspensions are centrifuged for 30 sec in the dark. The radioactive nucleotide content of an aliquot (0.1 ml) of the clear supernatant can be determined immediately. The supernatants can also be stored at 4°, but in order to avoid secondary interconversion of nucleotides by residual adenylate kinase, the addition of perchloric acid (4% final concentration) or 10 m$M$ EDTA, is recommended.

For the exchange reaction induced by an acid–base transition the prelabeled chloroplast pellets are resuspended in 1 ml of a solution containing 10 m$M$ sodium succinate, pH 4.0, and 25 $\mu M$ DCMU at a chlorophyll concentration of 0.5 mg/ml. After 15-sec incubation at the acid stage, 1 ml of a basic reaction mixture is added with mixing. The basic reaction mixture contains 110 m$M$ Tricine, pH 8.3, 50 m$M$ NaCl, 5m$M$ MgCl$_2$, and 0.1 m$M$ ADP. After 1 min at room temperature, the suspension is centrifuged at 15,000 $g$ and the radioactive nucleotide content of an aliquot (0.1 ml) of the clear supernatant is determined.

For the determination of the amount of the radioactive label incorporated into AMP, ADP, and ATP, conventional chromatographic procedures can be used. An easy and highly reproducible separation can be achieved by ion exchange chromatography on Dowex 1-X8, 200–400 mesh, Cl$^-$ form. The column length is 120 mm with a 3.5 mm diameter. Radioactive supernatant samples are mixed with 0.1 $\mu$mole of each carrier nucleotide AMP, ADP, and ATP. The samples are applied on the column by the use of a peristaltic pump. Using a multiple channel peristaltic pump permits parallel separation in less than 30 min. After washing with 1–2 ml distilled water, the nucleotides are discontinuously eluted with 8 ml of three different HCl solutions: 20 m$N$ AMP, 65 m$N$ 20 m$N$, and 200 m$N$ ATP. A flow rate of 1 ml/min should be maintained. To 2.5 ml fractions, 4.5 ml Unisolve 1 scintillator are added and the samples are counted.

*Phosphorylation of Membrane-Bound ADP.* Chloroplasts are isolated by conventional procedures, then washed three times by resuspension in 0.4 $M$ sucrose–1 m$M$ Tricine, pH 8.0, and centrifugation. The pellets are finally resuspended in the same medium, and the chlorophyll concentration is adjusted to about 1 mg/ml. Reaction mixture for phosphorylation contain in a final volume of 1 ml at pH 8.0: 20 m$M$ Tricine, 20 m$M$ KCl, 4 m$M$ MgCl$_2$, 0.5 $^{32}$P$_i$ (containing about 4 × 10$^7$ cpm); 60 $\mu M$ phenazine methosulfate or pyocyanine, and thrice-washed chlorplasts containing 100 $\mu$g chlorophyll. Samples are placed in small Pyrex test tubes (approximately 7.5 mm i.d.) and illuminated individually with a beam of strong white light from a 500 W projector lamp (Sylvania DAK) filtered through an infrared-absorbing lens at a light intensity of $4.5 \times 10^6$ erg/cm$^2$/sec. The duration of the light flash is controlled by an electromagnetic shutter opening of 30 msec or longer. The actual exposure time can be erated in conjunction with an electronic timer set to provide an effective shutter opening of 30 m or longer. The actual exposure time can be routinely checked with a photodiode connected to a storage oscilloscope. Upon extinction of illumination, the reaction is terminated by trichloroacetic acid addition to a final concentration of 6%. The estimated time for addition and mixing is about 200 msec. No significant ATP hydrolysis can be detected by introducing dark intervals (1 sec and longer) before the addition of trichloroacetic acid. Radioactivity incorporated into the organic phosphate fraction is determined as described.[2]

### Concluding Remarks

Light-induced incorporation of labeled ADP at saturating nucleotide concentrations ($K_m = 2$–$5$ $\mu M$) results in the incorporation of up to 1.5 nmoles ADP/mg chlorophyll. Using the molar ratio of CF$_1$/860 chlorophyll determined by Strotmann et al.,[10] maximal binding is therefore 1.2 moles/mole CF$_1$. It is advisable, however, to determine the ratio of CF$_1$ per chlorophyll in each preparation, since it is our experience that this ratio varies somewhat and depends on growth conditions, seasonal changes, and isolation procedure. The pattern of labeled bound nucleotides is 70–80% of ADP, 10–20% ATP, and less than 10% AMP. The distribution of label is largely independent of whether ADP $\pm$ P$_i$ or ATP are present during energization. Without energization, incorporation of label is practically zero, even after 1 hr of incubation. However, some incorporation is observed (about 0.2 nmole/mg chlorophyll) at higher nucleotide concentrations (100 $\mu M$). The amount of nucleotide exchanged by acid–base energization of the membranes is the same whether the nucleotide is added in the acid or basic stage. However, while no exchange is detectable in the base controls, a low but significant amount of nucleotide is exchanged in the acidification step, both in the forward and the back exchange reactions.[11] The light-induced release of labeled nucleotides requires the presence of ADP or ATP and is complete after 5 sec. The rate of release in the absence of added nucleotides is much slower and incomplete. No release occurs in the dark. Only 60–80% of

[10] H. Strotmann, H. Hesse, and K. Edelmann, *Biochim. Biophys. Acta* **314**, 202 (1973).
[11] B. Huchzermeyer and H. Strotmann, *Z. Naturforsch.* **32c**, 803 (1977).

the preloaded label is exchangeable due to partial inactivation of the membranes in the repeated washings of the pretreatment step. The method described for measuring ATP formation has the advantage of permitting kinetic analysis of the reaction product of a short, saturating, single flash. The increase in phosphorylation rate observed with increasing free ADP concentrations follows the usual Michaelis–Menten behavior. The product of photophosphorylation of membrane-bound ADP was shown to be $[\gamma\text{-}^{32}P]ATP$.[2] The lower initial rates of phosphorylation of this ADP suggest that upon energization it is released from the tight-binding site and then undergoes phosphorylation.[12,13] The energy-dependent exchange of membrane-bound ADP supports this conclusion.[3] Phosphorylation of membrane-bound ADP in the absence of added ADP indicates the formation of membrane-bound ATP. The ATP synthetase in deenergized membranes or the isolated enzyme appears to have bound ADP but not ATP.[13,14]

Acknowledgment

This work was supported in part by the Deutsche Forschungsgemeinschaft.

[12] N. Shavit, S. Lien, and A. San Pietro, *FEBS Lett.* **73**, 55 (1977).
[13] J. Rosing, D. J. Smith, C. Kayalar, and P. D. Boyer, *Biochem. Biophys. Res. Commun.* **72**, 1 (1976).
[14] R. P. Magnusson and R. E. McCarty, *Biochem. Biophys. Res. Commun.* **70**, 1283 (1976).

# [30] Crystallization and Assay Procedures of Tobacco Ribulose-1,5-bisphosphate Carboxylase–Oxygenase

By S. D. KUNG, R. CHOLLET, and T. V. MARSHO

Ribulose-1,5-bisphosphate (Rbu-$P_2$) carboxylase-oxygenase (EC 4.1.1.39) is the most abundant protein in all green plants. This bifunctional enzyme catalyzes both the carboxylation and oxygenation of Rbu-$P_2$ [Eq. (1) and (2)].

$$\text{D-Ribulose-1,5-bisphosphate} + CO_2 + H_2O \xrightarrow{Mg^{2+}} 2 \text{ D-3-phosphoglycerate} + 2\,H^+ \quad (1)$$

$$\text{D-Ribulose-1,5-bisphosphate} + O_2 \xrightarrow{Mg^{2+}} \text{2-phosphoglycolate}$$
$$+ \text{ D-3-phosphoglycerate} + 2\,H^+ \quad (2)$$

METHODS IN ENZYMOLOGY, VOL. 69

## I. Crystallization Procedure

*Principle.* The tobacco enzyme is soluble only in the presence of its substrate, Rbu-P$_2$, or high salt (NaCl).[1] Heat treatment is also required for maximum activity.[2] These properties provide the basis for the recent simplification of the crystallization procedure.[3] Crystalline enzyme can readily be prepared from many species of *Nicotiana*. The procedure involves three major steps: (1) breakage of chloroplasts in the presence of a high concentration of NaCl to release the enzyme, (2) heating the green filtrate to precipitate the undesirable material, and (3) removal of NaCl by gel filtration to yield crystalline enzyme.

*Reagents*

2.0 *M* NaCl

1.0 *M* Tris (unadjusted)

25 m*M* Tris-HCl buffer containing 0.2 m*M* EDTA (pH 7.4 at 25°) [Tris-EDTA]

10% (w/v) Na$_2$EDTA (pH 7.5)

2-Mercaptoethanol

5% (w/v) Trichloroacetic acid

Sephadex G-50 (coarse)

*Procedure.* The crystallization procedure shown in Fig. 1 is based on modification of the method developed by Lowe.[3]

Young tobacco leaves grown in a greenhouse or growth chamber (4–5 weeks after germination) are homogenized in 2.0 *M* NaCl plus 2-mercaptoethanol in a Waring blender. At first, leaves are added gradually as pieces with the blender at low speed. The slurry is blended at high speed for 30–50 sec. The resulting slurry is carefully squeezed through four layers of cheesecloth and one layer of Miracloth. The green filtrate is immediately heated in a 50° water bath with occasional stirring until the filtrate reaches 45°–48°. The filtrate is then rapidly cooled to 15°–20° in an ice bath. Two milliliters of 10% Na$_2$EDTA are added for every 100 ml of filtrate and the pH adjusted to 7.5 with 1.0 *M* Tris (about 4–5 ml/100 ml of filtrate). The green suspension is centrifuged for 10 min at 48,000 *g*. The supernatant is passed through a Sephadex G-50 column equilibrated with Tris-EDTA buffer. Two fractions are collected when protein starts to appear (as detected by precipitation with cold 5% trichloroacetic acid) in the effluent. A large fraction of effluent is collected first. Then a small fraction of effluent is collected separately. The protein concen-

[1] P. H. Chan, K. Sakano, S. Singh, and S. G. Wildman, *Science* **176**, 1145 (1972).

[2] S. Singh and S. G. Wildman, *Plant Cell Physiol.* **15**, 373 (1974).

[3] R. H. Lowe, *FEBS Lett.* **78**, 98 (1977).

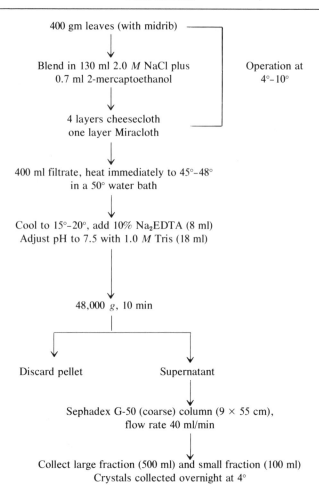

400 gm leaves (with midrib)

Blend in 130 ml 2.0 $M$ NaCl plus
0.7 ml 2-mercaptoethanol

Operation at
4°–10°

4 layers cheesecloth
one layer Miracloth

400 ml filtrate, heat immediately to 45°–48°
in a 50° water bath

Cool to 15°–20°, add 10% $Na_2EDTA$ (8 ml)
Adjust pH to 7.5 with 1.0 $M$ Tris (18 ml)

48,000 $g$, 10 min

Discard pellet                    Supernatant

Sephadex G-50 (coarse) column (9 × 55 cm),
flow rate 40 ml/min

Collect large fraction (500 ml) and small fraction (100 ml)
Crystals collected overnight at 4°

FIG. 1. Scheme for preparing approximately 1 gm of crystalline ribulose-1,5-bisphosphate carboxylase–oxygenase from tobacco leaves.

tration in the small fraction is rather low, and sometimes this fraction
fails to yield crystals. Usually crystals are formed in the column or will
be formed overnight at 4°. At 500× magnification of the light microscope
the crystals appear hexagonally shaped (Fig. 2). A yield of 2–3 mg of
crystals per gram leaf fresh weight can be expected. The column can be
cleaned by washing with distilled water to remove salt and pigments.

Recrystallization can be carried out as follows: Crystalline enzyme is
collected by centrifugation (10 min at 5000 g) or suctioning off the mother

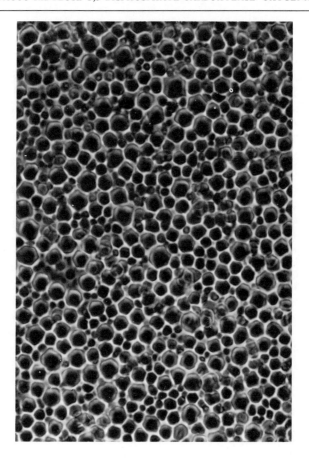

FIG. 2. Tobacco Fraction 1 protein prepared by this procedure will yield crystals of a hexagonal shape.

liquor and washing twice in Tris-EDTA. The crystals are resuspended in a small volume of Tris-EDTA and solubilized by adding saturated NaCl. Centrifugation at 48,000 $g$ for 10 min is sufficient to remove any residues. The solution is passed through the G-50 column and collected. Two recrystallizations are usually performed before subjecting the enzyme to further study. The column eluant containing crystals is stored at 4°. The recrystallization procedure may also be used to remove bacterial contamination from older preparations.

*Comments.* According to our results with leaves from young plants, only the NaCl solution and not Tris buffer is required in the homogenization medium. In some special cases, however, a buffered homogenization medium may be advisable. $NaHCO_3$ and $MgCl_2$ are also not necessary to induce crystallization. It is important to carry out the homogenization step at $4°-10°$ and to proceed as rapidly as possible. Failure to do so may result in enzyme degradation. In such a case, a fluffy slow-settling precipitate is obtained. The degraded enzyme has a reduced sedimentation coefficient as well as a reduced specific activity. Once the green filtrate has been adjusted to pH 7.5, the enzyme is stable for 2–3 hr. The Sephadex G-50 (coarse) column is more efficient than G-25. The column is washed periodically with $1.0 N$ NaOH to remove pigmented residues and any bacterial contamination.

*Physicochemical Properties.* Ribulose-1,5-bisphosphate carboxylase-oxygenase prepared by the direct crystallization procedure is homogeneous as determined by sedimentation velocity and by polyacrylamide gel electrophoresis.[4] The crystalline enzyme from tobacco contains no carbohydrate,[5] no tightly bound characteristic oxygenase prosthetic groups,[6] and approximately 100 sulfhydryl groups.[7] It has a sedimentation coefficient ($s_{20,w}^{°}$) of 18 S[8], corresponding to the most commonly reported molecular weight of $5.6 \times 10^5$ daltons. It is composed of eight large and eight small subunits[9], each of which has a molecular weight of $5.5 \times 10^4$ and $1.3 \times 10^4$, respectively. The eight large subunits, containing the carboxylase-oxygenase catalytic site,[10,11] are situated at the four corners of a two-layered structure with the eight small subunits on the surface.[9] The subunits can be dissociated under a variety of denaturing conditions, such as urea, sodium dodecyl sulfate, and extreme pH, and are separated by gel filtration.[8,12] The large and small subunits have different amino acid compositions and tryptic fingerprints.[4,12] The large subunit is much more hydrophobic than the small subunit.

Isoelectric focusing of $S$-carboxymethylated enzyme from *N. tabacum* in polyacrylamide gels has resolved the subunits into their component polypeptides,[13] the eight large subunits are resolved into three poly-

[4] S. D. Kung, K. Sakano, and S. G. Wildman, *Biochim. Biophys. Acta* **365**, 138 (1974).
[5] K. Sakano, J. E. Partridge, and L. M. Shannon, *Biochim. Biophys. Acta* **329**, 339 (1973).
[6] R. Chollet, L. L. Anderson, and L. C. Hovsepian, *Biochem. Biophys. Res. Commun.* **64**, 97 (1975).
[7] R. Chollet and L. L. Anderson, *Biochim. Biophys. Acta* **482**, 228 (1977).
[8] N. Kawashima and S. G. Wildman, *Annu. Rev. Plant Physiol.* **21**, 325 (1970).
[9] T. S. Baker, S. W. Suh, and D. Eisenberg, *Proc. Natl. Acad. Sci. U.S.A.* **74**, 1037 (1977).
[10] M. Nishimura and T. Akazawa, *Biochemistry* **13**, 2277 (1974).
[11] B. A. McFadden, *Biochem. Biophys. Res. Commun.* **60**, 312 (1974).
[12] S. D. Kung, C. I. Lee, D. D. Wood, and M. A. Moscarello, *Plant Physiol.* **60**, 89 (1977).
[13] S. D. Kung, *Science* **191**, 429 (1976).

TABLE I

PHYSICOCHEMICAL PROPERTIES OF RIBULOSE-1,5-BISPHOSPHATE CARBOXYLASE–
OXYGENASE

| Physicochemical property | | Plant source | Reference |
| --- | --- | --- | --- |
| Sedimentation coefficient, $s_{20,w}^{\circ}$ | 18.3 S | Tobacco | Kawashima and Wildman[8] |
| Molecular diameter (Å) | 112 | Tobacco | Kwok[16] |
| Molecular weight (daltons) | $5.6 \times 10^5$ | Tobacco | Baker et al.[9] |
| Partial specific volume (cm$^3$/gm) | 0.7 | Tobacco | Kwok[16] |
| Wet crystal density (gm/cm$^3$) | 1.058–1.095 | Tobacco | Kwok[16]; Baker et al.[9] |
| Water content of wet crystal (%) | 80 | Tobacco | Kwok[16] |
| $A_{280\ nm}/A_{260\ nm}$ | 1.92 | Tobacco | Singh[17]; Chollet et al.[6] |
| Diffusion constant $d_{20,w}^{\circ}$ (cm$^2$/sec) | $2.93 \times 10^{-7}$ | Spinach | Trown[18] |
| Frictional coefficient (f/f°) | 1.11 | Spinach | Paulsen and Lane[19] |
| Extinction coefficient $(E_{1\ cm}^{1\%})$ | 14.1 | Spinach | Trown[18] |

peptides each having a molecular weight of $5.5 \times 10^4$. The eight small subunits are resolved into two polypeptides each having a molecular weight of $1.3 \times 10^4$. The two polypeptides in the small subunit from *N. tabacum* have different N-terminal amino acid sequences.[14] The isoelectric points of the *S*-carboxymethylated large and small subunits are approximately 6.0 and 5.3, respectively.[15] Other physicochemical properties of the crystalline enzyme from tobacco and purified spinach enzyme are listed in Table I.[6,8,9,16−19]

## II. Carboxylase–Oxygenase Assays

*Principle.* Ribulose-1,5-bisphosphate carboxylase catalyzes the irreversible carboxylation of Rbu-P$_2$ to form 3-phosphoglycerate (3-P-glycerate) with the stoichiometry shown in Eq. (1). Besides functioning as a carboxylase, the enzyme can also act as an internal monooxygenase, catalyzing the oxygenation of Rbu-P$_2$ to yield 2-phosphoglycolate (2-P-glycolate) and 3-P-glycerate according to Eq. (2). Oxygen and CO$_2$ react as competitive inhibitor and substrate, respectively, in the carboxylase

[14] S. Strøbaek, G. C. Gibbons, B. Haslett, D. Boulter, and S. G. Wildman, *Carlsberg Res. Commun.* **41**, 335 (1976).

[15] S. D. Kung, *Bot. Bull. Acad. Sin.* **17**, 185 (1976).

[16] S. Y. Kwok, Ph.D. Thesis, University of California, Los Angeles (1972).

[17] S. Singh, Ph.D. Thesis. University of California, Los Angeles (1972).

[18] P. W. Trown, *Biochemistry* **4**, 908 (1965).

[19] J. M. Paulsen and M. D. Lane, *Biochemistry* **5**, 2350 (1966).

reaction and conversely as substrate and competitive inhibitor in the oxygenase reaction.[20,21] Likewise, both the carboxylase and oxygenase functions of the enzyme are reversibly activated by preincubation with $CO_2$ (rather than $HCO_3^-$) and $Mg^{2+}$ to form an activated ternary complex.[22-25]

$$Enzyme + CO_2 + Mg^{2+} \rightleftharpoons Enzyme-CO_2-Mg$$
$$\text{(Inactive)} \qquad\qquad \text{(Active)}$$

The carboxylase and oxygenase activities are dependent upon both the preincubation and assay pH, $pCO_2$, $pO_2$, and $Mg^{2+}$ concentrations;[22,24,25] therefore, care must be taken to distinguish the kinetics of activation from those of catalysis. This is especially important in kinetic studies [e.g., $K_m$ ($CO_2$)] where it is essential that the quantity of catalytically active enzyme, not simply the total quantity of enzyme protein, remains constant. This criterion can be satisfied by fully activating the enzyme before starting the reactions and limiting the assays to a 1- or 2-min duration to avoid inactivation effects. The assay procedures outlined below are based on those developed by Lorimer et al.[26] from studies of spinach Rbu-$P_2$ carboxylase–oxygenase. For a more complete description of these kinetic experiments, the reader is referred to refs. 22, 24, and 25.

*Assay for Carboxylase Activity.* Carboxylase activity is conveniently determined by following the rate of incorporation of $H^{14}CO_3^-$ into 3-P-glycerate (acid-stable radioactivity).[19] Alternatively, comparable activity can be measured by coupling through 3-P-glycerate kinase, ATP, an ATP-regenerating system, glyceraldehyde-3-phosphate dehydrogenase, and NADH by following the rate of NADH oxidation spectrophotometrically.[27-29]

*Reagents*

Activation medium consisting of Tris-HCl buffer, 0.1 $M$ (pH 8.6 at 25°); $MgCl_2$, 20 m$M$; $NaHCO_3$, 10 m$M$; and NaCl, 0.1 $M$
Assay medium consisting of Tris-HCl buffer, 0.11 $M$ (pH 8.2 at 25°) and $MgCl_2$, 5.5 m$M$

[20] W. A. Laing, W. L. Ogren, and R. H. Hageman, *Plant Physiol.* **54**, 678 (1974).
[21] M. R. Badger and T. J. Andrews, *Biochem. Biophys. Res. Commun.* **60**, 204 (1974).
[22] W. A. Laing and J. T. Christeller, *Biochem. J.* **159**, 563 (1976).
[23] W. A. Laing, W. L. Ogren, and R. H. Hageman, *Biochemistry* **14**, 2269 (1975).
[24] G. H. Lorimer, M. R. Badger, and T. J. Andrews, *Biochemistry* **15**, 529 (1976).
[25] M. R. Badger and G. H. Lorimer, *Arch. Biochem. Biophys.* **175**, 723 (1976).
[26] G. H. Lorimer, M. R. Badger, and T. J. Andrews, *Anal. Biochem.* **78**, 66 (1977).
[27] R. McC. Lilley and D. A. Walker, *Biochim. Biophys. Acta* **358**, 226 (1974).
[28] M. E. Delaney and D. A. Walker, *Biochem. J.* **171**, 477 (1978).
[29] V. Walbot, *Plant Physiol.* **59**, 107 (1977).

NaH$^{14}$CO$_3$ (New England Nuclear), 0.5 $M$ (approximately 0.2 Ci/
mole; the specific radioactivity must be accurately determined)
Aliquots of the stock solution are stored at $-20°$
Tetrasodium ribulosebisphosphate, 10 m$M$, dissolved in 10 m$M$
Tris-HCl, pH 8.2 (25°), just prior to use in order to minimize
inhibitor formation.[29a] Rbu-P$_2$ may be purchased from Sigma
Chemical Co. (St. Louis, Missouri) or synthesized enzymically
from ribose 5-phosphate[30] and subsequently purified by chroma-
tography on Dowex 1-X8 (Cl$^-$ form)[22] or DEAE-cellulose (Cl$^-$
form).[30a]
Acetic acid, 6 $M$
Handifluor (Mallinckrodt) or any other suitable liquid scintillation
(LSC) solution

*Procedure*. Sufficient crystalline enzyme is collected by centrifugation
(2000 $g$, 5 min), dissolved in activation medium, and recentrifuged (10,000
$g$, 15 min) to yield a protein solution of 5 mg/ml. Protein concentration
(as milligrams per milliliter) is calculated by the factor $A_{280 nm} \times 0.7$ (1-
cm light path). Protein content estimated by this method often exceeds
that determined by the Lowry method[31] using crystallized bovine albumin
as the standard. Just prior to use, the protein is heated for 20 min at 50°
to ensure complete heat reactivation of the enzyme[2,7,32] and held at 25°
until use. For routine assays the reactions are conveniently run directly
in glass LSC vials (25 × 60 mm) sealed with serum stoppers. The vials
are placed in a constant temperature water bath (25°) situated in a fume
hood. A 900-$\mu$l aliquot of the assay medium is added to each vial followed
by 50 $\mu$l of NaH$^{14}$CO$_3$ and 50 $\mu$l of Rbu-P$_2$. The vials are sealed and
placed in the water bath for temperature equilibration. Reactions are
initiated by injecting a small volume of activated enzyme (~5 $\mu$l). The
complete reaction mixture[33] thus contains the following components (in
$\mu$moles except as indicated) in a final volume of 1.0 ml: Tris-HCl, 100;
MgCl$_2$, 5; enzyme, 25 $\mu$g; NaH$^{14}$CO$_3$ (approximately 5 $\mu$Ci), 25; and Rbu-
P$_2$, 0.5. After a ≤90-sec incubation period at 25°, the reactions are ter-

---

[29a] C. Paech, J. Pierce, S. D. McCurry, and N. E. Tolbert, *Biochem. Biophys. Res.
Commun.* **83**, 1084 (1978).

[30] B. L. Horecker, J. Hurwitz, and A. Weissbach, *Biochem. Prep.* **6**, 83 (1958).

[30a] G. D. Kuehn and T.-C. Hsu, *Biochem. J.* **175**, 909 (1978).

[31] E. Layne, Vol. 3, p. 447.

[32] R. Chollet and L. L. Anderson, *Arch. Biochem. Biophys.* **176**, 344 (1976).

[33] The activity of dilute solutions (10 $\mu$g/ml) of the crystalline tobacco enzyme is stabilized
by including 0.5 mg/ml Fraction V bovine albumin in the reaction mixture (R. Chollet,
unpublished data, 1978).

minated by injecting 100 $\mu l$ of 6 $M$ acetic acid. The vials are unstoppered in the fume hood and the contents taken to dryness preferably in a forced-draft oven at 90°. Water (1 ml) is added to each vial followed by 12 ml of LSC solution. The vials are capped and vortexed, and the acid-stable radioactivity (as [$^{14}$C]3-P-glycerate) determined by liquid scintillation spectroscopy. Blanks are treated in an identical manner except that Rbu-$P_2$ is replaced with 10 m$M$ Tris-HCl, pH 8.2. The specific radioactivity of the NaH$^{14}$CO$_3$ stock is determined by adding a 30-$\mu l$ aliquot (15 $\mu$moles) to an LSC vial containing 1 ml of water previously made basic with 3 drops of ethanolamine, followed by 12 ml of LSC solution. Using this routine assay protocol, the incorporation of $^{14}$CO$_2$ by the enzyme is linear for at least 5 min.

*Assay for Oxygenase Activity.* In principle, the oxygenase activity of this bifunctional enzyme could be assayed by monitoring substrate utilization or accumulation of products according to the reaction depicted in Eq. (2). In practice, however, the enzyme must first be activated by preincubation with relatively high concentrations of CO$_2$ and Mg$^{2+}$. It is inevitable that some CO$_2$ will be carried over with the enzyme into the oxygenase reaction mixture, where it can act as both a competitive inhibitor and substrate. Thus, assay methods based upon the appearance of the common product 3-P-glycerate or the disappearance of the common substrate Rbu-$P_2$ are inadvisable. Oxygenase activity can be measured directly as Rbu-$P_2$-dependent O$_2$ uptake either polarographically[25,34,35] or manometrically.[11,36] Alternatively, an indirect spectrophotometric assay for 2-P-glycolate formation has been used.[20,32] In view of the complex kinetic properties of this enzyme,[24,25] and the laborious preparations required for the spectrophotometric assay, O$_2$ electrode measurements would appear to be the simplest method of choice. There are situations, however, where analysis of 2-P-glycolate formation may be desirable. For example, this method can be adapted for the *simultaneous* assay of both Rbu-$P_2$ oxygenase and Rbu-$P_2$ carboxylase activities in the same reaction vessel. Such a simultaneous assay is ideally suited for studies in which the objective is to critically evaluate the oxygenase and carboxylase reactions for possible differential regulation,[20,32] since it follows that the activation and assay conditions are identical for the two reactions.

### Reagents

Activation medium consisting of Tris-HCl buffer, 0.1 $M$ (pH 8.6 at 25°); MgCl$_2$, 20 m$M$; NaHCO$_3$, 10 m$M$; and NaCl, 0.1 $M$

[34] T. V. Marsho and S. D. Kung, *Arch. Biochem. Biophys.* **173**, 341 (1976).
[35] J. T. Bahr and R. G. Jensen, *Arch. Biochem. Biophys.* **164**, 408 (1974).
[36] G. H. Lorimer, T. J. Andrews, and N. E. Tolbert, Vol. 42, Part C, p. 484.

Assay medium consisting of Tris-HCl buffer, 0.1 $M$, and $MgCl_2$, 3 m$M$. Dissolved $CO_2$ is eliminated from the assay medium by adjusting the pH to $\leq 4$ with HCl and bubbling ($\geq 10$ min) with $N_2$. Subsequently, the medium is adjusted to pH 8.6 with carbonate-free NaOH, equilibrated with $CO_2$-free air (or the desired $O_2/N_2$ gas mixture) and kept at 25°

Rbu-$P_2$, 50 m$M$, prepared just prior to use as previously described

*Procedure for Polarographic Analysis.* Sufficient crystalline enzyme is dissolved in the activation medium to yield a protein solution of approximately 20 mg/ml. The protein solution is heated for 20 min at 50° and returned to room temperature before use. Oxygen uptake can be monitored with a Clark-type $O_2$ probe covered with a 1.0 mil Teflon membrane (Yellow Springs Instrument Company, Model 5331). Reactions are conveniently run in a lucite vessel having a reaction chamber volume of $\sim 1.0$ ml, magnetic stirring, and the capacity for thermoregulation (at 25°). The electrode is calibrated using the known concentration of $O_2$ (237 $\mu M$) in air-saturated water at 25°. The reaction vessel is filled with $\sim 1.0$ ml of assay medium, stoppered, and checked to ensure the absence of bubbles. Ten microliters of Rbu-$P_2$ are added through a small hole in the stopper and the mixture allowed to equilibrate ($\sim 1$ to 2 min). The reaction is initiated by the addition of a small aliquot ($\leq 25$ $\mu$l) of heat-activated enzyme to yield a final protein concentration of $\leq 0.5$ mg/ml. Additions should be kept to a minimum to avoid significant carryover of $CO_2$ and $Mg^{2+}$ from the activation medium. Rates of activity are calculated from the initial linear rate ($\sim 40$ to 60 sec) of Rbu-$P_2$-dependent $O_2$ uptake observed following the addition of protein. Rates of $O_2$ uptake subsequently decline reflecting the inactivation of the enzyme in the $CO_2$-free assay medium.

*Comments.* In most respects the enzymic properties of the activated crystalline Rbu-$P_2$ carboxylase–oxygenase from tobacco are similar to those reported for the electrophoretically pure enzyme obtained from spinach or soybean. Some kinetic constants which have been measured for the activated tobacco enzyme are as follows: The $K_m$ values determined for $O_2$, $CO_2$ (carboxylase), and Rbu-$P_2$ (oxygenase) are 245, 21[37] and 22[34] $\mu M$, respectively. Several apparent differences between the enzyme purified from spinach and tobacco should be noted. In particular, the specific activity for the oxygenase function of the tobacco enzyme is typically lower ($\leq 50\%$) than that reported for the spinach enzyme.[26] Whether this observation is a result of an inherent difference between the two proteins or varied activation and assay requirements for the two enzymes is unresolved. With regards to the last aspect we have observed

[37] R. G. Jensen and J. T. Bahr, *Annu. Rev. Plant Physiol.* **28**, 379 (1977).

that the carboxylase and oxygenase activities of the tobacco enzyme are markedly inhibited (40–50%) by the high $Mg^{2+}$ concentrations (20 m$M$) routinely used in carboxylase–oxygenase assays of the spinach enzyme.[26]

Acknowledgment

We thank Dr. Michael O'Neill for taking the photograph in Fig. 2. Grants from Tobacco Laboratory, United States Department of Agriculture (12-14-1001-976), and National Science Foundation (GB-38237) (PCM 78-06626 to R. C.) are acknowledged.

# [31] Bacteriochlorophyll a–Proteins of Two Green Photosynthetic Bacteria[1,2]

By JOHN M. OLSON[3]

The water-soluble, blue-green protein from *Prosthecochloris aestuarii* is known to have a molecular weight of 140,000 and to contain 21 molecules of bacteriochlorophyll (Bchl) *a*. (A very similar protein has been isolated from *Chlorobium limicola* f. sp. *thiosulfatophilum*.) *In vivo* bacteriochlorophyll *a* accepts excitation energy from Bchl *c* (*Chlorobium* chlorophyll) and transfers this energy to the reaction center chlorophyll, P840, which appears to be distinct from the Bchl *a*–protein *in vivo*.[4,5]

## Preparation[1,6]

*Prosthecochloris aestuarii* 2K in mixed culture[7,8] (formerly known as "*Chloropseudomonas ethylica*"[6] 2K) and *C. limicola* f. sp. *thiosulfato*-

[1] J. M. Olson, *in* "The Photosynthetic Bacteria" (R. K. Clayton, and W. R. Sistrom, eds.). p. 161. Plenum, New York, 1978.

[2] R. E. Fenna, and B. W. Matthews, *in* "The Porphyrins" (D. Dolphin, ed.). p. 473. Academic Press, New York, 1979.

[3] Research carried out at Brookhaven National Laboratory under the auspices of the United States Energy Research and Development Administration. By acceptance of this paper, the publisher and/or recipient acknowledges the United States Government's right to retain a nonexclusive, royality-free license in and to any copyright concerning this paper.

[4] J. M. Olson, T. H. Giddings, Jr., and E. K. Shaw, *Biochim. Biophys. Acta* **449**, 197 (1976).

[5] J. M. Olson, R. C. Prince, and D. C. Brune, *Brookhaven Symp. Biol.* **28**, 238 (1977).

[6] J. M. Olson, Vol. 23, Part A, p. 636.

[7] N. Pfennig, and H. Biebl, *Arch. Microbiol.* **110**, 3 (1976).

[8] Y. Shioi, K. Takamiya, and M. Nishimura, *J. Biochem. (Tokyo)* **79**, 361 (1976).

*philum* 6230 (Tassajara) are grown anaerobically in 20-liter carboys with illumination provided by two 60-W incandescent lamps (18 inch Lumiline type) and culture temperature maintained between 28° and 32°. Each culture is stirred continually with a magnetic bar. The cells are harvested by precipitation with alum [KAl(SO$_4$)$_2$·H$_2$O]: about 500 ml of saturated alum solution is added to 20 liters of culture, and the aggregated cells are collected by low speed centrifugation. All procedures are carried out at 20°–25° except as noted.

*Step 1.* The packed cells (~ 480 gm wet weight) are suspended in 1 volume of 0.2 *M* Na$_2$CO$_3$. Additional 2 *M* Na$_2$CO$_3$ is added if necessary to give pH ~ 10, and the cells are stored for 18 hr or more at 4°–5°. The cells are homogenized for 4 min at low speed in a blender and sonicated for 10 min at 1.2 A in a Ratheon 10-kHz oscillator. The sonicate is centrifuged at 14,000 *g* for 45 min.

*Step 2.* Solid ammonium sulfate (30 gm per 100 ml) is added to the supernatant, which is then stored at 4°–5° for 18 hr or longer. The resulting green precipitate containing both bacteriochlorophylls *a* and *c* is mixed with diatomaceous earth (0.5 kg of Celite 545) along with sufficient ammonium sulfate solution (35 gm per 100 ml in 10 m*M* Tris, pH 8.0) to make a slurry, which should be light gray-green. If too little Celite is used, the green (Bchl *c*) component will be washed off the column. Excess Celite is not harmful.

*Step 3.* The slurry is poured onto a 2.5 cm pad of dry Celite in a column (30 × 14 cm) and eluted with a constant gradient of decreasing ammonium sulfate (35 to 0 gm per 100 ml) in 10 m*M* Tris, pH 8.0 (~ 23 liter). Fractions are collected according to the color of the eluate. The first fraction is colorless; the second is straw colored (cytochromes); the third is blue-green (Bchl *a*–protein) and usually elutes between 12 and 5% ammonium sulfate; the fourth is yellow-green. The blue-green fraction (6 to 8 liter) is concentrated to ~ 150 ml by ultrafiltration and then dialyzed at 4°–5° verus 5–10 m*M* Tris, pH 8.0, to remove ammonium sulfate.

*Step 4.* The concentrated, salt-free Bchl *a*–protein solution is applied to a column (60 × 2.5 cm) of DEAE–cellulose (Schleicher & Schuell type 40), and is eluted with a constant gradient of increasing NaCl (0 to 0.3 *M*) in 5–10 m*M* Tris, pH 8.0 (3 liter). The Bchl *a*–protein is collected according to the spectral criterion that $A_{267}/A_{371} < 0.6$. (Material with a higher absorbance ratio may be rerun on DEAE–cellulose.) The Bchl *a*–protein may be stored in 30% ammonium sulfate as a slurry at −10°.

*Step 5.* Bchl *a*–protein may be prepared for crystallization by concentrating the Bchl *a*–protein to ~ 14 mg/ml by ultrafiltration in the presence of 1 *M* NaCl. Crystallization is effected by slow dialysis versus

5 to 10 gm ammonium sulfate per 100 ml in 1 $M$ NaCl in 5–10 m$M$ Tris buffer, pH 8.0, at ~ 4°. For storage as a noncrystalline precipitate, 30 gm ammonium sulfate is added per 100 ml of solution. Since the Bchl $a$-protein is slowly oxidized in the presence of light and oxygen, it should be isolated under green light (F40 green fluorescent lamps behind green celluloid or green 2092 Plexiglas) and stored in the dark.

### Properties

*Purity.*[1] After completion of step 4, the Bchl $a$-protein from strain 2K is at least 99% pure by the criterion of polyacrylamide gel electrophoresis, but the Bchl $a$-protein from strain Tassajara is only ~ 85% pure.

*Stability.* (See Olson[6].) The protein from strain Tassajara has essentially the same absorption spectrum in 2 $M$ guanidine HCl as in dilute buffer, but in 2 $M$ guanidine thiocyanate the narrow 808.5 nm band is replaced by two broad bands at ~ 795 and ~ 870 nm, and the 603.5 nm band is shifted to ~ 593 nm.[9] These data indicate that the protein is stable in 2 $M$ guanidine HCl but is denatured by 2 $M$ guanidine thiocyanate.

Although the protein from strain 2K appears to be unaffected by Triton X-100 (5%) or by 6 $M$ urea,[6] the combination of Triton X-100 (0.5%) and 8 $M$ urea at pH 7.8 does cause denaturation.[10] The protein is also denatured by 10 m$M$ hexadecyltrimethylammonium bromide at pH 7.9.[10] Sodium dodecylbenzenesulfonate (10 m$M$) denatures the protein and converts the Bchl $a$ to bacteriophephytin $a$ as does sodium dodecyl sulfate.[6]

*Molecular Size.* See Olson.[6]

*Isoelectric Point.*[11] Attempts were made to determine the isoelectric points for both Bchl $a$-proteins by isoelectric focusing. Since every preparation gave 5–13 bands, it was necessary to graph the distribution pattern for the bands from each protein. For the 2K protein there were 148 bands distributed between pH 5.3 and 7.4. The center of the distribution was at pH = 6.0. For the Tassajara protein there were 53 bands between pH 6.1 and 7.8, and the distribution was centered at pH = 7.0. The difference of one pH unit between the "average" isoelectric values for the two proteins may reflect the presence in the 2K protein of at least two more glutamic acid residues and two less arginine residues than in the Tassajara protein.

[9] J. M. Olson, unpublished results (1976).
[10] Y. D. Kim, *Arch. Biochem. Biophys.* **140,** 354 (1970).
[11] J. M. Olson, E. K. Shaw, and F. M. Englberger, *Biochem. J.* **159,** 769 (1976).

*Chemical Composition.*[11–13] The Bchl *a*–protein from strain 2K is a trimer of identical subunits containing only Bchl *a* and protein.[12,13] Each subunit contains seven Bchl *a* molecules surrounded by a polypeptide chain of ~ 354 amino acid residues.[13] The N-terminal amino acid is alanine.[6] The amino acid compositions of the proteins from strains 2K and Tassajara are compared in Table I on the assumption that both polypeptides contain ~ 354 residues. (This assumption is based on the observation that the two polypeptides are indistinguishable on polyacrylamide gel electrophoresis.[11]) Since the sum of positive differences is 8, and the sum of negative differences is −13, there must be at least 8–13 residues (~ 3%) in one polypeptide which differ from the corresponding residues in the other polypeptide. From the values of amino acid composition (Table I), the molecular weights of the two polypeptides (neglecting amide groups) are both calculated to be 38,600. With the addition of seven Bchl *a* molecules (6400), the total subunit weight becomes 45,000, and the trimer weight 135,000.[11]

*Spectral properties.*[14–16] (See Olson.[6]) The absorption characteristics of each Bchl *a*–protein at 20°–25° are summarized in Table II. For quantitative measurement of concentration, the absorbance value at 370.5 nm is recommended. The absorptivity at 808–809 nm is variable, depending on ionic strength, pH, temperature, and history of the sample. The absorbance ratio $A_{808-9}/A_{307.5}$ is a useful criterion of the state of the sample. Values of ~ 2.3 indicate trimers in "good conformation," whereas values below 2.0 suggest partial denaturation.

At 77°K, the 809-nm absorption band of the 2K Bchl *a*–protein is resolved into three sharp bands at 806, 814, and 825 nm and a broad shoulder at ~ 790 nm.[6,16] The 808-nm band of the Tassajara protein is similarly resolved into sharp bands at 806, 814, and 824 nm with a broad shoulder also at ~ 790 nm.[16] The circular dichroism (CD) spectra of the two proteins at 77°K, however, are quite different. The CD spectrum for strain 2K shows a band pattern of − (786 nm), + (798 nm), + (810 nm), − (814 nm), and − (821 nm), while the CD spectrum for strain Tassajara shows a pattern of − (786 nm), + (798 nm), − (806 nm), + (812 nm), and − (821 nm).[16] Clearly the exciton interactions between the Bchl *a* molecules in the two proteins must be different in order to account for the

[12] R. E. Fenna, and B. W. Matthews, *Nature (London)* **258,** 573 (1975).
[13] R. E. Fenna, and B. W. Matthews, *Brookhaven Symp. Biol.* **28,** 170 (1977).
[14] J. M. Olson, *in* "The Chlorophylls" (L. P. Vernon and G. R. Seely, eds.), p. 413. Academic Press, New York, 1966.
[15] J. M. Olson, K. D. Philipson, and K. Sauer, *Biochim. Biophys. Acta* **292,** 206 (1973).
[16] J. M. Olson, B. Ke, and K. H. Thompson, *Biochim. Biophys. Acta* **430,** 524 (1976).

TABLE I
COMPARISON OF BACTERIOCHLOROPHYLL a-PROTEINS: AMINO ACID COMPOSITION[a] OF
SUBUNITS[b]

| Amino Acid | Strain 2K | | Strain Tassajara | | Minimum difference nearest integer |
| | Moles | Nearest integers | Moles | Nearest integer | |
| --- | --- | --- | --- | --- | --- |
| Gly[c] | 40.5 | 39–42 | 37.0 | 36–38 | −1 |
| Asp | 39.0 | 38–40 | 39.4 | 38–41 | 0 |
| Val | 33.9 | 33–35 | 37.8 | 37–39 | +2 |
| Glu | 33.2 | 32–34 | 30.0 | 29–31 | −1 |
| Ser | 27.5 | 26–29 | 32.3 | 31–34 | +2 |
| Ile[c] | 23.3 | 23–24 | 14.1 | 14 | −9 |
| Ala[c] | 21.4 | 21–22 | 23.3 | 23–24 | +1 |
| Leu | 20.0 | 19–21 | 20.9 | 20–22 | 0 |
| Arg[c] | 19.5 | 19–20 | 22.4 | 22–23 | +2 |
| Pro[c] | 18.0 | 17–19 | 20.8 | 20–22 | +1 |
| Lys | 17.8 | 17–18 | 18.0 | 17–18 | 0 |
| Phe | 16.8 | 16–17 | 17.5 | 17–18 | 0 |
| Thr[c] | 13.9 | 13–14 | 11.8 | 11–12 | −1 |
| Tyr | 9.6 | 9–10 | 9.8 | 10 | 0 |
| His | 7.5 | 7–8 | 7.5 | 7–8 | 0 |
| Trp[c] | 6.6 | 6–7 | 4.6 | 4–5 | −1 |
| Met[c] | 3.5 | 3–4 | 4.3 | 4 | 0 |
| Cys | 2.0 | 2 | 2.1 | 2 | 0 |
| Total | 354.0 | | 353.6 | | −13 +8 |

[a] Calculated for 354 residues per subunit.[11]
[b] Calculated as moles/mole.
[c] Significant difference.

obvious differences in the CD spectra and the more subtle differences in the absorption spectra at 77°K.

*Crystal Properties.*[17] (See Olson[6]). Bchl a-protein from strain 2K forms crystals of two types: hexagonal[6] space group P6₃ or trigonal[17] space group P3₁ or P3₂. (The Tassajara protein also forms hexagonal crystals similar to those from strain 2K.) Dimensions of the unit cell in hexagonal crystals are $a = b = 112.4 \pm 0.4$ Å[17] and $c = 98.4 \pm 0.4$ Å.[6,17] The original value[6] given for $a$ and $b$ was erroneous.[17] Since each unit cell contains two macromolecules, the symmetry properties of the P6₃ unit cell require each macromolecule to consist of three subunits. Al-

[17] R. E. Fenna, B. W. Matthews, J. M. Olson, and E. K. Shaw, *J. Mol. Biol.* **84,** 231 (1974).

TABLE II
ABSORPTION CHARACTERISTICS[a] AT 20°–25°

| Strain 2K[14] | | | Strain Tassajara[d] | |
|---|---|---|---|---|
| $\lambda$ (nm) | $\epsilon$ (m$M^{-1}$ cm$^{-1}$)[b] | $\epsilon_\lambda/\epsilon_{370.5}$ [c] | $\lambda$ (nm) | $\epsilon_\lambda/\epsilon_{370.5}$ |
| 267 | 37 | 0.55 | 270 | 0.61 |
| 343 | 49 | 0.73 | 342.5 | 0.76 |
| 370.5 | 67 | 1.00 | 370.5 | 1.00 |
| 603 | 28.4 | 0.42 | 602.5 | 0.42 |
| 745 | 13.4 | 0.20 | 745 | 0.20 |
| 809 | 154[e] | 2.30[e] | 808 | 2.28[e] |
| | | 2.1–2.2[f] | | 2.0–2.1[f] |

[a] Samples dissolved in 10 m$M$ Tris, pH 8.0, 0.25 $M$ NaCl.
[b] Values based on Bchl $a$. Limit of error estimated to be ±4%.
[c] Average values. Standard deviation is about 2%.
[d] Olson et al.[15]
[e] Highest values observed.
[f] Typical values depending on pH, temperature, ionic strength, and history of sample.

though formation of hexagonal crystals is the usual mode of crystallization, trigonal crystals have been obtained once,[17] with cell dimensions $a$ = $b$ = 83.2 Å and $c$ = 165.8 Å. The trimers are much more densely packed in trigonal crystals than in hexagonal crystals.

*Structure*.[12,13,17a] The structure of the Bchl $a$–protein from strain 2K has been determined to a normal resolution of 2.8 Å by X-ray diffraction of single hexagonal crystals of native protein and of protein to which certain heavy atoms had been bound: mercury, mercury and platinum, platinum, and uranium.

The subunit (Fig. 1) contains seven Bchl $a$ molecules surrounded by a protein bag. The polypeptide chain forms a distorted hollow cylinder, one end of which is closed and the other open. (In the trimer the open end is covered by an adjacent subunit.) The cylinder wall facing the outside of the trimer is composed of 15 strands of $\beta$ sheet, 13 of which are antiparallel. The cylinder wall in contact with adjacent subunits consists of short stretches of $\alpha$-helix intermixed with regions of irregular conformation. The first 11 amino acid residues from the N-terminus are NH$_2$-Ala-Leu-Phe-Gly-Thr-Lys-Asx-Thr-Thr-Thr-Ala.[18]

The space occupied by the seven Bchl $a$ molecules is approximated by an ellipsoid with axial dimensions of 45, 35, and 15 Å. For nearest

[17a] B. W. Matthews, R. E. Fenna M. C. Bolognesi, M. F. Schmid, and J. M. Olson, *J. Mol. Biol.*, **131**, 259 (1979).
[18] C.-L. Tang, and C. H. W. Hirs, unpublished results (1976).

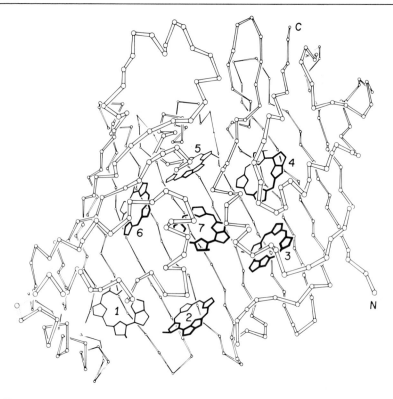

Fɪɢ. 1. Subunit of the Bchl $a$–protein from strain 2K, showing the arrangement of the polypeptide chain and the Bchl $a$ core. Circles indicate presumed $\alpha$-carbon positions. Some connections within the polypeptide chain are uncertain, and particularly ambiguous regions are indicated by broken lines. For clarity, Mg atoms, Bchl $a$ ring substituents, and phytyl chains are omitted. The direction of view is from the center of the trimer toward the exterior. The threefold axis of the trimer is horizontal. From Fenna and Matthews.[13]

neighbors the average center-to-center distance between porphine rings is 12 Å. Although no two porphine rings are exactly parallel, the seven rings do lie close to an "average" plane (defined as the plane for which the sum of direction cosines with respect to each of the rings is a maximum). The normal to the "average" plane makes an angle of 62° with the threefold axis of the trimer (Fig. 2), and the center of the Bchl $a$ aggregate in each subunit is about 20 Å from the center of the trimer. The three aggregates form a triangular funnel around the threefold axis of the trimer.

The phytyl chains of the Bchl $a$ molecules lie close together within each subunit. The phytyl tails of Bchl 4, 5, and 6 lie parallel between

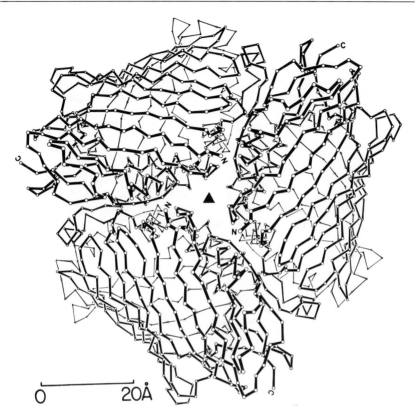

FiG. 2. Model of the Bchl *a*-protein trimer from strain 2K, showing the arrangement of subunits. The Bchl *a* molecules have been omitted for clarity. The threefold axis of the trimer is normal to the plane of the figure. From B. W. Matthews *et al.*[17a]

rings 5 and 6 and the β sheet of the outer wall. The tails of Bchl 2, 3, and 7 lie in extended conformation in the inner space between rings, and the tail of Bchl 1 is bent into a U-shaped loop.

Chemical interactions between neighboring Bchl *a* molecules are limited to hydrophobic associations between phytyl chains, but there is considerable evidence of interaction between each Bchl *a* molecule and the protein "bag." For each Bchl *a* molecule a fifth ligand to the Mg atom is indicated by electron density extending from the center of the porphine ring. For six Bchl *a* molecules the density extends as a continuum between the Mg atom and the polypeptide chain. For Bchl 3, 4, and 7, the ligands are the side chains of three histidine residues in an α-helix; for Bchl 1 and 6 the ligand are side chains of histidine residues in the β-sheet. The ligand for Bchl 5 seems to come directly from the polypeptide backbone, while the ligand for Bchl 2 appears to be a water molecule.

The Bchl $a$ ring substituents include several oxygen containing groups, which are potential hydrogen bond acceptors. Most of these groups are close to the periphery of the Bchl $a$ aggregate and are close enough to the protein wall to form hydrogen bonds with suitable donors in the polypeptide chain. Each Bchl $a$ molecule appears to be anchored to the protein through extensive hydrogen bonding and ligand bonding to the Mg atom, with additional hydrophobic interactions through the phytyl tail.

The subunits are tightly packed in the trimer, as shown in Fig. 2. The volume occpuied by the trimer is roughly equivalent to an oblate ellipsoid of revolution 57 Å alont the short axis and 83 Å along the long axis.[17] The closest distance[12] between Bchl $a$ molecules in adjacent subunits is 24 Å.

### Other Chlorophyll–Proteins[1]

The Bchl $a$-protein may be a model for certain light-harvesting chlorophyll–proteins such as the chlorophyll $a$-proteins of blue-green algae and chloroplasts and the chlorophyll $a/b$ proteins of chloroplasts.[19–21] According to the Bchl $a$-protein model each subunit of a light-harvesting chlorophyll–protein should consist of a protein bag with the chlorophyll molecules in fixed orientations inside the bag. Photochemical reaction centers would probably not follow this model, since in a reaction center at least one pair of chlorophyll molecules (the primary electron donor) needs to be accessible to the secondary electron donor (cytochrome or plastocyanin) outside the reaction center.

[19] J. P. Thornber, *Annu. Rev. Plant Physiol.* **26,** 127 (1975).

[20] J. P. Thornber, R. S. Alberte, F. A. Hunter, J. A. Shiozawa, and K.-S. Kan, *Brookhaven Symp. Biol.* **28,** 132 (1977).

[21] J. P. Thornber, this volume. [14].

# [32] NDP–NTP Kinase: Plant and Bacterial

*By* J. Yamashita and T. Horio

$$NDP + N'TP \rightleftharpoons NTP + N'DP$$

An enzyme capable of catalyzing the ADP–ATP exchange reaction (ADP–ATP exchange enzyme) was partially purified from green plants and photosynthetic bacteria.[1–3] In photosynthetically grown cells of *Rhodospirillum rubrum*, the enzyme is mostly present in the cytoplasmic fluid and partly in chromatophores. Although well-washed chromatophores hardly catalyze the exchange reaction, it is solubilized from chromatophores by sonication. It seems proper for this enzyme to be called nucleoside disphophate:nucleoside triphosphate phosphotransferase (NDP–NTP kinase, or NDP kinase) because of its broad substrate specificity towards various nucleoside phosphates. NDP–NTP kinase is also present in other plants and microorganisms as cited above.[4]

## Preparation

*Step 1.* Photosynthetically grown cells of *R. rubrum*[5] are centrifugally harvested, washed with water, and lyophilized. The lyophilized cells (1 kg) are suspended in 10 liters of 0.1 $M$ Tris-HCl buffer containing 0.5 $M$ NaCl (pH 8.0) and allowed to stand overnight while being stirred, followed by centrifugation at 100,000 $g$ for 1 hr. The precipitate is reextracted with the same volume of the buffer. To the combined supernatants, 176 gm/liter of ammonium sulfate is added. The mixture is supplemented with a minimum amount of celite, and filtered through filter paper covered with a thin layer of celite on a Büchner funnel. The filtrate is desalted by passing through a Sephadex G-25 column equilibrated with water.

*Step 2.* The desalted solution (approximately 15 liters) is passed through a DEAE–cellulose column (4.5 × 60 cm), previously equilibrated with 10 m$M$ Tris-HCl buffer (pH 8.0). The charged column is washed

[1] T. Horio, N. Yamamoto, Y. Horiuti, and K. Nishikawa, Vol. 23 [64].

[2] N. Yamamoto, Y. Horiuti, K. Nishikawa, and T. Horio, *J. Biochem.* (*Tokyo*) **72,** 599 (1972).

[3] J. S. Kahn, Vol. 23 [51].

[4] R. E. Parks, Jr. and R. P. Agarwal, *in* "The Enzymes" (P. D. Boyer, ed.), 3rd ed., Vol. 8, p. 307. Academic Press, New York, 1973.

[5] See Vol. 23 [63].

METHODS IN ENZYMOLOGY, VOL. 69

with 4 bed volumes of 0.1 $M$ Tris-HCl buffer (pH 8.0). The washed column is developed with 0.1 $M$ Tris-HCl buffer containing 0.3 $M$ NaCl (pH 8.0), and the eluate is divided into fractions. The fractions showing NDP–NTP kinase activity are mixed and desalted on Sephadex G-25.

*Step 3.* The enzyme solution is diluted with $\frac{1}{9}$ volume of 0.1 $M$ Tris-HCl buffer (pH 8.0), and then passed through a DEAE–cellulose column (4.5 × 60 cm), previously equilibrated with 0.1 $M$ Tris-HCl buffer (pH 8.0). The charged column is washed with 3 bed volumes of the same buffer and then eluted with 2 liters of a linear gradient of NaCl from 0.01 to 0.3 $M$ in 0.1 $M$ Tris-HCl buffer (pH 8.0). The eluate is divided into 10-ml fractions, those showing high specific activity are combined.

*Step 4.* To the combined enzyme solution, 350 gm/liter of ammonium sulfate are added to 0.55 saturation. After 0.5 hr, the precipitate is removed by centrifugation at 15,000 $g$ for 10 min. The supernatant is brought to 0.70 saturation by addition of 103 gm/liter of ammonium sulfate. After 1 hr, the precipitate is collected by centrifugation and dissolved in 10 ml of 50 m$M$ Tris-HCl buffer containing 0.1 M KCl (pH 7.5). The enzyme solution is dialyzed against to 50 mM Tris-HCl buffer containing 0.1 $M$ KCl (pH 7.5).

*Step 5.* The dialyzed solution is applied to a Sephadex G-75 column (2.5 × 9.0 cm), previously equilibrated with 50 m$M$ Tris-HCl buffer containing 0.1 $M$ KCl (pH 7.5). The charged column is then developed with the same buffer at a flow rate of 30 ml/hr and the eluate is divided into 3-ml fractions. Fractions showing the enzyme activity are collected and stored in frozen state. The yields at each purification step are summarized in Table I.

TABLE I

Summary for Purification Procedure of NDP–NTP Kinase

| Step | Total protein[a] | Specific activity[b] | Total Activity[c] | Yield (%) |
|------|------|------|------|------|
| 1. Extract | $7.9 \times 10^4$ | 17 | $13.4 \times 10^5$ | 100.0 |
| 2. First DEAE–cellulose | $2.7 \times 10^3$ | 290 | $78.6 \times 10^4$ | 59.0 |
| 3. Second DEAE–cellulose | $6.2 \times 10^2$ | 700 | $43.3 \times 10^4$ | 33.0 |
| 4. Ammonium sulfate | $7.6 \times 10$ | 2380 | $18.0 \times 10^4$ | 13.0 |
| 5. Sephadex G-75 | $1.7 \times 10$ | 5950 | $10.1 \times 10^4$ | 7.5 |

[a] $A_{280nm}$.
[b] Nanomoles of $P_i$ transferred or [$\beta$-$^{32}$P]ADP formed/$A_{280nm}$/min.
[c] Nanomoles of $P_i$ transferred or [$\beta$-$^{32}$P]ADP formed/min.

## Assay Method

Principally, spectrophotometric and isotopic methods are applicable for assay.

### Spectrophotometric Method

This method is based on the measurement of the ADP–GTP exchange activity of NDP–NTP kinase. In the presence of glucose, $NADP^+$, hexokinase, and glucose-6-phosphate dehydrogenase, the ATP (formed from ADP by NDP–NTP kinase) and glucose are converted to glucose-6-phosphate by the hexokinase, and the sugar phosphate and $NADP^+$ are then converted to 6-phosphogluconate and NADPH by the dehydrogenase. The reduction of $NADP^+$ is measured spectrophotometrically.

Reagents
  $MgCl_2$, 10 m$M$
  ADP, 1 m$M$
  GTP, 1 m$M$
  $NADP^+$, 1 m$M$
  Glucose, 10 m$M$
  Hexokinase, 1 unit
  Glucose-6-phosphate (G-6-P) dehydrogenase, 1 unit
  Tris-HCl buffer (pH 8.0), 50 m$M$
  Enzyme sample, appropriate amount of NDP–NTP kinase
  Total volume, 1 ml

Procedure. The reaction is started by adding the enzyme in a cuvette of 1-cm optical path. The reduction rate of $NADP^+$ is calculated from the absorbance change at 340 nm. The molar extinction coefficient of NADPH at the wavelength is regarded as $6.2 \times 10^3$ $M^{-1}$ $cm^{-1}$.[6] Hexokinase catalyzes the formation of glucose-6-phosphate from GTP and glucose to a lower extent than that from ATP and glucose. In addition, if adenylate kinase contaminates the enzyme sample, ATP is formed from ADP. Thus, two reactions for the control are carried out; in one, the enzyme sample and, in other, GTP is omitted from the complete reaction mixture.

### Radioisotopic method

Reagents
  0.1 $M$ $MgCl_2$
  10 m$M$ ATP

---

[6] A. Kornberg and B. L. Horecker, *Biochem. Prep.* **3**, 27 (1953).

10 m$M$ [$\beta$-$^{32}$P]ADP ($10^7$ cpm in excess)[1]

0.2 $M$ Tris-HCl buffer (pH 8.0)

*Procedure.* Five microliters of $MgCl_2$, 5 $\mu$l of ATP, 5 $\mu$l of [$\beta$-$^{32}$P]ADP, 25 $\mu$l of the buffer, and 10 $\mu$l of an enzyme sample are mixed in a total volume of 75 $\mu$l. The mixture without the enzyme sample is preincubated at 30° for 5 min. The reaction is started by adding the enzyme, carried out at 30° for 10 min, and stopped by adding 15 $\mu$l of cold 2.5 $M$ perchloric acid. The resulting mixture is supplemented with 5 $\mu$l each of 10 mA ATP, ADP, and AMP, and then allowed to stand in an ice water bath for 15 min. In order to remove perchloric acid, the resulting solution is neutralized by adding 2.5 $M$ potassium hydroxide, followed by centrifugation. The resulting supernatant is subjected to electrophoresis carried out at 2 kV, 150 mA in 50 m$M$ citrate buffer (pH 4.0) at 0°–4° for 1.5 hr, then dried in the air. Spots of nucleoside phosphates on the dried filter paper are detected under illumination from an ultraviolet light. The nucleoside phosphates are individually eluted with 2 ml of 0.1 $M$ HCl, and the radioactivities of the eluates are counted by a GM-counter.

## Properties

*Molecular Weight.* The molecular weight of the purified NDP–NTP kinase was determined by molecular sieve chromatography on Sephadex G-75 column to be approximately 30,000.

*Optimum pH.* The ADP–ATP and ADP–GTP exchange reactions have optimum pH's in a range 8.5–9.0.

*Activators and Inhibitors.* The ADP–ATP exchange reaction is activated by 10 m$M$ $Mg^{2+}$ to the highest extent. $Mg^{2+}$ is replaceable by $Mn^{2+}$ (66%) and $Ca^{2+}$ (23%). $K_m$ values for these cations are approximately 1 m$M$. The inhibitors and uncouplers for photophosphorylation of chromatophores from *R. rubrum* are not effective on the ADP–ATP exchange reaction, except that $p$-chloromercuribenzoate is slightly inhibitory.

*Substrate Specificity.* The specificity of nucleoside triphosphates are as follows: ATP (100), GTP (57), CTP (4), UTP (24) and ITP (41); of nucleosidediphosphates, ADP (100), dADP (100), GDP (81), dGDP (115), CDP (49), dCDP (27), UDP (106), dUDP (101), and IDP (110).

$K_m$ *Values for ADP and ATP.* The $k_m$ value for ADP is changed by ATP concentrations; 0.1, 0.25, and 1.4 m$M$ with 0.3, 1.0, and 5.0 m$M$ ATP, respectively. The $K_m$ values for ATP are 2.5 and 1.1 m$M$ with 2.5 and 5.0 m$M$ ADP, respectively.

# [33] Manganese Binding Sites and Presumed Manganese Proteins in Chloroplasts

## By GEORGE M. CHENIAE

At least three "classes" of Mn binding sites are identifiable in well-washed, $O_2$ evolving type II chloroplasts. One "class" is nonfunctional in electron transport, whereas the other two appear to be intimately involved within reactions of system II. Each "class" is presently identifiable only by estimates of its dissociation constants ($K_D$), the number of binding sites, and especially by the effects which the loss of each "class" imposes on the reactions of Fig. 1.

Figure 1 shows in an abbreviated manner, current thinking concerning the promotion of electrons from $H_2O$ to $X^-$ by the two photoacts connected in series.[1] Reaction 1 in Fig. 1 represents primary charge separation by the system II trap leading to the reduction of the primary acceptor, Q, and the oxidation of the primary donor, $P_{680}$; Reaction 2 in Fig. 1 denotes reduction ($< 1$ $\mu sec$)[2] of $P^+_{680}$[3] by the secondary electron donor, Z. Reaction 3 in Fig. 1 indicates the reduction of $Z^+$ by the S states of the $O_2$-evolving "enzyme"[4] in which, via successive steps, $(S_n \rightarrow S_{n+1})$ ($t_{1/2} \sim 0.1$-$1$ msec)[5,6] $S_0$ is raised to $S_4$ with resulting liberation of $O_2$ (reaction 4). Reaction 5 in Fig. 1 shows the oxidation of $Q^-$ ($t_{1/2} = 200$-$600$ $\mu sec$)[7,8] by the secondary electron acceptor, A, of system II. The rest of the abbreviated scheme shows the oxidation of $A^-$ by $P_{700}$ and ultimately the reduction of class I electron acceptors via $X^-$, the primary acceptor of System I. In green algae and chloroplasts each

[1] For the details of current thinking on the reaction steps within System II, see R. Radmer and B. Kok, *Annu. Rev. Biochem.* **44**, 409 (1975); P. Joliot, and B. Kok, *in* "Bioenergetics of Photosynthesis" (Govindjee, ed.), p. 387. Academic Press, New York 1975; and R. Radmer, and G. Cheniae, *in* "Topics in Photosynthesis" (J. Barber, ed.). p. 301. Elsevier, Amsterdam, 1977.

[2] G. A. Den Haan, L. N. M. Duysens, and D. J. N. Egberts, *Biochim. Biophys. Acta* **368**, 409 (1974).

[3] M. Gläser, C. Wolff, H. Buchwald, and H. Witt, *FEBS Lett.* **42**, 81 (1974).

[4] B. Kok, B. Forbush, and M. McGloin, *Photochem. Photobiol.* **11**, 457 (1970).

[5] R. E. Blankenship, G. T. Babcock, J. T. Warden, and K. Sauer, *FEBS Lett.* **51**, 287 (1975).

[6] G. T. Babcock, R. E. Blankenship, and K. Sauer, *FEBS Lett.* **61**, 286 (1976).

[7] D. Mauzerall, Proc. Natl. Acad. Sci. U.S.A. **69**, 1358 (1972).

[8] K. Zankel, *Biochim. Biophys. Acta* **325**, 138 (1973).

METHODS IN ENZYMOLOGY, VOL. 69

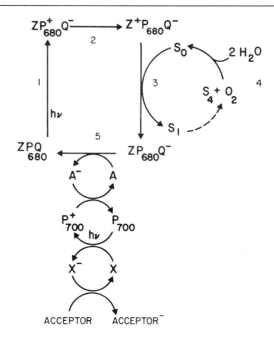

FIG. 1. Scheme for electron flow in photosystem II and abbreviated series connection to photosystem I.

trapping center occurs in an abundance of 1/200–400 Chl and provides a reference for Mn abundance in the three "classes" of Mn binding sites.

## Manganese Binding Sites in Chloroplasts

*Nonfunctional Mn.* Equilibration of type II spinach chloroplasts in isoosmotic media with $Mn^{2+}$ results in extensive binding of $Mn^{2+}$ to chloroplasts.[9,10] Such binding experiments (analyzed by Scatchard or Hughes-Klotz plots) reveal a single type binding site (80–160 Mn/400 Chl); however, vastly different $K_D$ values of $6.7 \times 10^{-6} \, M$[9] and $1.2 \times 10^{-4} \, M$[10] have been reported. Many proteins, nucleic acids, and anionic lipids bind $Mn^{2+}$, thus this apparent discrepancy in $K_D$ values may reflect differences in macromolecular composition of chloroplasts as isolated by different workers. For comparative purposes, both ribulose diphosphate (RuDP) carboxylase and bovine serum albumin can have two types of

[9] E. Gross, *Arch. Biochem. Biophys.* **150**, 324 (1972).
[10] R. E. Blankenship, and K. Sauer, *Biochim. Biophys. Acta* **357**, 252 (1974).

$Mn^{2+}$ binding sites: $K_D$ of $3 \times 10^{-3}$ and $1 \times 10^{-5}$ $M$, [11] and $3.0 \times 10^{-4}$ and $3.7 \times 10^{-5}$ $M$, [12] respectively. The latter values are two to three orders of magnitude less than $K_D$ values of $Mn^{2+}$–amino acid complexes.[13] These considerations suggest that the enormously high abundance of Mn (27–29 Mn/400 Chl) sometimes reported for isolated chloroplasts[14,15] may reflect Mn binding to chloroplast proteins and/or other macromolecules, which is nonspecific in contrast to the functional chloroplastic bound Mn. Such high abundances of chloroplastic Mn can be diminished to 5–8 Mn/400 Chl without affecting reactions of Fig. 1, and therefore Mn abundance greater than 5–8 Mn/400 Chl is considered to be nonfunctional.[1,16]

*Functional Mn in System II.* The pool of 5–8 Mn/400 Chl represents the minimum amount of bound Mn on the oxidant side of system II required for maximum quantum efficiency and high rates of $O_2$ evolution (quantum yields of about 10 $h\nu/O_2$ and uncoupled electron-transport rates in strong light of $\geq 1000$ Eq/Chl hr),[17] and normal kinetics of the reactions linking the S states to the reduction of $P^+_{680}$.[18] The bases for excluding functions of this Mn in reactions other than those on the oxidant side of system II and in a major structural role (lamellar stacking) are summarized elsewhere.[1,16]

In contrast to the nonfunctional Mn, the functional bound Mn (5–8 Mn/400 Chl) is not removed from chloroplasts by washes with grinding media or with chelating agents such as EDTA[19–21] even though this reagent readily penetrates thylakoid membranes and chelates any free $Mn^{2+}$ within the thylakoids.[18] Assuming that EDTA equilibrates with these Mn-binding sites, this implies that the sites binding the 5–8 Mn/400 Chl have an affinity for Mn greater than EDTA ($K_A$ for Mn–EDTA is $\sim 10^{14}$).[22] This conclusion is reinforced by the failure of EDTA to inhibit

[11] G. H. Lorimer, M. R. Badger, and J. T. Andrews, *Biochemistry* **15**, 529 (1976).

[12] A. S. Mildvan and M. Cohn, *Biochemistry* **2**, 910 (1963).

[13] B. L. Valee and W. C. Wacker, *in* "The Proteins" (H. Neurath, ed.), Vol. 5, p. 5. Academic Press, New York 1970.

[14] F. R. Whatley, L. Ordin, and D. I. Arnon, *Plant Physiol.* **26**, 414 (1951).

[15] M. Takahashi and K. Asada, *Eur. J. Biochem.* **64**, 445 (1976).

[16] G. M. Cheniae, *Annu. Rev. Plant Physiol.* **21**, 467 (1970).

[17] G. M. Cheniae and I. F. Martin, *Biochim. Biophys. Acta* **253**, 167 (1971).

[18] R. E. Blankenship, G. T. Babcock, and K. Sauer, *Biochim. Biophys. Acta* **387**, 165 (1975).

[19] G. M. Cheniae and I. F. Martin, *Brookhaven Symp. Biol.* **19**, 406 (1966).

[20] J. V. Possingham and D. Spencer, *Aust. J. Biol. Sci.* **15**, 58 (1962).

[21] P. Homann, *Plant Physiol.* **42**, 997 (1967).

[22] L. G. Sillen and A. E. Martell, eds., "Stability Constants of Metal-Ion Complexes," Spec. Publ. No. 25, p. 626. Chem. Soc., London, 1971.

the $Mn^{2+}$-dependent reactivation of $O_2$ evolution of Tris-treated chloroplasts, a process in which intrathylakoid free $Mn^{2+}$ is recomplexed at its binding sites,[18] and inactive $O_2$ centers are converted to $O_2$ evolving centers.[23]

Though the functionally bound Mn appears to have an extremely high affinity, a number of methods are available that rather specifically release the bound Mn in chloroplasts,[17,19,24-26] increase its abundance during development of $O_2$ evolution capacity in algae[27] or higher plants,[28] or decrease its abundance during differentiation of blue-green filamentous algal cells to non-$O_2$-evolving heterocysts.[29] The availability of methods to specifically alter the functionally bound Mn abundance has permitted correlative experiments between this Mn and various system II reactions. Correlative experiments have been required to date, since no definitive electron paramagnetic resonance (EPR) signal or absorbance change (with the possible exceptions of water proton spin relaxation[30,31] and the 310-nm absorbance change[32]) has been observed which can be specifically and directly related to the bound Mn. The results from such experiments have permitted the assignment of Mn function to specific reactions in the sequence shown by Fig. 1. Table I[2,5,10,17,19,24-27,30,33-40] summarizes some of the methodologies that alter the abundance of functional Mn and, where known, the consequence on the reactions of Fig. 1. This summary first indicates that a decrease of bound Mn abundance to <5-8 Mn/400 Chl results in decreased rates of $O_2$ evolution in any light regime without altering primary charge separation within the system II trap, as judged

[23] T. Yamashita, J. Tsuji, and G. Tomita, *Plant Cell Physiol.* **12**, 117 (1971).

[24] G. M. Cheniae and I. F. Martin, *Plant Physiol.* **47**, 568 (1971).

[25] K. Y. Chen and J. H. Wang, *Bioinorg. Chem.* **3**, 367 (1974).

[26] M. P. J. Pulles, H. J. Van Gorkom, and G. A. M. Verschoor, *Biochim. Biophys. Acta* **440**, 98 (1976).

[27] G. M. Cheniae and I. F. Martin, *Photochem. Photobiol.* **17**, 441 (1973).

[28] S. Phung-nu-hung, B. Houlier, and A. Moyse, *C. R. Hebd. Seances Acad. Sic.* **279**, 1669 (1974).

[29] E. Tel-Or and W. D. Stewart, *Nature (London)* **258**, 715 (1975).

[30] T. Wydrzynski and Govindjee, *Biochim. Biophys. Acta* **387**, 403 (1975).

[31] T. Wydrzynski, N. Zumbulyadis, P. G. Schmidt, H. S. Gutowski, and Govindjee, *Proc. Natl. Acad. Sci. U.S.A.* **73**, 1196 (1976).

[32] M. P. J. Pulles, H. J. Van Gorkom, and J. G. Willemsen, *Biochim. Biophys. Acta* **449**, 536 (1976).

[33] T. Yamashita and W. L. Butler, *Plant Physiol.* **43**, 1978 (1968).

[34] Govindjee, G. Doring, and R. Govindjee, *Biochim. Biophys. Acta* **205**, 303 (1970).

[35] S. Katoh, and A. San Pietro, *Arch. Biochem. Biophys.* **122**, 144 (1967).

[36] M. Kimimura, and S. Katoh, *Plant Cell Physiol.* **13**, 287 (1972).

[37] J. T. Warden, R. E. Blankenship, and K. Sauer, *Biochim. Biophys. Acta* **423**, 462 (1976).

TABLE I

EFFECTS OF Mn DEPLETION FROM SYSTEM II ON SYSTEM II ACTIVITIES

| Method of Mn depletion | Reaction 4 (O$_2$ evolution) | Reaction 3 (Signal II$_{vf}$) | Reaction 2 (Reduction of P$^+_{680}$) | Reaction 1 (Primary charge separation) |
|---|---|---|---|---|
| 0.8 $M$ Tris[33] | Abolished with loss of ~$\frac{2}{3}$ of functionally bound Mn[10,17,38] | Loss of ~$\frac{2}{3}$ of functionally bound Mn converts signal II$_{vf}$ to II$_f$, resulting in ~1000-fold slower rate of reaction 3 [5,10,39] | — | No effect[17,33,34] |
| NH$_2$OH ≤5 m$M$[17,24] | Abolished with loss of ~$\frac{2}{3}$ of functionally bound Mn[17,24] | — | Rate decreased[2] | No effect[2,24] |
| Mg$^{+2}$ ≥0.2 $M$[25] | Abolished with loss of ~$\frac{2}{3}$ of functionally bound Mn[25] (but see text) | Converted to signal II$_f$[25] | — | — |
| pH < 5[26] | Abolished with loss of Mn[26] | Converted to signal II$_f$; loss of all Mn results in loss of all signal II species[26] | Rate decreased[26] | No effect[26] |
| 35°–50°[19,30,35,40] | Abolished with loss of Mn[17,19,30] | Converted to signal II$_f$[37] | — | No effect[36] |
| Heterotrophic growth of green algae in darkness[27] | Abolished; only small Mn pool present[27] | — | — | No effect[27] |

either by direct measurements of $P^+_{680}$ or indirectly by measurements of the photooxidation of artificial electron donors to system II.

Detailed correlative experiments, employing at least some of the methodologies ($NH_2OH$, Tris) affecting functionally bound Mn abundance, indicate, however, that $O_2$ evolution is linearly correlated with only about two-thirds of the total bound Mn (two-thirds of the total bound Mn being readily released, but exhaustive treatment being required to remove the other third of the bound Mn).[10,17,30] Rates of water proton relaxation, which are dependent on the bound Mn and are closely associated with the $O_2$ evolving mechanism,[30,31] and the normal behavior of the species giving rise to signal $II_{vf}$[5,37] also show dependency on only about two-thirds of the total Mn abundance when Tris and/or $NH_2OH$ are employed to perturb the bound Mn. Current evidence suggests that signal $II_{vf}$ reflects the oxidized state of an electron donor to $Z^+$ and links the system II reaction center ($Z-P_{680}-Q$) to the $O_2$-yielding reactions.[39] The loss of $O_2$ evolution, decreased variable fluorescence, decreased rates of water proton relaxation, and the $\sim 1000$-fold slower rate of decay of signal $II_{vf}$ which accompany the loss of approximately two-thirds of the bound Mn have permitted assignment of the major fraction of bound Mn to the $O_2$-yielding reactions.

Such data imply that the total system II Mn pool is bound heterogeneously and has led to the concept of two functionally bound Mn pools: (1) one pool of $\sim 4$ Mn per system II trap associated with $O_2$ evolution and (2) the other pool of $\sim 2$Mn per system II trap, perhaps associated with reactions linking $P_{680}$ to the $O_2$-evolving "enzyme."[17]

Heterogeneous Mn binding also is observed when high $Mg^{2+}$ concentrations ($\geq 0.2\ M$) are used to affect release of bound Mn.[25] In this case linearity of rates of $O_2$ evolution with about two-thirds of the total bound Mn also is observed, but only with previously frozen chloroplasts. With freshly isolated chloroplasts, the curve describing the relation between $O_2$ evolution and total bound Mn indicates a correlation of $O_2$ evolution with the total Mn pool even though $O_2$ evolution is abolished with loss of only about two-thirds of the total bound Mn, the remaining one-third of the bound Mn being released with difficulty. Such results have led to the alternative hypothesis that the total bound Mn pool is required for $O_2$ evolution, but that loss of two-thirds of the total pool is accompanied by a conformational change of the Mn-containing $O_2$-evolving "enzyme" resulting in loss of $O_2$ evolution and a buried fraction of bound Mn which is inaccessible to $Mg^{2+}$ and further loss of bound Mn.[25]

Irrespective of interpretation, clear evidence for heterogeneity of functionally bound Mn is indicated. This heterogeneity, or two "classes" of functionally bound Mn, is observed also in studies of the development or loss of $O_2$ evolution in algae[27,29] and higher plants[28] and partially from the studies of the effects of pH $< 5$[26] and 35–50°[17] treatments on bound Mn and various system II parameters (Table I).

Exhaustive treatment of chloroplasts with Tris, $NH_2OH$, 35–50° shock, or pH $< 5$ results in release of even the smaller, more tightly bound Mn without affecting the photooxidation of $NH_2OH$[24] and DPC[36] by system II. However, decreased rates of reduction of $P^+_{680}$ (reaction 2, Fig. 1) are indicated in chloroplasts subjected to $NH_2OH$[2] and pH $< 5$.[26] Additionally, total loss of all signal II species has been observed following $\sim 83\%$ loss of the bound Mn (7.2 Mn/400 Chl) induced by pH 4.[26] Currently, it is not known that such behaviors are causally related to a specific depletion of the small Mn pool; thus, the possible function of the smaller, more tightly bound Mn in reactions linking $P_{680}$ to the $O_2$-evolving "enzyme" is unresolved.

## Determination of Functional Mn Abundance and Its Environment

The above relationship(s) between functionally bound Mn and system II activities have been derived from determinations[37a] of the abundance of bound and/or free Mn remaining in chloroplasts following exposure of chloroplasts to a given treatment along with measurements of the effects on the system II activities. The determination of the abundance of bound Mn remaining in chloroplasts after any specific treatment is not a trivial problem, since current evidence suggests that the $O_2$-evolving mechanism (with associated bound Mn) is located on the inner sides of the thylakoids and that the thylakoid membrane is a diffusion barrier to $Mn^{2+}$.[10] Thus, depending on the methodologies employed and, apparently, the integrity of thylakoid membranes, the bound Mn released with perturbation of system II may either be lost to the suspending medium[17,38] or retained as free $Mn^{2+}$ within the chloroplasts.[10,23] In the latter instance, no correlation between loss of bound Mn from chloroplasts and loss of $O_2$ evolution, for example, will be observed even through perturbation affected the release of bound Mn.

Electron paramagnetic resonance analyses of the environment of Mn in chloroplasts is the method of choice. However, indirect procedures have been employed[17,38] and have yielded conclusions very similar to those obtained using EPR spectroscopy.[10] The use of $^{54}$Mn uniformly

---

[37a] See Appendix for some of the many methods.
[38] P. H. Homann, *Biochem. Biophys. Res. Commun.* **33**, 229 (1968).
[39] R. E. Blankenship, and K. Sauer, *Biochim. Biophys. Acta* **459**, 617 (1977).
[40] W. Junge, and H. T. Witt, *Z. Naturforsch. Teil B* **23**, 244 (1968).

labeled chloroplasts in the indirect procedure allows determination of the washing procedures required to release into the suspending medium, as indicated by constant radioactivity in the chloroplast pellet, any $Mn^{2+}$ released from its binding sites by a specific treatment. Knowledge of the specific radioactivity of the chloroplastic $^{54}Mn$ permits calculation of gram atoms of bound Mn lost (See Appendix).

The analysis of the state of Mn by EPR spectroscopy is based on the following simplified considerations: (1) $Mn^{2+}$ in aqueous solution is in the form of a "free" hexaquo complex of high symmetry and yields, even at room temperature, a well-resolved EPR spectrum showing six hyperfine lines that are due to the interaction of unpaired manganese electrons ($3d^5$ electron configuration, $S = 5/2$) with the 5/2 spin of the $^{55}Mn$ nucleus.[41,42] (2) Oxidation states other than $Mn^{2+}$ usually do not yield an EPR signal in normal physiological solutions. (3) Complexing of the hexaquo $Mn^{2+}$ ion breaks up the high symmetry of the water molecules in the inner hydration shell resulting in broadening of the six hyperfine lines. Consequently, the interaction of $[Mn(H_2O)_6]^{2+}$ with simple organic ligands or complex biological macromolecules (phospholipids, proteins, nucleic acids) results in line broadening and decrease of the intensity of the $[Mn(H_2O)_6]^{2+}$ EPR signal. Line broadening often is sufficiently large in complexes of $[Mn(H_2O)_6]^{2+}$ with, e.g., proteins, that the signal is completely smeared out and invisible. (4) The peak-to-peak linewidth ($\Delta H_{pp}$) of the $g = 1.98$ line (fourth line from left when scan is from low to high magnetic field strength) reflects the correlation time of $Mn^{2+}$, whether free or complexed. $\Delta H_{pp}$ decreases with increasing temperature,[42] and the linewidth temperature dependence (assuming no large static field splitting which eliminates the spectrum) can be employed at temperatures $\leq 25°$ with chloroplasts to ascertain whether a $Mn^{2+}$ signal is derived from a hindered species or from freely rotating $[Mn(H_2O)_6]^{2+}$ in aqueous solution. These considerations predict that the concentration of "free" $Mn^{2+}$, released from native binding sites but retained within the chloroplasts, would show a peak amplitude and a $\Delta H_{pp}$ temperature dependency equivalent to the same concentration of freely rotating hexaquo $Mn^{2+}$ in aqueous solution.

Figure 2 shows, in part, one example of an application of EPR spectroscopy to the analysis of the effect of 0.8 $M$ Tris treatment of chloroplasts on the functional bound Mn of system II.[10] The figure shows the typical almost complete absence of a free $Mn^{2+}$ signal in $O_2$ evolving,

[41] B. B. Garrett and L. O. Morgan, *J. Chem. Phys.* **44**, 890 (1966).
[42] M. Cohn, *in* "Magnetic Resonance in Biological Systems" (A. Ehrenberg, B. G. Malmström, and T. Wanngard, eds.), p. 101. Pergamon, Oxford, 1967.

## SUCROSE – WASHED CHLOROPLASTS

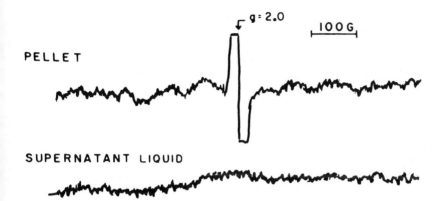

## TRIS–WASHED CHLOROPLASTS

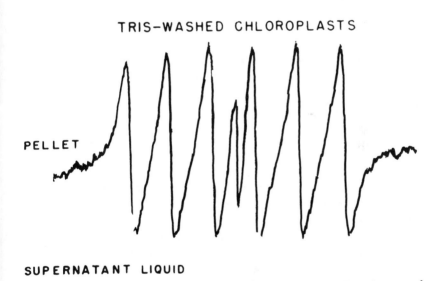

Fig. 2. Room temperature EPR spectra (first derivative) of chloroplast pellet and supernatant liquid from sucrose- or Tris-washed chloroplasts.[10] (The single line at $g$ 2.0 is signal II; the six-line pattern observed in Tris-washed chloroplasts is a typical signal for $[Mn(H_2O)_6]^{2+}$.

STN washed chloroplasts. Additionally, Fig. 2 shows that Tris-washing resulted in the appearance of the six-line pattern typical for $[Mn(H_2O)_6]^{2+}$, although the bound Mn released was not lost from the thylakoids to the supernatant liquid. Quantitation of the EPR-detectable Mn accompanying the loss of $O_2$ evolution showed: (1) the properties of the $Mn^{2+}$ released inside the thylakoids were those of uncomplexed, freely rotating hexaquo $Mn^{2+}$ (at least presumably in the presence of 0.8M Tris) and (2) the signal represented about 60% of the total chloroplast bound Mn (5.5–7.6 Mn/400 Chl). By comparison, the previously mentioned indirect procedure yielded the conclusion that loss of $O_2$ evolution was accompanied with loss of about two-thirds of total chloroplast bound Mn (5.5–6 Mn/400 Chl).[17]

The relatively high sensitivity and precision of EPR detection of $[Mn(H_2O)_6]^{2+}$ (2% variation of amplitude of signal[10] from sample to sample with 36 $\mu M$ $Mn^{2+}$)[42a] coupled with the above considerations and the fact that all chloroplastic functionally bound Mn is released at pH $< 4.5$[26] show, in part, the considerable utility of EPR spectroscopy for analyses of the properties of thylakoid-bound Mn.

Information on the biochemical nature of the chloroplast constituents binding the three identifiable classes of chloroplast Mn is minimal. The few reported studies have been directed toward the functionally bound Mn and have yielded some general, but few specific, properties of the moeities involved. A complex of Mn with a "special" chlorophyll has been ruled out.[19]

### Chloroplast Mn–Proteins

Most workers believe that the functionally bound Mn, particularly the Mn correlating with $O_2$ evolution, is bound to an easily denatured protein. This supposition is based on the following. (1) The Mn is retained with the protein phase after extraction of pigments and lipids with organic solvents.[19,43,44] (2) Many of the reagents listed in Table I generally affect protein structure. Additionally, pronase digestion,[45,46] high concentrations of Triton X-100 (but not digitonin),[19] chaotropic agents (urea, guanadine, perchlorate, thiocyanate),[19,47] and high alkalinity[19,33] also release bound Mn and are deleterious to many proteins. (3) Attempts to solubilize

---

[42a] A concentration equivalent to that present in a chloroplast suspension of 2.4 mg Chl/ml (Mn abundance of 6 Mn/400 Chl).

[43] B. Lagoutte, and J. Duranton, *FEBS Lett.* **51**, 21 (1975).

[44] F. Henriques, and R. B. Park, *Biochim. Biophys. Acta 430*, 312 (1976).

[45] P. Donnat and J. M. Briantais, **C. R. Hebd. Seances Acad. Sci. 265**, 21 (1967).

[46] P. H. Homann, *Plant Physiol.* **42**, 997 (1967).

and purify Mn-containing protein(s) have been unsuccessful because of the instability of the protein(s) and loss of Mn during purification.[19]

Separation of proteolipids from hydrophylic proteins of organelles can be accomplished by extraction with water–methanol–chloroform or with a weakly protic solvent such as $n$-butanol. Employing such procedures, a few attempts have been made to identify Mn binding to these broad protein categories and to individual proteins within a category.[43,44] Lagoutte and Duranton have observed a Mn-containing 25 kilodaltons band following SDS–polyacrylamide gel electrophoresis of proteolipids isolated from *Zea mays* chloroplasts.[43] This protein contained tightly bound Mn (1 Mn per eight 25 kilodalton units) with a Mn content enriched some six times more than whole chloroplast lamellae. Considering the rather drastic conditions used to isolate this Mn-containing protein, Lagoutte and Duranton postulated that this complex is identifiable with the smaller, more tightly bound Mn pool and was involved in a structural role. In this context, we note that the 25 kilodalton protein accounts for about 25% of the total chloroplast lamellar protein and has been identified as the main protein component of the light-harvesting chlorophyll–protein complex.[48,49]

In similar studies, but with spinach chloroplasts, Henriques and Park[44] found, however, that the 25 kilodalton protein was largely depleted in Mn and that the hydrophylic protein fraction, containing photosynthetic cytochromes and water-soluble proteins, was enriched in Mn. To date, there is no cogent explanation for this apparent discrepancy. Some of the discrepancy is possibly attributable to differences in total Mn abundance originally present in the corn versus spinach chloroplasts. Generally, chloroplasts isolated from corn, in contrast to spinach, show rather low rates of $O_2$ evolution and possibly diminished abundance of the Mn correlating with $O_2$ evolution.

A Mn-containing, low-molecular-weight ($\sim$ 1000 daltons), heat-stable (100°), and acid-stable (1–4 $N$ HCl) substance has been isolated from the blue-green alga, *Phormidium luridum*.[50] The properties of this substance are therefore drastically different from the functional Mn of system II. Nevertheless, the addition of a fraction containing this purified "Hill factor" to washed spheroplast fragments from *Phormidium* resulted in about 2- to 15-fold increases in rates of BQ or FeCN mediated rates of $O_2$ evolution.[50]

[47] R. Lozier, M. Baginski, and W. L. Butler, *Photochem. Photobiol.* **14**, 323 (1971).

[48] S. Genge, D. Pilger, and R. Hiller, *Biochim. Biophys. Acta* **347**, 22 (1974).

[49] J. Thornber and H. Highkin, *Eur. J. Biochem.* **41**, 109 (1974).

[50] E. Tel-Or, and M. Avron, *Proc. Int. Congr. Photosynth. Res., 3rd, 1974* Vol. 1, p. 569 (1975).

This "factor" was not required in the photooxidation of artificial electron donors to either system II or I, thus suggesting that the site of action of the factor was in close proximity to the $O_2$-yielding reactions. It is not clear, however, that the stimulatory effect on $O_2$ evolution by this "factor" can be specifically attributed to the Mn-containing substance. Much of the effect may be attributable simply to salt within fractions containing the "factor" (M. Avron, personal communication), thus reflecting the high salt requirement generally observed for system II electron flow in lamellar-fragments of blue-green algae.

As yet no Mn-containing protein of chloroplasts has been clearly demonstrated, using conventional assays, to possess enzymatic activity involving any of the various oxidation states of $O_2$. Lumsden and Hall,[51] however, have indicated indirectly that Triton subchloroplast particles (but not intact chloroplast lamellae) possess a superoxide dismutase-like activity. This activity, like $O_2$ evolution in intact chloroplasts, was abolished by Tris washing and partially by heat ($\sim 60\%$ at 85°, but not at 50°). Unlike $O_2$ evolution in chloroplasts, this activity was abolished by EDTA but not by $Mg^{2+}$. Additional dissimilarities exist between the properties of this superoxide dismutase-like activity and the $O_2$-evolution process. Most notably, the superoxide dismutase-like activity can be restored in the Tris-washed particles by simple equilibration of the particles with either $Mn^{2+}$ or $Cu^{2+}$. In contrast, $O_2$ evolution of Tris-treated chloroplasts is not restored by this simple treatment[23]; moreover, $O_2$ evolution is specific for Mn. Also, in contrast, the Mn of the mangano-superoxide dismutase of *E. coli* cannot be reversibly dissociated.[52] Thus, some reservations are indicated before identifying the superoxide dismutase-like activity of Triton-subchloroplast particles with a Mn-containing superoxide dismutase and postulating its involvement in the $O_2$ evolution process.

### Appendix–Chemical Analyses of Chloroplast Mn

*Colorimetric Formaldoxime Method.*[53] Manganese standards (0 to 10 $\mu$g) and chloroplasts (equivalent to 2–5 mg Chl) are digested in test tubes at boiling temperature with 2–5 ml concentrated $HNO_3$ with precautions to avoid excessive charring of the samples. After nearly all of the acid is evaporated, the samples are cooled and 0.8 ml of 60% $HClO_4$ is added. Samples then are boiled until white fumes appear, after which digestion

[51] J. Lumsden, and D. O. Hall, *Biochem. Biophys. Res. Commun.* **64,** 595 (1975).
[52] B. B. Keele, J. M. McCord, and I. Fridovich, *J. Biol. Chem.* **245,** 6176 (1970).
[53] G. Bradfield, *Analyst* **82,** 254 (1957).

is continued at incipient boiling for about 15 min. Digestions are conveniently carried out in a sandbath.

Following digestion, 3–5 ml of $H_2O$ are added, then 1 ml of 10% trisodium $N$-hydroxyethylethylenediaminetriacetate (to preclude subsequent precipitation of $Ca^{2+}$ and $Mg^{2+}$ as phosphates). After neutralization with 10% NaOH ($\sim$ 1–1.5 ml), 0.2 ml of formaldoxime reagent[53a] is added, followed immediately by 0.2 ml of 10% NaOH. Volumes are adjusted to 10 ml and the solutions heated at 65° for 2 hr to destroy interfering formaldoxime complexes of iron and copper. Glass "blob" air condensors on the tubes prevent excessive evaporation. After cooling, the sample volume is adjusted volumetrically to 10 ml with $H_2O$, and the absorbance is measured against a blank at 450 nm. In a cell of 2.5-cm light path, 4 $\mu$g of Mn typically yields an absorbancy of 0.2. The absorbance is linear with Mn concentration up to at least 20 $\mu$g Mn.[17]

*Atomic Absorption Analyses.* Standards and chloroplast samples (equivalent to 1–5 mg Chl in 1–2 ml) may be either dry-ashed (500°) in quartz tubes, platinum, or unscratched porcelain crucibles, or wet-ashed at 150°–170° overnight in centrifuge tubes containing 2 ml of a mixture of 17 parts concentrated $HNO_3$ and 3 parts 70% $HClO_4$ (v/v).[54] Wet digestion for prolonged durations at elevated temperatures may not be required,[55] at least when the solvent extraction procedure described below is used. The supernatant obtained following addition of $HNO_3$–$HClO_4$ to chloroplasts has been extracted directly and yields results equivalent to those obtained following complete digestion and extraction.

If dry ashed, the ashed samples are dissolved with 3–4 drops concentrated HCl, 1 ml of water is added, and the samples are heated at 80° for 5–10 min, then evaporated almost to dryness. This latter step is repeated, then 3–4 ml of $H_2O$ is added and the contents and washings transferred to a centrifuge tube and centrifuged to remove any undissolved material. The supernatant and a 2-ml water wash of the pellet are combined into a test tube for extraction (see later). Alternately, the volumes of samples are adjusted to 10 ml with water and aspirated directly to the spectrophotometer.

If wet-ashed, 3–4 ml $H_2O$ is added to the digested samples, centrifuged, and the supernatant and a 2 ml water wash of any precipitate are combined into a test tube for solvent extraction of the manganese.

[53a] Two grams of paraformaldehyde and 5.5gm of $NH_2OH$ sulfate are dissolved in 5 ml of boiling $H_2O$, then diluted to 10 ml. This solution is diluted ten times for use as above.

[54] E. B. Sandell, "Colorimetric Determination of Traces of Metals," p. 620. Wiley (Interscience), New York, 1959.

[55] R. L. Heath, and G. Hind, *Biochim. Biophys. Acta* **189**, 222 (1969).

Solvent extraction[56,57] prior to aspiration into the spectrophotometer offers at least two distinct advantages: (1) the chelated Mn (or other metals) can be easily concentrated into a small volume of solvent and (2) compared to aqueous solutions, an increase in absorption of three to five times is obtained for a given concentration of metal. Extraction of the Mn from either the dry-ashed or wet-ashed prepared samples is carried out as follows: (1) The pH is adjusted to 3.0–3.5 with 10 and 0.01 $M$ NaOH (bromphenol blue may be used as indicator); (2) 1 ml of 2% aqueous ammonium 1-pyrrolidine dithiocarbamate is added; (3) 2 ml of water saturated methyl isobutyl ketone (MIBK) are added, mixed thoroughly with the aqueous phase, and the Mn complex in the MIBK drawn off. Extraction is repeated two more times, and (4) the combined extractions are centrifuged briefly, the volume adjusted to 10 ml with MIBK, and the sample aspirated into the spectrophotometer. Samples (aqueous or MIBK) are read at 2795 Å on the spectrophotometer equipped with an acetylene–air burner head. The absorbance is linear with respect to Mn up to at least 2 $\mu$g Mn/ml in the MIBK system.

*Radioassays.* Either of the above analytical procedures has sufficient sensitivity to determine the Mn abundance (5–8 Mn/400 Chl) normally occuring in well-washed, class II chloroplasts. In principle, these procedures also are applicable for analysis of Mn abundance in chloroplasts sufficiently depleted of Mn to abolish $O_2$ evolution and/or to reveal partial reactions in Fig. 1. However, additional sensitivity and conservation of chloroplasts can be obtained by radioassay of $^{54}$Mn-labeled chloroplasts or particles from higher plants or algae cultured in media containing $^{54}$Mn$^{2+}$ (halflife = 313 days). The high sensitivity of radioassays, coupled with the necessity of only one chemical determination of Mn (to establish the specific activity of chloroplastic Mn), permits use of small quantities of chloroplasts in experiments correlating effects of Mn depletion from chloroplasts and any resulting alterations in parameters of system II activities.[17]

With use of highly purified ("Specpure", Johnson, Matthey and Co., Limited, London, England) salts used for "trace elements" in media (iron salts often contain high amounts of Mn), glass-distilled, deionized water, and HCl-washed containers for the cultures, it is possible to obtain chloroplasts with high $^{54}$Mn specific activity with a rather minimum amount of carrier-free $^{54}$Mn in the cultures.[17] Even with the above precautions, variable and often large amounts of manganese in other reagents, particularly salts of phosphate used in growth media, often limit

[56] R. E. Mansell, *At. Absorpt. Newsl.* **4**, 276 (1965).
[57] C. E. Mulford, *At. Absorp. Newsl.* **5**, 88 (1966).

the accuracy of sometimes reported chloroplastic Mn abundance determinations based only on a presumed specific activity of Mn in the growth media.

Acknowledgment

This work was supported, in part, by a grant from the National Science Foundation (PCM 76-02536).

## [34] Chloroplast Membrane Polypeptides

*By* SHOSHANA BAR-NUN and ITZHAK OHAD

The separation and identification of chloroplast membrane polypeptides was one of the major goals of many laboratories involved in the study of various aspects of chloroplast membrane structure, function, and development. During the last decade spectacular progress has been made in this field of research due to the development of SDS (sodium dodecyl sulfate)–polyacrylamide gel electrophoresis techniques and their application to the separation of chloroplast membrane polypeptides derived from a variety of sources.

Today, this method provides a rapid and accurate means for the analysis of purity of membrane fractions or isolated active complexes, of changes in membrane composition during development, or, as a result of mutations, identification of subcellular origin of membrane polypeptides. In addition, large-scale gel electrophoresis has enabled the isolation of specific polypeptides for analysis of their amino acid composition or to serve as antigens.

The wide application of this technique has inevitably resulted in a large number of modifications of the basic method to suit the material obtained from different sources. Despite the fact that the general pattern of chloroplast membrane polypeptides from higher plants and algae has many common features, a general nomenclature for the different polypeptides (accepted by all laboratories) is not yet available. As a result, a certain degree of confusion exists, and difficulties are encountered when polypeptides from various sources, with apparently similar migration properties, are compared.

These difficulties are not entirely artificial since, as will be mentioned below, the electrophoretic mobility of an SDS–polypeptide complex can hardly be, by itself, a sufficient criterion for considering two polypeptides as identical. Indeed, when additional criteria, such as binding of chlorophyll, correlation with biochemical activities, or intracellular origin are available, identification is facilitated and the nomenclature is generally accepted. This is the case for the chlorophyll–protein complexes I and II, polypeptide constituents of CFI, and the reaction center of PSI.

The present article describes the general methodology that can be applied to the separation and identification of different membrane polypeptides as well as the possible use of this technique, in combination with other techniques, for the study of chloroplast membranes.

## Theoretical Considerations

The underlying principles for the SDS–polyacrylamide gel electrophoresis technique have been discussed in detail in the past and only a brief outline will be given here.

The mobility of a polypeptide in the electrophoretic field is a function of its charge per unit mass, its shape, and its size. In the presence of SDS, a relatively stable complex is formed between the polypeptide and the detergent. Ideally, a constant weight of SDS is bound per weight of polypeptide (1.4 gm/gm)[1] and thus, at a basic pH, the bound SDS confers a constant negative charge per unit mass of polypeptide. Due to the binding of the detergent and resulting charge repulsion, the polypeptide unfolds. Opening of S–S bonds by reducing agents will then cause a complete unfolding of the polypeptide molecule. According to Reynolds and Tanford,[2] the unfolded SDS–polypeptide complexes assume a rodlike configuration of constant width and a length dependent on their molecular weights. Thus, the mobility of a polypeptide complexed with SDS is a function only of its size, which is directly proportional to its molecular weight.

According to Ferguson,[3] the electrophoretic mobility ($R_f$) of a polypeptide is

$$\log R_f = \log Mo - Kr \times T \tag{1}$$

where $T$ is the acrylamide concentration or any other parameter representing the porousity or viscosity of the medium. Thus, in a viscous or porous medium, the larger the polypeptide, the greater the retardation of its mobility ($Kr$) will be. On the other hand, the mobility in a nonviscous

---

[1] J. A. Reynolds and C. Tanford, *Proc. Natl. Acad. Sci. U.S.A.* **66**, 1002 (1970).

[2] J. A. Reynolds and C. Tanford, *J. Biol. Chem.* **245**, 5161 (1970).

[3] K. A. Ferguson, *Metab., Clin. Exp.* **13**, 985 (1964).

medium ($Mo$, free mobility) is only slightly dependent on the molecular weight and is almost identical for all polypeptides.

Any departure from this ideal behavior will cause a drastic difference between the expected and the observed migration of a polypeptide of a known molecular weight. Such departure will occur for polypeptides having a high intrinsic charge or undergoing incomplete unfolding due to residual S–S bonds or binding of ligands to the polypeptide (chlorophyll, lipids, carbohydrates) which, as a consequence, will alter the shape of the complex as well as the degree of SDS binding and, thus, the charge per unit mass. Such complexes will show abnormal retardation coefficients (either smaller or larger than expected) as well as free mobility different from unity. Thus, the molecular weights of such complexes, estimated by their electrophoretic mobility, will be erroneous.

Several simple techniques have been used to verify whether the polypeptides in an electrophoretogram have retardation coefficients and free mobility values expected for a fully unfolded SDS–polypeptide complex having 1.4 gm SDS per 1 gm of polypeptide.

a. Any change in the mobility of a certain polypeptide band, induced by increasing the concentration of the reducing agents, the concentration of SDS in the solution, the time of solubilization and, most effectively, the temperature during solubilization,[4,5] will indicate that the polypeptide(s) present in the above-mentioned band is not ideally complexed with SDS and/or not completely unfolded.

b. Anomalies in the retardation coefficient ($Kr$) or free mobility ($Mo$) values can be detected by separating the same polypeptide mixture on gels of various polyacrylamide concentrations and then plotting the electrophoretic mobility (log $R_f$) of a certain polypeptide versus the different polyacrylamide concentrations ($T$). The free mobility is calculated from the intersect with the $Y$ axis, while $[-Kr]$ is the slope of the line.[6]

c. A more direct method for the detection of such anomalies can be obtained when a mixture of polypeptides layered along a linear gradient of polyacrylamide is run with the electric field perpendicular to the gradient. The polypeptides migrate simultaneously through several acrylamide concentrations. Since the electrophoretic mobility is a logarithmic function of the gel concentration, a series of exponential, sequential lines is observed. Any band representing a certain polypeptide changing

[4] N. H. Chua, K. Matlin, and P. Bennoun, *J. Cell Biol.* **67**, 361 (1975).
[5] S. Bar-Nun, R. Schantz, and I. Ohad, *Biochim. Biophys. Acta* **459**, 451 (1977).
[6] G. A. Banker and C. W. Cotman, *J. Biol. Chem.* **247**, 5856 (1972).

its slope and crossing other bands, would indicate anomalies in the migration of the polypeptide(s).

## Techniques

### Electrophoresis

The analysis of chloroplast membrane polypeptides using SDS–polyacrylamide gel electrophoresis was first carried out by Ogawa et al.[7] and by Thornber et al.[8] who resolved the chloroplast membrane into three to four polypeptides, including two chlorophyll–protein (CP) complexes.

The chloroplast membrane of Chlamydomonas reinhardi was analyzed by several SDS–polyacrylamide gel electrophoresis techniques and can serve as a model system to describe these techniques.

This membrane was resolved by Hoober[9] into seven to ten polypeptides using the method described, as follows: Gels (0.6 × 16 cm) are prepared in glass tubes, 18 cm in length, after degassing a solution containing 8% (w/v) acrylamide, 0.28% (w/v) $N,N'$-methylenebisacrylamide, 0.1 $M$ Tris-acetate (pH 8.4), 0.1% (w/v) SDS, 0.01% (w/v) EDTA, and 0.5 $M$ urea. The gels are polymerized chemically by the addition of 0.04% (v/v) TEMED ($N,N,N',N'$-tetramethylethylenediamine) and 0.06% (w/v) ammonium persulfate. The electrode buffer has the same composition as the gel buffer, but is prepared with Tris-acetate at pH 9.0. The buffer is degassed under vacuum, and just before use 0.019 $M$ 2-mercaptoacetate is added, providing a final pH of 8.4. The 2-mercaptoacetate, which moves into the gel ahead of the proteins by preliminary electrophoresis for 1½ hr at 5 V/cm, prevents oxidation of sulfhydryl groups during the electrophoresis. Lipids and pigments are extracted from the chloroplast membranes with 90% (v/v) acetone at room temperature. Since some membrane polypeptides can be extracted by the acetone, the soluble phase is treated with several grains of NaCl, which causes precipitation of solubilized protein.[10] This precipitate is added to the insoluble protein phase and the combined precipitates are solubilized in 2% (w/v) SDS, 0.1 $M$ Tris-acetate (pH 9.0), 0.01% (w/v) EDTA, and 0.5 $M$ urea to give a protein concentration of about 8 mg/ml. The membranes are treated under nitrogen with 2% (v/v) 2-mercaptoethanol for 1½ hr at 37° and then

[7] T. Ogawa, F. Obata, and K. Shibata, Biochim. Biophys. Acta 112, 223 (1966).
[8] J. P. Thornber, J. C. Stewart, M. W. C. Hatton, and J. L. Bailey, Biochemistry 6, 2006 (1967).
[9] J. K. Hoober, J. Biol. Chem. 245, 4327 (1970).
[10] J. K. Hoober, J. Cell Biol. 52, 84 (1972).

are dialyzed under nitrogen against 0.1% (w/v) SDS, 0.002 $M$ Tris-acetate (pH 9.0), 0.001 $M$ EDTA, and 0.5 $M$ urea. The 2-mercaptoethanol is again adjusted to 2% (v/v), and the membrane solution is incubated under nitrogen for 2 hr at 37° or over night at 25°. Sucrose is added up to 0.2 $M$, and 10 to 50 $\mu$l of a sample are layered on each gel. Electrophoresis is performed at 1.8 V/cm for 3 min and 6 V/cm (about 2.5 mA/gel) for 6½ to 7 hr at room temperature.

The method described by Hoober[8] has also been used by Levine *et al.*[11,12]

Improved resolution of the chloroplast membrane of *Chlamydomonas reinhardi* was obtained by Bar-Nun and Ohad[13] using a modification of the disc SDS–polyacrylamide gel electrophoresis technique described by Laemmli.[14] Gels (0.6 × 8 cm) are prepared in glass tubes, 14 cm in length, from a stock solution of 30% (w/v) acrylamide and 0.8% (w/v) $N,N'$-methylenebisacrylamide. The separation gel contains 10% (w/v) acrylamide, 0.375 $M$ Tris-HCl (pH 8.9), and 0.1% (w/v) SDS. The stacking gel, 1 cm long, contains 3% (w/v) acrylamide, 0.0625 $M$ Tris-HCl (pH 6.8), and 0.1% (w/v) SDS. The gels are polymerized chemically by the addition of 0.03 to 0.05% (v/v) TEMED, and 0.05 to 0.1% (w/v) ammonium persulfate. The electrode buffer (pH 8.3) contains 0.05 $M$ Tris, 0.38 $M$ glycine, and 0.1% (w/v) SDS. Membranes are solubilized in 2% (w/v) SDS, 0.05 $M$ Tris-HCl (pH 6.8), 10% (v/v) glycerol, either 1% (v/v) 2-mercaptoethanol or 0.1 $M$ dithiothreitol, and 0.001% (w/v) bromphenol blue, to give a protein concentration of 2 to 5 mg/ml, and SDS–protein ratio of between 10:1 and 4:1. The membranes are solubilized either at room temperature or in a boiling water bath for 2 to 3 min, and 20 to 200 $\mu$l of sample are layered on each gel. In certain experiments, lipids and pigments were extracted from the membranes prior to electrophoresis by 80% (v/v) acetone at room temperature. Electrophoresis is carried out at room temperature with a current of 0.5 mA/gel for 15 min, 1 mA/gel for an additional 15 min, and then 2 mA/gel until the bromphenol blue marker reaches the bottom of the gel (about 3 hr).

The slab-gel technique, which was designed by Studier,[15] was applied to *Chlamydomonas reinhardi* by Chua and Bennoun.[16] Their method is

[11] R. P. Levine, W. G. Burton, and H. A. Duram, *Nature (London), New Biol.* **237,** 176 (1972).
[12] D. P. Beck and R. P. Levine, *J. Cell Biol.* **63,** 759 (1974).
[13] S. Bar-Nun and I. Ohad, *Plant Physiol.* **59,** 161 (1977).
[14] U. K. Laemmli, *Nature (London)* **227,** 680 (1970).
[15] F. W. Studier, *J. Mol. Biol.* **79,** 237 (1973).
[16] N. H. Chua and P. Bennoun, *Proc. Natl. Acad. Sci. U.S.A.* **72,** 2175 (1975).

essentially the discontinuous alkaline buffer system of Neville[17] with a stacking gel of 1 to 2 cm and a separating gel of about 20 cm. The stacking gel is made up of 6% (w/v) acrylamide, whereas the separating gel is made up of a linear concentration gradient of acrylamide, 7.5 to 15% (w/v), as described by Alvares and Siekevitz,[18] accompanied by a 5 to 17.5% (w/v) sucrose gradient in the gel. The ratio of acrylamide to $N,N'$-methylenebisacrylamide for both gels is 30:0.8. The following buffers are used: upper reservoir buffer—0.04 M boric acid, 0.041 $M$ Tris, 0.1% (w/v) SDS (pH 8.64); stacking gel buffer—0.0267 $M$ $H_2SO_4$, 0.0541 $M$ Tris, 0.1% (w/v) SDS (pH 6.10); separating gel buffer—0.0308 $M$ HCl, 0.4244 $M$ Tris, 0.1% (w/v) SDS (pH 9.18); lower reservoir buffer—same as the separating gel buffer, except that the SDS is omitted. Membranes are solubilized in 0.05 $M$ $Na_2CO_3$, 0.05 $M$ dithiothreitol, 2% (w/v) SDS, 12% (w/v) sucrose, and 0.04% (w/v) bromphenol blue, to a final chlorophyll concentration of 1 mg/ml and a ratio of SDS to chlorophyll of 20:1. The membranes are solubilized either at room temperature[4] or in a boiling water bath for 1 min.[16] In certain experiments the pigments and lipids were extracted from the chloroplast membranes with either 90% (v/v) acetone or a chloroform–methanol 2:1 mixture at room temperature, and then the membranes were solubilized as described above. Electrophoresis is performed at a constant current of 17.5 mA for about 12 hr at room temperature.

A combination of the slab-gel technique and the methods of Laemmli[14] was described by Gershoni and Ohad.[19] The stacking gel is made up to 6% (w/v) acrylamide and contains 10% (v/v) glycerol, whereas the separating gel is made up of a linear concentration gradient of acrylamide, 7 to 15% (w/v), accompanied by a 0 to 25% (v/v) glycerol gradient. The buffers used are essentially those used in the method of Bar-Nun and Ohad[13]: stacking gel—0.0625 $M$ Tris-HCl (pH 6.8); separating gel—0.375 $M$ Tris-HCl (pH 8.9); electrode buffer—0.05 $M$ Tris and 0.38 $M$ glycine (pH 8.3) (both upper and lower reservoirs); all buffers contain 0.1% (w/v) SDS. Chloroplast membranes are solubilized in 0.05 $M$ Tris-HCl (pH 6.8), 1% (v/v) 2-mercaptoethanol, 2% (w/v) SDS, 10% (v/v) glycerol, and 0.001% (w/v) bromphenol blue, to give a protein concentration of 5 mg/ml and a ratio of SDS to protein of 10:1. The membranes are solubilized either at room temperature or in a boiling water bath for 2 to 3 min. Electrophoresis is performed at a constant current of 20 mA for about $3\frac{1}{2}$ hr at room temperature.

[17] D. M. Neville, Jr., *J. Biol. Chem.* **246**, 6328 (1971).

[18] A. P. Alvares and P. Siekevitz, *Biochem. Biophys. Res. Commun.* **54**, 923 (1973).

[19] J. M. Gershoni and I. Ohad, Colloques Internationaux C.N.R.S. N°261, p. 447 (1976).

## Staining of Gels

Several different methods have been used for staining of the polypeptides. Essentially, either Coomassie brilliant blue or Amido black serve as dyes. Coomassie brilliant blue is more sensitive, but the staining intensity obtained is not well correlated to the concentration of the polypeptide. Amido black gives staining intensity closely related to the polypeptide concentration. The polypeptides are fixed either prior to or during the staining, and washing out of the SDS is essential for staining.

The most common method was published by Fairbanks et al.[20] The gels are fixed and stained overnight in 0.05% (w/v) Coomassie brilliant blue (R-250) in 25% (v/v) isopropanol and 10% (v/v) acetic acid, then transferred for 6 to 9 hr to 0.005% (w/v) Coomassie brilliant blue in 10% (v/v) isopropanol and 10% (v/v) acetic acid, and destained completely in 10% (v/v) acetic acid. The second solution may be omitted. A modification of this method, using Amido black instead of Coomassie brilliant blue, can also be applied. Staining with 0.25% (w/v) Coomassie brilliant blue in 50% (v/v) methanol and 7% (v/v) acetic acid for 3 to 5 hr, and destaining in 30% (v/v) methanol and 7% (v/v) acetic acid have also been described.[16]

Gels can be fixed overnight with 5% (w/v) sulfosalicylic acid in 10% (w/v) trichloroacetic acid, rinsed for 1 hr in 7% (v/v) acetic acid, stained for 20 hr in 0.2% (w/v) Coomassie brilliant blue in 7% (v/v) acetic acid, and destained in 7% (v/v) acetic acid.[9] When using this method, the fine details and the low molecular weight polypeptides are apparently lost.[9] Glycoproteins might be poorly stained by both of the above-mentioned stains. Their detection requires the periodic acid–Schiff procedure.[20] Stained gels can be photographed and/or scanned at 550 to 600 nm with a spectrophotometer equipped with a linear transport gel scanner.

## Trouble Shooting

The most common difficulties encountered when using the SDS–polyacrylamide gel electrophoresis technique are poor separation of the polypeptides and incomplete staining or destaining of the gel. These difficulties may be due to the following reasons.

a. Incomplete solubilization of the sample or presence of polymers, such as nucleic acids or starch, in relatively large quantities will accumulate and progressively plug the interface between the stacking and separating gels. The solubilized polypeptides will penetrate the gel at a

[20] G. Fairbanks, T. L. Steck, and D. F. H. Wallach, *Biochemistry* 10, 2606 (1971).

progressively slower rate, as the plug is formed, and thus be retarded during separation. This difficulty can be overcome by centrifuging the solubilized sample at 15,000–25,000 $g$ for 5 min and using only the clear soluble supernatant. Solubilization of the sample by sonication is not advisable since it might cause local heating and dissociation of protein–ligand complexes. Since it is possible that incomplete solubilization is specific to certain polypeptides, it is essential to ensure (through repeated solubilization of the sediment and electrophoresis) that the initially un-solubilized material is not enriched in certain polypeptides.

b. Uneven polymerization or distribution of acrylamide concentration, impurities, leaks, and air bubbles between the gel and the glass container will cause uneven migration, bending of the migration front, and eventual plugging of the gel.

c. Extensive heating developing progressively during the electrophoresis might cause dissociation of polypeptide–ligand complexes.

d. Use of unsuitable acrylamide or SDS preparations might cause diffuse migration due either to uneven polymerization and cross-linking of the gel or formation of inappropriate detergent–polypeptide, detergent–lipid, or detergent–detergent complexes, especially when using samples which have not been delipidized. This might occur if the SDS preparation contains unsulfonated molecules or sulfonated alcohols, the carbon chains of which are longer or shorter than 12. These difficulties can be overcome by changing the source of reagents or crystallizing the acrylamide and SDS repeatedly.

## Chlorophyll–Protein Complexes

Chloroplast membranes solubilized with SDS are usually resolved into three chlorophyll-containing bands. The fastest migrating band contains chlorophyll $a$ as a free pigment, while the two others are chlorophyll-protein (CP) complexes. The slowest migrating band, CPI, contains chlorophyll $a$ and the middle band, CPII, contains almost equal amounts of chlorophyll $a$ and chlorophyll $b$.[7,8,21]

The chlorophyll-containing bands, which are recognized on nonstained gels, can be obtained by all the SDS–polyacrylamide gel electrophoresis techniques described above if, and only if, the membranes are solubilized with SDS at room temperature. Heating the solubilized membranes, even for a short time, causes dissociation of the chlorophyll from the protein and all the chlorophyll migrates as free pigment. The methods for obtaining chlorophyll–protein complexes were described by Shibata.[22]

[21] J. P. Thornber, *Annu. Rev. Plant Physiol.* **26,** 117 (1975).
[22] K. Shibata, Vol. 23, p. 296.

Chlorophyll–protein complexes have free mobility much higher than unity and $Kr$ values higher than expected from their estimated molecular weights.[4,5] These complexes are not completely unfolded, as the chlorophyll is still bound to the polypeptide. The high $Mo$ values may indicate a higher charge per unit mass, and the high $Kr$ values may indicate a configuration different from the rodlike configuration assumed for completely unfolded polypeptides.

*Radioactive Labeling*

Chloroplast membrane polypeptides radioactively labeled can be detected by SDS–polyacrylamide gel electrophoresis. Gels cast in glass tubes can be sliced into 1-mm slices. The slices are dissolved overnight with 4% (w/v) SDS in 10% (v/v) $H_2O_2$ at 60° in tightly closed vials, and then counted for radioactivity. When using the slab-gel technique, autoradiography can be applied for determining radioactivity. The gels are dried down on a Whatman 3 MM chromatography paper and then placed in contact with an X-ray film for the required time. Equipment for drying slab-gels is commercially available. Detection of $^3H$-labeled polypeptides by autoradiography is possible through the use of the fluorography technique.[23] The gels are thoroughly washed with $H_2O$, kept in dimethyl sulfoxide overnight, soaked for 3 hr in 22% (w/v) PPO solubilized in dimethyl sulfoxide, then washed thoroughly again with $H_2O$, and dried as above.

During exposure of the PPO-containing gel in close apposition with the sensitive film, light emitted by the PPO present in the vicinity of radioactively labeled bands causes sensitization of the film. In order to improve the resolution, the exposure is carried out at −70°.

In both fluorgraphy and autoradiography techniques, about $5 \times 10^4$ cpm loaded on a slot in the slab gel can be detected after about 1 week of exposure. Gels obtained using the glass tubes technique can be sliced into 1-mm layers along the midpart of the gel, the slices dried on a Whatman 3 MM paper, as above, and then further processed as described for the slab gel.

Determination of the relative radioactive content of labeled polypeptides, using the autoradiography technique, can be obtained by scanning the developed autoradiogram. This estimation is accurate only within a limit of exposing conditions (cpm × time) in which the response of the film to the radiation is proportional to the concentration of radioactivity per unit area. Since the range of the linear response is not large enough, one should expose the same autoradiogram for increasing lengths of time

[23] W. M. Bonner and R. A. Laskey, *Eur. J. Biochem.* **46,** 83 (1974).

so as to be able to detect the various bands following exposure time within the linearity range of each of them. The techniques described above were used for analyzing chloroplast membrane polypeptides of *Chlamydomonas reinhardii*. Basically, the same techniques have been used for other algae[24-26] and higher plants.[11,27-29]

### Treatments Influencing the Polypeptide Pattern

Different polypeptide patterns are obtained from chloroplast membranes solubilized under different conditions.

*Temperature of Solubilization.* Several polypeptides, especially those which bind ligands such as chlorophyll, lipids, and carbohydrates, are not completely unfolded when solubilized with SDS at room temperature. Chlorophyll–protein complexes, for instance, which are not completely unfolded at room temperature and thus migrate as complexes, dissociate upon heat treatment. The chlorophyll migrates as free pigment and the polypeptides involved in the complexes are now completely unfolded and migrate as real SDS–polypeptide complexes.[4,5]

*Concentration of SDS.* Complete unfolding and constant amount of SDS bound to constant weight of protein are obtained when the polypeptides are saturated with SDS. Lower concentrations of SDS allow partial unfolding and certain ligands, such as chlorophyll or lipids, remain bound to the partially unfolded polypeptide. Thus, the ratio of SDS to protein is much more crucial than the SDS concentration.

*Reducing Agents.* Complete unfolding of the polypeptides is dependent on breaking of S–S bonds by reducing them to sulfhydryl groups. Complete solubilization of chloroplast membranes requires the presence of reducing agents such as 2-mercaptoethanol or dithiothreitol. However, the concentration of these reagents is crucial; at lower concentrations (up to 0.1 $M$) the chlorophyll remains bound to the polypeptides, while at concentrations above 0.2 $M$ the chlorophyll–protein complexes dissociate.[5]

*Use of Organic Solvents.* Lipids and pigments can be extracted from chloroplast membranes with organic solvents such as acetone, 80 to 90% (v/v) or chloroform:methanol (1:1 or 2:1).[4,9,10,13] Several polypeptides,

[24] K. Apel, L. Bogorad, and C. L. F. Woodcock, *Biochim. Biophys. Acta* **387,** 568 (1975).

[25] M. Gurevitz, H. Kratz, and I. Ohad, *Biochim. Biophys. Acta* (in press).

[26] G. Galling, *in* "Genetics and Biogenesis of Chloroplast and Mitochondria" (T. Bücher *et al.,* eds.), p. 53. North-Holland Publ., Amsterdam, 1976.

[27] O. Machold, *Biochim. Biophys. Acta* **238,** 324 (1971).

[28] R. J. Ellis, *in* "Membrane Biogenesis—Mitochondria, Chloroplasts and Bacteria" (A. Tzagoloff, ed.), p. 247. Plenum, New York, 1975.

[29] J. M. Anderson and R. P. Levine, *Biochim. Biophys. Acta* **333,** 378 (1974).

probably the most hydrophobic ones, which are eventually bound to the pigments, are also extracted from the membranes by these solvents.[4,10] The extracted polypeptides can be precipitated and added back to the membrane fraction by addition of salt to the extract.[10]

*Ionic Strength.* Certain polypeptides might be solubilized if various ions are washed out of the membranes. For instance, washing of chloroplast membranes with EDTA changes the polypeptide composition of the membranes and several polypeptides are solubilized.[24] Pretreatment of chloroplast membranes, which may change the bound ion concentration and composition, should be carried out under controlled conditions.

### Resolution of Polypeptide Bands

Usually a distinct band on an electrophoretogram is considered as a distinct polypeptide. Considering all the parameters involved in electrophoretic separation of polypeptides, the possibility that each band contains more than one polypeptide cannot be excluded. Two polypeptides with different molecular weights may comigrate if the smaller one, for instance, is not completely unfolded and thus more retarded.

Elucidation of this problem has been attempted by the use of additional techniques. Several techniques have already been published.

*Two-Dimensional Gel Electrophoresis.* Two-dimensional gel electrophoresis has been applied to many systems, such as ribosomes,[30] endoplasmic reticulum,[31] and mitochondria,[32] but not yet to chloroplast membranes. The principles of the separation are different in the two dimensions, and thus the possibility that two polypeptides will comigrate in both dimensions is greatly reduced.

*Second Run.* As has been mentioned above, the polypeptide pattern is dependent on the conditions of the solubilization. Slicing of a polypeptide band from a polyacrylamide gel, and separating it on an identical gel after treating the slice differently, may reveal additional polypeptides included in the first slice. For instance, slices containing chlorophyll–protein complexes can be heated before the second run, and the several polypeptides involved in the complex can now be separated.[5]

*Peptide Mapping.* Polypeptides which have been isolated from gels containing SDS, may be cleaved by partial enzymatic proteolysis in the

---

[30] E. Kaltschmidt and H. G. Wittmann, *Anal. Biochem.* **36,** 401 (1970).
[31] H. H. Czosnek and A. A. Hochberg, *Mol. Biol. Rep.* **2,** 19 (1975).
[32] F. Cabral, J. Saltzgaber, W. Birchmeier, D. Deters, T. Frey, C. Kohler, and G. Schatz, *in* "Genetics and Biogenesis of Chloroplasts and Mitochondria" (T. Bücher *et al.,* eds.), 215. North-Holland Publ., Amsterdam, 1976.

presence of SDS, and the cleavage products analyzed by polyacrylamide gel electrophoresis. The pattern of peptide fragments produced is characteristic of the polypeptide substrate and the proteolytic enzyme.[33]

Additional criteria may be applied for improved resolution of polypeptides. Among them amino acid analysis[34] and immunological studies.[35]

Acknowledgments

The preparation of this chapter is based partly on experimental work carried out in our laboratory and supported during the last five years by United States–Israel Binational Science Foundation (Grant No. 184), Deutsche Forschungsgemeinschaft (Grant No. DR 29/17), and The Israel Commission for Basic Research (Grant No. D-2).

[33] D. W. Cleveland, S. G. Fischer, M. W. Kirschner, and U. K. Laemmli, *J. Biol. Chem.* **252,** 1102 (1977).
[34] J. Kyte, *J. Biol. Chem.* **246,** 4157 (1971).
[35] E. Lazarides and K. Weber, *Proc. Natl. Acad. Sci. U.S.A.* **71,** 2268 (1974).

# [35] Plastoquinones in Algae and Higher Plants

## By R. BARR and F. L. CRANE

Plastoquinone was discovered by Kofler[1] from alfalfa in 1946 and rediscovered in chloroplasts by Crane[2] in 1959. In the decade from 1960 to 1970 plastoquinone research flourished, resulting in the discovery of the additional plastoquinone series B and C, isolation methods for which have previously been described in the series.[3] There is also the possibility that esterified hydroxy plastoquinones (the PQ Z series) exist, along with postulated epoxy and dihydroxy derivatives.[4,5]

[1] M. Kofler, *in* "Festschrift Herrn Emil Christopher Barell" (H. M. Wuest, ed.), p. 199. Hoffman-La Roche AG, Basel, 1946.
[2] F. L. Crane, *Plant Physiol.* **34,** 546 (1959).
[3] R. Barr and F. L. Crane, Vol. 23, p. 372.
[4] J. C. Wallwork and J. F. Pennock, *Chem. Ind.* (*London*) p. 1571 (1968).
[5] J. C. Wallwork and J. F. Pennock, *Prog. Photosynth. Res., Proc. Int. Congr.* [*1st*], *1968* Vol. I, p. 315 (1969).

## New Forms of Plastoquinones

Three new forms of naturaliy occurring plastoquinone have recently been identified in *Euglena gracilis*.[6,7] They include phytylplastoquinone (2,3-dimethy-5-phytyl-1,4-benzoquinone), phytylplastoquinol monomethyl ether [2,3-dimethyl-4-methoxy-5(or 6)-phytylphenol], and 1-*o*-methyl-2-demethylphytylplastoquinol. Methods for their characterization are summarized in Table I.

## Plastoquinone Biochemistry

Plastoquinone and other chloroplast quinone biochemistry has been reviewed extensively by Wallwork and Pennoc,[5] Wallwork and Crane,[8] and Morton.[9] Plastochromanols, cyclized derivatives of PQ A, have been described by Peake, Dunphy, and Pennock,[10] and will be omitted from further discussion. An integrated view of how chloroplast quinones fit into a general electron transport scheme is provided by Trebst.[11]

## Plastoquinone Distribution

### In Higher Plants

Plastoquinone distribution studies in higher plants and algae at various stages continue to appear in the literature. Most productive have been Lichtenthaler and associates (general distribution),[12] *Fagus silvatica*,[13] *Nicotiana tabacum*,[14] *Ficus elastica*,[15] *Raphanus sativum*,[16–18] *Cereus*

[6] A. Law and D. Threlfall, Phytochemistry 11, 829 (1972).
[7] G. R. Whistance and D. R. Threlfall, *Phytochemistry* 9, 213 (1970).
[8] J. C. Wallwork and F. L. Crane, *Prog. Phytochem.* 2, 267 (1970).
[9] R. A. Morton, *Biol. Rev. Cambridge Philos. Soc.* 46, 47 (1971).
[10] I. R. Peake, P. J. Dunphy, and J. F. Pennock, *Phytochemistry* 9, 1345 (1970).
[11] A. Trebst, *Annu. Rev. Plant Physiol.* 25, 423 (1974).
[12] H. K. Lichtenthaler, *Planta* 81, 140 (1968).
[13] H. K. Lichtenthaler, *Z. Naturforsch. Teil B* 26, 832 (1971).
[14] H. K. Lichtenthaler, V. Straub, and K. H. Grumbach, *Plant Sci. Lett.* 4, 61 (1975).
[15] H. K. Lichtenthaler, *Z. Naturforsch., Teil B* 24, 1461 (1969).
[16] V. Straub and H. K. Lichtenthaler, *Z. Pflanzenphysiol.* 70, 34 (1973).
[17] V. Straub and H. K. Lichtenthaler, *Z. Pflanzenphysiol.* 70, 308 (1973).
[18] C. Buschmann and H. K. Lichtenthaler, *Proc. Int. Congr. Photosynth. Res., 3rd, 1974* p. 753 (1975).

TABLE I

Properties of Phytylplastoquinone, Phytylplastohydroquinone Monomethyl Ether, and 10-Methyldemethylphytylplastoquinol from *Euglena gracilis*[a,b]

| Properties | Phytylplastoquinone | Phytylplastohydroquinone monomethyl ether | 10-Methyldemethylphytylplasto-quinol |
|---|---|---|---|
| Amount ($\mu$g/gm dry wt.) | 40 | 37 | 29 |
| Spectral characteristics | $\lambda_{max}$ 254 and 261 nm in cyclohexane and 255 nm in ethanol | $\lambda_{max}$ 287 nm with a shoulder at 295 nm in cyclohexane | $\lambda_{max}$ 290 nm in cyclohexane |
| Mass spectral characteristics | $m/e$ = 416 (M+2) $m/e$ = 414 (M) and 189 (base peak) | high resolution mass at $m/e$ = 430 | M$^+$ at $m/e$ 416 with major fragment ions at $m/e$ 205, 191, and 151 |
| Oxidation with HAuCl$_4$ | — | Yields phytylplastoquinone | Yields 2-demethylphytylplasto-quinone |
| $R_F$ on silica gel TLC in benzene-petroleum ether (2:3, v/v) | 0.40 | 0.45 | 0.56 in benzene |
| $R_F$ on reverse-phase TLC[c] in 95% aqueous acetone | 0.52 | 0.81 | 0.82 in 98% aqueous acetone |

[a] Law and Threlfall.[6]
[b] Whistance and Threlfall.[7]
[c] TLC, thin layer chromatography.

*peruvianus,*[19] *Hordeum vulgare,*[20–27] nongreen tissues,[28]chromoplasts,[29] and plastoglobuli.[30,31] Williams[32] has studied plastoquinones in *Vicia faba;* Baszynski[33] in *Zea mays;* Wild *et al.*[34] in *Zea mays* and *Atriplex;* Griffiths *et al.*[35] in maize and barley shoots; Mercer and Pughe[36] in maize; Hall and Laidman[37] in germinating *Triticum vulgare;* Barr and Crane[38] in maize, oats, peas, and horse beans; and Barr *et al.*[39] in mineral-deficient maize leaves.

## In Algae

Plastoquinones in algae have been studied by Takamiya *et al.,*[40] Carr *et al.*[41] (Myxophyceae), Allen *et al.*[42] (*Anacystis*), Oku *et al.*[43] (*Chlorella*),

[19] H. K. Lichtenthaler, *Planta* **87**, 304 (1969).
[20] H. K. Lichtenthaler, *Z. Pflanzenphysiol.* **56**, 273 (1967).
[21] H. K. Lichtenthaler, *Biochim. Biophys. Acta* **184**, 164 (1969).
[22] H. K. Lichtenthaler and K. Becker, *Photosynth., Two Centuries Its Discovery Joseph Priestley, Proc. Int. Congr. Photosynth. Res., 2nd, 1971* p. 2451 (1972).
[23] K. H. Grumbach and H. K. Lichtenthaler, *Z. Naturforsch., Teil C* **28**, 439 (1973).
[24] H. K. Kleudgen and H. K. Lichtenthaler, *Z. Naturforsch., Teil C* **20**, 142, (1974).
[25] H. K. Lichtenthaler, and K. H. Grumbach, *Proc. Int. Congr. Photosynth. Res., 3rd, 1974* p. 2017 (1975).
[26] H. K. Kleudgen and H. K. Lichtenthaler, *Proc. Int. Cong. Photosynth. Res., 3rd, 1974* p. 2017 (1975).
[27] H. K. Grumbach and H. K. Lichtenthaler, *Z. Naturforsch., Teil C* **30**, 337 (1975).
[28] H. K. Lichtenthaler, *Z. Pflanzenphysiol.* **59**, 195 (1968).
[29] H. K. Lichtenthaler, *Ber. Dtsch. Bot. Ges.* **82**, 483 (1969).
[30] H. K. Lichtenthaler, *Prog. Photosynth. Res., Proc. Int. Congr. [1st], 1968* Vol. I, p. 304 (1969).
[31] K. H. Grumbach and H. K. Lichtenthaler, *Proc. Int. Congr. Photosynth. Res., 3rd, 1974* p. 515 (1975).
[32] J. P. Williams, *J. Chromatogr.* **36**, 504 (1968).
[33] T. Baszynski, *Polonia* **26**, Sect. C, 187 (1971).
[34] A. Wild, G. Conrad, and H. O. Zickler, *Planta* **103**, 181 (1972).
[35] W. T. Griffiths, *Biochem. J.* **103**, 589 (1967).
[36] E. T. Mercer and J. E. Pughe, *Phytochemistry* **8**, 115 (1969).
[37] G. S. Hall and D. L. Laidman, *Biochem. J.* **108**, 495 (1968).
[38] R. Barr and F. L. Crane, *Plant Physiol.* **45**, 53 (1970).
[39] R. Barr, J. D. Hall, F. L. Crane, and H. Al-Abbas, *Proc. Indiana Acad. Sci.* **80**, 130 (1971).
[40] K. M. Takamiya, M. Nishimura, and A. Takamiya, *Plant Cell Physiol.* **8**, 79 (1967).
[41] N. G. Carr, G. Exell, V. Flynn, M. Hallaway, and S. Talukdar, *Arch. Biochem. Biophys.* **120**, 503 (1967).
[42] C. F. Allen, H. Franke, and O. Hirayama, *Biochem. Biophys. Res. Commun.* **26**, 562 (1967).
[43] T. S. Oku, S. Okayama, I. Aiga, and T. Sasa, *Plant Cell Physiol.* **9**, 599 (1968).

TABLE II
THE DISTRIBUTION OF PLASTOQUINONES IN PHOTOSYSTEM I AND II PARTICLES ISOLATED FROM SPINACH CHLOROPLASTS

| Type of particle | Whole chloroplast quinones | | PS I | Quinone | PS II | Quinone | Quinone units | Reference |
|---|---|---|---|---|---|---|---|---|
| Digitonin | PQ A | (0.13) | 144,000 *g* | PQ A (0.05) | 10,000 *g* | PQ A (0.01) | μmoles/mg Chl | *a* |
| | PQ C | (0.057) | 144,000 *g* | PQ C (0.015) | 10,000 *g* | PQ C (0.006) | | |
| Triton X-100 | PQ A | (0.1) | 50,000 *g* | PQ A (0.08) | 10,000 *g* | PQ A (0.11) | | |
| | PQ C | (0.05) | 50,000 *g* | PQ C (0.02) | 10,000 *g* | PQ C (0.085) | | |
| Sonicated | PQ A | (0.08) | 144,000 *g* | PQ A (0.04) | 10,000 *g* | PQ A (0.04) | | |
| | PQ C | (0.018) | 144,000 *g* | PQ C (0.012) | 10,000 *g* | PQ C (0.046) | | |
| SDBS (sodium dodecyl-benzene-sulfonate) | PQ A | (6μg) | Complex I | PQ A (<2μg) | Complex II | PQ A (<2μg) | μg/ml | *b* |
| Combination digitonin-Triton X-100 and sonication | PQ A | (85.8) | 14 G | PQ A (116.5) | 1.2 G-T | PQ A (14.9) | nanomoles/mg Chl | *c* |
| | PQ B | (12.2) | 14 G | PQ B (22.4) | 1.2 G-T | PQ B (2.5) | | |
| | PQ C+D | (20.7) | 14 G | PQ C+D (42.9) | 1.2 G-T | PQ C+D (1.7) | | |
| Particles isolated from the supernatant after treating frozen-thawed chloroplasts with NaCl | | | CPC I | Present but unidentified | CP II | Present but unidentified | No units | *d* |

| Method | Quinone | | | Quinone | | | Quinone | | Units |
|---|---|---|---|---|---|---|---|---|---|
| Triton X-100 | PQ A | (7.2) | TSF I | PQ A | (1.8) | TSF II | PQ A | (5.5) | moles quinone/ 100 moles Chl [e] |
| | PQ B | (4.2) | TSF I | PQ B | (0.2) | TSF II | PQ B | (4.5) | |
| | PQ C+D | — | TSF I | PQ C+D | (0) | TSF II | PQ C+D | (3.3) | |
| Digitonin | PQ A | (5.6) | S 150 | PQ A | (3.6) | S 10 | PQ A | (6.2) | moles/100 moles Chl [f] |
| | PQ AH₂ | (2.9) | S 150 | PQ AH₂ | (0.7) | S 10 | PQ AH₂ | (3.1) | |
| Digitonin | PQ A | (51) | S 150 | PQ A | (51.6) | S 10 | PQ A | (53) | % of total chloroplast quinone [g] |
| | PQ AH₂ | (32.2) | S 150 | PQ AH₂ | (14) | S 10 | PQ AH₂ | (46) | |

[a] M. D. Henninger, L. Magree, and F. L. Crane, *Biochim. Biophys. Acta* **131**, 119 (1967).

[b] J. P. Thornber, J. C. Stewart, M. W. C. Hatton, and J. L. Bailey, *Biochemistry* **6**, 2006 (1967).

[c] H. Huzisige, H. Usiyama, T. Kikuti, and T. Azi, *Plant Cell Physiol.* **10**, 441 (1969).

[d] K. H. Tachiki, P. R. Parlette, and N. G. Pon, *Biochem. Biophys. Res. Commun.* **34**, 162 (1969).

[e] L. P. Vernon, B. Ke, H. H. Mollenhauer, and E. R. Shaw, *Prog. Photosynth. Res., Proc. Int. Congr. [1st], 1968* Vol. I, p. 137 (1969).

[f] M. Tevini and H. Lichtenthaler, *Z. Pflanzenphysiol.* **62**, 17 (1970).

[g] H. K. Lichtenthaler and M. Tevini, *Z. Pflanzenphysiol.* **62**, 3

Melandri *et al.*[44] (*Euglena*), Senger and Frickel-Faulstich[45] (*Scenedesmus* and *Chlamydomonas*), and Hanigk and Lichtenthaler[46] (*Scenedesmus*).

*In Photosystem I or II Particles*

Plastoquinone distribution in photosystem I or photosystem II particles is summarized in Table II.

### Plastoquinone Mutants

Mutants having lesser amounts of PQ A have been described by Lavorel and Levine[47] (*Chlamydomonas*), Wild and Zickler[48,49] (*Antirrhinum, Chlorella*), Bishop and Wong[50] (*Scenedesmus*), Keck *et al.*[51] (soybeans), and Shahak *et al.*[52] (*Lemna*).

### Biosynthesis of Plastoquinones

The biosynthesis of isoprenoid quinones has been reviewed by Threlfall and Whistance.[53] Their efforts were concentrated on incorporation of radioactive precursors into maize or tobacco shoots and subsequent analysis of the various radioactive quinone fractions. It was found that $^{14}CO_2$, and [2-$^{14}$C]mevalonic acid were incorporated into the side chain of PQ A,[54–56] L-[U-$^{14}$C]phenylalanine and L-[U-$^{14}$C]tyrosine or various other labeled forms of these compounds into the nucleus of PQ A.[57–59] [U-

[44] B. A. Melandri, A. Baccarini-Melandri, and A. San Pietro, *Arch. Biochem. Biophys.* **138,** 598 (1970).

[45] H. Senger and B. Frickel-Faulstich, *Proc. Int. Congr. Photosynth. Res., 3rd, 1974* p. 715 (1975).

[46] H. Hanigk and H. K. Lichtenthaler, *Proc. Int. Congr. Photosynth. Res., 3rd, 1974* p. 2021 (1975).

[47] J. Lavorel and R. P. Levine, *Plant Physiol.* **43,** 1049 (1968).

[48] A. Wild and H.-O. Zickler, *Planta* **97,** 208 (1971).

[49] A. Wild and H.-O. Zickler, *Photosynth., Two Centuries Its Discovery Joseph Priestley, Proc. Int. Congr. Photosynth. Res., 2nd, 1971* Vol. III, p. 2461 (1972).

[50] N. I. Bishop and J. Wong, *Biochim. Biophys. Acta* **234,** 433 (1971).

[51] R. W. Keck, R. A. Dilley, C. F. Allen, and S. Biggs, *Plant Physiol.* **46,** 692 (1970).

[52] Y. Shahak, H. B. Posner, and M. Avron, *Plant Physiol.* **57,** 577 (1976).

[53] D. R. Threlfall and G. R. Whistance, Vol. 23 p. 369.

[54] D. R. Threlfall, W. T. Griffiths, and T. W. Goodwin, *Biochem. J.* **103,** 831 (1967).

[55] O. A. Dada, D. R. Threlfall, and G. R. Whistance, *Eur. J. Biochem.* **4,** 329 (1968).

[56] W. T. Griffiths, D. R. Threlfall, and T. W. Goodwin, *Eur. J. Biochem.* **5,** 124 (1968).

[57] G. R. Whistance, D. R. Threlfall, and T. W. Goodwin, *Biochem. J.* **105,** 145 (1967).

[58] G. R. Whistance and D. R. Threlfall, *Biochem. Biophys. Res. Commun.* **28,** 295 (1967).

[59] G. R. Whistance and D. R. Threlfall, *Biochem. J.* **109,** 577 (1968).

[14]C]Homogentisic acid, [α-[14]C]homogentisic acid,[60,61] or shikimic acid[62] were also incorporated into the nucleus of PQ A, whereas L-[Me-[14]C,[3]H]methionine[63] lead to nuclear methyl substituents. Subsequent work with chloroplast-rich particles of sugar beet and *Euglena gracilis* has demonstrated the biosynthesis of nonaprenyltoluquinol[64] (2-deme-thylplastoquinol-9, a postulated PQA precursor[65]) and an octaprenylto-luquinol from [U-[14]C]homogentisic acid and other forms of labeled hom-ogentisic acid in presence of protein-bound polyprenylpyrophosphates. Cell-free homogenates of *Euglena* can also carry out the biosynthesis of nonaprenyltoluquinol and an octaprenyltoluquinol.[66]

Lichtenthaler[67] touches upon aspects of regulation of plastid quinone biosynthesis. Janizowska *et al.*,[68] using [14]$CO_2$, [1-[14]C] acetate, and [2-[14]C] mevalonate, confirm that biosynthesis of PQ A occurs in chloroplasts of *Calendula officinalis*. Bickel and Schultz[69] are the first to show that isolated chloroplasts incorporate the [14]$CO_2$ label into PQ A under pho-tosynthetic conditions as do excised barley shoots.[70]

[60] G. R. Whistance and D. R. Threlfall, *Biochem. J.* **109,** 482 (1968).
[61] G. R. Whistance and D. R. Threlfall, *Biochem. J.* **117,** 593 (1970).
[62] G. R. Whistance and D. R. Threlfall, *Phytochemistry* **10,** 1533 (1971).
[63] D. R. Threlfall, G. R. Whistance, and T. W. Goodwin, *Biochem. J.* **106,** 107 (1968).
[64] G. Thomas and D. R. Threlfall, *Biochem. J.* **142,** 437 (1974).
[65] D. R. Threlfall and G. R. Whistance, *in* "Aspects of Terpenoid Chemistry and Biochem-istry" (T. W. Goodwin, ed.), p. 357. Academic Press, New York, 1971.
[66] G. Thomas and D. R. Threlfall, *Phytochemistry* **14,** 2607 (1975).
[67] H. K. Lichtenthaler, *Ber. Dsch. Bot. Ges.* **86,** 313 (1973).
[68] W. Janiszowska, W. Michalski, and Z. Kasprzyk, *Phytochemistry* **15,** 125 (1976).
[69] H. Bickel and G. Schultz, *Phytochemistry* **15,** 1253 (1976).
[70] G. Y. Schultz, Y. Huchzermeyer, B. Reupke, and H. Bickel, *Phytochemistry* **15,** 1383 (1976).

# [36] Thioredoxin and Ferredoxin–Thioredoxin Reductase of Spinach Chloroplasts

By RICARDO A. WOLOSIUK, PETER SCHÜRMANN, and BOB B. BUCHANAN

The ferredoxin/thioredoxin system is a newly found mechanism that links light to enzyme regulation in chloroplasts.[1-4] In the light, electrons from chlorophyll are transferred to ferredoxin and then via the enzyme ferredoxin–thioredoxin reductase (earlier called ARP$_a$)[3] to thioredoxin[5] (earlier called ARP$_b$).[3] Reduced thioredoxin, in turn, activates regulatory enzymes that include members of the reductive pentose phosphate cycle (fructose-1,6-diphosphatase, sedoheptulose-1,7-diphosphatase, NADP-glyceraldehyde-3-phosphate dehydrogenase, phosphoribulokinase)[3,4] and enzymes not associated with the cycle (NADP-linked malate dehydrogenase and phenylalanine ammonia lyase)[6,6a] (Fig. 1). The rate of activation of each of these enzymes is slow relative to the rate of catalysis.

Thioredoxin can also be reduced with the nonphysiological sulfhydryl reagent dithiothreitol and, in this case, chloroplast membranes, ferredoxin, and ferredoxin–thioredoxin reductase are not needed. Monothiols tested (cysteine, reduced glutathione, 2-mercaptoethanol) cannot replace dithiothreitol (DTT) in this reaction.[2]

In the dark, the activated (reduced) chloroplast enzymes are deactivated (oxidized) by, depending on the enzyme, either a soluble oxidant (e.g., oxidized glutathione or dehydroascorbate[4]) or an as yet unidentified membrane oxidant.[6] In chloroplasts, the "oxidation–reduction" mechanism appears to act in conjunction with ion-mediated[7] and effector-me-

[1] B. B. Buchanan, P. P. Kalberer, and D. I. Arnon, *Biochem. Biophys. Res. Commun.* **29**, 74 (1967).

[2] B. B. Buchanan, P. Schürmann, and P. P. Kalberer, *J. Biol. Chem.* **246**, 5952 (1971).

[3] P. Schürmann, R. A. Wolosiuk, V. D. Breazeale, and B. B. Buchanan, *Nature (London)* **263**, 257 (1976).

[4] R. A. Wolosiuk and B. B. Buchanan, *Nature (London)* **266**, 565, (1977).

[5] A. Holmgren, *J. Biol. Chem.* **252**, 4600 (1977).

[6] R. A. Wolosiuk, B. B. Buchanan, and N. A. Crawford, *FEBS Lett.* **81**, 253 (1977).

[6a] A. N. Nishizawa, R. A. Wolosiuk, and B. B. Buchanan, *Planta* **145**, 7 (1979).

[7] H. W. Heldt, K. Werdan, M. Milovancev, and G. Geller, *Biochem. Biophys. Acta* **314**, 224 (1973).

METHODS IN ENZYMOLOGY, VOL. 69

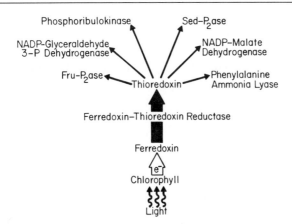

FIG. 1. Role of thioredoxin in the light-dependent activation of chloroplast enzymes (reprinted with permission from "Trends in Biochemical Sciences").

diated[8–12] mechanisms of enzyme control. The effector-mediated and ferredoxin–thioredoxin mechanisms seem to function also in enzyme development in greening seedlings.[13]

This article describes the isolation from spinach leaves of the two newly found components of the chloroplast ferredoxin/thioredoxin system, viz, ferredoxin–thioredoxin reductase, thioredoxin $f$ (the thioredoxin specific for Fru-$P_2$ase and certain other chloroplast enzymes) and thioredoxin $m$ (the thioredoxin specific for NADP–malate dehydrogenase).[13a]

## Thioredoxins $f$ and $m$

### Assay Method for Thioredoxin $f$

*Principle.* The assay for thioredoxin $f$ is based on its capacity to activate fructose-1,6-diphosphatase (Fru-$P_2$ase) when reduced either by ferredoxin (in the presence of ferredoxin–thioredoxin reductase) or by dithiothreitol. The original work that led to the finding of thioredoxin in chloroplasts[2] was based on the assay with reduced ferredoxin. More

[8] B. Müller, I. Ziegler, and H. Ziegler, *Eur. J. Biochem.* **9,** 101 (1969).
[9] H. P. Ghosh and J. Preiss, *J. Biol. Chem.* **241,** 4491 (1966).
[10] P. Pupillo and G. G. Piccari, *Arch. Biochem. Biophys.* **154,** 324 (1973).
[11] R. A. Wolosiuk and B. B. Buchanan, *J. Biol. Chem.* **251,** 6456 (1976).
[12] G. H. Lorimer, M. R. Badger, and T. J. Andrews, *Biochemistry* **15,** 529 (1976).
[13] B. B. Buchanan, N. A. Crawford, and R. A. Wolosiuk, *Plant Sci. Lett.* **12,** 257 (1978).
[13a] R. A. Wolosiuk, N. A. Crawford, B. C. Yee, and B. B. Buchanan, *J. Biol. Chem.* **254,** 1627 (1979).

recently, because of convenience, we have used the assay with dithiothreitol. In this assay, Fru-P$_2$ase can be followed by measuring either released P$_i$ colorimetrically or newly formed fructose-6-phosphate spectrophotometrically (by following the reduction of NADP in the presence of excess glucose-6-phosphate isomerase and glucose-6-phosphate dehydrogenase). In either case, the assay is executed in the presence of a limiting concentration of Mg because higher concentrations activate Fru-P$_2$ase in the absence of reduced thioredoxin.

The assay is an excellent indicator of the presence of thioredoxin. However, because of contaminating Fru-P$_2$ase present during the early purification steps, it has been difficult to quantitate thioredoxin activity in absolute terms. Accordingly, no unit of activity has been defined for thioredoxin.

*Method I. Colorimetric Assay*

*Reagents for Assay*

Tris-HCl buffer (pH 7.9). 1$M$

MgSO$_4$, 10 m$M$

Dithiothreitol (DTT), 50 m$M$

Sodium fructose-1,6-diphosphate (FDP), 60 m$M$

Chloroplast fructose-1,6-diphosphatase (purified as given by Buchanan et al.[14]), 1.7 mg/ml

*Assay Procedure.* The reaction is carried out in a 1 × 10 cm test tube containing 0.005 ml of Fru-P$_2$ase and 0.05 ml each of Tris-HCl buffer (pH 7.9), MgSO$_4$, DTT, and the thioredoxin fraction to be assayed. Total volume is made up to 0.5 ml with water. After the assay components stand for 5 min, the reaction is initiated by adding 0.05 ml of FDP and is continued for 15 min at room temperature. The reaction is stopped by the addition of 2.0 ml of the mixture for P$_i$ analysis[1] described below. The samples are allowed to stand for 10 min and the absorbance is determined spectrophotometrically at 660 nm.

*Reagents for P$_i$ Analysis*

H$_2$SO$_4$, 9$N$

Ammonium molybdate [(NH$_4$)$_6$Mo$_7$O$_{24}$·4H$_2$O], 6.6 gm/100 ml

FeSO$_4$·7H$_2$O, 2.0 gm/25 ml of a solution containing 24 ml H$_2$O + 1 ml 9 $N$ H$_2$SO$_4$

*Procedure for P$_i$ Analysis.*

To 0.5 ml of sample is added 2.0 ml of the mixture used for P$_i$ analysis (prepared daily by combining 125 ml of H$_2$O, 25 ml of H$_2$SO$_4$, 25 ml of ammonium molybdate, and 25 ml of FeSO$_4$). After the mixture stands

[14] B. B. Buchanan, P. Schürmann, and R. A. Wolosiuk, *Biochem. Biophys. Res. Commun.* **69**, 970 (1976).

for 5 min at room temperature, absorbance at 660 nm is measured in a cuvette (3-ml capacity, 1-cm light path); 0.5 umole of $P_i$ gives an absorbance of about 0.6 under these conditions.

*Method II. Spectrophotometric Assay*

*Reagents for Assay*

   Tris-HCl buffer (pH 7.9), 1 $M$

   DTT, 50 m$M$

   Fru-$P_2$ase, 1.7 mg/ml

   MgSO$_4$, 10 m$M$

   Glucose-6-phosphate dehydrogenase, 270 units/ml

   Glucose-6-phosphate isomerase, 2100 units/ml

   NADP, 0.1 $M$

   FDP, 60 m$M$

*Assay Procedure.* The following are added to a test tube (0.1 × 7 cm): 0.01 ml of Tris-HCl buffer (pH 7.9), 0.02 ml of DTT, and a limiting amount of thioredoxin. The total volume is made up to 0.09 ml with water. After the addition of 0.01 ml of Fru-$P_2$ase the mixture is incubated at room temperature for 10 to 20 min. Following preincubation, the mixture is injected into a cuvette of 1-cm light path containing, in a total volume of 0.9 ml: 0.005 ml of glucose-6-phosphate dehydrogenase, 0.002 ml of glucose-6-phosphate isomerase, 0.04 ml of Tris-HCl buffer (pH 7.9), 0.01 ml of NADP, and 0.1 ml each of MgSO$_4$ and FDP. NADP reduction is followed spectrophotometrically at 340 nm with a Cary 14 recording spectrophotometer.

*Assay Method for Thioredoxin m*

*Principle.* The assay for thioredoxin *m* is based on the capacity to activate NADP-malate dehydrogenase (NADP–MDH) when reduced either by ferredoxin (in the presence of ferredoxin–thioredoxin reductase) or by dithiothreitol. Because of convenience, we routinely use the assay with dithiothreitol. In this assay, NADP–MDH is followed spectrophotometrically by measuring NADPH oxidation in the presence of oxalacetate. For the reasons described above for thioredoxin *f*, no unit of activity has been defined for thioredoxin *m*.

*Reagents for Assay*

   Tris-HCl buffer (pH 7.9), 1 $M$

   DTT, 50 m$M$

   NADP–MDH (purified as described by Wolosiuk *et al.*[6]), 1.2 mg/ml

   NADPH, 25 m$M$

   Oxalacetic acid, 50 m$M$

*Assay Procedure.* The reaction is carried out in air at 22°. To a 1 × 7 cm test tube is added 0.03 ml Tris-HCl buffer pH 7.9, 0.05 ml DTT, 0.02 ml NADP–MDH, and $H_2O$, to a final volume of 0.1 ml. After preincubation for 10 min, the mixture is transferred with a Lang-Levy pipette to a 1-cm cuvette of 1-ml capacity that contains 0.09 ml Tris-HCl buffer (pH 7.9), 0.01 ml NADPH and 0.65 ml $H_2O$. The reaction is started by the immediate addition of 0.05 ml of oxalacetic acid. The change in absorbance at 340 nm is measured with a Cary 14 M spectrophotometer.

## Purification of Thioredoxins

Spinach leaves, purchased from a local market, are de-stemmed, rinsed in distilled water, and stored overnight at 4° prior to use in enzyme preparations. Each of the proteins is purified at 4°. Buffers are adjusted to the indicated pH at 22°. The purification methods described below have been used repeatedly for the isolation of ferredoxin–thioredoxin reductase and the leaf thioredoxins. However, because of contaminating Fru-$P_2$ase and ferredoxin during the early stages of purification and because of the cross-reactivity of the different thioredoxins in the Fru-$P_2$ase assay (see below), it is not feasible to quantitate in absolute terms the activity of ferredoxin–thioredoxin reductase or any one of the thioredoxins throughout the purification procedure. Accordingly, no specific activity measurements are included below.

## Isolation of Thioredoxins f and m from Chloroplasts

*Step 1: Preparation of Chloroplast Extract.* Spinach leaves (in 75-gm lots) are homogenized for 10 sec in a Waring blender (2-qt capacity, model HBG-100) containing 200 ml of homogenizing buffer solution [25 m$M$ Tris-HCl buffer (pH 7.6) and 0.35 $M$ sucrose]. The homogenate is filtered through 8 layers of filtering silk (124 threads warp, 82 threads filling per in²; Precision Cells, Inc., Hicksville, New York). The residue is discarded. The filtrate is centrifuged for 1 min at 3000 $g$ and the supernatant fraction is discarded. The pellet, containing whole chloroplasts, is resuspended gently in the preparative solution to a volume about 1/20 that of the original. The resuspended whole chloroplasts are collected by centrifugation for 3 min at 3000 $g$, the supernatant fraction is discarded, and the dark-green pellet, containing once-washed chloroplasts, is resuspended in 50 m$M$ Tris-HCl buffer (pH 7.9). The chlorophyll content of the chloroplast suspension is adjusted to 1.5 mg/ml and the green membrane fraction removed from the osmotically ruptured chloroplast suspension by centrifugation (10 min, 35,000 $g$). The supernatant fraction ("chloroplast extract") is used as a source of thioredoxins $f$ and $m$.

*Step 2: pH 4.5 Fractionation.* Chloroplast extract (250 ml) is adjusted to pH 4.5 with 2 $M$ formic acid and centrifuged (5 min, 27,000 $g$). The greenish precipitate is discarded, and the pH of the supernatant fraction is adjusted to 7.1 with 1 $M$ NH$_4$OH.

*Step 3: Acetone Precipitation.* Acetone, cooled to $-15°$, is added to the neutralized pH 4.5 supernatant fraction, with constant stirring, to a final concentration of 75%. The suspension is left for 1 hr at $-15°$ to allow the precipitate to settle. The bulk of the yellow supernatant fraction is decanted and discarded. The flocculent precipitate is collected by centrifugation of the remainder of the suspension (5 min, 27,000 $g$) and resuspended in 9.0 ml of 30 m$M$ Tris-HCl buffer (pH 7.9) (subsequently designated "Buffer A"). The suspension is dialyzed overnight against 2 liters of Buffer A and then clarified by centrifugation for 30 min at 105,000 $g$. The precipitate is discarded and the supernatant fraction, containing the two chloroplast thioredoxins, is used.

*Step 4: DEAE–Cellulose Chromatography.* The fraction obtained by acetone precipitation is applied to a DEAE–Cellulose column (3.0 × 15 cm) equilibrated beforehand with Buffer A. After application of the sample, the column is eluted with 150 ml of Buffer A and then with 500 ml of a linear gradient 0 to 0.2 $M$ in NaCl added to Buffer A. Finally, the column is eluted with 200 ml of Buffer A supplemented with 0.5 $M$ NaCl. Fractions (7.3 ml) are collected and analyzed for thioredoxin activity with the Fru-P$_2$ase and NADP–malate dehydrogenase assays in the presence of dithiothreitol. The two chloroplast thioredoxins are separated in this step. The fractions showing either preferential Fru-P$_2$ase-linked thioredoxin activity (thioredoxin $f$) or preferential NADP–malate dehydrogenase-linked thioredoxin activity (thioredoxin $m$) are pooled separately. Each of the two thioredoxin fractions is then treated separately through the remainder of the purification procedure.

*Step 5: Concentration by Dialysis vs. Solid Sucrose.* The thioredoxin $m$ and thioredoxin $f$ fractions are added to several strips of dialysis tubing (2.5-cm wide) which are placed in a glass dish, covered with solid sucrose, and left overnight. In this step, the volume of each component is reduced from 80 ml to 5–10 ml. The concentrated solution is clarified by centrifugation (1 hr, 105,000 $g$).

*Step 6: Sephadex G-75 Chromatography.* Each of the thioredoxins concentrated on sucrose is applied separately to a Sephadex G-75 column (2.6 × 100 cm) equilibrated beforehand with Buffer A. The column is eluted with the same buffer solution at a flow rate of 12 ml/hr. Fractions (5.3 ml) are collected and assayed for thioredoxin $f$ and thioredoxin $m$

activity. The peak thioredoxin fractions are pooled, concentrated by dialysis vs. solid sucrose as before, and dialyzed overnight vs. Buffer A.

*Step 7: Hydroxyapatite Chromatography.* The pooled Sephadex G-75 fractions of either thioredoxin *f* or thioredoxin *m* are concentrated by dialysis vs. solid sucrose as described above and dialyzed overnight vs. 50 m$M$ Tris-HCl buffer (pH 7.9). The dialyzate is clarified by centrifugation (1 hr, 105,000 $g$) and the supernatant fraction (about 60 ml) is applied to a 1.8 × 3 cm hydroxyapatite column (Bio-Gel HTP, Bio-Rad Labs, Richmond, California) that is equilibrated beforehand with 50 m$M$ Tris-HCl buffer (pH 7.9). The column is subsequently washed with 40 ml of 50 m$M$ Tris-HCl buffer (pH 7.9), containing 0, 25, 100, and 300 m$M$ potassium phosphate buffer (pH 7.7). Fractions (6.6 ml) are collected at a flow rate of 60 ml/hr. The absorbance at 280 nm is determined, and those fractions containing 280 nm-absorbing material are pooled and dialyzed overnight vs. Buffer A. The combined fractions are then assayed for thioredoxin *f* and thioredoxin *m* activity. Thioredoxin *m* is recovered mainly in the 25 m$M$ potassium phosphate buffer eluate, whereas thioredoxin *f* is recovered mainly in the 100 m$M$ potassium phosphate buffer eluate.

### General Comments

*Assay.* All of our purification work has been accomplished by assaying thioredoxin *f* with Fru-P$_2$ase and thioredoxin *m* with NADP–MDH. Although not yet developed, it would seem that assays for thioredoxin *f* based on other thioredoxin *f*-dependent enzymes could be used equally well (NADP–glyceraldehyde-3-phosphate dehydrogenase,[11,13a] phosphoribulokinase,[3] phenylalanine ammonia lyase[14a]). Each of the former two enzymes can be assayed spectrophotometrically by classical assays based on the measurement of oxidation–reduction of NAD(P)H/NAD(P). Because the rate of activation is slower than the rate of catalysis, assays for thioredoxin with these enzymes would be most successful with a two-stage assay of the type described above for the spectrophotometric Fru-P$_2$ase or NADP–MDH assay.

*Purification.* During the first DEAE–cellulose chromatography (Step 6), a thioredoxin peak elutes with the front. Recent work[6a] has shown that this fraction contains two extrachloroplastic thioredoxins—one that reacts preferentially with chloroplast Fru-P$_2$ase (thioredoxin $c_f$) and one that reacts preferentially with chloroplast NADP–MDH (thioredoxin $c_m$).

---

[14a] N. A. Crawford, B. C. Yee, A. N. Nishizawa, and B. B. Buchanan, *FEBS Lett.* **104,** 141 (1979).

Neither the function nor the physiological mechanism of reduction is known for the extrachloroplast thioredoxins. Certain properties of the chloroplast[13a] as well as of the extrachloroplast[14b] thioredoxins have been reported.

*Stability.* The stability of thioredoxins $f$ and $m$ seems to be variable. In general, concentrated thioredoxin solutions retain activity for several months when stored at $-15°$. Stability seems to be greater when thioredoxins $f$ and $m$ are not resolved and are stored as the $f$–$m$ mixture.

### Ferredoxin–Thioredoxin Reductase

#### *Assay Method*

*Principle.* Ferredoxin–thioredoxin reductase is assayed by measuring the stimulation in the Fru-$P_2$ase-mediated release of $P_i$ from fructose-1,6-diphosphate activated in the presence of photoreduced ferredoxin and thioredoxin. As noted above for thioredoxin, the ferredoxin–thioredoxin reductase assay is executed with thioredoxin and a limiting concentration of $Mg^{2+}$.

*Reagents for Assay*
    Tris-HCl buffer, 1 $M$ pH 7.9
    MgCl$_2$ or MgSO$_4$, 10 m$M$
    2,6-Dichlorophenolindophenol (DPIP), 2 m$M$
    Sodium ascorbate, 0.2 $M$
    Chloroplast Fru-$P_2$ase (purified as described by Buchanan
        *et al.*[14]), 1.7 mg/ml
    Spinach ferredoxin, 1.2 mg/ml
    Chloroplast thioredoxin $f$, 2.5 mg/ml
    Spinach chloroplast fragments ($P_1S_2$ prepared in sucrose as
        given by Buchanan and Arnon[15]), 1 mg/ml
    Fructose-1,6-diphosphate, sodium salt, 60 m$M$
    Ferredoxin–thioredoxin reductase
    Trichloroacetic acid

*Assay Procedure.* The reaction is carried out at 20° in Warburg vessels. The main compartment contains: 0.1 ml each of Tris-HCl buffer (pH 7.9), $Mg^{2+}$, spinach ferredoxin, and chloroplast fragments; 0.05 ml each of sodium ascorbate and DPIP; 0.01 ml each of thioredoxin and Fru-$P_2$ase; the ferredoxin–thioredoxin reductase as needed; and $H_2O$ to a final volume of 1.4 ml. The sidearm contains 0.1 ml of fructose-1,6-diphosphate. After equilibration for 5 min with $N_2$, vessels are preincu-

[15] B. B. Buchanan and D. I. Arnon, Vol. 23, p. 413.

bated 5 min in the light (20,000 lux). The reaction is started by adding fructose-1,6-diphosphate from the sidearm and is continued for 30 min under illumination. The reaction is stopped by the addition of 0.5 ml of 10% trichloroacetic acid. The precipitate is centrifuged off, and 1.0 ml of the supernatant solution is analyzed for $P_i$.

As noted above for thioredoxin, we have not been able to quantitate the assay for ferredoxin–thioredoxin reductase and therefore have not defined a unit of activity for the enzyme.

## Purification of Ferredoxin–Thioredoxin Reductase

*Step 1: Preparation of Leaf Extract.* Sixteen kilograms of leaves (1-kg lots) are blended for 3 min in a gallon-size Waring blender containing 1000 ml of $H_2O$, 1 ml of 2-mercaptoethanol (14 $M$), and 30 ml of 1 $M$ Tris-HCl buffer (pH 7.9). The homogenate is filtered through four layers of cheesecloth and the fibrous retentate is discarded. The green filtrate is used in the next step.

*Step 2: Ammonium Sulfate Fractionation.* Solid ammonium sulfate is added to the filtrate (18.0 1) to give 50% saturation. The suspension is centrifuged (10 min, 13,000 $g$) and the green precipitate is dicarded. Ammonium sulfate is added to the yellow supernatant fraction to 90% saturation. The solution is stirred for 15 min and then centrifuged as described above. The precipitate is resuspended in 330 ml of a solution 50 m$M$ in Tris-HCl buffer (pH 7.9) with 0.1% 2-mercaptoethanol (14 m$M$.

*Step 3: Acetone Fractionation.* Acetone (0.99 liter) cooled to $-15°$ is added, with constant stirring, to the resuspended 50–90% ammonium sulfate fraction. The suspension is centrifuged immediately at $-15°$ (5 min, 20,000 $g$) and the supernatant fraction is discarded. The precipitate is resuspended in 280 ml of a solution containing 50 m$M$ Tris-HCl buffer (pH 7.9) and 14 m$M$ 2-mercaptoethanol. The suspension is then dialyzed overnight vs. a solution containing Buffer A supplemented with 14 m$M$ 2-mercaptoethanol. The dialyzate is centrifugated at 40,000 $g$ for 20 min and the precipitate is discarded. The supernatant fraction is frozen and then thawed after 4 hr. The suspension is clarified by centrifugation (20 min, 20,000 $g$).

*Step 4: Concentration with Solid Sucrose.* This step is as described in Step 5 for the isolation of thioredoxins $f$ and $m$. The dialyzed acetone precipitate from Step 3 (485 ml) is concentrated to about 50 ml with sucrose. The dark-brown solution is dialyzed overnight against Buffer A containing 14 m$M$ 2-mercaptoethanol and then clarified by centrifugation (1 hr, 105,000 $g$).

*Step 5: Sephadex G-100 Chromatography.* The supernatant fraction from Step 4 is applied to a Sephadex G-100 column (5 × 150 cm) equilibrated beforehand with Buffer A containing 14 m*M* 2-mercaptoethanol. The column is eluted with the same buffer solution at a flow rate of 55 ml/hr. Fractions (10 ml) are collected. Those fractions showing ferredoxin–thioredoxin reductase activity are pooled (590 ml).

*Step 6. DEAE–Cellulose Chromatography.* The combined reductase fraction from Step 5 is applied to a DEAE–cellulose column (1.5 × 35 cm) previously equilibrated with Buffer A supplemented with 14 m*M* 2-mercaptoethanol. The column is eluted with 250 ml of this solution 0.08 *M* in NaCl. The bulk of the reductase activity is then eluted with a linear gradient of 0.08 to 0.3 *M* NaCl in the same buffer (total volume 1000 ml). Finally, the reductase remaining in the column is eluted with 250 ml of the same buffer solution containing 0.5 *M* NaCl. Fractions (8 ml) are collected and those containing reductase activity are pooled (290 ml) and concentrated to about 50 ml with solid sucrose as in Step 4.

*Step 7. Hydroxyapatite Chromatography.* The pooled fractions from Step 6 are dialyzed vs. a solution containing Buffer A supplemented with 14 m*M* 2-mercaptoethanol. The dialyzate is clarified by centrifugation (30 min, 105,000 *g*) and the supernatant fraction (95 ml) is applied to a 1.8 × 3.5 cm hydroxyapatite column equilibrated beforehand with the same solution as that used for dialysis. The column is successively eluted with 85 ml each of 25, 50, 100, and 300 m*M* potassium phosphate buffer (pH 7.7) containing 14 m*M* 2-mercaptoethanol. Fractions (4 ml) are collected at a flow rate of 70 ml/hr and the effluent is monitored at 280 nm. The fractions showing absorbance at 280 nm are pooled and dialyzed overnight vs. Buffer A supplemented with 14 m*M* 2-mercaptoethanol. The fractions that elute with 25 m*M* and 300 m*M* potassium phosphate solutions both show ferredoxin–thioredoxin reductase activity. However, only the 300 m*M* phosphate fraction (which contains about 30% of the reductase activity) is free of thioredoxin.

*Stability.* The enzyme is labile in the absence of a sulfhydryl reagent. Stabilization is achieved by supplementing the 30 m*M* Tris-HCl buffer (pH 7.9) used during purification with 14 m*M* 2-mercaptoethanol. In this buffer, the enzyme can be stored at −15° (or −70°) for two months with no appreciable loss of activity. Incubation for 30 min at 30° in air results in complete loss of activity.

*Distribution.* Sources other than spinach chloroplasts have not been tested systematically for ferredoxin–thioredoxin.reductase activity. The enzyme has been detected in crude preparations from the blue-green alga *Nostoc muscorum* and from dark- and light-grown barley seedlings.

## [37] Purification and Properties of Flavodoxins

### By GORDON TOLLIN and DALE E. EDMONDSON

### I. Biological Functions of Flavodoxins

In general, flavodoxins mediate the transfer of electrons at a low redox potential between the prosthetic groups of other microbial proteins. In some cases (e.g., *Clostridium pasteurianum*), flavodoxin acts as a replacement for ferredoxin under low iron growth conditions; in others (e.g., *Azotobacter vinelandii*), it is a constitutive protein which plays a unique biochemical role. A large number of ferredoxin-dependent enzymes have been identified in bacteria.[1] Flavodoxin will replace ferredoxin in most of these. Among the most extensively studied systems are the following.

### A. Pyruvate Catabolism (Phosphoroclastic Reaction)

$$\text{Pyruvate} + \text{CoA} \rightarrow \text{acetyl CoA} + CO_2$$

The enzyme system catalyzing this reaction, as found in *C. pasteurianum* and other anaerobes, consists of pyruvate dehydrogenase, phosphotransacetylase, and hydrogenase. Electrons derived from pyruvate reduce a nonheme iron sulfur moiety in the dehydrogenase.[2] These electrons are then used to reduce ferredoxin (or flavodoxin), which in turn is reoxidized by hydrogenase with the accompanying formation of molecular hydrogen. In *C. acidi-urici,* the reduced ferredoxin is involved as an electron source in uric acid fermentation.[3] Ferredoxin-dependent hydrogen evolution is also essential to the growth of the ethanol fermenting organism *C. kluyverii.*[4]

### B. CO₂ Fixation (Pyruvate Synthase)

$$\text{Acetyl CoA} + CO_2 \rightarrow \text{Pyruvate} + \text{CoA}$$

This is essentially the reverse of the process described in Section I,A. It has been found in *C. pasteurianum,* where its significance is uncertain,

---

[1] D. C. Yoch and R. C. Valentine, *Annu. Rev. Biochem.* **26**, 139 (1972).

[2] K. Uyeda and J. C. Rabinowitz, *J. Biol. Chem.* **246**, 3111 (1971).

[3] R. C. Valentine, R. C. Jackson, and R. S. Wolfe, *Biochem. Biophys. Res. Commun.* **7**, 453 (1962).

[4] W. W. Fredericks and E. R. Stadtman, *J. Biol. Chem.* **240**, 4065 (1965).

METHODS IN ENZYMOLOGY, VOL. 69

and in *C. kluyverii,* where it is an essential metabolic component. In the latter organism, reduction of ferredoxin occurs via hydrogenase, or by reduced pyridine nucleotides utilizing an NADH or NADPH–ferredoxin oxidoreductase.[5,6]

## C. Dissimilatory Sulfate Reduction

In *Desulfovibrio,* ferredoxin (or flavodoxin) mediates electron transfer between hydrogen and sulfite.[7] The details of this pathway have not yet been elucidated.

## D. Nitrogen Fixation

In the obligate anaerobes, the endogenous donor of electrons to nitrogenase is ferredoxin or flavodoxin.[8–11] In *A. vinelandii,* a constitutive flavodoxin fulfills this function.[12] It has been suggested[13] that the system for generating reducing equivalents for nitrogenase in this organism is localized in the cytoplasmic membrane, and is mediated by an NADH–flavodoxin oxidoreductase operating at a site of low pH (~5), at which the semiquinone–hydroquinone potential is high enough to allow NADH to function as a reductant. The pH gradient which is generated by membrane energization is then used to bring the flavodoxin hydroquinone to a low enough potential so that it can serve as the donor to nitrogenase. Among the miscellaneous activities found for flavodoxins is the ability to function in the light-dependent reduction of $NADP^+$ by plant chloroplasts.[14] It is not known whether or not this represents a biologically occurring process.

[5] K. Jungermann, E. Rupprecht, C. Ohrloff, R. K. Thauer, and K. Decker, *J. Biol. Chem.* **246,** 960 (1971).

[6] R. K. Thauer, E. Rupprecht, C. Ohrloff, K. Jungermann, and K. Decker, *J. Biol. Chem.* **246,** 954 (1971).

[7] E. C. Hatchikian, J. LeGall, M. Bruschi, and M. Dubourdieu, *Biochim. Biophys. Acta* **258,** 701 (1972).

[8] A. J. D'Eustachio and R. W. Hardy, *Biochem. Biophys. Res. Commun.* **15,** 319 (1964).

[9] L. E. Mortenson, *Biochim. Biophys. Acta* **81,** 473 (1964).

[10] J. R. Benemann, D. C. Yoch, R. C. Valentine, and D. I. Arnon, *Proc. Natl. Acad. Sci. U.S.A.* **64,** 1079 (1969).

[11] D. C. Yoch, J. R. Benemann, R. C. Valentine, and D. I. Arnon, *Proc. Natl. Acad. Sci. U.S.A.* **64,** 1404 (1969).

[12] G. Scherings, H. Haake, and C. Veeger, *Eur. J. Biochem.* **77,** 621 (1977).

[13] H. Haake and C. Veeger, *Eur. J. Biochem.* **77,** 1 (1977).

[14] R. M. Smillie, *Biochim. Biophys. Res. Commun.* **20,** 621 (1965).

## II. General Comments on Flavodoxin Purification

All known flavodoxins have low molecular weights (14,500–23,000 gm/mole) and are strongly anionic. Because of these properties, DEAE-cellulose chromatography, $(NH_4)_2SO_4$ precipitation, and Sephadex chromatography are techniques usually used in all published purification procedures.

This discussion will deal with those flavodoxins which function in either photosynthesis or nitrogen fixation. The properties and purification procedures for the flavodoxins from C. *pasteurianum*[15] and *Anacystis nidulans*[16] have been described in previous volumes of this series and thus will not be described here.

The flavodoxin purification procedures described below will include those from the nitrogen-fixing bacteria: A. *vinelandii* and *Klebsiella pneumoniae*, and from the photosynthetic organisms: *Rhodospirillum rubrum*, *Synechoccus lividus*, and *Chlorella*.

## III. Purification of *Azotobacter vinelandii* Flavodoxin

The procedure described is an adaptation of that published by Hinkson and Bulen.[17] *Azotobacter vinelandii* (strain OP) cells are harvested from cultures grown under $N_2$-fixing conditions using sucrose as the sole carbon source as described by Bulen and LeComte[18] The wet cell paste may be used directly or stored in the freezer before use.

Wet cell paste (400–500 gm) is suspended at a ratio of 22 gm/60 ml of cold 25 m$M$ phosphate buffer, pH = 7.0. The cells are broken either by passage through a French pressure cell at 16,000–18,000 psi or by sonication of 100-ml aliquots of cell suspension at full power for 4 min. Unbroken cells and large cell debris are removed by centrifugation for 10 min at 12,000 $g$. The resulting supernatant is then centrifuged for 120 min at 105,000 $g$ to sediment membranous material. Nucleic acids are removed by the addition of 5 ml (dropwise with stirring) of 2% protamine sulfate (Eli-Lilly) solution, pH = 6.0, per 25 ml of the supernatant. After standing for 30 min at 0°, the precipitated nucleic acids are separated by centrifugation at 12,000 $g$ for 15 min. If desired, the preparation may be interrupted at this point and the crude supernatant stored overnight at 0°. Any precipitate formed during the overnight storage should be eliminated by centrifugation before proceeding to the next step.

[15] E. Knight, Jr. and R. W. F. Hardy, Vol. 23, p. 552.
[16] R. M. Smillie and B. Entach, Vol. 23, Part A, p. 504.
[17] J. W. Hinkson and W. A. Bulen, *J. Biol. Chem.* **242**, 3345 (1967).
[18] W. A. Bulen and J. R. LeComte, Vol. 24, p. 456.

The crude extract is applied to a DEAE–cellulose column (3 × 20 cm) (either Bio-Rad Cellex D or Whatman DE-52 is suitable) which has previously been equilibrated in 25 m$M$ phosphate, pH = 7.0 at 4°. After all of the crude extract has been applied, the column is washed with 25 m$M$ phosphate, pH = 7.0, until the $A_{280}$ is less than 0.1. The blue-green flavodoxin band (a mixture of the oxidized and semiquinone forms) is eluted with 0.4 $M$ NaCl–25 m$M$ phosphate, pH = 7.0. Solid $(NH_4)_2SO_4$ is added to the pooled flavodoxin fraction to 75% saturation, the pH adjusted to 7.0 with dilute $NH_4OH$, and after 30 min at 0°, centrifuged at 15,000 $g$ for 30 min. The greenish-yellow supernatant, containing pure flavodoxin is then either dialyzed or chromatographed on Sephadex G-25 to remove the $(NH_4)_2SO_4$. The flavodoxin is concentrated by ultrafiltration (Amicon, PM-10 membrane) and stored at −20°. The purified material in its oxidized form has a ratio of absorbance at 274 to 452 nm of 4.7. Approximately 30 mg of flavodoxin is obtained per 100 gm wet cell paste.

The dimerization of *Azotobacter* flavodoxin via intermolecular disulfide formation has been shown by Yoch[19] and has also been observed in the authors' laboratories. Dimerization results in a decreased activity in donation of reducing equivalents to nitrogenase.[19] To prevent dimerization, it is suggested that flavodoxin preparations be stored in the presence of mercaptoethanol or dithiothreitol.

### IV. Purification of *Rhodospirillum rubrum* Flavodoxin

This procedure is essentially that described by Cusanovich and Edmondson.[20] *R. rubrum* (strain 2.1.1, Van Niel) is grown on a media consisting of the following reagents (per liter): succinate, 4 gm; 2-(hydroxymethyl)-1,3-propanediol, 2.4 gm; $(NH_4)_2SO_4$, 1 gm; Mg $Cl_2$, 0.05 gm; casamino acids (Difco), 1 gm; biotin 7 $\mu$g; and Larsen's trace elements without iron, 1 ml.[21] The iron content of the media should be determined and should not exceed 350 $\mu$g/liter. The organism is grown, using a 20% inoculum, in an illuminated water bath at 35° for 48 hr. The cells are then harvested and stored at −10° until used.

*Rhodospirillium rubrum* cells are suspended in 2 volumes of 0.1 $M$ Tris, pH 7.3, at 4°. The cells are broken by passage through a Ribi Cell

[19] D. C. Yoch, *Arch. Biochem. Biophys.* **170,** 326 (1975).

[20] M. A. Cusanovich and D. E. Edmondson, *Biochem. Biophys. Res. Commun.* **45,** 327 (1971).

[21] S. K. Bose, *in* "Bacterial Photosynthesis" (H. Gest, A. San Pietro, and L. P. Vernon, eds.), p. 501. Antioch Press, Yellow Springs, Ohio, 1963.

Fractionator (Ivan Sorvall Co.) at 20,000 psi. The broken cell suspension is centrifuged at 30,000 $g$ for 30 min, the resulting pellet resuspended in an equal volume of 0.1 $M$ Tris, pH 7.3, centrifuged again under the same conditions and the supernatant combined with the supernatant from the initial centrifugation. The slightly turbid, orange solution is then centrifuged at 100,000 $g$ for 1 hr. The resulting supernatant is applied to a DEAE–cellulose column (4 × 10 cm) (Type 40, Brown and Co. or Whatman DE-23) which has been previously equilibrated with 0.1 $M$ Tris, pH 7.3. The column is washed with the equilibration buffer until the $A_{280}$ is less than 0.1. The yellow flavodoxin band is eluted from the column with 0.5 $M$ NaCl–0.02 $M$ Tris, pH 7.3. The pooled flavodoxin fractions are desalted by either Sephadex G-25 chromatography or by dialysis and then applied to a DEAE–cellulose column (4 × 10 cm) (Type 20, Brown and Co. or Whatman DE-52) equilibrated with 0.01 $M$ Tris, pH 7.3. The column is washed with 2 volumes of equilibration buffer and 40 volumes of 0.1 $M$ Tris, pH 7.3, and the flavodoxin band is eluted with 0.2 $M$ Tris, pH 7.3. The pooled flavodoxin fractions are concentrated by ultrafiltration (Amicon PM-10 membrane) and chromatographed on a Sephadex G-100 column (4 × 80 cm) with 0.02 $M$ Tris–0.5 $M$ NaCl, pH 7.3, as the eluting buffer. The flavodoxin is eluted as a strongly retarded symmetrical peak with an $A_{272}/A_{460}$ spectral ratio of 4.84. The pure flavodoxin is concentrated by ultrafiltration (Amicon PM-10 membrane) and stored at −20°. A yield of approximately 30 mg of flavodoxin can be expected to be obtained from 240 gm (wet weight) of cells.

## V. Purification of *Synechoccus lividus* Flavodoxin

This procedure is taken from that described by Crespi *et al.*[22] The thermophilic blue-green algae *S. lividus* is grown in nutrient media according to the procedure of Daboll *et al.*[23] The harvested cells may be stored either as a frozen cell paste or as lyophilized cells.

The cells (800 gm wet cell paste or 200 gm freeze-dried cells) are suspended in 1.5 liters of 10 m$M$ phosphate, pH 6.9. The suspended cells are sonicated for a few minutes and then allowed to stand at 5° for 3 hr with occasional stirring. Unbroken cells and cell debris are removed by centrifuging twice for 30 min at 15,000 $g$. The combined precipitates from both centrifugations are resuspended in an equal volume of 10 m$M$ phosphate, pH 6.9, and recentrifuged. The supernatants from the above cen-

[22] H. L. Crespi, U. Smith, L. Gajda, T. Tisue, and R. M. Ammeraal, *Biochim. Biophys. Acta* **256,** 611 (1972).

[23] H. F. Daboll, H. L. Crespi, and J. J. Katz, *Biotechnol. Bioeng.* **4,** 281 (1962).

trifugations are combined and brought to 50% saturation with solid $(NH_4)_2SO_4$, which is added slowly with stirring. After standing for 1 hr at 0°, the suspension is centrifuged for 2 hr at 15,000 $g$. The resulting supernatant is golden brown or only slightly green. If it is heavily colored, the supernatant is brought to 52.5% of saturation with $(NH_4)_2SO_4$ and centrifuged as above. The supernatant is then brought to 90% saturation with $(NH_4)_2SO_4$, stirred overnight at 4°, and centrifuged (2 hr, 15,000 $g$). The precipitate is suspended in a minimum volume of 10 m$M$ phosphate, pH 6.9, and dialyzed to remove residual $(NH_4)_2SO_4$.

The dialyzed solution is clarified by centrifugation (15,000 $g$) and applied to a DEAE–cellulose column (5 × 13 cm) (Cellex D, Bio-Rad) equilibrated in 10 m$M$ phosphate, pH 6.9. The column is washed with 0.3 $M$ NaCl–10 m$M$ phosphate, pH 6.9, until the $A_{280}$ is less than 0.1. Elution with 0.5 $M$ NaCl–10 m$M$ phosphate, pH 6.9, brings down a crude fraction containing both flavodoxin and ferredoxin. This fraction is diluted with 10–20 volumes of cold $H_2O$ and absorbed onto Whatman DE-52, poured into a column and eluted with 0.5 $M$ NaCl–10 m$M$ phosphate, pH 6.9. The eluted material is desalted on a Sephadex G-10 column equilibrated and eluted with 1 m$M$ phosphate, pH 6.9. The flavodoxin eluate is applied to a hydroxylapatite column and washed (under 2–4 lb pressure of $N_2$) with 10 m$M$ phosphate, pH 6.9, until the $A_{280}$ is less than 0.1. Ferredoxin is eluted with 40 m$M$ phosphate, pH 6.9, and the column is washed with the same buffer until the $A_{280}$ is less than 0.1. The pure flavodoxin fraction is then eluted with 0.1 $M$ phosphate, pH 6.9. This flavodoxin has a molecular weight of 17,000, and the purified protein shows an $A_{273}/A_{464}$ ratio of 6.67.[22] Yields of flavodoxin are in the range of 0.1–0.3 mg/gm of freeze-dried cells.

## VI. Purification of *Chlorella* Flavodoxin

The purification of flavodoxin from the green alga *Chlorella* has been published only schematically by Zumft and Spiller.[24] The procedure is quite similar to those given above for other flavodoxins. *Chlorella fusca* (strain 211–15) is grown as described by Kessler and Czygan[25] with an iron concentration in the growth medium of 0.03 mg/liter.

After cell breakage, the resulting suspension is centrifuged at high speed to remove particulate material. The supernatant is chromatographed on DEAE–cellulose (Whatman DE-52), and the eluted flavodoxin fraction treated with 2% protamine sulfate solution to remove

[24] W. G. Zumft and H. Spiller, *Biochem. Biophys. Res. Commun.* **45**, 112 (1971).
[25] E. Kessler and F. C. Czygan, *Arch. Mikrobiol.* **70**, 211 (1970).

nucleic acids and rechromatographed on DEAE–cellulose. Final purification is achieved by chromatography on Sephadex G-75. The purified flavodoxin (approximately a 200-fold purification of mediated $NADP^+$ reduction activity in the chloroplast assay) is electrophoretically homogeneous and shows an $A_{275}/A_{464}$ spectral ratio of 5.46.[24] The flavodoxin is of particular interest in that it has been reported to be composed of two subunits[26] rather than a single polypeptide chain as found in all other known flavodoxins.

### VII. Purification of *Klebsiella pneumoniae* Flavodoxin

The purification procedure for this flavodoxin was originally published by Yoch.[27] *Klebsiella pneumoniae* (strain M 5al) is grown as described by Yoch and Pengra.[28] As with *A. vinelandii*, flavodoxin is present in this organism when it is grown in the presence of iron salts.

*Klebsiella* cell paste (500 gm) is suspended in 2 volumes of 20 m*M* phosphate, pH 7.4. Cells are ruptured by sonication for 5 min with a 20 kHz Bronson sonifier at full power. The broken cell suspension is then brought to 30% (v/v) with cold (−20°) *n*-butanol and stirred for 30 min at 5°. After centrifugation for 30 min at 10,000 *g*, the supernatant is applied to a DEAE–cellulose column equilibrated in 20 m*M* phosphate, pH 7.4. After washing the column with several volumes of buffer, the crude flavodoxin is eluted with 0.5 *M* NaCl–20 m*M* phosphate, pH 7.4. The eluted flavodoxin fraction is desalted by dialysis overnight against 20 m*M* phosphate, pH 7.4. This dialyzed protein solution is applied to a DEAE–cellulose column (1.5 × 45 cm) equilibrated with 20 m*M* phosphate, pH 7.4. The column is washed with 150 ml of 0.1 *M* NaCl–20 m*M* phosphate, pH 7.4, followed by the same volume of 0.2 *M* NaCl–20 m*M* phosphate, pH 7.4. A flavoprotein with no known activity is eluted in the 0.2 *M* NaCl fraction. Flavodoxin is eluted with 0.32 *M* NaCl–20 m*M* phosphate followed by a blue protein and a brown protein. The flavodoxin fraction is concentrated and chromatographed on Sephadex G-50. After this step, gel electrophoresis shows the presence of contaminating proteins which, however, do not have any visible absorbance.

The yield of *Klebsiella* flavodoxin is quite low as compared with the yields from other organisms (see above). This flavodoxin has a molecular weight of 21,000 gm/mole and is active in coupling reducing equivalents from illuminated chloroplasts to either *Klebsiella* or *Azotobacter* nitrogenase.[27]

---

[26] H. Spiller, W. G. Zumft, and L. Gürtler, Paper Presented at Fourth International Symposium on Flavins and Flavoproteins, Konstanz, Germany, 1972.

[27] D. C. Yoch, *J. Gen. Microbiol.* **83**, 153 (1974).

[28] D. C. Yoch and R. M. Pengra, *J. Bacteriol.* **92**, 618 (1966).

Values for the extinction coefficients of several flavodoxins are given in the review by Mayhew and Ludwig.[29]

## VIII. General Chemical Properties of Flavodoxins

Flavodoxins consist of a single polypeptide chain (with the possible exception of *Chlorella* flavodoxin) to which is noncovalently attached one mole of FMN. No redox active component other than flavin has been found. Polypeptide chain lengths generally fall into two groups: 14,500–17,000 and 20,000–23,000. There does not seem to be any correlation between this size distinction and any of the biochemical properties of the flavodoxins.[30] The amino acid compositions display several general similarities.[31] Thus, an excess of acidic amino acids is always present (resulting in low isoelectric points); the content of histidine is usually low (nine proteins have either no histidine or only a single residue); and no disulfides are found (cysteine content varies from one to five residues).

The flavin cofactors are bound rather tightly to the apoproteins; association constants are the order of $10^8$ $M^{-1}$ or greater.[29,32,33] The FMN can be reversibly removed by exposure to low pH (<4) or by dialysis against concentrated KBr,[32,34] although certain of the flavodoxins appear to be resistant to one or the other of these procedures.[30,32] Structurally modified flavin analogs are able to bind to apoflavodoxins,[29,32,33] and investigation of this property has led to significant insights into the nature of the flavin–protein interaction. Some of the results obtained with *A. vinelandii* flavodoxin[32,33,35] are shown in Table I. Although the hydroxyl groups in the ribityl phosphate side chain are not essential to binding (deoxy-FMN binds almost as well as does FMN), substitution of these groups with acetyl functions leads to complete loss of binding ability. This is most likely due to steric effects and suggests that the side chain is buried within the protein. Substitution of a methyl group at the 3-position, or transfer of the 8-methyl group to the 6-position (as in iso-FMN), results in a significant reduction in binding strength, suggesting that the 3- and 6-positions are in contact with the protein. Removal of

[29] S. G. Mayhew and M. L. Ludwig, *in* "The Enzymes" (P. D. Boyer, ed.), 3rd ed., Vol. 12, p. 57. Academic Press, New York, 1975.

[30] J. A. D'Anna and G. Tollin, *Biochemistry* **11,** 4073 (1972).

[31] J. L. Fox, S. S. Smith, and J. R. Brown, *Z. Naturforsch.,* *Teil B* **27,** 1096 (1972).

[32] D. E. Edmondson and G. Tollin, *Biochemistry,* **10,** 124 (1971).

[33] B. G. Barman and G. Tollin, *Biochemistry* **11,** 4746 (1972).

[34] S. G. Mayhew, *Biochim. Biophys. Acta* **235,** 289 (1971).

[35] K. Shiga, G. Tollin, M. C. Falk, and D. B. McCormick, *Biochem. Biophys. Res. Commun.* **66,** 227 (1975).

TABLE I

KINETICS AND THERMODYNAMICS OF FLAVIN ANALOG BINDING TO APOFLAVODOXIN
FROM *A. vinelandii*[a]

| Analog | $K_a(M^{-1})$ | $k_{on}(M^{-1} sec^{-1})$ | $k_{off}(sec^{-1})$ |
|---|---|---|---|
| FMN | $2.2 \times 10^8$ | $2.0 \times 10^5$ | $9.1 \times 10^{-4}$ |
| Deoxy-FMN | $1.3 \times 10^8$ | — | — |
| 3-Methyl FMN | $3.7 \times 10^7$ | — | — |
| Iso-FMN | $4.8 \times 10^7$ | — | — |
| Riboflavin | $1.6 \times 10^6$ | $8.5 \times 10^5$ | 0.5 |
| FAD | $1.3 \times 10^6$ | — | — |
| Lumiflavin | $2.2 \times 10^5$ | $4.0 \times 10^7$ | $1.8 \times 10^2$ |
| 8α[S-(N-Acetyl)-L-Cys-L-Tyr]-FMN | $>10^5$ | — | — |
| Tetraacetylriboflavin | Not bound | — | — |

[a] The structure of FMN and its numbering system is as follows:

the ribityl phosphate group (as in riboflavin), or the first four carbon atoms of the ribityl side chain (as in lumiflavin), leads to a large reduction in association constant. The rate at which the cofactor binds to the protein is actually increased in these last two cases, and the off-rate is increased to a still greater extent. This suggests that the rate-determining step in the binding process involves the phosphate–protein interaction. A large, positive entropy change has been shown to result from this interaction,[33] and a study of the pH dependence of the binding kinetics[36] leads to the conclusion that the phosphate can bind in both the singly and doubly charged anionic forms. That FAD binds with about the same strength as does riboflavin indicates that the bulky adenylate moiety can be positioned away from the protein. Finally, the fact that an FMN analog with a bulky peptide moiety attached to the 8α-position is bound relatively strongly indicates that the edge of the benzenoid ring of the flavin is oriented away from the protein. All of these results are consistent with the X-ray crystal structures found for the flavodoxins from *D. vulgaris* and *Clostridium* MP (see below). This points to the possibility

[36] M. L. MacKnight, J. M. Gillard, and G. Tollin, *Biochemistry* **12,** 4200 (1973).

that all of the flavodoxins have structurally analogous coenzyme binding sites. Still further evidence in support of this concept derives from amino acid sequence studies[37] which indicate that the ribityl phosphate binding regions in six of the flavodoxins are similar in structure.

Another cofactor binding property which is of interest is that whereas some of the flavodoxins bind riboflavin rather strongly (e.g., *A. vinelandii* and *D. vulgaris*),[30] others (e.g., *Clostridium* MP and *P. elsdenii*) do not bind this analog at all.[34] It has been shown[30] that this selectivity correlates well with a variety of other properties of the flavodoxins, such as circular dichroism (CD) spectra and protein fluorescence maxima, suggesting some common structural features which serve to distinguish one group from the other. As will be described below, the X-ray results are consistent with this in that they show that the coenzyme binding sites of the *Clostridium* MP and *D. vulgaris* flavodoxins are significantly different.

The redox potentials of the bound FMN are considerably modified from those of the free coenzyme. In particular, the second one-electron potential ($FH\cdot + e^- \rightarrow FH^-$) is shifted from $-0.24$ V for free FMN to $-0.37$ to $-0.5$ V in the flavodoxins.[29] The low potential for the semiquinone to hydroquinone conversion is clearly related to the biological ability of the flavodoxins to participate in such strongly reductive electron transfer processes as nitrogen fixation and hydrogen formation. Another consequence of this low redox potential is that the semiquinone form of the flavodoxins is considerably more stable than for the free coenzyme. As will be discussed below, the X-ray results allow at least a partial explanation of these properties to be given in structural terms.

Complete amino acid sequence determinations have been reported for three flavodoxins, and partial sequences are available for several others.[31,37−41] The homology which exists in the vicinity of the ribityl phosphate binding site region has already been noted (see above). In addition, homology is found in several other parts of the proteins as well, in particular those chain segments which are, according to the X-ray results, in or near the coenzyme binding site. The evolutionary relatedness of the flavodoxins seems to be clearly established by these data. It has been

[37] M. L. MacKnight, W. R. Gray, and G. Tollin, *Biochem. Biophys. Res. Commun.* **59,** 630 (1974).

[38] M. Dubourdieu, J. LeGall, and J. L. Fox, *Biochem. Biophys. Res. Commun.* **52,** 1418 (1973).

[39] M. Tanaka, M. Haniu, K. T. Yasunobu, and S. G. Mayhew, *J. Biol. Chem.* **249,** 4393 (1974).

[40] M. Tanaka, M. Haniu, K. T. Yasunobu, S. G. Mayhew, and V. Massey, *J. Biol. Chem.* **249,** 4397 (1974).

[41] M. Tanaka, M. Haniu, K. T. Yasunobu, and D. C. Yoch, *Biochemistry* **16,** 3525 (1977).

suggested[41] that they have all arisen from a common ancestor and that the larger molecular weight flavodoxins have an additional peptide chain at the carboxyl-terminal end.

## IX. X-Ray Structure Analysis of Flavodoxins

X-ray structure determinations have been done with both the *D. vulgaris*[42] and *Clostridium* MP flavodoxins.[29,43] In both cases, the most detailed analyses have been performed on the oxidized forms, although for the latter protein, some structural information is available for both the semiquinone and hydroquinone forms as well. The overall polypeptide folding patterns of the two flavodoxins are highly similar. Both contain a high proportion of secondary structure with a central parallel β-sheet flanked on either side by pairs of helices. In both proteins, the isoalloxazine ring of the flavin is planar and is found at the periphery of the molecule, with the dimethylbenzene end accessible to the solvent and the pyrimidine portion enclosed by polypeptide chain segments. The ribityl side chain extends toward the interior of the protein and is largely buried.

The orientation of the protein around the ribityl phosphate group appears to be identical in the two flavodoxins. Four hydroxyamino acids (two serines and two threonines) and five backbone NH groups are close to the phosphate oxygens and are presumed to be involved in hydrogen bond interactions. No countercharged groups are present to balance out the negative charges on these oxygens. The ribityl side chain hydrogen bonding interactions are similar though not identical in the two proteins. In both cases, the isoalloxazine rings are sandwiched between hydrophobic residues, but these differ in the two flavodoxins. In the *Clostridium* MP protein, these are Met-56, toward the interior of the molecule, and Trp-90, which partially shields the ring from solvent. In *D. vulgaris* flavodoxin, Trp-60 occupies the inside position and Tyr-98 the outside position. In neither case are the indole and isoalloxazine rings coplanar, but in the *D. vulgaris* protein, the phenol ring and the flavin ring are parallel. All of the hydrogen bond interactions between the protein and the flavin ring appear to be different in the two structures. In both cases, two acidic groups are in the vicinity of the flavin. It is of considerable interest that, despite what appears to be major differences in the detailed

[42] K. D. Watenpaugh, L. C. Sieker, and L. H. Jensen, *Proc. Natl. Acad. Sci. U.S.A.* **70,** 3857 (1973).

[43] R. M. Burnett, G. D. Darling, D. S. Kendall, M. LeQuesne, S. G. Mayhew, W. W. Smith, and M. L. Ludwig, *J. Biol. Chem.* **249,** 4383 (1974).

flavin environments in the two flavodoxins, particularly in the side chain groups which adjoin the isoalloxazine ring, their redox properties are in general quite similar.

The major differences between the oxidized and semiquinone forms of the *Clostridium* MP protein seem to involve in the latter a small movement of the indole ring of Trp-90 and a reorientation of the peptide connecting residues 57 and 58, allowing the formation of a hydrogen bond between N-5 and the peptide oxygen 57. Inasmuch as the N-5 position is protonated upon going from the oxidized form to the neutral semiquinone, this latter interaction may be significant in terms of radical stabilization. Gillard and Tollin[44] have provided evidence from model studies that the indole–isoalloxazine interaction may also be important in this regard.

There appears to be little or no further structural changes in going from semiquinone to hydroquinone, which is consistent with solution properties.[33,45] It is important to note that the flavin ring conformation is also planar in the fully reduced protein. Inasmuch as a bent structure is energetically more favorable for a reduced flavin,[46] this constraint may play a large role in establishing the low potential of the hydroquinone form.[29]

## X. Spectroscopic Properties of Flavodoxins

Inasmuch as flavins absorb light in the visible region of the spectrum, optical probes have proved quite useful in the study of flavoproteins, both as a determinant of flavin redox state and of flavin environment. In the oxidized form, flavodoxins exhibit two broad bands in the visible and a third in the near UV ($\lambda_{max}$ = 272–275 nm). The latter overlaps with the absorption of the protein aromatic residues. The positions, shapes and intensities of the visible bands vary with the flavodoxin and are different from free FMN. Thus, the 445-nm band of free flavin is shifted to the red by 5–22 nm in many flavodoxins, blue-shifted to 443 nm in *C. pasteurianum* flavodoxin, and unchanged in position in *P. elsdenii* and *Clostridium* MP flavodoxins. The extinction coefficient is usually smaller (~15%) than for the free coenzyme and vibrational fine structure is

[44] J. M. Gillard and G. Tollin, *Biochem. Biophys. Res. Commun.* **58**, 328 (1974).

[45] M. Dubourdieu, M. L. MacKnight, and G. Tollin, *Biochem. Biophys. Res. Commun.* **60**, 649 (1974).

[46] P. Kierkegaard, R. Norrestam, P. E. Werner, I. Csöregh, M. von Glehn, R. Karlsson, M. Leijonmarck, O. Rönnquist, B. Stensland, O. Tillberg, and L. Torbjörnsson, *in* "Flavins and Flavoproteins" (H. Kamin, ed.), p. 1. Univ. Park Press, Baltimore, Maryland, 1971.

apparent. The position and shape of the 373-nm flavin band are also changed in the flavodoxins. These observations have led to the conclusion that the flavin is in a nonpolar milieu, perhaps in the vicinity of aromatic side chains.[47] This is confirmed by the X-ray work (see above).

Upon the addition of one reducing equivalent, the FMN adds one electron and one proton to form the semiquinone. This species is blue in color (as contrasted to the yellow color of the oxidized form) and has broad absorption bands in the visible between 400 and 700 nm. Addition of a second equivalent leads to a one-electron reduction to the hydroquinone, which is pale yellow in color and has weak bands at around 450 and 365 nm.

Flavins are intrinsically optically active (due to the asymmetric centers in the ribityl side chain) and also gain an induced optical activity as a consequence of their environment in the protein. Thus, CD spectroscopy has proved useful in the study of flavodoxins.[47] One of the major conclusions that has come out of these studies[30] is that although the flavin moieties in the flavodoxins all seem to be in generally similar environments, specific differences exist which group these proteins into two classes which correlate with the binding site environments found by X-ray analysis, as well as other physicochemical properties (see above).

Flavins are also highly fluorescent in the free state. However, when bound to the flavodoxin apoproteins, this fluorescence is almost completely quenched, probably as a consequence of the proximity of aromatic side chains. In this regard, protein tryptophan fluorescence is quenched as well. These properties have proved useful in the study of the kinetics and thermodynamics of flavin binding.[32-34]

Magnetic resonance spectroscopy (ESR, ENDOR, NMR) has also been applied to the study of flavodoxins.[48-55] The ESR and ENDOR results have indicated that the spin density distribution in flavodoxin semiquinone is similar to that found in free flavins and that the flavin

[47] D. E. Edmondson and G. Tollin, Biochemistry 10, 113 (1971).
[48] J. S. Hyde, L. E. G. Eriksson, and A. Ehrenberg, Biochim. Biophys. Acta 222, 688 (1970).
[49] G. Palmer, F. Müller, and V. Massey, in "Flavins and Flavoproteins" (H. Kamin, ed.), p. 123. Univ. Park Press, Baltimore, Maryland, 1971.
[50] C. C. MacDonald and W. D. Phillips, in "Fine Structure of Proteins and Nucleic Acids" (G. D. Fasman and S. N. Timasheff, eds.), p. 1. Dekker, New York, 1971.
[51] H. L. Crespi, J. R. Norris, J. P. Bays, and J. J. Katz, Ann. N.Y. Acad. Sci. 222, 800 (1973).
[52] J. Fritz, F. Müller, and S. G. Mayhew, Helv. Chim. Acta 56, 2250 (1973).
[53] H. L. Crespi, J. R. Norris, and J. J. Katz, Nature (London) 236, 178 (1972).
[54] T. L. James, M. L. Ludwig, and M. Cohn, Proc. Natl. Acad. Sci. U.S.A. 70, 3292 (1973).
[55] L. E. G. Eriksson and A. Ehrenberg, Biochim. Biophys. Acta 293, 57 (1973).

radical is probably in a planar conformation. The latter conclusion is verified by the X-ray structure work. The NMR studies have argued against any large conformational changes upon change in redox state, have demonstrated that aromatic protons are shifted upfield when flavin is bound to apoprotein and disappear upon semiquinone formation, and have shown that the linewidths of some of the flavin resonances are consistent with rigid attachment of the flavin to the protein. All of these conclusions are in agreement with the X-ray crystallography.

## XI. Redox Reactivity of Flavodoxins

Free flavins readily participate in the following equilibrium.

$$F + FH_2 \rightleftharpoons 2 FH \cdot \tag{1}$$

The rate constants for both the forward (comproportionation) and reverse (disproportionation) reactions are in the range of $10^7-10^9$ $M^{-1}$ sec$^{-1}$.[33,56] In contrast, the corresponding rates for the flavodoxins are much slower. For example, the comproportionation rate constant for *P. elsdenii* flavodoxin (at pH 8.5) is approximately $10^5$ $M^{-1}$ sec$^{-1}$ and that for disproportionation (at pH 8.3) is about 1 $M^{-1}$ sec$^{-1}$.[57] For *A. vinelandii* flavodoxin, the disproportionation rate is immeasurably slow below about pH 10.[58] The very slow rate of the latter reaction is in keeping with the large shift in the equilibrium of Eq. (1) in favor of the semiquinone form.

Several other redox reactions of flavodoxins have been studied. These include photoreduction, dithionite reduction, and several oxidation reactions of the semiquinone ($O_2$, ferricyanide, cytochrome $c$). Photoreduction (usually with EDTA as reductant) produces the semiquinone form of the flavodoxin, and is a convenient method for preparing this redox state. The rates vary from species to species, but are generally quite slow compared to the free flavin.[59] This is undoubtedly a consequence of the high level of quenching of flavin excited states upon binding to apoprotein.

Dithionite reduction is more complex. In general, both the semiquinone and hydroquinone forms are produced, but the ratio between these is pH dependent. With the *Azotobacter* protein, below pH 7 only semiquinone is formed.[58] As the pH is raised above 7, more and more hydroquinone is generated. This follows a one proton dependence, with a p$K$ of about 7, which is ascribed to the following equilibrium.

[56] S. P. Vaish and G. Tollin, *J. Bioenerg.* **2,** 33 (1971).
[57] S. G. Mayhew and V. Massey, *Biochim. Biophys. Acta* **315,** 181 (1973).
[58] D. E. Edmondson and G. Tollin, *Biochemistry* **10,** 133 (1971).
[59] K. Shiga and G. Tollin, *in* "Flavins and Flavoproteins" (T. P. Singer, ed.), p. 422. Elsevier, Amsterdam, 1976.

$$PFH_2 \rightleftharpoons PFH^- + H^+$$

This p$K$ value appears to vary with the flavodoxin. Thus, for the *D. vulgaris* protein, it has a value of 6.6.[45] For *P. elsdenii* flavodoxin, it appears to be around 5.[57] The most extensive study of the kinetics of dithionite reduction were carried out by Mayhew and Massey[57] at pH 8.3 with *P. elsdenii* flavodoxin. They interpreted the results according to the following scheme.

$$S_2O_4{}^{2-} \overset{K_1}{\rightleftharpoons} 2\ SO_2{}^-\cdot$$

$$H^+ + PF + SO_2{}^-\cdot \overset{k_1}{\rightarrow} PFH\cdot + SO_2$$

$$PFH\cdot + SO_2{}^-\cdot \overset{k_2}{\rightarrow} PFH^- + SO_2$$

The ratio $k_2/k_1$ is approximately 450. The faster rate of reaction of PFH· also seems to be the case for *A. vinelandii* flavodoxin.[58]

Studies of the oxidation of flavodoxin semiquinones by oxidants such as $O_2$, ferricyanide, and cytochrome $c$,[58-60] have provided some insights into the mechanism of electron transfer. The rates of oxidation by $O_2$ are quite slow and vary considerably among the flavodoxins (rate constants range from $8.3 \times 10^{-2}$ sec$^{-1}$ for *A. vinelandii* to 3.2 sec$^{-1}$ for *C. pasteurianum*). This has been attributed to the inaccessibility of the 5-position of the flavin ring. Another factor might be the different potentials for the semiquinone to oxidized conversion. Oxidations by ferricyanide and cytochrome $c$ proceed considerably faster, and the rate constants are less dependent on the flavodoxin, particularly for the cytochrome reaction. In this case, the oxidation probably occurs via the partially exposed benzenoid ring of the cofactor. Studies in which the 7- and 8-methyl groups of the flavin were replaced by chlorine atoms are in agreement with this concept.[59] It is not known whether the hydroquinone to semiquinone conversion (which is the physiologically important reaction) can also proceed by electron transfer via the benzene ring of the flavin.

Unfortunately, very little is known concerning the details of the redox chemistry of flavodoxin interaction with physiological electron donors and acceptors. This is largely due to the complexity of the biochemical systems involved (see above).

Acknowledgments

The following grants provided support for this work: National Institutes of Health Grants AM 15057 (to G. T.), NIH HL 16251 (to D.E.E.), and National Science Foundation Grant PCM 74-03264 (to D.E.E.).

[60] S. G. Mayhew, G. P. Foust, and V. Massey, *J. Biol. Chem.* **244**, 803 (1969).

# Section IV
## Methodology

# [38] Amperometric Hydrogen Electrode

## By RICHARD T. WANG

This paper describes the use and care of a commercially available Clark-type electrode for measurement of hydrogen in aqueous solution. The electrode is applicable to the determination of biological hydrogen production.[1-4]

At a platinum electrode maintained at proper potential, hydrogen is oxidized.

$$H_2 \rightarrow 2\ H^+ + 2e^-$$

Therefore $5.4 \times 10^{16}$ electrons (equivalent to 2.4 $\mu$A hr of electricity can be extracted from 1 $\mu$ liter of molecular hydrogen at STP. With currently available operational amplifiers employing field effect transistors (FET), current to output transfer ratio as small as one nA/V can easily be achieved. The advantage of electrochemical detection of hydrogen is obvious. Wang, Healey, and Myers[1] introduced the use of the amperometric hydrogen electrode, which is sensitive, fast and specific for hydrogen measurement. In order to achieve satisfactory performance, certain precautions, given herein, must be taken.

An apparatus, such as the one shown in Fig. 1, consisting of a OX 700 Clark-type electrode (equivalent to YSI 5331, platinum electrode about 0.5 mm) and OX 705 water-jacketed cuvette from Gilson Medical Electronics (Middleton, Wisconsin, 53562) may be used. This apparatus was intended for use as an oxygen electrode. The electrode (E) contains a small platinum wire at the center surrounded by a much larger reference silver electrode. For hydrogen measurement, the following reaction occurs at the silver electrode.

$$AgCl + e^- \rightarrow Ag + Cl^-$$

This reaction complements the oxidation of hydrogen at the platinum electrode. Both reactions occur in the thin film of half-saturated KCl

---

[1] R. Wang, F. P. Healey, and J. Myers, *Plant Physiol.* **48**, 108 (1971).

[2] L. W. Jones and N. I. Bishop, *Plant Physiol.* **57**, 659 (1976).

[3] A. C. McBride, S. Lien, R. K. Tagasaki, and A. San Pietro, *in* "Biological Solar Energy Conversion" (A. Mitsui, S. Miyachi, A. San Pietro, and S. Tamura, eds.), p. 77. Academic Press, New York, 1977.

[4] R. H. Burris, *in* "Biological Solar Energy Conversion" (A. Mitsui, S. Miyachi, A. San Pietro, and S. Tamura, eds.), p. 275. Academic Press, New York, 1977.

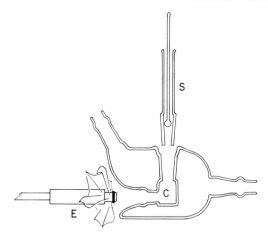

FIG. 1. The cuvette assembly. Two ports of the water-jacketed cuvette received the electrode, E, and the stopper, S. Liquid level in the cuvette was maintained slightly above the point of seal between glass ball and bottom of the ground glass taper.

solution covering both electrodes. It is necessasary to plate AgCl onto the silver electrode by applying a positive voltage to the silver while the tip of the electrode is immersed in dilute HCl. For hydrogen measurement, the platinum electrode is polarized +0.6 V relative to the reference Ag/AgCl electrode.

A simple voltage divider consisting of two properly chosen resistors can be used to obtain a fixed polarizing voltage from a mercury battery. A potentiometer should be used if a change in polarizing voltage is desired. The resistance in the divider chain should be chosen so that the current flowing through the divider chain is at least 100 times greater than the maximum anticipated electrode current. For example, if the maximum electrode current anticipated is 0.1 $\mu$A, then the divider chain current should be at least 10 $\mu$A. An equivalent statement is that for a 1.35-V mercury battery the total divider chain resistance should not exceed 135 k$\Omega$. The electrode current can be amplified by any of a number of instruments in which the voltage drop across the input is small (<1%) compared to the polarizing voltage. Output voltage of the amplifier can be recorded on a stripchart recorder. As will be noted below, the signal is proportional to the concentration of hydrogen in the cuvette. The slopes of recorded signals versus time are used to compute the rates of hydrogen evolution.

Rate of hydrogen consumption by the electrode and hence electrode current, is determined by several factors: most importantly; size of the platinum electrode, thickness and composition of the membrane, stirring, and hydrogen concentration. A larger surface of the platinum gives greater sensitivity but also greater consumption of hydrogen and sensitivity to stirring. As with the oxygen electrode, the major contribution to noise seems to arise from variation in movement of fluid across the membrane surface, i.e., "stirring" noise.

For constant stirring my experience recommends the following. Stirring bars are handmade by sealing into glass capillaries short (4 to 6 mm) lengths cut from a magnetized sewing needle. Constant stirring speeds are provided best by a driving magnet on the shaft of a 300 or 450 rpm synchronous motor.

Several kinds of membrane (polyethylene, polypropylene, Teflon) are available. We elected to use 19-$\mu$m polypropylene for its thickness and availability. The membrane chosen must be permeable to gases but not to ions. It serves to separate the KCl electrolyte on the electrode side from the aqueous solution on the cuvette side. The membrane also allows hydrogen molecules to diffuse through to the electrode surface where electrochemistry occurs. The thicker the membrane the longer the hydrogen molecules need to travel before reaching the platinum surface, resulting in lower sensitivity and slower response time.

Calibration of the electrode sensitivity is accomplished by injecting into the cuvette a small known volume of aqueous solution previously equilibrated with pure hydrogen gas. The dissolved hydrogen can be calculated from its solubility at the temperature and pressure of the experiment. The dilution factor, i.e., the injected volume divided by the total volume in the cuvette, is used to obtain the dissolved hydrogen concentration in the cuvette. A typical calibration curve is given in Fig. 2. The linearity of response is good. However, a small residual current is usually present even when there is no hydrogen in the system. This is probably due to some nonspecific electrochemistry independent of the hydrogen and can always be substracted to give the true hydrogen current.

Some technical problems were observed in experiments utilizing this apparatus for hydrogen measurement. The apparatus proved to be more leaky for hydrogen than for oxygen. The top port of the cuvette is normally sealed by a ground glass stopper with a capillary open to air. Except at oxygen concentrations far removed from that at equilibrium with air, diffusion leakage through the capillary was negligible. However, even at very low concentrations of hydrogen, leakage was appreciable as evidenced by a drift toward lower concentrations with time. To help correct this, the glass stopper was redesigned. A 6 mm inner diameter glass tube was sealed to a standard taper joint of 5-mm inner diameter at

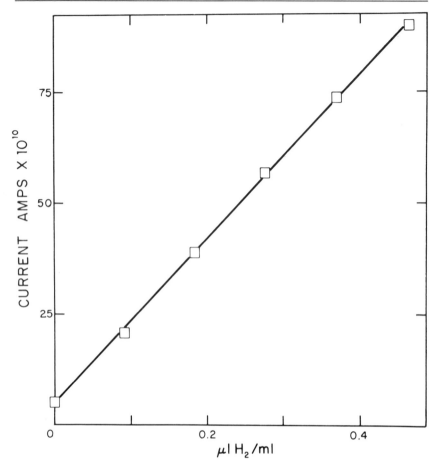

Fig. 2. Current signal versus hydrogen concentration. Polarizing voltage was +0.6 V. Water in the cuvette (1.8 ml) was initially aerated with nitrogen. Thereafter, 10-$\mu$l aliquots of hydrogen-saturated water were introduced by microsyringe. The sensitivity shown (1.9 $\times$ 10$^{-8}$ A/$\mu$l H$_2$/ml) and the current for zero hydrogen concentration (5 $\times$ 10$^{-10}$ A) were typical but not exactly reproducible day to day.

the lower opening. A spherical glass ball at the end of a 2-mm glass rod was hand-ground to seal against the inside bottom of the tapered joint. The liquid level in the cuvette was maintained slightly above the sealing point. Raising the ball to an upper part of the taper gave sufficient clearance for insertion of a 20-gauge needle for aeration or for injection of added microliter quantities of solutions. With this modification the only leakage path remaining was that through the membrane and electrode assembly.

A second technical difficulty arose in the stability of electrode sensitivity, which drifted with time after imposition of a chosen polarizing potential. Polarograms of current signal versus potential showed a marked hysteresis when traversed first toward higher and then toward lower potentials. For example, the current at 0.6 V after a period of 0.7 V was almost twice that observed after a period of 0.5 V. The following precedures suggested by Gilman.[5] We conditioned the electrode surface through a treatment consisting of about 10 min of timed 50/min alternations of polarizing potential between 0.2 and 0.8 V. This can be accomplished by utilizing a timer coupled to a single-pole double-throw microswitch which alternately connects the electrodes to two points in the divider chain to provide the desired voltages.

Practical experiences has shown that $H_2S$, CO, and HCN behave as electrode poisons and greatly reduce sensitivity. Photoflo, a wetting agent, that is routinely added to KCl solutions by the manufacturers can degrade electrode performance. It is, therefore, suggested that the KCl solution be prepared from purified salt. A poisoned electrode can generally be reconditioned by thorough cleaning. A polishing compound (No. 600 or finer) can be used to polish the platinum electrode. Dilute ammonium hydroxide can be used to remove the silver chloride from the silver electrode, followed by recoating.

Acknowledgments

I am grateful to Dr. L. W. Jones and Dr. S. Lien for sharing with me their recent experiences with the electrode.

[5] S. Gilman, in "Electroanalytical Chemistry" (A. J. Gard, ed.), Vol. 2, p. 111. Dekker, New York, 1967.

# [39] Acceptors and Donors for Chloroplast Electron Transport

By S. Izawa

The methodology of chloroplast electron-transport studies using exogenous electron acceptors and donors was last reviewed in 1972 by Trebst[1] in this series. As a sequel to it, this article is concerned mainly

[1] A. Trebst, Vol. 24, p. 146.

with the new developments that have been made in the field since around 1970.[1a]

One of the notable recent developments is the introduction of a new group of Hill oxidants which are capable of intercepting electrons from photosystem II (PS-II) at very high rates even in unfragmented chloroplasts.[2] Represented by oxidized $p$-phenylenediamines, these PS-II electron acceptors (or "class III acceptors") are all lipid-soluble oxidants with high redox potentials ($E_0' > + 0.1$ V). Together with the introduction of a number of new electron-transport inhibitors,[3,4] the introduction of class III acceptors has opened up a new approach toward the mechanisms of photosynthetic electron transport and associated phosphorylation. Furthermore, the recognition of lipophilicity as an essential character of PS-II electron acceptors has stimulated the interests of investigators in the membrane topology of the electron-transport system and thereby contributed much to our understanding of the structure–function of the thylakoid membrane. Recent review articles by Trebst[5] and by Hauska[6] include discussions of these and other new studies involving the use of artificial electron donors and acceptors.

This article deals only with experiments which utilize the envelope-free "class II" chloroplasts[7] (or "type D" chloroplasts according to Hall's terminology[8]), the standard material for electron-transport and photophosphorylation studies. These chloroplasts consist simply of sheets of thylakoids (lamellae) which usually retain all the components of the photosynthetic electron-transport chain except for ferredoxin which is leached out during chloroplast isolation. Figure 1 represents a simplified model of photosynthetic electron transport in isolated chloro-

---

[1a] Abbreviations: DAD, diaminodurene (2,3,5,6-tetramethyl-$p$-phenylenediamine); $DAD_{ox}$, oxidized form of DAD (duroquinonediimide); DBMIB, 2,5-dibromo-3-methyl-6-isopropyl-$p$-benzoquinone (dibromothymoquinone); DCIP, 2,6-dichlorophenolindophenol; $DCIPH_2$, reduced form of DCIP; DCMU, 3-(3,4-dichlorophenyl)-1,1-dimethylurea; PD, $p$-phenylenediamine; $PD_{ox}$, oxidized PD ($p$-benzoquinonediimide); PMS, phenazine methosulfate ($N$-methylphenazonium methosulfate); TMPD, $N,N,N',N'$×tetramethyl-$p$-phenylenediamine.

[2] S. Saha, R. Ouitrakul, S. Izawa, and N. E. Good, $J.$ $Biol.$ $Chem.$ **346**, 3204 (1971).

[3] S. Izawa, $in$ "Photosynthesis I" (A. Trebst and M. Avron, eds.), Encycl. Plant Physiol., New Ser., Vol. 5, p. 266. Springer-Verlag, Berlin and New York, 1977.

[4] A. Trebst, this volume, p. 675.

[5] A. Trebst, $Annu.$ $Rev.$ $Plant$ $Physiol.$ **25**, 423 (1974).

[6] G. Hauska, $in$ "Photosynthesis I" (A. Trebst and M. Avron, eds.), Encycl. Plant Physiol., New Ser., Vol. 5, p. 253. Springer-Verlag, Berlin and New York, 1977.

[7] D. Spencer and H. Unt, $Aust.$ $J.$ $Biol.$ $Sci.$ **18**, 197 (1965).

[8] D. O. Hall, $Nature$ ($London$), $New$ $Biol.$ **235**, 125 (1972).

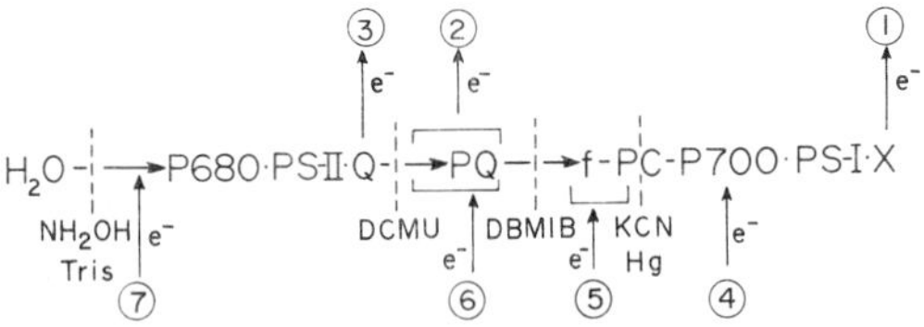

Fig. 1. A model of photosynthetic electron transport in isolated chloroplasts showing sites of electron donation and acceptance (arrows) and sites of inhibition (broken lines). The symbols and abbreviations used are: P680, the reaction center chlorophyll of photosystem II (PS-II); Q, the primary electron acceptor of PS-II; PQ, the plastoquinone pool; f, cytochrome f; PC, plastocyanin; P700, the reaction center chlorophyll of photosystem I (PS-I); X, the primary electron acceptor of PS-I; DCMU, 3-(3,4-dichlorophenyl)-1,1-dimethylurea; DBMIB, 2,5-dibromo-3-methyl-6-isopropyl-$p$-benzoquinone (dibromothymoquinone).

plasts (for details, see Trebst[5] and Golbeck[9]). The numbered arrows indicate regions where electron donation and acceptance by exogenous redox agents are believed to take place. (In the following sections, electron donation and acceptance will be discussed in reference to these regions.) The broken lines indicate sites where electron transfer can be blocked by specific inhibitors.[4]

## Acceptors of Electrons from Photosystem I (Region 1)

Of all parts of the thylakoid membrane-bound photosynthetic electron-transport chain, the reducing end of PS-I (X in Fig. 1) seems to be the part which is most easily accessible to exogenous electron acceptors. Presumably, this region is exposed on the outer surface of the membrane.[2,5,6] Furthermore, the primary electron acceptor of PS-I (X), when reduced, is an extremely strong reductant ($E_0' \geq -0.5$ V).[10,11] One must, therefore, assume that any known Hill oxidant, regardless of its standard redox potential and of its solubility properties, should be able to accept electrons from PS-I. All but a few of the PS-I electron acceptors listed below are already covered by the previous review,[1] some of them under "PS-II acceptors." The regrouping here is based on information which

[9] J. H. Golbeck, S. Lien, and A. San Pietro, in "Photosynthesis I" (A. Trebst and M. Avron, eds.), Encycl. Plant Physiol. New Ser., Vol. 5, p. 94. Springer-Verlag, Berlin and New York, 1977.

[10] R. Malkin, in "Photosynthesis I" (A Trebst and M. Avron, eds.), Encycl. Plant. Physiol. New Ser., Vol. 5, p. 179. Springer-Verlag, Berlin and New York, 1977.

[11] B. Ke, Biochim. Biophys. Acta 301, 1 (1973).

has become available in the past several years. Some additional comments are also given to well-established PS-I electron acceptors.

## NADP/Ferredoxin

This reconstituted natural electron-acceptor system is completely PS-I specific, but it tends to be inefficient and requires a relatively large amount of ferredoxin (50–100 $\mu$g/ml) for rate saturation.[12] Whether or not the terminal NADP reduction step is rate-limiting may be examined by comparing the rate of NADP reduction or of associated $O_2$ production with the rate of methyl viologen reduction as observed by $O_2$ consumption (see next section). If it is rate-limiting, the electron flow does not respond properly to exogenous agents such as inhibitors, uncoupling agents, and ADP/phosphate, which is misleading.[12]

*Methods.* Besides the method given previously[1] the NADP Hill reaction can be measured as $O_2$ evolution or by directly observing the absorbance increase (340 nm) of a reaction mixture containing a relatively small amount of chloroplasts (e.g., 5–10 $\mu$g chlorophyll/ml). An example of a reaction mixture for $O_2$ assay under phosphorylating conditions is 0.1 $M$ sucrose, 40 m$M$ tricine–NaOH buffer (pH 8.0), 5 m$M$ MgCl$_2$, 1 m$M$ ADP, 10 m$M$ Na$_2$H$^{32}$PO$_4$, 0.2 m$M$ NADP, 100 $\mu$g ferredoxin/ml, and chloroplasts equivalent to 25 $\mu$g chlorophyll/ml.

## Autoxidizable Electron Carriers

$\gamma,\gamma'$-Dipyridyl salts such as methyl viologen ($E_0'$ −0.44 V) and low-potential quinones such as anthraquinone sulfonates ($E_0' \approx -0.2$ V) are representatives of this group of electron acceptors. They accept electrons exclusively from PS-I, and pass them on to ambient $O_2$ to reduce it to the level of superoxide radical, which dismutates to $H_2O_2$ and $O_2$. Thus, with catalase-free chloroplast preparations or in the presence of a catalase inhibitors (e.g., 1 m$M$ KCN or NaN$_3$), the transport of electrons from water to the end of PS-I can be measured as $O_2$ uptake (Mehler reaction). The rate of $O_2$ uptake corresponds exactly to the rate of concurrent (but masked) $O_2$ evolution.[13] For more details, see the previous review.[1] These low potential acceptors are also very convenient to use with artificial electron donors, but their use for this purpose requires caution (see section for PS-II electron donors).

---

[12] H. E. Davenport, *in* "Biological Structure and Function" (T. W. Goodwin and O. Lindberg, eds.), Vol. 2, p. 449. Academic Press, New York, 1961.

[13] In the absence of any of these artificial low potential acceptors, washed chloroplasts normally show a Mehler reaction activity up to 50 $\mu$equiv/hr·mg chlorophyll. See, for example, J. M. Gould and S. Izawa, *Eur. J. Biochem.* **37**, 185 (1973).

*Methods.* The reaction mixtures described for ferricyanide (next section) can be modified for methyl viologen experiments by simply replacing the ferricyanide with 0.1 m$M$ methyl viologen.

## Potassium Ferricyanide

Thermodynamically, ferricyanide ($E_0'$ +0.42 V) has a capacity to accept electrons directly from PS-II, and indeed it does so efficiently in subchloroplast particles.[14] These facts led many workers to the belief that ferricyanide reduction is always a pure PS-II reaction. However, recent inhibition experiments with plastocyanin inhibitors[15–17] and with DBMIB[18,19] strongly indicate that in unfragmented thylakoid membranes ferricyanide in fact acts more as a PS-I electron acceptor than as a PS-II acceptor. PS-II reduction may account for only ~5–10% of the total ferricyanide reduction in chloroplasts with high membrane integrity[15,19] and 20–40% in average chloroplast preparations.[17,18] The accessibility of the PS-II reduction sites (region 2 in Fig. 1) to highly polar oxidants such as ferricyanide seems to be very limited unless the thylakoid membrane is disrupted. This appears to be also true of the cytochrome/plastocyanin region.[20]

*Methods.* The ferricyanide Hill reaction may be measured as $O_2$ evolution or by following the absorbance decrease of the reaction mixture at 420 nm (absorption peak of ferricyanide; millimolar extinction, 1.00 m$M^{-1}$ cm$^{-1}$) but the optical assay tends to be inaccurate when large changes in the light-scattering properties of chloroplasts occur during electron transport. An example of a reaction mixture useful for both $O_2$ assay and optical assay under phosphorylating conditions is 0.1 $M$ sucrose, 40 m$M$ tricine–NaOH buffer (pH 8.3), 3 m$M$ MgCl$_2$, 10 m$M$ Na$_2$H$^{32}$PO$_4$, 1 m$M$ ADP, 0.4 m$M$ potassium ferricyanide, and chloroplasts equivalent to 25 $\mu$g chlorophyll/ml. The maximum rate of electron transport can be measured at pH 7.0 using phosphate or MOPS buffer[21] and methylamine hydrochloride as an uncoupling agent.

[14] J. M. Anderson and N. K. Boardman, *Biochim. Biophys. Acta* **112**, 403 (1966).

[15] R. Ouitrakul and S. Izawa, *Biochim. Biophys. Acta* **305**, 105 (1973).

[16] M. Kimimura and S. Katoh, *Biochim. Biophys. Acta* **325**, 167 (1973).

[17] W. G. Nolan and D. G. Bishop, *Arch. Biochem. Biophys.* **166**, 323 (1975).

[18] H. Böhme, S. Reimer, and A. Trebst, *Z. Naturforsch., Teil B* **26**, 341 (1971).

[19] S. Izawa, J. M. Gould, D. R. Ort, P. Felker, and N. E. Good, *Biochim. Biophys. Acta* **305**, 119 (1973).

[20] P. Horton and W. A. Cramer, *Biochim. Biophys. Acta* **368**, 348 (1974).

[21] N. E. Good and S. Izawa, Vol. 24, Part B, p. 53.

## 2,6-Dichlorophenolindophenol (DCIP) and Related Oxidants

DCIP ($E_o' = +0.22$ V, $pK_a'$ 5.7) is another oxidant which was often treated as a PS-II electron acceptor and which has recently been shown to behave more as a PS-I acceptor when used with unfragmented chloroplasts.[16,18,19] This situation was already predicted by the kinetic experiments of Kok et al.[22] The blue, ionized form of the dye, which predominates at neutral to basic pH's is hydrophilic and, like ferricyanide, seems to have only limited access to the PS-II reduction sites. However, in fragmented chloroplasts[14] or at acidic pH's where the red, lipid-soluble undissociated form predominates, the dye will certainly intercept electrons from PS-II more freely. It is also this lipid-soluble form of the dye that seems to be responsible for the well-known uncoupler action of DCIP.[2] In line with these observations, it has been reported that the highly hydrophilic sulfonated DCIP does not uncouple and behaves as a typical PS-I electron acceptor.[23] One of the quinonimide dyes synthesized and tested by Hill,[24] compound XVI or reduced dichroin, also showed properties of a nonuncoupling PS-I electron acceptor.

Methods. DCIP photoreduction can be easily assayed spectrophotometrically by following the absorbance decrease of the reaction mixture at or near 600 nm (absorption peak of DCIP at pH > 7; millimolar extinction, 21 m$M^{-1}$ cm$^{-1}$). An example of a reaction mixture is 0.1 $M$ sucrose, 30 m$M$ HEPES–NaOH buffer (pH 7.5), 30 $\mu M$ DCIP ($A_{600} \approx$ 0.6) and chloroplasts equivalent to 5 $\mu$g chlorophyll/ml. $O_2$ assay is possible but not very practical.

### Acceptors of Electrons from Photosystem II (Region 2)

The Hill reaction supported by lipid soluble oxidants, such as oxidized phenylenediamines and quinones, show properties which deviate from those of the standard Hill reaction with water-soluble electron acceptors such as ferrianide. The electron transport is distinctively more rapid, less responsive to addition of ADP and phosphate, and is less efficient in supporting phosphorylation. Extreme deviations are found with oxidized p-phenylenediamine and oxidized diaminodurene. They are photoreduced at rates which are even faster than the rate of fully uncoupled

[22] B. Kok, S. Malkin, O. Owens, and B. Forbush, Brookhaven Symp. Biol. 19, 446 (1966).
[23] G. Hauska, A. Trebst, and W. Draber, Biochim. Biophys. Acta 305, 632 (1973).
[24] R. Hill, Bioenergetics 4, 432 (1972).

ferricyanide reduction. The electron flow, which is almost completely independent of the presence or absence of ADP and phosphate, does support phosphorylation but with only half the efficiency of conventional noncyclic photophosphorylation (Fig. 2). To explain these findings, Saha et al.[2] proposed a model of chloroplast electron transport which postulated that the electron transport chain contained two energy coupling sites and an intermediate reduction site. The intermediate reduction site (X' in the scheme below), which was placed between the two sites of energy coupling (~), was assumed to be buried in the lipid membrane and therefore only accessible to lipid-soluble Hill oxidants. The terminal reduction site (X) was assumed to be exposed on the external surface of the membrane and therefore accessible to all Hill oxidants.

$$H_2O \rightarrow PS \cdot II \xrightarrow{\text{fast}} X'(\text{buried}) \xrightarrow{\text{slow}} PS \cdot I \rightarrow X (\text{exposed})$$

Lipophilic oxidants ↑      All oxidants ↗

This model acquired strong support when it was demonstrated that the photoreduction of lipid-soluble oxidants and associated phosphorylation are only partially inhibited by the plastocyanin inhibitor KCN[15] and by the plastoquinone antagonist DBMIB.[19,25] These inhibitors not only abolished strictly PS-I-requiring reactions, such as the methyl viologen Hill reaction; in well-coupled chloroplasts they nearly abolished ferricyanide photoreduction as well, indicating the relative inaccessibility of the intermediate reduction site (X') to hydrophilic oxidants. Evidently the lipophilic quinoid compounds do have access to, and accept electrons efficiently from the site(s). However, the fact that their photoreduction is partially inhibited by KCN and DBMIB (20 to 50% inhibition depending on the oxidant) strongly suggests that substantial portions of the reduction take place at PS-I when the pathway of electrons from the intermediate reduction site(s) to PSI-I is open. (This two-site reduction model has been discussed in some detail.[15])

Oxidized p-phenylenediamines and substituted quinones are now in frequent use for investigations of various PS-II-associated phenomena. In such experiments DBMIB-poisoned or KCN-treated chloroplasts are routinely employed to ensure that PS-I is not involved in the reaction under study (see Yocum, this volume.[26])

[25] A. Trebst and S. Reimer, Biochim. Biophys. Acta 305, 129 (1973).
[26] C. F. Yocum, this volume, Article [54].

## Oxidized p-Phenylenediamines and Related Oxidants

At neutral to basic pH's, $p$-phenylenediamine (PD) ($E_0' = +0.36$ V) and 2,3,5,6-tetramethyl-$p$-phenylenediamine (diaminodurene or DAD) ($E_0' = +0.22$ V) can be oxidized quantitatively or nearly quantitatively by ferricyanide to $p$-benzoquinonediimide (PD$_{ox}$) and duroquinonediimide (DAD$_{ox}$), respectively.

PD$_{ox}$ and DAD$_{ox}$ are both excellent PS-II electron acceptors.[2] PD$_{ox}$ can support vast rates of electron transport (up to 2000 $\mu$Eq/hr mg chlorophyll) which, according to inhibition data,[15] is mostly (70–80%) due to PS-II reduction. Unfortunately, PD$_{ox}$ is chemically highly unstable. The relatively stable DAD$_{ox}$ is, therefore, the preferred oxidant for routine work even though DAD$_{ox}$-supported PS-II electron transport and phosphorylation are appreciably slower than those supported by PD$_{ox}$. Interestingly, photoreduction of PD$_{ox}$ and DAD$_{ox}$ is markedly inhibited, rather than stimulated, by uncoupling agents.[27–30] (An indication of this uncoupler effect is seen in Fig. 2.) The phenomenon has been interpreted to mean that energy-linked proton translocation is contributing to the high rates of photoreduction in coupled chloroplasts, either by facilitating the shuttling of the amine/imine forms of the acceptor across the thylakoid membrane,[28] or by inducing the internal accumulation of the imine form.[30] Other useful PS-II electron acceptors of this category include oxidized forms of 2,5-diaminotoluene, 4,4'-diaminodiphenylamine, TMPD,[2] $N,N$-dimethyl-$p$-phenylenediamine, $N,N$-diethyl-$p$-toluidine, 2-methyl-5-methoxy-$p$-phenylenediamine, etc.[31] Not surprisingly, many of these and other amines which act in their oxidized forms as PS-II electron acceptors are known to donate electrons to PS-I.[31]

*Methods.* Because of their chemical instability, oxidized $p$-phenylenediamines (PD$_{ox}$, DAD$_{ox}$, etc.) are prepared in a buffered reaction me-

[27] J. M. Gould and D. R. Ort, *Biochim. Biophys. Acta* **325**, 157 (1973).
[28] A. Trebst and S. Reimer, *Biochim. Biophys. Acta* **325**, 546 (1973).
[29] W. S. Cohen, D. E. Cohen, and W. Bertsch, *FEBS Lett.* **49**, 350 (1975).
[30] J. M. Guikema and C. F. Yocum, *Biochemistry* **15**, 362 (1976).
[31] A. Trebst and S. Reimer, *Z. Naturforsch., Teil C* **28**, 710 (1973).

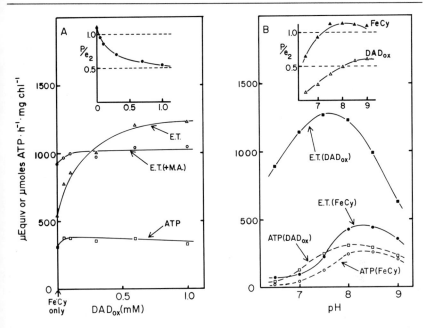

FIG. 2. The Hill reaction and associated phosphorylation with oxidized diaminodurene (DAD$_{ox}$) as the electron acceptor. The reaction mixture for (A) contained 40 m$M$ Tricine–NaOH buffer (pH 8.2), 1 m$M$ ADP, 5 m$M$ K$_2$H$^{32}$PO$_4$, 3 m$M$ MgCl$_2$, 5 m$M$ methylamine–HCl (MA) (if added), 0.4 m$M$ potassium ferricyanide plus varied concentrations of DAD$_{ox}$, and chloroplasts equivalent to 20 $\mu$g chlorophyll/ml. At DAD$_{ox}$ = 0, only ferricyanide was present as the electron acceptor. In the pH profile experiments of (B), the data for the regular ferricyanide Hill reaction are given for comparison. The buffers used are 2-($N$-morpholino) ethanesulfonic acid (MES)–NaOH for pH 6.5, $N$-2-hydroxyethylpiperazine-$N'$-ethanesulfonic acid (HEPES)–NaOH for pH 7 and 7.5, Tricine–NaOH for pH 8, and $N$-tris(hydroxymethyl)-3-aminopropanesulfonic acid (TAPS)–NaOH buffer for pH 8.5 and 9. All buffers were at 40 m$M$. The ferricyanide concentration was 0.4 mM. When present, DAD$_{ox}$ (see text) was 0.6 m$M$. Other conditions were the same as in (A). Electron transport was assayed as ferricyanide reduction observing the absorbance changes of the reaction mixture at 420 nm. The intensity of actinic light, approximately 500 kergs sec$^{-1}$ cm$^{-2}$ (>600 nm). (Adapted from S. Saha, R. Ouitrakul, S. Izawa, and N. E. Good, *J. Biol. Chem.* **246**, 3204 (1971).)

dium, immediately before the reaction, by mixing dihydrochloride salts of the amines (typical final concentration, 0.4 m$M$) and excess ferricyanide (1.2 m$M$). After mixing, the medium should remain colorless except for the pale yellow due to the part of ferricyanide which has remained unreduced. (Impure $p$-phenylenediamines tend to form deep-brown oxidation by-products.) For details of the experimental procedure and purification of $p$-phenylenediamines, see Yocum, this volume.[26]

*Substituted Benzoquinones*

The ability of $p$-benzoquinone ($E_0' = +0.29$ V) to penetrate biological membranes was already recognized by Warburg[32] who introduced this Hill oxidant more than 30 years ago. It is still used quite often, especially for the purpose of investigating the photosynthetic electron-transport activity of plant and algal cells. However, in terms of the ability to intercept electrons from PS-II in isolated chloroplasts, 2,5-dimethyl-$p$-benzoquinone ($E_0' = +0.18$ V) is far superior to $p$-benzoquinone.[2] Furthermore, the dimethylquinone is chemically much more stable than the unsubstituted quinone and can be used freely at pH 8 (standard pH for phosphorylation experiments) where the latter will denature in seconds. As a PS-II electron acceptor, 2,5-dimethylquinone is also much more convenient to use than oxidized $p$-phenylenediamines (PD$_{ox}$, DAD$_{ox}$, etc.) although the latter oxidants do support faster rates of PS-II electron transport and phosphorylation. 2,5-Dichloro-$p$-benzoquinone is also a good PS-II acceptor.[2] Trebst's group, who routinely use 2,6-dimethyl-$p$-benzoquinone ($E_0' = +0.18$ V)[25] rather than the 2,5-dimethyl analog, have recently introduced an autooxidizable PS-II electron acceptor, dimethylmethylenedioxy-$p$-benzoquinone,[33] which allows one to assay its reduction by PS-II as $O_2$ consumption. No doubt many more interesting PS-II electron acceptors will be found in the future, for instance among those various quinones and their derivatives which Trebst and his associates used in their early work on chloroplast photophosphorylation.[34]

*Methods.* This author uses 2,5-dimethylquinone routinely for PS-II electron transport and phosphorylation studies. The quinone can be used alone (rate-saturating concentration, 0.5 m$M$) or in combination with ferricyanide (0.5 m$M$) to observe the reaction as $O_2$ evolution or as ferricyanide reduction (see section for ferricyanide under PS-I electron acceptors).

*DBMIB as PS-II Electron Acceptor*

The plastoquinone analog DBMIB ($E_0' \approx +0.17$ V), now widely used as an electron-transport inhibitor,[4] acts as a PS-II electron acceptor when

---

[32] O. Warburg and W. Lüttgens, *Naturwissenschaften* **38**, 301 (1944).

[33] A. Trebst, S. Reimer, and F. Dallcker, *Plant Sci. Lett.* **6**, 21 (1976).

[34] A. Trebst and H. Eck, *Z. Naturforsch., Teil B* **16**, 44 (1961); see also A. Trebst, H. Eck, and S. Wagner, *in* "Photosynthetic Mechanisms of Green Plants" (B. Kok and A. T. Jagendorf, eds.), p. 174. Natl. Acad. Sci.—Natl. Res. Counc., Washington, D.C., 1963.

used at relatively high concentrations (optimum, 10–20 $\mu M$).[19,35] The advantage of using DBMIB as the oxidant is that the reaction observed is a pure PS-II reaction because electron transfer between PS-II and PS-I is completely blocked by the DBMIB itself. The disadvantage is that the reaction tends to be slow (<200 $\mu$Eq/hr mg chlorophyll) presumably because of the side effects which high concentrations of DBMIB exert on PS-II. The reaction supports phosphorylation as well as proton translocation.[35]

*Methods.* DBMIB reduction can be measured as $O_2$ evolution (pH < 8), $O_2$ consumption (pH > 8.2) or spectrophotometrically using ferricyanide as the terminal acceptor.

## Ferricyanide and DCIP

As already discussed, these ionic oxidants accept electrons mostly from PS-I in undisturbed thylakoid membrane but they can be reduced in large part by PS-II when the membrane is disturbed (see discussion of PS-I electron acceptors).

## Electron Acceptors for the DCMU-Insensitive Hill Reaction (Region 3)

Photoreduction of heteropoly compounds such as silicotungstate, phosphotungstate, phosphomolybdate, and silicomolybdate, has been shown to be totally or partially insensitive to DCMU.[36–38] Fluorescence experiments suggest that these oxidants may be capable of accepting electrons from the primary electron acceptor of PS-II (Q in Fig. 1).[39,40] Since silicomolybdate-washed chloroplasts can also photoreduce such common oxidants as DCIP and ferricyanide, it seems that silicomolybdate perturbs the thylakoid membrane in some way and that this membrane modification renders the primary acceptor accessible to exogenous oxidants including silicomolybdate itself.[41] In line with this notion, it has recently been shown that mildly trypsin-treated chloroplasts can photo-

[35] J. M. Gould and S. Izawa, *Eur. J. Biochem.* **37**, 185 (1973); see also J. M. Gould and S. Izawa, *Biochim. Biophys. Acta* **333**, 509 (1974).

[36] G. Girault and J. M. Galmiche, *Biochim. Biophys. Acta* **333**, 314 (1974).

[37] R. T. Giaquinta, R. A. Dilley, F. L. Crane, and R. Barr, *Biochem. Biophys. Res. Commun.* **59**, 985 (1974).

[38] R. Barr, F. L. Crane, and R. T. Giaquinta, *Plant Physiol.* **55**, 460 (1975).

[39] R. T. Giaquinta and R. A. Dilley, *Biochim. Biophys. Acta* **387**, 288 (1975).

[40] B. Zillinskas and Govindjee, *Biochim. Biophys. Acta* **387**, 306 (1975).

[41] G. Ben-Hayyim and J. Neumann, *FEBS Lett.* **56**, 240 (1975).

reduce ferricyanide in a DCMU-insensitive reaction, although in this case rather high concentrations of ferricyanide ($>10$ m$M$) is required.[42] The section below describes the chemical properties and the reaction characteristics of silicomolybdate, the most frequently used oxidant of this category.

### Silicomolybdic Acid (12-Molybdosilicic acid)

In 1 $N$ HCl, silicomolybdic acid ($H_4Mo_{12}O_{40}$) accepts up to 8 electrons in a four-step reaction ($E_0' = +0.56, +0.43, +0.19, +0.02$ V).[43] The redox properties of this substance in neutral pH regions (where the complex is gradually hydrolyzed) are not clear, except for the fact that the fully oxidized form (usual form) is still a strong enough oxidant to be visibly reduced by comparable concentrations of ferrocyanide ($E_0' = +0.42$ V). This strong poly acid forms insoluble complexes with various organic bases, such as amines, quarternary ammonium salts, amides, etc. It also tends to form insoluble complexes and salts with various metals (including $K^+$) and their chelates.[43] It precipitates such commonly used cationic oxidants as methylviologen and $N$-methylphenazonium ion ("PMS")[44] and seems to aggregate gramicidin. Clearly the use of silicomolybdate for electron transport and phosphorylation experiments calls for special caution. Silicomolybdate is also a multiple-character Hill oxidant. Its behavior apparently depends on the extent of the damage it causes to the chloroplast membrane. Thus in the presence of high concentrations of bovine serum albumin (a membrane protective agent) silicomolybdate behaves exactly like ferricyanide, supporting well-coupled, completely DCMU-sensitive electron transport. Only when the protection is weakened, does it begin to extract electrons from PS-II in a DCMU-insensitive reaction which may[44–47] or may not[37,39] support phosphorylation depending on the extent of protection and on other as yet unclarified conditions. In the total absence of protective agents, 0.1 m$M$ silicomolybdate strongly uncouples and quickly abolishes all electron transport activities.[44]

*Methods.* The following are modifications of the reaction mixture and procedure which have been used in this author's laboratory to demon-

---

[42] G. Renger, *FEBS Lett.* **69**, 225 (1976).

[43] G. A. Tsigdinos and C. J. Hallada, *J. Less-Commun. Met.* **36**, 79 (1974); see also G. A. Tsigdinos, *Bull.* Cdb-12a. Climax Molybdenum Co., Greenwich, Connecticut, 1969.

[44] S. P. Berg and S. Izawa, *Biochim. Biophys. Acta* **460**, 206 (1977).

[45] S. Izawa and S. P. Berg, *Biochem. Biophys. Res. Commun.* **72**, 1512 (1976).

[46] L. Rosa and D. O. Hall, *Biochim. Biophys. Acta* **449**, 23 (1976).

[47] G. Ben-Hayyim and J. Neumann, *Eur. J. Biochem.* **72**, 57 (1977).

strate the phosphorylation associated with DCMU-insensitive silicomolybdate reduction. Reaction mixture (2 ml): 0.1 $M$ sucrose, 40 m$M$ HEPES–NaOH buffer,[21] 2 m$M$ MgCl$_2$, 5% (v/v) glycerol, 5 m$M$ Na$_2$H$^{32}$PO$_4$, 0.75 m$M$ ADP, and tobacco chloroplasts containing 50 $\mu$g chlorophyll/ml. The reaction is initiated by simultaneously adding 10 $\mu$l of silicomolybdic acid solution (50 mg/ml in dimethyl sulfoxide–water, 1:1, v/v) and turning on the light. The reaction is measured as O$_2$ evolution. Barr et al.[38] have assayed silicomolybdate reduction following the absorbance increase at 750 nm ("molybdate blue" formation).

### Donors of Electrons to Photosystem I (Regions 4–6)

Electron donors which support DCMU-insensitive, PS-I-mediated reactions are covered extensively in the previous review.[1] These PS-I electron donors can now be subdivided into at least three groups based on the sensitivity of the reaction (electron transport or phosphorylation) to KCN or other plastocyanin inhibitors and to the plastoquinone antagonist DBMIB.

1. Reactions insensitive to KCN: the probable site of electron donation at P700 (region 4 of Fig. 1).
2. Reactions sensitive to KCN but insensitive to DBMIB: the probable site of electron donation in the cytochrome $f$/plastocyanin region (region 5).
3. Reactions sensitive to DBMIB: the probable site of electron donation in the plastoquinone region (region 6).

Reaction systems for PS-I-mediated noncyclic electron transport require three basic components besides the electron donor to be tested: (a) a PS-II blocking agent, (b) an electron reservoir for the donor, and (c) a low-potential electron acceptor. DCMU (1–5 $\mu M$) is usually employed as the PS-II blocking agent, although any one of well-established PS-II inhibitors[4,47a] may be used. As for the electron reservoir, ascorbate (D- or L-; 0.5–2 m$M$) is practically the only choice. Ascorbate is satisfactory both in terms of its high enough reducing potential ($E_0' = +0.06$ V) to keep most of PS-I electron donors nearly completely in their reduced forms (which prevents or minimizes electron cycling) and in terms of its poor ability to serve itself as a direct electron donor to the transport chain. Because of its susceptibility to superoxide oxidation, ascorbate does pose a problem when it is used with autoxidizable electron acceptors (e.g., viologens and anthraquinones) to assay electron flow as O$_2$ uptake.

[47a] S. Izawa and N. E. Good, Vol. 23, p. 335.

This problem can be circumvented by adding excess superoxide dismutase to the reaction mixture (see discussion of PS-II electron donors) or by using NADP/ferredoxin as the electron acceptor couple. However, the NADP/ferredoxin couple tends to permit a cycling of electrons around PS-I when artificial electron donors are present. Since this hidden cyclic electron flow is usually coupled to phosphorylation, one must exert caution in interpretating phosphorylation data from donor reactions involving NADP/ferredoxin.

### Donors of Electrons to P700 (Region 4)

Of known electron donors to PS-I, the reduced forms of PMS and of DCIP (DCIPH$_2$) seem to have the easiest access to P700. This is suggested by (a) the relative insensitivity of PMS-mediated cyclic phosphorylation and DCIPH$_2$-supported electron transport to KCN[15] and (b) electron paramagnetic resonance (EPR) experiments which showed a rapid dark reduction of photooxidized P700 by reduced PMS and DCIPH$_2$ in KCN-blocked chloroplasts.[48] However, PMS-mediated cyclic photophosphorylation is largely sensitive to KCN when the PMS concentration is low ($<50$ $\mu M$). Furthermore, although DCIPH$_2$-supported PS-I noncyclic electron flow is only partially (50%) inhibited, concurrent phosphorylation is completely blocked by KCN.[15] These results suggest that the primary site of electron donation by reduced PMS and DCIPH$_2$ is not P700 but is probably in the cytochrome $f$/plastocyanin region (see next section). Presumably, however, given at very high concentrations any PS-I electron donor will gain sufficient access to P700 to support measurable electron flow, as indicated by the fact that even ferrocyanide does so when given at 0.2 $M$.[49] The inaccessibility of P700 to low concentrations of ferricyanide has been noted by Kok.[50] Clearly, however, the accessibility of P700 to exogenous redox agents is a function of the membrane integrity.[51]

*Methods.* A typical reaction mixture designed for assay of KCN-resistant DCIPH$_2$ oxidation by PS-I (O$_2$ uptake) is 0.1 $M$ sucrose, 40 m$M$ HEPES-NaOH buffer (pH 7.5), 0.2 m$M$ DCIP, 1 m$M$ ascorbate, 2 $\mu M$ DCMU, 0.1 m$M$ methyl viologen, and KCN-treated chloroplasts[15,26] equivalent to 50 $\mu$g chlorophyll/ml.

[48] S. Izawa, R. Kraayenhof, E. K. Ruuge, and D. DeVault, *Biochim. Biophys. Acta* **314**, 328 (1973).

[49] D. Rosen, R. Barr, and F. L. Crane, *Biochim. Biophys. Acta* **408**, 35 (1975).

[50] B. Kok, *Biochim. Biophys. Acta* **48**, 527 (1961).

[51] J. M. Gould and S. Izawa, *Biochim. Biophys. Acta* **314**, 211 (1973).

Donors of Electrons to Cytochrome $f$/Plastocyanin Region (Region 5)

The best known of PS-I electron donors and mediators of PS-I-dependent cyclic photophosphorylation all belong to this category: DAD ($E_0' = +0.22$ V), TMPD ($E_0' = +0.22$ V), DCIPH$_2$ ($E_0' = +0.22$ V), PMS ($E_0' = +0.08$ V), and pyocyanine ($E_0' = -0.04$ V). Of these, PMS and pyocyanine are only useful as mediators of cyclic photophosphorylation. Reactions supported by these substances are all insensitive to DBMIB but are largely or almost completely inhibited by KCN, indicating a site (or sites) of electron donation in the cytochrome $f$/plastocyanin region. There is spectroscopic evidence which suggests that the main site of electron donation by DCIPH$_2$ may be cytochrome $f$ or some component before the cytochrome.[52,53] Besides DAD and TMPD, various other C- or N-substituted phenylenediamines,[54] 3,3'-diaminobenzidine (a histochemical reductant),[55] and indamines (4,4'-diaminodiphenylamines)[56] have been shown to be useful as electron donors of this category. Among these amines, completely N-substituted compounds such as TMPD, $N$-phenyl-$N',N',N'$-trimethyl-$p$-phenylenediamine, and pentamethylindamine, are unique in two ways: (a) PS-I noncyclic electron transport supported by them is not coupled to phosphorylation, and (b) their oxidation, which is a monovalent oxidation to relatively stable free radicals, does not release protons. This structure–function relation has been in-

(TMPD)                    (Oxidized TMPD or "Wurster s blue")

terpreted as indicating the involvement of a chemiosmotic energy coupling mechanism in PS-I-mediated photophosphorylation. The highly hydrophilic sulfonated PMS, sulfonated pyocyanine,[57] and sulfonated DCIP(H$_2$)[23] have been shown to be ineffective as mediators of PS-I-

[52] S. Izawa, in "Comparative Biochemistry and Biophysics of Photosynthesis" (K. Shibata, A. Takamiya, A. T. Jagendorf, and R. C. Fuller, eds.), p. 140. Univ. Park Press, State College, Pennsylvania, 1968.

[53] A. W. D. Larkum and W. D. Bonner, Biochim. Biophys. Acta 267, 149 (1973).

[54] G. Hauska, W. Oettmeier, S. Reimer, and A. Trebst, Z. Naturforsch., Teil C 30, 37 (1975).

[55] J. Goffer and J. Neumann, FEBS Lett. 36, 62 (1973).

[56] W. Oettmeier, S. Reimer, and A. Trebst, Plant Sci. Lett. 2, 267 (1974).

[57] G. Hauska, FEBS Lett. 28, 217 (1972).

supported reactions. This suggest that the component or components of the electron transport chain which accept electrons from PS-I donors are located near or on the inner surface of the lipid membrane.[23,57]

*Methods.* An example of reaction mixtures for measurement (as $O_2$ consumption) of very fast rates of electron transfer from DAD to methyl viologen and associated phosphorylation is given in the legend for Fig. 3. The reaction mixture contained excess superoxide dismutase to suppress superoxide oxidation of ascorbate (see discussion of PS-II electron donors).

### Donors of Electrons to Plastoquinone (Region 6)

Tetramethyl-*p*-hydroquinone (durohydroquinone; $E_0' = +0.06$ V) has recently been shown to donate electrons preferentially to the plastoquinone region of the electron transport chain in the presence of a PS-II inhibitor, DCMU. Although this reductant is susceptible to air oxidation, below pH 8 the oxidation rate is sufficiently slow to allow measurement of electron flow as $O_2$ uptake using methyl viologen as the electron acceptor.[58,58a] No secondary donor (electron reservoir) is needed in short term measurements (3–4 min). In fact, no reducing agent presumably exists which may be useful as an electron reservoir for this system. The durohydroquinone-supported electron flow is quite fast and well coupled ($P/e_2$ 0.6). It is highly sensitive to DBMIB, indicating that electron donation takes place at or very close to the plastoquinone pool. Unfortunately durohydroquinone is sensitive to the superoxide radical and this necessitates the use of exogenous superoxide dismutase when the electron flow rate is to be determined critically[58] (see also section below). Other electron donors of this category include reduced forms of: 9,10-phenanthrenequinone, 2-hydroxy-1,4-naphthoquinone, 2-methyl-1,4-naphthoquinone (menadione), 1,4-naphthoquinone and, interestingly, ferredoxin.[58b] To date, however, the ability of these latter substances to donate electrons to plastoquinone has only been shown in terms of DBMIB-sensitive, anaerobic cyclic photophosphorylation.

*Methods.* Durohydroquinone solution can be prepared by adding approximately 2 mg of $NaBH_4$ to 1 ml of ice-chilled alcoholic solution of duroquinone (20 m$M$) and subsequently acidifying the solution with 10 $\mu$l of 5 $N$ HCl.[58] The rate-saturating concentration of durohydroquinone is about 1 m$M$, but in routine work 0.5 m$M$ will suffice. A typical reaction mixture for phosphorylation experiments with durohydroquinone: 0.1 $M$

[58] S. Izawa and R. L. Pan, *Biochem. Biophys. Res. Commun.* **83**, 1171 (1978).
[58a] C. C. White, R. K. Chain and R. Malkin, *Biochim. Biophys. Acta* **502**, 127 (1978).
[58b] G. Hauska, S. Reimer, and A. Trebst, *Biochim. Biophys. Acta* **357**, 1 (1974).

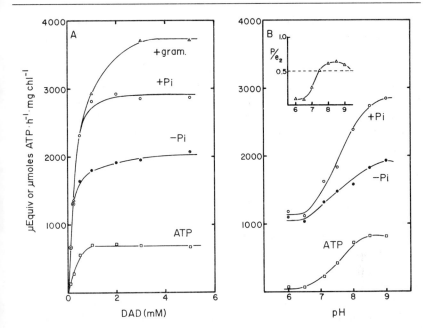

FIG. 3. Effects of DAD concentration and of pH on the rate of electron transport and phosphorylation in the ascorbate/DAD → PS-I → methyl viologen/$O_2$ reaction. In the experiment of (A) the reaction mixture contained 0.1 $M$ sucrose, 2 m$M$ MgCl$_2$, 50 m$M$ Tricine–NaOH buffer (pH 8.1), 0.75 m$M$ ADP, 5 m$M$ Na$_2$H$^{32}$PO$_4$ (P$_i$), 2.5 $\mu M$ DCMU, 2.5 m$M$ D-ascorbate, 0.1 m$M$ methyl viologen, indicated concentrations of DAD, chloroplasts equivalent to 5 $\mu$g chlorophyll/ml and bovine eruthrocyte superoxide dismutase at 0.4 mg/ml (approximately 1200 units/ml). When added, gramicidin (gram) was 4 $\mu$g/ml. In the pH experiments of (B) the buffers used were the same as in Fig. 2B except that TAPS buffer was replaced by tricine buffer. In these experiments, rates of electron transport were calculated from $O_2$ uptake rates based on the relation $O_2 = 2e^-$, which is valid in experiments such as this with excess superoxide dismutase. Although not shown in these figures, the $O_2$ consumption rates were approximately 50% higher when the dismutase was omitted from the reaction mixture (see section for PS-II electron donors). The intensity of actinic light, 500 kergs sec$^{-1}$ cm$^{-2}$ (>600 nm). [Adapted from J. M. Gould, *Biochim. Biophys. Acta* **387**, 135 (1975).]

sucrose, 40 m$M$ tricine–NaOH buffer (pH 7.8), 3 m$M$ MgCl$_2$, 5 m$M$ NaH$^{32}$PO$_4$, 0.8 m$M$ ADP, 2 $\mu M$ DCMU, 0.1 m$M$ methyl viologen, 0.5 m$M$ durohydroquinone and chloroplasts containing 12 $\mu$g chlorophyll.

## Donors of Electrons to Photosystem II (Region 7)

During the last several years a number of new reductants have been added to the list of PS-II electron donors given in the previous review.[1]

Some of these new donors are now in frequent use for investigations of energy coupling at PS-II. Another recent development is the recognition of the problems of superoxide radical which are associated with the use of $O_2$ as the terminal electron acceptor in the presence of certain reducing agents. This latter topic is discussed below because it is an important one, bearing upon the quantitation of donor-supported electron flow observed as $O_2$ consumption. It should be noted that the use of ambient $O_2$ as the terminal electron acceptor is a very convenient and the most widely applicable way of measuring both PS-I-mediated and PS-II-mediated electron-donor reactions.

### Superoxide and the Stoichiometry of Electron Flow Assayed as $O_2$ Uptake

It is well known that addition of ascorbate greatly stimulates the $O_2$ uptake by the Mehler reaction, i.e., the Hill reaction supported by an autoxidizable electron acceptor such as viologen and anthraquinone sulfonate. Since the enhanced $O_2$ uptake is sensitive to PS-II inhibitors and is accompanied by an oxidation of ascorbate, the phenomenon has been interpreted as indicating that ascorbate contributes additional electrons to PS-II or replaces water as the electron donor.[1] However, it has been shown recently that the ascorbate-enhanced part of $O_2$ uptake can be abolished by the addition of superoxide dismutase.[59-61] This demonstrates that ascorbate photooxidation by normal, water-oxidizing chloroplasts has nothing to do with electron donation to PS-II. It is a pure chemical oxidation by the superoxide radical $O_2^-$, the product of monovalent reduction of $O_2$ by the photoreduced methyl viologen or anthraquinone. $O_2$ uptake in the Mehler reaction is stimulated by ascorbate because $O_2^-$, which normally dismutates to $H_2O_2$ and $O_2$, is mostly reduced to $H_2O_2$ by the ascorbate. $Mn^{2+}$ also induces a similar "false" PS-II donor reaction.[59] Some reductants, such as hydrazine, sulfite, dopamine,[62-64] are subject to chain oxidation when exposed to $O_2^-$, the result of which is a vast amplification of $O_2$ uptake.

Superoxide inevitably poses a problem when one is to deal with a donor-supported Mehler reaction, since it usually requires ascorbate (as an electron reservoir) which is one of the most $O_2^-$-sensitive reductant.

[59] B. L. Epel and J. Neumann, *Biochim. Biophys. Acta* **325**, 520 (1973).
[60] J. F. Allen and D. O. Hall, *Biochem. Biophys. Res. Commun.* **52**, 856 (1973).
[61] D. R. Ort and S. Izawa, *Plant Physiol.* **53**, 370 (1974).
[62] K. E. Mantai and G. Hind, *Plant Physiol.* **48**, 5 (1971).
[63] K. Asada and K. Kiso, *Eur. J. Biochem.* **33**, 253 (1973).
[64] E. F. Elstner and A. Heupel, *Z. Naturforsch., Teil C* **29**, 559 (1974).

Some donors (e.g., catechol) are also $O_2^-$ sensitive. In many cases the problem is a relatively simple one, ascorbate just scavenging all $O_2^-$ and thereby doubling the rate of $O_2$ uptake. However, as noted above, further amplification of $O_2$ uptake can readily occur depending on the nature and the purity of the direct donor used. Thus, the only reliable solution to the problem is to add large amounts of superoxide dismutase to the reaction mixture to abolish whatever superoxide reactions that may be going on besides dismutation. It is also only under these conditions that the long-known and long-misused equation does apply:

$$2\ e^- = O_2$$

that is, one molecules of $O_2$ taken up for a pair of electrons transported through the photosynthetic chain (see formulae in footnote[65]).

*Inhibition of Water Oxidation as Prerequisite to PS-II Donor Reactions*

As has been noted by many workers, PS-II does not usually oxidize exogenous reductants unless forced to do so by inhibition of water oxidation. What appeared to be cases of normal chloroplasts preferentially oxidizing artificial reductants through PS-II (with $O_2$ as acceptor) have turned out to be nonbiological superoxide oxidation (see section above). Selective blocking of water oxidation can be achieved in a variety of ways,[1,3] but the most often used are Tris treatment and hydroxylamine treatment. These two are also the only methods that have been shown to

---

[65] The reaction sequence is as follows (symbols: $AH^-$, ascorbate; A, dehydroascorbate; MV, methyl viologen):

$$AH^- \xrightarrow[\text{chloroplasts}]{2e-\text{ transport}} A + 2\ e^- \quad H^+ + (1)$$

$$2\ e^- + 2\ O_2 \xrightarrow{MV} 2\ \dot{O}_2^- \quad (2)$$

$$2\ \dot{O}_2^- + AH^- + 3\ H^+ \xrightarrow{\text{scavenging}} 2\ H_2O_2 + A \quad (3)$$

---

$$2\ AH^- + 2\ O_2 + 2\ H^+ \xrightarrow[\text{chloroplasts}]{2e-\text{ transport}} 2\ A + 2\ H_2O_2 \quad (O_2 = e^-\ \text{without excess dismutase})$$

When excess superoxide is added to the reaction mixture, Eq. (3) will be replaced by

$$2\ O_2^- + 2\ H^+ \xrightarrow[\text{(spontaneous)}]{\text{dismutase}} H_2O_2 + O_2$$

The overall reaction would then become

$$AH^- + O_2 + H^+ \xrightarrow[\text{chloroplasts}]{2e-\text{ transport}} A + H_2O_2 \quad (O_2 = 2\ e^-\ \text{with excess dismutase})$$

TABLE I

NEWLY INTRODUCED PHOTOSYSTEM II ELECTRON DONORS AND THEIR PROPERTIES

| Electron donor | $E_0'$ (pH 7) (V) | $e^-$ or $e^- + H^+$ release | Rate saturating concentration (mM) | Optimal pH (or pH used) | Applicable $e^-$ acceptor[a] | Typical rate ($\mu$Eq/hr mg chl) | Phosphorylation ($P/e_2$) |
|---|---|---|---|---|---|---|---|
| Catechol[b] | 0.36 | $2e^- + 2E^+$ | 0.5[i] | 8.0 | MV/O$_2$, (NADP/Fd) | 350 | 1.0 |
| 3,3'-Diaminobenzidine[c] | — | $2e^- + 2H^+$ | 0.3 | (8.0) | MV/O$_2$, (NADP/Fd) | 300 | 0.8 |
| Hydrazobenzene[d] | — | $2e^- + 2H^+$ | > 0.1 | (7.8) | DCIP, (MV/O$_2$) | 300 | Not known |
| I$^-$[e] | 0.53 | $e^-$ | 20 | 8.0 | MV/O$_2$ FeCy | 250 | 0.5 with MV/O$_2$ |
| Vanadyl sulfate[f] | — | $2e^- + 4H^+$ | 30 | 6.5 | MV/O$_2$ | 250 | Not known |
| H$_2$O$_2$[g] | 0.27 | $2e^- + 2H^+$ | 10 | 8.0 | FeCy, DCIP, DMQ[j] | 150 | 0.3 with DMG[j] |
| N,N,N',N'-tetramethylbenzidine[h] | — | $e^-$ | 0.05[i] | (8.0) | NADP/Fd, (MV/O$_2$) | 100 | 0.6 |
| Ferrocyanide[e] | 0.42 | $e^-$ | 30 | 8.0 | MV/O$_2$ | 100 | 0.5 |

[a] Abbreviations: DCIP, 2.6-dichlorophenolindophenol; DMQ, 2,5-dimethyl-$p$-benzoquinone; Fd, ferredoxin; FeCy, ferricyanide; MV, methyl viologen. Electron acceptors in parentheses are those which have not been tested but which are presumably applicable.

[b] D. R. Ort and S. Izawa, *Plant Physiol.* **53**, 370 (1974).

[c] G. Ben-Hayyim, Z. Drechsler, J. Goffer, and J. Neumann, *Eur. J. Biochem.* **52**, 135 (1975). (This compound is also a PS-I electron donor according to these authors.)

[d] J. Haveman and M. Donze, *Proc. Int. Congr. Photosynth. Res., 2nd, 1971* p. 81 (1972).

[e] S. Izawa and D. R. Ort, *Biochim. Biophys. Acta* **357**, 127 (1974).

[f] D. Rosen, R. Barr, and F. L. Crane, *Biochim. Biophys. Acta*, **408**, 35 (1975).

[g] H. Inoue and M. Nishimura, *Plant Cell Physiol.* **12**, 739 (1971).

[h] E. Harth, W. Oettmeier, and A. Trebst, *FEBS Lett.* **43**, 231 (1974).

[i] Ascorbate is required as the electron reservoir.

[j] R. L. Pan and S. Izawa, *Biochim. Biophys. Acta*, **547**, 311 (1979).

abolish the water oxidation reaction without appreciably impairing the energy-coupling efficiency of the membrane.[66,67]

## Donors of Electrons to PS-II

Of the various PS-II electron donors listed in the previous review,[1] 1,5-diphenylcarbazide (DPC)[68] now seems to have established itself as the standard PS-II donor. It is routinely used to test the photochemical activity of PS-II and to locate the site of action of PS-II inhibitors. The main reason for the popularity of this donor appears to be the applicability of DCIP as an electron acceptor, and the relatively fast rate of electron transport it supports. The behavior of various carbazides and azobenzenes as PS-II donors has been studied in some detail. The oxidation of hydrazobenzene by "PS-II particles" is reported to be rather insensitive to DCMU.[69] Unique among the newly introduced PS-II electron donors are ferrocyanide ($E_0' = +0.42V$), iodide ion ($E_0' = +0.56$ V)[70] and $N,N,N',N'$-tetramethylbenzidine ($E_0' = +0.5$ V).[71] Electron transport from these non-proton-producing donors to PS-I acceptors (methyl viologen, NADP) generates ATP with only half the efficiency of electron transport from proton-releasing donors, such as hydroquinone, $p$-aminophenol, benzidine. Like the case of TMPD for PS-I, the result strongly points to a chemiosmotic coupling mechanism associated with PS-II.[70,71] Another notable new PS-II electron donor is hydrogen peroxide,[72] the well known end product of electron transport reaction in which $O_2$ is utilized as the terminal electron acceptor. Given at high concentrations (>5 m$M$), $H_2O_2$ is rapidly oxidized to $O_2$ in the presence of standard electron acceptors, such as ferricyanide, DCIP, and quinones. Furthermore, dimethylquinone reduction supported by $H_2O_2$ is coupled to phosphorylation with a near-normal efficiency (P/e$_2$ = 0.3).[73] Thus, there are intriguing similarities between $H_2O$ and $H_2O_2$ as electron donors to PS-II.

*Methods.* Some technical data pertaining to selected new PS-II electron donors are summarized in Table I.

---

[66] T. Yamashita and W. L. Butler, *Plant Physiol.* **44**, 435 (1969); see also *Ibid.* **43**, 1978 (1968).

[67] D. R. Ort and S. Izawa, *Plant Physiol.* **52**, 595 (1973); see also Ort and Izawa.[61]

[68] L. P. Vernon and E. R. Shaw, *Plant Physiol.* **44**, 1645 (1969).

[69] J. Haveman and M. Donz, *Proc. Int. Congr. Photosynth. Res., 2nd,* 1971 p. 81 (1972).

[70] S. Izawa and D. R. Ort, *Biochim. Biophys. Acta* **357**, 127 (1974).

[71] E. Harth, W. Oettmeier, and A. Trebst, *FEBS Lett.* **43**, 231 (1974).

[72] H. Inoue and M. Nishimura, *Plant Cell Physiol.* **12**, 739 (1971).

[73] R. L. Pan and S. Izawa, *Biochim. Biophys. Acta.* **547**, 311 (1979)

Acknowledgments

The author wishes to thank Patrick M. Kelley for reading through the manuscript and for his assistance in art work. This work was supported by a grant (PCM76-19887) from the National Science Foundation.

# [40] Electrophoretic Analysis of Chloroplast Proteins

By Nam-Hai Chua

Electrophoresis in sodium dodecyl sulfate (SDS) polyacrylamide gels has become a standard technique for structural and biosynthetic studies of chloroplast proteins. A number of electrophoretic systems with different degrees of resolution have been employed in such investigations.[1-12] This article describes a method[13-15] which combines the alkaline SDS-discontinuous buffer system of Neville[16,17] and, in the resolving gel, a linear acrylamide concentration gradient.[18] The discontinuous buffer system is capable of stacking SDS–protein complexes over a wide range of molecular weights, thus providing very sharp bands, while the linear pore gradient[18,19] allows the separation and resolution of polypeptides with widely different molecular weights. The combination of these two meth-

[1] J. K. Hoober, *J. Biol. Chem.* **245**, 4327 (1970).
[2] R. P. Levine, W. G. Burton, and H. A. Durham, *Nature (London)* **237**, 176 (1972).
[3] S. M. Klein and L. P. Vernon, *Photochem. Photobiol.* **19**, 43 (1974).
[4] J. P. Thornber and H. R. Highkin, *Eur. J. Biochem.* **41**,109 (1974).
[5] A. R. J. Eaglesham and R. J. Ellis, *Biochim. Biophys. Acta* **335**, 396 (1974).
[6] W. Bottomley, D. S. Spencer, and P. R. Whitfeld, *Arch. Biochem. Biophys.* **164**, 106 (1974).
[7] O. Machold, *Biochim. Biophys. Acta* **382**, 494 (1975).
[8] W. G. Nolan and R. B. Park, *Biochim. Biophys. Acta* **375**, 406 (1975).
[9] K. Apel, L. Bogorad, and C. L. F. Woodcock, *Biochim. Biophys. Acta* **387**, 568 (1975).
[10] J. J. Morgenthaler and L. Mendiola-Morgenthaler, *Arch. Biochem. Biophys.* **172**, 51 (1976).
[11] A. R. Cashmore, *J. Biol. Chem.* **251**, 2848 (1976).
[12] S. Bar-Nun, R. Schantz, and I. Ohad, *Biochim. Biophys. Acta* **459**, 451 (1977).
[13] N.-H. Chua and P. Bennoun, *Proc. Natl. Acad. Sci. U.S.A.* **72**, 2175 (1975).
[14] N.-H. Chua, K. Matlin, and P. Bennoun, *J. Cell Biol.* **67**, 361 (1975).
[15] L. Y. W. Bourguignon and G. E. Palade, *J. Cell Biol.* **69**, 327 (1976).
[16] D. M. Neville, Jr., *J. Biol. Chem.* **246**, 6328 (1971).
[17] D. M. Neville, Jr. and H. Glossmann, Vol. 32 p. 92.
[18] J. Margolis and K. G. Kendrick, *Anal. Biochem.* **25**, 347 (1968).
[19] D. Rodbard, G. Kapadia, and A. Chrambach, *Anal. Biochem.* **40**, 135 (1971).

ods gives consistently high resolution of thylakoid membrane polypeptides and chloroplast stromal proteins.

## Preparation of Slab Gels with a Linear Concentration of Acrylamide

*Chemicals*

Acrylamide (Eastman; practical grade)
$N,N'$-Methylene bisacrylamide (Eastman; practical grade)
Sodium dodecyl sulfate (SDS) (Pierce; sequanal grade)
Tris(hydroxymethyl)aminomethane (Tris) (Sigma; trizma base)
Ammonium persulfate (Fisher; reagent grade)
$N,N,N',N'$-Tetramethylethylenediamine (TEMED) (Sigma; 99% pure)
Sucrose (Schwarz/Mann; ultra pure)
Activated charcoal, neutralized (Sigma)
Boric acid (Mallinckrodt, analytical grade)
Dithiothreitol (Sigma)

*Stock solutions*

(1) 30% (w/v) acrylamide–0.8% (w/v) bisacrylamide
Acrylamide, 300 gm
Bisacrylamide, 8 gm
$H_2O$ to 1 liter
To decolorize solution add 2–3 gm of neutralized, activated charcoal and stir for 30 min. Pass solution through two layers of filter paper (Schleicher & Schuell No. 597) and then through 0.45 $\mu$m Millipore filter. Store solution in amber bottle at 4°.

(2) 10% (w/v) SDS
SDS, 10 gm
$H_2O$ to 100 ml. Store at room temperature.

(3) 10% (w/v) Ammonium persulfate
Ammonium persulfate, 10 gm
$H_2O$ to 100 ml
Prepare fresh weekly. Store at 4°.

(4) 60% (w/v) Sucrose
Sucrose 600 gm
$H_2O$ to 1000 ml
Pass solution through 1.2 $\mu$m Millipore filter. Store at 4°.

(5) 20x Upper reservoir buffer (0.04 $M$ Tris–0.04 $M$ borate, pH 8.64)
Trizma base, 198.4 gm
SDS, 40 gm
Boric acid, 56 gm

H₂O to 1800 ml
Adjust with saturated solution of boric acid to pH 8.64 at 25°.
Make up to 2 liters with $H_2O$. Store at room temperature.
(6) 4x Stacking gel buffer (0.0541 $M$ Tris–0.0267 $M$ $H_2SO_4$, pH 6.1)
Trizma base, 26.2 gm
$H_2O$, 980 ml
Adjust with concentrated sulfuric acid (about 4.4 ml) to pH 6.1
at 25°. Make up to 1 liter with $H_2O$. Store at room temperature.
(7) 5x Lower reservoir and resolving gel buffer (0.4244 $M$ Tris–
0.0308 $M$ HCl, pH 9.18)
Trizma base, 514 gm
$H_2O$, 1800 ml
Adjust with concentrated HCl (about 28.6 ml) to pH 9.18 at 25°.
Make up to 2 liters with $H_2O$. Store at room temperature.

*Resolving Gel Solution (45 ml)*

|  | Dense (15% acrylamide) | Light (7.5% acrylamide) |
|---|---|---|
| 30% Acrylamide–0.8% bisacrylamide | 22.5 ml | 11.25 ml |
| 5X Resolving gel buffer | 9.0 ml | 9.0 ml |
| 60% Sucrose | 12.9 ml | 3.8 ml |
| 10% SDS | 0.45 ml | 0.45 ml |
| H₂O | — | 20.35 ml |
| TEMED | 10 μl | 20 μl |
| 10% Ammonium persulfate (add last) | 0.15 ml | 0.15 ml |

*Stacking Gel Solution (20 ml; 6% Acrylamide)*

| | |
|---|---|
| 30% Acrylamide–0.8% bisacrylamide | 4.0 ml |
| 4X Stacking gel buffer | 5.0 ml |
| 10% SDS | 0.2 ml |
| H₂O | 10.6 ml |
| TEMED | 20 μl |
| 10% Ammonium persulfate (add last) | 0.2 ml |

Any slab gel apparatus constructed according to the design of Studier[20] may be employed. Assemble the glass plates and spacers to form a gel mold as described.[20] Prepare the dense and light resolving gel solutions. Establish the linear acrylamide concentration gradient by mixing the two solutions using a Buchler density gradient mixer (No. 2-5102)

[20] F. W. Studier, *J. Mol. Biol.* **79**, 237 (1973).

and stirrer assembly (No. 2-5070A). The volumes are determined by gel dimensions; for slab gels 0.1 × 24 × 30 cm, use 35 ml of the dense and 38 ml of the light resolving gel solutions. The slightly smaller volume of the dense solution prevents its initial backflow into the light solution. Feed the resolving gel solution into the mold either by gravity or by means of a peristaltic pump. In either case, maintain the flow rate at approximately 3 ml/min so that the gradient is poured in 25 min. After the resolving gel solution has been poured, overlay it with a few millimeters of isobutanol. Complete polymerization of the gel, which should occur about 60 min after addition of ammonium persulfate, is marked by formation of a water–isobutanol interface. Decant the isobutanol layer and rinse the top of the resolving gel with distilled water. Pour in the stacking gel solution and insert slot former taking precautions to exclude air bubbles. After the stacking gel has set, store the mold containing slab gel in a moistened chamber at 4° for at least 1 day before use. Gels may be stored under these conditions for as long as 2 weeks without any apparent effect on their resolving power.

Before use, fasten mold containing aged slab gel to electrophoresis apparatus and pour in the upper and lower reservoir buffers. Remove slot former slowly, taking care not to tear the stacking gel projections which partition the slots. Generally it is easier to remove slot former from aged gels since the latter are slightly dehydrated. Straighten the gel projections with a spatula before applying samples into slots.

## Sample Preparation

*Stock Solutions*
    1.0 $M$ $Na_2CO_3$
    1.0 $M$ Dithiothreitol (DTT)
    1.0 $M$ Tris–HCl (pH 7.5 at 25°)
    10% (w/v) Sodium dodecyl sulfate (SDS) (Pierce; sequanal grade)
    60% (w/v) Sucrose. Pass solution through 1.2 $\mu$m Millipore filter.

Solution A (0.1 $M$ $Na_2CO_3$–0.1 $M$ DTT)
    1 $M$ $Na_2CO_3$, 1 ml
    1 $M$ DTT, 1 ml
    $H_2O$ to 10 ml

Solution B (5% SDS–30% sucrose–0.1% bromphenol blue)
    10% SDS, 50 ml
    60% Sucrose, 50 ml
    Bromphenol blue, 0.1 gm

*Unextracted Thylakoid Membranes.* Suspend purified thylakoid membranes[13,14] in solution A to a chlorophyll concentration of 1.67 mg/ml. To

0.3 ml (500 $\mu$g chlorophyll or 2.5 mg protein) of the membrane suspension add 0.2 ml of solution B and mix thoroughly. The final concentrations of the components in the mixture are: thylakoid membranes, 1 mg chlorophyll or 5 mg protein/ml; $Na_2CO_3$, 50 m$M$; DTT, 50 m$M$; SDS, 2%; sucrose, 12%; and bromphenol blue, 0.04%. The SDS–chlorophyll weight ratio is 20:1 and the SDS–protein weight ratio 4:1. Divide the SDS-solubilized membrane preparation into two equal aliquots. Incubate one aliquot in a boiling water bath for 1–2 min to dissociate chlorophyll-protein complexes.[14] This aliquot is referred to as "heated," whereas the untreated aliquot is designated "nonheated."

*Chloroform: Methanol (C/M) Extract and Residue of Thylakoid Membranes.* C/M extractions of thylakoid membranes[14,21] are carried out in 12 ml conical centrifuge tubes. To 0.25 ml of thylakoid membranes (1.5 mg chlorophyll/ml solution A) add 5.25 ml of a 2:1 (v/v) mixture of C/M and mix vigorously with a Vortex mixer. Keep the tube on ice for about 5 min and collect the precipitate by centrifugation at 5000 $g$ for 10 min. Transfer the C/M extract to another centrifuge tube and evaporate the extract to dryness under a stream of nitrogen at room temperature. Carefully aspirate away residual supernatant from the C/M residue, and dry the latter under nitrogen. With the sharp end of a stainless steel spatula, disperse precipitates in the C/M residue or the dried C/M extract in 0.15 ml of solution A. Add 0.1 ml of solution B, and heat the mixture at 100° for 1–2 min.

*Chloroplast Stromal Proteins.* To 0.28 ml of chloroplast stromal proteins in 50 mM Tris–HCl (pH 7.5), add 0.02 ml 1 $M$ DTT and 0.2 ml solution B. The protein concentration of the final mixture should not exceed 5 mg/ml. Incubate the SDS–protein mixture at 100° for 1–2 min before loading onto gels.

Dilute protein solution may be concentrated by precipitation at 10% (w/v) TCA in the cold for 1 hr. Pellet the TCA precipitate by centrifugation at 3000 $g$ for 10 min and aspirate away the supernatant. To remove residual TCA, wash the pellet with a 1:1 (v/v) mixture of ethanol–ether. Repeat the wash with ether only and airdry the pellet. With the sharp end of a spatula disperse the dried pellet (1–1.3 mg protein) in 0.15 ml of solution A. Add 0.1 ml of solution B, mix, and heat the final mixture at 100° for 1–2 min.

### Sample Application and Electrophoresis

Samples are placed into gel slots by means of a Hamilton syringe. The volume used depends on the gel dimensions and on the number

[21] N.-H. Chua and N. W. Gillham, *J. Cell Biol.* **74**, 441 (1977).

and relative amounts of proteins to be separated. In general, the amount of protein loaded should be varied systematically over a wide range in order to resolve polypeptides of similar electrophoretic mobility and to ensure the detection of minor components. For 1-mm slab gels with 9.4-mm slots, load 20 $\mu$l of the unextracted membrane samples, or of the samples derived from C/M residue and extract. These volumes give a protein load of about 70–100 $\mu$g/slot and provide optimal resolution of the thylakoid membrane polypeptides.

Gels are run at room temperature under conditions of constant electrophoretic current. It is convenient to adjust the amperage such that the electrophoresis is carried out overnight (14–16 hr). In the case of thylakoid membranes, the bromphenol blue dye front and the lower half of the "free" pigment zone are allowed to run off, since there are no low molecular weight polypeptides migrating in this region.

*Staining, Destaining, and Drying of Gels*

    *Staining Solution*
       Coomassie brilliant blue R250 (Sigma), 2.5 gm
       Methanol, 500 ml
       Glacial acetic acid, 70 ml
       $H_2O$ to 1000 ml
    *Destaining Solution*
       Methanol, 400 ml
       Glacial acetic acid, 70 ml
       $H_2O$ to 1000 ml

The compositions of the staining and destaining solutions are adapted from Weber and Osborn.[22] Gels are stained and destained in clear polystyrene plastic boxes available from Althor Products (202 Bay 46 Street, Brooklyn, New York). After completion of electrophoresis, place gel in the staining solution and shake gently for at least 3 hr. Pour off the staining solution and elute the excess dye with three or four changes of the destaining solution, until the gel background is clear. Destained gels may be dried under vacuum onto Whatman 3 MM chromatographic paper as described by Maizels.[23] Before drying, equilibrate gels with the destaining solution containing 3% glycerol. Gels infiltrated with glycerol are more pliable and do not crack upon drying or after prolonged storage. Such gels, when tightly wrapped with Saran Wrap, may be stored as permanent records.

---

[22] K. Weber and M. Osborn, *J. Biol. Chem.* **244**, 4406 (1969).
[23] J. V. Maizels, Jr., *Methods Virol.* **5**, 179 (1971).

### Densitometric Tracings and Autoradiography

If densitometric tracings of polypeptide bands are desired, scan gel strips at 550 mm with any commercially available spectrophotometer equipped with a linear transport device (e.g., Gilford Model 2400). For autoradiography of $^{14}$C or $^{35}$S samples, place gels directly in contact with X-ray film (Kodak or DuPont Cronex 2DC). Gels containing $^{3}$H samples may be processed for fluorography, as described by Bonner and Laskey.[24]

### General Comments

1. For optimal resolution use approximately 20 cm of resolving gel and 1.5–2.0 cm of stacking gel. The acrylamide concentration of the stacking gel may be decreased to 4 or 5% depending on the molecular weight distribution of the polypeptide mixture to be analyzed.

2. The 7.5–15% acrylamide concentration gradient is stabilized by the simultaneous establishment of a 5–17% sucrose gradient, as recommended by Margolis and Kendrick.[18] To avoid mixing of gradient by thermal convection current generated during polymerization, the resolving gel contains an inverted gradient of TEMED so that gelation proceeds from the top to the bottom of the poured gel.[18]

3. The system of Neville[16,17] is adapted so that both the stacking and resolving gels contain 0.1% SDS. Polymerization of the resolving gel is carried out at room temperature instead of 15° as originally suggested by Neville[16,17]

4. The use of recrystallized acrylamide and bisacrylamide is not necessary for good resolution as long as the stock solution of acrylamide-bisacrylamide is decolorized with activated charcoal before use.

5. The resolution of the polypeptide bands as well as their relative electrophoretic mobilities in the gels are dependent upon the quality of SDS used. Satisfactory results are obtained consistently with the sequanal grade SDS from Pierce Chemical Co.

6. No difference in the polypeptide profile is detected when 0.1 $M$ $Na_2CO_3$ in solution A is replaced by 50 m$M$ HEPES–NaOH (pH 7.5) or 50 m$M$ Tris-HCl (pH 7.5 or 8.6). Similar results are also obtained when DTT is substituted by 2-mercaptoethanol; however, the polypeptide bands are not distinct when sulfhydryl reagent is totally omitted from the solubilization mixture.

7. Thylakoid membranes or chloroplast stromal proteins can be stored

---

[24] W. M. Bonner and R. A. Laskey, *Eur. J. Biochem.* **46,** 83 (1974).

in solution A at $-20°$, and are stable for at least 3 years under these conditions. Samples solubilized in SDS are stable for only 2 weeks at $-20°$ and 3 months at $-80°$.

8. Vertical streaking of protein in the gel is mostly due to the presence of incompletely solubilized protein aggregates. Recalcitrant aggregates may be dispersed by brief sonication of the sample in the presence of SDS.

9. The resolution of low-molecular-weight (<15,000) polypeptides in the unextracted membrane sample is not optimal due to the presence of interfering SDS-pigment complexes which migrate in that gel region. To resolve these polypeptides thylakoid membranes are extracted with C/M followed by electrophoretic analysis of the C/M-soluble and insoluble fractions. Extraction of photosynthetic pigments with 80% or 90% acetone should be avoided since such treatment results in the aggregation of specific polypeptides on top of the stacking gels.[13,14,21]

10. C/M extraction of thylakoid membranes may be carried out at neutral or alkaline pH with no quantitative or qualitative difference in the polypeptide profiles of the soluble and insoluble fractions. In contrast to mitochondrial inner membrane proteins[25] all thylakoid membrane polypeptides are extracted by acidic C/M. Since thylakoid membrane polypeptides of similar electrophoretic mobility may have differential solubility in nonacidic C/M, the organic solvent extraction of thylakoid membranes followed by electrophoretic analysis of the soluble and insoluble fractions provide an excellent method for simplifying polypeptide banding patterns, thereby facilitating their resolution.

11. The choice of the starting and limiting acrylamide concentrations for the resolving gels is determined by the molecular weight distribution of the polypeptide mixture and the relative abundance of the components. Since different acrylamide concentration gradients provide optimal resolution at different molecular weight ranges, it is desirable to analyze the sample in several gradient systems. For example, we have separated thylakoid membrane polypeptides from a number of sources on gels containing the following linear acrylamide concentration gradients: 5–10, 7.5–10, 7.5–12.5, 7.5–15, and 10–15%, and found the 7.5–15% system to be optimal.[13] This gradient system also gives good resolution of the chloroplast coupling factor, ribulose-1,5-diphosphate carboxylase, and total chloroplast stromal proteins.

12. In the characterization of any polypeptide mixture it is important to eliminate possible proteolytic artifacts that may have occurred during

[25] A. Tzagoloff and A. Akai, *J. Biol. Chem.* **247**, 6517 (1972).

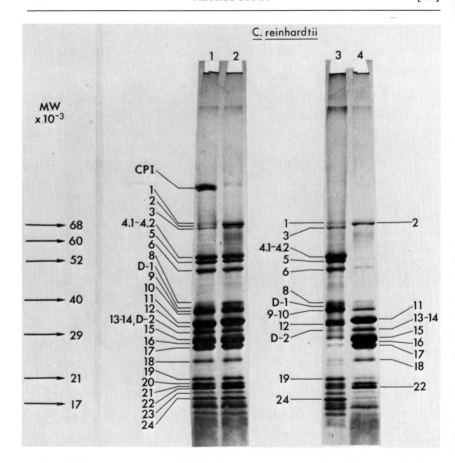

FIG. 1. Electrophoretogram of thylakoid membrane polypeptides of *C. reinhardii*. Thylakoid membranes were purified[13,14] and analyzed on the 7.5–15% SDS gradient gel system described in this article. (1) Nonheated sample, 20 μg chlorophyll. (2) Heated sample, 20 μg chlorophyll. (3) C/M residue, equivalent to 30 μg chlorophyll. (4) C/M extract, equivalent to 30 μg chlorophyll.

cell fractionation. The use of protease inhibitors to avoid such artifacts has been discussed extensively by Pringle.[26]

13. The 7.5–15% gradient gel system generally shows a linear relationship between $R_f$ and log molecular weight, if the latter is in the 15,000 to 70,000 range.[13] Caution should be exercised, however, in applying this relation to the case of thylakoid membrane polypeptides and chlorophyll–

[26] J. R. Pringle, *Methods Cell Biol.* **12**, 149 (1975).

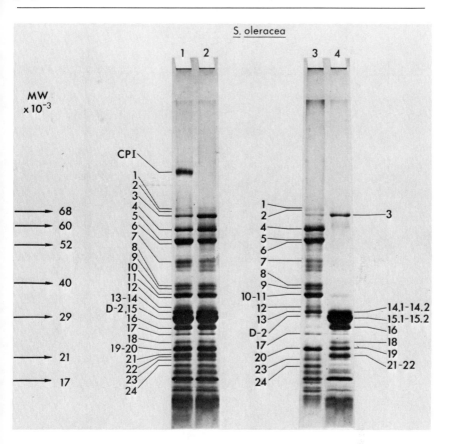

Fig. 2. Electrophoretogram of thylakoid membrane polypeptides of *S. oleracea* (spinach). Thylakoid membranes were purified and washed successively with distilled $H_2O$ and 0.5 m$M$ EDTA (pH 7.5) as described.[4] Membrane polypeptides were separated on the 7.5–15% gradient gel system described in this article. (1) Nonheated sample, 20 $\mu$g chlorophyll. (2) Heated sample, 20 $\mu$g chlorophyll. (3) C/M residue, equivalent to 30 $\mu$g chlorophyll. (4) C/M extract, equivalent to 30 $\mu$g chlorophyll.

protein complexes, since their free electrophoretic mobilities may differ significantly from those of the protein standards used for calibration.[14] Apparent molecular weight assigned to thylakoid membrane polypeptides should, therefore, be considered as tentative until verified by techniques[27] other than SDS gel electrophoresis.

[27] C. Tanford, Y. Nozaki, J. A. Reynolds, and S. Makino, *Biochemistry* **13**, 2369 (1974).

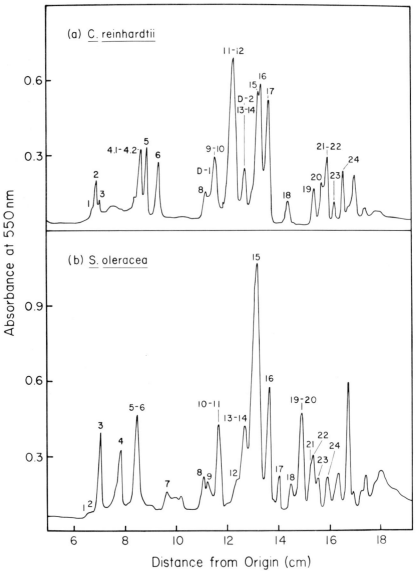

Fig. 3. Densitometric tracings of polypeptide profiles of heated thylakoid membrane samples. (a) *C. reinhardii*. (b) *S. oleracea*.

## SDS Gradient Gel Electrophoresis of Thylakoid Membrane Polypeptides

The 7.5–15% gradient gel system described in this article has been used for the separation of thylakoid membrane polypeptides of several algae and higher plants. Figures 1 and 2 show the membrane polypeptide profiles of *Chlamydomonas reinhardii* and *Spinacea oleracea,* respectively, and Fig. 3 presents densitometric tracings of the protein staining patterns obtained with unextracted membrane samples. In both plants, more than 30 polypeptide bands are resolved, ranging in molecular weight from 68,000 to less than 15,000. For convenience, all the polypeptides, except D-1 and D-2, are numbered consecutively starting from the high molecular weight region. Some polypeptides which migrate as a single band in the 7.5–15% gradient gel system can be resolved into two bands, either by decreasing the protein load, or by changing the range of the acrylamide gradient. In those cases where band splitting has been observed, the newly resolved bands are designated N.1 and N.2, where N refers to the unresolved bands in the 7.5–15% gradient gel system. Those polypeptide bands which are not split by changing the separation procedures are assumed to represent single polypeptides. Polypeptide bands with variable recovery are not numbered; neither are several low molecular weight bands which fall in the region of sub-optimal resolution.

Approximately 60% (w/w) of thylakoid membrane protein is soluble in a 2:1 (v/v) mixture of C/M. Electrophoretic analysis of the C/M residues (Figs. 1 and 2, slot 3) and extracts (Figs. 1 and 2, slot 4) reveals that polypeptide patterns of the two fractions are simplified as compared to that of the unextracted sample (Figs. 1 and 2, slot 2). In many cases, polypeptides (e.g., D-2 and 13-14 in Fig. 1, 19 and 20 in Fig. 2), which migrate closely together in the unextracted sample, can be separated and resolved because of their different solubility in the organic solvent.

## Structural and Biosynthetic Studies of Thylakoid Membrane Polypeptides from C. reinhardii

Structure–function correlative studies of nonphotosynthetic mutants of *C. reinhardii* suggest that polypeptides 2 and 6 are associated with photosystem I and II reaction center activities, respectively.[14,13] Pulse labeling of thylakoid membrane polypeptides in the presence of specific inhibitors reveals that at least nine of the polypeptides (2, 4.1, 4.2, 5, 6, D-1, D-2, and two low-molecular-weight components) are made on chloroplast ribosomes.[21] Among them, polypeptide 5 is synthesized as a high-molecular-weight variant (polypeptide 5′) in a non-Mendelian mutant, thm-u-1.[28] Another non-Mendelian mutant, $C_1$, has been shown to be deficient in photosystem I activity as well as in chlorophyll–protein complex I and polypeptide 2.[29]

Acknowledgments

I thank R. G. Piccioni and G. W. Schmidt for reading the manuscript and for helpful suggestions and Sally Chao for expert technical assistance. The work described in this article was supported in part by National Institutes of Health Grant GM-21060. I am a recipient of a National Institutes of Health Research Career Development Award GM-00223.

[28] N.-H. Chua, in "Genetics and Biogenesis of Chloroplasts and Mitochondria" (T. Bücher et al., eds.), p. 323. Elsevier, Amsterdam, 1976.
[29] P. Bennoun, J. Girard, and N.-H. Chua, Mol. Gen. Genet. 153, 343 (1977).


# [41] Isolation of Thylakoid Proteins

By W. MENKE and F. KOENIG

The thylakoid membrane contains 55–65% proteins,[1] and these consist of a large number of different polypeptides.[2-5] Most of them are insoluble in water. It has long been known that the thylakoids can be brought into solution by means of different detergents;[6] from the detergents investigated so far, sodium dodecyl sulfate (SDS) has the strongest dissociating effect. However, this detergent does not dissociate all proteins into individual polypeptides, even under conditions in which disulfide bonds are reduced by mercaptoethanol. A disadvantage of the use of SDS arises from the fact that the polypeptides are denatured and that removal of the detergent causes and additional alteration of the secondary and tertiary structure.[7]

Upon removal of SDS, aggregation of the polypeptides occurs, and the molecular weight of the aggregates depends on the pH of the medium.[8,9] For direct investigation of their function in photosynthesis these

[1] F. Koenig, Z. Naturforsch. Teil B 26, 1180 (1971).
[2] W. Menke and E. Jordan, Z. Naturforsch. Teil B 14, 234 (1959).
[3] W. Menke and E. Jordan, Z. Naturforsch. Teil B 14, 393 (1959).
[4] W. Menke and E. Schölzel, Z. Naturforsch. Teil B 26, 378 (1971).
[5] K. Apel, K. R. Miller, L. Bogorad, and G. J. Miller, J. Cell Biol. 71, 876 (1976).
[6] E. L. Smith and E. G. Pickels, J. Gen. Physiol. 24, 753 (1941).
[7] W. Menke, A. Radunz, G. H. Schmid, F. Koenig, and R.-D. Hirtz, Z. Naturforsch. Teil C 31, 436 (1976).
[8] H. Craubner, F. Koenig, and G. H. Schmid, Z. Naturforsch. Teil C 30, 615 (1975).
[9] H. Craubner, F. Koenig, and G. H. Schmid, Z. Naturforsch. Teil C 32, 384 (1977).

denatured polypeptides are hardly suitable. However, it was shown that antisera to the denatured polypeptides contain antibodies to native antigenic determinants because they agglutinate stroma-free chloroplasts and inhibit photosynthetic electron transport in a specific way at discrete sites.[10-14] Therefore, the antisera can be used for localization of the polypeptides in the thylakoid membrane as well as for functional characterization. In addition, the primary structure of isolated polypeptides can be determined by the usual methods.

The polypeptide mixture solubilized by means of SDS can be fractionated by gel permeation chromatography. The separation works in principle, as the SDS–polyacrylamide gel electrophoresis, according to the size of the polypeptide–detergent micelles. Due to the large number of components to be separated and due to unavoidable losses, it is necessary to start with large amounts of chloroplasts if one wants to end up with pure polypeptides. This requires columns of correspondingly large diameters and considerable length. The results depend on the obedience to certain working conditions. Therefore, in the following, the method is described in detail.

For the isolation of certain polypeptides, modifications may be advantageous. A separation of polypeptide fractions with the same or very similar molecular weights is possible by adsorption chromatography on hydroxylapatite. Attempts to separate mixtures after the removal of SDS by ion exchange chromatography fail, since the polypeptides form mixed aggregates in detergent-free solutions. According to our experience, a prefractionation with other detergents, such as the fractionation into photosystem I and II particles is not generally applicable, since after this procedure these particles only partially dissolve in SDS-containing buffers. For the isolation of some membrane proteins such as cytochromes, special preparation procedures are described in the literature.[15-17]

The scale of the procedure is chosen as to yield sufficient amounts of the major polypeptides in one preparation step.

[10] W. Menke, F. Koenig, A. Radunz, and G. H. Schmid, *FEBS Lett.* **49**, 372 (1975).

[11] F. Koenig, G. H. Schmid, A. Radunz, B. Pineau, and W. Menke, *FEBS Lett.* **62**, 342 (1976).

[12] G. H. Schmid, W. Menke, F. Koenig, and A. Radunz, *Z. Naturforsch. Teil C* **31**, 304 (1976).

[13] G. H. Schmid, G. Renger, M. Gläser, F. Koenig, A. Radunz, and W. Menke, *Z. Naturforsch. Teil C* **31**, 594 (1976).

[14] F. Koenig, W. Menke, A. Radunz, and G. H. Schmid. *Z. Naturforsch. Teil C* **32**, 817 (1977).

[15] D. S. Bendall, H. E. Davenport, and R. Hill, Vol. 23, [33].

[16] A. R. Wasserman, Vol. 32, [37].

[17] G. H. Schmid, A. Radunz, and W. Menke, *Z. Naturforsch. Teil C* **32**, 271 (1977).

## Preparation of the Starting Material

*Isolation of Chloroplasts.* It is useful to isolate chloroplasts from plants which permit fast preparation of sufficient amounts of stroma-free chloroplasts. It is important that the chloroplasts preserve their morphological structure since the yield will be higher. A plant which fullfills these requirements is *Antirrhinum majus.* It is advantageous to minimize isolation time since the solubility of the proteins decreases with time and, because proteases attack the proteins.

Leaves (1.2 kg) are ground with 4 liters of 0.4 $M$ sucrose in water. The homogenate is filtered and centrifuged for a few minutes at 400 $g$. From the supernatant the chloroplasts are spun down (20 min at 2300 $g$). In order to remove the stroma proteins, the sediment is washed three times with distilled water and suspended in 100 ml of water. From this suspension 0.5 ml is used for dry weight determination. The average yield is 4.5 gm of stroma-freed chloroplasts. Centrifugation in a sucrose density gradient yields cleaner preparations. All steps are carried out at $0°–5°$.

*Dissolution of the Chloroplasts.* To the suspension of stroma-free chloroplasts, the same volume of the following solution is added:

0.02 $M$ Sodium phosphate buffer pH 7.2, 1000 ml
Sodium dodecyl sulfate, 44 gm
$\beta$-Mercaptoethanol, 20 ml

Part of this buffer is diluted 1 : 1 (v/v) with water. To the chloroplast solution an amount of the 1 : 1 diluted buffer is added so as to give a final chloroplast concentration of 1%. In order to accelerate the dissolution of the chloroplasts, the solution is stirred for approximately 20 hr at room temperature. Undissolved residues are removed by centrifugation (60 min at 30,000 $g$). All operations with SDS—containing solutions must be carried out at 22°.

## Gel Chromatography

*Fractionation at Neutral pH.* The solution is pumped via a membrane filter (pore width 0.45 $\mu$m, Sartorius) onto the column which has a diameter of 21.5 cm. The optimal column length is 6 to 8 m. It has proven to be useful to take columns of $l$-m length (Pharmacia) and to connect these in series. The columns are filled with CL-6B Sepharose (Pharmacia). The elution buffer has the following composition:

0.01 $M$ Sodium phosphate buffer, pH 7.2, 1000 ml
Sodium dodecyl sulfate, 2.5 gm
$\beta$-Mercaptoethanol, 1 ml

The use of Tris instead of phosphate buffer offers a certain advantage, since the probability of bacterial infections is lower with Tris buffer. By addition of sodium chloride the adsorption of polypeptides onto the gel is reduced and the aggregation of some polypeptides is prevented. The Tris buffer used has the following composition:

0.05 $M$ Tris-HCl buffer, pH 7.5, 1000 ml
Sodium dodecyl sulfate, 2.5 gm
Sodium chloride, 5.0 gm
$\beta$-Mercaptoethanol, 2.5 ml

The flow rate is 2 cm/hr, and 250-ml fractions are collected. For the fractionation we use a combination of two multiport valves equipped with appropriate steering device (WFN, Köln) together with a flow-through photometer. The photometer curve gives only insufficient information concerning the success of the separation. Therefore, every fraction is checked by SDS–polyacrylamide gel electrophoresis.[18] The electropherograms show one, two, or three bands. Fractions with the same composition are pooled. From fractions, which contain several components, pure preparations may be obtained after rechromatography. For polypeptides with smaller molecular weights Sephacryl S-200 Superfine (Pharmacia) is suitable.

It should be noted that fractions which exhibit in the gel electrophoresis only one sharp band may still contain several components. In this case gel chromatography in alkaline medium occasionally leads to success.

*Fractionating at Alkaline pH.* A volume, which contains 200–250 mg polypeptides, is concentrated by ultrafiltration, (Amicon, H1DP10 hollow fiber and PM 10 membrane) to about 70 ml and dialyzed for 48 hr against flowing buffer of the following composition:

0.1 $M$ Tris–HCl buffer, pH 9.2, 1000 ml
Sodium dodecyl sulfate, 2.5 gm
$\beta$-Mercaptoethanol, 1 ml

During dialysis the volume increases to about 100 ml. The solution is applied onto 6 to 8 columns each of 10 cm diameter and 100 cm length, connected in series, and eluted with the above described buffer. Again, the columns contain CL-6B Sepharose. Fractionation is achieved with the same equipment described above. In this case, fractions of 100 ml are collected.

By chromatography in alkaline medium, from a polypeptide fraction with apparent molecular weight 66,000, a component is isolated which

[18] K. Weber and M. Osborn, *J. Biol. Chem.* **244**, 4406 (1969).

belongs to reaction center of photosystem II. In addition, from the electrophoretically uniform preparation two components of lower molecular weights are separated.

## Chromatography on Hydroxylapatite

Under certain circumstances, fractions with apparently uniform molecular weight are separated into components by the method of Moss and Rosenblum.[19] The preparations are transferred by dialysis into the following buffer.

0.1 $M$ Sodium phosphate buffer, pH 6.4, 1000 ml
Sodium dodecyl sulfate, 1 gm
$\beta$-Mercaptoethanol, 1 ml

If the diluted buffer as indicated by the authors is used, a considerable portion of the polypeptides is irreversibly adsorbed onto hydroxylapatite. One milliliter of settled bed volume of hydroxylapatite (BioGel HT, Bio-Rad) per milligram of polypeptide is used with the above described buffer, the phosphate concentration of which increases in an approximately linear fashion to 0.5 $M$. The eluate is led through a conductivity and a photometer cell. It is advantageous to use a two-beam photometer.

By hydroxylapatite chromatography, two components were isolated from the polypeptide fraction of the apparent molecular weight 66,000. Both components play a role in the region of photosystem I. Antibodies directed toward one component inhibit electron transport if absorbed onto the outer surface of the thylakoid membrane; antibodies to the other component inhibit only if absorbed onto the inner surface.[14]

## Removal of Sodium Dodecyl Sulfate

For the immunological investigation of the isolated polypeptides it is in most cases necessary to remove the detergent. This is achieved by anion exchange chromatography according to Weber and Kuter.[20] The anion exchange resin AG1-X8 (Bio-Rad) is more suitable than the AG1-X2 proposed by these authors because it retains less protein on the column. The sample is placed on the column in 0.01 $M$ sodium phosphate buffer, containing 6 $M$ urea besides SDS and mercaptoethanol. The column volume depends on the SDS content of the sample. It is advisable to adjust the desired protein concentration by ultrafiltration prior to the removal of the SDS. Urea is removed from the SDS-free eluate by

[19] B. Moss and E. N. Rosenblum, *J. Biol. Chem.* **247,** 5194 (1972).
[20] K. Weber and D. J. Kuter, *J. Biol. Chem.* **246,** 4504 (1971).

diafiltration (Amicon, PM 10 membrane). The solution contains the poly-peptides in an aggregated form; the molecular weight depends on the pH.[8,9] The occurrence of turbidity and aggregation leading to sedimenting particles cannot be avoided sometimes. If the preparations are to be used for immunization, they are transferred by diafiltration or dialysis to $\beta$-mercaptoethanol-free 0.06 $M$ potassium–sodium phosphate buffer, pH 7.8. For the immunization of one rabbit 8–10 mg polypeptides are required. This amount is relatively high, which is probably due to the fact that the antigens are more or less denatured.

### Extraction of the Lipids

In the course of the listed operations, the major part of the lipids are removed from the proteins. Extraction of the remaining lipids with organic solvents causes the formation of aggregates, which hardly can be brought back into solution. If, however, the polypeptides are precipitated from the SDS-containing solution with five times the volume of acetone, followed by acetone washing of the precipitate until the supernatant is colorless, the major part of the precipitate can be redissolved in SDS-containing phosphate buffer. In order to avoid aggregation of the polypeptides, it is advantageous to apply the acetone extraction as late as possible.

### Selective Preseparations

For the isolation of certain polypeptides additional procedures are sometimes necessary. Thus, it may facilitate the separation if coupling factor and other less tightly bound proteins are removed from stroma-free chloroplasts prior to dissolution by washing with EDTA. This is also advised if, for example, the polypeptide fraction of MW 11,000 is to be prepared free of the $\epsilon$ component of the coupling factor. The EDTA treatment, however, has the disadvantage that a considerable portion of the proteins cannot be dissolved to polypeptides.

For gel chromatography also chloroplast extracts which one obtains with buffers of different pH and lower SDS concentrations may be used. These contain, comparable to the EDTA extract, the easier removable components of the thylakoids.

Finally, it should be noted, that according to Henriques and Park one can separate prior to gel chromatography the polypeptides in a hydrophobic and a hydrophilic portion.[21] However, also after this treatment,

[21] F. Henriques and R. B. Park, *Biochim. Biophys. Acta* **430**, 312 (1976).

the protein aggregates can only partially be dissolved in SDS–containing solutions.

## Cleaning of the Columns

After prolonged use it is necessary to clean and desinfect the separation plant. For this purpose washing with increasing concentrations of ethanol is used (15–70% ethanol in water). Due to their stability in organic solvents CL-6B Sepharose and Sephacryl S-200 Superfine are superior to Sepharose 6B and Sephadex. Some lots of SDS contain impurities which sediment in the lines and columns and which reduce the flow rate. These impurities must be removed by recrystallization.

## Comments

As the polypeptides are used primarily for the preparation of antisera, verification of the purity of the fractions is important. Unequivocal results are obtained in the immunological tests only if the antisera are monospecific. The best way to achieve this goal is to immunize with pure polypeptides. A purity test is possible only by determination of the terminal amino acid sequences. With the above described separation methods, polypeptides can be isolated whose purity is sufficient for sequence determination and for preparation of monospecific antisera. On the other hand, antisera may be obtained which are monospecific with respect to their action even though the uniformity of the preparation is doubtful. For example, besides antibodies which inhibit electron transport, the serum may also contain antibodies which do not. In addition, the terminal amino acid residue determination may show nonuniformity of the preparation, if partial decomposition by bacterial and cell proteases has occurred.

Finally it should be noted, that chlorophyll–protein complexes with intact chlorophyll can be isolated in SDS–containing buffers only if the reduction of the disulfide bonds with mercaptoethanol is omitted. Chlorophyll, not protected by protein, is transformed by SDS into pheophytin and other derivatives.

Acknowledgments

The authors wish to thank Dr. G. H. Schmid for assistance in the preparation of the manuscript.

## [42] Measurement of Photorespiratory Activity and the Effect of Inhibitors

*By* ISRAEL ZELITCH

"Photorespiration" is defined as the rapid evolution of $CO_2$ in light by photosynthetic tissues. It is brought about mainly by biochemical reactions associated with the synthesis and metabolism of glycolate. The existence and importance of photorespiration was first clearly described by Decker and Tió[1] based on the $CO_2$ burst they observed upon darkening leaves of many $C_3$ species after a period of steady state photosynthesis. The postillumination burst was attributed to an overshoot resulting because the primary substrate of photorespiration (later recognized as being glycolate[2]) was synthesized only in the light and had a small pool size). Thus synthesis of the substrate was promptly cut off when photosynthesis ceased in darkness and its oxidation was observed as a burst of $CO_2$ release.

The use of an electrical analog to explain gaseous diffusion into and within a leaf has been helpful in understanding the relation between photosynthesis and photorespiration.[3-7] Such models assist in perceiving the difficulties associated with obtaining precise measurements of rates of photorespiration. The evolution of photorespiratory $CO_2$ in photosynthetic systems is perceived as occurring while the main flux of $CO_2$ is from the atmosphere surrounding the leaf through a series of diffusive resistances (boundary layer, stomata, physical, photochemical, and enzymatic) into the chloroplast. Because of these diffusive resistances and fluxes, all assays of photorespiration will underestimate the rate because a portion of the released $CO_2$ will be refixed by the chloroplasts and will not escape to the atmosphere where it can be measured.

[1] J. P. Decker and M. A. Tió, *J. Agric. Univ. P. R.* **43**, 50 (1959).

[2] I. Zelitch, *Annu. Rev. Plant Physiol.* **15**, 121 (1964).

[3] P. Gaastra, *Meded. Landbouwhogesch. Wageningen* **59**, 1 (1959).

[4] I. Zelitch, "Photosynthesis, Photorespiration, and Plant Productivity." Academic Press, New York, 1971.

[5] M. M. Ludlow and P. G. Jarvis, *in* "Plant Photosynthetic Production. Manual of Methods" (Z. Šesták, J. Čatský, and P. G. Jarvis, eds.), p. 294. Junk, The Hague, 1971.

[6] D. T. Canvin and H. Fock, Vol. 24, p. 246.

[7] C. Schnarrenberger and H. Fock, *in* "Encyclopedia of Plant Physiology, New Series" (C. R. Stocking and U. Heber, eds.), Vol. 3, p. 185. Springer-Verlag, Berlin and New York, 1976.

METHODS IN ENZYMOLOGY, VOL. 69

In spite of these inherent handicaps, measurements show that in $C_3$ species $CO_2$ release in the light is three to five times faster than dark respiration. Photorespiration is much more oxygen dependent than dark respiration, and photorespiration increases greatly with increasing $O_2$ levels in the atmosphere up to 100%, while dark respiration usually reaches a maximal rate at 2 to 3% $O_2$. Photorespiration is more rapid at higher temperatures (about a threefold increase is observed from 25° to 35°) than dark respiration. Photorespiration is strongly inhibited at $CO_2$ concentrations above 0.2%, while dark respiration is usually unaffected by $CO_2$ concentration. Photorespiration can also be blocked by biochemical inhibitors that affect the synthesis and metabolism of glycolate, and these inhibitors do not affect dark respiration.

Various assays, as discussed later, show that at about 25° photorespiration in a number of $C_3$ species occurs at rates at least 50% of net $CO_2$ assimilation.[8] In $C_4$ species, such as maize, however, photorespiration is slow and difficult to detect. A number of photorespiration assays have been described, and their advantages and disadvantages discussed.[4-7] Only the more commonly used assays are evaluated here. Descriptions of flow diagrams and apparatus used for the measurement of $CO_2$ gas exchange in leaves by means of infrared $CO_2$ gas analyzers in open or closed systems abound in the literature[3,5-7,9] and do not require repeating here. All systems consist of a transparent chamber usually constructed of Plexiglas, a light source, air pumps to move the gas rapidly over the leaf or leaves, flow meters to monitor the air movement, temperature and humidity controls for the leaf chamber, driers to prevent water vapor from entering the infrared gas analyzer, and sometimes $^{14}CO_2$ detectors. The advantages and limitations of four commonly used assays are discussed below.

### Postillumination $CO_2$ Outburst in Leaves

This assay was first described in tobacco by Decker[10] (Fig. 1). Leaves are placed in an illuminated chamber until the rate of net $CO_2$ uptake is constant as determined with an infrared $CO_2$ analyzer in a closed system. The chamber is then darkened, and a rapid rate of $CO_2$ evolution is observed lasting from several minutes to as long as 9 min in some species. After the $CO_2$ outburst, the dark respiration resumes its normal constant rate. Decker estimated the initial rate of $CO_2$ release by extrapolation to

[8] I. Zelitch, *Annu. Rev. Biochem.* **44**, 123 (1975).
[9] D. N. Moss, *Conn., Agric. Exp. Stn. Bull.*, New Haven, **664**, 86 (1964).
[10] J. P. Decker, *Plant Physiol.* **30**, 82 (1955).

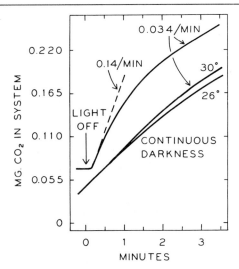

FIG. 1. The postillumination $CO_2$ outburst by a tobacco leaf (from Decker[10]). The leaf was illuminated at 3800 ft-c in a transparent chamber while the leaf temperature was kept at 24° to 26°. The air surrounding the leaf in the closed system was rapidly recycled through an infrared $CO_2$ analyzer. When the $CO_2$ concentration reached the steady $CO_2$ compensation point, the light was turned off (upper curve). The initial rate of the $CO_2$ outburst was determined by extrapolation to zero time of the outburst and is compared with the rate of $CO_2$ released in continuous darkness.

zero time in darkness. In the experiment shown in Fig. 1, for example, photorespiration was 4.1 times faster than dark respiration. He found that photorespiration increased with increasing irradiance during the photosynthetic period immediately preceding the outburst, it increased with higher leaf temperatures between 14.5° and 33.5°, and it was unchanged when the leaf was kept either at the $CO_2$ compensation point (45 ppm $CO_2$) or in normal air.[11] The latter observation is especially important, because it demonstrates that rates of photorespiration must be similar at low $CO_2$ levels (such as $CO_2$-free air) and in normal air. At high $CO_2$ levels (1200 ppm) the $CO_2$ outburst was completely eliminated.[12]

Krotkov[13] later confirmed Decker's observations and extended them by showing high $O_2$ levels were required to obtain the $CO_2$ outburst[14] and

[11] J. P. Decker, *Plant Physiol.* **34**, 100 (1959).
[12] K. Egle and H. Fock, *in* "The Biochemistry of Chloroplasts" (T. W. Goodwin, ed.), Vol. 2, p. 79. Academic Press, New York, 1967.
[13] G. Krotkov, *in* "Photosynthetic Mechanisms in Green Plants," Publ. No. 1145, p. 452. Natl. Acad. Sci.—Natl. Res. Counc., Washington, D. C., 1963.
[14] M. L. Forrester, G. Krotkov, and C. D. Nelson, *Plant Physiol.* **41**, 422 (1966).

it was abolished in 2% $O_2$. Maize leaves do not show a typical outburst,[15] an indication that photorespiration is slow in $C_4$ species, and no outburst was observed in maize even in 100% $O_2$.

This assay is rapid and relatively easy to perform, but there are several possible problems associated with it. Since the rate of $CO_2$ release decreases with time (Fig. 1), the determination of the maximal rate is subjective.[5] If the stomata should decrease in width during the assay, photorespiration will be underestimated even more. This assay is probably less sensitive to smaller changes in photorespiration than other assays, since two tobacco varieties that appeared similar by this method differed in photorespiration when assayed by the rate of $CO_2$ efflux in $CO_2$-free air.[16]

Rates of photorespiration determined by the postillumination outburst with leaves of several species, as a percentage of net photosynthesis in normal air, are soybean, 75% [17]; tobacco at 25.5°, 45% [11]; tobacco at 33.5°, 66%.[11]

### Inhibition of Net $CO_2$ Uptake by Oxygen in Leaves

Large increases in net photosynthesis (33 to 50% are usually observed in $C_3$ species when the $O_2$ content surrounding a leaf is lowered from 21 to 1–3% (Table 8.5 in Zelitch[4]), while there is no change in $CO_2$ uptake under the same conditions in $C_4$ species even when the $O_2$ level is increased to 50%.[15] The oxygen inhibition of photosynthesis was first observed by Warburg in *Chlorella* in 1920, and this inhibition has often been called the Warburg effect. Both the synthesis of glycolate and its oxidation to produce photorespiratory $CO_2$ are highly dependent on $O_2$ concentration; hence it is often assumed that the decrease in net photosynthesis with increasing $O_2$ levels is a function of the rate of photorespiration. It has been suggested that this assay overestimates photorespiration because some of the inhibition is caused by an inhibition of the carboxylation reaction.[18] Although this is true, it seems more likely that the assay underestimates photorespiration because glycolate metabolism occurs even at low concentrations of $O_2$,[19] and oxygen is always produced by chloroplasts during photosynthesis so that photorespiration cannot be completely inhibited by decreasing atmospheric $O_2$ levels.

[15] M. L. Forrester, G. Krotkov, and C. D. Nelson, *Plant Physiol.* **41,** 428 (1966).
[16] G. H. Heichel, *Plant Physiol.* **51S,** 42 (1973).
[17] N. R. Bulley and E. B. Tregunna, *Can. J. Bot.* **49,** 1277 (1971).
[18] A. L. D Aoust and D. T. Canvin, *Can. J. Bot.* **51,** 457 (1973).
[19] J. D. Eickenbusch and E. Beck, *FEBS Lett.* **31,** 225 (1973).

Recent results of rates of photorespiration determined by the inhibition of net photosynthesis in 21% $O_2$ with leaves of several species, as a percentage of net photosynthesis in normal air, are wheat, 53% [20]; potato, 50% [21]; tall fescue, 47 and 36% for two different varieties.[22]

## $CO_2$ and $^{14}CO_2$ Release in $CO_2$-Free Air by Leaves

As indicated above, photorespiration is not greatly affected by $CO_2$ levels close to "zero," and glycolate synthesis is similar in "zero" $CO_2$ and normal air. When a $C_3$ leaf is placed in light in a rapid stream of $CO_2$-free air, the rate of $CO_2$ released depends upon the rate of photorespiration, the diffusive resistances to $CO_2$ fixation by the chloroplasts, and the stomatal diffusive resistance.[23] This method of assay, like all others, therefore, reveals only a portion of the photorespiratory $CO_2$ even when stomata are wide open. The method demonstrates that rapid rates of photorespiration occur in $C_3$ species compared with dark respiration, that photorespiration is twice as rapid at 100% $O_2$ compared with 21% $O_2$, and that no detectable photorespiration occurs in maize.[4] Some failures to detect rapid photorespiration by this method are described in the literature, and such failures may be attributed to lack of sufficient irradiance, slow flow rates of $CO_2$-free air, or stomata being closed.

This assay is a convenient one, but the rate of $CO_2$ efflux is determined in an open system, and it requires rapid flow rates to keep the external $CO_2$ concentration close to zero. With rapid gas streams at concentrations of $CO_2$ close to zero, the sensitivity of the measurement with infrared $CO_2$ gas analyzers and the accuracy are limited.

To increase the sensitivity a modification was introduced whereby leaf disks are first allowed to assimilate $^{14}CO_2$ (initial concentration 0.2% $CO_2$) in a closed system at high irradiance at 30°,[24,25] The $^{14}CO_2$ is completely assimilated in about 15 min by six leaf disks (1.6 cm diameter) floating on a thin layer of water in large Warburg flasks. The $^{14}C$-labeled products are then allowed to recycle at the $CO_2$ compensation concentration for an additional 30 to 45 min. Then the $^{14}CO_2$ released in a rapid stream of moist $CO_2$-free air (3 to 7 flask volumes per minute) is bubbled through ethanolamine solution and the radioactivity measured (Fig. 2).

[20] A. J. Keys, E. V. S. B. Sampaio, M. J. Cornelius, and I. F. Bird, *J. Exp. Bot.* **28,** 525 (1977).

[21] S.-B. Ku, G. E. Edwards, and C. B. Tanner, *Plant Physiol.* **59,** 868 (1977).

[22] C. J. Nelson, K. H. Asay, and C. D. Patton, *Crop. Sci.* **15,** 629 (1975).

[23] B.-A. Bravdo, *Plant Physiol.* **43,** 479 (1968).

[24] A. Goldsworthy, *Phytochemistry* **5,** 1013 (1966).

[25] I. Zelitch, *Plant Physiol.* **43,** 1829 (1968).

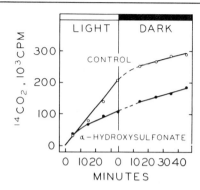

FIG. 2. $^{14}CO_2$ released by tobacco leaf disks in the $^{14}C$ assay of photorespiration and the effect of a sulfonate (from Zelitch[25]). Leaf disks (1.6 cm diameter) were floated on water at 35° in air for 45 min at 1000 ft-c. The flasks (75 ml) were closed and 5 $\mu$mole of $^{14}CO_2$ (2.95 × 10⁶ cpm) were released. After an additional 45 min, at zero time the fluid was replaced with water or 10 m$M$ $\alpha$-hydroxy-2-pyridinemethanesulfonic acid and $CO_2$-free air (7 flask volumes/min) was passed through the vessels. The $^{14}CO_2$ evolved was collected in ethanolamine solution and the radioactivity determined. After 35 min, the flasks were darkened.

The results are usually expressed as the ratio of the $^{14}CO_2$ released in the light to that in darkness during 30 min periods when the rates are constant. Photorespiration was three to five times faster than dark respiration in leaves of $C_3$ species and very slow in maize.[25,26] The specific radioactivity of $^{14}CO_2$ released in the light was not greatly different from that in darkness (70 versus 50% of the initial specific radioactivity, respectively),[24] and in sunflower leaves the rate of $^{14}CO_2$ evolution and the specific radioactivity became constant after about 10 min of supplying $^{14}CO_2$ in the light.[27]

The $^{14}C$-assay of photorespiration was used to demonstrate the importance of having open stomata, fast flow rates, higher temperatures, higher $O_2$ levels, and $CO_2$-free air in order to observe rapid rates of photorespiration.[25] The $^{14}C$ assay is usually carried out under conditions where the leaf disks completely fix the $^{14}CO_2$ within 15 min in a closed system, and a further 30 min is allowed to elapse before $CO_2$-free air is passed over the disks at zero time. If the collection of the released $^{14}CO_2$ is started immediately after the first 15 min, the ratio of the $^{14}CO_2$ released in the light/dark is about twice as great as the ratio observed after an

[26] I. Zelitch and P. R. Day, *Plant Physiol.* **43**, 1838 (1968).
[27] A. L. D'Aoust and D. T. Canvin, *Photosynthetica* **6**, 150 (1972).

incubation of 30 or the usual 45 min. The ratio is constant, however, after 30 min incubation, and the release of $^{14}CO_2$ is linear with time. This suggests that after about 15 min of recycling within the leaf disks the specific radioactivity of the $^{14}CO_2$ released into the $CO_2$-free air is constant.

The $^{14}C$ assay does not measure photorespiration in absolute units because the specific radioactivity of the $^{14}CO_2$ is usually not determined, and this is a limitation of the method. It measures photorespiration relative to dark respiration. In spite of its limitations, the $^{14}C$-assay has been found useful in studies on the genetic variation of photorespiration[26] and the biochemical control of this process as discussed below.

Rates of photorespiration determined by measurement of $CO_2$ released in $CO_2$-free air with leaves of several $C_3$ species, as a percentage of net photosynthesis in normal air, are soybean, 46%[28]; soybean, 42%[29]; sugar beet, 43%[30]; tobacco, 55%[31]; wheat, 69%.[20]

### Short-Time Uptake of $^{14}CO_2$ and $^{12}CO_2$ in Leaves

If a leaf is allowed to carry out photosynthesis under steady state conditions and the $^{12}CO_2$ is suddenly replaced with a stream of $^{14}CO_2$ at the same concentration, the initial rate of $^{14}CO_2$ uptake will represent the gross photosynthesis (sometimes called "true" photosynthesis). This rate will be greater than the previous rate of $^{12}CO_2$ uptake, which is a measure of the net photosynthesis (sometimes called "apparent" photosynthesis). The difference between the $^{14}CO_2$ and net $^{12}CO_2$ uptakes should equal the $CO_2$ evolved by photorespiration. This method of assaying photorespiration was first reported in principle by Bidwell et al.[32] In practice, however, recently fixed $^{14}CO_2$ is respired very rapidly. The $CO_2$ released within a leaf can be detected as a decreased specific radioactivity in the air outside sunflower leaves within 15 to 45 seconds.[33] Therefore, considerable internal $CO_2$ release and refixation must occur in periods less than 15 seconds within the leaf, making the precise determination of gross photosynthesis very difficult. This method then also underestimates photorespiration because this recycling will cause gross photosynthesis

[28] Y. B. Samish, J. C. Pallas, Jr., G. M. Dornhoff, and R. M. Shibles, *Plant Physiol.* **50**, 28 (1972).
[29] G. Hofstra and J. D. Hesketh, *Planta* **85**, 228 (1969).
[30] N. Terry and A. Ulrich, *Plant Physiol.* **54**, 379 (1974).
[31] T. Kisaki, *Plant Cell Physiol.*, **14**, 505 (1973).
[32] R. G. S. Bidwell, W. B. Levin, and D. C. Shepard, *Plant Physiol.* **44**, 946 (1969).
[33] L. J. Ludwig and G. Krotkov, *Plant Physiol.* **42S**, 47 (1967).

to be underestimated even in the short times used to make the necessary measurements.

The type of apparatus required to simultaneously measure net $CO_2$ gas exchange and $^{14}CO_2$ uptake in short times has been described in the papers cited below. The method as initially described in Canvin's laboratory required 30 seconds for the simultaneous measurement of $^{14}CO_2$ and $^{12}CO_2$ uptake.[34] Later the apparatus was modified and the times decreased to 20 seconds,[6] and finally to 15 seconds.[35] The specific radioactivity of the $^{14}CO_2$ available to the leaf is calculated from the average of the specific radioactivity of the gas stream entering the leaf chamber and the specific radioactivity of the exiting gas stream.[6] Large errors may be introduced by this calculation, especially when high $CO_2$ concentrations (and low specific radioactivity) are used. This may explain why by this method of assay the unusual result was obtained that photorespiration *increased* at higher $CO_2$ concentration.[35,36] This result is inconsistent with considerable data in the literature showing that photorespiration *decreases* at high $CO_2$ levels. By this method results were also obtained showing that photorespiration was unaffected by increasing the temperature from 15° to 35°C in sunflower and tobacco leaves,[36] another result contrary to most other reports in the literature.

Rates of photorespiration determined by measurement of the difference between short time uptake of $^{14}CO_2$ and $^{12}CO_2$ with leaves of several $C_3$ species, as a percentage of net photosynthesis in normal air, are: sunflower, 31% [35]; sunflower, 33% [37] tobacco, 47% [36]; wheat, 17%.[20]

### Assay of Photorespiration in Algae and Submerged Aquatic Plants

The assay of photorespiration in aqueous systems is subject to even greater underestimation than for leaves of land plants because the diffusion of $CO_2$ in water is several orders of magnitude slower than in air and thus the refixation of photorespired $CO_2$ will be even greater. The $^{14}C$-assay applied to *Chlorella* and *Chlamydomonas* suspensions grown in normal air showed a faster $^{14}CO_2$ release in the light than in darkness.[28] Lower values for photorespiration for these algae were obtained by the same assay procedure[38] and still lower rates for blue-green algae. An active photorespiration was demonstrated in seven species of marine

[34] L. J. Ludwig and D. T. Canvin, *Can J. Bot.* **49,** 1299 (1971).
[35] B. Bravdo and D. T. Canvin, *Proc. Int. Congr. Photosynth. 1974* p. 1277 (1975).
[36] H. Fock and K.-R. Przybylla, *Ber. Dtsch. Bot. Ges.* **89,** 643 (1976).
[37] D. W. Lawlor and H. Fock, *Planta* **126,** 247 (1975).
[38] K. H. Cheng and B. Colman, *Planta* **115,** 207 (1974).

algae and one freshwater species by showing the inhibition of net photosynthesis in 21% $O_2$ compared with an atmosphere of $N_2$ and the
existence of a postillumination $CO_2$ outburst.[39]

The [14]C assay of photorespiration has been applied to submerged
aquatic angiosperms, and, in a spite of extensive refixation of $CO_2$ in
such environments, the presence of photorespiration was established by
the strong dependence of the rate of [14]$CO_2$ release in the light on the
dissolved $O_2$ concentration.[40,41]

### Biochemical Inhibitors of Glycolate Oxidation

In order to be metabolized glycolate must first be oxidized to glyoxylate by the flavoprotein glycolate oxidase. Long-term inhibition of the
glycolate oxidase reaction would not, therefore, seem a practical solution
for the regulation of photorespiration, since glycolate would continue to
accumulate and ultimately reach toxic concentrations. Biochemical in-
hibitiors of glycolate oxidase are useful to show the importance of the
glycolate pathway as a source of photorespiratory $CO_2$.

α-Hydroxysulfonates, aldehyde–bisulfite addition compounds, are effective competitive inhibitors of glycolate oxidase.[42] When α-hydroxy-2-
pyridinemethanesulfonic acid was supplied to tobacco leaf disks in the
light, the enzymatic oxidation in the tissue was blocked and glycolate
accumulated at initial rates (70 μmole/g fresh weight/hr) sufficiently rapid
to account for photorespiration in tobacco and sunflower.[43] In maize leaf
disks, glycolate synthesis was 10% of that in the $C_3$ species.[43,44] This
sulfonate inhibited photorespiration, but not dark respiration, in the [14]C
assay[25] (Fig. 2), and under suitable conditions of temperature and short
times of exposure, large increases in photosynthetic uptake were obtained[45] (Table I).

Jewess et al.[46] described a class of irreversible inhibitors of glycolate
oxidase. These are acetylenic substrate analogs of glycolate. One of
these, 2-hydroxy-3-butynoate, was supplied to pea leaf disks in the light
and glycolate accumulation was measured. The action of this inhibitor
was much slower than that observed with sulfonates. Maximal rates of

[39] J. E. Burris, *Mar. Biol.* **39,** 371 (1977).
[40] R. A. Hough and R. G. Wetzel, *Plant Physiol.* **49,** 987 (1972).
[41] R. A. Hough, *Limnol. Oceanogr.* **19,** 912 (1974).
[42] I. Zelitch, *J. Biol. Chem.* **224,** 251 (1957).
[43] I. Zelitch, *Plant Physiol.* **51,** 299 (1973).
[44] I. Zelitch, *Arch. Biochem. Biophys.* **163,** 367 (1974).
[45] I. Zelitch, *Plant Physiol.* **41,** 1623 (1966).
[46] P. J. Jewess, M. W. Kerr, and D. P. Whitaker, *FEBS Lett.* **53,** 292 (1975).

TABLE I
Compounds that Inhibit Glycolate Synthesis, Glycolate Oxidation, and Photorespiration and their Effect on Net Photosynthesis in Tobacco Leaf Disks

| Solution supplied to disks in light | Inhibition of glycolate synthesis (%) | Inhibition of photorespiration in $^{14}C$-assay (%) | Increase in net $^{14}CO_2$ assimilation (%) | Reference |
|---|---|---|---|---|
| Isonicotinic acid hydrazide, 10 mM | | 69 | | 24 |
| Isonicotinic acid hydrazide, 9 mM | | 49[a] | | 51 |
| Isonicotinic acid hydrazide, 10 mM | 35 | | | 43 |
| α-Hydroxy-2-pyridinemethane-sulfonic acid, 10 mM | | 60[a] | | 25 |
| α-Hydroxy-2-pyridinemethane-sulfonic acid, 10 mM | | | >200 | 45 |
| 3-(4-Chlorophenyl)-1,1-dime-thylurea (CMU), 0.1 mM | 43 | 48[a] | | 43 |
| 2,3-Epoxypropionate (glyci-date), 10 mM | 40–50 | 40 | 40–50 | 44 |
| Phosphoenolpyruvate, 10 mM | 72 | | | 43 |
| Phosphoenolpyruvate, 30 mM | 24 | | | 49 |
| L-Aspartate, 30 mM | 20 | | | 49 |
| L-Glutamate, 30 mM[b] | 40 | 60 | 15–25[c] | 49 |
| Glyoxylate, 20 mM[d] | 50 | 57 | 100[a] | 50 |

[a] Under the conditions used in these experiments the inhibitors did not close the leaf stomata.

[b] The average glutamate concentration in leaf disks floated on water was 2.2 mM and in disks floated on glutamate was 5.4 mM.

[c] Treatment of leaf disks with glutamate, and particularly glyoxylate, stimulated $CO_2$ fixation rates much more than the rates of glutamate or glyoxylate metabolism on a molar basis.

[d] The average glyoxylate concentration in leaf disks floated on water was 0.61 mM and in disks floated on glyoxylate was 0.87 mM.

glycolate accumulation were 1.8 $\mu$mole compared with the 70 $\mu$mole/g fresh weight/hr often observed with a sulfonate in $C_3$ tissues.

## Inhibitors of Glycolate Synthesis in Leaves

Inhibitors of photosynthetic electron transport such as 3-(4-chloro-phenyl)-1,1-dimethylurea (CMU) also strongly inhibit glycolate synthesis

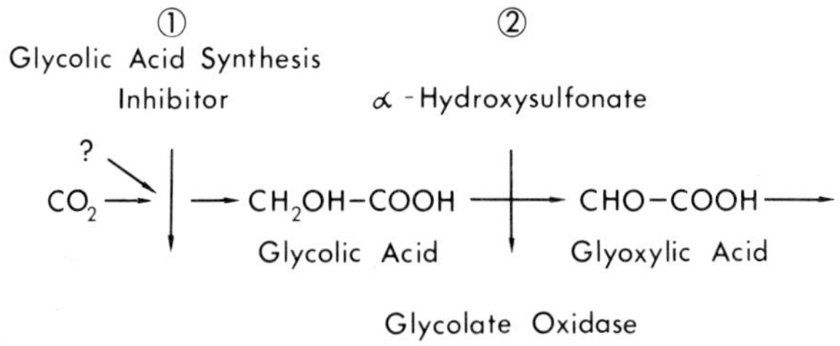

①                                    ②
Glycolic Acid Synthesis
        Inhibitor                α -Hydroxysulfonate

?
CO₂ ─→ | ─→ CH₂OH–COOH ─┼─ CHO–COOH ─→
                Glycolic Acid          Glyoxylic Acid

Glycolate Oxidase

FIG. 3. Diagrammatic representation of an assay for the activity of inhibitors of glycolate synthesis in leaf disks (from Zelitch[44]). Leaf discs were floated on water for 60 min in the light. Then during the first experimental period, the water was replaced with solutions of the assumed inhibitor of glycolate synthesis for 60 min. This solution was removed, and during the second period the disks were floated for 3 min on 10 m$M$ α-hydroxy-2-pyridinemethanesulfonic acid, an inhibitor of glycolate oxidase. The initial rate of glycolate accumulation (glycolate synthesis) was determined and compared with glycolate accumulation in leaf disks floated on water during the first experimental period.

(Table I). Inhibitors of glycolate synthesis were sought that would specifically block photorespiration in $C_3$ species and increase $CO_2$ assimilation in leaf disks. The concentration of glycolate in leaf tissue is low (usually less than 0.5 m$M$), and its turnover is exceedingly rapid (at least 50% of net photosynthesis). An assay was developed[44] in which leaf disks were floated on solutions of the inhibitor to be tested for 60 min in the light (Fig. 3). The inhibitor solution was then removed and replaced with solution of a sulfonate for 3 min. Glycolate accumulation was measured colorimetrically in extracts of killed leaf disks after first isolating the glycolate by column chromatography with Dowex-1 acetate anion exchange resin.

With such an assay 20 m$M$ glycidate, 2,3-epoxypropionate, inhibited glycolate synthesis and photorespiration about 50% and increased photosynthetic $^{14}CO_2$ in leaf disks about 50% [44] (Table I). The products of $^{14}CO_2$ assimilation were examined. As expected if glycolate synthesis were inhibited, the pool sizes of products of the glycolate pathway, glycine and serine, were decreased. The pool sizes of aspartate and glutamate were, unexpectedly, about twice as great in leaf disks treated with glycidate. As discussed later, these changes in pool size may largely account for the regulatory effect of glycidate.

[1-$^{14}$C]Glycidate was synthesized, and its binding to proteins in leaf disks was investigated under conditions where glycolate synthesis in the tissue was inhibited at least 50%.[47] Glycidate did not combine with or

[47] I. Zelitch, *Plant Physiol.* **61**, 236 (1978).

inhibit ribulosediphosphate carboxylase or affect the inhibition of this enzyme by oxygen. Glycidate also had no effect on the activities of glycolate oxidase, phosphoglycolate phosphatase, and NADH–glyoxylate reductase, and slightly inhibited NADPH–glyoxylate reductase. A strong inhibition of glutamate:glyoxylate aminotransferase activity by glycidate was found in particulate preparations of tobacco leaf and callus.[48] The pool sizes of glyoxylate as well as those of aspartate and glutamate increased under conditions where net photosynthesis increased in leaf discs in the presence of glycidate.

Using the above assay (Fig. 3) common metabolites were also supplied to leaf disks to determine whether changing the pool sizes of metabolites would aslo regulate glycolate synthesis and photorespiration. These studies revealed that floating leaf disks on about 30 m$M$ solutions of L-glutamate, L-aspartate, phosphoenolpyruvate, or glyoxylate effectively inhibited glycolate synthesis.[49,50] With glutamate, for example, glycolate synthesis was inhibited about 40%, photorespiration in the [14]C assay was inhibited about 60%, and there was no substantial effect on dark respiration. Fixation of [14]CO$_2$ by the leaf discs was increased about 25% by glutamate treatment (Table I). It is believed that the regulation of glycolate synthesis by supplying glutamate is probably brought about by a metabolite of glutamate rather than by glutamate itself.[49]

Floating leaf disks on 20 m$M$ glyoxylate solution inhibited glycolate synthesis about 50% and inhibited photorespiration similarly.[50] Net photosynthesis was increased twofold under optimal conditions (Table I). Treating leaf disks with glyoxylate increased the total glyoxylate concentration in the disks from 0.61 to 0.87 m$M$. Thus modest alterations in the pool size of a common metabolite can increase photosynthesis by inhibiting glycolate synthesis and photorespiration probably by a feedback-type mechanism. The effect of glycidate on inhibiting glycolate synthesis is thus probably an indirect one brought about by increasing the pool sizes of glutamate, aspartate, and glyoxylate. Compounds that inhibit glycolate synthesis, glycolate oxidation, and photorespiration in a relatively specific manner when supplied to leaf disks (1.6-cm diameter) are summarized in Table I.[24,25,43–45,49–51]

[48] A. L. Lawyer and I. Zelitch, *Plant Physiol.* **61**, 242 (1978).
[49] D. J. Oliver and I. Zelitch, *Plant Physiol.* **59**, 688 (1977).
[50] D. J. Oliver and I. Zelitch, *Science* **196**, 1450 (1977).
[51] I. Zelitch, *Plant Physiol.* **50**, 109 (1972).

# [43] Cycling Assay for Nicotinamide Adenine Dinucleotides

By HISAKO MATSUMURA and SHIGETOH MIYACHI

The methods currently available for the quantitative estimation of nicotinamide adenine dinucleotides are classified into spectrophotometric (I), fluorometric (II), and enzyme cycling (III). The spectrophotometric method is dependent on the measurement of the absorption of the reduced coenzyme at 340 nm or that of the cyanide addition complex of the oxidized coenzyme at 325 nm.[1,2] The fluorometric method is dependent on the fluorescence of the reduced coenzymes or that of the fluorescent derivatives made by adding methyl ethyl ketone[1,2] to the oxidized coenzymes. Table I[1-11] shows that the spectrophotometric and fluorometric methods require coenzyme concentrations higher than $2 \times 10^{-4}$ and $1 \times 10^{-6}$ to $2 \times 10^{-7}$ $M$, respectively. On the other hand, the enzyme cycling method can be applied to concentrations as low as $3 \times 10^{-7}$ to $1 \times 10^{-9}$ $M$. Since the nicotinamide adenine dinucleotide levels in plant tissues are generally extremely low, it is necessary to use one of the enzyme cycling methods. Among the several variants of the enzyme cycling method cited in Table I, the use of 3-(4,5-dimethylthiazolyl-2)-2,5-diphenyltetrazolium bromide (MTT) as a terminal electron acceptor[10,11] (III, f in Table I) is described below.

*Extraction of Nicotinamide Adenine Dinucleotides from Plants.* An equal amount of isolated chloroplasts or unicellular algae is transferred to the preheated NaOH and HCl solutions (final concentration, 0.1 $N$),

---

[1] M. M. Ciotti and N. O. Kaplan, Vol. 3 [128].

[2] M. Klingenberg, *in* "Methoden der Enzymatischen Analyse" (H.-U. Bergmeyer, ed.), 2nd ed., p. 2045. Verlag Chemie, Weinheim, 1974.

[3] O. H. Lowry and J. V. Passonneau, Vol. 6 [111].

[4] J. V. Passonneau and O. H. Lowry, *in* "Methoden der Enzymatischen Analyse" (H.-U. Bergmeyer, ed.), 2nd ed., p. 2059. Verlag Chemie, Weinheim, 1974.

[5] G. E. Glock and P. McLean, *Biochem. J.* **61**, 381 (1955).

[6] Y. Yamamoto, *Plant Physiol.* **38**, 45 (1963).

[7] T. F. Slater and B. Sawyer, *Nature (London)* **193**, 454 (1962).

[8] P. J. C. Smith, *Nature (London)* **190**, 84 (1961).

[9] O. H. Cartier, *Eur. J. Biochem.* **4**, 247 (1968).

[10] J. S. Nisselbaum and S. Green, *Anal. Biochem.* **27**, 212 (1969).

[11] C. Bernofsky and M. Swan, *Anal. Biochem.* **53**, 452 (1973).

METHODS IN ENZYMOLOGY, VOL. 69

TABLE I
METHODS FOR ASSAYING NICOTINAMIDE ADENINE DINUCLEOTIDES

| | Sensitivity | |
|---|---|---|
| Principle[a] | $M$ | moles/assay |

I. Spectrophotometric method[1,2]

  a. Absorption of reduced coenzyme at 340 nm  $2 \times 10^{-4}$  $1 \times 10^{-8}$

  b. Absorption of cyanide complex of oxidized form at 325 nm  $2 \times 10^{-4}$  $1 \times 10^{-8}$

II. Fluorometric method[1,2]

  a. Fluorescence of reduced form  $2 \times 10^{-6}$  $2 \times 10^{-9}$

  b. Fluorescence of the derivative of oxidized form produced by treating with strong alkali  $2 \times 10^{-7}$  $2 \times 10^{-10}$

  c. Fluorescence of the derivative formed by the reaction between oxidized form and methyl ethyl ketone  $1 \times 10^{-6}$  $1 \times 10^{-8}$

    Note: Standard curve must be made for each assay

III. Enzyme cycling method

  a. [3,4]

NAD(H):

Glu ⤸ ⤷NAD ⤵ ⤹Lactate
αKG ⤸ ⤸NADH ⤵ ⤷Pyruvate
   GluDH     LDH

Pyruvate + NADH $\xrightarrow{\text{LDH}}$ Lactate + NAD⁺  $1 \times 10^{-9}$  $1 \times 10^{-13}$

NAD⁺ + strong alkali → fluorescent substance

NADP(H):

Glu ⤸ ⤷NADP⁺ ⤵ ⤹G6P
αKG ⤸ ⤸NADPH ⤵ ⤷6PG
   GluDH    G6PDH

6PG + NADP⁺ → Ru5P + NADPH
       6PGDH

Note: Assay system is rather complicated. Also highly purified enzymes are required

  b. [5]

NAD(H):

Ethanol ⤵ ⤹NAD⁺ ⤸ ⤷Cytochrome $c^{2+}$
Acetaldehyde ⤸ ⤷NADH ⤵ ⤸Cytochrome $c^{3+}$
    ADH   NAD⁺–Cyt $c$ reductase  $4 \times 10^{-8}$  $1 \times 10^{-8}$

NADP(H):

G6P ⤵ ⤹NADP⁺ ⤸ ⤷Cytochrome $c^{2+}$
6PG ⤸ ⤸NADPH ⤷Cytochrome $c^{3+}$
   G6PDH    NADP⁺Cyt $c$ reductase

Note: Follow the increase in absorbancy at 550 nm. Very sensitive. However must use enzymes which are commercially unavailable

TABLE I (*Continued*)

| | Sensitivity | |
|---|---|---|
| Principle[a] | M | moles/assay |

**c. [6]**

NAD(H):

Ethanol   NAD$^+$   DCPIP$_{red}$     $3 \times 10^{-7}$   $1 \times 10^{-9}$
Acetaldehyde   NADH   DCPIP$_{ox}$
     ADH      NADH–diaphorase

NADP(H):

Isocitrate   NADP$^+$   DCPIP$_{red}$
$\alpha$KG   NADPH   DCPIP$_{ox}$
     IDH      NADPH–diaphorase

Note: Follow the change in absorbancy at 600 nm, which is due to oxidized DCPIP. Therefore, the amount of DCPIP applicable to the assay system is limted. This limitation results in a shortening of the time period during which the cycle is linear at higher levels of coenzyme

**d. [7,8]**

NAD(H):

Ethanol   NAD$^+$   PMS$_{red}$   DCPIP$_{ox}$    $1 \times 10^{-8}$   $1 \times 10^{-11}$
Acetaldehyde   NADH   MS$_{ox}$   DCPIP$_{red}$
     ADH      PMS$_{ox}$

NADP(H):

G6P   NADP$^+$   PMS$_{red}$   DCPIP$_{ox}$
6PG   NADPH   PMS$_{ox}$   DCPIP$_{red}$
  G6PDH

Note: PMS may be subject to oxidation at higher pH[11]

**e. [9]**

NAD(H):

Ethanol   NAD$^+$   PMS$_{red}$   resazurine    $2 \times 10^{-8}$   $1 \times 10^{-11}$
Acetaldehyde   NADH   PMS$_{ox}$   resorufine
     ADH

NADP(H):

G6P   NADP$^+$   PMS$_{red}$   resazurine
6PG   NADPH   PMS$_{ox}$   resorufine
  G6PDH

Note: Follow the initial rate of flourescence increase at 580 nm which is induced by illuminating 540-nm light

**f. [10,11]**

NAD(H):

Ethanol   NAD$^+$   PES$_{red}$   MTT$_{ox}$    $1 \times 10^{-9}$   $1 \times 10^{-12}$
Acetaldehyde   NADH   PES$_{ox}$   MTT$_{red}$
     ADH

*(Continued)*

TABLE I (*Continued*)

| Principle[a] | Sensitivity | |
|---|---|---|
| | $M$ | moles/assay |

NADP(H):

$$\text{G6P} \searrow \quad \text{NADP}^+ \quad \searrow \quad \text{PES}_{red} \quad \searrow \quad \text{MTT}_{ox}$$
$$\text{6PG} \nearrow \quad \text{NADPH} \quad \nearrow \quad \text{PES}_{ox} \quad \nearrow \quad \text{MTT}_{red}$$
G6PDH

Note: PES and reduced MTT are stable. Therefore, the reaction proceeds linearly for a long time

[a] Abbreviations: DCPIP, 2,6-dichlorophenol indophenol; Glu, glutamate; GluDH, glutamate dehydrogenase; G6PDH, glucose-6-phosphate dehydrogenase; IDH, isocitrate dehydrogenase; $\alpha$KG, $\alpha$-ketoglutarate; MTT, 3-(4,5-dimethylthiazolyl-2)-2,5-diphenyltetrazolium bromide; PES, phenazine ethosulfate; 6PG, 6-phosphogluconate; PMS, phenazine methosulfate; Ru5P, ribulose-5-phosphate; Cyt, cytochrome; ADH, alcohol dehydrogenase; LDH, lactate dehydrogenase.

respectively. The suspensions are kept at 100° for 2 min, then cooled to 0° and centrifuged.[12] NADH and NADPH are specifically extracted in the supernatant obtained after the alkaline treatment, while NAD$^+$ and NADP$^+$ are extracted after acid treatment.

Larger plant tissues, such as leaf segment, may be quickly transferred to light petroleum (b.p. 40°–60°) at $-120°$[13] and lyophilized. Reduced and oxidized coenzymes are extracted by alkali (0.1 $N$ NaOH[13] or boiling 0.1 $N$ NaHCO$_3$[14]) or acid (0.1 $N$ perchloric acid[13] or 5% trichloroacetic acid[14]), respectively.

*Principle.* This method utilizes a cycling mixture composed of MTT, phenazine ethosulfate (PES), and ethanol and alcohol dehydrogenase for the determination of NAD$^+$ (or NADH) or glucose-6-phosphate (G6P) and G6P dehydrogenase for the determination of NADP$^+$ (or NADPH). NAD$^+$ or NADP$^+$ is reduced by the respective dehydrogenase system, and the NADH or NADPH formed reduces MTT through the mediation of PES (see III, f in Table I). The rate of reduction of MTT is proportional to the concentration of coenzyme.

*Reagents*

Bicine–NaOH buffer, 1.0 $M$, pH 8.0

[12] T. Oh-hama and S. Miyachi, *Biochim. Biophys. Acta* **34**, 202 (1959).
[13] U. W. Heber and K. A. Santarius, *Biochim. Biophys. Acta* **109**, 390 (1965).
[14] W. L. Ogren and D. W. Krogman, *J. Biol. Chem.* **240**, 4603 (1965).

NaOH, 0.1 $N$

HCl, 0.1 $N$

Ethanol, 5.0 $M$

Glucose-6-phosphate (disodium salt), 50 m$M$ [sometimes contains NADP, which can be removed by adding active charcoal (2%)]

EDTA (disodium salt), 40 m$M$

Phenazine ethosulfate, 16.6 m$M$

3-(4,5-Dimethylthiazolyl-2)-2,5-diphenyltetrazolium bromide, 4.2 m$M$

Alcohol dehydrogenase, 500 units/ml of 0.1 $M$

Bicine–NaOH buffer, pH 8.0

Glucose-6-phosphate dehydrogenase, 35 units/ml of 0.1 $M$

Bicine–NaOH buffer, pH 8.0

Standard solution of nicotinamide adenine dinucleotides, 5 m$M$ [Prepare and store according to Burch.[15] Dilute to 0.5 $\mu M$ before the assay.]

*Procedures.* Reaction should be carried out in the dark or in a brown test tube. The blank without enzyme addition is higher if the reaction is carried in light.

The acid as well as alkaline extract (0.1–0.5 ml) from the plant material is placed in a small test tube containing 0.1 ml Bicine–NaOH buffer. The extract is neutralized by adding an equivalent amount of NaOH or HCl, followed by addition of 0.1 ml each of EDTA, MTT, PES, and ethanol for the determination of NAD(H), or 0.05 ml of glucose-6-phosphate for the determination of NADP(H). After adjusting the total volume to 1 ml by adding $H_2O$, the test tubes are kept at 37° for 5 min. The reaction is started by adding 0.02 ml of alcohol dehydrogenase [for NAD(H)] or G6P dehydrogenase [for NADP(H).]. After a predetermined reaction time, the reaction medium is transferred to a microcuvette having a light path of 10 mm, and the absorbancy at 570 nm is determined. The proper reaction time will be 30–60 min. With each extract, a blank measurement must also be carried out. This may be done by adding an equal amount of Bicine–NaOH buffer instead of enzyme.

The concentration of the coenzyme in each extract can be determined according to the standard curves which are determined with the cycling systems and known amounts of nicotinamide adenine dinucleotides. The presence of the compounds in the extract which might affect the rate of enzyme cycling must also be checked.

---

[15] H. B. Burch, Vol. 17 [100].

*Comments and Variations.* The above cycling reaction proceeds linearly when the coenzyme content is lower than 120 pmoles/test tube. When the coenzyme content is higher, the initial velocity of the cycling reaction can be compared. The reaction with alcohol dehydrogenase may be stopped by adding 12 m$M$ iodoacetate.[16] However, the reaction with G6P dehydrogenase cannot be stopped by this inhibitor. G6P and G6P dehydrogenase can be replaced with isocitrate and isocitrate dehydrogenase,[16] which can be stopped by PCMB.

[16] E. L. Jacobson and M. K. Jacobson, *Arch. Biochem. Biophys.* **175**, 627 (1967).

# [44] Enhancement

## By JACK MYERS

Enhancement (the Emerson effect) describes the synergistic or more than additive effect of two light beams of properly chosen wavelengths. The wavelengths are chosen to give a light 1 and a light 2 for the experimental material. The effect is observed in terms of a photochemical rate measured as $O_2$ evolution, as $CO_2$ uptake, or as a rate of some partial process requiring operation of two photoreactions. Presumably the synergism results from some interdependence or interaction between two photoreactions with different action spectra.

As a practical matter, enhancement is conveniently assayed as an increment in rate $(V_{12} - V_1)$ arising when a measuring beam $I_m$ of $\lambda_2$ (a light 2, as 650 nm) is added to a background beam $I_b$ of $\lambda_1$ (a light 1, as 700 nm). The increment $(V_{12} - V_1)$ divided by rate $V_2$ ($\lambda_2$ without background) becomes an instructive measure of enhancement

$$E_2 = (V_{12} - V_1)/V_2$$

$E_2$ approaches a maximum value with increasing ratio $V_1/V_2$. A more complete and technically advantageous modification is to measure $V_2$ also as an increment in rate observed when $\lambda_2$ is added to a background beam at identical wavelength $\lambda_2$. Then

$$E_2 = (V_{12} - V_1)/(V_{22} - V_2)$$

The procedure also can be inverted to obtain

$$E_1 = (V_{12} - V_2)/V_1$$

which approaches a maximum value with increasing $V_2/V_1$ ratio. Both $E_1$ and $E_2$ depend upon the ratio between, but not the absolute values of, $V_1$ and $V_2$.

Several particular rate measurements for demonstration of enhancement were described by Ben Hayyim and Avron.[1] Of these the most elegant is the rate-measuring $O_2$ electrode of Haxo and Blinks with modulation of the measuring light beam $\lambda_2$ and extraction of the modulated amperometric signal.[2] When added to a steady background beam, the modulated signal measures directly and separately the rate attributable to the labeled measuring beam. In order to achieve further analysis of interaction, it has been possible to label two beams at different frequencies of modulation.[3] However, there is no special restriction on method of measurement. This article will be directed toward considerations of interpretation with updating from past treatments.[1,4]

Demonstration and measurement of enhancement require that, except for the interaction effect, all rates are strictly proportional to intensities. Approach to light saturation and the Kok effect are conditions known to obscure the phenomenon.[4] Deactivation within the oxygen-generating system[5] gives rise to more than linear rates at very low intensities[6]; hence use of too low an intensity for a measuring beam $I_m$ could exaggerate or give false indication of enhancement. Large errors arising from any of the above can be guarded against by comparing rate increments with those obtained when $I_m$ and $I_b$ are purposely made identical in wavelength (see above).

*Interaction between Photoreactions.* The interaction giving rise to enhancement can be shown from the three equations needed for kinetic analysis of the Z scheme.

Photoreaction II

$$\frac{1}{2} H_2O + Q \rightarrow Q^- + H^+ + \frac{1}{4} O_2 \tag{1}$$

with rate

$$v_q = I\alpha q \tag{1a}$$

Photoreaction I

$$P + X \rightarrow P^+ + X^- \tag{2}$$

[1] G. Ben Hayyim and M. Avron, Vol. 24, p. 293.
[2] P. Joliot and A. Joliot, *Biochim. Biophys. Acta* 153, 625 (1968).
[3] R. T. Wang, C. L. R. Stevens, and J. Myers, *Photochem. Photobiol.* 25, 103 (1977).
[4] J. Myers, *Annu. Rev. Plant Physiol.* 22, 289 (1971).
[5] B. Kok, B. Forbush, and M. McGloin, *Photochem. Photobiol.* 11, 457 (1970).
[6] R. T. Wang and J. Myers, *Biochim. Biophys. Acta* 347, 134 (1974).

with rate

$$v_p = I (1 - \alpha) p \tag{2a}$$

and

$$Q^- + P^+ \leftrightarrows Q + P \tag{3}$$

such that

$$K = qp/[(1 - q)(1 - p)] \tag{3a}$$

Notation used is conventional. Fractions of open reaction centers, Q and P, are $q$ and $p$. $I$ is the rate of total quantum absorption. The critical term $\alpha(\lambda)$ is the fraction of absorbed quanta giving excitations available to the centers of photoreaction II ($Q + Q^-$); the fraction $1 - \alpha$ is available to centers of photoreaction I ($P + P^+$). This definition differs from early usage in which absorption was viewed in terms of particular pigments and assumptions were made about mechanism of energy transfer.[7] Enhancement data alone contain no information on pathway or mechanism of energy transfer (e.g., spillover) and reflect only the distribution of excitation energy as viewed by reaction centers.

The above equations provide an abridged framework for consideration of enhancement. The nonlinearity in rate of photoreaction II[9] is ignored and the quantum yield of each photoreaction is assumed to be one. A more detailed treatment has been made.[8] Equation (3a) is included only to support the following observation. Early interpretations of enhancement used a simple bottleneck principle applied to interaction. Practically this meant that $K$ was assumed to be high (>1000) so that $q \cong 1.0$ for *any* light 1 ($\alpha < 0.5$) and $p \cong 1.0$ for *any* light 2 ($\alpha > 0.5$). It appears that $K$ is not sufficiently high to allow this simple rationale.

All wavelength-dependent features of enhancement are embodied in $\alpha$. At steady state the two photoreactions proceed at equal rates so that

$$\alpha q = (1 - \alpha)p \tag{4}$$

It is readily shown that throughput quantum yield is maximum at $\alpha = 0.5$, where $q = p$ and decreases as $\alpha$ departs from 0.5.[8] Low values of $\alpha$ at longer wavelengths give rise to the red drop in quantum yield. The adding together of two light beams as 690 nm with $\alpha < 0.5$ and 640 nm with $\alpha > 0.5$ gives an effective $\alpha$ closer to 0.5 and improves quantum yield as reckoned for quanta of both beams. As a practical matter it

[7] J. Myers, *in* "Photosynthesis Mechanisms in Green Plants," Publ. No. 1145, p. 301. Natl. Acad. Sci.—Natl. Res. Counc., Washington, D.C., 1963.

[8] R. T. Wang and J. Myers, *Photochem. Photobiol.* **23**, 405 (1976).

[9] P. Joliot, A. Joliot, and B. Kok, *Biochim. Biophys. Acta* **153**, 635 (1968).

appears that, even in plants of very different pigmentation, $\alpha$ has limits of about 0.1 to 0.6.

Analysis of enhancement requires consideration of two effects. These will be discussed for measurement of the incremental oxygen rate caused by addition of a weak measuring beam $I_m$ to a strong background beam $I_b$. A *direct* effect on rate arises from excitations produced by $I_m$ operating upon $q$ as poised by $I_b$. There is also an *indirect* effect of $I_m$ in perturbing $q$ from its value as poised in $I_b$ alone. Identification of the two effects is shown by Fig. 1 of Joliot *et al.*[9]

When $I_b$ is a light 1 (as 700 nm) with low $\alpha$, $q$ is well poised at $q \cong 1.0$. Then an increment of oxygen rate produced by $I_m$ is almost entirely a direct effect; $\Delta q$ is negative and small. Hence an action spectrum obtained upon a sufficient background of light 1 closely approximates the action spectrum of photoreaction II. However, the reverse is not true. When $I_b$ is a light 2 (as 650 nm) with $\alpha$ never much greater than 0.5, $q$ is not well poised. Now an increment in rate produced by $I_m$ may contain a large component due to the indirect effect, arising from excitations provided by $I_b$ acting upon the small positive $\Delta q$ produced by $I_m$. Hence, action spectra obtained upon any background of light 2 cannot be described as action spectra for photoreaction I.

The original Emerson experiment was an observation of increase in oxygen rate for $I_m$ at 690 nm when an $I_b$ at 644 nm was added.[10] The interpretation, excitingly important at the time, was that $I_b$ at 644 nm increased the quantum yield of $I_m$ at 690 nm. This interpretation, as originally given and persistently repeated, actually is in error. A perturbation of $+\Delta q$, caused by $I_m$ at 690 nm, is acted upon by the (uncounted) quanta of $I_b$ to give the observed increment in rate.

*State 1 – State 2 Phenomenon.* A second kind of interaction which contributes to enhancement can be explained in terms of (slow) variation in $\alpha$. Apparently, at any wavelength, $\alpha$ shows some adjustment toward $\alpha = 0.5$, toward the condition $q = p$, and toward greater quantum yield.[6,8] Contributions to enhancement by the slower adjustments in $\alpha$, as compared to more rapid adjustment of reaction center conditions, is shown by Fig. 8 of Bonaventura and Myers.[11] The state 1 – state 2 phenomenon in whole cells behaves qualitatively as the effects of $+Mg^{2+}$ and $-Mg^{2+}$ in chloroplasts.[12]

*Information Derivable from Enhancement.* Spectral characteristics of

[10] R. Emerson, R. Chalmers, and C. Cederstrand, *Proc. Natl. Acad. Sci. U.S.A.* **43**, 133 (1957).

[11] C. Bonaventura and J. Myers, *Biochim. Biophys. Acta* **189**, 366 (1969).

[12] P. Bennoun, *Biochim. Biophys. Acta* **368**, 141 (1974).

enhancement must depend upon and contain information about the action spectra for photoreactions I and II. Action spectrum II is easily obtained because it can be observed as a direct effect on oxygen rate. Action spectrum I is more difficult to extract because it must be measured from the indirect effect. Analysis and separation of the direct and indirect effects has allowed simultaneous measurement of action spectra I and II in Chlorella[3] and in Anacystis[13] and demonstration that at least one previous action spectrum I[14] was in error.

In study of chloroplast reactions, the occurrence or absence of enhancement has been set forth as evidence for[15] or against[16] participation and interaction of two photoreactions. If all measurements are proper and significant enhancement is observed, a case for interaction is made. If significant enhancement is not observed, the argument against interaction is less compelling. The difficulty is as follows: In whole cells enhancement has been observed only when the two wavelengths used were a light 1 and a light 2.[17] In the model used above to describe interaction, significant enhancement ($>1.05$) is predicted only for use of a light 1 ($\alpha < 0.5$) and a light 2 ($\alpha > 0.5$). If photoreaction II were damaged in chloroplast preparation, no wavelength could act as a light 2 with $\alpha > 0.5$; enhancement then would not be significant regardless of potential interaction. Hence lack of enhancement seems to require additional information before it becomes an argument against interaction between photoreactions.

[13] R. T. Wang and J. Myers, *Photochem. Photobiol.* **23**, 411 (1976).

[14] L. W. Jones and J. Myers, *Plant Physiol.* **39**, 938 (1964).

[15] M. Avron and G. Ben Hayyim, *Prog. Photosynth. Res., Proc. Int. Congr.* [1st], *1968* p. 1185 (1969).

[16] B. D. McSwain and D. I. Arnon, *Proc. Natl. Acad. Sci. U.S.A.* **61**, 989 (1968).

[17] J. H. Eley and J. Myers, *Plant Physiol.* **42**, 598 (1967).

# [45] Measurement of Spillover

## By Elizabeth L. Gross

At wavelengths greater than 680 nm, where the light absorbed by photosystem II is rate-limiting, a drop in the quantum yield for $O_2$ evolution is observed (the red drop phenomenon). This effect does not occur when photosystem I is rate limiting. It is thought that, under these conditions, the excess excitation energy is transferred from photosystem

METHODS IN ENZYMOLOGY, VOL. 69

II to photosystem I. The term "spillover"[1] has been coined to describe this process. However, the term spillover applies strictly to the case in which the energy absorbed by photosystem II is in excess and the photosystem II traps are closed. I prefer the term "excitation energy distribution" since it can also apply to other conditions. Most of the work in recent years has involved the cation regulation of spillover.[2,3]

We have used measurements of both chlorophyll *a* fluorescence and the quantum yields for electron transport reactions of the individual photosystems to monitor spillover. In addition, measurements of enhancement have also been used.[4,5]

*Chloroplast Isolation.* Class II chloroplasts which have lost their outer membranes and stroma proteins are used.[6] [See Telfer *et al.*[7] for experiments involving chloroplasts with intact outer membranes.] Market spinach (40–50 gm) is placed in a blender with 200 ml of an isolation medium containing 50 m$M$ Tris-HCl (pH 7.6) + 350 m$M$ sucrose.[8] More spinach can be used without affecting the results. The mixture is blended for 20–30 sec after which it is filtered through two layers of cheesecloth. Blending for longer times tends to decrease activity. The filtrate is centrifuged for 10 min at 3000 $g$ to sediment the chloroplasts. All operations are performed in the cold. Originally, we used an initial centrifugation for 1 min at 3000 $g$ but we omitted this step after we found that we discarded two many active chloroplasts. After the centrifugation step, the chloroplasts are resuspended in 40–50 ml unbuffered sucrose and recentrifuged for 15 min at 12,000 $g$. This step removes any remaining outer membranes, all but the most tightly bound mono- and divalent cations, and approximately 56% of the divalent cation-binding sites (i.e., those which are not membrane bound).[9] After centrifugation, the chloroplasts are resuspended in 100 m$M$ sucrose. The suspension is filtered through two layers of Kim-wipes to remove large particles which interfere with optical measurements. The chlorophyll concentration is then determined after which the suspension is diluted to 0.2 mg/ml chlorophyll. Chlorophyll concentrations are determined according to the method of Arnon.[10] The

[1] J. Myers and J. R. Graham, *Plant Physiol.* **38,** 105 (1963).

[2] N. Murata, *Biochim. Biophys. Acta* **189,** 171 (1969).

[3] G. Papageorgiou, *in* "Bioenergetics of Photosynthesis" (Govindjee, ed.), p. 319. Academic Press, New York, 1975.

[4] A. S. K. Sun and K. Sauer, *Biochim. Biophys. Acta* **256,** 409 (1972).

[5] W. P. Williams and Z. Salamon, *Biochim. Biophys. Acta* **430,** 282 (1976).

[6] D. O. Hall, *Nature (London) New Biol.* **235,** 125 (1972).

[7] A. Telfer, J. Barber, and J. Nicolson, *Biochim. Biophys. Acta* **396,** 301 (1975).

[8] E. L. Gross, *Arch. Biochem. Biophys.* **147,** 77 (1971).

[9] E. L. Gross and S. C. Hess, *Biochim. Biophys. Acta* **339,** 334 (1974).

[10] D. I. Arnon, *Plant Physiol.* **24,** 1 (1949).

chloroplasts isolated in this manner continue to show monovalent cation-induced decreases in chlorophyll a fluorescence for 4–5 hr after isolation. The decrease in activity after this time is probably due to spontaneous unstacking of membranes.[11] The pH of the chloroplast suspension is approximately 6.5. The chloroplasts lose activity more rapidly if stored in buffer at higher pH.

## Measurements of Chlorophyll a Fluorescence at Room Temperature

*Rationale.* There are four possible fates for the excitation energy absorbed by a chlorophyll molecule in the antenna of photosystem II (Fig. 1). These are photochemistry, fluorescence, radiationless deexcitation, and transfer to photosystem I. If photochemistry is inhibited by a poison such as DCMU [3-(3,4-dichlorophenyl)-1,1-dimethylurea], then a decrease in fluorescence is thought to reflect an increase in spillover. Changes in radiationless deexcitation are thought to be proportional to the fluorescence changes and, therefore, can be ignored. However, this may not be a very good assumption (see below).

*Equipment.* An Aminco-Bowman spectrofluorometer was used for most measurements of room temperature fluorescence. However, a more sophisticated fluorometer can be used. A red-sensitive photomultiplier such as the S-1 is recommended.

*Procedure.* We have used either 435 or 470 nm as the excitation wavelength. Chlorophyll *a* is excited by 435-nm light, and 470-nm light excites chlorophyll *b*. The results obtained are qualitatively the same. The major emission peak is observed at 680–682 nm[12] and is due to fluorescence from photosystem II. A decrease in fluorescence at this wavelength corresponds to an increase in spillover from photosystem II to photosystem I. There is also a small peak at 740 nm which is due to fluorescence from photosystem I. We have never been able to observe it with our equipment, but others have.[13] An increase in spillover will cause an increase in fluorescence of this peak.

*Assay Mixture.* The assay mixture we use contains 100 m$M$ sucrose + sufficient Tris base (usually 0.3 m$M$) to titrate the suspension to pH 8.0 ± 0.2. DCMU(10 $\mu M$) is added to prevent photosystem II photochemistry. The sucrose provides osmotic support and prevents the swell-

[11] E. L. Gross and S. H. Prasher, *Arch. Biochem. Biophys.* **14,** 460 (1974).
[12] E. L. Gross and S. C. Hess, *Arch. Biochem. Biophys.* **159,** 832 (1973).
[13] S. Murakami and L. Packer, *Arch. Biochem. Biophys.* **146,** 337 (1971).

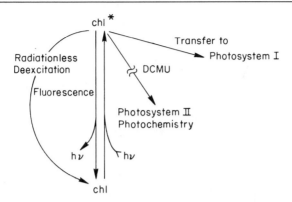

FIG. 1. Four possible fates for the excitation energy absorbed by a chlorophyll molecule (chl) in the antenna of photosystem II.

ing and membrane disorganization, which occurs under low ionic strength conditions in its absence (compare Murakami and Packer[13] with Gross and Prasher[11]).

For most experiments, we used a chlorophyll concentration of 6.7 $\mu$g/ ml. This can be varied, but it is important that the fluorescence versus chlorophyll concentration curve is linear.

*Results.* Addition of low concentrations (3–10 m$M$) of $Na^+$ or $K^+$ salts causes a decrease in fluorescence indicative of an increase in spillover (Fig. 2). Half-maximal effects are observed between 1 and 2 m$M$. Subsequent addition of 1 m$M$ $MgCl_2$ or $CaCl_2$ or 100 m$M$ Na or KCl reverses the effects of the low concentrations of monovalent cations.[12] Under our conditions 0.1 m$M$ $CaCl_2$ is sufficient to produce half-maximal effects. Increasing the cation content of the medium will increase the $CaCl_2$ concentration required for half-maximal effects (compare Murata[2] and Gross and Hess[12]) due to cation competition for the $Ca^{2+}$ binding sites.[9] Addition of divalent cations initially to spinach chloroplasts incubated under low ionic strength conditions caused only a slight decrease in the fluorescence level. However, in other species, such as peas and oats, divalent cations cause an increase in fluorescence under these conditions.[14] Also when chloroplasts are incubated in a zwitterionic buffer, such a Tricine,[2] divalent cations cause an increase in fluorescence. This is because the membranes are already unstacked in Tricine buffer.[15] See Table I for a comparison of the various systems.

[14] D. L. Vander Muelen and Govindjee, *Biochim. Biophys. Acta* **368,** 61 (1974).
[15] S. Izawa and N. E. Good, *Plant Physiol.* **41,** 544 (1966).

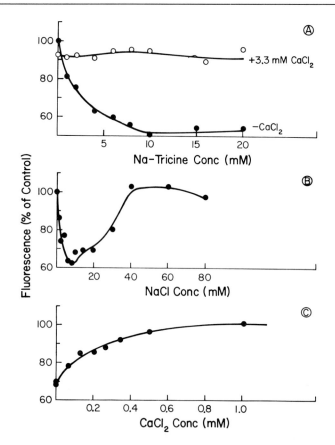

Fig. 2. Concentration dependence of cation effects on chlorophyll *a* fluorescence. (A) The effect of Na Tricine (pH 8) on chlorophyll *a* fluorescence in the presence and absence of 3.3 m*M* CaCl$_2$. (B) The effect of NaCl on chlorophyll *a* fluorescence. (C) CaCl$_2$-induced reversal of the Na Tricine-induced decreases in flourescence Na Tricine (10 m*M*) (pH 8) was added to the reaction mixture for (C). Other conditions are described in the text discussion of assay mixture. Taken from E. L. Gross and C. Hess, *Arch. Biochem. Biophys.* **159**, 832 (1973).

## Measurements of Chlorophyll *a* Fluorescence at 77°K

*Rationale.* The advantages of measuring chlorophyll *a* fluorescence at 77°K rather than at room temperature is that additional emission bands, particularly from photosystem I, are observed. In addition to a band at 685 nm which is due to the light-harvesting antenna of photosystem II, there is a shoulder at 695 nm and an intense band at 735 nm. The shoulder at 695 nm is attributed to either the photosystem II reaction center or to

TABLE I
The Effect of Cations on Spillover

| Assay medium | Addition | Results | | |
|---|---|---|---|---|
| | | Chlorophyll $a$ fluorescence | Quantum yields | |
| | | | Photosystem II | Photosystem I |
| 100 m$M$ sucrose | 10 m$M$ NaCl | Decrease[a] | Decrease | Increase[b] |
| | 1 m$M$ CaCl$_2$ or 100 m$M$ NaCl added after 10 m$M$ NaCl | Increase back to zero salt level | No effect | No effect |
| | 1 m$M$ CaCl$_2$ alone | No effect or slight decrease | Decrease | Increase |
| 10 m$M$ NaCl or 50 m$M$ Tricine buffer | 100 m$M$ NaCl | Increase[d] | Increase | Decrease[c,d] |
| | 1–5 m$M$ CaCl$_2$ or MgCl$_2$ | Increase[e] | Increase | Decrease[c,e] |

[a] E. L. Gross and S. C. Hess, Arch. Biochem. Biophys. **159**, 832 (1973).
[b] E. L. Gross, R. J. Zimmerman, and G. F. Hormats, Biochim. Biophys. Acta **440**, 59 (1976).
[c] M. Avron and G. Ben Hayyim, Prog. Photosynth. Res., Proc. Int. Congr. [1st], 1968 Vol. 3, p. 1185 (1969).
[d] N. Murata, Biochim. Biophys. Acta **226**, 422 (1971).
[e] N. Murata, Biochim. Biophys. Acta **189**, 171 (1969).

chlorophyll molecules very close to it.[16] The band at 735 nm is attributed to photosystem I. The intensity of this band should increase if spillover is increased. The disadvantages of this method are that the path length is uncertain, making comparisons between different samples difficult. Also, it is impossible to study kinetics.

*Equipment.* We have used the Aminco-Bowman spectrofluorometer for some measurements. However, we have not been able to detect the 695 nm shoulder with our equipment. The systems described by Cho *et al.*,[17] Butler,[18] and Mayne *et al.*[19] are better than ours.

*Procedure.* The excitation wavelength was 435 nm, and the emission was scanned from 650 to 770 nm. Some authors correct their data for the spectral sensitivity of the photomultiplier but others do not.

*Assay Medium.* The reaction medium is the same as that described above for the room temperature fluorescence studies.

*Results.* Addition of monovalent cations causes an increase in the 740-nm emission band relative to the 685-nm emission band.[12] Addition of divalent cations reverses the effect. Divalent cations also cause an increase in the 695 nm shoulder.

### Quantum Yield Measurements

*Rationale.* The quantum yield is the ratio of electrons transported to photons absorbed. The difficult part of the measurement is determining the number of photons absorbed. In the case of chloroplasts, this involves using an integrating sphere to minimize light scattering. However, relative quantum yields are sufficient for spillover measurements. The relative quantum yield ($\phi$) is the slope of a plot of reaction rate ($v$) versus light intensity ($I$) [Eq. (1)].

$$v = \phi I \tag{1}$$

If the reaction center is irradiated directly, the quantum yield will depend only on the efficiency of the primary photochemistry. On the other hand, if the light harvesting antenna is irradiated, the quantum yield will also depend both on the efficiency of the transfer of the excitation energy from the antenna to the trap. Thus, an increase in the efficiency of energy transfer will result in an increase in the quantum yield. A comparison of the results obtained when the antenna is irradiated compared to when

[16] N. Murata, M. Nishimura, and A. Takamiya, *Biochim. Biophys. Acta* **126,** 174 (1966).
[17] F. Cho, J. Spencer, and Govindjee, *Biochim. Biophys. Acta* **126,** 174 (1966).
[18] W. L. Butler, Vol. 24, p. 3.
[19] B. C. Mayne, G. E. Edwards, and C. C. Black, Jr., *Plant Physiol.* **47,** 600 (1971).

the reaction center is irradiated will determine whether energy transfer is affected.

*Equipment.* We use an Aminco-Chance spectrophotometer in the split beam mode of operation. However, any sensitive spectrophotometer can be used provided it has been adapted for actinic illumination.

*Procedure.* The actinic light was supplied by a projector with a 750-W bulb. Interference filters of 10-nm half-bandwidth were used to provide reasonably monochromatic light at sufficient intensities to make reliable measurements. We use a 650 nm filter to irradiate the chlorophyll *b* of the light-harvesting chlorophyll *a/b* complex[20] which serves as the antenna of photosystem II. Filters (680 and 710 nm) are used to irradiate the reaction centers of photosystem II and photosystem I, respectively. The light intensity is varied by adjusting the voltage to the projector and is measured using a Kettering-Yellow Springs Instruments Radiometer. Once the light-intensity dependence of the reaction has been determined, an intensity on the linear region of the *v* versus *I* plot can be used for further measurements.[21] Under these conditions the reaction rate will be proportional to the quantum yield.

*Assay Mixture.* The basic reaction mixture contains 100 m$M$ sucrose + 0.3 m$M$ Tris base (to titrate the chloroplast suspension to pH 8) + the electron donor or acceptor of choice. For photosystem II, we measure dichlorophenolindophenol (DCIP) reduction. DCIP (40 $\mu M$) is added to each reaction mixture. For photosystem I, we monitor diphenylcarbazone disproportionation at 485 nm.[22,23] We add 0.15 m$M$ diphenylcarbazone to each reaction mixture + 10 $\mu M$ DCMU to prevent photosystem II photochemistry. Higher concentrations of diphenylcarbazone are inhibitory. Diphenylcarbazone should be recrystallized from a methanol–$H_2O$ mixture prior to use. The diphenylcarbazone solutions should not be exposed to white light.

We have found diphenylcarbazone disproportionation to be a good choice for photosystem I measurements, since it can be used under very low ionic strength conditions. NADP reduction cannot be used under low ionic strength conditions because $Mg^{2+}$ ions are required for ferredoxin binding to the membranes.[24] Also, the addition of ferrodoxin usually involves adding some salt to the system (i.e., the zero salt case will contain considerable salt).

[20] J. P. Thornber, *Annu. Rev. Plant Physiol.* **26**, 127 (1975).
[21] E. L. Gross, R. J. Zimmermann, and G. F. Hormats, *Biochim. Biophys. Acta* **440**, 59 (1976).
[22] L. P. Vernon and E. R. Shaw, *Plant Physiol.* **49**, 862 (1972).
[23] A. Shneyour and M. Avron, *Biochim. Biophys. Acta* **253**, 412 (1971).
[24] G. Harnischfeger and N. Shavit, *FEBS Lett.* **45**, 286 (1974).

See Murata[2] or Avron and Ben-Hayyim[25] for the details of doing quantum yield measurements in Tricine buffer.

*Results.* Ten millimolar concentrations of salts of monovalent cations caused a decrease in the quantum yields for photosystem II and a corresponding increase in those for Photosystem I[21] (see Table I). These results agree with those obtained for chlorophyll *a* fluorescence described above. However, higher concentrations of monovalent cations had no effect. This contrasts with the increases in chlorophyll *a* fluorescence observed under the same conditions. We feel that the increases in fluorescence may be due to decreases in the rate constant for radiationless decay (see also Malkin and Siderer[26]). Consequently, we feel that the quantum yield measurements are a more direct and reliable measure of energy transfer. Divalent cations at 1 m$M$ concentrations caused a decrease in the quantum yields for photosystem II as well as an increase in those for photosystem I. Again, these results disagree with those obtained for chlorophyll a fluorescence.

If the experiments are conducted in 50 m$M$ Tricine buffer[2,25,27] instead of our system, both mono- and divalent cations cause an increase in the quantum yields for photosystem II and a corresponding decrease in those for photosystem I. Thus, in 50 m$M$ tricine buffer, the results obtained from quantum yield and fluorescence measurements agree with each other. See Table I for a comparison of the various systems and types of measurements.

[25] M. Avron and G. Ben Hayyim, *Prog. Photosynth. Res., Proc. Int. Congr. [1st], 1968* Vol. 3, p. 1185 (1969).

[26] S. Malkin and Y. Siderer, *Biochim. Biophys. Acta* **368**, 422 (1974).

[27] N. Murata, *Biochim. Biophys. Acta* **226**, (1971).

# [46] Binding of Fluorescent Nucleotide Analogs to Chloroplast Coupling Factor

*By* V. Shoshan and N. Shavit

Substrate analogs of ADP and ATP represent potentially valuable probes of enzymatic mechanisms, the substrate analog resembling the natural substrate and capable to interact with the active site of the enzyme. Fluorescent nucleotide analogs can provide further information on the conformation of an enzyme active site. In cases in which the ligand itself undergoes a change in fluorescence on binding, the analog can be used to monitor the formation of the enzyme–ligand complex. Studies on the interaction between adenine nucleotide analogs and isolated or membrane-bound ATPase ($CF_1$), the only protein known to participate in ATP synthesis and hydrolysis in chloroplast membranes, have provided information on the nature, affinity, and number of nucleotide binding sites and on kinetic parameters of the binding process. The activity of ADP and ATP analogs in ATP formation, exchange and ATPase reactions (Table I) suggests the participation of two different sites or conformations of the ATPase, in the forward and back reactions of photophosphorylation.[1,1a]

Several types of nucleotide analogs are known, although not all of them are yet commercially available. As summarized in Table I, the replacement of a nonbinding oxygen in the $\alpha$-phosphate group of ADP by sulfur results in two diastereoisomers, one of which (ADP$\alpha$S, A form) binds to $CF_1$ and is phosphorylated as well as ADP, while the B form is almost inactive.[2] Replacing the $\alpha,\beta$ bridge oxygen by a methylene group strongly reduces the phosphorylation rate of this analog.[3] From the activity measured with the ribose-modified ADP analogs, it appears that deoxidation in the 3' or 2' position does not seriously affect the capacity to replace ADP. However, chemical cleavage of the ribose ring between the 2' and 3' position gives an analog (rroADP) which neither binds nor is phosphorylated. Modification of the base moiety shows that any alteration results in loss of activity indicating that the site on the enzyme is highly specific for this part of the molecule. Nevertheless, two fluorescent analogs of ADP ($\epsilon$ADP and FDP) were phosphorylated rather satisfactorily allowing their use as substrate analogs in photophosphorylation.

The triphosphate nucleotide analogs tested were modified either in the bridge oxygen of the phosphate chain or in the base moiety of the molecule. With one exception, so far, all of them were very poor substi-

[1] Y. Shahak, D. M. Chipman, and N. Shavit, in "Photosynthesis III" (M. Avron. ed.), Vol. 2, p. 859. Elsevier, Amsterdam, 1975.

[1a] Abbreviations: ADP$\alpha$S, adenosine-5'-(1-thiodiphosphate); 2'dADP, 9-($\beta$-D-2'-deoxyribofuranosyl)adenine diphosphate; 3'dADP, 9-($\beta$-D-3'-deoxyribofuranosyl)adenine diphosphate; rroADP, 2,2'[1-(9-adenyl)-1'-diphosphoryloxymethyl]dihydroxydiethyl ether; $\epsilon$ADP, 1,$N^6$-ethenoadenosine 5'-diphosphate; $\epsilon$ATP, 1,$N^6$-ethenoadenosine 5'-triphosphate; $\epsilon$CTP, 3,$N^4$-ethenocytidine 5'-triphosphate; Formycin [7-amino-3($\beta$-D-ribofuranosyl)pyrazolo [4,3-$d$]pyrimidine]; FMP, FDP and FTP, formycin mono-di-, and triphosphates, respectively; $CF_1$, chloroplast coupling factor one or ATPase.

[2] H. Strotmann, S. Bickel-Sandkötter, K. Edelmann, E. Schlimme, K. S. Boss, and J. Lüstorff, in "Structure and Function of Energy-Transducing Membrane" (J. M. Tager, ed.), p. 307. Elsevier, Amsterdam, 1977.

[3] A. Horak and Z. Zalik, Biochim. Biophys. Acta 430, 135 (1976).

TABLE I
NUCLEOTIDE DI- AND TRIPHOSPHATE ANALOGS AS SUBSTRATES IN REACTIONS
CATALYZED BY THE ATP SYNTHETASE OF CHLOROPLASTS

| Reaction | Substrate analog | Relative activity[a] (%) | Reference[b] |
|---|---|---|---|
| ATP formation | Phosphate chain modification | | |
| | $\alpha,\beta$-Methylene-ADP | 11 | a |
| | ADP$\alpha$S, A form | 100 | b |
| | ADP$\alpha$S, B form | 4 | b |
| | Ribose modification | | |
| | 2'dADP | 40 | c |
| | 3'dADP | 95 | c |
| | rroADP | 0 | c |
| | Base modification | | |
| | $\epsilon$ADP | 52 | d |
| | 1-$N$-Oxido-ADP | 24 | b |
| | 8-Bromo-ADP | 2 | b |
| | FDP | 36 | e |
| P-ATP exchange | $\epsilon$ATP | 0.2 | d |
| | FTP | 0.2 | e |
| | $\alpha,\beta$-Methylene-ATP | 4.2 | a |
| | $\beta,\gamma$-Methylene-ATP | 0 | a |
| Light-triggered ATPase | $\epsilon$ATP | 15 | d |
| | $\alpha,\beta$-Methylene-ATP | 13 | a |
| | $\epsilon$CTP | 40 | c |
| CF$_1$-ATPase | $\epsilon$ATP | 5 | d |
| | $\epsilon$CTP | 5 | d |
| | FTP | <0.2 | f |

[a] Activity relative to ADP or ATP.
[b] Key to references: a, A. Horak and S. Zalik, *Biochim. Biophys. Acta* **430,** 135 (1976);   b, H. Strotmann, S. Bickel-Sandkötter, K. Edelmann, E. Schlimme, K. S. Boos, J. Lüstorff, *in* "Structure and Function of Energy-Transducing Membranes" (J. M. Tager, ed.), p. 307 Elsevier, Amsterdam, 1977;   c, K. S. Boos, J. Lüstorff, E. Schlimme, H. Hesse, and H. Strotmann, *FEBS Lett.* **71,** 124 (1976);   d, Y. Shahak, D. M. Chipman, and N. Shavit, *in* "Photosynthesis III" (M. Avron, ed.), Vol. 2, p. 859. Elsevier, Amsterdam, 1975;   e, M. Avron, personal communication;   f, V. Shoshan, N. Shavit, and D. M. Chipman, *Biochim. Biophys. Acta* **504,** 108 (1978).

tutes of ATP in the exchange or ATPase reactions of the membrane-bound or isolated $CF_1$.

## Binding Studies Using Fluorescent Nucleotides

The binding of $\epsilon$ADP and $\epsilon$ATP to $CF_1$ was investigated by measuring the decrease of the fluorescence intensity[4] or the increase in the polarization of fluorescence of the ligand upon binding.[5] A binding system amenable to fluorescence intensity measurements developed recently consists of the fluorescent nucleotide analog, formycin di- and triphosphate and $CF_1$. We found that the binding of FDP or FTP to $CF_1$ is accompanied by a two- to threefold enhancement in fluorescence intensity of the protein–ligand complex and a markedly increased polarization of the nucleotide fluorescence.[6]

Formycin[7] contains a carbon–carbon glycosyl bond rather than the carbon–nitrogen bond present in adenosine. The nucleoside and the nucleotides of formycin are fluorescent having an excitation maximum at 295 nm and an emission maximum at 340 nm.[8] The relative fluorescence intensity is not changed in the range of pH between 5.8 and 8.5. Protonation of the ring results in a strong quenching of the 340 nm emission band. A reduced fluorescence intensity and a shift toward the red, both in the excitation and emission maximum, are characteristic for the anionic form of formycin. The fluorescence lifetime of both fluorescent forms of formycin is less than 1 nsec. The spectral and fluorescent properties of formycin nucleotides together with the finding that FTP is not hydrolyzed to any significant extent by the $CF_1$-ATPase, are particularly favorable to study enzyme–substrate interactions by fluorescence measurements. The method detailed henceforth enables calculation, using the fluorescence intensity measurements, of the dissociation constant and the association and dissociation rate constants of the $CF_1$–formycin nucleotide complex.[6] The mono-, di-, and triphosphate nucleotides of formycin can be synthesized from formycin according to the procedure of Ward *et al.*[9]

[4] G. Girault and J. M. Galmiche, *Eur. J. Biochem.* **77,** 501 (1977).

[5] D. L. Vandermeulen and Govindjee, *FEBS Lett.* **57,** 272 (1975).

[6] V. Shoshan, N. Shavit, and D. M. Chipman, *Biochim. Biophys. Acta* **504,** 108 (1978).

[7] M. Hori, E. Ito, R. Takita, G. Koyama, T. Takeuthi, and H. Umezawa, *J. Antibiot., Ser. A* **17,** 96 (1964).

[8] D. C. Ward, E. Reich, and L. Stryer, *J. Biol. Chem.* **244,** 1228 (1969).

[9] D. C. Ward, A. Cerami, E. Reich, G. Acs, and L. Altwerger, *J. Biol. Chem.* **244,** 3243 (1969).

## Fluorescence Measurements

Fluorescence emission spectra of formycin nucleotides are determined in any appropriate fluorescence spectrophotometer. We have used a Hitachi-Perkin Elmer MPF-2A. The measurement is performed at room temperature, using an exciting Xenon lamp source of 150 W, an excitation wavelength of 296 nm, and a 6-nm slit opening. The emission spectra between 320 and 400 nm are recorded in the ratio mode with a slit opening of 8 nm. Many types of fluorimetric cells are available; we have used 3-ml cells (10 × 10 mm) or microcells, 0.2–0.4 ml (5 mm in diameter). Microcells make it possible to economize on materials or to use high concentrations without having high total absorbance. Larger cells, on the other hand, may be necessary to obtain accurate results with very low concentrations. An important requirement for successful fluorescence measurements is that the solutions have low light scattering (turbidity). Aggregation of protein preparations under many conditions make this a condition difficult to meet. We chose to solve this problem by using a $CF_1$ preparation released from chloroplasts by chloroform treatment. This preparation is a latent ATPase containing only four type subunits lacking the δ-subunit.[6] Typical spectra of the fluorescence of free FMP and FTP in the presence of $CF_1$ are given in Fig. 1. A low level of fluorescence peaking at 330 nm is observed with the soluble $CF_1$ protein alone. The mixture of $CF_1$ and FMP gives a fluorescence intensity value corresponding to about the sum of the individual fluorescence intensity of each compound. The mixture of FTP and $CF_1$ gives a significant enhancement in the fluorescence intensity as compared with the FMP + $CF_1$ mixture. Direct binding experiments corroborate that AMP is not bound to $CF_1$, while one mole of ATP binds to one mole of this protein. Taking into consideration the dissociation constant for the ATP–$CF_1$ complex determined by direct binding experiments, the lack of AMP binding and the additivity of the fluorescence intensity of FMP and $CF_1$ when combined, one can calculate a two- to threefold enhancement in fluorescence intensity for the $CF_1$-bound FTP. The enhancement of emission fluorescence intensity is similar at all wavelengths up to 400 nm. In order to minimize the contribution of the protein fluorescence to the total fluorescence, it is advisable to determine $\Delta F$ (fluorescence intensity of D minus that of C, Fig. 1) at 380 nm rather than at 340 nm.

Fluorescence polarization measurements support the formation of the $CF_1$–FTP complex. FMP alone shows a fluorescence polarization of 0.09, a value also obtained for the free FTP. The binding of FTP to $CF_1$ gives a markedly increased polarization of the nucleotide fluorescence (0.42), indicating considerable rotational restriction of the base.

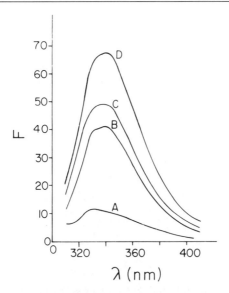

FIG. 1. Fluorescence emission spectra of formycin nucleotides and $CF_1$. Reaction mixtures contained the following components at pH 8.0: tricine-NaOH, 50 m$M$; NaCl, 25 m$M$; $MgCl_2$, 1.25 m$M$; FMP or FTP, 2.2 $\mu M$; $CF_1$, 0.55 $\mu M$. A, $CF_1$ alone; B, FMP or FTP; C, FMP + $CF_1$; D, FTP + $CF_1$.

## Measurement of the Dissociation Constant ($K_d$), the Association ($k_1$) and Dissociation ($k_{-1}$) Rate Constants of the $CF_1$–FTP Complex

*Reagents*
Tricine–NaOH, 1 $M$, pH 8.0
NaCl, 1 $M$
$MgCl_2$, 0.1 $M$
FMP, $10^{-4}$ and $2 \times 10^{-5}$ $M$
FTP, $10^{-4}$ and $2 \times 10^{-5}$ $M$
ATP, $10^{-3}$ $M$
Desalted $CF_1$, 1–2 mg/ml

*Procedure for $K_d$ Determination.* $K_d$ is obtained from the titration curve of $CF_1$ with increasing FTP concentration. A reaction mixture of 0.12 ml tricine, 0.06 ml NaCl, 0.03 ml $MgCl_2$, 0.030–0.060 ml $CF_1$, and $H_2O$ to a final volume of 2.4 ml is added to each of two fluorimetric cells. The initial fluorescence intensity in each cell is recorded as described in fluorescence measurements. The titration curve is obtained by successive addition, to both cells, of aliquots of 5 $\mu$l of the appropriate stock solution of FTP and FMP, respectively. It is essential that the stock solutions of FTP and FMP be at exactly identical concentrations. At the enzyme concentration given (5 to $8 \times 10^{-8}$ $M$), the range of concentrations of FTP and FMP should be between $4 \times 10^{-8}$ and $10^{-6}$ $M$. With each addition, the fluorescence intensity in each cell is recorded after 1–2 min. A plot of $\Delta F$ (the difference between the fluorescence intensity of the cuvette containing FTP and that containing FMP) against the total FTP concentration permits calculation of the enzyme–ligand complex [EL] concentration and $K_d$.

*Procedure for $k_1$ Determination.* $k_1$ is determined by following the fluorescence intensity change with time. If the reaction is slow enough, as in our case with the latent enzyme, the mixing and response time are not limiting factors in the measurement. Where shorter reaction times are involved, as with the activated enzyme, a stopped-flow fluorimeter technique should be used. A reaction mixture of 0.02 ml tricine; 0.01 ml NaCl; 0.005 ml $MgCl_2$, 0.02–0.03 ml $CF_1$, and $H_2O$ to a final volume of 0.4 ml is added to each of two fluorimetric microcells. The initial fluorescence intensity in each cell is recorded. With the recorder operating and photomultiplier shutter kept in the open position, 0.01–0.02 ml of FTP or FMP stock solutions are placed carefully on a thin Teflon rod, which is used for mixing the cell contents. The lid of the instrument is closed immediately and the fluorescence intensity change is recorded. This procedure enables measuring the response after 3–5 sec.

*Procedure for $k_{-1}$ Determination.* The displacement of FTP from the $CF_1$-FTP complex by ATP is used for measurement of $k_{-1}$. The assay is carried out as described for the measurement of $k_1$. After the binding of FTP to $CF_1$ reaches equilibrium, 5 $\mu$l of the ATP stock solution are added to the microcell using a thin Teflon rod. The decrease in fluorescence intensity when the bound FTP is displaced by ATP is recorded every 2–5 min until no further change in fluorescence is observed.

*Calculation of Enzyme–Ligand complex [EL].* Characterization of binding equilibria involves the determination of concentrations of components of the equilibrium system. The concentration of the enzyme–ligand complex and that of free enzyme and free ligand are calculated from the $\Delta F$ data as follows: since the absorbance of the solutions employed is small (<0.05 O.D.), the total fluorescence for the cuvette containing FMP and $CF_1$ is equal to

$$F = f_E[E_0] + f_L[L_0] \tag{1}$$

and for the one containing FTP and $CF_1$

$$F = f_E[E] + f_L[L] + f_{EL}[EL] \tag{2}$$

where $[E_0]$ and $[L_0]$ are total concentrations of enzyme and ligand; $[E]$, $[L]$, and $[EL]$ are concentrations of free enzyme, free ligand, and enzyme–ligand complex, respectively; and $f_E$, $f_L$, and $f_{EL}$ are characteristic fluorescence parameters for each species. Considering that $[E_0 - E] = [L_0 - L] = [EL]$, we obtain

$$\Delta F = -f_E[E_0 - E] - f_L[L_0 - L] + f_{EL}[EL]$$

$$= [EL]\,(f_{EL} - f_E - f_L) \quad (3)$$

we can define $R = (f_{EL} - f_E - f_L)$ and then,

$$[EL] = \Delta F(1/R) \quad (4)$$

$R$ can be calculated from fluorescence titration of $CF_1$ with saturating concentrations of FTP (at least 5–10 times $[E]$ and 5–10 times $K_d$)+ then the $\Delta F$ obtained is divided by the $CF_1$ concentration to give $R$.

*Calculation of $K_d$.* In the case of a protein with one nucleotide binding site and when measurements are made at protein concentrations close to those of the ligand, the dissociation constant becomes

$$K_d = \frac{([E_0 - EL])\,([L_0 - EL])}{[EL]} \quad (5)$$

$[EL]$ is calculated from $\Delta F$ as described above. When $[EL] = [E_0]/2$ (50% saturation), $K_d = [L_0] = [E_0]/2$.

Alternatively, the dissociation constant of the complex $CF_1$–FTP can be calculated by nonlinear squares fit[6] to the quadratic equation for the dependence of fluorescence on enzyme and ligand concentration which is obtained by combining Eqs. (4) and (5).

*Calculation of the Number of Binding Sites and of $K_d$.* The calculation is done as described[10] for the case of ligand binding to identical noninteracting sites. For a protein with $n$ binding sites, Eq. (5) becomes

$$\frac{1}{K_d} = \frac{b}{(n[E_0] - b)\,([L_0] - b)} \quad (6)$$

where $b$ is the concentration of bound ligand. We can define a saturated fraction as $b/b_\infty$ or $b/n[E_0]$ which is equal to $\Delta F/\Delta F_\infty$. Substituting $b$ in

[10] H. Gutfreund, "Enzymes Physical Principles." Wiley (Interscience), New York, 1972.

Eq. (6) and rearranging, we obtain

$$K_d \frac{1}{1 - \Delta F/\Delta F_\infty} = \frac{[L_0]}{\Delta F/\Delta F_\infty} - n[E_0] \qquad (7)$$

A plot of $1/(1 - \Delta F/\Delta F_\infty)$ versus $[L_0]/(\Delta F/\Delta F_\infty)$ is linear. The intercept on the abscissa is $n[E_0]$ and the slope is equal to $1/K_d$. Thus the number of binding sites and $K_d$ can be calculated.

*Calculation of $k_1$.* If one assumes that the kinetic processes observed are due to simple bimolecular association:

$$E + L \underset{k_{-1}}{\overset{k_1}{\rightleftarrows}} EL \qquad (8)$$

and that $k_{-1}/k_1 = K_d$, one can write the differential equation for the appearance of [EL]:

$$\frac{d[EL]}{dt} = k_1[E][L] - k_{-1}[EL]$$

$$= k_1\{(E_0 - [EL])(L_0 - [EL]) - K_d[EL]\} \qquad (9)$$

If experiments are carried out under conditions for which one can assume $[L_0] \gg [EL]$, Eq. (9) can be simplified to

$$\frac{d[EL]}{dt} = k_1\{(E_0 - [EL])[L_0] - K_d[EL]\}$$

$$= k_1([E_0][L_0] - (K_d + [L_0])[EL]) \qquad (10)$$

The solution of the equation is

$$\ln\left(1 - \frac{K_d + [L_0]}{[E_0][L_0]}[EL]\right) = -k_1(K_d + [L_0])t \qquad (11)$$

The concentration of [EL] at time $t$ is determined from the fluorescence change using Eq. (4) as above. A plot of $\ln(1 - \Delta F Q)$, where $Q = (K_d + [L_0])/[E_0][L_0]R$, against time gives a straight line of slope $-k_1(K_d + [L_0])$. Assuming that $K_d$ has been determined, this method can be used to calculate $k_1$.

When it is not possible to carry out experiments with $[L_0] \gg [EL]$, Eq. (9) must be used without simplification. The solution of Eq. (9) cannot be linearized and can only be fit using a nonlinear least squares approach.[6]

*Calculation of* $k_{-1}$. The simplest interpretation of a displacement experiment is that described by Eq. (12)

$$E + FTP \underset{k_{-1}}{\overset{k_1}{\rightleftharpoons}} E \cdot FTP$$

$$E + ATP \underset{k'_{-1}}{\overset{k'_1}{\rightleftharpoons}} E \cdot ATP \tag{12}$$

If experiments are carried out using a very high concentration of ATP (or other displacing ligand), so that one can assume that the reaction of free enzyme with ATP is essentially instantaneous on the time scale of the experiments and that [E] drops to nearly zero at once, it is reasonable to treat the process as a first-order disappearance of [E–FTP] with the rate constant $k_{-1}$. For very high ATP concentration, a good straight line is obtained upon plotting $\ln(\Delta F)$ against time, and the slope $(-k_{-1})$ is independent of the concentration of ATP.

For cases in which the displacement is not complete, the treatment of the data becomes more complex.[6]

### Concluding Remarks

Fluorescence intensity measurements made with FTP or FDP and a chloroplast coupling factor released from lettuce chloroplasts by chloroform treatment were used to determine the nucleotide binding properties of $CF_1$. In the presence of 1 m$M$ $Mg^{2+}$, 1 mole of FTP is bound to $CF_1$, with a $K_d$ of $2 \times 10^{-7}$ $M$. The binding of FTP behaves as a bimolecular process with an association constant $k_1 = 2.4 \times 10^4$ $M^{-1}$ sec$^{-1}$. FTP is displaced by ATP in a first-order process with a dissociation constant $k_{-1} = 3 \times 10^{-3}$ sec$^{-1}$. This method can be used to study the binding of FTP to the activated ATPase enzyme by a stopped-flow fluorescence technique and could be further developed to measure the nucleotide binding characteristics of the membrane-bound ATPase.

Acknowledgments

This work was supported in part by the Deutsche Forschungsgemeinschaft.

# [47] Antibody Approach to Membrane Architecture

## By RICHARD J. BERZBORN

Antibodies are valuable tools for biochemical investigations, since their reaction with soluble antigens is highly specific, fast, and strong.[1] Precipitation and inhibition of enzymes[2,3] are useful for the detection of the interaction of antigen and antibodies. When applied to membranes, the immunological methods partially have to be revised, but additional conclusions can be drawn concerning the location of the antigens as constituents of membranes.

This approach to the architecture of the chloroplast thylakoid membrane system has been reviewed[4-6] together with considerations on the correlation of structure and function, especially the asymmetric distribution of components perpendicular to the thylakoid surface,[7] their distribution along this membrane system,[8] and the action of inhibiting antibodies on photosynthetic electron transport.[9]

In this article, the basis of the approach, the experimental possibilities, the conclusion patterns, and the inherent limitations are considered. A complementary treatise on the application of antibodies to studies on the mitochondrial membrane will appear.[10]

### Antibodies as Membrane Surface Probes

Antibody molecules are found in the $\gamma$-globulins in the serum of

---

[1] E. A. Kabat and M. M. Mayer, "Experimental Immunochemistry." Thomas, Springfield, Illinois, 1967.

[2] B. Cinader, *Ann. N. Y. Acad. Sci.* **103**, 495 (1963).

[3] R. Arnon, *Curr. Top. Microbiol. Immunol.* **54**, 47 (1971).

[4] A. Trebst, *Annu. Rev. Plant Physiol.* **25**, 423 (1974).

[5] J. M. Anderson, *Biochim. Biophys. Acta* **416**, 191 (1975).

[6] C. J. Arntzen and J. M. Briantias, *in* "Bioenergetics of Photosynthesis" (Govindjee, ed.), Academic Press, New York, p. 51. 1975.

[7] K. Mühlethaler, *in* "Photosynthesis I" (A. Trebst and M. Avron, eds.), Encycl. of Plant Physiol., New Ser., Vol. 5, p. 503, Springer-Verlag, Berlin and New York, 1977.

[8] P. V. Sane, cf. Mühlethaler,[7] p. 522 and R. J. Berzborn, *Z. Naturforsch., Teil B* **24**, 436 (1969).

[9] R. J. Berzborn and W. Lockau, cf. Mühlethaler,[7] p. 283.

[10] S. H. P. Chan and G. Schatz, *in* "Methods in Enzymology" Vol. 56, p. 223. Academic Press, New York, 1979.

immunized animals, i.e., they are hydrophilic macromolecules with a partially characterized structure.[11,12]

It is generally assumed that they do not penetrate biological membranes.[2] Since a reaction can occur only if the matching antigen determinant group is accessible to the antibody added from the outside, an observed reaction indicates that the antigen was at least partially localized at the membrane surface.[13-15]

The basic assumption was corroborated by investigations with black lipid membranes.[16-18] In some cases, however, a reaction between antigens and antibodies added to the two different sides of the bilayer was observed; this was due to a penetration of the antigens used. Transport of antibodies through the plasma membrane of lymphocytes, during secretion or uptake after capping, as well as passage through the placenta are probably achieved via pinocytosis and exocytosis vesicles, respectively. This basic assumption is supported also by the agreement of the conclusions on the localization of mitochondrial components from antibody studies[14] with the conclusions from other techniques (e.g., resolution and reconstitution, labeling with impermeable probes). For the chloroplast thylakoid, the picture from the antibody approach seemed at first to be in good agreement with spectroscopic[19] and chemical[20] studies on vectorial photosynthetic electron transport. However, some recent contradictory conclusions seem to cast doubt on the reliability of the antibody approach.

### Experimental Possibilities

Antiserum can be produced by either injecting an isolated component into rabbits (see schedule below), or by challenging the animal with complex particles, e.g., the entire thylakoid membrane. Even after injection of hydrophobic proteins, isolated in the presence of SDS, antibodies

---

[11] R. C. Valentine and N. M. Green, *J. Mol. Biol.* **27**, 615 (1967).

[12] D. R. Davies, E. A. Padlan, and D. M. Segal, *Annu. Rev. Biochem.* **44**, 639 (1975).

[13] R. J. Berzborn, Ph.D Thesis, University of Köln (1967).

[14] E. Racker, C. Burstein, A. Loyter, and R. O. Christiansen, *in* "Electron Transport and Energy Conservation" (J. M. Tager, S. Papa, E. Quagliariello, and E. C. Slater, eds.), p. 235. Adriatica Editrice, Bari, 1970.

[15] O. J. Bjerrum, *Biochim. Biophys. Acta* **472**, 135 (1977).

[16] J. Del Castillo, A. Rodrigues, C. A. Romero, and V. Sanchez, *Science* **153**, 185 (1966).

[17] P. Barfort, E. R. Arquilla, and P. O. Vogelhut, *Science* **160**, 1119 (1968).

[18] D. Wobshall and C. McKeon, *Biochim. Biophys. Acta* **140**, 152 (1970).

[19] W. Junge, cf. Mühlethaler,[7] p. 59.

[20] G. Hauska, cf. Mühlethaler,[7] p. 253.

are sometimes found which cross-react with the protein in its native conformation in the membrane.[21,21a]

The reaction of an antibody with the membrane is established by *inhibition* of a membrane associated reaction, by *agglutination* of the suspension, or by experimental proof of just *binding* of antibodies to the membrane. Methods which allow the latter are standard procedures and are not described in detail here. Binding (absorption) of antibodies can be shown with anti-γ-globulin serum, i.e., antiserum against the γ-globulin of the animal which produced the antibody,[22] or if the antibody had been labeled with [125]I, fluorescent dyes, ferritin, or lactoperoxidase.[23,24] An indirect proof for antibody binding to the membrane is the reaction of the treated membranes with labeled immunoglobulin from antiglobulin sera. Binding of antibodies to the thylakoid can also be shown after centrifugation by measuring the increase of protein in the pellet or decrease of protein (or specific antibody) in the supernatant.

Binding of antibodies to determinant groups in crevices of the membrane surface will hinder antibody combination with an antigen in a crevice of a second particle; agglutination is then very weak or absent. Since antibodies are bivalent, the binding can be shown by addition of the soluble antigen (indirect or mixed antigen agglutination).

If a serum against the entire membrane agglutinates the membrane suspension or inhibits a reaction, it is worthwhile to identify the reacting antigen in the membrane by isolating a component which neutralizes specifically the antibody.

### Patterns of Conclusions

Under conditions of the above assumption, that antibodies do not penetrate membranes, the following conclusions can be drawn.

1. If an antiserum inhibits a membrane associated reaction specifically, antibodies must have reacted with a determinant group in the membrane; thus the antigen or cross-reacting (i.e., extremely similar) determinant group must have been accessible to the antibody during the time of the reaction.

[21] W. Menke, F. Koenig, A. Radunz, and G. H. Schmid, *FEBS. Lett.* **49**, 372–375 (1975).
[21a] D. H. Chua and F. Blomberg, *J. Biol. Chem.* **254**, 215 (1979).
[22] R. R. A. Coombs, M. H. Gleeson-White, and J. C. Hall, *Br. J. Exp. Pathol.* **32**, 195 (1951).
[23] L. A. Sternberger, "Immunocytochemistry." Prentice-Hall, Englewood Cliffs, New Jersey, 1974.
[24] W. C. Davis, Vol. 32, p. 60.

2. If antibodies against a component isolated from the membrane inhibit a reaction sequence carried out by the membrane, the component was accessible and probably involved in the reaction as a constituent of the membrane surface. It does not follow, however, that the site of the reaction is at the surface, since the protein might span the membrane.[5]

3. If an antiserum against the entire membrane (thylakoid) contains inhibiting antibodies against a reaction sequence carried out by the membrane and if an isolated component neutralizes the inhibition and can be shown to react with the antibody (e.g., by precipitation), the inhibition was probably due to the reactions of the antibody with the neutralizing component as a constituent of the membranes (surface). This does not necessarily mean, however, that the component itself is involved in the reaction.[25] If no reaction of the neutralizing component with the antibody is detected in solution, the reactivation of the inhibited reaction might be due to a "bypass," i.e., the component might introduce a new reaction sequence. In this case, no conclusions concerning accessibility of the neutralizing component in the membrane can be drawn with certainty.

4. If a stimulation of a reaction occurs by an antigen mixture, added to the membrane suspension, and if this stimulation is absent after the mixture was precipitated with a monospecific antiserum, the precipitated antigen was responsible for the stimulation and must have reacted with the membrane surface somehow.

5. If an antiserum against an isolated component does not inhibit a membrane-associated reaction sequence, one cannot conclude that the antigen is not involved in the reaction. The antigen could be inaccessible. However even accessibility is not sufficient for inhibition, because not all precipitating or agglutinating antibodies are inhibiting.[2] Also it is obvious that a certain reaction sequence can only be inhibited if the particular step, where the antibody interferes, was rate limiting or was at least made rate limiting by the antibody. If insufficient antibody was added or if the preincubation was not allowed to proceed long enough, inhibition might not be detected.

6. If an antiserum against an isolated component agglutinates the suspension of the membrane particles, it follows that antibodies did react with determinant groups at the surface, even if no inhibition of any reaction occurs. There is high probability that the component itself is an accessible membrane constituent.

7. Antisera against a membrane (thylakoid) contain many agglutinating antibodies of different quality. It is improbable that a single isolated component could neutralize all agglutinating antibodies; but from quan-

[25] W. Lockau, Ph.D. Thesis, University of Bochum (1976).

titative titration experiments it might be possible to show that a certain component neturalizes most of the agglutinating antibodies. In this case it follows, that the neutralizing isolated component is an accessible membrane constituent at the surface.

8. If no direct agglutination of a membrane suspension occurs by an antiserum against a single component, even if all conditions are correct, it does not follow that the antigen is inaccessible, since not all binding antibodies are agglutinating.[26]

9. If no binding of antibodies against a component with the membrane surface can be detected, the conformation of the component in solution and as a constituent of the membrane may be different as sensed by a highly specific antibody.[27] Thus it does not follow with certainty that the component is inaccessible.

10. If no reaction of an antibody occurs with membrane vesicles, but a reaction can be established by opening the vesicles, it is concluded that the reacting antigen was localized at the other (inner) side of the membrane.

11. If an antiserum against a membrane vesicle preparation contains antibodies against an isolated component, it does not follow that the component is localized at the surface, because the vesicle (thylakoid) might have been damaged before recognition of the antigen pattern by the immune system, and the animal might have synthesized antibodies against the inside of the vesicle. To our knowledge this has never been observed. This approach can only be taken seriously, if agglutination or inhibition neutralization titrations are performed.[13]

## Experimental Procedures

### Production of Antisera

For immunization, soluble proteins, larger particles, or entire membrane systems can be used. If a protein was purified by electrophoresis in the presence of SDS, it can be cut out from the gels, eluted, or used together with the polyacrylamide for the primary injection. For an intraveneous booster injection, a physiological solution or suspension is necessary, e.g., thylakoid systems in physiological buffer can be injected intraveneously.

*Procedures of Immunization.*[10,13,28] First about 20 ml of control blood

---

[26] R. J. Berzborn, *Prog. Photosynth. Res., Proc. Int. Congr. [1st], 1968.* **I,** (1969).

[27] C. J. Lazdunski, *Trends Biochem. Sci.* **1,** 231 (1976).

[28] D. H. Campbell, J. S. Garvey, N. E. Cremer, and D. H. Sussdorf, "Methods in Immunology," Benjamin, New York, 1963.

are taken from the marginal vein of the rabbit's right ear. After clotting at room temperature, the blood is carefully loosened from the glass, stored at 4° over night, decanted, clarified by centrifugation (10 min at 4000 $g$), and filtered through a 0.45 $\mu$m filter (Millipore, Sartorius) and stored at $-20°$. For some purposes decomplementation is recommended, i.e., heating for 30 min at 56°.

The antigen (0.2 to 2 mg protein), dissolved or suspended in 1.0 ml of 150 m$M$ NaCl or any physiological buffer with as little detergent present as possible (dialyzed), is thoroughly mixed with 1.5 ml Freund's complete adjuvant (Difco) and emulsified in a clear plastic tube by a short (5–10 sec) ultrasonication (microtip) under ice cooling. The paste is administered to a rabbit with a 2-ml syringe (14 gauge) intradermally, 0.5 ml in two portions in the footpads of one leg, and four to six portions into the back. It is also possible to inject a thylakoid suspension intravenously without the use of adjuvant, increasing amounts (0.1 to 0.3 mg chlorophyll) at weekly intervals.[13]

After 4–6 weeks, 10–20 ml of blood is taken from the marginal ear vein to check for antibodies. If agglutination (see below) or ring test is strong,[28a] no more than 0.5 mg of antigen is injected at once for the booster injection to avoid anaphylactic shock. Another 0.5 mg can be injected the next day. Usually the titer of antibodies is low after the primary injection only, and about 1 mg of antigen in physiological buffer is injected into the marginal vein of the left ear (18 gauge needle).

At days 7,9, and 12 after the booster, up to 20, 40, and 20 ml of blood can be taken from the marginal vein of the right ear, depending on the size of the rabbit; a trace of xylene on the tip of the ear leads to good sanguination of the ear and has to be washed off later with alcohol. An alternative booster can be administered intradermally[10] with the antigen mixed with incomplete Freund's adjuvant. Additional booster injections can be given at 4–6 week intervals. In a final bleeding 10–12 days after a booster injection, after narcosis by an intravenous injection of 40–60 mg Nembutal, 80–100 ml of blood can be recovered from the carotid. As antigens isolated from membranes are often weak immunogens, they can be polymerized to enhance the response.[10] Antisera can be stored for years at $-20°$, but lyophilization facilitates storage. Lyophilized sera are reconstituted with distilled water, conveniently to double the original

---

[28a] Put 10–15 $\mu$l serum into a glass tube of 1-mm diameter standing in lute (e.g., Seal-Ease, Clay Adams) and add antigen solution slowly to get a sharp boundary; the antigen solution should contain 0.1–1 mg protein/ml, 20–150 m$M$ salt, pH 6.5–8.5; detergent, if necessary, Triton X-100 or deoxycholate below 1%; urea $< 1$ $M$. After 5–10 min at 22° a specific precipitin ring indicates the presence of antibodies.

volume. Tests for the quality of the antiserum include double diffusion in gels and immunoelectrophoretic analysis[15,28] with crude antigen mixtures, to test for the number of antigen antibody systems, and quantitative precipitation[1,28] and agglutination to determine the amount of antibody (titer). For immunization with chloroplast lipids, see Radunz and Berzborn.[29]

### Inhibition by Antisera

For inhibition of photosynthetic reactions appropriate amounts of thylakoid systems are preincubated with increasing amounts of antiserum and, for comparison, of control serum. The reaction has to be titrated. The buffer must allow the antibody reaction and preserve the thylakoid structure. Without salt antibodies are insoluble; their reaction is optimal between pH 7 and 8.5. Sucrose at high concentrations slows down the reaction. Detergent should be avoided, but may not inhibit the antibody completely.[10,15] The amount of serum to be used depends on the titer. With a good undiluted antiserum against the photosynthetic ATP synthetase $(CF_1)$ 10 $\mu$l/10 $\mu$g chlorophyll will give complete inhibition of photophosphorylation. Antisera can be diluted in 150 m$M$ NaCl or in control sera. Control sera might stimulate or inhibit a reaction, if higher amounts are used. Less unspecific interference of serum constituents are observed if the serum is dialysed or the $\gamma$-globulin fraction used only.[28] After incubation (2–15 min at 4° or 22°), the thylakoid systems are transfer to the assay medium or first washed by centrifugation.

### Agglutination by Antisera

Agglutination of thylakoid systems or smaller membrane fragments leads to macroscopic aggregates,[13] analogous to the agglutination of red blood cells. This can be observed on white porcelain plates with shallow molds (about 1 cm diameter), on microscope slides, or in small glass test tubes, but not in the commercially available plastic plates with several holes, neither U nor V shaped.

For the test on plates 10 $\mu$l aliquots of diluted serum (e.g., a 1 : 2 dilution series up to 1 : 8000 or more) is mixed with 10 $\mu$l of a thylakoid suspension (0.1 mg chlorophyll/ml) in buffer (e.g., 10 m$M$ tricine, pH 8, 50 m$M$ NaCl) and moved about by tilting the plate slowly five to ten times.

For the test on microscope slides, 5 $\mu$l of the diluted serum and 5 $\mu$l of the thylakoid suspension are mixed and the droplet moved around by tilting the slide five to ten times. In both cases, agglutination occurs

---

[29] A. Radunz and R. J. Berzborn, Z. Naturforsch., Teil B **25**, 412 (1970).

within 10–20 sec at room temperature. By comparing the effect of aliquots of diluted control sera it is quite easy to decide whether an aggregation of particles is specific or not. For very small vesicle preparations a phase contrast microscope or dark field illumination is used (magnification × 200–400, no cover glass).

For agglutination in glass tubes (4-mm diameter, round bottom) 40 $\mu$l aliquots of diluted serum and 40 $\mu$l of thylakoid suspension (0.1 mg chlorophyll/ml) are mixed, and after 30 min at room temperature the rack with the series of tubes is incubated overnight at 4°. Analogous to the agglutination of red blood cells, the appearance of sediments indicate agglutination; without agglutinating antibodies the entire bottom (look from underneath with the aid of a mirror) is covered with a green homogeneous layer. With increasing reaction the sediment becomes coarse and displays overlapping rims. With large antibody excess the sediment looks more like the negative sample (zone phenomenon). It requires experience to decide which tube in comparison with a dilution series of control sera is the last positive one. An error of one tube makes a 100% error in titer in the 1 : 2 dilution series. Therefore, less dilution should be used at each step.

For indirect agglutination the slide test with microscopic examination is recommended. The Coombs test uses anti-$\gamma$-globulin.[22] For mixed antigen agglutination[30] the thylakoid suspension is preincubated for 20–30 sec with the antiserum under investigation, as for direct agglutination, then 5–10 $\mu$l of soluble antigen is added; the concentration has to be titrated since this·test exhibits a two-dimensional zone phenomenon.[26] The order of addition is rather important; preincubation of the soluble antigen with the serum will neutralize the antibodies within seconds, and no agglutination occurs upon addition of the thylakoids. If the serum is added to a mixture of soluble and membranous antigens, the antibody might react faster with the soluble antigen, leading to false negative results, or the soluble antigen might have been absorbed to the membrane (analogous to the so-called passive agglutination of red blood cells) leading to false positive results. It is better to incubate larger amounts of membranes and antiserum for 2 min, to wash by centrifugation, and to add the soluble antigen later as a test for antibody binding.

*Neutralization of Antibody Reactions*

For inhibition or agglutination neutralization tests, the serum is first absorbed with increasing amounts of the antigen suspected to be the responsible membrane constituent for combination of the antibody with

---

[30] G. Uhlenbruck and O. Prokop, *Dtsch. Med. Wochenschr.* **92,** 940 (1967).

the membrane. After incubation for 2–10 min at 4°, the sample is tested for residual antibody activity. If the absorption was done with different membrane preparations and the test with soluble antigen, quantitative results can be obtained for the amount of reacting antigen exposed at the membrane surface.[31]

### Reproducibility and Limitations

The immunization of an animal sometimes leads to a unique antiserum. At least three animals (rabbits) should be treated according to identical immunization procedures. The immunogenicity of an antigen is defined as the number of responding animals per treated animals, by the amount of antigen needed to elicit the antiserum, and by the amount of antibody per milliliter of antiserum achieved. The specificity of the serum is tested with a crude antigen mixture; membrane proteins can be solubilized in 5% Triton X-100, and the agarose gel may contain up to 1% Triton.[10,15] The observed number of lines is a minimum number of proteins present in the antigen mixture, since, due to tolerance and/or immunocompetition, the animals do not synthesize antibodies to all antigens in a mixture.

Qualitatively different antibodies are produced in different amounts by individual animals; thus the titer of precipitating, agglutinating, inhibiting, or binding antibodies may vary independently. Due to a large excess of either antigen or precipitating antibody, i.e., if all binding sites are saturated, the combination reaction is not detected, and no macroscopic precipitation occurs. If, therefore, a precipitation test shows only one arc, i.e., about equivalent amounts of one antigen and its antibody did react, other antigen–antibody systems might not be present in equivalent concentrations and are not documented unless the concentrations are varied. For agglutinating antibodies this saturation effect explains the zone phenomenon.

If no precipitation occurs at all, it does not follow that no antibody is present, since precipitations requires a three-dimensional lattice formation, which becomes improbable if the antigen is smaller than 10,000 dalton.[15]

Unspecific inhibition can occur just by high concentrations of serum proteins. This effect is eliminated, if the influence of control sera from the same rabbit is compared, or only the $\gamma$-globulin fraction is used. Control sera often stimulate a reaction. Thus, a reaction can be inhibited

---

[31] G. H. Schmid, A. Radunz, and W. Menke, *Z. Naturforsch., Teil C* **32,** 271 (1977).

by antisera, if compared to the proper control, even if the rate is not changed, if compared to a sample without any serum.

Inhibition of a membrane reaction can occur even when the antibody reacts with a membrane constituent which is not involved in the reaction. Any agglutinating antibody may inhibit a reaction by steric hindrance,[13,25] although with soluble antigens the contribution of precipitation to inhibition is believed to be small.[2] But inhibitions of 10 or 15% may not be significant.

More specifically, some antisera against $CF_1$ inhibit photosynthetic electron transport via photosynthetic control.[9] It can be argued that an antibody will induce a conformational change in an integral membrane protein and via allosteric interaction influence the reaction of a different component; in this case the inhibited enzyme (electron transport carrier) and the reacting antigen would not be identical.

Nonspecific agglutination has also been observed. Every undiluted serum does agglutinate class I chloroplasts because it contains components neutralizing the charge of the envelope and thus diminishes repulsion.[32] Thus agglutination may depend on the residual amount of envelopes. However, diluted control sera also sometimes agglutinate thylakoid systems, presumably due to natural antibodies against cross-reacting substances (microorganisms?).

Specific and unspecific binding (absorption) of serum proteins has been described.[33] This effect is not understood. The serum complement system would lead to such increased binding of proteins from antisera, but the conditions for complement activity[1] are very different from the incubation conditions described here. The serum complement does uncouple photophosphorylation, if started properly.[34] To avoid uncertainties the serum should be decomplementalized.

The described semiquantitative agglutination in glass tubes is almost essential if several sera or supernatants from absorption tests are to be titrated. However, this test should be avoided if conclusions on the membrane architecture are the purpose of the experiment, since one has no control over changes in the mutual orientation of membrane constituents during the necessary 12–16 hr of incubation. For this purpose the short time agglutination on slides or plates should be carried out.

The first combination of an antibody with its antigen is finished in seconds.[1] Therefore, a reaction of an inhibiting antibody should be complete after a few minutes, since no lattice formation is required for this.

---

[32] R. J. Berzborn, C. Larsson, and G. Hauska, unpublished observation (1976).
[33] A. Radunz, Z. Naturforsch., Teil C **30**, 484 (1975).
[34] R. J. Berzborn and M. Tobien, unpublished data (1977).

Some inhibiting antibodies display their ability only after 20 min of incubation, however, or even during the measurement of photosynthetic electron transport in the light.[31] On the other hand, there is also an increase in inhibition after 2 to 10 min with antisera against $CF_1$, probably due to bad accessibility of the corresponding antigen in the labyrinth of the stacked, although not tightly compressed, stroma membranes. With digitonin vesicles the inhibition is complete within 3 min. If inhibition develops only in the light, it may be that the energization of the thylakoid exposes new determinant groups, either by a conformational change in a surface protein or by a change in the orientation of components, i.e., exposure of previously inaccessible components. Further studies in this direction may resolve the discrepancies in results and conclusions from antibody studies concerning the location of plastocyanin[35,36] and cytochrome $f$.[37,31]

Finally, there is the possibility that for agglutination or inhibition an aggregation of antigens in the plane of the membrane is necessary (analogous to capping at the lymphocyte membrane). This is only possible above the transition temperature of chloroplast lipids, i.e., when the membrane is in fluid state. Thus false conclusions may be drawn if the antibody was allowed to react at 0° only.

Nevertheless antibodies are valuable hydrophilic and macromolecular probes for membrane research, because they are highly specific.

[35] G. A. Hauska, R. E. McCarty, R. J. Berzborn, and E. Racker, *J. Biol. Chem.* **246**, 3524 (1971).
[36] G. H. Schmid, A. Radunz, and W. Menke, *Z. Naturforsch., Teil C* **30**, 201 (1975).
[37] E. Racker, G. A. Hauska, S. Lien, R. J. Berzborn, and N. Nelson, *Proc. Int. Congr. Photosynth. 2nd, 1971* Vol. 2, p. 1097 (1972).

# [48] Analysis of Membrane Structure–Function Relationships: A Chemical Approach

*By* L. J. PROCHASKA *and* R. A. DILLEY

## Introduction

The molecular structure of the chloroplast membrane is unresolved and the location of specific electron transport components within the membrane is still unknown. Furthermore, how polypeptides interact in the chloroplast membrane to synthesize ATP is not understood. One approach to the study of chloroplast membrane structure and function is chemical modification of membrane components. Chemical modification exploits the differences in chemical reactivity of amino acid side chains

METHODS IN ENZYMOLOGY, VOL. 69

in proteins. Protein chemists have used the differences in the physical organic chemistry of amino acid moieties to study protein function and structure for many years.[1,2] More recently, chemical modification has begun to be applied to red blood cell,[3] mitochondrial,[4] and chloroplast[5] structure and function.

Methods to measure chemical modification into membrane proteins and lipids include radiochemical techniques, fluorescence, and absorbance spectroscopy. Combinations of the above techniques can give quantitative measurement of the amount of reagent incorporated and the location of the reagent in the membrane. The pattern of inhibition of membrane function induced by the chemical modifier can provide additional information on the locus of membrane–chemical probe interaction.

The amino acid specificities of chemical modification reagents vary due to the chemical nature of the reagent and the characteristics of amino acid groups on proteins. Reagents such as diazonium compounds have broad amino acid specificities, whereas other modification processes such as reductive alkylation with formaldehyde and sodium borohydride seem to be quite specific.[1] The pH value during the modification process also effects the manner in which chemical probes react with amino acid functional groups of chloroplast membrane functional groups.

### Inhibitors of Electron Transport

Chemical modifiers have been used as electron-transport inhibitors in chloroplast membranes. Reagents such as diazonium benzenesulfonate,[6] 1-cyclohexyl-3-(2-morpholino-4-ethyl)carbodiimide plus glycine ethyl ester,[6] and p-nitrothiophenol[7] all appear to inhibit chloroplast electron transport at the water oxidation level. All of these reagents appear to cause more inhibition of water oxidation activity when chloroplasts are modified in the light than in the dark. Electron transport thorough photosystem II can be restored to diazonium and carbodiimide-treated chlo-

[1] G. E. Means, and R. E. Feeney, "Chemical Modification of Proteins." Holden-Day, San Francisco, California, 1971.
[2] A. N. Glazer, R. J. Delange, and D. S. Sigman, "Chemical Modification of Proteins." North-Holland Publ., Amsterdam, 1975.
[3] K. L. Carraway, Biochim. Biophys. Acta 415, 379 (1975).
[4] D. L. Schneider, Y. Kagawa, and E. Racker, J. Biol. 247, 4074 (1972).
[5] R. T. Giaquinta, R. A. Dilley, B. R. Selman, and B. J. Anderson, Arch. Biochem. Biophys. 162, 200 (1974).
[6] R. T. Giaquinta, R. A. Dilley, and B. J. Anderson, Biochem. Biophys. Res. Commun. 52, 1410 (1973).
[7] Y. Kobayashi, Y. Inoue, and K. Shibata, Biochim. Biophys. Acta 423, 80 (1976).

roplasts upon the addition of diphenylcarbazide.[6] Lactoperoxidase catalyzed iodination of chloroplast membranes inhibits photosystem II at the reaction center.[8] 1-Ethyl-3-(3-dimethylaminopropyl)carbodiimide in the absence of added nucleophile has been shown to inhibit electron transport between plastoquinone and cytochrome $f$.[9] Heavy metal ions such as $Ag^+$ [10] and $Hg^{2+}$ [11] inhibit electron transport in the plastocyanin region. Examples of chemical modifiers that inhibit photosystem I electron transport activity are diazonium benzenesulfonate,[12] diazonium-1,2,4-triazole,[13] and glutaraldehyde.[14]

### Inhibitors of Phosphorylation

Many different chemical probes have been used to study the structure and function of chloroplast coupling factor. Dicyclohexylcarbodiimide in the absence of a nucleophile has been used as an energy transfer inhibitor in chloroplasts for many years.[15] It can also inhibit electron transport under certain conditions, and it can restore proton uptake in EDTA-treated (coupling factor devoid) chloroplasts.[16] The site of dicyclohexylcarbodiimide binding is an 8000 dalton lipoprotein that appears to be involved in proton permeability in chloroplasts.[17] N-Ethylmaleimide has been used to capture a light-activated, uncoupler-sensitive conformational change in coupling factor. The $\gamma$-subunit of $CF_1$ is involved in the conformational change.[18] The mercurials, mercury and p-chloromercuribenzonate, are energy transfer inhibitors.[19] Reversible binding of reagents, such as 2,2-dithio-bis-5-nitropyridine,[20] o-iodosobenzonate,[20] and

[8] C. J. Arntzen, C. Vernotte, J. M. Briantais, and P. Armond, *Biochim. Biophys. Acta* **368**, 39 (1974).

[9] R. E. McCarty, *Arch. Biochem. Biophys.* **161**, 93 (1974).

[10] D. P. O'Keefe, and R. A. Dilley, unpublished results.

[11] M. Kimimura, and S. Katoh, *Biochim. Biophys. Acta* **283**, 279 (1972).

[12] B. R. Selman, R. T. Giaquinta, and R. A. Dilley, *Arch. Biochem. Biophys.* **162**, 210 (1974).

[13] I. Maruyama, K. Nakaya, K. Ariga, F. Obata, and Y. Nakamura, *FEBS Lett.* **47**, 26 (1974).

[14] H. Hardt, and B. Kok, *Biochim. Biophys. Acta* **449**, 125 (1976).

[15] R. E. McCarty, and E. Racker, *J. Biol. Chem.* **242**, 3435 (1967).

[16] E. G. Uribe, *Biochemistry* **11**, 4228 (1972).

[17] N. Nelson, E. Eytan, B.-E. Notsani, H. Sigrist, K. Sigrist-Nelson, and C. Gitler, *Proc. Natl. Acad. Sci. U.S.A.* **74**, 2375 (1977).

[18] R. E. McCarty, and J. Fagan, *Biochemistry* **12**, 1503 (1973).

[19] S. Izawa, and N. E. Good, *Prog. Photosynth. Res., Proc. Int. Congr.* [1st], 1968 Vol. 2, p. 1288 (1969).

[20] R. H. Vallejos, R. A. Ravizzini, and C. S. Andreo, *Biochim. Biophys. Acta* **459**, 205 (1977).

2,3-butanedione,[21] have been used to measure both conformational changes and to study $Ca^{2+}$-ATPase activity in coupling factor. Trinitrobenzenesulfonate reaction with chloroplasts after methyl acetimidate modification led to $Ca^{2+}$-ATPase inhibition due to a decrease in affinity of ATP binding.[22] One trinitrophenyl molecule was incorporated into the $\alpha$- and $\beta$-subunits, while two molecules were incorporated into the $\gamma$-subunit. Diazonium benzenesulfonate appears to uncouple chloroplasts by inactivating coupling factor.[23] Cross-linking chemical modifiers, such as 1,5-difluoro-2,4-dinitrobenzene, dimethyl-3,3'-dithiobispropionimidate, and dimethylsuberimidate, have been used to determine the minimal stoichiometry of $CF_1$ to be $\alpha_2\beta_2\gamma\delta\epsilon_2$.[24] Absorbance spectroscopy has determined the amino acid functional group that is modified by 7-chloro-4-nitrobenzo-2-oxa-1,3-diazole in $CF_1$ [25] and the molecular distance between subunits of $CF_1$ has also been determined by using fluorescence energy transfer spectroscopy using the same chemical probe.[26]

### Structural Studies with Chemical Modifiers

Chemical modification has also been used to elucidate chloroplast membrane structure. Diazonium benzenesulfonate has been used to test the location of photosystem I and II in the chloroplast membrane.[27] Photosystem I was labeled much more heavily than photosystem II with the nonpenetrating diazonium compounds leading to the interpretation that photosystem I is more externally localized than photosystem II. Plastocyanin is labeled by both diazonium benzenesulfonate and 1-cyclohexyl-3-(2-morpholino-4-ethyl)carbodiimide plus glycine ethyl ester and is thought to be in a hydrophillic cleft in the chloroplast membrane.[28] Lactoperoxidase-catalysed iodination of chloroplasts showed that the reducing side of photosystem I was externally located[29] and that photosystem II under mild iodination conditions was inhibited at the

[21] C. S. Andreo, and R. H. Vallejos, *FEBS Lett.* **78**, 207 (1977).

[22] D. Oliver and A. Jagendorf, *J. Biol. Chem.* **251**, 7168 (1976).

[23] R. T. Giaquinta, B. R. Selman, C. L. Bering, and R. A. Dilley, *J. Biol. Chem.* **249**, 2873 (1974).

[24] B. A. Baird, and G. G. Hammes, *J. Biol. Chem.* **251**, 6953 (1976).

[25] D. W. Deters, E. Racker, N. Nelson, and H. Nelson, *J. Biol. Chem.* **250**, 1041 (1975).

[26] L. C. Cantley, and G. G. Hammes, *Biochemistry* **14**, 2976 (1975).

[27] R. A. Dilley, G. A. Peters, and E. R. Shaw, *J. Membr. Biol.* **8**, 163 (1972).

[28] D. D. Smith, B. R. Selman, K. K. Voegeli, G. Johnson, and R. A. Dilley, *Biochim. Biophys. Acta* **459**, 468 (1977).

[29] C. J. Arntzen, P. A. Armond, C. S. Zettinger, C. Vernotte, and J.-M. Briantais, *Biochim. Biophys. Acta* **347**, 329 (1974).

reaction center. This suggested that the reaction center of photosystem II was also externally located.[8] 1-Cyclohexyl-3-(2-morpholino-4-ethyl)carbodiimide plus glycine methyl ester has been used to study the salt-induced membrane association of chloroplasts.[30] The reagent prevented polycation inhibition of photosystem I electron transfer and inhibited low salt-induced grana unstacking. The low affinity divalent cation binding site in chloroplasts was located in the light harvesting chlorophyll $a$, $b$, pigment protein complex by the use of 1-ethyl-3-(3-dimethylaminopropyl)carbodiimide plus glycine ethyl ester.[31]

### Methods of Chemical Modification of Chloroplasts

Chemical modification techniques used to study chloroplast structure and function can take two approaches. One approach is exhaustive reaction of chloroplast membrane functional groups with saturating concentrations of chemical modifier. This type of experiment gives a quantitative method for estimating the number of reactive groups in the chloroplast membrane. Inhibition of chloroplast function can result. Prochaska and Gross[31] used the carboxyl group modifying reagents, 1-ethyl-3-(3-dimethylaminopropyl)carbodiimide plus glycine ethyl ester, to locate a divalent cation binding site of the chloroplast in the light harvesting chlorophyll $a,b$ pigment protein complex.

Another approach for studying chloroplast structure and function is to limit the modification of the chloroplast membrane to maintain part of the chloroplast function. These types of experiments can lead to the identifying of proteins or amino acids are involved in a functional process. Davis and San Pietro[32] have used this technique to study how ferredoxin interacts with NADP reductase in solution as a model for *in vivo* function. Upon trinitophenylation of a single amino group in spinach ferredoxin (measured spectrophotometrically), the diaphorase activity of NADP reductase was inhibited, while the NADP reductase catalysed, ferredoxin-linked, cytochrome $c$ reductase activity was uneffected. Another example is the work of Dilley *et al.*[33,34] who have used diazonium benzenesulfonate, iodoacetate, and acetic anhydride to detect a photosystem II electron transport-dependent conformational change in chloroplasts. They found that for diazonium benzenesulfonate and iodoacetate an increase in

[30] S. Berg, S. Dodge, D. W. Krogmann, and R. A. Dilley, *Plant Physiol.* **53**, 619 (1974).
[31] L. Prochaska and E. L. Gross, *J. Membr. Biol.* **36**, 13 (1977).
[32] D. J. Davis, and A. San Pietro, *Arch. Biochem. Biophys.* **182**, 266 (1977).
[33] R. T. Giaquinta, D. R. Ort, and R. A. Dilley, *Biochemistry* **14**, 4392 (1975).
[34] L. Prochaska, and R. A. Dilley, *Arch. Biochem. Biophys.* **187**, 61 (1978).

incorporation of radioactive reagent into chloroplasts occurred in the light as compared to the dark or DCMU-inhibited state. However, acetic anhydride showed a decrease in incorporation upon photosystem II electron transport that is sensitive to DCMU. They later showed that this effect is specific for protolytic events in photosystem II.[33,34] Upon electrophoresis of chloroplast membranes treated with radioactive acetic anhydride on sodium dodecyl sulfate polyacrylamide gels, they observed that a low-molecular-weight component of the chloroplast membrane underwent the greatest change in radioactive labeling during photosystem II protolytic events.[35,36] This protein was identified as the dicylohexyl-carbodimide-binding protein, the proposed proton channel of the chloroplast ATP synthetase.[36] These results suggested that protons liberated from water oxidation interact with the chloroplast membrane in a different manner than those generated from photosystem I reactions.

### Detecting a Photosystem II-Dependent Chloroplast Membrane Conformational Change

This section will present a detailed example of the use of chemical modification agents to measure a membrane conformational change.

Chloroplasts are isolated and resuspended in a highly buffered, isosmotic medium. Buffer selection is important in that it should not react with the modification reagent. The buffer concentration should be tenfold higher than the chemical modifier due to fact that many chemical modifiers liberate or take up protons upon reaction. The pH value greatly affects both the kinetics and concentration dependence of the reaction of the chemical probe and chloroplast membrane functional groups. Chloroplasts are diluted to a concentration of 0.3 mg/ml in the presence of electron transport cofactors. The high chlorophyll concentration is important because it enables one to recover enough membranes by centrifugation to assay for radioactive probe incorporation. The reaction takes place in a water jacketed cell attached to a constant temperature water circulator. Since temperature affects the chemical reaction rate, constant temperature is important for comparison of radioactive probe incorporation levels from day to day. Two actinic light sources (500-W lamps) are projected onto the water jacketed cell, and broad band red filters are used if the chemical modifier has any visible or ultraviolet absorption bands. The light intensity for the experiment should be from 300 to 1000

[35] L. Prochaska, and R. A. Dilley, *Biochem. Biophys. Res. Commun.* **83**, 644 (1978).
[36] L. Prochaska and R. A. Dilley *in* "Frontiers of Biological Energetics" (P. L. Dutton, J. Leigh, and A. Scarpa, eds.) Vol. 1, p. 265. Academic Press, New York, 1978.

kergs cm$^{-2}$ sec$^{-1}$. Photochemical assays of chloroplast electron transport are measured prior to or during the modification process with either the oxygen electrode or pH meter.

The concentration of reagent used for the chemical modification experiment should be estimated by consulting Means and Feeney[1] and Glazer *et al.*[2] and should be based upon the reagent to protein stoichiometry. Usually, these ratios are much higher than necessary for chloroplasts studies, but they provide a basis for preliminary experiments. The pH of the modification experiment should be adjusted in the region from pH 5 to 9 due to the fact that chloroplast electron transport is labile to extremes of pH.

The effect of the reagent upon electron transport activity should be investigated. Since the photosystem II conformational change is dependent on electron transport, the chemical modification should not entirely inhibit electron transport. Also, experiments that test if the chemical modifier interfers with the cofactor's electron donor or acceptor efficiency must be performed. The reagent should be made just before the modification experiment to minimize decomposition of the reagent in water or other solvents. If radioactive detection techniques are used, the radioactive probe should be added to the carrier at this time.

The chloroplast suspension (2 ml) is temperature equilibrated for 1 min, and then a 15–30 sec light activation step is used to initiate electron transport. Uncouplers should be present to avoid detection of coupling factor conformational changes. The chemical modifier is then added, and the modification process occurs for 15 to 60 sec, depending on the reaction rate of the chemical modifier. The modification is quenched by the addition of a buffer that can readily react with the reagent. The buffer concentration should be in a tenfold molar excess to chemical modifier, and a large volume (30 ml) of the buffer should be used. The modified chloroplasts should then be washed three times with large volumes of either the quenching buffer or resuspension media and resuspended in distilled water until homogeneous. Small aliquots of the chloroplasts are transferred into scintillation vials (40–120 $\mu$g chlorophyll) and bleached in 30% hydrogen peroxide for 4–8 hr at 60°. Ten milliliters of Tritosol,[37] a water-soluble liquid scintillator, are added to each vial and the vials are stored in the dark for 8–12 hr to allow for the decay of chemilumenscence. Other aliquots are used to determine chlorophyll concentrations, and the counts are adjusted for differences in chlorophyll.

An alternative method for quenching the modification process is precipitating the chloroplast membrane proteins with 20% trichloroacetic

[37] U. Fricke, *Anal. Biochem.* **63**, 555 (1975).

TABLE I
The Effects of Light and DCMU upon Chemical Modifier Incorporation into
Chloroplasts

| Treatment[a] | Iodoacetate bound (nmoles/mg Chl) |
| --- | --- |
| Light | 1.65 ± 0.15 |
| Light + 20 $\mu M$ DCMU | 1.35 ± 0.12 |
| Dark | 1.29 ± 0.19 |

| Treatment[b] | Acetic anhydride bound (nmoles/mg protein) |
| --- | --- |
| Light | 1.34 ± 0.13 |
| Light + 12.5 $\mu M$ DCMU | 2.36 ± 0.10 |
| Dark | 2.14 ± 0.16 |

| Treatment[c] | DABS bound (nmoles/mg Chl) |
| --- | --- |
| Light | 31.0 |
| Light + 20 $\mu M$ DCMU | 7.7 |
| Dark | 7.0 |
| Light + 20 $\mu M$ DCMU + 30 $\mu M$ PMS | 8.2 |
| Light + 10 $\mu M$ nigericin | 36.0 |

[a] Chloroplasts were reacted with 5 m$M$ iodoacetic acid for 30 sec in 25 m$M$ NaCl, 20 m$M$ Na$_2$HPO$_4$, pH 7.6, 0.5 m$M$ MV and 2.5 m$M$ Na ascorbate.[34]

[b] Chloroplasts were reacted with 1 m$M$ acetic anhydride in 50 m$M$ KCl, 2 m$M$ MgCl$_2$, 50 m$M$ HEPPS NaOH, pH 8.6, 0.5 m$M$ NaN$_3$, 5 $\mu M$ valinomycin, 15 $\mu M$ nigericin, and 0.5 m$M$ MV.[34]

[c] As published in Giaquinta et al.[6]

acid. After two subsequent washes with 10% trichloroacetic acid in acetone, the chloroplast proteins are resuspended in 1 $M$ NaOH and aliquots are taken for scintillation counting and protein determination by the Lowry method.[38] After the liquid scintillator is added, 6 hr in darkness is required to allow for the decay of chemilumenescence. The counts from the vials are then corrected for differences in protein content.

Table I shows representative data of three reagents which capture the photosystem II chloroplast membrane conformational change, using water as the electron donor and methyl viologen as the electron acceptor.

[38] D. H. Lowry, N. J. Rosebrough, A. L. Farr, and R. J. Randall, J. Biol. Chem. 193, 265 (1951).

[³H]Iodoacetate and [³⁵S]diazonium benzenesulfonate both show an increase in incorporation of reagent into chloroplasts undergoing electron transport, while [³H]acetic anhydride shows a decrease in radioactive label incorporation. The final concentrations of iodoacetate, diazonium benzenesulfonate, and acetic anhydride were 5, 2, and 1 m$M$, respectively. The following reaction times were used: iodoacetate, 30 sec; diazonium benzenesulfonate, 60 sec; and acetic anhydride, 15 sec. The diazonium benzenesulfonate reacts the most extensively with the chloroplast while the iodoacetate reacts with the least amount of membrane functional groups. This technique was used to test what electron transport and protolytic events are required to induce the change in labeling (see Giaquinta et al.[33] and Prochaska and Dilley[34]).

# [49] Analysis of Membrane Architecture: Fluorimetric Approach

By Ruud Kraayenhof

The fluorescence method offers a powerful tool in attempts to unravel membrane structure and function in addition to other techniques. The possibility to measure different fluorescence parameters (i.e., fluorescence intensity, quantum yield, spectral position, and fluorescence polarization and lifetime) makes this approach an extremely versatile and dynamic one. In principle, one can derive structural information from both intrinsic and extrinsic fluorescent molecules, but only the latter type of probe has been applied in the analysis of photosynthetic membrane structure thus far, and the results are largely qualitative and indicative.

An obvious reason for the marginal success obtained with fluorescence probes in photosynthetic membranes is the interference of the pigments with the fluorescence of added probes. Another is that binding of many fluorophores to membranes affects their electronic configuration in such a way that the fluorescence is quenched so that little remains to be measured. However, the few useful results obtained justify further investigation toward a meaningful application of this technique in photosynthesis. In particular, the advantage of most fluorescence probes is that they can be used in the functioning system and should be exploited in studying the dynamic features of photosynthetic membrane structure.

It is not the aim of this contribution to survey all relevant literature but rather to introduce the basic methodology and to exemplify a few types of fluorescent reporter groups that have been used in isolated chloroplasts and chromatophores. These may provide a basis for further research of membrane structure.

For theoretical considerations of fluorescence phenomena and applications in a more general context the reader is referred to some textbooks[1-3] and reviews.[4-8]

## Adaptation of Fluorimeters for Photosynthesis Research

Most commercially available filter or monochromator fluorimeters can either be used directly or adapted easily for studies with photosynthetic membrane preparations. A number of investigators prefer to assemble the apparatus themselves because of simultaneous measurements of fluorescence and other phenomena, such as absorption, pH, or oxygen content changes. In the last case it is recommended to design a modular configuration that allows a maximum of experimental flexibility. It is impossible to give a generalized design, because each worker attaches primary importance to different features, but a few remarks will be made here that may help investigators in setting up a device for fluorescence measurements.

It is always preferable to use monochromators rather than filters in both excitation and emission beams in order to enable the measurements of spectral position changes. If the excitation light used is absorbed by the photosynthetic pigments, one wants to keep the incident light intensity at a minimum to avoid activation of the system. Photodecomposition of the probes is kept to a minimum by exciting in the longer wavelength region of the spectrum and by avoiding intense and focused excitation light.

The actinic light should be evenly distributed over the whole cuvette with its axis of incidence away from the direction of the emission light path. In practically all cases one has to protect the detector from scattered

[1] S. Udenfriend, "Flourescence Assay in Medicine and Biology," Vols. 1 and 2. Academic Press, New York, 1962 and 1969, resp.

[2] A. A. Thaer, and M. Sernetz, eds., "Fluorescence Techniques in Cell Biology." Springer-Verlag, Berlin and New York, 1973.

[3] R. F. Chen, and H. Edelhoch, eds., "Biochemical Fluorescence—Concepts," Vols. 1 and 2. Dekker, New York, 1975.

[4] A. S. Waggoner, and L. Stryer, *Proc. Natl. Acad. Sci. U.S.A.* **67,** 579 (1970).

[5] G. K. Radda, and J. Vanderkooi, *Biochim. Biophys. Acta* **265,** 509 (1972).

[6] C. Gitler, *Annu. Rev. Biophys. Bioeng.* **1,** 51 (1972).

[7] G. Weber, *Annu. Rev. Biophys. Bioeng.* **1,** 553 (1972).

[8] A. Azzi, Vol. 32, p. 234.

actinic light by appropriate guard filters (available from Schott, Kodak, Corning, etc.). It is recommended to use a constant-stirring device in the experiments with chloroplasts. We use a 1 ml cylindrical cuvette suited with a small dc motor which drives a black Delrin stirring rod; this system enables a mixing time of less than 80 msec. Since fluorescence is highly temperature-sensitive, thermostatic control is a must.

There are three possible geometries of the excitation and emission light paths: straight through (180°), right-angle, and front-face. The resulting different concentration–fluorescence relationships are discussed by Udenfriend.[1] The front-face geometry gives least problems with absorption and scattering by photosynthetic membrane particles, and this is the preferred method to use. For fluorescence polarization measurements a 90° geometry is required. The 90° geometry as it is situated in most commercial fluorimeters (Aminco, Perkin-Elmer, etc.) can be modified to a front-face system; it is important that the cuvette is positioned so that no direct reflection from the cuvette wall enters the emission path. It is preferable to use a small angle (30° or 0°) geometry such as applied in the Eppendorf filter fluorimeter. For cases where the cuvette has to be located at variable sites with respect to light source and detector (for instance in combined measurements), a flexible bifurcated light guide is convenient. Schott (Mainz, GFR) makes these in glass or quartz fibers according to the investigator's design. A 0° geometry can be made with a neutral beam splitter or a dichromatic mirror. The latter method, invented by Ploem[9] for fluorescence microscopy, combines a maximum of efficiency in fluorescence detection and a minimum of scattering and reflection problems. The dichromatic mirrors, mounted in cubical holders with or without additional cut-off and interference filters, are available from Leitz (Wetzlar, GFR) in four types, giving separation of excitation and emission at 400, 455, 510, and 580 nm. We use these mirrors in combination with solid quartz light guides (available from Schott in diameters of 2, 5, and 10 mm and variable lengths) for routine experiments on the time-dependent fluorescence changes in chloroplasts and algae at fixed excitation and emission wavelength. For spectral analysis one has to correct for the mirror characteristics or, alternatively, replace them by a neutral beam splitter. In all our fluorimeters the cylindrical cuvettes are illuminated with actinic light from the bottom by flexible glassfiber guides (diameter 10 mm), at right angles with excitation and emission beams.

The ultimate in versatility of fluorescence measurements is offered by

[9] J. S. Ploem, Z. Wiss. Mikrosk. Mitrosk. Tech. **68**, 129 (1967).

the instrument developed by Spencer and Weber[10] which is now commercially available from SLM Instruments Inc. (Urbana, Illinois). We use a modified type SLM-480, capable to produce quantum efficiency data, corrected spectra, polarization, and lifetime data. This instrument is of the modular type so that it can be easily changed to suit the special requirements of photosynthetic studies. It is often important to carry out absorption measurements of extrinsic or intrinsic probes simultaneously with fluorescence readout. An instrument for this purpose is described by Chance et al.[11]

### Selection of Suitable Fluorophores for Investigation of Photosynthetic Membrane Phenomena

The way one arrives at the use of fluorescence probes for certain purposes may be varied in nature. The most obvious and least time-consuming one is to try out a number of probes that have been proven useful in other systems, such as proteins, lipid bilayers, or other membranes. Alternatively, or in addition to this approach, one may more purposely select or synthesize fluorophores that are expected to respond to a specific membrane phenomenon, e.g., changes of "local" pH or electric charges, dielectric constant, "microviscosity," or changes of proximity between fluorescence probes and membrane components.

In order to arrive at unambiguous conclusions it is very important to investigate the structure–activity relationship of a family of probe analogues under a variety of conditions. Very often an expected response is masked by larger unexpected effects (e.g., changes of membrane absorption, of light scattering, and probe binding changes). Moreover, many fluorophores respond to more than one environmental change at the same time. The latter problem often becomes evident and can be "corrected" for if one tries out a series of probes in which one group or property is gradually modified, keeping the other groups the same.

Most fluorescent reporter groups used thus far can be added to suspensions of intact or broken chloroplasts, algal cells or particles, or chromatophores in their usual incubation media. There is no need for a recipe in this respect. One must be conscious of possible chemical interaction between the probe and compounds present in the medium (e.g., thiol or amino group-containing chemicals and polymers).

[10] R. D. Spencer, and G. Weber, *Ann. N.Y. Acad. Sci. U.S.A.* **158**, 361 (1969).
[11] B. Chance, V. Legallais, J. Sorge, and N. Graham, *Anal. Biochem.* **66**, 498 (1975).

Examples of Fluorescence Probes Used in Chloroplasts and
Chromatophores

*Probes for the Hydrophobic Membrane Domain; Microviscosity
Probes.* Waggoner and Stryer[4] have developed probes that orient themselves in the hydrocarbon region of lipid bilayer membranes. An example
is 12-(9-anthroyl) stearic acid (AS). This probe has been tested in chloroplasts.[12,13] The light-induced fluorescence change observed has a half-
life in the order of seconds and may reflect a hydrophobicity change in
the thylakoid membrane which goes hand in hand with osmotic volume
changes. We found the effect to be badly reproducible and difficult to
distinguish from direct effects of medium pH. Similar results were obtained with the typical viscosity probes N-phenylnaphthylamine[5] and
perylene.[14] On the other hand, our recent (unpublished) experiments with
the superior viscosity probe 1,6-diphenyl-1,3,5-hexatriene (DPH)[15] are
more reproducible, but we cannot draw any conclusion on the nature of
the responses as yet. The major difference with the other compounds
seems to lie in the faster and stronger binding of DPH to the thylakoid
membrane. We may tentatively conclude that this family of fluorescence
probes reports on relatively slow structural changes as result of energization.

The cationic probe auramine-O is sensitive to solvent viscosity.[16]
Kobayashi and Nishimura[17] found its fluorescence to become enhanced
by the presence of chromatophores of *Rhodospirillum rubrum*. Illumination causes a biphasic fluorescence quenching interpreted as a monitor
of membrane protonation and, in part, a local hydrophobicity change.
This work indicates the existence of a mutual influence of ionic environment and membrane viscosity, as emphasized by Träuble and Eibl.[18]
Further research with the aid of viscosity probes may produce significant
contributions in the study of this interesting phenomenon.

*Electric Field Probes: Merocyanine and Oxonol Dyes.* There has been
an extensive search for molecular probes of the action potential in squid

[12] D. L. Vandermeulen and Govindjee, *FEBS Lett.* **45**, 186 (1974).
[13] R. Kraayenhof, J. R. Brocklehurst, and C. P. Lee, *in* "Biochemical Fluorescence—
Concepts" (R. F. Chen and H. Edelhoch, eds.), Vol. 2, p. 767. Dekker, New York, 1975.
[14] M. Shinitzky, A. C. Dianoux, C. Gitler, and G. Weber, *Biochemistry* **10**, 2106 (1971).
[15] M. Shinitzky, and Y. Barenholtz, *J. Biol. Chem.* **249**, 2652 (1974).
[16] M. Nishimura, *in* "Probes of Structure and Function of Macromolecules and Membranes" (B. Chance, C. P. Lee, and J. K. Blasie, eds.), Vol. 1, p. 227. Academic Press,
New York, 1971.
[17] Y. Kobayashi, and M. Nishimura, *J. Biochem. (Tokyo)* **71**, 275 (1972).
[18] H. Träuble, and H. Eibl, *Proc. Natl. Acad. Sci. U.S.A.* **71**, 214 (1974).

giant axon,[19] which produced several useful fluorescent potential probes, among which the merocyanine dyes were superior. This development is quite important for the research of photosynthetic energy conservation which is in urgent need for an independent method to monitor the membrane potential in addition to the work with microelectrodes and electrochromic absorption changes of pigments. We have tested merocyanine I[19] (available from Eastman Kodak as merocyanine 540) in chloroplasts but did not find any fluorescence or absorption change.[13] The reason for this may be the inability of this negatively charged fluorophore to orient itself properly in the thylakoid membrane which is negatively charged itself. Chance and Baltscheffsky[20] observed interesting but complex responses with merocyanine V, more recently identified as oxonol V[21] [bis(3-phenyl-5-oxoisoxasol-4-yl)pentamethine oxonol] in *Rhodospirillum rubrum* chromatophores. Energization causes a red shift of the absorption spectrum and a decrease of the fluorescence. However, valinomycin-induced $K^+$ influx causes a fluorescence increase together with an absorption red shift; $K^+$ efflux causes a fluorescence decrease and absorption blue shift. This is not in accord with a purely electrochromic behavior of the probe, and interactions with carotenoids may also occur.[21] At present we are studying oxonol VI[22] (oxonol V with the phenyl groups replaced by propyl groups) in chloroplasts. We found that in spinach thylakoids, oxonol VI undergoes a spectral shift in response to a membrane potential formation, induced by either illumination or a gradient of $K^+$ activity.[22a] The probe response is sufficiently rapid to allow its use in single-turnover experiments, so that one can compare the kinetics with those of the intrinsic potential-probe response at 515 nm. Oxonol VI can be studied by both absorption and fluorescence changes. It is evident that good candidates for trans-membrane potential monitors may be found among this group of fluorophores.[19,22]

*N-Arylaminonaphthalene Sulfonates: Polarity Probes.* These anionic fluorophores show a polarity-dependent shift of the emission spectrum. The blue shift occurring upon energization of submitochondrial particles with 1-anilino-8-naphthalene sulfonate (ANS) is therefore interpreted by a decrease of dielectric constant in the membrane. This conclusion is supported by $D_2O$ isotope effects and fluorescence lifetime measure-

[19] L. B. Cohen, B. M. Salzberg, H. V. Davila, W. N. Ross, D. Landowne, A. S. Waggoner, and C. H. Wang, *J. Membr. Biol.* **19**, 1 (1974).

[20] B. Chance, and M. Baltscheffsky, *Biomembranes* **7**, 33 (1975).

[21] J. C. Smith, P. Russ, B. S. Cooperman, and B. Chance, *Biochemistry* **15**, 5094 (1976).

[22] C. L. Bashford, and W. S. Thayer, *J. Biol. Chem.* **252**, 8459 (1977).

[22a] J. J. Schuurmans, R. P. Casey, and R. Kraayenhof, *FEBS Lett.* **94**, 405 (1978).

ments.[23] X-ray diffraction studies showed that ANS orients itself preferentially in the membrane–medium interface.[23] The aromatic ring intercalates in a hydrophobic region, while the negative sulfonate group protudes into the aqueous medium, thereby affecting the electric surface potential.[24] Although this picture of ANS behavior seems to be the most satisfactory one, there are deviations in different biological membranes caused by energy-dependent changes of the ANS binding constant.[8]

Gromet-Elhanan[25] showed that the ANS responses in chromatophores are very similar to the ones in submitochondrial particles.[8,23] She suggested that this probe monitors an energized state, leading to ATP synthesis, but independent of $H^+$ uptake. In chloroplasts ANS has a very poor light response in terms of relative increase of the quantum yield and emission blue shift. Its kinetics were very similar to those of the light-induced pH rise and increase of light scattering.[26] We found that in general negatively charged probes have a poor response in chloroplasts,[26] presumably due to electrostatic repulsion by the thylakoid surface.

*9-Amino-Substituted Acridines; ΔpH Probes versus Surface Potential Probes.* A large number of positively charged fluorescence probes show a strong interaction and concomitant response in chloroplasts, chromatophores, submitochondrial, and bacterial "inside-out" membranes.[13] Among these the 9-aminoacridines, notably the uncouplers atebrin and 9-aminoacridine, have been found useful in studying energy-dependent phenomena in photosynthetic and respiratory membranes. Their energy-induced binding is usually associated with a proportional fluorescence quenching without a shift of the emission spectrum. The responses of these probes are particularly useful in studying the dynamics of light-induced membrane changes.

We have observed that the aminoacridines that show the largest and most rapid responses in chloroplasts are all potent uncouplers. Variation of the degree of energization (by changing the rate of electron transfer or ATP hydrolysis) leads to a proportional shift of the acridine binding titration curve.[27] A titration with a suitable acridine such as atebrin or, better, 9-amino-6-chloro-2-methoxyacridine (ACMA) thus enables a comparison of the magnitude of the "energized state" under different metabolic conditions (e.g., at different temperatures, in the presence of certain

[23] See the review by J. Vanderkooi and A. McLaughlin, *in* "Biochemical Fluorescence—Concepts" (R. F. Chen and H. Edelhoch, eds.), Vol. 2, p. 737. Dekker, New York, 1975

[24] S. McLaughlin, *Curr. Top. Membr. Transport* **9**, 71 (1977).

[25] Z. Gromet-Elhanan, *Eur. J. Biochem.* **25**, 84 (1972).

[26] R. Kraayenhof, and M. B. Katan, *Photosynth., Two Centuries After Its Discovery by Joseph Priestley, Proc. Int. Congr. Photosynth. Res., 2nd, 1971,* Vol. 2, p. 937 (1972).

[27] R. Kraayenhof, S. Izawa, and B. Chance, *Plant Physiol.* **50**, 713 (1972).

inhibitors, and in reconstitution experiments). The titration point usually coincides with the probe concentration giving complete inhibition of ATP synthesis.[28]

The interpretation of the acridine behavior in energy-conserving membranes is a matter of continuing debate. On overlooking differences in detail one can distinguish two proposed mechanisms of interaction. One is proposed by Rottenberg et al.[29,30] and interprets the response by a penetration of the acridine in its uncharged form through the bulk membrane and distribution of the charged species proportional with an energy-generated $\Delta$pH. Formulas were derived to calculate $\Delta$pH from the percentage of fluorescence quenching. The other envisages the responses as a consequence of electrostatic binding of the positive acridines to negative sites on the thylakoid membrane,[13,28] the density of which increases upon energization. In this mechanism the acridine probes are proposed to monitor the local negative surface potential rather than a transmembrane phenomenon. It should be realized, however, that under certain circumstances there can be a mutual effect of transmembrane and localized potentials and, therefore, that the responses can be easily accommodated in either mechanism.

Although this is not the place to sum up all arguments *pro* and *con* the two proposals, this author feels that a serious warning against gratuitous use of the acridine fluorescence response for routine $\Delta$pH measurement has its place. We have done extensive research on this interesting phenomenon in spinach chloroplasts in order to locate the site of acridine interaction and interpret the nature of its response. Over thirty different acridine analogs were synthesized (according to the recipes given by Albert[31]), including species with different number of charges, $pK_a$ values, hydrophobic groups, and paramagnetic or nonpermeant groups bound to the 9-amino side chain,[13,32,33] More recently, we have studied the binding to differently charged lipid vesicles and the upward $pK_a$ shift of the acridines upon binding to these vesicles and chloroplasts, the light-induced acridine dimer formation, the responses of covalently

[28] R. Kraayenhof and, J. W. T. Fiolet, *in* "Dynamics of Energy-Transducing Membranes" (L. Ernster, R. W. Estabrook, and E. C. Slater, eds.), BBA Libr. Vol. 13, p. 355. Elsevier, Amsterdam, 1974.

[29] H. Rottenberg, T. Grunwald, and M. Avron, *Eur. J. Biochem.* **25,** 54 (1972).

[30] S. Schuldiner, H. Rottenberg, and M. Avron, *Eur. J. Biochem.* **25,** 64 (1972).

[31] A. Albert, "The Acridines," Arnold, London, 1966.

[32] R. Kraayenhof, *in* "Fluorescence Techniques in Cell Biology" (A. A. Thaer and M. Sernetz, eds.), p. 381. Springer-Verlag, Berlin and New York, 1973.

[33] R. Kraayenhof, and E. C. Slater, *in* "Photosynthesis III" (M. Avron, ed.), Vol. 2, p. 985. Elsevier, Amsterdam, 1974.

bound acridines, and the light-induced increase of electrophoretic mobility of chloroplasts.[34,35] These studies provide a large body of evidence in favor of the proposal that the aminoacridine probes respond to a membrane surface potential change which is directly associated with energy-dependent structural reorientations of membrane-bound groups.

*Covalently Bound Fluorophores; Conformation Probes.* Much of the confusion in the interpretation of the responses of "free" or loosely bound probes is the frequently observed variability of the affinity toward the membrane. Obviously, covalent binding to the membrane or specific components therein circumvents this problem.

Aminoacridines, covalently bound to broken chloroplasts, show a 50- to 100-fold enhanced light-response. These acridines have either an isothiocyanato, a nitrogen mustard, or an azido group in their 9-amino side chain.[34] The isothiocyanato and mustard analogs are incubated with the chloroplasts which are brought to pH 9.0 for 30 sec. The azido compound can be coupled by a 30 to 60 sec illumination with a mercury arc. The concentrations of these compounds must be kept as low as possible; uncoupling activity as well as inhibition of electron transfer is observed. It appears that these acridine probes have an orientation with respect to the membrane that allows for a much more direct response to the energy-driven structural changes.

De Kouchkovsky[36] used covalently bound fluoresceine isothiocyanate (FITC) as a probe for the light-induced pH rise at close proximity to the chloroplast thylakoid membrane. It would be very interesting to extend this type of approach with fluorophores, covalently bound but with different proximity to the membrane surface. This may enable a "calibration" of the pH or potential profile in the membrane–medium interface.

The fluorimetric approach using covalent fluorescent labels seems to be of particular use in attempts to study conformational reorientations in membrane-bound proteins. We have made a first step along this line by labeling chloroplast ATPase with the purpose of detecting light-induced structural changes in the reconstituted system.[33] A number of labels for carboxyl and amino groups have been tested. Only the latter ones were found to be useful because the inhibitions of enzymic activities were either small or absent under the proper conditions. The amine-specific

[34] R. Kraayenhof, and J. C. Arents, *in* "Electrical Phenomena at the Biological Membrane Level" (E. Roux, ed.), p. 493. Elsevier, Amsterdam, 1977.

[35] R. Kraayenhof, *in* "Structure and Function of Energy-Transducing Membranes" (K. van Dam and B. F. van Gelder, eds.), p. 223. Elsevier, Amsterdam, 1977.

[36] Y. de Kouchkovsky, *in* "Photosynthesis III" (M. Arron, ed.), Vol. 2, p. 1013 Elsevier, Amsterdam, 1974.

fluorogenic label fluorescamine[37] was found to be far superior over FITC and other dyes. We have labeled and studied the fluorescence parameters and activities before and after reconstitution of the membranes with ATPase. Only if membrane-bound ATPase is modified with fluorescamine, followed by isolation according to the procedure of Strotmann *et al.*[38,39] and reconstitution with freshly depleted membranes,[39] we obtained a preparation with fully recovered ATP synthesis and reproducible fluorescence response. Modification of isolated ATPase leads to partial and irreversible loss of ATP hydrolysis and synthesis activities as well as reconstitution capacity.

Harnischfeger and Schopf[40,41] have labeled the isolated ATPase with fluorescamine and FITC for estimating the retention of ATPase in reconstitution studies and effects of ATPase activation on the reconstitution capacity. They also showed that dithiothreitol, often used in activation of the ATPase, interacts directly with fluorescamine, which could at least partly explain our observation of an activation-dependent difference in fluorescamine labeling. However, chloroplasts that are activated and then washed free of thiol compound retain their activated state and still show a decreased reactivity toward fluorescamine.

Although this approach with covalent probes is still in a developmental stage, it seems to be quite promising. If employed for the more deeply buried membrane proteins, this method will undoubtedly encounter severe hindrance by the unavoidable modification of the desired component as well as the membrane residue. This problem may be partly circumvented by taking specifically binding inhibitors as the basis for designing fluorescent labels. A combination of this approach with that of chemical modification[42] is obviously significant.

### Final Remarks

From the above survey it is evident that we are not yet in a position to present a standard methodology for the application of fluorescence in the analysis of photosynthetic membranes. We are repeatedly confronted with the finding that theory and practice do not go hand in hand in applying fluorescence probes. The extrapolation of fluorescence phenom-

[37] S. Udenfriend, S. Stein, P. Böhlen, W. Dairman, W. Leimgruber, and M. Weigele, *Science* **178**, 871 (1972).
[38] H. Strotmann, H. Hesse, and K. Edelman, *Biochim. Biophys. Acta* **314**, 202 (1973).
[39] H. Hesse, R. Jank-Ladwig, and H. Strotmann, *Z. Naturforsch., Teil C* **31**, 445 (1976).
[40] G. Harnischfeger and R. Schopf, *Z. Naturforsch., Teil C* **32**, 392 (1977).
[41] R. Schopf and G. Harnischfeger, *Z. Naturforsch., Teil C* **32**, 398 (1977).
[42] L. J. Prochaska, and R. A. Dilley, this vol. [48].

ena in simple model system (i.e., different solvents, proteins, and lipid membranes) to complicated biological membranes often turns out to be more than dubious. One has to pay a great deal of attention to the specific properties of the membrane studied in order to understand apparent deviations from expected probe responses.

Hydrophobic and strongly negatively charged probes seem to have difficulties in reaching the "expected" target in the membrane, whereas positively charged probes usually show spectacular responses. Proper pretreatment of the membrane will possibly facilitate the entrance of probes in the desired domain.

The use of covalent fluorophores seems to offer a most challenging method, since it is almost the only way to unravel specific structural events in the photosynthetic membrane by means of fluorescence measurements.

## [50] Analysis of Dynamic Changes in Membrane Architecture: Electron Microscopic Approach

*By* Charles J. Arntzen and John J. Burke

### Introduction

Electron microscopy has now been used for more than 30 years to evaluate the structural organization of photosynthetic membranes. Initial studies were largely descriptive in nature; the overall morphology of photosynthetic bacteria, algae, and higher plant chloroplasts have been defined and terminology for various structural features has been established.[1] Since the photosynthetic pigments and the enzymatic components which catalyze electron transport and photophosyphorylation are constituents of the internal chloroplast membranes, these lamellae have received intense study over the last 15 years. The use of electron microscopy has been coupled to many other biochemical or biophysical techniques to examine submembrane fragments, enzymatically digested membranes, genetically altered membranes, etc. These combined studies have greatly broadened our understanding of chloroplast membrane ar-

[1] J. T. O. Kirk, and R. A. E. Tilney-Basset, "The Plastids." Freeman, San Francisco, California, 1967.

METHODS IN ENZYMOLOGY, VOL. 69

chitecture and have led to the knowledge that the structure of the chloroplast lamellae is not static; dynamic macrostructural and microstructural changes are intimately involved in regulation of photochemical activity.[2]

This article will deal with the various general types of electron microscopic techniques that are used to examine chloroplast lamellae, the terminology used to define structural components, and current interpretations of structural data with respect to membrane function. We will primarily focus attention upon higher plant chloroplasts. We will not attempt to discuss electron microscopic techniques in depth; there are now many basic methodology manuals for this purpose (see, for example, Dawes[3] and Glauert[4]). Examples of procedures which have been found to work successfully with chloroplast lamellae will be described.

## Chloroplast Membrane Morphology: Use of Thin Sectioning for Electron Microscopy

The morphological organization of chloroplasts can be studied either *in vivo* or *in vitro*. The organization of internal lamellae into grana (stacked membranes) is usually the most characteristic feature of the plastids; the reversible nature of the stacking process can be demonstrated by altering the salt concentration around isolated plastid lamellae.[2,5,6]

### Fixing and Embedding Intact Leaf Tissue

All steps are carried out at room temperature.

*Fixation.* Leaf segments are cut into 1-mm$^2$ pieces in the primary fixing solution (3% glutaraldehyde in 0.1 $M$ Na cacodylate buffer, pH 7.4); these may be vacuum infiltrated to enhance fixation penetration. Fixation continues for 60 min.

*Wash.* The primary fixative is removed by five washes with 0.1 $M$ Na cacodylate buffer (pH 7.4) over a 60 min period.

*Postfixation.* Tissue pieces are treated for 60 min with 2% osmium tetroxide in 50 m$M$ Na cacodylate (pH 7.4).

---

[2] C. J. Arntzen, *Curr. Top. Bioenerg.* **8**, 111–160 (1978).
[3] C. J. Dawes, "Biological Techniques in Electron Microscopy." Barnes & Noble, New York, 1971.
[4] A. M. Glauert, "Fixation, Dehydration and Embedding of Biological Specimens." North-Holland Pub., Amsterdam, 1975.
[5] S. Izawa, and N. E. Good, *Plant Physiol.* **41**, 544 (1966).
[6] E. L. Gross, and S. H. Prasher, *Arch. Biochem. Biophys.* **164**, 460 (1974).

*Wash.* Osmium tetroxide is removed by two washes in deionized water.

*Dehydration.* The tissue is transferred through a series of graded acetone solutions (10 min per step in 30, 50, 75, 95, 100, and 100% repeated).

*Embedding.* The tissue is transferred into a plastic such as Spurr's low viscosity resin.[7]

### Fixing and Embedding Isolated Plastid Membranes

*Fixation.* An isolated chloroplast preparation is centrifuged to yield a pellet which is resuspended in a medium containing the desired osmoticum, salts, and buffers plus 3% glutaraldehyde. The sample is held on ice for 30 min.

*Wash.* The fixed membranes are pelleted by centrifugation and resuspended in 0.1 $M$ Na cacodylate buffer (pH 7.4). This wash is repeated three times.

*Postfixation.* The washed chloroplast pellet is gently cut into small segments and is transferred into small volumes of warm 2% agar held in the hemispherical concavity of a porcelain spot plate. After cooling, the agar is trimmed into smaller pieces such that the clumped chloroplasts are left encased in agar. The agar/chloroplast samples are fixed for 1 hr at room temperature in 2% osmium tetroxide in 50 $mM$ Na cacodylate (pH 7.4). Dehydration and embedding are conducted as described above.

### Preparation of Stroma-Free Stacked and Unstacked Chloroplast Lamellae

Fifty grams of washed, chilled leaf tissue are homogenized for 15 sec at maximum speed in a blender containing 100 ml of cold grinding solution (0.1 $M$ Na Tricine, pH 7.8, 0.4 $M$ sorbitol, 0.01 $M$ Na ascorbate, pH 7.8, and 5 mg/ml bovine serum albumin). The resulting brei is squeezed through 4 layers of cheesecloth and then poured through 12 layers of cheesecloth. The filtered solution is centrifuged in two or more tubes at 1000 $g$ for 10 min. The resultant supernatants are discarded, and the chloroplast pellets are dispersed with small camel's hair brushes. Equal portions of the samples are then diluted with either (a) 30 ml with 0.1 $M$ sorbitol, 0.01 $M$ Na Tricine, pH 7.8, and 0.01 $M$ NaCl or (b) an identical solution which also contains 10 m$M$ MgCl$_2$. After 15 min, the resuspended chloroplasts are centrifuged at 5000 $g$ for 10 min to yield pellets of (a) unstacked and (b) stacked lamellar membranes. This procedure has been

[7] A. R. Spurr, *J. Ultrastruct. Res.* **26**, 31 (1969).

found to work successfully on spinach, pea, lettuce, and barley chloroplasts.

## Terminology

The chloroplast of a typical higher plant is lense-shaped with a long diameter of 3–10 $\mu$m. An example of a chloroplast of *Phleum pratense* is shown in Fig. 1. This sample was chemically fixed *in vivo*, dehydrated, embedded in plastic, and thin sectioned for examination by electron microscopy. The characteristic features of the plastid are a double outer (limiting) membrane called the chloroplast envelope (E), a dense, particulate matrix called the stroma (S), and a complex lamellar system embedded within the stroma. The lamellae characteristically consist of disk-like membranes (called thylakoids) which fuse or stack to form grana (G). The most common other stroma inclusions are starch grains (not in this figure) or osmophillic globules (O).

An individual grana stack is shown in Fig. 2. The large, flat membrane surfaces of the thylakoids exposed to the stroma at the top and bottom of each stack are termed end membranes (EM), whereas the end portion (in contact with stroma) of those thylakoids which are within the grana stack are called the margins (M). Unstacked membranes extending out between grana stacks are stroma lamellae (SL). A region where two

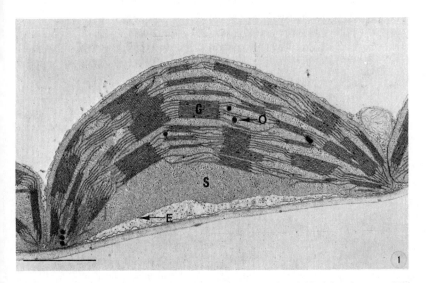

FIG. 1. A thin-sectioned sample showing a chloroplast in a leaf of timothy grass (*Phleum pratense*). E, chloroplast envelope; S, stroma; G, grana stack; O, osmiophilic globule. Scale, 1 $\mu$m. Figure provided by E. Newcomb and W. Wergin.

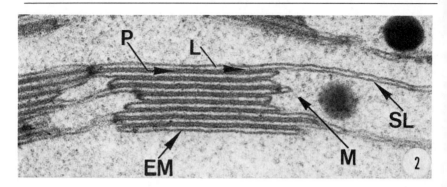

FIG. 2. Descriptive definition of nomenclature used for chloroplast membranes. P, partition; L, loculus (or lumen); M, margin; EM, end membrane; SL, stroma lamellae.

membranes come in contact within the stack is a partition (P). The internal space within a thylakoid is called the loculus or lumen (L). The relative ratio of thylakoid membrane surface exposed to the stroma (end membranes + margins + stroma lamellae) with respect to the membrane surfaces within the partitions (not readily accessible to soluble stroma constituents) is usually near 1 : 2 for normally developed, mature plastids of $C_3$ plants or mesophyll chloroplasts of $C_4$ plants.

*Reversible Grana Stacking*

The involvement of salts in regulating the morphological organization of chloroplast thylakoids was first demonstrated by Izawa and Good.[5] The fact that divalent cations such as $Mg^{2+}$ have a dramatic effect on stacking is clearly evident in Fig. 3. In the absence of $Mg^{2+}$ (Fig. 3a) there is no evidence of membrane fusion; the sample prepared in the presence of $Mg^{2+}$ was tightly appressed lamellae (Fig. 3b). Many attempts have been made to identify the membrane component (s) which regulate the cation-induced stacking process. At present it is thought that a light-harvesting pigment protein is essential for the stacking.[2,8,9] There is evidence that reversible stacking–unstacking occurs *in vivo*. These changes may in part be in response to cation fluxes in the plastid. It has been suggested that the stacking process per se is involved in the regulation of excitation energy distribution within the membranes.[2] Combined use

[8] J. M. Anderson, *in* "International Cell Biology" (B. R. Brinkley and K. R. Porter, eds.), p. 183. Rockefeller Univ. Press, New York, 1977.
[9] C. J. Arntzen, P. A. Armond, J.-M. Briantais, J. J. Burke, and W. P. Novitzky, *Brookhaven Symp. Biol.* **28**, 316, (1976).

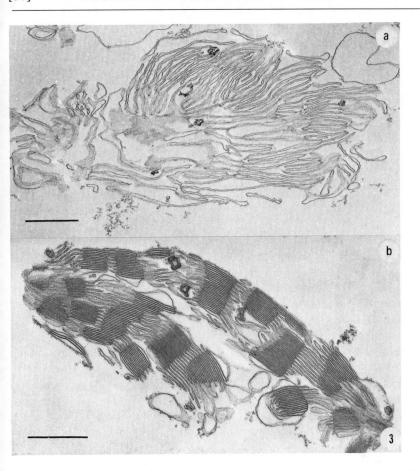

FIG. 3. The effects of cations on grana stacking in isolated, stroma-free chloroplast lamellae (see text for methods). (a) Isolated membranes suspended in the absence of $MgCl_2$ which show no regions of membrane fusion. (b) Well-defined grana stacking in the presence of $MgCl_2$. Scale, $1\mu m$.

of both electron microscopic and biochemical/biophysical assay techniques are needed to further resolve this problem.

### Analysis of Intrinsic and Extrinsic Thylakoid Components

The most important preparative procedure for the analysis of internal membrane structural organization is freeze fracturing. In this procedure, samples are frozen in liquid Freon 12 or Freon 22 at approximately $-150°$, fractured at $-100°$ or lower, and a platinum-carbon shadowed replica of the exposed frozen surface is prepared. Following chemical cleaning (Chlorox, 1 hr; $H_2O$ wash; 30% chromic acid, 1 hr) and mounting on an electron microscope grid, the replica can be examined in the electron microscope. It is now generally accepted that the fracture follows the hydrophobic interior of the membrane so that intrinsic membrane components are revealed.[10–12] A closely correlated procedure which can be used to examine membrane surfaces is the freeze etch (or deep etch) technique; this entails fracturing a frozen sample and then allowing some of the solution surrounding the membranes to sublime away (under vacuum at low temperature). The resultant exposed true membrane surfaces are replicated and the replica is transferred to grids for examination.

## Preparation of Samples for Freeze Fracturing of Freeze Etching

For freeze fracture, isolated chloroplasts prepared as described above are resuspended to a concentration of approximately 0.5 mg chlorophyll per ml in 0.4 $M$ sorbitol, 10 m$M$ Na Tricine (pH 7.8), 10 m$M$ NaCl, and 10 m$M$ $MgCl_2$. This solution is stirred as glycerol is slowly added over a 30-min period until a final glycerol concentration of 30% is achieved. The chloroplasts are then centrifuged at 10,000 $g$ for 15 min to yield a pellet. Small aliquots of the pellet are frozen on sample supports by plunging them into the liquid Freon for 3 sec and then quickly transferring them to liquid nitrogen for storage. To examine particle distributions in unstacked lamellae, the 10 m$M$ $MgCl_2$ is omitted from the resuspension solution above and the sample is allowed equilibrate for 1 hr at 4° before glycerol addition. Prefixation of samples with glutaraldehyde can result in modification of the fracture plane (L. A. Staehelin, personal communication).

For deep-etching experiments, the samples must be frozen in solutions containing less than 10 m$M$ total solute concentration.[13] A representative procedure would be to wash isolated membranes once in a solution containing 1 m$M$ Na Tricine (pH 7.8) and 5 m$M$ $MgCl_2$, pellet this sample by centrifugation at 3000 $g$ for 5 min, and then freeze portions of the sample. $MgCl_2$ can be replaced by an equivalent concentration of NaCl to observe unstacked membranes. Samples can be prefixed in 0.5% glutaraldehyde.

[10] A. J. Verkleij, and P. H. J. T. Ververgaert, *Annu. Rev. Phys. Chem.* **26**, 101 (1975).

[11] D. Branton, *Annu. Rev. Plant Physiol.* **20**, 209 (1969).

[12] K. Mühlethaler, *in* "Photosynthesis I" (A. Trebst and M. Avron, eds.) p. 503. Springer-Verlag, Berlin and New York, 1977.

[13] L. A. Staehelin, *J. Cell. Biol.* **71**, 136 (1976).

*Preparation of Samples for Negative Staining or Shadowing*

The chloroplast membrane sample used for negative staining or shadowing must be obtained free of sugars and in a very low salt concentration for maximal clarity of the specimen. The washing procedure described above for deep etching experiments should be repeated twice to obtain a washed membrane sample; for negative staining, a small portion of the pellet obtained should be diluted to approximately 10–20 $\mu$g chlorophyll/ml with neutralized 2% phosphotungstic acid or 2% uranyl acetate. Small aliquots are then either atomized onto carbon-coated electron microscopy grids or, alternatively, small drops of the suspended chloroplasts are placed upon the grid and all fluid except for a very thin film is removed by touching the grid edge to a filter paper. For shadowing, the washed membranes are diluted with either distilled water or 5 m$M$ MgCl$_2$ to 5–20 $\mu$g chlorophyll/ml and are applied to coated grids by aspiration or the droplet method.

*Nomenclature for Freeze Fracturing and Freeze Etching*

The currently accepted nomenclature for freeze fractured and freeze etched membranes[14] is based on a general acceptance of the fluid mosaic membrane model.[15] It is recognized that all biological membranes have an exoplasmic leaflet (E) and a protoplasmic leaflet (P). Each leaflet has a true surface (S) and a fracture face (F); this leads to the use of either PF or EF designations for fracture faces or PS or ES for membrane surfaces. Finally, since chloroplast membranes can be either stacked or unstacked, subscripts "s" or "u" are used with the fracture face or surface designation.[13,16] The use of this nomenclature is indicated diagrammatically in Fig. 4. The drawing represents portions of three thylakoids forming a grana stack. The partition region where the central thylakoid makes contact with the other thylakoids is a stacked membrane region; the surface and fracture faces occurring in these stacked regions are indicated. The fracture faces and surfaces of end membranes (or stroma lamellae, not shown in this diagram) are indicated as PS$_u$, EF$_u$, etc.

[14] D. Branton, S. Bullivant, N. B. Gilula, M. J. Karnovsky, H. Moor, K. Mühlethaler, D. H. Northcote, L. Packer, B. Satir, P. Satir, V. Speth, L. A. Staehelin, R. L. Steere, and R. S. Weinstein, *Science* **190**, 54 (1975).

[15] S. J. Singer, and G. L. Nicolson, *Science* **175**, 720 (1972).

[16] L. A. Staehelin, P. A. Armond, and K. R. Miller, *Brookhaven Symp. Biol.* **28**, 278 (1976).

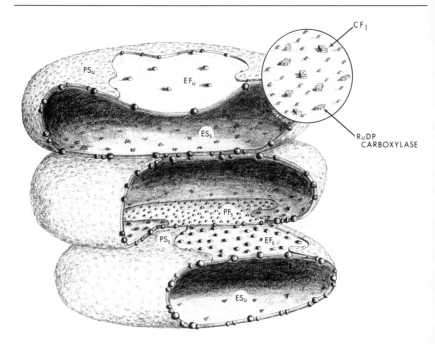

FIG. 4. A diagrammatic representation of three thylakoids forming a grana stack. Freeze fracture reveals inner membrane faces; these are indicated as half-membrane regions labeled either EF or PF. True membrane surfaces can be seen by deep etching; these are labeled as ES or PS. The enlarged inset portrays extrinsic protein complexes. Further discussion of terminology is detailed in the text.

### Descriptive Characteristics of Freeze Fracture Preparations

An example of freeze-fractured pea chloroplast membranes is shown in Fig. 5. The four types of fracture faces are indicated. The greater density of the $EF_s$ subunits as compared to the $EF_u$ subunits has often been noted; it is commonly reported that there are three times as many $EF_s$ particles as $EF_u$ particles on a unit area basis in mature chloroplast lamellae.[16,17] In contrast, the particle densities of $PF_s$ and $PF_u$ regions are reported to be nearly identical.

One of the most important advances in recent years in the evaluation of freeze fractured images has been the introduction of careful quantitation of particle sizes by Staehelin and co-workers.[16,18] A particle size histogram for the four fracture faces of spinach chloroplast thylakoids is

[17] P. A. Armond, L. A. Staehelin, and C. J. Arntzen, *J. Cell Biol.* **73**, 400 (1977).
[18] U. W. Goodenough and L. A. Staehelin, *J. Cell. Biol.* **48**, 594 (1971).

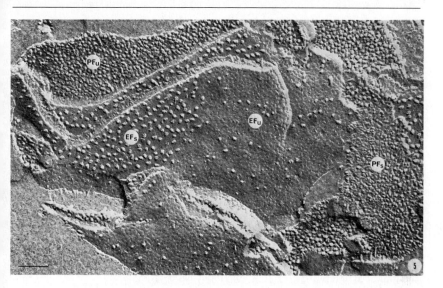

FIG. 5. Characteristic fracture faces of stacked chloroplast membranes isolated from pea (*Pisum sativum*) leaves. Two thylakoids forming a portion of a grana stack are fractured on the left portion of the figure. The protoplasmic leaflet of the end membrane is indicated as a $PF_u$ face, since this is an unpaired membrane. A $PF_s$ face is shown on the right side of the figure; the particles of this face are characteristically less distinct than the $PF_u$ face. Note the demarcation between $EF_s$ and $EF_u$ faces due to a change in particle density; the area occupied by $EF_s$ particles defines a partition (i.e., the point of membrane contact between the thylakoids). Scale, 0.1 μm. (Figure provided by P. Armond.)

shown in Fig. 6. On the EF faces, particle sizes range from 60 to more than 200 Å in diameter, with major size classes centered at approximately 110 and 160 Å. The $PF_s$ face has relatively uniform-sized particles of 80 Å average diameter, whereas the $PF_u$ face has an additional size class centered near 110 Å. The segregation of various sized particles into distinct regions of the membrane is lost in unstacked chloroplasts. Figure 7, a freeze-fractured "low-salt" chloroplast prepared in the absence of $MgCl_2$, shows the uniform EF and PF particle distribution pattern.

*Functional Interpretations of Freeze Fracture Data*

The particulate structures visible in freeze fractured membranes are generally believed to be protein complexes embedded in a lipid bilayer which appears as a smooth matrix background.[10,19] Various attempts have

[19] D. W. Deamer, R. Leonard, A. Tardieu, and D. Branton, *Biochim. Biophys. Acta* **219**, 47 (1970).

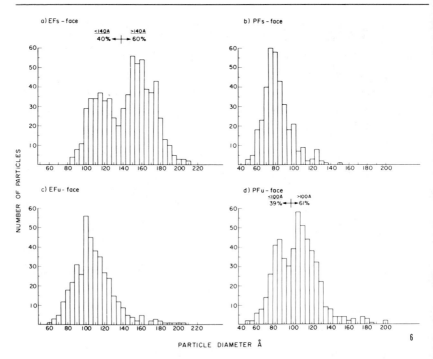

Fig. 6. Particle size histograms for membrane subunits observed on freeze fracture faces of mature spinach chloroplast thylakoids. In histograms showing more than one size class maximum, the histogram was arbitrarily divided into two subsets for which the percentage distribution of size classes is indicated. (Figure provided by L. A. Staehelin.)

been made to correlate particulate substructures with functional units of the membrane. Arntzen[2] has recently summarized much of the existing data into the membrane model shown in Fig. 8. The essential point of this scheme is that five different classes of protein complexes have been isolated from chloroplast lamellae by mild detergent disruption and subsequent fractionation procedures. It is thought that the five complexes represent "native" protein aggregates which are simply solubilized out of the lipid matrix of the membrane by detergent action and must, therefore, exist in the membrane as distinct structural entities. The five complexes are (a) a photosystem II–reaction center complex which also contains light-harvesting chlorophyll $a$; (b) a photochemically inactive light-harvesting pigment–protein complex, which contains chlorophyll $b$ and which acts primarily as a sensitizer of photosystem II centers; (c) a nonpigmented enzymatic complex containing cytochromes $b_6$ and $f$, a nonheme iron and a copper-containing protein (plastocyanin?) which

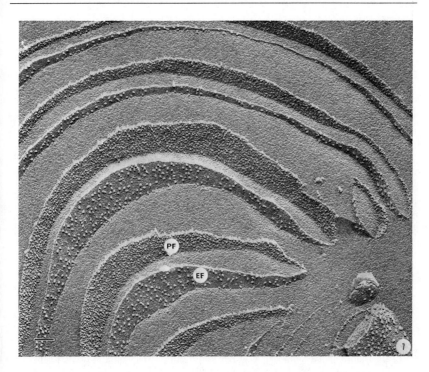

FIG. 7. The membrane substructure of unstacked pea chloroplast lamellae as revealed by freeze fracture. The relatively uniform distribution of EF particles in this figure should be compared with particle density differences in $EF_s$ and $EF_u$ faces of Fig. 5. Scale 0.1 $\mu$m. (Figure provided by P. Armond.)

serves in the electron transport chain; (d) a photosystem I reaction center complex that contains P700 and the primary electron acceptor plus light-harvesting chlorophyll $a$; and (e) a nonpigmented hydrophobic protein complex that acts as a proton channel and which binds the extrinsic portion of the coupling factor.

All known light-harvesting and energy-coupling characteristics of chloroplast lamellae can be accounted for by these complexes. Each can be correlated with one of the types of freeze fracture particles observed by electron microscopic observation. The different size classes of EF particles can be explained by the observations that different amounts of light-harvesting complexes bind to photosystem II centers.[9,17] The number of EF particles observed can be correlated to known photosystem II distribution in grana and stroma lamellae[20] and in bundle-sheath chloro-

[20] P. A. Armond, and C. J. Arntzen, *Plant Physiol.* **59**, 398 (1977).

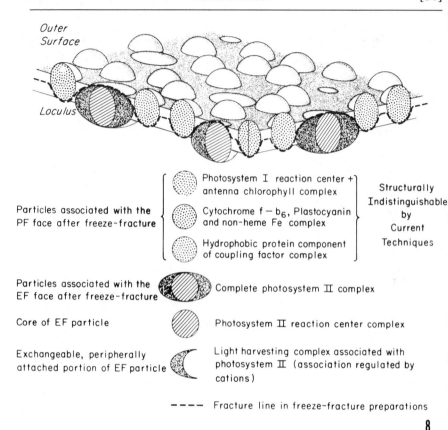

*Outer Surface*

*Loculus*

Particles associated with the
PF face after freeze-fracture
{
Photosystem I reaction center +
antenna chlorophyll complex

Cytochrome f – b$_6$, Plastocyanin
and non-heme Fe complex

Hydrophobic protein component
of coupling factor complex
}
Structurally
Indistinguishable
by
Current
Techniques

Particles associated with the
EF face after freeze-fracture
Complete photosystem II complex

Core of EF particle
Photosystem II reaction center complex

Exchangeable, peripherally
attached portion of EF particle
Light harvesting complex associated with
photosystem II (association regulated by
cations)

– – – – Fracture line in freeze-fracture preparations

8

FIG. 8. A diagrammatic interpretation of the structural organization of functional components of the chloroplast membranes. Intrinsic protein complexes are thought to be embedded in a lipid bilayer continuum. A freeze fracture of the lipid bilayer results in segregation of the asymmetrically distributed complexes to either half of the membrane, resulting in distinct EF and PF faces. Five different classes of protein complexes are drawn to be consistent with the number of different types of complexes that can be isolated as membrane subunits from chloroplast lamellae by mild detergent techniques.

plasts of certain C$_4$ plants[21]. The relative abundance of PF and EF particles is compatible to measured amounts of the corresponding functional complexes. In addition, the fact that the various particles have fluid mobility within the membrane (compare EF particle distribution in Fig. 5 and 7) is compatible with the model of Fig. 8. Further details of the model have previously been discussed.[2]

[21] K. R. Miller, G. J. Miller, and K. R. McIntyre, *Biochim. Biophys. Acta* **459**, 145 (1977).

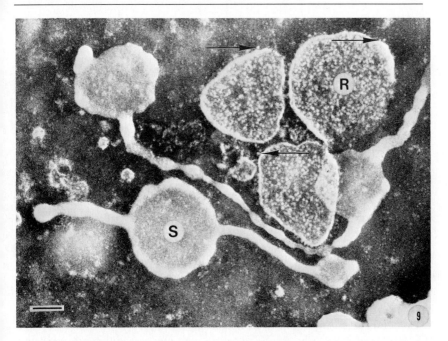

FIG. 9. A preparation of membranous vesicles derived from plastids isolated from dark-grown maize leaves. The sample was negatively stained with 2% phosphotungstic acid. This sample demonstrates the usefulness of negative staining in comparing membrane surface characteristics. S, Smooth membranes; R, rough membranes. Arrows indicate surface-bound particles which clearly protrude out from the membrane surface. Scale, 0.1 μm.

## Descriptive Characterization of Chloroplast Membrane Surfaces

In many early studies of chloroplasts, thylakoids were spread upon electron microscope grids and shadowed before examination; it was routinely observed that there were subunits either upon or within the membrane. With the advent of the rapid sample preparation used in negative staining procedures, membrane surface analysis was greatly facilitated. This technique allows the visualization of surface bound particles, since the staining involves deposition of an electron dense material around a particle resulting in what appears to be a white object in a field of black. An example of the usefulness of negative staining in resolving differences in smooth and rough membranes is shown in Fig. 9. In this figure, membranous vesicles from an etiolated corn mesophyll plastid appear either with a smooth surface (S) (fragments derived from the prolamellar body during sample preparation) or a rough surface (R) (the lamellae extending out from the prolamellar body *in situ*). The presence of coupling factor particles ($CF_1$; see arrows in Fig. 9) on the lamellae is consistent with biochemical data which has described the distribution of $CF_1$ in developing chloroplasts.[22]

In recent years, use of deep etching procedures has been extensively utilized for high resolution examination of membrane surfaces. Examples of the ES and PS surfaces of chloroplast lamellae are shown in Figs. 10 and 11. The $ES_s$ surface is characterized by the presence of relatively large particles, each of which contain two to six subunits.[13,16] Under certain conditions, the particles aggregate into crystalline lattices. In this configuration, the particles always appear to have four subunits which are sometimes called tetramers.[23] (These lattices are equivalent to the particles previously described by Park[24] and co-workers as "quantasomes" in shadowed membrane preparations.) The $ES_u$ surfaces also have a particulate surface, although the particles are more widely spaced than in the $ES_s$ region and usually have only one or two subunits. In unstacked chloroplast lamellae, the differences between $EF_s$ and $EF_u$ particle regions are lost, as all particles diffuse and intermix randomly in the lipid phase.[13]

The $PS_s$ surface of chloroplast membranes is characterized by the presence of both large (150 Å average diameter) and small (90 Å average diameter) particles (Fig. 11). The larger size class particles are removed by washing procedures that extract extrinsic membrane proteins.[16,25,26] The 90-Å particles of the PS surface are not affected by washing procedures that remove extrinsic membrane proteins; they occur in approximately the same density as the PF fracture particles and therefore appear to correspond to external protrusions of the proteins that make up the intrinsic membrane complexes.[16] Occasionally, a third type of subunit can be detected as slight protrusions on PF surfaces. These occur only in unstacked membranes and appear as regularly spaced linear and/or crystalline arrays. The spacing of these arrays is very similar to that observed for $EF_s$ or $ES_s$ arrays.[16,26]

*Interpretations of Membrane Surface Structural Analysis with Respect to Lamellar Function*

Two general classes of subunits can be detected by analysis of the chloroplast membrane surface: (1) the extrinsic protein complexes which

[22] A. R. Wellburn, *Planta* **135**, 191 (1977).

[23] K. R. Miller, G. J. Miller, and K. R. McIntyre, *J. Cell Biol.* **71**, 624 (1976).

[24] R. B. Park, *in* "Plant Biochemistry" (Y. Bonner and J. E. Varner, eds.), p. 124. Academic Press, New York, 1965.

[25] M. P. Garber, and P. L. Steponkus, *J. Cell. Biol.* **63**, 24 (1974).

[26] K. R. Miller, and L. A. Staehelin, *J. Cell Biol.* **68**, 30 (1976).

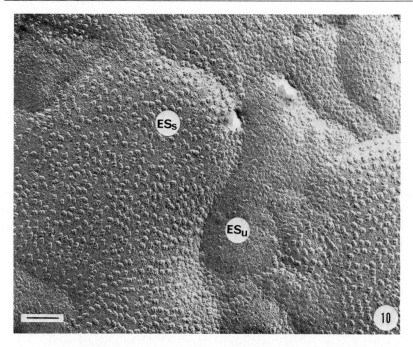

FIG. 10. The true inner surface of chloroplast thylakoid membranes as shown by freeze etching. The smooth-textured membrane matrix is interrupted by particles, each of which are made up of 2–6 subunits. The area of high particle density ($ES_s$) corresponds to that portion of the membrane which was in the partition of a grana stack. There are fewer particles in the $ES_u$ regions, and these particles have a lower average number of subunits than the $ES_s$ particles. These latter observations are correlated with smaller $EF_u$ as compared to $EF_s$ membrane subunits and a reduced light-harvesting pigment–protein complex content in the unstacked membranes. Scale, 0.1 $\mu$m. (Figure provided by L. A. Staehelin).

actually reside outside the lipid matrix of the membrane and (2) portions of intrinsic protein complexes which are intercalated within the lipid matrix but protrude to varying extents outside that matrix.

The extrinsic membrane protein complexes have been most carefully identified by biochemical techniques. The pioneering work in this area was by Moudrianakis[27] and co-workers, who used negative staining techniques to visualize surface-bound proteins. They demonstrated that washing thylakoids with dilute ethylenediaminetetraacetic acid caused a parallel removal of ribulosebisphosphate carboxylase (RuBP carboxylase) and coupling factor ($CF_1$) and the appearance of smooth membranes. The $CF_1$ could be reconstituted back onto smooth membranes with the con-

[27] E. N. Moudrianakis, *Fed. Proc. Fed. Am. Soc. Exp. Biol.* **27**, 1180 (1968).

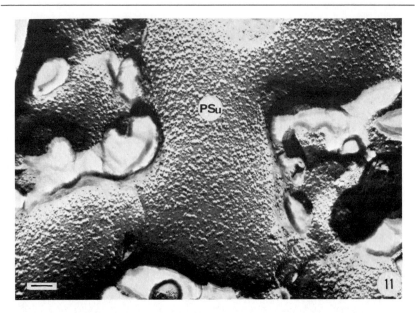

Fig. 11. The true outer surface of a chloroplast membrane as revealed by freeze etching. Large particles averaging 150 Å in diameter and small particles averaging 90 Å are visible on the surface. Only the 150-Å particles are removed by washing treatments that release extrinsic protein complexes. The 90-Å particles represent portions of the intrinsic protein complexes that protrude out of the lipid matrix of the membrane. Scale, 0.1 μm. (Figure provided by L. A. Staehelin.)

comitant appearance of a particulate surface and restored phosphorylation activity. These findings have subsequently been verified in several other laboratories using both negative staining and deep etching techniques.[16,25,26,28] Oleszko and Moudrianakis[29] have also found that the protruding $CF_1$ particles can be visualized in thin-sectioned samples by a uranyl acetate staining procedure.

Berzborn et al.[30] have shown that the membrane subunits corresponding $CF_1$ can be caused to clump by $CF_1$ antibody. These data indicate lateral mobility of the subunits along the membrane; the antibody studies also suggest that most $CF_1$ is exposed in unstacked membranes. This latter viewpoint is also the conclusion of Miller and Staehelin[26] based on combined structural analysis and $CF_1$ removal and reconstitution exper-

[28] C. J. Arntzen, R. A. Dilley, and F. L. Crane, J. Cell Biol. **43**, 16 (1969).
[29] S. Oleszko, and E. N. Moudrianakis, J. Cell. Biol. **63**, 936 (1974).
[30] R. J. Berzborn, F. Kopp, and K. Mühlethaler, Z. Naturforsch., Teil B **29**, 694 (1974).

iments. The question of where $CF_1$ is localized along chloroplast membranes has been reviewed.[8] The mechanism(s) of association of the externally bound $CF_1$ complex (see Fig. 4) with intrinsic hydrophobic protein component of the $CF_1$ complex (see Fig. 8) has not been established.

Particles with a 90-Å diameter can be observed by deep etching on the outer surface of thylakoids from which extrinsic proteins have been removed.[16,26] This suggests that the intrinsic PF particles protrude out of the lipid matrix of the membrane; this is consistent with the model of Fig. 8. Based on various lines of biochemical and biophysical analysis, it seems highly probable that the protein complexes making up the PF particles must extend across the lipid phase even though they are not visible on the ES surface with our present techniques.[2]

The particles observed on the ES surface of chloroplasts have the same density and spacing patterns as the EF particles; both are concentrated in partitions in stacked membranes and diffuse out uniformly when the membranes are experimentally unstacked, and both can occur in paracrystalline arrays. It is thought that the ES particles represent a luminally protruding portion of the intrinsic protein complexes which make up the EF particle. Particles which form low ridges in paracrystalline arrays on the PS surface have been suggested to be due to an external protrusion of the EF particles.[16,23] A documented transmembrane organization of the photosystem II complex as shown in the model of Fig. 8 is consistent with the trans-membrane structure of the EF particles (see Arntzen[2]).

When organized into regular arrays, the ES particles have 4 subunits; this has been termed a "quantasome" or "tetramer".[23,24] When the ES particles are in a random organization pattern, however, there are 2 to 6 subunits per particle.[13,16] Based on the recent evidence that EF particles are composed of a photosystem II–reaction center complex plus varying amounts of associated light-harvesting pigment–protein complexes,[17,20] it now appears that the subunits of the EF particles actually represent the separate light-harvesting complexes bound around a "core" particle.

## Summary

We have summarized examples of techniques which can be used to show dynamic changes in the morphological organization and/or substructural organization of chloroplast lamellae. These techniques, coupled with existing biochemical and biophysical analysis are currently being utilized for refinement of a membrane model which will adequately describe our understanding of the photochemical properties of chloroplast lamellae.

Acknowledgments

We are indebted to Dr. L. A. Staehelin for providing us with Fig. 6, 10, and 11, and for his invaluable discussions on chloroplast structure. We thank Dr. E. Newcomb for Fig. 1 and Dr. P. Armond for Fig. 5 and 6. This work was supported in part by DOE Contract No. EE77502-4475. J. J. B. is supported by a National Institutes of Health Predoctoral Fellowship (NIH Grant No. 6M7283-1); his current address is Crop Science Dept., North Carolina State Univ., Raleigh, N.C. 27650, USA.

## [51] Measurement of Membrane ΔpH

### By Uri Pick and Richard E. McCarty

Illuminated suspensions of chloroplast thylakoids carry out the uptake of protons from the medium associated with electron transport.[1] In this article, we describe three procedures which estimate the magnitude of the proton concentration gradient (ΔpH) across thylakoid membranes.

### General Principles of the Assay of ΔpH

The internal pH of illuminated thylakoids can not be calculated from the extent of proton uptake, since most of the protons entering the thylakoids are buffered inside, and the buffering capacity of the internal space cannot be determined readily. Moreover, the internal space of the thylakoids is too small to allow the use of micro pH electrodes. Thus, indirect methods must be used to estimate ΔpH. Amines are accumulated by illuminated thylakoids.[2] At the steady state, the transmembrane concentration gradient of amine cation will be equivalent to that of protons provided certain conditions hold true. This relationship may be readily derived from considerations of acid–base equilibria. For a monobasic amine, one may write:

$$pH_o = pK + \log \frac{[A]_o}{[AH^+]_o} \quad \text{and} \quad pH_i = pK + \log \frac{[A]_i}{[AH^+]_i}$$

where A stands for the basic form of the amine; $AH^+$, for the amine cation; o, for the external compartment, and i for the internal compart-

[1] J. Neumann and A. T. Jagendorf, *Arch. Biochem. Biophys.* **107**, 109 (1964).

[2] A. R. Crofts, *J. Biol. Chem.* **242**, 3352 (1967).

ment. $\Delta$pH equals $pH_o - pH_i$ and since the $pK$ in the two compartments is probably equal, $\Delta$pH would be

$$\Delta pH = \log \frac{[A]_o}{[AH^+]_o} - \log \frac{[A]_i}{[AH^+]_i} = \log \frac{[AH^+]_i}{[AH^+]_o} + \log \frac{[A]_o}{[A]_i}$$

The amine base is likely to be much more permeable than the amine cation. Assuming that the amine base is in concentration equilibrium, $\Delta pH = \log[AH^+]_i/[AH^+]_o$. In principle, therefore, $\Delta$pH may be estimated if the extent of the uptake of an amine and the thylakoid internal volume are known. As long as the pK of the amine is at least one pH unit higher than the pH of the external medium, essentially all of the amine will be in the cation form and $\Delta$pH may be determined from the total amine uptake.

Amines uncouple phosphorylation from electron flow[3] and lead to marked swelling of the thylakoids. Therefore, low amine concentrations must be used in the determination of $\Delta$pH. Moreover, if the amine cation binds to the internal surface of the thylakoid membrane in response to internal acidification, $\Delta$pH will be overestimated. On the other hand, $\Delta$pH will be underestimated if the amine base is not maintained at concentration equilibrium.[4] In this article, we describe three different procedures for estimating the extent of the uptake of amines for the assay of $\Delta$pH.

## Silicone Fluid Centrifugation[5,6]

*Principle.* Thylakoids are illuminated in the presence of radioactive amines within a microcentrifuge. The thylakoids are then rapidly centrifuged through a layer of silicone fluid to separate them from the suspending medium. Internal volumes may be estimated from the uptake of $^3H_2O$.

*Materials.* A Coleman model TM 14K30 microcentrifuge fitted with a transparent plexiglass top is used. The Beckman or Eppendorf microcentrifuges are also satisfactory. A 650 W quartz-halide lamp (lamp DVY) positioned 40 cm over the top of the microcentrifuge provides about 2.5 $\times$ $10^5$ ergs $cm^{-2}$ $sec^{-1}$ at the rotor surface after filtration through 25 cm of 0.5% (w/v) of aqueous $CuSO_4 \cdot 5H_2O$ and through an infrared absorbing filter (Corning glass No. 4602). Fans are used to cool the lamp and the top of the centrifuge. The temperature of reaction mixtures increases less

[3] N. E. Good, *Biochim. Biophys. Acta* **40**, 502 (1960).

[4] A. R. Portis, Jr. and R. E. McCarty, *Arch. Biochem. Biophys.* **156**, 621 (1973).

[5] R. E. Gaensslen and R. E. McCarty, *Anal. Biochem.* **48**, 504 (1972).

[6] A. R. Portis, Jr. and R. E. McCarty, *J. Biol. Chem.* **251**, 1610 (1976).

than 1° in 3 min of continuous illumination. Polyethylene microcentrifuge tubes of 0.4 ml capacity are used (A. H. Thomas, Philadelphia, Pennsylvania). Polypropylene tubes may also be used, but are more difficult to cut than polyethylene ones.

The silicone fluids, Versilube F-50 and SF-96(50) may be obtained by writing Mr. Harold Sober, Silicone Products Division, General Electric Company, Waterford, New York. The Dow-Corning 702 diffusion pump silicone oil may be purchased from scientific supply houses.

*Procedure.* Microcentrifuge tubes are prepared for assay by adding 100 $\mu$l of 8.5% glycerol in 2% trichloroacetic acid followed by 100 $\mu$l of a silicone fluid mixture. Generally, a mixture (by weight) of 2.2 parts of Versilube F(50) to 1 part of SF96(50) is used. In cases where the density of the incubation mixture is high, a ternary mixture, consisting of (by weight) 1 part SF96(50) to 0.9 part Versilube F(50) to 0.6 part Dow-Corning 702, is more satisfactory. The tubes are briefly centrifuged to remove air pockets. The temperature of the room should be no higher than about 23°. At temperatures greater than 23°, centrifugation can lead to inversion of the upper two layers.

Assays are conducted at room temperature in a dark room with dim green safelight. The incubation mixtures contain in 0.25 ml, 20 or 40 m$M$ tris(hydroxymethyl)methylglycine–NaOH (pH 8.0), 50 m$M$ NaCl, 5 m$M$ MgCl$_2$, a mediator of electron transport [0.05 m$M$ pyocyanine, for cyclic electron flow or 1 m$M$ K$_3$Fe(CN)$_6$, for noncyclic flow], spinach chloroplast thylakoids equivalent to 25 $\mu$g of chlorophyll, radioactive amine, a radioactive external space indicator, and other additions as needed. The thylakoids are suspended to 2.5 mg of chlorophyll/ml in a medium which contains 0.4 $M$ mannitol, 0.01 $M$ NaCl, and 0.02 $M$ tris/(hydroxymethyl)methylglycine–NaOH (pH 8.25). In most experiments, the amine used is 25 $\mu M$ [$^{14}$C]hexylamine (1.3 $\mu$Ci/$\mu$mole) and [$^3$H]sorbitol (0.25 or 0.5 $\mu$Ci/0.25 ml) is the external space indicator. [$^{14}$C]Hexylamine is no longer commercially available, but large amounts of [$^3$H]hexylamine can be prepared by catalytic tritation at relatively modest cost. The [$^3$H]hexylamine obtained must be purified and is not as stable as [$^{14}$C]hexylamine. If $^3$H-labeled amines are used, [$^{14}$C]sorbitol (about 0.2 $\mu$Ci/0.25 ml) is used as the external space indicator. [$^{14}$C]Methylamine, which is commercially available (New England Nuclear Co., Boston, Massachusetts), may be used as long as the external pH of the assay mixture is greater than 8.0.

As soon as possible after the thylakoids are added to the incubation mixture, duplicate 100 $\mu$l aliquots are withdrawn and added to the top of microcentrifuge tubes which contain the silicone fluid and the glycerol–trichloroacetic acid mixture. The tubes are placed in the microcentrifuge.

To equalize the light intensity, the tubes are spun at low speed. The light is switched on and, after 30 to 120 sec, the centrifuge is rapidly brought to top speed (about 9000 $g$ at the center of the tubes). Centrifugation with the light still on is continued for 15 sec. The steady state of hexylamine uptake is reached within 15 sec. When all the samples have been illuminated, the incubation mixture and most of the silicone fluid are removed by aspiration. The tubes are then cut a few millimeters above the glycerol–trichloroacetic acid–silicone fluid interphase, and the top parts are discarded. The denatured thylakoid pellets are resuspended in the glycerol–trichloroacetic acid with a slender metal rod. The tubes are placed in 15-ml conical centrifuge tubes and briefly centrifuged to sediment the thylakoids. Using a Pedersen-type micropipet (A. H. Thomas Co.) 50 $\mu$l aliquots of the supernatant fluids are taken and added directly to scintillation vials. The pipette must be cleaned to remove traces of silicone fluids. Rinsing the pipette using an aspirator, in the following solvents is effective: water, acetone, benzene, acetone. Automatic pipettors with disposable tips may also be used, but the accuracy is not as great as with a Pedersen micropipette.

Nine milliliters of a scintillation fluid consisting of 0.5% 2,5-diphenyloxazole and 0.01% 1,4-bis[2-(diphenyloxazole)]benzene in toulene–Triton X-100 (2:1, v/v) are added. The radioactivity in aliquots (10 $\mu$l) of the reaction mixture is also determined. Double label counting is carried out in a liquid scintillation spectrometer.

Thylakoid internal volumes are determined by the same procedure, except that nonradioactive hexylamine is substituted for the [$^{14}$C]hexylamine and [$^{14}$C]sorbitol and $^3$H$_2$O (2.5 $\mu$Ci per 0.25 ml) are used.

*Calculations.* The counts are first corrected for background and spillovers. The total cpm of [$^{14}$C]hexylamine and [$^3$H]sorbitol present in 50 $\mu$l of the incubation mixture are calculated. To estimate the amount of hexylamine carried through the silicone fluid with the thylakoids, the [$^3$H]sorbitol counts are used. Usually only 0.5 to 0.8% of the [$^3$H]sorbitol is carried through. This correction is made by dividing the $^3$H counts per minute observed by the total $^3$H and multiplying by the total $^{14}$C counts per minute. This yields the $^{14}$C in the glycerol layer which is carried through with the thylakoids and is subtracted from the $^{14}$C counts per minute observed. The corrected $^{14}$C counts per minute are converted to nanomoles hexylamine per aliquot of the lower layer. The internal and external concentrations of hexylamine are then calculated if the internal volume is known. Internal volumes are calculated from the uptake of $^3$H$_2$O.

*Validity of the Method and Some Drawbacks.* A number of experiments suggest that silicone fluid centrifugation may be used for the quantitative estimation of ΔpH. The uptake of aniline determined by this method agrees well with that determined by an entirely independent procedure.[4] ΔpH is independent of the concentration of thylakoids over a tenfold range (20 to 200 µg chlorophyll/ml). Increasing the osmolarity of the incubation mixtures also had no effect on ΔpH. Since the binding of [$^{14}$C]triethylmethylammonium cation to thylakoids was negligible, it is unlikely that hexylamine cation binds significantly.[6]

Although amine uptake by illuminated thylakoids may be estimated to ±1 to 2%, the precision of internal volume measurements is ±10% or more. This is because the internal volume is small and corrections for trapped suspending medium are high. Under the conditions reported here, the internal volume is about 10µl/mg chlorophyll, and no significant change in the internal volume was detected when the light intensity was altered or if nucleotides or an uncoupler was added. Assuming a fixed internal volume of 10µl/mg chlorophyll, ΔpH may be estimated to ±0.05 pH. Even a 20% variation in internal volumes has relatively little effect on ΔpH. For example, a ΔpH of 3.0 is obtained for a certain extent of amine uptake using 10 µl/mg chlorophyll, whereas if the internal volume is 8 µl/mg chlorophyll, ΔpH is 3.10 and if it is 12 µl/mg chlorophyll, ΔpH is 2.92.

Generally 85 to 90% of the thylakoids are recovered in the lower layer after centrifugation. For most precise work, the amount of thylakoids recovered in the lower layer should be determined and the results calculated on the basis of the recovered amount rather than the total amount added. Recoveries may be assayed by omitting the trichloroacetic acid from the lower layer and assaying the chlorophyll content of the layer. If a ΔpH of 3.0 is calculated assuming full recovery, a ΔpH of 3.10 is calculated for 90% recovery, and 3.15 for 85% recovery.

Photophosphorylation and electron transport may be assayed[6] under the same conditions used to determine ΔpH.

### Fluorescent Amines[7]

A suspension of thylakoids including a fluorescent amine (9-aminoacridine or N-1-naphthylethylenediamine) contained in a glass cuvette is placed in a fluorimeter and the changes in the fluorescence of the amine are continuously followed. Illumination of the thylakoid suspension will result in uptake of the amine by the thylakoid which will be manifested as a reversible decrease of the fluorescence emitted from the cuvette.

*Materials.* Fluorescence can be measured in any fluorimeter (Eppen-

---

[7] S. Schuldiner, H. Rottenberg, and M. Avron, *Eur. J. Biochem.* **25,** 64 (1972).

dorf, Zeiss, Perkin-Elmer) provided with an opening for side (or top) illumination of the cuvette. The exciting light used for measuring the fluorescence of acridine derivatives should be filtered through a 405/436 filter when provided by a mercury lamp (Eppendorf fluorimeter). When a monochromatic light source is available, the recommended exciting light for 9-aminoacridine is 400 nm, 5 nm bandwidth. The exciting light intensity should be reduced to about 20% with the aid of a neutral glass filter when the 405/436 light is used to avoid energization of the thylakoids by the exciting light. Emission can be measured either front face at an angle of 30°–40° or at 90° relative to the exciting light. The emission light should be passed to the photomultiplier through a filter that will cut off the red actinic light and the short wavelength light but transmit most of the light between 460–500 nm. The following filter combination can be used: Corning CS 4-96 filter, Strand Electric Co. cinemoid filter No. 62, and Wratten No. 58 filter. Similarly, for naphthylamine fluorescence the exciting light should be filtered through a 313/366 filter (Eppendorf) and the emission can be measured through Corning CS 4-96 and CS 3-73 filters. Actinic light can be provided by a 24 V halogen projector lamp filtered through a red cutoff filter such as Schott RG-645 and a heat filter made of glass or 5 cm water. It will provide an incident light intensity of 5 to $7 \times 10^5$ erg cm$^{-2}$ sec$^{-1}$ (650–750 nm).

*Procedure.* To a clean 3-ml cuvette the following reaction mixture is added: KCl, 30 m$M$; Tricine–NaOH (pH 8.0), 15 m$M$; MgCl$_2$, 5 m$M$; and an electron carrier such as pyocyanine, 30 $\mu M$; K$_3$F$_e$(CN)$_6$, $2 \times 10^{-4}$ $M$ or diquat 15 $\mu M$ plus NaN$_3$, 1 m$M$. $N$-Methylphenazonium methosulfate (PMS) is not recommended as an electron carrier since it interferes with the fluorescence measurements. Thylakoids are added next (15–30 $\mu$g chlorophyll in 3 ml) and mixed with a glass rod. The actinic light is turned on for a few seconds (with the photomultiplier open) to make sure that no light leaks reach the photomultiplier. The fluorescent probe is added on a glass rod to a final concentration of 1 $\mu M$. The deflection of the pen will mark the 100% fluorescence. Illumination results in a decrease in fluorescence which is reversed when the light is turned off. To assure that the exciting light is low enough and does not energize the chloroplasts, an excess of uncoupler should be added in the dark (1 $\mu M$ SF-6847 or 5 $\mu M$ nigericin). If an increase in fluorescence is obtained, the exciting light should be further attenuated. When measurements are carried out in the presence of adenine nucleotides, care should be taken to keep the Mg always in excess over the nucleotide[8] to avoid direct quenching of the amine fluorescence by the nucleotide.

[8] U. Pick and M. Avron, *FEBS Lett.* **32**, 91 (1973).

*Calculations.* The $\Delta$pH can be calculated from the following equation[7]

$$\Delta pH = pH_o - pH_i = \log \frac{V}{v} \frac{Q}{1 - Q}$$

where $v$ is the internal thylakoid volume ($\mu l$), $V$ the external volume, practically equals total volume ($\mu l$), and $Q$ the fraction of the fluorescence quenched in the light. This equation can be used as long as (a) the external pH is significantly lower than the higher $pK$ of the fluorescent amine (up to pH 9 with 9-aminoacridine and 8.5 with $N$-1-naphthylethylenediamine). For higher external pH the following equation can be used

$$-pH_i = \log (H_o^+ + K) + \log \frac{V}{v} \frac{Q}{1 - Q}$$

where $K$ is the higher dissociation constant of the amine (10 for 9-aminoacridine, 9.5 for $N$-naphthylethylenediamine). (b) The internal pH should be higher than the *lower* $pK$ of the amine ($-2$ for 9-aminoacridine, 4.9 for $N$-naphthylethylenediamine). When such a correction is needed Eq. (7) in Schuldiner *et al.*[7] should be used. This will be the case when $N$-naphthylethylenediamine is used at low pH. The internal thylakoid volume should be determined independently using the silicone method (see Section I) or an equivalent method.[9] The values regularly obtained are 10–30 $\mu l$ water/mg chlorophyll.

*Validity of the Method and Drawbacks.* The method described above utilizing fluorescent amines to estimate pH gradients is convenient and attractive for kinetic measurements and it is becoming very popular also in other systems. Nevertheless, the method does have some limitations which should be mentioned. One of them is that the method relies on values of internal volumes determined separately usually at different conditions (of light intensity and chloroplast concentration). But as mentioned above even a twofold over- or underestimation of the internal volume will shift the calculated $\Delta$pH by no more than 0.3 pH units. Another limitation is the sensitivity of the method—it is suitable only for $\Delta$pH values higher than 1.5 pH units.

A basic assumption of the model on which the calculation of $\Delta$pH is based is that no binding or aggregation of the dye occur inside the thylakoids. This assumption was challenged by some workers[10] based on lack of response of the acridine fluorescence quenching to the medium osmolarity, dependence on the acridine concentration, partial competi-

---

[9] H. Rottenberg, T. Grunwald, and M. Avron, *Eur. J. Biochem.* **25**, 54 (1972).
[10] J. W. T. Fiolet, E. P. Bakker, and K. VanDam, *Biochim. Biophys. Acta* **368**, 432 (1974).

tion with other amines,[11,12] and differences in the calculated ΔpH values obtained with different probes. Internal binding and aggregation will lead to overestimation of the calculated ΔpH as mentioned above. Internal aggregation can be avoided by keeping the acridine concentration low so that it will not exceed its critical micelle concentration inside illuminated thylakoids. The upper concentration limits determined experimentally[12] are 0.5 $\mu M$ at pH 9.0, 1 $\mu M$ at pH 8.0, 3 $\mu M$ at pH 7.0, and 10 $\mu M$ at pH 6.0. Internal binding is hard to control or evaluate. Indirect estimates from competition with other amines[11,12] suggest that although significant internal binding may take place, its impact on the calculated ΔpH will be an overestimation of about 0.3 pH units only. Therefore, although the mechanism of response of fluorescent amines to a pH gradient is not yet unequivocally known, the experimental results fit well the predicted values and are in reasonable agreement with other methods.

### Stimulation of Proton Uptake by Amines[13]

*Principle.* The pH increase in the medium of an illuminated thylakoid suspension is measured. Following addition of an amine having a p$K$ lower than the external pH, an increase in the extent of proton uptake is observed.[14] This increment in proton uptake resulting from internal buffering of protons by the amine is used to calculate the internal pH.

*Materials and Methods.* Proton uptake can be measured with any pH recording system composed of a combination glass pH electrode, a pH meter, and a recorder.[15] Illumination can be provided by an 24-V halogen projector lamp filtered through a red cutoff filter (Schott-RG 645) and a heat filter (glass filter or 5–10 cm of water). The reaction mixture should contain KCl, 30 m$M$; Tricine–NaOH, 0.5 m$M$ for pH 7.0–9.0 or MES–NaOH for lower pH values; a mediator of electron flow, preferably pyocyanine, 30 $\mu M$, and thylakoids suspended in a medium of low buffering capacity added to a final concentration of 20 $\mu$g of chlorophyll per milliliter. Illumination is provided until steady state is reached (15–30 sec); the light is turned off and after a complete decay the process is repeated. Calibration of the signal is made by addition of standardized NaOH or HCl. The pH should not change by more than 0.05 pH unit. Illumination is repeated for a third cycle as before. Next the buffered

[11] J. E. Tillberg, C. Giersch, and U. Heber, *Biochim. Biophys. Acta* **461**, 31 (1977).
[12] U. Pick and M. Avron, unpublished results.
[13] U. Pick and M. Avron, *Eur. J. Biochem.* **70**, 569 (1976).
[14] N. Nelson, H. Nelson, Y. Naim, and J. Neuman, *Arch. Biochem. Biophys.* **145**, 263 (1971).
[15] R. Dilley, Vol. 24, p. 68.

amine is added, and illumination is repeated for another two to three steady state cycles. At the end, a second calibration should be made, since the amine may increase the buffering capacity of the mixture. If in the presence of the amine the pH increase does not reach a steady state within 30 sec, or increases progressively from one illumination cycle to the next, the experiment should be repeated using a lower amine concentration. The recommended amines are for pH 8.5–9.3, morpholine (0.03 m$M$) and imidazole (0.1 m$M$); for pH 7.5–8.0, imidazole and $p$-phenylenediamine (1 m$M$); and for pH 7.0 $p$-phenylenediamine and aniline (10 m$M$).

*Calculation.* The internal pH of the illuminated chloroplast thylakoids can be calculated by using the following equation[13]

$$H_i^+ = \frac{\Delta H^+ (H_o^+ + K)^2}{A v K}$$

where $H_i^+$ is the internal proton concentration ($M$), $H_o^+$ the external proton concentration in the light ($M$), $K$ the amine dissociation constant, $A$ the total amine concentration ($M$), $\Delta H^+$ the increment in extent of proton uptake induced by the amine ($\mu$Eq H$^+$/mg chlorophyll), and $v$ the internal thylakoid volume determined separately or estimated ($\sim$20 $\mu$l/mg chlorophyll).

*Validity of the Method and Drawbacks.* The method described here is simple, does not require sophisticated apparatus, and is not limited to a specific amine probe. One of the limitations of the method is that the calculation relies on values for internal volume that have to be measured separately and under different conditions (see p. 541). Another limitation is that the amine used should be restricted to relatively low concentrations to avoid uncoupling and swelling, both of which will affect the calculated internal pH: uncoupling will lead to an underestimation of the $\Delta$pH, while swelling will lead to overestimation. For each of the amines recommended here, the highest concentration that can be used without interference of either uncoupling or swelling was mentioned. Another limitation encountered is that several amines that give reliable results comparable to other amines at a certain pH give an underestimation of the $\Delta$pH at a lower pH (morpholine at pH < 8.0).[13] Pyridine and to a smaller extent also aniline show a concentration dependence—at low concentrations they stimulate proton uptake to a larger extent than would be expected perhaps due to internal binding. Therefore, pyridine is not recommended for quantitative measurements, and aniline should be used only at relatively high concentration (around 10 m$M$).

[52] Measurement of the Oxygen Cycle: The Mass
Spectrometric Analysis of Gases Dissolved in a Liquid
Phase

*By* R. RADMER and O. OLLINGER

Introduction

The oxygen cycle is a light-driven reaction occurring in algae in which $O_2$ is directly reduced, i.e., without the intervention of carbon metabolism. Electrons, derived from $H_2O$ by photosystem II (PS-II), are diverted to the reduction of $O_2$ rather than $CO_2$ under conditions in which the $CO_2$ fixation cycle cannot keep pace with the production of strong reductant by photosystem I (PS-I).[1]

The measurement of the $O_2$ cycle is really a special case of the more general problem of measuring simultaneous $O_2$ evolution and $O_2$ uptake (as well as $CO_2$ uptake). Since, in some cases, the $O_2$ cycle is only a transient phenomenon (e.g., during the "lag" following a dark–light transition) a technique with relatively good time resolution (a few seconds) is required.

No radioactive isotope of oxygen is sufficiently long-lived to be used in labeling experiments.[2] The most useful stable isotope is $^{18}O$, which has an atom percent abundance of 0.204. Since the oxygen atoms of $O_2$ do not appreciably exchange with those of $H_2O$ under normal conditions, labeled $O_2$ can be dissolved in $H_2O$ without complicating exchange reactions.

The general technique used to measure the oxygen cycle is based on a system developed by Hoch and Kok.[3] A mass spectrometer (MS) is used to determine the identity and isotopic composition of gases admitted via the MS inlet. [A mass spectrum, obtained by bombarding the sample gas stream with an electron beam (usually 70–100 eV) and separating the resultant ions, is expressed as a series of mass to charge ($m/e$) ratios, e.g., for $^{16}O_2$, one would observe peaks at $m/e = 32$ ($O_2^+$), 16 ($O_2^{2+}$, $O^+$) etc; for $^{12}C^{16}O_2$, the main peaks would occur at $m/e = 44$ ($CO_2^+$), 28 ($CO^+$), 16 ($O^+$), 12 ($C^+$), etc. For most common gases, the largest peak

[1] R. Radmer and B. Kok, *Plant Physiol.* **58**, 336 (1976).

[2] M. Kamen, "Isotopic Tracers in Biology." Academic Press, New York, 1957.

[3] G. Hoch and B. Kok, *Arch. Biochem. Biophys.* **101**, 160 (1963).

METHODS IN ENZYMOLOGY, VOL. 69

is the $m/e$ value corresponding to the removal of a single electron, e.g., $^{16}O_2^+$, $CO_2^+$, etc.]

The mass spectrometer inlet system is centered around a semipermeable membrane that allows dissolved gases, but not the liquid phase, to enter the mass spectrometer. Because the dissolved gases are monitored directly, the time response is much faster than that obtained by "manometric" methods, in which the gas phase composition is used to infer the gas concentrations in the liquid phase.

This mass spectrometer inlet system is complemented by a specially designed mass selector data handling system. This apparatus sequentially tunes the mass spectrometer to the mass peaks of interest. Simultaneously, the mass spectrometer output is channeled through preselected attenuators before being fed in parallel to a fast stripchart recorder and a minicomputer.

In the following sections, we describe various aspects of this technique. Some recent data obtained using these methods are presented in Radmer et al.[1,4]

Theory

The measurement of the oxygen cycle, which involves the concomitant evolution and uptake of $O_2$, is based on the different isotopic compositions of the substrates of $O_2$ evolution ($H_2O$) and uptake ($O_2$). The oxygen in naturally occurring $H_2O$ has an isotopic composition (to three decimal places) of $^{16}O = 0.998$, $^{18}O = 0.002$. (The other naturally occurring stable isotope, $^{17}O$, has an abundance of $\approx 0.0004$; since it does not appreciably contribute to the $m/e$ values discussed herein, we will neglect it.) If the fractional abundance of $^{16}O$ [$^{16}O/(^{16}O + ^{18}O)$] in the $H_2O$ is denoted $\alpha$, then the fractional abundance of $^{18}O$ is $(1 - \alpha)$, and the relative abundances of $^{16}O_2$, $^{16,18}O_2$, and $^{18}O_2$ in the photosynthetically-evolved oxygen are given by $\alpha^2$, $2\alpha(1 - \alpha)$, and $(1 - \alpha)^2$, respectively. Thus, the $O_2$ derived from this water will have the isotopic composition (to three places) $^{16}O_2$ ($m/e = 32$) $= 0.996$, $^{16,18}O_2$ ($m/e = 34$) $= 0.004$, and $^{18}O_2 = 0.000004$ (which we can safely neglect).

$^{18}O$-Labeled oxygen gas is available in high isotopic purity from several sources[5]; we routinely use 99 atom% $O_2$. Since the $^{18}O$ label is usually well randomized, these isotope preparations are composed of 0.98 $^{18}O_2$ and 0.02 $^{16,18}O_2$; 99.5 atom % $^{18}O_2$, which is also commercially available, is composed of 99% $^{18}O_2$. The assumption that the added

[4] R. Radmer, B. Kok, and O. Ollinger, *Plant Physiol.* **61**, 915 (1978).
[5] Bio-Rad Laboratories, Richmond, California; Prochem/BOC Limited, London, England.

isotope is reflected solely at $m/e = 36$, thus results in an error of only 1 to 2%. Since this is on the order of the isotope discrimination of the membrane and the mass discrimination of the MS, and since all parameters are calibrated end-to-end (see below), it is unnecessary to monitor $m/e = 34$ ($^{16,18}O_2$); the monitoring of $m/e = 32$ for evolution and $m/e = 36$ for uptake is sufficient.

At any given time during the course of an experiment, the reaction suspension will contain a certain amount of $^{16}O_2$ (from photosynthesis, nonideal buffer equilibration procedures, etc.) in addition to $^{18}O_2$. Since the $O_2$ uptake reaction does not appreciably discriminate between isotopes, a correction must be made for the amount of $^{16}O_2$ consumed. The rates of $^{16}O_2$ and $^{18}O_2$ uptake will be proportional to their concentrations, i.e.

$$\frac{\Delta^{16}O_2}{\Delta^{18}O_2} \sim \frac{^{16}O_2}{^{18}O_2}$$

since the total uptake of oxygen

$$U_0 = \Delta^{16}O_2 + \Delta^{18}O_2$$

$$U_0 = \Delta^{18}O_2 (1 + {^{16}O_2}/{^{18}O_2})$$

The "true" evolution of oxygen $E_0$ is equal to the observed evolution plus the $^{16}O_2$ consumed, i.e.

$$E_0 = \Delta^{16}O_2 - \Delta^{18}O_2 ({^{16}O_2}/{^{18}O_2})$$

(note the sign conventions).

The smaller the $^{16}O_2/^{18}O_2$ ratio, the closer the experiment approaches the ideal conditions in which the two isotopes uniquely reflect the two opposing processes. It is advantageous, then, to start with as high an $^{18}O_2/^{16}O_2$ ratio as possible (see methods below). The $^{16}O_2$ produced during the course of the experiment will cause a continuous decline in the precision and accuracy of this determination.

### Apparatus

Figure 1 is a diagram of the mass spectrometer inlet system used in these measurements. The reaction vessel is a 2-ml cylindrical cavity made of plexiglass that is completely filled with the algal suspension. The suspension is stirred with a rotating Teflon-coated magnetic bar.[6]

---

[6] Best results have been obtained with rare-earth magnetic bars obtained from Markson Science, Inc., Del Mar, California.

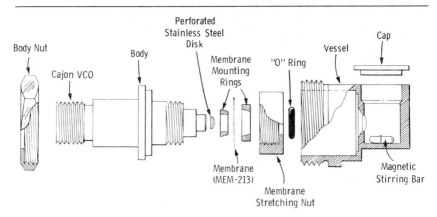

Body Nut
Cajon VCO
Body
Perforated
Stainless Steel
Disk
Membrane
Mounting
Rings
"O" Ring
Vessel
Cap
Membrane
(MEM-213)
Membrane
Stretching Nut
Magnetic
Stirring Bar

FIG. 1. Exploded diagram of mass spectrometer inlet system.

Gases are introduced into the mass spectrometer vacuum system through an approximately 16 mm² circular window of 1 mil silicone rubber membrane[7] that forms part of the wall of the cavity. The membrane is held in place by pinching it between two concentric stainless steel rings of slightly differing tapers; this membrane ring unit can be handled conveniently during assembly of the inlet. Once installed, the membrane is quite sturdy, provided that sharp implements are not allowed to come in contact with it. We routinely use a membrane at least 1 month before it requires replacement.

The mass spectrometer currently being utilized in our apparatus is a quadrupole mass spectrometer[8]; we use a thoria-coated iridium filament because of its oxidation-resistant properties. Actually, the MS requirements for these measurements are rather minimal; only limited mass range and resolution are needed. The main requirement is that the mass number to mass number switching time (peak to peak) be rather rapid; for our quadrupole MS, this time is about 20 msec.

In our apparatus, the mass spectrometer scan voltage and output are programmed using a peak selector stepper module developed in house. In this system, the mass spectrometer can monitor up to eight $m/e$ values, stepped cyclically. The output signals are also routed through this system; the signal at each $m/e$ value is amplified by a preselected factor before being fed in parallel to a fast-running stripchart recorder and a PDP-8 minicomputer. In this way, the progress of the experiment

---

[7] MEM 213, obtained from General Electric Co.

[8] The MS analyzer and ancillary electronics were obtained from Extranuclear Laboratories, Inc., Pittsburgh, Pennsylvania; the rest of the system was constructed in house.

can be monitored in real time (on the recorder) and averaged data obtained through the use of the minicomputer. A detailed description of this apparatus is given in the Appendix.

Illumination of the algal suspension is obtained from a 1000-W projector lamp. The light beam is passed through a heat filter [30 cm of water and either a dilute copper sulfate solution or a KG3 (Schott) filter] and an OG-3 (Schott) filter before being focused on the reaction vessel.

## Methods

### Experimental Protocol

In the usual assay procedure, the algae are concentrated to about 0.1 ml and transferred as quantitatively as possible to the reaction vessel using a Pasteur pipette. Although this procedure is not particularly precise, it allows the addition of a maximum amount of $^{18}O_2$-equilibrated buffer to the vessel. [After the measurement, the algae can be quantitatively transferred from the reaction vessel (with repeated washing, if necessary) for a chlorophyll determination.] The reaction vessel is then filled with $^{18}O_2$-equilibrated buffer and other reagents, e.g., a $CO_2$ source. The vessel is closed, the stirrer turned on, and the cyclical measurement of the masses of interest initiated. After an appropriate dark period (long enough for the system to stabilize), the vessel is illuminated. In order to observe the so-called "lag," in which there is a pronounced transient gulp of $O_2$ (balanced by $O_2$ evolution), algae such as *Scenedesmus* require at least 10 min of darkness.

To facilitate the observation of the $CO_2$ exchange, we routinely use pH 6.0 phosphate buffer. Since the p$K$ of the $CO_2$-$HCO_3^-$ reaction is 6.3, at pH 6.0 about two-thirds of the "total $CO_2$" is present as $CO_2$ gas, which is transmitted by the membrane (the membrane is not permeable to ions, such as $HCO_3^-$). Provided the algae can tolerate this low pH (a question that can be answered beforehand), $CO_2$ exchange can be monitored with a sensitivity about equal to that of $O_2$ exchange. If this measurement is to accurately reflect the status of "total $CO_2$" (i.e., $CO_2$ + $HCO_3^-$ + $CO_3^{2-}$), equilibrium between these species must be maintained by the addition of carbonic anhydrase.

$^{18}O_2$-Equilibration of the buffer is effected by taking up a deaerated aliquot in a 10-ml syringe and shaking vigorously (using a Vortex mixer) with a bubble of $^{18}O_2$ for about 10 min. Figure 2 shows the gas handling system used to provide the $^{18}O_2$. The heart of the system is a gas chromatograph septum[9] which can be repeatedly pierced by the syringe hy-

---

[9] Available from Pierce Chemical Co., Rockford, Illinois.

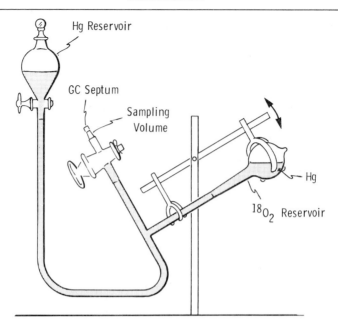

FIG. 2. Schematic diagram of gas handling system for $^{18}O_2$. Small aliquots of $^{18}O_2$ are transferred from the $^{18}O_2$ reservoir to the sampling volume by rotating the apparatus (so that the reservoir is below the sampling volume) and then pumping the gas by raising and lowering the Hg reservoir.

podermic needle. The syringe itself is equipped with a syringe stopcock and a stripped-down Cornwall pipetter.[10] After admission of the $^{18}O_2$ bubble to the buffer, the pipetter is locked and the stopcock closed before shaking.

*Calibrations and Other Considerations*

The sensitivity of the overall mass spectrometer inlet system is a function of the permeability of the membrane, the characteristics of the vacuum system, and the performance characteristics of the mass spectrometer. Although magnetic mass spectrometers show relatively little mass discrimination (i.e., changes in the relative sensitivity to two different $m/e$ values), the electron multipliers used to detect the ion current do display a mass discrimination, which is a function of factors such as multiplier voltage and aging. Quadrupole mass spectrometers themselves can show pronounced mass discrimination, which can vary with changes in mass separation parameters and aging of components. Thus, for quan-

[10] Both available from Fisher Scientific Co.

titative work, an end-to-end calibration of the system is required. For precise work the standardization and background determination procedures may be required after each experiment.

*Oxygen-16.* The usual standard solution for $^{16}O_2$ is air-saturated $H_2O$ or buffer. At room temperature air-saturated $H_2O$ contains about 0.25 $\mu$mole/ml $O_2$; precise values as a function of temperature and some salt concentrations are available.[11] Because the permeability of the membrane can be affected by the buffer composition, it is judicious to determine standard values in a buffer identical to (or closely approximating) the buffer being used in the experiment.

The background value of $m/e = 32$ can be determined by monitoring the signal in the presence of an $O_2$ getter, such as a few crystals of sodium hydrosulfite (dithionite). In using this procedure, one must guard against the generation of a reaction product that can appear at $m/e = 32$ (e.g., $SO_2$). This background measurement can be checked by comparing it to the $m/e = 32$ signal observed with a buffer that has been exhaustively deaerated by bubbling with argon, preferably in the reaction vessel. Oxygen background determinations can be quite important; a system of this type can have a substantial $m/e = 32$ signal derived from the recombination of MS-generated $H_2O$ fragments ($H_2O$ is usually the primary species passing through the membrane).

*Oxygen-18.* The prohibitive expense of large quantities of $^{18}O_2$ precludes standardization procedures based on usual gas handling techniques. In our laboratory, we determine the $^{18}O_2$ sensitivity of the system by referring it to $^{16}O_2$. For example, a system using isolated chloroplasts, the electron acceptor methyl viologen (MV), and catalase drives the mole for mole exchange of $^{16}O_2$ and $^{18}O_2$ according to the following sequence of reactions

$$H_2^{16}O + 2\ MV \xrightarrow{\ h\nu\ } \tfrac{1}{2}\ ^{16}O_2 + 2\ MVH$$

$$2\ MVH + ^{18}O_2 \longrightarrow MV + H_2^{18}O_2$$

$$H_2^{18}O_2 \xrightarrow{\ \text{catalase}\ } H_2^{18}O + \tfrac{1}{2}\ ^{18}O_2$$

$$\overline{\qquad\qquad\qquad\qquad\qquad\qquad\qquad}$$

$$H_2^{16}O + \tfrac{1}{2}\ ^{18}O_2 \longrightarrow H_2^{18}O + \tfrac{1}{2}\ ^{16}O_2$$

A typical reaction mixture for this standardization contains 50 m$M$ Tris or Tricine, pH 7.4, 30 m$M$ methylamine (an uncoupler), 0.1 m$M$ methyl viologen, catalase (sufficient to keep the $H_2O_2$ concentration near zero),

[11] "International Critical Tables of Numerical Data." Published for the National Research Council by McGraw-Hill, New York, 1926–1933.

and isolated chloroplasts.[12] The concentration of chloroplasts is not critical; it should be high enough so that the catalyzed $^{16}O_2$–$^{18}O_2$ exchange rate is fast compared to the withdrawal rate (see below).

Since the stoichiometry of this reaction is such that one $^{18}O_2$ is taken up for each $^{16}O_2$ evolved, the ratio $R$ of the two *measured* changes (neglecting withdrawal corrections) will be proportional to the relative sensitivities. Thus, the absolute sensitivity for $^{18}O_2$ will be the product: i.e., $S_{36} = RS_{32}$.

Other reactions can also be used to obtain this sensitivity ratio. For example, in the presence of iodoacetamide, *Scenedesmus* catalyzes the mole for mole exchange of $^{16}O_2$ and $^{18}O_2$ similar to the chloroplast-driven reaction described above.[1]

The background at $m/e = 36$ is usually quite low. It can be determined along with the $m/e = 32$ background by the use of dithionite.

*Carbon Dioxide.* Carbon dioxide standardization can be obtained by measuring known concentrations of carbonate or bicarbonate ($^{12}C$ or $^{13}C$) at a known pH. Since the concentration of free $CO_2$ in solution is a strict function of pH, care must be taken to provide sufficient buffering capacity.

Most mass spectrometers have a substantial background at $m/e = 44$ due to outgassing of carbon (from the MS housing, filament, etc.) and subsequent combustion. Thus, the background at $m/e = 44$ can be quite significant. The primary $CO_2$ signal can be moved to $m/e = 45$ (a position in the mass spectrum with relatively low background) by using $^{13}CO_2$ as the $CO_2$ source, or by the use of $^{13}C$-labeled algae.[4]

The backgrounds at $m/e = 44$ and 45 can be conveniently determined by monitoring these mass numbers in the presence of a dilute solution of KOH. In some mass spectrometers, there is also a contribution to $m/e = 44$ from $N_2O$, formed from MS-generated fragments of $N_2$ and $O_2$. The magnitude of this (usually rather small) contribution can be ascertained by changing the amount of $N_2$ (and/or $O_2$) admitted to the MS, e.g., by depleting the buffer of $N_2$ by bubbling with argon.

*Use of Argon as a Monitor of Stability.* In the course of depleting the buffer of $^{16}O_2$, substantial quantities of argon are dissolved. This carried-over argon can be monitored and the time course used to evaluate the stability of stirring, magnitude of heat effects, etc. In our laboratory, we have not used the argon time course to correct experimental results; instead, we reject any experiment in which an unperturbed, monotonic depletion of argon is not obtained.

---

[12] M. Schwartz, *Biochim. Biophys. Acta* **112**, 204 (1966).

*Withdrawal Factor*

The membrane inlet constitutes a significant fraction of the wall area of the reaction vessel. Since there is a high vacuum on the other side of this membrane, the result is a one-way leak path by which the membrane-transmissible contents of the reaction vessel are constantly being depleted. Although this withdrawal is negligible for many experiments in which the rates of gas exchange are orders of magnitude higher than the rate of MS withdrawal, it can be a significant perturbation in experiments in which low rates of gas exchange are being monitored.

The MS withdrawal is proportional to the concentration of the species of interest; i.e., $dC/dt = -wC$; thus the withdrawal factor $w = (1/t) \ln (C_0/C_t)$, where $C_0$ and $C_t$ are the concentrations at time zero and $t$, respectively (time zero can any arbitrarily chosen time). The withdrawal constant $w$ is a function of membrane permeability (which can be altered by stretching, deposition of foreign substances, etc.), membrane area, stirring efficiency, etc. It must be empirically determined for each apparatus.

For each species, the net change per measuring interval is equal to the observed change minus that due to MS withdrawal. Thus

$$\Delta^{16}O_2 \text{ (net)} = \Delta^{16}O_2 \text{ (observed)} + w'^{16}O_2$$

and

$$\Delta^{18}O_2 \text{ (net)} = \Delta^{18}O_2 \text{ (observed)} - w'_{36}{}^{18}O_2$$

where $w'$ is the withdrawal factor normalized to the measuring interval. (The difference in signs occurs because $\Delta^{16}O_2$ and $\Delta^{18}O_2$ have opposite signs.) To check the validity of this correction, it should be tested using gas-equilibrated buffer (no algae); the calculated uptake and production should be equal (or very close) to zero.

*Time Constant*

Because the MS inlet (the membrane) is in intimate contact with the liquid being monitored, many of the time delays inherent in the more usual MS sample-handling procedures are eliminated. The main time-response limitations in the membrane direct-sampling technique are due to the gas transmission characteristics of the membrane and its associated vacuum system. The end-to-end time response of the apparatus can be readily determined; the instrument currently in use in our laboratory has a response ($\tau$) time of about 3 sec.

This time response can be determined in several ways. Probably the most straightforward technique (although it does not take into account limitations associated with stirring) is to monitor the appearance and disappearance of the $m/e$ peak with the application and removal of a jet of gas (e.g., argon) directed onto the membrane; the time it takes the $m/e$ signal to reach 63% $(1 - e^{-1})$ of its final value is equal to its $\tau$ time.

In another method the transient response of a chloroplast suspension to a high intensity light beam is monitored. (Since chloroplasts show very minimal induction, the $O_2$ evolution rate can be taken as constant from light-on; this is not true of algae.) The time response at light-on can then be calculated by determining the $x$ intercept of the extrapolated steady-state rate.

A more intuitive feeling for the time constant can be acquired by subjecting the chloroplast suspension to equal light–dark intervals of varying duration; as the alternation time $t$ is decreased to $<\tau$, the apparent rate of $O_2$ evolution will reflect the light and dark periods less and less, and at $t \ll \tau$ will attain an apparent linear steady state.

### Representative Data

Figure 3A shows the recorder output from an experiment using *Scenedesmus*. Gas exchange was monitored at a rate of 10 cycles/min; during each cycle the amplitude of five $m/e$ values (32, $^{16}O_2$; 36, $^{18}O_2$; 40, Ar; 44, $^{12}CO_2$; and 45, $^{13}CO_2$) was monitored. These data were simultaneously fed to a minicomputer and stored.

Computation was carried out as described above. Gas exchange values (e.g., $U_o$, $E_o$, $U_c$) were calculated for each measuring interval. The results were then summed to provide a concentration versus time profile (Fig. 3B). We should note that the values for $E_o$ and $U_o$ obtained by these techniques represent minimum values, since there is no readily available technique to correct for the preferential reduction of nascent oxygen due to the possible juxtaposition of the evolving and consuming sites.[13]

### Appendix

The apparatus used in our laboratory employs a mass selection stepping system in conjunction with a PDP8/E minicomputer[14] to program

---

[13] A proposed method to determine the extent of this "internal cycling" requires the assumption that the rate of light-induced $O_2$ uptake is independent of $O_2$ concentration [N. E. Good and A. H. Brown, *Biochim. Biophys. Acta* **50**, 544 (1961)]. However, this does not appear to be the case for the $O_2$ cycle.[4]

[14] Digital Equipment Corporation, Maynard, Massachusetts.

A

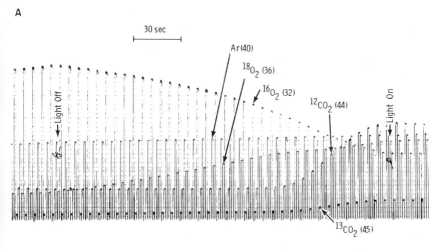

FIG. 3. (A) Recorder output of the time course of gas exchange by *Scenedesmus* $D_3$ under conditions of $CO_2$ limitation. The relative attenuations were $^{16}O_2 \times 32$, $^{18}O_2 \times 32$, Ar $\times 16$, $^{12}CO_2 \times 4$, $^{13}CO_2 \times 1$. The reaction medium contained 100 m$M$ phosphate buffer, pH 6.0, and 2.5 m$M$ KNO$_3$. No $CO_2$ was added; the $CO_2$ in the reaction vessel originated from carry over from the culture medium and respiration during the preceding 4 min dark period. Final cell concentration was 1% (v/v).

the MS and process the output. In this system only the $m/e$ values of interest are monitored; no time is wasted scanning undesired $m/e$ values.

The minicomputer is equipped with 8K of memory and an analog multiplexer (MUX) in series with an analog-to-digital converter (A/D). The analog multiplexer has eight input channels to which an analog signal of $-1$ to $+1$ V may be applied. Under software control, any one of these eight inputs can be selected and its analog voltage applied to the A/D for digitizing. The process of selection and digitizing requires approximately 30 $\mu$sec. The A/D has a dynamic range of only $\pm 2^9$. Therefore, gain factors of the MS stepper volume are in multiples of two (rather than 10) to avoid a possible loss of significant digits. This particular minicomputer configuration has no accessible output other than the teletype and high-speed tape punch. Thus, control functions are of necessity one way.

Figure 4 is a block diagram of the MS computer control unit, constructed using readily available digital (TTL) and analog integrated circuits and package reed relays. A software package (developed in-house) operates in conjunction with the control unit. The unit supplies the necessary control signals to the MS and to the computer, utilizing a coding of the MUX inputs.

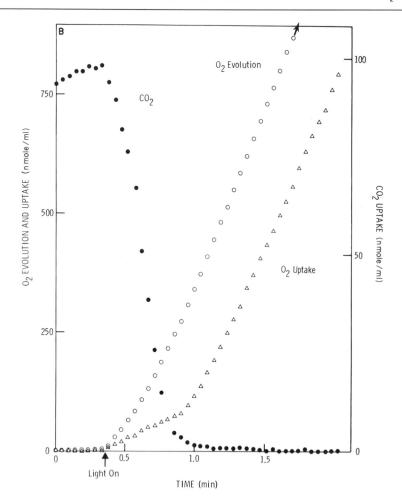

FIG. 3. (B) Results of the same experiment, calculated according to the methods described in the text. The points represent the cumulative value of $O_2$ evolution ($E_o$), $O_2$ uptake ($U_o$), and $CO_2$ uptake ($U_c$). The initial lag in $O_2$ evolution of ~10 points seems to have a biological basis, and does not reflect the time response of the system.

The allocations for the MUX channels are as follows.

CH∅: the 0 to +1 V MS preamp signal output.

CHl, "sync" signal. With a "high" level (> +0.25 V) computer remains in, or goes to, the MS signal sampling mode. With a low level (< +0.25 V) the computer remains in the software waiting loop.

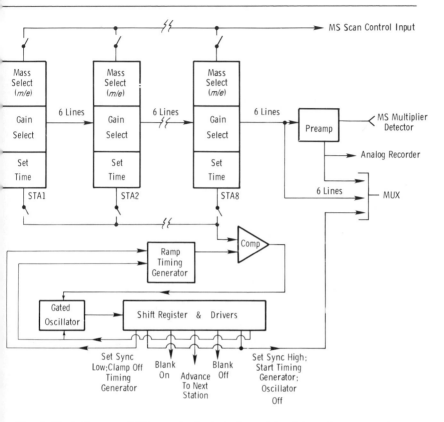

FIG. 4. Block diagram of MS control unit. Only three of the eight stations (STA1, STA2, and STA8) are shown.

CH2 through 7, gain code. Each input is associated with a stored gain factor and corresponding feedback resistor in the preamp. A "high" level on one input indicates the appropriate factor.

The control unit itself has eight stations which can be stepped either manually or automatically. Each station has access to the timing bus, the gain control, and the mass selection bus.

The desired $m/e$ value is selected by applying a dc voltage, derived from a coarse range switch and fine tuning potentiometer, to the external input of the mass spectrometer scan control circuit. A rotary switch selects one (of six) gain resistors in the preamplifier and provides the corresponding signal to the MUX. The computer then links the analog signal from the mass spectrometer to the proper gain factor.

The "dwell time" for each $m/e$ value is adjusted by varying a dc voltage applied to one side of a comparator; a linear ramp voltage is applied to the other side. When the ramp voltage reaches the level set by the particular station "dwell time" potentiometer, the comparator output changes state and initiates the advance to the next station.

At the beginning of each measurement the stations are advanced manually; each $m/e$ value is fine tuned, and the appropriate gain factor chosen. The cyclic observation of the masses is then begun by manually starting the computer, which enters a software waiting loop, continuously monitoring the MUX sync input. When the step button is depressed on the control unit, the sync level goes high and the ramp generator begins the dwell timing cycle. The computer, upon detecting the high sync level, exits its waiting loop and sequentially samples the six multiplexer inputs corresponding to the gain factors of the preamp. After selecting the activated gain line and the corresponding gain factor, the computer proceeds to sample the analog signal from the mass spectrometer. As long as the sync level remains high, the MS analog signal is repeatedly sampled, with a tally kept on the total number of samplings. Present program execution allows for a sampling rate of approximately 2000/sec.

To advance to the next station, the gated oscillator and associated circuitry provide a series of outputs. These sequential signals (1) set the sync level low and turn off the timing ramp, (2) blank all analog circuits to prevent switch overloads, (3) advance control of buses to the next station, (4) remove blanking, and (5) return sync level high and turn on the timing ramp.

During switching time between stations, the computer processes the data from the $m/e$ just measured by averaging the total samplings and storing the result. Upon completing this process, the computer enters a timing-waiting loop; if the sync level returns high within 2 sec, the computer reenters the measuring routine; if not, the computer exits to the data output routine.

The data are read out via a Teletype® or a high-speed paper tape punch. Calculated values of $E_0$ and $U_0$ can be obtained, either immediately after the data printout, or alternately, at a later time, after loading the data via the paper tape reader.

### Acknowledgments

This work was supported by National Science Foundation Grant No. AER73-03291 and United States Energy Research and Development Administration Contract No. E(11-1)-3326. We thank B. Kok for helpful suggestions.

# [53] Measurement of Translocation

## By DONALD R. GEIGER

The term translocation is applied to the movement of materials, primarily organic substances, through phloem from a source region to sink regions. A variety of measurements are commonly used to describe the various aspects of translocation. Included are the following.

a. *Mass transfer rate,* the rate of export of material from a source region into sink regions, expressed in mass units per unit time,[1] e.g., micrograms of carbon per minute or nanomoles of sucrose per minute.

b. *Specific mass transfer rate,* the mass transfer rate through a unit of cross sectional area of sieve tube or of phloem,[2,3] e.g., grams of sugar per hour per square centimeter of phloem or grams of sugar per hour per square centimeter of sieve tube.

c. *Phloem loading rate,* the rate of entry of sugar into minor vein endings in a source region,[4] e.g., nanomoles of sugar per minute per square centimeter of membrane.

d. *Translocation velocity,* an average velocity of some component of the material being translocated,[3] e.g., centimeters per hour.

e. *Pattern of distribution,* the pattern of movement of some component of the materials being translocated into sink regions. Generally qualitative or perhaps expressed as a set of ratios.

Because of the variety of ways of describing translocation, it is usually ambiguous to speak simply of translocation rate. Some measurements focus on the quantity of material moved, others on the speed along the path, on the flux through a conducting region or on the distribution into specific organs. In addition to addressing these specific purposes, all of the methods for measuring translocation are based on certain assumptions about the mechanism of translocation as well as about the methods themselves. The values of these ways of describing translocation depends on the validity of the assumptions. One common area of difficulty is the method of supplying the radioactive tracer in labeling experiments.

[1] D. R. Geiger and C. A. Swanson, *Plant Physiol.* **40,** 942 (1965).
[2] M. J. Canny, *Bot. Rev.* **35,** 507 (1960).
[3] M. H. Zimmermann, *Planta (Berl)* **84,** 272 (1969).
[4] B. R. Fondy and D. R. Geiger, *Plant Physiol.* **59,** 958 (1977).

Steady State Labeling

Since radioisotopes became readily available, translocation measurements have relied largely on radiotracer methods. While the methods appear to be reasonably straightforward, there are problems in these tracer studies which are not always realized or acknowledged. In addition to knowing the total amount of radioisotope transferred in a period of time, it is valuable and, in some cases, essential to know the specific radioactivity of the compound being traced, i.e., the ratio of the number of labeled atoms to the total number of atoms of that species. Because of differences in turnover time between various pools of intermediates, not all compounds which are labeled arrive simultaneously at the specific activity of the source of the radioisotope being supplied.

Bassham and Calvin[5] reported that hexose monophosphates become saturated with $^{14}C$ after 2 to 3 min of labeling with $^{14}CO_2$, while sucrose requires a much longer time to reach isotopic saturation. When $^{14}CO_2$ is supplied to photosynthesizing leaves, the newly formed starch arrives at isotopic saturation within a few minutes, while sucrose requires over 90 min to attain a steady ratio of $^{14}C$ to $^{12}C$.[1] As a consequence, the relative contribution of starch, for example, is emphasized over sucrose in short periods of labeling. To determine the actual amount of material present in a pool or transferred to or from a region, the amount of the radioisotope present in the compound of interest is divided by its specific radioactivity to give the total amount of the labeled compound. To obviate the need for measuring the specific radioactivity of all of the labeled compounds we wish to study, it is possible to supply $^{14}CO_2$ or [$^{14}C$]sucrose to leaves at a steady concentration over a sufficiently long period of time so that the compounds of interest, such as those being translocated, reach isotopic saturation.[1,4] The time course of label present in the pool or compartment being studied provides evidence for the attainment of isotopic saturation. At steady state, the label content of an intermediate pool becomes constant at isotopic saturation, while a terminal pool accumulates label at a constant rate if the pool supplying label to it has reached isotopic saturation.[1] For sugar beet, bean, and squash plants in air, under light of 250 $\mu E$ $m^{-2}$ $sec^{-1}$, the transport sugar pools require approximately 90 min to reach isotopic saturation. After this time, the pool size of the sugars being translocated and the rate of export can be computed from the radioactivity of the item of interest and the specific radioactivity of the $^{14}CO_2$ or [$^{14}C$]sucrose being supplied. This method of applying $^{14}CO_2$

[5] J. Bassham and M. Calvin, *Proc. Int. Congr. Biochem., 5th, 1961* Vol. 6, p. 285 (1963).

at a constant specific radioactivity and concentration is termed steady state labeling of photosynthate.

## The Labeling System

The system for supplying $^{14}C$-labeled $CO_2$, which will be described, can be used both for pulse labeling and steady state labeling. The components of the basic system are shown in Fig. 1.

*Leaf Chamber.* A variety of leaf chambers have been constructed either from brass or from Plexiglas assembled with dichloroethane. The dimensions and shape are adapted to the size and shape of the leaves used. The lid is sealed by an O ring seated in a channel cut out to half the diameter of the ring. The lid is not channeled. The petiole is sealed with caulking compound rope. Luggage fasteners clamp the lid tightly and permit rapid access to the leaf for sampling. Manifolds direct air both above and below the surface of the leaf which is held between two nylon monofilament networks. Air flow is approximately 1000 ml min$^{-1}$. Thermocouples or other devices for monitoring the status of the leaf or its surroundings can be incorporated into the chamber, with electrical connectors sealed into the side walls to provide convenient connections for sensors. A method for measuring export as well as accumulation in sinks has been published recently.[6]

$^{14}CO_2$ *Measurements.* $^{14}CO_2$ is measured by means of an ion chamber electrometer. Both the Nuclear Chicago model 6000 Dynacon and the Victoreen model 725 are satisfactory but are no longer manufactured. Both the Cary model 401 (Varian Associates, Palo Alto, California) and the Kiethley model 642 (Kiethley Instruments, Cleveland, Ohio) are well suited for the system. A 250-ml ion chamber is appropriate for the system which has a total volume of approximately 1500 ml. A $10^{12}$ $\Omega$ standard resistor allows the ion current for less than 1 $\mu Ci$ per total system volume to be measured. $^{14}CO_2$ monitors based on a GM tube can also be used.[7]

$CO_2$ *Measurements.* In the system described, carbon dioxide is measured by a Mine Safety Appliances Co. (201 N. Braddock Ave., Pittsburgh, Pennsylvania) infrared gas analyzer, LIRA model 200 or 300. The reference cell is sealed while the measuring cell has 1/8 inch ID inlet and outlet ports. The infrared gas analyzer is standardized daily with zero gas and span gas. For zeroing, air is passed through Ascarite (Arthur H. Thomas Co., Philadelphia, Pennsylvania) to remove $CO_2$. The standardizing gases pass through a cold-finger, flow meter, pressure meter, and

---

[6] D. R. Geiger and B. F. Fondy, *Plant Physiol.* **64**, 361 (1979).
[7] A. G. Swan and H. M. Rawson, *Photosynthetica* **7**, 325 (1973).

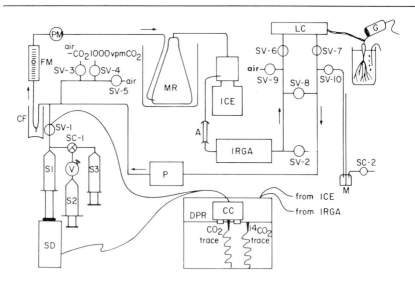

Fig. 1. Schematic diagram of the components of a closed system for supplying $^{14}CO_2$ for steady state labeling experiments: A, removable Ascarite cartridge: DPR, dual pen recorder with carbon dioxide concentration controller (CC) which activates syringe drive (SD) and solenoid valve (SV-1); CF, cold finger for humidity control; FM, flow meter; G, G-M detector; ICE, ion chamber electrometer; IRGA, infrared gas analyzer; LC, leaf chamber; MR, mixing reservoir in water bath; P, tubing pump; PM, pressure meter; S-1, 50-ml syringe with $^{14}CO_2$ to be metered into closed system; S-2, 50-ml syringe for mixing $^{14}CO_2$ released in vial (V) by injection of $Na_2$ $^{14}CO_3$ followed by phosphoric acid. Syringe can fill the 50-ml reserve syringe, S-3 or the working syringe, S-1; SC-1, stopcock for distributing $^{14}CO_2$ mixture; SV-2, solenoid valve for venting standardizing gas or for releasing pressure in system; SV-3, SV-4, and SV-5, solenoid valves for admitting standardizing gases for IRGA; SV-6, SV-7 and SV-8, solenoid valves for bypassing leaf chamber, LC; SV-9 and SV-10, solenoid valves and SC-2 stopcock for passing air over source leaf and performing leak test; M, manometer. The entire flow system is contained in a fume hood for handling radioactive carbon dioxide.

mixing reservoir prior to entering the LIRA and are vented to the air by opening solenoid valve SV-2. Flow is adjusted to 1000 ml min$^{-1}$. Following zeroing, span gas of a known $CO_2$ concentration of approximately 1000 volumes per million (vpm) $CO_2$ is passed through the LIRA and the span control is adjusted. The meter settings are adjusted at zero and full scale by alternately purging the LIRA with zero and span gas until correct settings are attained.

*Standard Curves.* It is necessary to make standard curves of infrared gas analyzer readings as a function of the $CO_2$ content and ion current readings as a function of $^{14}C$ content of a system of a specified, known volume. The bypass system (SV-6 and SV-7 closed, SV-8 open) is used for the initial standardizing procedure. A measured amount of $^{14}C$-labeled $CO_2$ of a known radioactivity of approximately 10 $\mu Ci$ is injected into the system with a gas-tight syringe. A known volume of $CO_2$, corrected for temperature and pressure, is selected by trial and injected to give nearly full-scale deflection of the infrared gas analyzer. After the first readings for $CO_2$ and $^{14}C$ are recorded, the pump is stopped, the Ascarite cartridge A is inserted and a small amount of $^{14}CO_2$ is removed by running the pump briefly. After removal of the Ascarite, another tracing is made. The amounts of carbon in $CO_2$ and radioactivity in $^{14}CO_2$ are calculated by multiplying the original values by the factor $G$

$$G = \frac{m}{M}$$

where $G$ is the proportion of $CO_2$ or $^{14}CO_2$ remaining, $M$ is the original ion current reading, and $m$ is the reading after some $CO_2$ is removed. Repetition of this procedure gives two standard curves: micrograms of carbon as $CO_2$ as a function of LIRA recorder readings and microcuries of $^{14}C$ as a function of the ion chamber electrometer recorder readings. Repeating this procedure several times and averaging the data gives a basic set of standard curves for micrograms of carbon as $CO_2$ and for microcuries as $^{14}CO_2$ in the bypass system.

System volume also must be known to perform certain calculations. Both the $CO_2$ concentration ($\mu l/l$) and the volume of $CO_2$ at known temperature and pressure which must be added to give a full-scale LIRA reading are known. The former is known directly from the $CO_2$ concentration of the standardized span gas and the latter is obtained from volume of $^{14}CO_2$ which needed to be injected to give a full scale LIRA reading. The latter may require a slight extrapolation if the volume of $^{14}CO_2$ injected does not give a reading of exactly 100% scale. Dividing the volume injected (corrected for 25° and 745 mm Hg, the usual operating conditions) by the vpm $CO_2$ gives the system volume

$$V = \frac{i}{C}$$

where $V$ is the system volume, $i$ is the volume of $CO_2$ injected corrected to 25° and 745 mm Hg pressure, and $C$ is the concentration of $CO_2$ in vpm.

A check on this volume which gives an approximate value is the following

$$V = \frac{vA}{a}$$

where $V$ is the system volume, $v$ is the nominal sensitive volume of the ion chamber (250 ml in the system shown; the accuracy of this value probably limits the accuracy of this method), and $A$ is the radioactivity injected, and $a$ is the radioactivity found to be present in the ion chamber by computation from the ion current.

The volumes of the various leaf chambers can be found by filling them with a measured volume of water. Once the standard curves are made for micrograms of carbon and microcuries in the bypass system, values for other systems can be gotten by multiplying by a factor, larger than one, obtained by dividing the larger system volume by the bypass system volume. The basic value for microcuries or micrograms of carbon is read from the appropriate standard curve and converted by multiplying by the correct factor. For example, if the bypass system volume is 1500 ml and the added volume of the sugar beet chamber is 280 ml, then:

$$F = \frac{(1500 + 280)}{1500}$$

$$F = 1.187$$

where $F$ is the factor used to adjust standard curve readings for $CO_2$ and $^{14}CO_2$. If the standard curve for the 1500-ml bypass system gives a value of 1.20 $\mu Ci$, then the total system including the larger system would contain 1.42 $\mu Ci$ for the same percent full-scale reading of the ion chamber electrometer. Similarly, if the LIRA standard curve for the bypass system shows 450 $\mu g$ of carbon then the larger system will contain 534 $\mu g$ of carbon. However, the LIRA reading for $CO_2$ concentration is independent of the system volume and no correction should be applied.

*Specific Radioactivity of $CO_2$.* The readings from the LIRA and the ion chamber electrometer provide a means of calculating the specific radioactivity of the $^{14}CO_2$ being supplied. To prevent serious changes in specific radioactivity, it is advisable to insert the Ascarite into the system at A to remove unlabeled $CO_2$ just prior to labeling. With a flow rate of 1000 ml min$^{-1}$ and a system volume of 1500 ml, circulating the air for 3 min with the Ascarite in place will significantly lower the content of unlabeled $CO_2$. Labeling can begin once the Ascarite cartridge is removed. For labeling periods of several hours duration, it is advisable to average specific radioactivity calculations from three sets of measurements, for instance after 1 hr, in the middle, and at the end of the labeling period.

Readings at the same point in time should be used for the $CO_2$ and $^{14}CO_2$ levels. It is not necessary to correct for the system volume because the correction factor needs to be applied to both numerator and denominator of the specific radioactivity fraction. Specific radioactivity is calculated as follows

$$S = \frac{L}{U}$$

where $S$ is the specific radioactivity, for example, $\mu Ci/\mu g$ of carbon, $L$ is the $^{14}C$ contained in the system (e.g., the bypass volume), and $U$ is the carbon contained in the $CO_2$ in the same system volume.

It has been observed that there is an isotopic discrimination against $^{14}C$ with the result that the specific radioactivity of compounds derived from the $^{14}CO_2$ is less than that of the $^{14}CO_2$ supplied.[8] For sucrose, the specific radioactivity of the sucrose was found to be approximately 85% of that of the $^{14}CO_2$ supplied.[9]

*Controlled $^{14}CO_2$ Supply.* Steady state labeling over a long period is most conveniently done using a reservoir of $^{14}CO_2$ of known specific radioactivity and a means of regulating the concentration at a specific level. In the system described, the $^{14}CO_2$ is generated by release from labeled carbonate solution. Labeled $Na_2^{14}CO_3$ is made by releasing the $BaCO_3$ with acid and absorbing it in 0.1 $N$ NaOH to give approximately 0.25 $\mu Ci/\mu l$. A mixture of 4 volumes of 6.7% $K_2CO_3$ as unlabeled carrier to 1 volume of $^{14}C$-labeled sodium carbonate gives a specific radioactivity of approximately 10 $\mu Ci/\mu g$ of carbon, a good value for initial experiments. The specific activity can be adjusted to fit the needs for particular experiments. Alternatively, $^{14}CO_2$ of appropriate specific activity may be purchased.

The delivery system for $^{14}CO_2$, shown in Fig. 1, includes a solenoid valve (SV-1), syringe pump (SD), delivery syringe (SI), and generating and reservoir syringes, S2 and S3, respectively. A controller (CC) activates the syringe drive and the solenoid valve on demand as indicated by the level of $CO_2$ desired. The $^{14}CO_2$ is generated in container V by injecting carbonate and concentrated phosphoric acid through a serum-vial stopper in the sidearm and mixing them thoroughly by moving the plunger of S2 repeatedly. The released $CO_2$ is moved to the glass syringes S1 or S3 for use or storage.

The delivery of $^{14}CO_2$ is controlled by the microswitch-activated controller shown in Fig. 2. This controller, which responds to the recorder pen position, is able to operate in either of two modes. In "control" mode, microswitches MS-1 and MS-2 slow or speed the entry of $CO_2$, respectively, by modifying the frequency of entry of $CO_2$ pulses. In

[8] O. D. Bykov, *Photosynthetica* **4**, 195 (1970).
[9] D. R. Geiger and J. W. Batey, *Plant Physiol.* **42**, 1743 (1967).

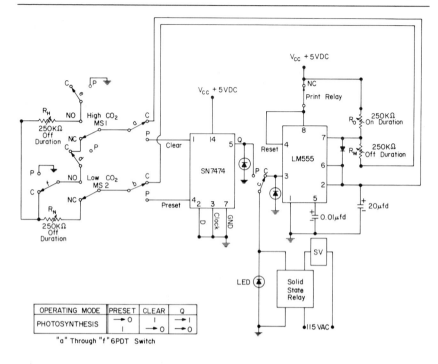

FIG. 2. Schematic circuit diagram for carbon dioxide concentration controller. Components: LM555, linear integrated circuit, timer (National Semiconductor, Santa Clara, California); SN7474, digital integrated circuit, D-type edge-triggered flip-flop (Texas Instruments, Dallas, Texas); Solid State Relay, model 601-1401 (Teledyne, Hawthorn, California); Vcc, 5 V dc power supply, model 81-05-230 (Sola Electric Div. Elk Grove Village, Illinois); LED, light emitting diode; SV, syringe drive model 341 (Sage Instruments Divn, Orion Research, Inc., Cambridge, Massachusetts) and solenoid valve, ASCO Red Hat Model 8262C1, 3/64 in orifice, normally closed (Automatic Switch Co., Florham Park, New Jersey); $R_M$, $R_N$, $R_H$ and $R_O$, 250K potentiometers for setting minimum, normal and high off duration and on duration, respectively; MS-1 and MS-2, microswitches for sensing pen position; print relay, relay for disabling LM555 during printing cycle. Schematic of printer/timer/counter available from author.

"photosynthesis" mode, switches MS-1 and MS-2 alternately close and open the solenoid valve and stop and start the syringe drive to produce a tracing of the $CO_2$ utilization rate. To avoid protracted entry of $^{14}CO_2$ after the solenoid valve shuts, it is necessary to reduce the tube between the valve and the entry to the flow system to 1/16 inch ID or less. In the "control" mode, the controller is adjusted by first setting the off duration potentiometers $R_M$, $R_H$, and $R_N$ to medium settings of approximately 2 sec each and then adjusting the on duration potentiometer $R_0$ to keep the

$CO_2$ concentration steady. Thereafter, all adjustment are normally made by adjusting all three off duration potentiometers slightly, if necessary. At low $CO_2$ level, only $R_M$ contributes to the off period, at the control point $R_M$ plus $R_N$ control the off, while at high $CO_2$ all three determine off duration. The rate of photosynthesis can be determined in arbitrary units from the number of on cycles recorded by a counter during a fixed period. The rates can be converted to micrograms of carbon per minute by using the slopes of $CO_2$ utilization from operation in "photosynthesis" mode. Counting and printing can be done with a modified printer calculator such as TI model 5025 (Texas Instruments, Dallas, Texas). The $CO_2$ measuring system lacks some of the refinements found in systems designed specifically for measurement of carbon fixation rates and this limitation must be kept in mind.[10]

### Rate Measurements

*Photosynthetic Carbon Fixation Rate.* The rate of net photosynthetic carbon fixation is calculated by determining the carbon dioxide content of the system at points at the upper and lower ends of the near-linear portions of the slopes. The regions of transition at either end of the slope tracing should be avoided. After the carbon dioxide content is calculated for the two levels, the total change in carbon dioxide is divided by the time interval for the change to give net carbon fixation rate. Calculation of rate in the "control" mode has been described in the previous paragraph.

*Mass Transfer Rate.* The measurement of the mass transfer rate depends upon a continuous record of the arrival of $^{14}C$-labeled material translocated from the source region. To obtain this, a portion of the sink region such as a small leaf is held securely in front of the window of a GM tube throughout the course of the experiment. The accumulation of $^{14}C$ in this sink region is used to monitor the arrival of $^{14}C$ in the sinks generally. Obviously the method is dependent for its validity on the basic assumption that the specific radioactivity of the material translocated is the same as that of the $^{14}CO_2$ supplied. Because of discrimination against $^{14}CO_2$, the specific radioactivity of the sucrose in sugar beet leaves approaches approximately 85% of the specific radioactivity of the $CO_2$ supplied.[9] This factor can be used to correct the specific radioactivity in calculating mass transfer rate.

Another assumption is that the proportion of material translocated to the monitored sink, such as a young sink leaf, represents a constant

[10] D. T. Canvin and H. Fock, Vol. 24, p. 246.

proportion of the material translocated to the sinks as a whole. While this assumption is true under many circumstances, there are probably some treatments which would change the partition between the monitored sink and the rest of the sinks. The assumption should be tested.

It is further assumed that isotopic saturation of the labeled material being translocated exists as soon as the rate of arrival of label in the monitored sink region becomes steady. Once this condition exists, the rate of export out of the source leaf can be calculated as follows

$$T = \frac{RE}{AS}$$

where $T$ is the mass transfer rate out of the source leaf into the sink regions, micrograms of carbon per minute per square decimeter, $R$ is the rate of accumulation of $^{14}C$ in the monitored sink, counts per minute per minute, $E$ is the efficiency of detection of label in the sink; it is the ratio of microcuries found in the sink regions to the count rate in the monitored sink region at the end of the experiment. The former is determined by oxidation of the sink regions and assay for the $^{14}CO_2$, while the latter is the final count rate for the monitored sink region, microcuries per counts per minute, $A$ is the area of the source leaf, square decimeter; $S$ is the specific radioactivity of the translocated sugar; in sugar beet it is 0.85 of the $^{14}CO_2$ specific radioactivity, microcuries per microgram of carbon.

The rate of accumulation of $^{14}C$-labeled compounds in the monitored sink region is obtained by recording the ratemeter output from the GM detector positioned against the sink leaf being monitored. If the count rates are such that deadtime corrections are minimal and the mass transfer rate is steady, the slope may be read directly from the ratemeter tracing. Otherwise, dead-time corrections must be made. If the mass transfer rate is not linear, a polynomial, least-squares curve fit of the deadtime corrected output of the GM detector can be made. The slope or the first derivative of the curve is multiplied by the appropriate factors to calculate the mass transfer rate.

Certain treatments of the plant may change the water status which could conceivably change the amount of self-absorption by the sink leaf and thus change the counting efficiency. Significant growth by the sink leaf could also conceivably change the counting geometry and the count rate. Experience has shown that these factors usually do not present serious difficulties but that they must be considered for any experimental design.

The most satisfactory method for determining the total $^{14}C$ accumulated is the oxidation of the entire sink region to $CO_2$ and assay of the $^{14}CO_2$ released. Presently tissue of no more than 1.5 gm fresh weight is oxidized by the method of the Van Slyke, Plazin, and Weisiger and swept into an evacuated ion chamber with $CO_2$.[11] An apparatus formerly available from Nuclear Chicago Corp. is used to oxidize the samples and fill the ion chamber. Systems for oxidation followed by liquid scintillation counting which are available commercially are discussed by Rapkin.[12]

The method estimates the mass transfer rate by measuring the accumulation of [14]C-labeled material in the sink regions. Labeled carbon lost by respiration in the sink regions is not taken into account. Recovery data indicate that there is a 15 to 20% loss of label, presumably by respiration in the sink regions. Where this loss is of importance, the experimental design should include steps to trap and measure the respired $CO_2$.

The methodology of steady-state labeling can be combined with anatomical measurements to obtain specific mass transfer rates and to determine rates of various processes related to translocation.[1,4,9] Combined with experimental treatments, the method can be used to study the kinetics of the response of translocation to the treatment.[13] The labeling system shown can also be used for pulse labeling experiments to study translocation velocity and response to various treatments.[14,15]

Acknowledgment

The preparation of this section was supported in part by National Science Foundation Grant BMS 71-01572 AO1.

[11] D. D. Van Slyke, J. Plazin, and J. R. Weisiger, *J. Biol. Chem.* **191**, 299 (1951).
[12] E. Rapkin, "Digitechniques," Vol. 6 Tech. Bull. Teledyne Intertechnique, Westwood, New Jersey, 1972.
[13] D. R. Geiger, *Ohio J. Sci.* **69**, 356 (1969).
[14] R. T. Giaquinta and D. R. Geiger, *Plant Physiol.* **51**, 372 (1973).
[15] D. R. Geiger and S. A. Sovonick, *Plant Physiol.* **46**, 847 (1970).

# [54] System for Long-Term Exposure of Aerial Part of Plants to Altered Oxygen and/or Carbon Dioxide Concentration

*By* BRUNO QUEBEDEAUX and R. W. F. HARDY

Oxygen as a physiological factor controlling long-term growth in whole plants is poorly defined and deserves further consideration. It has not been explored in detail largely because of the complexity and diversity of experimental systems and plant enclosure chambers utilized for growing plants at altered $O_2$ concentrations. The procedures described here have been greatly simplified and standardized, and they enable experimentation with altered $O_2$ or $CO_2$ concentration in controlled environments at all plant growth stages of development from early seedling to maturity and provide the most defined system available. In our laboratory this system has been used mainly to elucidate the mode of action of the $O_2$ process regulating seed growth, but, in addition, we have used it to explore oxygen effects on a number of plant processes including, photosynthesis, photorespiration, respiration, assimilate partitioning and $^{14}C$ translocation, senescence, ATP and energy levels, and symbiotic $N_2$ fixation.[1-3]

### Equipment and Methods

Experimental plants are normally grown in a controlled environment growth room (4000–6000 ft-c, 12-hr photoperiod, 75% relative humidity, 25°–30° day, 18°–20° night) in 15–20 cm plastic pots containing a 50:50 mixture of Jiffy-Mix (Jiffy Products of America, West Chicago, Illinois) and sand with Hoagland's nutrient solution applied daily. Oxygen growth chambers are made of Lucite acrylic resin measuring 45 × 45 × 72 cm and a 145-liter capacity with the aerial plant portion enclosed by the chamber and urethane foam gaskets around the stem entrance with roots exposed to air as shown schematically in Fig. 1. Altered $O_2$ gas mixtures are humidified and filtered in sterilized $H_2O$ and purged continuously in an open system at 15 liters/min and stirred continuously by a small fan with flow rates calibrated and regulated by flow meters and needle valves. Nitrogen serves as the carrier gas with $O_2$ and $CO_2$ adjusted to the desired concentrations. Subambient $O_2$ mixtures are prepared by diluting air with $N_2$ and adjusting $CO_2$ additions to 320 ppm with small micro needle valves (Hoke Incorporated; Cresskill, New Jersey) connected to compressed gas cylinders of either pure $CO_2$ or 10% $CO_2$ in $N_2$. The partial pressure of $O_2$ and $CO_2$ in the oxygen growth chambers is calibrated by adjusting flow rates and gas analysis with $O_2$ monitored by a Beckman $O_2$ electrode and $CO_2$ by a Beckman Model 215 infrared gas analyzer. For example, to prepare a 5% $O_2$ mixture with ambient $CO_2$ levels purged at 15 liters/min in the oxygen growth chamber, air is first precalibrated to flow at 3.57 liters/min and diluted with $N_2$ flowing at 11.43 liters/min with $CO_2$ addition adjusted to 320 ppm. The oxygen and $CO_2$ concentra-

[1] B. Quebedeaux, and R. W. F. Hardy, *Nature (London)* **243**, 277 (1973).
[2] B. Quebedeaux, and R. W. F. Hardy, *Plant Physiol.* **55**, 102 (1975).
[3] B. Quebedeaux and R. W. F. Hardy, *in* "$CO_2$ Metabolism and Plant Productivity (R. H. Burris and C. C. Black, eds.), p. 185. Univ. Park Press, Baltimore, Maryland, 1976.

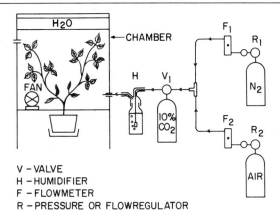

FIG. 1. Diagrammatic representation of the continuous-flow system for growing plants at modified $O_2$ and/or $CO_2$ concentrations. From Quebedeaux and Hardy.[2]

tions are monitored throughout the course of the experiment. Elevated $O_2$ mixtures are prepared by diluting air or $N_2$ with purified $O_2$ from compressed gas cylinders or liquid $O_2$ in cylinders adapted with valve connections to form gaseous $O_2$ generated from the liquid system. Carbon dioxide additions are adjusted to 320 ppm in the continuous flow system similar to that described for subambient $O_2$ mixtures.

Plant Responses to Altered $O_2$ Concentrations

The growth and development of plants is sensitive to oxygen concentration. Our recent plant growth studies with altered $O_2$ concentrations in controlled environments have established $O_2$ as an important physiological factor regulating plant growth and reproductive development. High levels of $O_2$ are required to facilitate photosynthate transfer from leaves to developing seeds. The high $O_2$ requirement for seed growth occurs in both $C_3$ and $C_4$ plants, and this $O_2$ process is a major factor in the facilitation of assimilate translocation and accumulation in reproductive structures. In contrast to the high $O_2$ requirement for reproductive growth, low $O_2$ increases net photosynthesis and vegetative growth in $C_3$ plants, but not in $C_4$ plants. This low $O_2$ stimulation of net photosynthetic $CO_2$ fixation is due to a decrease in photorespiration and a reduction of the direct $O_2$ inhibition of photosynthesis. In the carbon cycle, oxygen effectively competes with $CO_2$ for the $CO_2$ fixing enzyme, ribulose-1,5-diphosphate (RuDP) carboxylase, catalyzing the reactions for both photosynthesis and photorespiration. Low $O_2$ and high $CO_2$ concentrations

FIG. 2. Effect of subatmospheric $O_2$ concentrations on vegetative and reproductive growth of soybean plants. From Quebedeaux and Hardy.[1]

TABLE I

CONCLUSIONS OF EFFECTS OF $O_2$ CONCENTRATION ON THE REGULATION OF CROP
GROWTH AND PRODUCTIVITY

1. Oxygen concentration controls total growth in most crops and reproductive growth in all crops tested
2. Root and vegetative growth—oxygen limitation increases in inefficient ($C_3$) plants with greatest effect on root
   Exposure of aerial portion of legume increases $N_2$ fixation
3. Reproductive Growth
   Essential role of $O_2$ for both $C_3$ and $C_4$ plants occurs at all stages of reproductive growth
   Seed Development more sensitive than pods
   Early exposure arrests pod while latter arrests seed development
   Localized to the part exposed
   Irreversible except for short exposures of $\leq 3$ days
   Independent of $pCO_2$
4. Oxygen concentration for maximal reproductive growth
   Approaches or exceeds 21 versus 1–3% to saturate dark respiration in leaves
   Dark respiration not altered in intact plants at 5% $O_2$
   ATP content constant in seeds at 5% $O_2$
5. Physiological and biochemical effect of $O_2$ on reproductive growth
   Fertilization is normal
   Indirectly related to a light reaction
   Altered translocation and accumulation of photosynthate through altered source-sink relationship
   Altered level or activity of endogenous hormones may mediate $O_2$ effect
   An enzyme with low affinity for $O_2$ may be involved
   A physical versus a chemical phenomenon is also a possibility
6. Altitude Effects
   Seed yield at high altitudes may be limited by amount of $O_2$
8. Crop Yield
   Understanding role of $O_2$ may provide information to control sink activity and thereby optimize harvest index, i.e. maximize seed yield of cereals and vegetative growth of forage and root crops

favor carboxylation and photosynthesis, while low $CO_2$ and high $O_2$ favor oxygenation, glycolate synthesis and photorespiration.

The vegetative and reproductive growth rates of plants grown at subambient $O_2$ during a complete growth cycle are most dramatic. The most striking effects are on the morphology and reproductive character of the plant (Fig. 2). With soybeans there is an increase in lateral development, diameter of stems, and leaf thickness. Some delay in senescence occurs, while the time of flowering is essentially unchanged. The number of fully developed pods and seeds is decreased by all subatmospheric $O_2$ levels so that at the time of normal maturation there are neither completely developed pods nor seeds for plants exposed to 5% $O_2$ and only 50% as many developed seeds in 15% $O_2$ as there are in air. The final dry weight of the aerial portion of the plant is increased 74% by 5% versus 21% $O_2$, while an increase of over sixfold was found for the roots. In sorghum, a $C_4$ species with low rates of photorespiration, reproductive growth also is decreased by all subatmospheric levels of $O_2$, whereas the morphology, flowering, and dry weights of the aerial portion and roots are not altered significantly. Oxygen alters reproductive growth of both $C_3$ and $C_4$ plants. An exposure to 40% $O_2$ reduces both total vegetative and reproductive growth of $C_3$ species but does not affect the balance of vegetative to reproductive growth, while $C_4$ species are not altered significantly. Conclusions about the effect of oxygen on the regulation of crop growth and productivity are summarized in Table I.

# [55] Measurement of Photophosphorylation Associated with Photosystem II

*By* CHARLES F. YOCUM

## I. Introduction[1]

Noncyclic and cyclic electron transport catalyzed by photosystem II is coupled to the synthesis of ATP.[2-6] Assay of the photophosphorylation activity unique to photosystem II-catalyzed electron transport requires the elimination of electron-transport reactions associated with photosystem I. The plastoquinone antagonist dibromothymoquinone (DBMIB) is a potent inhibitor of electron transport between photosystem II and photosystem I, but this compound does not inhibit the donation of elec-

---

[1] Abbreviations used are DBMIB, 2,5-dibromo-3-methyl-6-isopropyl-$p$-benzoquinone; DCMU, 3-(3,4-dichlorophenyl)-1,1-dimethylurea; PD, $p$-phenylenediamine; $PD_{ox}$, $p$-phenylenediimine; DAD, 2,3,5,6-tetramethyl-$p$-phenylenediamine; TMPD, $N,N,N',N'$-tetramethyl-$p$-phenylenediamine; DAT, 2,5-diaminotoluene; DMQ, 2,5 dimethyl-$p$-benzoquinone; $P/e_2$, ratio of ATP synthesis to electron pairs transferred; MV, methyl viologen; Asc, ascorbate; Chl, chlorophyll; DNP, 2,4-dinitrophenol.

[2] S. Saha, R. Ouitrakul, S. Izawa, and N. E. Good, *J. Biol. Chem.* **346**, 3204 (1971).
[3] R. Ouitrakul and S. Izawa, *Biochim. Biophys. Acta* **305**, 105 (1973).
[4] A. Trebst and S. Reimer, *Biochim. Biophys. Acta* **325**, 546 (1973).
[5] C. F. Yocum, *Biochem. Biophys. Res. Commun.* **68**, 828 (1976).
[6] C. F. Yocum and J. A. Guikema, *Plant Physiol.* **59**, 33 (1977).

METHODS IN ENZYMOLOGY, VOL. 69

trons to photosystem I by artificial mediators.[7] In addition, DBMIB possesses oxidation–reduction activity[8] and the electron-transport block imposed by this inhibitor can be by-passed under certain conditions.[4] A method for inhibiting electron donation to photosystem I, which avoids the difficulties associated with DBMIB, involves destruction of plastocyanin function with KCN[3,9] or the combination of KCN and Hg.[6] Chloroplasts subjected to KCN/Hg inhibition retain the ability to transfer electrons via photosystem II from water to $p$-phenylenediimines or $p$-benzoquinones with an accompanying photophosphorylation efficiency $(P/e_2)$ of about 0.3. Treatment of KCN/Hg-inhibited chloroplasts with $NH_2OH$ plus EDTA[10] to inhibit oxygen evolution creates a condition whereby a photosystem II-dependent cyclic photophosphorylation reaction can be observed. This section describes the methods for preparation of inhibited chloroplasts and conditions for assay of photosystem II-catalyzed photophosphorylation reactions.

## II. Preparation of KCN/Hg- and KCN/Hg/NH₂OH-Inhibited Chloroplasts

### A. Isolation of Chloroplasts from Spinach

All preparative operations described in this section are carried out at $0°$–$4°$. Chlorophyll concentrations are determined by Arnon's procedure.[11] Depetiolated spinach leaves (125 gm) are homogenized in a chilled Waring blender for 10 sec. The homogenizing medium (250 ml) contains Tricine (20 m$M$, pH 8), NaCl (15 m$M$), sucrose (0.4 $M$), and 2 mg/ml bovine serum albumin (Sigma Catalogue No. A-4378). The homogenate is filtered through four layers of cheesecloth and centrifuged at 1000 $g$ for 1 min. The pellets are discarded and the supernatant is centrifuged at 4000 $g$ for 10 min. The pellets from this step are resuspended in homogenizing medium and recentrifuged at 4000 $g$ for 10 min.

### B. Preparation of KCN/Hg-Inhibited Chloroplasts

Chloroplast pellets from Section II,A are resuspended in a medium containing HEPES buffer (50 m$M$, pH 8), sucrose (0.1 $M$), NaCl (15

[7] H. Bohme, S. Reimer, and A. Trebst, Z. Naturforsch, Teil B **26,** 341 (1971).
[8] S. Izawa, J. M. Gould, D. R. Ort, P. Felker, and N. E. Good, Biochim. Biophys. Acta **305,** 119 (1973).
[9] S. Izawa, R. Kraayenhof, E. K. Ruuge, and D. DeVault, Biochim. Biophys. Acta **314,** 328 (1973).
[10] D. R. Ort and S. Izawa, Plant Physiol. **52,** 595 (1973).
[11] D. I. Arnon, Plant Physiol. **24,** 1 (1949).

m$M$), and bovine serum albumin (2 mg/ml). Inhibition of photosystem I activity is initiated by adding to the suspension sufficient 0.3 $M$ KCN (neutralized to pH 8 with 1 $N$ HCl just before use), 20 m$M$ HgCl$_2$, 20 m$M$ ferricyanide, and resuspending medium to produce final concentrations of 300 $\mu$g/ml Chl, 50 m$M$ KCN, 100 $\mu M$ HgCl$_2$, and 50 $\mu M$ ferricyanide. The reaction vessel (an Erlenmeyer flask) is stoppered and stirred gently in the dark. The progress of KCN/Hg inhibition is monitored at 15–20 min intervals using two assays. Inhibition of ferricyanide reduction is assessed by measuring oxygen evolution with ferricyanide as the terminal oxidant. The stirred, thermostatted (25°) reaction cuvette (1.6 ml) is fitted with an O$_2$ electrode (Yellow Springs Instrument Co.) and contains a reaction mixture consisting of Tricine (12 m$M$, pH 8), NaCl (15 m$M$), ferricyanide (2.5 m$M$), and 100 $\mu$l of the incubating chloroplast suspension (30 $\mu$g Chl). The oxygen evolution rate observed during 1 min of illumination with heat-filtered white light (10$^6$ ergs/cm$^2$/sec) is compared to a control value obtained with a sample of chloroplasts not exposed to KCN plus HgCl$_2$. Inhibition of photosystem I electron transport is assayed by following the rate of oxygen uptake catalyzed by DAD in the presence of MV. The reaction medium is that described above, but ascorbate (3m$M$), DAD (0.6m$M$) recrystallized as described in Section III,A, and MV (0.125 m$M$) replace ferricyanide, and DCMU (6 $\mu M$) is present to inhibit photosystem II.

Incubation of chloroplasts under the conditions described above for 90 min results in strong (90% or greater) inhibition of ferricyanide photoreduction and DAD photooxidation. Incubation is terminated by centrifuging the chloroplast suspension at 10,000 $g$ for 10 min. The KCN/Hg-inhibited preparations are then resuspended and subjected to further inhibition with NH$_2$OH plus EDTA (see below) or stored at $-70°$ for later assay (Section II,D).

### C. Treatment of KCN/Hg-Inhibited Chloroplasts with NH$_2$OH plus EDTA

The pelleted chloroplasts from Section II,B are resuspended in a medium containing sucrose (0.4 $M$), Tricine (20 m$M$, pH 8), NaCl (15 m$M$), MgCl$_2$ (2 m$M$), and bovine serum albumin (2 mg/ml). Sufficient 0.1 $M$ NH$_2$OH in 0.01 $N$ HCl (neturalized to pH 8 with 1 $N$ NaOH just before use), 0.5 $M$ EDTA, and resuspending medium are added to the gently stirred chloroplast suspension to produce final concentrations of 1–1.5 mg Chl/ml, 5 m$M$ NH$_2$OH, and 1 m$M$ EDTA. Inhibition of ferricyanide reduction is assayed as described in Section II,B, but the reaction mixture also contains recrystallized PD (0.25 m$M$), purified as described in Section II,A, and the Chl concentration (per ml) used for assay is 50–75 μg instead of 30 μg. Incubation of KCN/Hg-inhibited chloroplasts under these conditions for 20–25 min is sufficient to destroy more than 95% of the oxygen evolution activity supported by oxidized PD. The NH₂OH/EDTA treatment is terminated by centrifuging the chloroplast suspension for 10 min at 10,000 g. The resulting pellets are resuspended in homogenizing medium (Section II,A) and centrifuged again for 10 min at 10,000 g.

## D. Storage of KCN/Hg- and KCN/Hg/NH₂OH-Inhibited Chloroplasts at −70°

The inhibited chloroplasts collected by centrifugation at the end of the inhibition treatments described in Section II,B and C can be resuspended in the original homogenizing medium (Section II,A) and used immediately. Alternatively, these chloroplast suspensions (1.5–2 mg Chl/ml) can be dispensed into vials (0.5 ml aliquots) and frozen for later use. There appear to be no special conditions for freezing the chloroplasts; the stoppered vials are placed in a convenient container, such as a glass beaker or ice cream carton and placed directly in a −70° freezer. This storage procedure stabilizes electron transport and photophosphorylation activities for periods of several months. The electron-transport properties of freshly prepared and −70° preserved KCN/Hg- and KCN/Hg/NH₂OH-inhibited chloroplasts are shown in Table I. The activities

TABLE I

ELECTRON-TRANSPORT PROPERTIES OF KCN/HG- AND KCN/HG/NH₂OH-INHIBITED CHLOROPLASTS

| Treatment | $H_2O \rightarrow Fe(CN)_6^{3-}$ | Activity[a] $H_2O \rightarrow PD_{ox}$ | $Asc/DAD \rightarrow MV$[b] |
|---|---|---|---|
| None | | | |
| a. Freshly prepared | 600–800 | 1000–1200 | 1400–1600 |
| b. Stored at −70°, 6 weeks | 600–800 | 1000–1200 | 1400–1600 |
| KCN plus Hg | | | |
| a. Freshly prepared | 40–60 | 700–800 | 80–150 |
| b. Stored at −70°, 6 weeks | 40–60 | 700–800 | 80–150 |
| KCN/Hg plus NH₂OH/EDTA | | | |
| a. Freshly prepared | 0 | 0 | 80–150 |
| b. Stored at −70°, 6 weeks | 0 | 0 | 80–150 |

[a] Expressed as μEq e⁻ transferred/hr mg Chl.
[b] Corrected for superoxide ion formation.

presented are representative of the results obtained from a number of experiments in the author's laboratory.

## III. Assay of Photophosphorylation Associated with Photosystem II

### A. Recrystallization of Electron Transport Catalysts

Lipophilic catalysts of photosystem II electron transport may be obtained from the following sources: Eastman Organics, Eastman Kodak Co., Rochester, New York: [p-phenylenediamine (PD), N,N,N',N',-tetramethyl-p-phenylenediamine (TMPD), and 2,5-dimethyl-p-benzoquinone (DMQ)]; Research Organics/Inorganics, 507-519 Main St., Belleville, New Jersey: [2,3,5,6-tetramethyl-p-phenylenediamine (DAD)]; and Aldrich Chemical Co., Milwaukee, Wisconsin: [2,5-diaminotoluene (DAT)]. The material obtained from these suppliers contains impurities which must be removed before the catalysts are suitable for use in assays of electron transport activity. Purification of PD, DAD, and DAT is accomplished by dissolving 3 gm of the crude material in 50 ml of 1 $N$ HCl at 25°. One gram of activated charcoal is added to the stirred solution, the suspension is filtered, and the filtrate is cooled to 0°-4° in an ice bucket. Crystallization is initiated by addition of 35 ml of chilled concentrated HCl to the cooled solution. The acid solution is allowed to stand for 1-2 hr, and then filtered to remove the white crystals, which are dried on a Büchner funnel and stored at −20°. Purification of TMPD is accomplished by dissolving 10 gm of the crude material in 50 ml of hot (80°) absolute ethanol to which 1 ml of concentrated HCl has been added. The solution is maintained at 80° while 1 gm of activated charcoal is added with stirring, and the suspension is quickly filtered. The light-blue color in the filtrate is removed by addition of a small grain of dithionite, and the colorless solution is then transferred to a −20° freezer (2-3 hr) for cooling to promote crystallization. The colorless crystals are collected, dried on a Büchner funnel and stored at −20°. DMQ (3-5 gm) is dissolved in 30-50 ml of 95% ethanol, treated with charcoal as described above, and recrystallized by addition of 3-4 volumes of cold (0°-4°) water to the ethanolic solution. The bright yellow crystals are collected and dried on a Büchner funnel and stored at −20°.

### B. Assay of Photosystem II Noncyclic Photophosphorylation

Analysis of photosystem II noncyclic electron transport and associated ATP synthesis requires measurement of electron transport catalyzed by KCN/Hg-inhibited chloroplasts under phosphorylating conditions. This may be accomplished either by measurement of oxygen evolution (with an oxygen electrode) or by following the photoreduction of the terminal oxidant, ferricyanide (see Vol. 24, p. 159). Conditions are given below for measurement of electron flow as oxygen evolution. The reaction mixture, in a stirred, thermostatted (25°) 1.6-ml cuvette fitted with an oxygen electrode, contains 50 m$M$ Tricine (pH 8), 50 m$M$ NaCl, 3 m$M$ MgCl$_2$, 1 m$M$ ADP, 5 m$M$ NaH$_2{}^{32}$PO$_4$($10^6$ cpm/ml), and KCN/Hg-inhibited chloroplasts (25–40 $\mu$g Chl), which have been thawed in an ice bucket just prior to assay. The complete reaction mixture also contains ferricyanide (2.5 m$M$) and a lipophilic catalyst (0.25 m$M$). The latter is added to the reaction mixture from a 20-m$M$ stock solution in water just prior to illumination with heat-filtered white light ($10^6$ ergs/cm$^2$/sec). Electron transport activity is monitored by observing the evolution of oxygen on a chart recorder during a 1-min illumination period. At the end of the illumination period, the reaction mixture is acidified with 0.2 ml of 30% trichloroacetic acid and transferred to an ice bucket. At the end of the series of experiments, unreacted $^{32}$P is extracted from the centrifuged (2000 $g$, 5 min) reaction mixtures by Avron's procedure,[12] and AT$^{32}$P synthesis is determined by gas-flow counting of dried aliquots of the extracted material or by Cerenkov counting[13] of liquid samples. The ratio of ATP synthesis, determined as above, to the rate of electron transport may then be used to calculate the efficiency (P/$e_2$) of photosystem II noncyclic photophosphorylation.

Care should be taken in conducting these assays to avoid prolonged exposure of lipophilic acceptors to ferricyanide prior to illumination. Failure to observe this precaution may result in the formation of secondary oxidation products of $p$-phenylenediimines in the reaction mixture. Fresh solutions of $p$-phenylenediamines should be prepared prior to each series of assays, since these compounds may degrade upon aging in solution. KCN/Hg-inhibited chloroplasts may be used for up to 1 hr after thawing without observable decreases in electron transport of photophosporylation activities, provided that the chloroplasts are kept at 0°–4° in an ice bucket.

Table II summarizes the properties of photosystem II electron transport and photophosphorylation catalyzed by KCN/Hg-inhibited chloroplasts in the presence of several different lipophilic acceptors. The electron transport and photophosphorylation rates obtained depend on the acceptor used, with PD and DAT producing the highest activity, and DMQ and TMPD producing considerably lower activities. Table II shows that the electron transport activity catalyzed by $p$-phenylenediimines in

[12] M. Avron, *Biochim. Biophys. Acta* **40**, 257 (1960).
[13] J. M. Gould, R. Cather, and G. D. Winget, *Anal. Biochem.* **50**, 540 (1972).

TABLE II
PROPERTIES OF PHOTOSYSTEM II NONCYCLIC ELECTRON TRANSPORT AND ATP
SYNTHESIS CATALYZED BY KCN/HG-INHIBITED CHLOROPLASTS

| Electron acceptor[a] system | Electron transport[b] | ATP synthesis[c] | $P/e_2$ | Inhibition of electron transport by uncouplers (%) |
|---|---|---|---|---|
| PD + ferricyanide | 920 | 164 | 0.36 | 60 |
| DAD + ferricyanide | 829 | 90 | 0.22 | >90 |
| DAT + ferricyanide | 1082 | 145 | 0.27 | 70 |
| TMPD + ferricyanide | 448 | 65 | 0.29 | 40 |
| DMQ + ferricyanide | 488 | 55 | 0.23 | 0 |

[a] The lipophilic acceptor concentration was 250 $\mu M$; the ferricyanide concentration was 2.5 m$M$.

[b] Expressed as $\mu$Eq. $e^-$ transferred/hr mg Chl.

[c] Expressed as $\mu$moles ATP synthesized/hr mg Chl.

KCN/Hg-inhibited chloroplasts is sensitive to inhibition by uncouplers (methylamine or gramicidin). This phenomenon has been observed by a number of investigators,[4,14-16] and although there is a disagreement about its cause, one should be aware that factors affecting chloroplast energy conservation may also affect electron transport supported by $p$-phenylenediimines. This problem may be avoided by the use of DMQ as an acceptor; although this compound catalyzes lower rates of electron transport than the $p$-phenylenediimines, the $P/e_2$ obtained with DMQ is similar to that observed with other catalysts, and the electron transport activity of DMQ is unaffected by uncouplers.

## C. Assay of Photosystem II-Catalyzed Cylic Photophosphorylation

The assay medium (1.6 ml) contains 50 m$M$ Tricine (pH 8), 50 m$M$ NaCl, 3 m$M$ MgCl$_2$, 1 m$M$ ADP, 5 m$M$ NaH$_2$ $^{32}$PO$_4$, and KCN/Hg/NH$_2$OH-inhibited chloroplasts (20–30 $\mu$g Chl) which have been thawed in an ice bucket just prior to assay. The artificial catalyst (PD or TMPD) is added from a freshly prepared 20 m$M$ solution to water to a final concentration of 250 $\mu M$, and the complete reaction mixture is mixed

[14] J. A. Guikema and C. F. Yocum, Biochemistry 15, 362 (1976).
[15] W. S. Cohen, D. E. Cohn, and W. Bertsch, FEBS Lett. 49, 350 (1975).
[16] A. Trebst, S. Reimer, and F. Dallacker, Plant Sci. Lett. 6, 21 (1976).

thoroughly and incubated in a circulating 25° water bath for 10 min in the dark. At the end of the incubation period, the reaction mixtures are illuminated with heat-filtered white light ($10^6$ ergs/cm²/sec) for 1 min. After illumination, the reaction mixtures are immediately acidified with 0.2 ml of 30% trichloroacetic acid, denatured material is removed by centrifugation (2000 $g$, 5 min), and unreacted $^{32}P$ is extracted by Avron's procedure.[12] AT$^{32}P$ synthesis is determined as described in Section III,B.

The purpose of the incubation step described above is to permit oxidation of the artificial catalyst to occur. Omission of this incubation period produces rates of photosystem II cyclic photophosphorylation approximately half those obtained after incubation. The catalyst concentration used in these experiments (250 $\mu M$) is sufficient to produce saturated rates of ATP synthesis in excess of 100 $\mu$moles/hr mg Chl.

Table III presents a summary of the properties of photosystem II cyclic photophosphorylation reactions obtained in KCN/Hg/NH$_2$OH-inhibited chloroplasts with PD or TMPD as the artificial catalyst of activity. Note that cyclic photophosphorylation by these chloroplast preparations is not dependent upon the ability of the artificial catalyst to liberate protons upon oxidation. The photosystem II cycle is strongly inhibited by DCMU. In addition, both DBMIB and antimycin A inhibit ATP syn-

TABLE III
PROPERTIES OF PHOTOSYSTEM II-CATALYZED CYCLIC PHOTOPHOSPHORYLATION

| Property | Catalyst | |
| --- | --- | --- |
| | PD | TMPD |
| Catalyst liberates H$^+$ upon oxidation | Yes | No |
| Catalyst concentration ($\mu M$) for optimum ATP synthesis | 250 | 250 |
| Maximum rate of ATP synthesis ($\mu$moles ATP/hr mg Chl) | 95–110 | 95–120 |
| pH optimum of ATP synthesis | 8 | 8 |
| Light intensity for optimum ATP synthesis (ergs/cm²/sec) | $10^6$ | $8 \times 10^5$ |
| Extent of proton uptake (nmoles/mg Chl) | 600 | 840 |
| Inhibition (%) of ATP synthesis by | | |
|   DCMU (6 $\mu M$) | 100 | 100 |
|   DBMIB (12.5 $\mu M$) | 60 | 90 |
|   Antimycin A (18 $\mu M$) | 70 | 95 |
|   Valinomycin (1.25 $\mu M$) | 65 | 63 |
|   DNP (250 $\mu M$) | 67 | 73 |
|   Uncouplers (saturating concentrations) | 100 | 100 |
|   Energy transfer inhibitors (saturating concentrations) | 100 | 100 |

thesis, but the extents of inhibition obtained with these compounds is dependent upon the artificial catalyst used to elicit ATP synthesis; TMPD-catalyzed activity is more sensitive to these inhibitors than is the activity supported by PD.[17]

A variety of uncouplers (methylamine, gramicidin, atebrin, CCCP) and energy transfer inhibitors (DCCD, phlorizen, triphenyltin chloride) inhibit photosystem II cyclic activity. The extents of these inhibitions of activity are independent of whether PD or TMPD are present as catalysts of ATP synthesis. Valinomycin (1.25 $\mu M$) and DNP (250 $\mu M$) have also been demonstrated to inhibit photosystem II cyclic activity.[18] The nature of these inhibitions is ascribed to interference with electron transport rather than uncoupling, since the inhibition by valinomycin does not require the presence of KCl, and neither compound by itself inhibits photosystem II catalyzed proton uptake. As found in other systems,[19] the combination of valinomycin, DNP, and KCl will uncouple ATP synthesis and proton uptake by the photosystem II cycle, but the uncoupling activity of this combination is complicated owing to the additional inhibitions of electron transport by both valinomycin and DNP.

Compounds other than PD and TMPD may be used to catalyze photosystem II cyclic photophosphorylation. Among the *p*-phenylenediamines tested thus far, 250 $\mu M$ DAD or DAT will produce rates of ATP synthesis equivalent to those shown in Table III. The activity catalyzed by these compounds is only 95–98% inhibited by DCMU, however, indicating that a small amount of photosystem I electron transport may be contributing to phosphorylation. Ferrocyanide is also a catalyst of photosystem II cyclic ATP synthesis; low rates (20–25 $\mu$moles ATP/hr mg Chl) are obtained, and much higher catalyst concentrations (600–650 $\mu M$) are required to saturate ATP synthesis.

[17] C. F. Yocum, *Plant Physiol.* **60**, 592 (1977).
[18] C. F. Yocum, *Plant Physiol.* **60**, 597 (1977).
[19] S. J. D. Karlish, N. Shavit, and M. Avron, *Eur. J. Biochem.* **9**, 291 (1969).

# [56] Techniques for Studying Ionic Regulations of Chloroplasts

*By* J. Barber and H. Y. Nakatani

## Introduction

There are two special features of higher plant chloroplasts which must be recognized before carrying out a detailed study to understand their ionic relations. First, they are two compartment systems, having a stromal and an intrathylakoid space. Second, they are peculiar in that they contain an enormous amount of internal surface due to the thylakoid membrane system.[1] The thylakoid membranes are negatively charged at physiological pH as indicated by particle electrophoresis studies,[2] and the charge density has been estimated to be about 2.5 $\mu$coulombs/cm².[3-5] The existence of this extensive negatively charged surface has important implications in terms of the way ions distribute themselves in this organelle.[5] Cations are attracted into a diffuse layer adjacent to the membrane, and, in general, there will be a preference for multicharged species over monovalent. Since the thylakoids can in a sense be considered as an ion exchange system, the nature of the medium they are suspended in will greatly influence the cations associated with them. It was the failure to realize this in much of the earlier work reported in the literature which has led to confusion and conflicting observations. Of course, in the intact chloroplast the thylakoids are bathed in the aqueous stromal phase, and the nature of this solution will govern the cations attracted to the fixed negative charges of the membrane. The outer chloroplast membranes, which forms an envelope around the intrathylakoid and stromal phases, represents an additional barrier to inorganic ions and in this sense is also an important ion transport site. Chloroplasts contain a large range of organic ions and molecules as well as many inorganic elements firmly bound to protein and membrane constituents. This article, however,

[1] J. Barber, *in* "Bioenergetics of Membranes" (L. Packer, G. C. Papageorgiou, and A. Trebst, eds.), p. 459. Elsevier, Amsterdam, 1977.

[2] H. Y. Nakatani, J. Barber, and J. A. Forrester, *Biochim. Biophys. Acta* **504**, 215 (1978).

[3] J. Barber and J. Mills, *FEBS Lett.* **68**, 288 (1976).

[4] J. Barber, J. Mills, and A. Love, *FEBS Lett.* **74**, 174 (1977).

[5] J. Barber, *Proc. Colloq. Int. Potash Inst.* **13**, 67 (1977).

focuses attention on measurement of inorganic species which are likely to be the major thermodynamically active inorganic ions, in particular $H^+$, $K^+$, $Na^+$, $Mg^{2+}$, $Ca^{2+}$, and $Cl^-$.

### Chloroplast Isolation and Characterization

There are many methods of isolating chloroplasts, and most yield envelope free preparations. As explained above, the ionic properties of these isolated thylakoid systems will be governed by the preparation procedure and the nature of the medium in which they are finally suspended. For this reason, any ion studies with isolated thylakoids should be carried out in well-defined media. In any case, such studies may have little value if they cannot be related with ionic control mechanisms which take place in the intact organelle.[6] For this reason, it is recommended that whenever possible studies should begin with chloroplasts isolated such that their outer envelope is intact. The intactness of the isolated organelle (i.e., class A chloroplasts defined by Hall[7]) can be tested using the "ferricyanide test.[8,9]" Potassium ferricyanide cannot penetrate the outer chloroplast membrane, and the degree of intactness of the preparation can be determined by comparing the ferricyanide-dependent $O_2$ evolution before and after subjecting the preparation to osmotic shock (see Fig. 1). A method of isolating intact chloroplasts has already been described by Walker in this series,[10] and more recently an improvement on this technique was reported by Nakatani and Barber.[11] In this way, preparations can be obtained which can be suspended in a cation-free medium and still maintain their main biochemical functioning. Such preparations are, therefore, free of ionic contamination resulting from the isolation procedure and presumably contain their physiological complement of ions.

### Flame Emission Photometry and Atomic Absorption Spectrometry

Flame photometry is ideally suited to the analysis of $K^+$ and $Na^+$ because of their low excitation energies, excellent detection limits and precisions are readily obtained. While $Ca^{2+}$ and $Mg^{2+}$ can be analyzed by

[6] J. Barber, in "The Intact Chloroplast" (J. Barber, ed.), Vol. 1 of Topics in Photosynthesis, p. 89. Elsevier, Amsterdam, 1977.

[7] D. O. Hall, Nature (London) 235, 125 (1972).

[8] U. Heber and K. A. Santarius, Z. Naturforsch., Teil B 25, 718 (1970).

[9] W. Cockburn, C. W. Baldry, and D. A. Walker, Biochim. Biophys. Acta 143, 614 (1967).

[10] D. A. Walker, Vol. 23, p. 211.

[11] H. Y. Nakatani and J. Barber, Biochim. Biophys. Acta 461, 510 (1977).

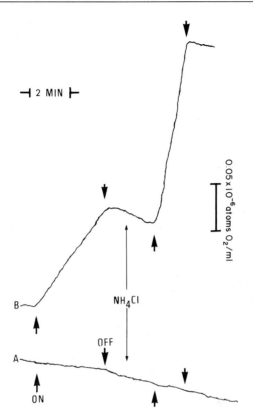

FIG. 1. Ferricyanide-dependent $O_2$ evolution by intact (A) and envelope-free (B) chloroplasts. During the dark period, 2.5 m$M$ $NH_4Cl$ added. Assay medium contained 0.33 $M$ sorbitol, 1 m$M$ $MgCl_2$, 1 m$M$ $MnCl_2$, 50 m$M$ HEPES-KOH, pH 7.6, 2 m$M$ EDTA, 1 m$M$ potassium ferricyanide, and 20 μg chl/ml.

flame emission superior performance is attained from atomic absorption techniques. For both of these procedures a digestion step prior to analysis is recommended; this is usually carried out by taking a known quantity of chloroplasts, specified in terms of total chlorophyll content and subjecting it to acid digestion. Digestion mixtures consisting of nitric and perchloric acids are commonly used, for example, Chow et al.[12] used 50% concentrated nitric and 50 or 70% perchloric acid. Slight warming, in a fume cupboard, aids digestion, and the resulting clear solutions can be diluted to the appropriate concentration range for the optimal sensitivity of the instrument.

[12] W. S. Chow, G. Wagner, and A. B. Hope, *Aust. J. Plant Physiol.* **3**, 853 (1976).

Until relatively recently, atomic absorption has relied on the use of a flame for the atomization (i.e., formation of neutral atoms) process, but during the last few years another option has become available. The flame can, if required, be replaced by a resistively heated atomizer, such as the graphite furnace[13] or the carbon rod.[14] These devices use very small discrete samples (5–100 $\mu$l of solution or 1–2 mg of solid sample), and because of their absence of flame noise, their relatively long residence times for the analyte atoms and their rapid ability to pulse the sample into the optical path, they achieve excellent improvements in detection limits. Absolute detection limits of $10^{-10}$ to $10^{-13}$ gm are achieved for many elements. In addition to these two obvious advantages they can, because of the long residence time and the absence of a nebulization requirement, handle solid samples and suspensions. However, as with any new advance some disadvantages are also present. To protect these carbon devices from becoming oxidized during the high temperature portion of the analytical cycle, they are operated in an inert atmosphere, such as argon. While this protection prevents oxidation of the elements, it also prevents oxidation of any organic material in the sample being analyzed, which often results in the evolution of a smokey carbon vapor which will act as a nonatomic absorber and give an apparent analytical signal. Simultaneous background correction will overcome this problem, and it is recommended that no analysis be carried out by these furnace techniques without background correction. In addition to this disadvantage, the technique is considerably slower than flame atomic absorption (3 min per sample versus 15 sec), and the precision is somewhat lessened. However, one of the major manufacturers of atomic absorption equipment has recently started to sell an autosampler for the resistively heated atomizer which improves the precision of the analytical measurements to such an extent that it equals those obtained by flame techniques. Although the analytical cycle time is not reduced, automation decreases the amount of technical time required during the analysis.

*Radiotracer Techniques*

Incubation of chloroplasts, especially isolated thylakoids, in solutions containing known specific activity of a particular radioisotope can be used to estimate ionic levels.[15] As a guide Chow et al.[12] have used the following activities (in $\mu$Ci/ml): $^{36}$Cl, 0.7; $^{22}$Na, 0.5; $^{86}$Rb, 1.0 ($^{86}$Rb is

[13] G. Buttgereit, in "Methodicum Chimicum" (F. Korte, ed.), Vol. 1, p. 710. Academic Press, New York, 1974.
[14] G. F. Kirkbright, *Analyst* **96**, 609 (1971).
[15] B. A. Winocur, R. I. Macey, and A. B. Tolberg, *Biochim. Biophys. Acta* **150**, 32 (1968).

used for K$^+$ determinations, $^{42}$K can be used, but has a half-life of only 12.5 hr), with a chlorophyll level of about 100 $\mu$g/ml. After incubating isolated thylakoids with the isotope for 5-10 min. they separated the thylakoids from the radioactive solution by using the silicone oil layer technique (see Heldt[16]). A suitable oil mixture consists of 4 parts Dow Corning MS 550 and 3 parts MS 510. The Beckman 152 microfuge was found to be ideal for carrying out the rapid separation. Separation procedures using Millipore membrane filters have been used,[17] but in our laboratory we have not found this approach to be satisfactory.[18] Calcium-45, a pure $\beta$ emitter, can also be used. Gimmler et al.[19] and Pflüger[20] have used $^{28}$Mg for studies with chloroplasts, however, it has a half-life of only 21.3 hr and is not readily available.

Short-term treatments with radioisotopes to determine ionic levels may not be suitable with intact class A chloroplasts, since it is unlikely that complete isotopic exchange will occur across the envelope membranes. In this case, it is probably necessary to grow plants hydroponically in the particular radioisotope and to isolate the labeled intact chloroplasts. This also has the drawback of assuming that the specific activity in the tissue is identical to that in the external medium.

*Neutron Activation*

In our opinion, this is a very valuable approach for determining the levels of K, Na, Mg, Ca, and Cl in chloroplasts. This technique allows the simultaneous determination of all these elements without the need for liquid digestion. The sample is placed in a thermal neutron beam, and the subsequent analyses of the radioactive nuclides formed by $\gamma$-ray spectrometry permits the identification and quantitative determination of the elements present. Consider an element X of atomic mass A and atomic number Z, then

$$_{Z}^{A}X + {_0^1}n \rightarrow ({_Z^{A+1}}X)^* \rightarrow \text{radioactive decay by particle or electromagnetic emission}$$

Neutron activation, therefore, occurs by the transformation of an isotope to an excited state with an increased mass (absorption of a neutron) which may then deexcite by various pathways, involving $\alpha$ and

[16] H. W. Heldt, in "The Intact Chloroplast" (J. Barber, ed.), Vol. 1, p. 215. Elsevier, Amsterdam, 1976.

[17] R. A. Dilley, Vol. 14, p. 68.

[18] W. J. Varley, Ph.D. Thesis, University of London (1972).

[19] H. Gimmler, G. Schafer, and U. Heber, in "Photosynthesis III" (M. Avron, ed.), p. 1381. Elsevier, Amsterdam 1975.

[20] R. Pflüger, Z. Naturforsch., Teil C 28, 779 (1973).

$\beta$ particles and $\gamma$-ray emissions. With thermal neutrons (average kinetic energy at 20° of 0.038 eV) the most common reaction is the $(n,\gamma)$ reaction involving the emission of gamma rays usually having energies of several MeV. Analytical determinations are thus made possible by high resolution $\gamma$-ray spectrometry. This technique is particularly useful for biological material since the matrix elements (C, H, O, and N) are relatively insensitive. For the elements mentioned above, short irradiation times can be used, for example, we have irradiated chloroplast samples (200–400 $\mu$g chl) for 8–15 min with a thermal neutron flux of ~1.5 × 10$^{12}$ neutrons/sec cm$^2$ and reactor power of 100 kW. The optimal procedure for irradiation times and subsequent $\gamma$-ray analysis is dependent on several factors, including the relative abundance of the parent nuclide, its thermal neutron capture cross section, its half-life, and $\gamma$-ray energy. The physical properties of the elements of interest are listed in Table I. Other elements such as Mn, V, Cu, Ti, and Al can also be determined with relatively short irradiation times. Longer irradiation times greatly extends the list of detectable elements. A useful description of this technique is de Soete et al.[21]

We have used this method to study the ionic levels in isolated intact chloroplasts and the ions associated with isolated thylakoid membranes.[5,22] Samples of chloroplast material are placed in small polythene capsules which had been thoroughly washed in 10% $HNO_3$. The samples are dried in an oven at 60°. To avoid possible contamination to the reactor core itself, the capped sample capsules are placed into larger polythene tubes and sealed. The sample capsule is positioned in the bottom of the larger capsule with folded filter paper. Table II shows the level of ions in class A pea chloroplasts, while Fig. 2 demonstrates the use of neutron activation in determining the relationship between mono- and divalent cations associated with the thylakoid membranes.

The thylakoids carry a net negative charge (see Barber[1] and Nakatani et al.[2]) and the nature of the co-ion balancing this charge in the diffuse layer adjacent to the membrane is determined by the ionic composition of the suspending medium.[3,4,23] Note in Fig. 2 the preference for divalents over monovalents, an observation which can be explained by the Gouy–Chapman theory for the electrical double layers (see Barber et al.[2−4]) and emphasized in the calculated theoretical curves shown in Fig. 3.

[21] D. de Soete, R. Gijabel, and J. Hoste, "Neutron Activation Analysis." Wiley (Interscience), New York, 1972.

[22] H. Y. Nakatani, J. Barber, and Minski, M. J., *Biochim. Boiphys. Acta* **545**, 24 (1979).

[23] J. Mills and J. Barber, *Biophys. J.* **21**, 257 (1978).

TABLE I
PROPERTIES OF APPROPRIATE NUCLIDES FOR NEUTRON ACTIVATION ANALYSIS

| | Parent nuclide | | | Activated nuclide | |
|---|---|---|---|---|---|
| Isotope | Relative Abundance (%) | Thermal neutron capture cross-section $(10^{-24}cm^2)$ | Isotope | $\gamma$-Ray energy (keV) | Half-life (min) |
| $^{23}_{11}Na$ | 100 | 0.53 | $^{24}_{11}Na$ | 1368,2754 | 900 |
| $^{26}_{12}Mg$ | 11.3 | 0.27 | $^{27}_{12}Mg$ | 1014 | 9.45 |
| $^{37}_{17}Cl$ | 24.6 | 0.56 | $^{38}_{17}Cl$ | 1642,2168 | 37.29 |
| $^{41}_{19}K$ | 6.9 | 1.1 | $^{42}_{19}K$ | 1525 | 751.2 |
| $^{48}_{20}Ca$ | 0.18 | 1.1 | $^{49}_{20}Ca$ | 3083 | 8.8 |
| $^{55}_{25}Mn$ | 100 | 13.3 | $^{56}_{25}Mn$ | 1811,2113 | 154.8 |
| $^{65}_{29}Cu$ | 31.0 | 2.2 | $^{66}_{29}Cu$ | 1039 | 5.1 |

*Plasma Emission Spectroscopy*

This new and powerful technique, like neutron activation, can be used for multielemental analyses on a single sample, although as yet we know of no reports of its use with chloroplasts.

Greenfield and co-workers[24] in the early 1960's established a new approach to the production and use of plasmas for atomic spectroscopy. They demonstrated that a unique plasma could be formed when argon was blown through the center of a quartz tube which was placed in the center of a load coil through which a radio frequency current was passing. Under this condition large amounts of radio frequency power can be inductively coupled into the argon, whereupon a high temperature (7000°–10,000°K) plasma is formed. Generally the energy used is in the 3.5 to 4.8 MHz frequency range at power levels of 1–7 kW. The key features of this inductively coupled plasma (ICP) technique are the absence of electrodes and the injection of the sample as an aerosol into a very narrow channel at the centre of the plasma. These features result in all of the sample entering the plasma experiencing an extremely uniformly high temperature which gives intense atomic and ionic emission of the analyte over an extended linear dynamic concentration range (typically from 1–10 ng/ml to 1000–10,000 $\mu$g/ml of the analyte atoms). Further, the high temperature results in a more complete fragmentation of molecular

[24] S. Greenfield, L. L. Jones, and C. T. Berry, *Analyst* **89**, 713, (1964).

TABLE II
IONIC CONTENT OF INTACT PEA CHLOROPLASTS AS DETERMINED BY NEUTRON
ACTIVATION[a]

| Element | $\mu$moles/mg Chl ($\pm$ S.E.M. of 17 sets) |
|---------|---------------------------------------------|
| Na | 0.22 $\pm$ 0.02 |
| K | 2.56 $\pm$ 0.06 |
| Mg | 1.72 $\pm$ 0.04 |
| Ca | 0.76 $\pm$ 0.03 |
| Cl | 0.21 $\pm$ 0.01 |
| Mn | 0.017 $\pm$ 0.0003 |

[a] Mean intactness, 91%.

species, which coupled with the massive excess concentrations of free electrons greatly reduces the chemical and ionization interferences found in flame spectroscopy. Most elements, including those of interest to this article, are capable of being detected at concentration levels of less than 1$\mu$g/ml. The analytical process that is quantified electrooptically is that of emission and thus when used in conjunction with a polychromator

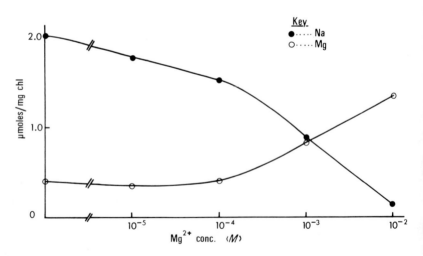

FIG. 2. Neutron activation analyses of ions associated with thylakoid membranes isolated from peas. The membranes were washed with 100 m$M$ NaCl before being treated with various levels of MgCl$_2$ with constant background NaCl level of 10 m$M$. The results demonstrate the differential of ability of Mg$^{2+}$ to displace Na$^+$ initially associated with the membranes.

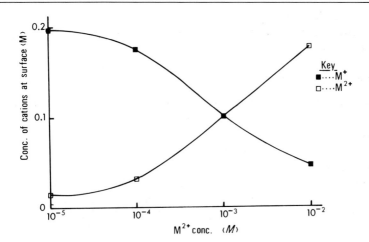

FIG. 3. Calculated curves of the ability of divalent cations to displace monovalent cations from surface of a membrane having a fixed negative charge density ($\sigma$) of 2.5 $\mu$coulombs/cm$^2$. The surface potential $\psi_0$ has been calculated from the following expression for mixed electrolytes derived from the Gouy–Chapman theory (see Barber *et al.*[3,4]).

$$2C_\alpha'' \cosh^2\left(\frac{F\psi_0}{RT}\right) + C_\alpha' \cosh\left(\frac{F\psi_0}{RT}\right) - \left(2C_\alpha'' + C_\alpha' + \frac{\sigma^2}{2A^2}\right) = 0$$

where $C_\alpha'$ and $C_\alpha''$ are the concentrations of the mono- and divalent salts respectively, $F$ is the Faraday, $R$ is the gas constant, $T$ is the absolute temperature, and $A = (RT\epsilon/2\pi)^{1/2}$ (where $\epsilon$ is the permittivity of water). The concentrations at the surface ($C_\alpha'$ and $C_\alpha''$) for various mixtures has been calculated ($C_\alpha'$ constant at 10 m$M$) using the Boltzmann equation:

$$C_0 = C_\alpha \exp\left(-ZF\psi_0/RT\right)$$

where $Z$ is the charge on the cation.

simultaneous multielement analysis is achieved. Precisions of less than 1% (relative S.D.) are readily attained.

Since the appearance of the first commercial multielemental system in 1974 many different applications of the ICP have been investigated. Silvester and co-workers[25,26] have been one of the most active groups in applying this technique to the analysis of organic samples.

Like flame atomic absorption or emission spectroscopy, the technique normally requires the sample to be in solution, and thus digestion pro-

[25] J. A. C. Fortesque, M. D. Silvester, and F. A. Abercrombie, *Trace Subst. Environ. Health—9, Proc. Univ. Mo. Annu. Conf.*, (1975).
[26] M. D. Silvester, F. A. Abercrombie, and R. B. Cruz, *174th Natl. Meet., Am. Chem. Soc.* Paper 9 (1977).

cedures are needed. However, recent work by Silvester and co-workers[27] has demonstrated that a laser can be used to vaporize solid samples prior to injection into the central core of the ICP.

## Electrometric Titrations

Chloride level in chloroplasts could be determined by an electrometric technique involving titration of extracts with $AgNO_3$. This approach is often applied by animal and plant physiologists (see, for example, Ramsey et al.[28]). However, care in preparing the extract is required since acid digestion could tend to convert chloride to chlorine gas. Hind et al.[29] adapted this technique for analyzing chloride levels in chloroplasts by using the Aminco–Cotlove chloride titrator. The method has detection limits of $10^{-7}$ gm.

## Ion Selective Electrodes

A whole range of electrodes are now commercially available (having both solid and liquid membranes) which could be used for determining activities of elements present in extracts. However they are perhaps more useful for monitoring ionic fluxes across chloroplast membranes (see pp. 595–596).

## Electron Probe X-Ray Microanalysis

This is an attractive technique which has as yet not been extensively used on chloroplasts. A brief report of its use to locate bound Mn and Fe in chloroplasts has been given.[30] The principle of the technique is to detect characteristic X-ray emissions brought about by irradiating a particular element with high-energy electrons. This treatment displaces electrons from the $K$, $L$, or $M$ shells and characteristic X rays are produced when electron replacement occurs during relaxation. With elements of atomic numbers less than 56, X-ray emission due to electron replacement of vacancies in $K$ shells are monitored. Micro areas analyzed can be as low as 100–200 Å with thin sections depending on the type of X-ray spectrometer employed, and, therefore, it should be possible to obtain information about levels and location of elements within chloroplasts.

[27] M. D. Silvester, F. A. Abercrombie, and G. Stoute, *Pittsburg Anal. Conf.*, *19* p. (1977).
[28] J. A. Ramsey, R. H. J. Brown, and P. C. Croghan, *J. Exp. Biol.* **32**, 822 (1955).
[29] G. Hind, H. Y. Nakatani, and S. Izawa, *Biochim. Biophys. Acta* **172**, 274 (1969).
[30] A. Janossy, L. Mustardy, and A. Faludi-Daniel, *Proc. Int. Biophy. Congr.*, *5th*, *19* Abstract, p. 64 (1975).

The drawback with the technique is sample preparation, since sectioning of material can give rise to disturbances in ion distribution. For further information see Reed[31] and Hall et al.[32]

## Ion Chromatography (IC)

A new analytical technique developed by Small et al.[33] combines the principles of the chromatographic separation of ions through ion exchange with conductimetric detection for the determination of trace quantities of ions. Ion chromatography can be applicable to the separation and detection of anions and cations ($< 1$ $\mu g/ml$), including the alkali and alkali earth metals and halides as well as organic anions. Thus, this method appears to have great potential for the routine analysis of chloroplast material. Anderson[34] recently reported on the application of IC to the ionic analysis of biological fluids of clinical interest: serum, urine, cerebrospinal, and tissue extracts.

## Ion Fluxes—Direct Methods

### Selective Ion Electrodes

To date, most reports of ion fluxes in chloroplasts have involved the use of selective electrode systems. The most comprehensive work on isolated thylakoids using these detectors was carried out by Hind et al.[35] They used a multielectrode set-up and simultaneously measured the light induced fluxes of several ions; $K^+$, $Na^+$, $Cl^-$, $Mg^{2+}$, and $H^+$. An excellent description of the electrode and electronic systems used and problems encountered in this type of work has been given by Hind and McCarty.[36] Some of the traces obtained by Hind et al.[35] are shown in Fig. 4. It is worth mentioning that the findings of these workers demonstrated the importance of the ionic composition of the suspending medium in determining which ions are available for exchange with the light-induced proton pump. These findings, also seen by Bulychev and Vredenberg,[37] are in line with the ionic analyses presented in Fig. 2, assuming that the $H^+$

[31] S. J. B. Reed, "Electron Probe Analysis," Cambridge Monograph on Physics. Cambridge Univ. Press, London and New York, 1975.

[32] T. A. Hall, P. Echlin, and R. Kaufmann, "Microprobe Analysis as Applied to Cells and Tissue." Academic Press, New York, 1974.

[33] H. Small, T. S. Stevens, and W. C. Bauman, Anal. Chem. 47, 1801 (1975).

[34] C. Anderson, Clin. Chem. 22, 1424 (1976).

[35] G. Hind, H. Y. Nakatani, and S. Izawa, Proc. Natl. Acad. Sci. U.S.A. 71, 1484 (1976).

[36] G. Hind and R. E. McCarty, Photophysiology 8, 113 (1973).

[37] A. A. Bulychev and W. J. Vredenberg, Biochim. Biophys. Acta 449, 48 (1976).

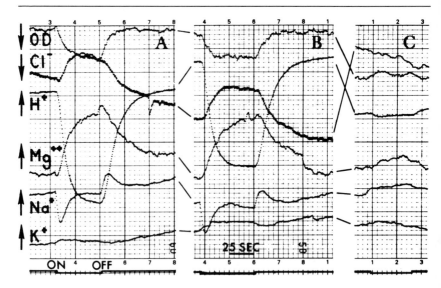

Fig. 4. Changes in the external ion activities of illuminated thylakoids using ion selective electrodes. Also shown are changes in optical density (O.D.) due to scattering changes. (A) and (B) represent control recordings, and (C) shows the effect of adding the uncoupler carbonyl cyanide *m*-chlorophenylhydrazone. From Hind *et al.*[35]

taken into the intrathylakoid space exchange with cations at the inner surface of the thylakoid membrane.[6]

Selective ion electrodes have also been used with suspensions of intact chloroplasts,[19,38] where it has been possible to monitor light-induced exchanges across the envelope membrane and cation effluxes induced on adding specific ionophores.

### Radiotracer Approach

In principle ion fluxes could be measured by adding a radioisotope of known specific activity at zero time and following its uptake by using the rapid oil centrifugation technique described on pp.    –    . Effluxes could be measured by suspending chloroplasts pretreated with radiotracers in a nonradioactive medium. Measurements of this type have not been reported with the exception of Winocur *et al.*[15] who have studied $Na^+$, $K^+$, and $Cl^-$ exchange with cold, dark treated spinach chloroplasts. The problem with this technique, when isolated thylakoids are used, is that the exchange processes occur rapidly relative to the slow centrifu-

[38] W. J. Vredenberg, *Proc. Int. Congr. Photobiol., 4th* Abstracts, p. 76 (1977).

gation times required for the separation procedures. We suggest, however, that the radiotracer technique may be useful for studying ionic
fluxes across the outer chloroplast membranes where the exchange is
much slower (see, for example, Pflüger[20]).

*Direct Chemical Analyses*

The existence of a light-induced net flux of $Mg^{2+}$ from the intrathylakoid to the stromal space of isolated intact chloroplasts has been demonstrated by Portis and Heldt.[39] Although these workers could not determine actual rates of movement, they were able to detect increase in
stromal $Mg^{2+}$ levels using atomic absorption measurements coupled with
the silicone oil centrifugation technique. They related the $Mg^{2+}$ levels
measured with stromal volume by monitoring the release of $^{32}P$-labeled
carbon compounds which reflected the degree of chloroplast breakage
due to the centrifugation procedures used. Another interesting approach
adopted by these workers was to use EDTA to determine the "free"
$Mg^{2+}$ levels associated with illuminated or dark treated thylakoids and
thus avoid the background correction required due to the $Mg^{2+}$ content
of chlorophyll.

*Use of Ion Selective Dyes*

Recently, Krause[40] reported the use the metallochromatic indicator
Eriochrome Blue S.E. for monitoring $Mg^{2+}$ fluxes across chloroplast
membranes. This dye undergoes an absorbancy change at 554 nm when
$Mg^{2+}$ binds to it and does not readily bind $Ca^{2+}$.[41] In agreement with the
direct measurements of Portis and Heldt,[39] Krause was able to use this
approach to detect a $Mg^{2+}$ flux between the intrathylakoid and stromal
compartments of isolated intact chloroplasts and estimate the light-induced rise in the stromal concentration of this element. To detect the
changes he had to use a sensitive double beam instrument (reference 592
nm) and observed that an increase of 1 $\mu M$ $Mg^{2+}$ gave an absorbancy
change at 554 nm of $1.4 \times 10^{-3}$.

Another metallochromatic indicator which has been used extensively
with a wide range of biological systems is murexide (see Scarpa[42]). It is
ideal for detecting $Ca^{2+}$ when an absorption increase occurs at 470 nm
and a decrease at 540 nm (isosbestic point at 507 nm). Although this

[39] A. R. Portis and H. W. Heldt, *Biochim. Biophys. Acta* **449**, 434 (1976).
[40] G. H. Krause, *Biochim. Biophys. Acta* **460**, 500 (1977).
[41] A. Scarpa, *Biochemistry* **13**, 2789 (1977).
[42] A. Scarpa, Vol. 24, p. 343.

indicator does not respond to $Mg^{2+}$, it does also complex $Sr^{2+}$, $Ba^{2+}$, and $Mn^{2+}$. As far as we know, there are no published reports of the use of murexide with chloroplasts.

An obvious word of warning when using indicators of this type with chloroplasts is to be aware of any interferring absorbance changes which may also occur due to the various pigments present. However, bearing this in mind, metallochromatic indicators have the distinct advantage of fast response time as well as being relatively sensitive and requiring only small volumes.

In addition to the above type of indicator, changes in $H^+$ levels in various photosynthetic tissue have been studied extensively using a variety of pH indicating dyes, e.g., bromcresol purple, bromphenol red, and bromthymol blue (see Chance and Scarpa[43]). With the reservations mentioned above, these indicators are useful for measuring kinetics while the extent of light-induced $H^+$ gradients across the thylakoid membrane has been estimated using the fluorescent compound 9-aminoacridine.[44] In the latter case, fluorescence is quenched due to the inward movement of the amine in response to the pH gradient developed due to electron transport.

## Ion Fluxes—Indirect Methods

*Osmotically Induced Volume Changes.*

A whole range of techniques are available for determining volume changes of chloroplasts and have been previously reviewed.[45,46] Swelling and shrinking studies of chloroplasts using different ions for changing the osmotic strength coupled with the use of specific ionophores can give valuable information about the permeability properties of chloroplast membranes. For example, Gimmler et al.[19] have followed volume changes of isolated intact chloroplasts by monitoring scattering changes recorded as changes in apparent absorption at 535 nm. They found that KCl and NaCl, like sorbitol, caused volume changes which followed the Boyle-van't Hoff relationship. Thus, it was concluded that the chloroplast envelope has a low permeability to these salts, although, of course, exchange processes could have occurred. When valinomycin was present, the intact chloroplasts no longer responded to KCl, indicating that normally the outer membranes are relatively impermeable to $K^+$ but

[43] B. Chance and A. Scarpa, Vol. 24, p. 336.
[44] S. Schuldiner, H. Rottenberg, and M. Avron, *Eur. J. Biochem.* **25**, 64 (1972).
[45] K. Shibata, Vol. 14, p. 171.
[46] L. Packer and S. Murakami, Vol. 14, p. 181.

permeable to Cl⁻. In this way, it should be possible to obtain information on relative permeabilities of the membrane to various ions, for example, Gimmler et al.[19] showed the outer membrane to be impermeable to the gluconate anion. Similar experiments have been carried out with isolated thylakoids to determine anion permeabilities and the reader is referred particularly to the work of Schuldiner and Avron.[47]

*Slow Chlorophyll Fluorescence Yield Changes*

Slow changes in the yield of chlorophyll fluorescence measured with isolated intact chloroplasts or with "reconstituted" broken chloroplasts are indicative of cation fluxes across the thylakoid membrane.[6] The nature of the cations involved can be determined by using specific ionophores which do or do not inhibit the fluorescence change. For example, Barber et al.[48] showed that with intact chloroplasts suspended in a cation-free medium, the fluorescence change was not inhibited by the K⁺ mediating ionophores valinomycin or nigericin, by the Na⁺ ionophore monensin, or by the Ca⁺ ionophore beauvaricin. Only the divalent specific ionophore A23187 was effective, indicating that the fluorescence changes reflected the movement of $Mg^{2+}$ between the intrathylakoid and stromal compartments.

*Salt Stimulation of Delayed Light*

The intensity of delayed light emission can be enhanced by generating an electrical gradient across the thylakoid membrane.[49] A convenient way to create such a gradient is to give a salt pulse to a preilluminated sample where the size of the electrical potential difference created is a function of the differential ionic permeability of the membrane as well as being dependent on the concentration gradient created. For a salt $C^+A^-$ the appropriate expression is

$$\Delta \psi = \frac{RT}{F} \ln \frac{P_C[C]_o + P_A[A]_i}{P_C[C]_i + P_A[A]_o} \tag{1}$$

or

$$\Delta \psi = \frac{RT}{F} \ln \frac{[C]_o + \alpha [A]_i}{[C]_i + \alpha [A]_o} \tag{2}$$

where $\Delta \psi$ is the electrical potential, $[C]_o$ and $[A]_i$ are outside and inside

[47] S. Schuldiner and M. Avron, *Eur. J. Biochem.* **19**, 227 (1971).
[48] J. Barber, J. Mills, and J. Nicolson, *FEBS Lett.* **49**, 106 (1974).
[49] J. Barber and G. P. B. Kraan, *Biochim. Biophys. Acta* **197**, 49 (1970).

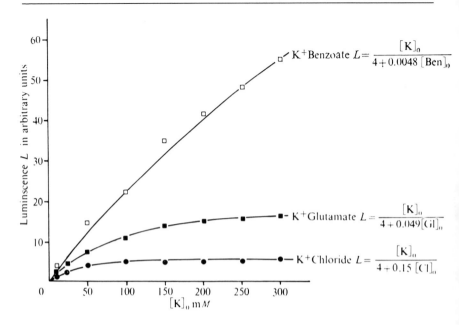

FIG. 5. Effect of adding various levels of potassium salts on the intensity of 10 sec delayed light emission from isolated spinach thylakoids. The curves have been drawn according to the equations shown [see Eq. (4) in text] and give estimated values of the relative permeability coefficients. (From Barber and Varley.[50])

concentrations, respectively, $P_C$ and $P_A$ are the permeability coefficients for $C^+$ and $A^-$, $\alpha = P_C/P_A$, $R$ is gas constant, $T$ is absolute temperature, and $F$ is the Faraday (see Barber and Varley.[50,51])

Since the action of the electrical gradient seems to be to decrease the activation energy giving rise to the emission process then the observed stimulated luminescence intensity ($L$) will be exponentially related to the electrical gradient. That is

$$L \propto \exp \Delta\psi \tag{3}$$

therefore

$$L \propto \frac{[C]_o + \alpha [A]_i}{[C]_i + \alpha [A]_o} \tag{4}$$

Thus, by changing the nature of salt, it is possible to obtain values of

[50] J. Barber and W. J. Varley, *J. Exp. Bot.* **23**, 216 (1972).
[51] J. Barber, *Biochim. Biophys. Acta* **275**, 105 (1972).

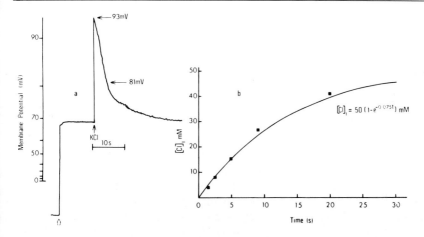

FIG. 6. (a) The kinetics of a 50 m$M$ KCl induced increase in the intensity of 1 msec delayed light emission measured with a phosphoroscope. (b) The influx of Cl⁻ estimated from the decay of the signal shown in (a). The closed squares are experimental points and the curve has been drawn according to the first-order law shown where $t$ = time in seconds and [Cl]$_i$ is the inside Cl⁻ concentration in mmoles/liter. (From Barber.[51])

the relative permeability coefficient by monitoring the intensity of the salt-induced emission $L$. Barber and Varley[50] adopted this approach by recording salt-stimulated emission after 10 sec of darkness, and some of their results using different K⁺ salts are shown in Fig. 5. Barber extended this approach by using a phosphoroscope to record continuously the intensity of 1 msec delayed light.[51] In this way, the salt-induced signals can be used not only to estimate relative permeabilities but also calculate absolute permeabilities. Figure 6 gives some data which can be used to emphasize this approach. By using different KCl pulses the value of $\alpha$ was calculated to be 0.47, and then using Eq. (2) the electrical potential scale determined. The zero value corresponds to the emission intensity when 0.5 $\mu M$ valinomycin was present such that no light-induced potential existed and $\alpha = 0.04$. In this type of experiment it was argued that the decay of the salt-induced signal reflected the diffusion of the salt across the thylakoid membrane. Since the membrane was shown to be more permeable to K⁺ than Cl⁻ ($\alpha$ less than 1, see Fig. 5) then the decay is controlled by the rate of Cl⁻ movement. Since both the concentration gradient and the electrical gradient is known, then by calculating the initial Cl⁻ influx from the decay it is possible to estimate the absolute permeability of the thylakoid membrane to this anion using the appropriate form of the Goldman flux equation

$$\theta = \frac{F\Delta\psi}{RT} P_{Cl} \frac{[Cl^-]}{1 - \exp(-F\Delta\psi/RT)} \qquad (5)$$

where $\theta$ is the $Cl^-$ flux in moles/cm$^2$ sec and $P_{Cl}$ is the permeability coefficient for $Cl^-$ (see Barber[51]).

In Fig. 6 the salt pulse was 50 m$M$ KCl, and the decay follows the first-order law shown assuming that the internal KCl level rises from 0 to 50 m$M$. The rate constant of 0.075 sec$^{-1}$ corresponds to an initial influx of 3.75 m$M$/sec. Assuming 1 $\mu$g chl is equivalent to 16.7 cm$^2$ of membrane surface and the intrathylakoid volume is 3.3 $\mu$l/mg chl[61] then the above influx is $7.4 \times 10^{-11}$ moles/cm$^2$ sec. Taking $\Delta\psi = +93$ mV (see Fig. 6) and $[Cl]_o = 50 \times 10^{-6}$ moles/cm$^3$ then Eq. (5) gives $P_{Cl} = 3.4 \times 10^{-7}$ cm sec$^{-1}$. With $\alpha = 0.47$, $P_K$ is equal to $7.2 \times 10^{-6}$ cm sec$^{-1}$, while after valinomycin treatment ($\alpha = 0.04$) $P_K = 8.5 \times 10^{-5}$ cm/sec. It should be noted that the values in the original paper[51] were incorrectly calculated.

### Flash-Induced 515 nm Decay

The flash-induced 515 nm signal is due to an electrochromic shift of the absorption spectra of the thylakoid membrane pigments thought to be brought about by the establishment of an electrical gradient generated by primary charge separation at the reaction centers (see Junge[52]). As emphasized by Junge and Schmid,[53] its dark decay is a reflection of charge dissipation across the thylakoid membrane and therefore a measure of ion fluxes. For isolated thylakoids treated and suspended in different salts, the decay of the 515 nm signal could be used to estimate the relative permeability of the membrane. This is emphasized by the experiments of Junge and Witt,[54] where they showed that with thylakoids suspended in NaCl or KCl the decay of the 515 nm signal was considerably faster with the latter when valinomycin was present. This approach, could be extended to recording decays after subjecting thylakoids to a range of different salts. For example, Nickson and Barber[55] have shown that changing the anion of various $K^+$ salts can significantly influence the dark decay rate finding that $Cl^-$ and $NO_3^-$ are more effective than $SO_4^{2-}$.

[52] W. Junge, *Annu. Rev. Plant Physiol.* **28**, 503 (1977).
[53] W. Junge and R. Schmid, *J. Membr. Biol.* **4**, 179 (1971).
[54] W. Junge and H. T. Witt, *Z. Naturforsch. Teil B* **23**, 244 (1968).
[55] D. Nickson and J. Barber, *Proc. Int. Congr. Photobiol., 4th* Abstracts p. 270 (1977).

## Use of Selective Ionophores

As already indicated above specific ionophores such as valinomycin ($K^+$), nigericin ($K^+/H^+$), monensin ($Na^+/H^+$), A23187 ($Mg^{2+}$ and $Ca^{2+}$), beauvaricin ($Ca^{2+}$), gramicidin (monovalent cations), nystatin ($Cl^-$), and many others can be used to gain understanding about the nature of ion fluxes and gradients in chloroplasts. These compounds may or may not exert their influence on such phenomena as slow chlorophyll fluorescence quenching (see Barber et al.[48]), 515 nm decay (see Junge et al.[52-54] or uncoupling of electron transport (see Telfer et al.[56,57]) indicating the possible involvement of specific ionic species.

### Chloroplast Parameters

Ionic contents and fluxes can be conveniently related to chlorophyll concentration determined by the well-established methods.[58,59] However, many workers have attempted to transform chlorophyll levels into stromal and intrathylakoid volumes or water content and into membrane surface area parameters. The total water content of a pellet is often determined using $^3H$, while the nonosmotic space is given as the space corresponding to the distribution of a $^{14}C$-labeled sugar such as sorbitol (see Rottenberg et al.[60]) or sucrose.[61] The difference between the two measurements gives the osmotic space. This approach is satisfactory for thylakoid preparations but has less meaning for isolated intact chloroplasts where the osmotic space includes both the intrathylakoid and stromal compartments. To overcome this, Heldt et al.[61] used electron micrographs to establish the size of the stromal and intrathylakoid volumes of intact spinach chloroplasts. These workers concluded that stromal volume is about 23 $\mu l$/mg chl, while the intrathylakoid volume is 3.3 $\mu l$/mg chl. Other estimates of intrathylakoid volumes have been made (see Hall[62]), but as emphasized by the work of Rottenberg et al.[60] the value obtained will depend on the osmolarity of the medium. The object

[56] A. Telfer, J. Barber, and J. Nicolson, Biochim. Biophys. Acta 386, 301 (1975).
[57] A. Telfer and J. Barber, Biochim Biophys. Acta 501, 94 (1978).
[58] D. I. Arnon, Plant Physiol. 24, 1 (1949).
[59] J. Bruinsma, Biochim. Biophys. Acta 52, 576 (1961).
[60] H. Rottenberg, T. Grunwald, and M. Avron, Eur. J. Biochem. 25, 54 (1972).
[61] H. W. Heldt, K. Werdan, M. Milovancev, and G. Geller, Biochim. Biophys. Acta 314, 224 (1973).
[62] D. O. Hall, in "The Intact Chloroplast." (J. Barber, ed.), Vol. 1, p. 135. Elsevier, Amsterdam, 1976.

of obtaining the volumes of the chloroplast compartments is to convert levels of ions in terms of chlorophyll into true concentration units (i.e., moles/liter). However, because the chloroplasts have an extensive negatively charged membrane system and a host of membrane-associated and soluble charged proteins, it is probably unfair to quote concentrations since this does not take into account the heterogenous distribution of ions within the chloroplast compartments nor their activity coefficients. For this reason, it is probably undesirable to quote ionic levels in chloroplasts in normal concentration terms, and we feel that chlorophyll is the best reference parameter. However, knowledge of the surface area of the membranes is more meaningful when comparing ionic fluxes across chloroplast membranes with similar studies on other natural or artificial membrane systems. For the thylakoid membranes, Barber[51] has used the conversion that 1 $\mu$g chl corresponds to 16.7 cm$^2$, a value calculated from figures given by Junge and Witt.[54]

Acknowledgments

We wish to thank the SRC and the EEC Solar Energy Research and Development Programme for financial support.

# [57] Measurement of Metabolite Movement across the Envelope and of the pH in the Stroma and the Thylakoid Space in Intact Chloroplasts

By H. W. HELDT

## Silicone Layer Filtering Centrifugation

This method, originally introduced by Werkheiser and Bartley[1] and later applied with great success to investigations on metabolite transport in mitochondria by Klingenberg and Pfaff,[2] enables a rapid separation of mitochondria and chloroplasts from the suspension medium by centrifugation through a layer of silicone oil into a denaturing agent, e.g., $HClO_4$ or NaOH. It is advantageous to carry out the centrifugation in capillary tubes from polypropylene (length: 47 mm, inner diameter: 4 mm) of 400 $\mu$l content. We found that tubes made from polyethylene sometimes

---

[1] W. C. Werkheiser and W. Bartley, *Biochem. J.* **66**, 79 (1957).
[2] M. Klingenberg and E. Pfaff, Vol. 10, p. 680.

ruptured during centrifugation. The centrifugation tube is filled at the bottom with 20 $\mu$l of 1 $M$ HClO$_4$, followed by a layer of 70 $\mu$l of silicone oil and 200–300 $\mu$l of chloroplast suspension on top of this. If the uptake of bicarbonate is to be measured, the bottom layer consists of 20 $\mu$l of 2.5 $M$ NaOH.

The density of the silicone layer is of crucial importance. There has to be a density gradient between the three layers. If this gradient between the chloroplast suspension and the silicone layer is too large, the chloroplasts remain on the interface, whereas with a too low gradient the sedimented chloroplasts carry large amounts of adhering medium through the silicone layer. The optimal oil is selected empirically from testing mixtures of different densities and that oil is chosen which yields a maximum amount of chloroplasts and a minimum of medium carried through the silicone layer. As discussed later, the medium carried through is determined as the sorbitol-permeable space, whereas the sorbitol-impermeable $^3$H$_2$O space is taken as a measure for the filtered chloroplasts. It may be noted that the optimal silicone oil mixture is very dependent on temperature. We employ in our experiments with intact chloroplasts suspended in 0.33 $M$ sorbitol at 20° a mixture of AR 100:AR 150 (1:3) (Wacker Chemie, München, Germany). Under these conditions only the intact chloroplasts are sedimented, the broken chloroplasts remain on the interface between the medium and the oil. For the sedimentation of osmotically shocked chloroplasts suspended in 100 m$M$ NaCl a mixture of AR 50:AR 100 (2:1) is used. In order to measure the uptake of a substance into chloroplasts, the radioactively labeled substance is added in a volume of 5–10 $\mu$l to a stirred chloroplast suspension, and the incubation is stopped by centrifugation. Using a rapidly accelerating microcentrifuge (Microfuge, No. 154, Beckman, 16,000 rpm) the sedimentation time of intact chloroplasts can be reduced to 2 sec after starting the centrifuge. The sedimentation of osmotically shocked chloroplasts takes much longer (30–60 sec).

For the measurement of radioactivity in the sedimented chloroplasts, the tip of the centrifugation tube containing HClO$_4$ or NaOH is cut slightly above this fraction in the silicone layer and is then transferred into a lidded polypropylene microreaction vessel (Netheler and Hinz, Hamburg, Germany) containing 300 $\mu$l H$_2$O (Fig. 1).[3] The pellet is homogenized by vigorous shaking, and the protein precipitate is separated by centrifugation. Two hundred microliters of the clear supernatant are usually withdrawn for radioactivity measurement by liquid scintillation counting.

[3] H. W. Heldt, in "The Intact Chloroplast" (J. Barber, ed.), p. 215. Elsevier, Amsterdam, 1976.

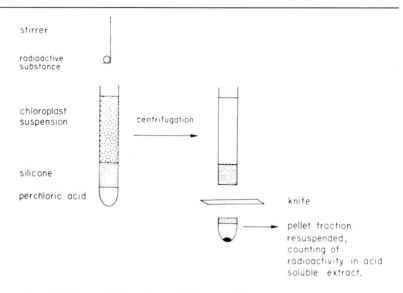

FIG. 1. Scheme of silicone layer filtering centrifugation. (From Heldt.[3])

### Determination of Solute Spaces

The intact chloroplast is surrounded by an envelope consisting of the inner and the outer envelope membrane (Fig. 2). Furthermore, there are the thylakoid membranes within the chloroplast. The outer envelope membrane is permeable to all compounds of low molecular weight which have been tested, such as nucleotides, inorganic phosphate, phosphate derivates, carboxylic acids, and sorbitol, but it is not penetrated by macromolecules such as dextran.[4] The inner envelope membrane, being the functional border between the external space and the stroma, is impermeable to sorbitol. With osmotically shocked chloroplasts, [$^{32}$P]orthophosphate was found to be a suitable impermeable agent.[5]

The determination of solute space is based on the measurement of that portion of the volume of the radioactively labeled supernatant which is found, after centrifugation, in the pellet fraction. The sorbitol space is equivalent to the volume of the external medium present in the intermembrane space between the two envelope membranes and adhering to the outer surface of the chloroplasts. The $^{3}H_2O$ space, corrected by subtraction of the sorbitol space, is taken as a measure for the space

[4] H. W. Heldt and F. Sauer, Biochim. Biophys. Acta 234, 83 (1971).
[5] C. J. Chon,    Portis, and H. W. Heldt, unpublished.

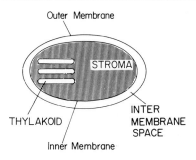

Outer Membrane

STROMA

INTER
MEMBRANE
SPACE

THYLAKOID

Inner Membrane

Fig. 2. Schematic representation of chloroplast structure.

surrounded by the inner envelope membrane and consists of the stroma and the thylakoid spaces. Using double isotope radioactivity measurement, this space can be measured in a single experiment by adding 1 $\mu$Ci $^3$H$_2$O and 0.2 $\mu$Ci [$^{14}$C]sorbitol to 200 $\mu$l chloroplast suspension equivalent to 20 $\mu$g of chlorophyll. The spaces are calculated as follows

$$S = \frac{\text{dpm}_p \, \text{Vol}}{\text{dpm}_U \, \text{Chl}} \tag{1}$$

where $S$ is the solute space (in microliters per milligram of chlorophyll), $\text{dpm}_P$ and $\text{dpm}_U$ are total radioactivity in pellet and supernatant, respectively, Vol is the total volume of chloroplast suspension (in microliters), and Chl is the amount of chlorophyll in the sample (in milligrams). With intact spinach chloroplasts in 0.33 $M$ sorbitol a $^3$H$_2$O space of 60 $\mu$l and a sorbitol space of 35 $\mu$l are usually observed.

Measurement of the Uptake into the Sorbitol-Impermeable $^3$H$_2$O Space

In order to measure the uptake of a labeled compound, e.g., [$^{32}$P]orthophosphate, 10 $\mu$l is added to the chloroplast suspension, and the incubation is terminated after a certain time (e.g., 20 sec) by centrifugation. From measurement of the total radioactivity in the pellet and the supernatant (dpm $^{32}$P$_P$, dpm $^{32}$P$_U$), the [$^{32}$P]phosphate space ($S_{P_i}$) is determined according to Eq. (1). In a parallel experiment the $^3$H$_2$O space ($S_{H_2O}$) and the [$^{14}$C]sorbitol space ($S_{Sorb.}$) are also determined. The phosphate concentration in the sorbitol impermeable $^3$H$_2$O space, [P$_i$]$_C$, is calculated as follows

$$[\text{P}_i]_C = \left( \frac{S_{P_i} - S_{Sorb}}{S_{H_2O} - S_{Sorb}} \right) [\text{P}_i]_U \tag{2}$$

Combined with Eq. (1)

$$[P_i]_C = \left( \frac{dpm(P_i)_P/dpm(P_i)_U - dpm(Sorb)_P/dpm(Sorb)_U}{dpm(H_2O)_P/dpm(H_2O)_U - dpm(Sorb)_P/dpm(Sorb)_U} \right) [P_i]_U \quad (3)$$

$[P_i]_U$ represents the corresponding concentration in the supernatant. If the amount taken up is neglegible with regard to the total amount added to the chloroplast suspension, the initial concentration in the medium $[P_i]_M$ may be introduced into Eq. (3) instead of $[P_i]_U$. Otherwise, $[P_i]_U$ is calculated from $C_{P_iM}$ in the following way

$$[P_i]_U = \frac{dpm(P_i)_U \, [P_i]_M}{dpm(P_i)_U + \left[ dpm(P_i)_P - \left( \dfrac{dpm(Sorb)_P \, dpm(P_i)_U}{dpm(Sorb)_U} \right) \right]} \quad (4)$$

It has been estimated that about 13% of the sorbitol-impermeable space of spinach chloroplasts belongs to the thylakoid space[6] and the rest to the stroma. The thylakoid membrane is impermeable to phosphate and phosphorylated compounds. Furthermore, some of the stromal $^3H_2O$ will be nonosmotic (i.e., hydrate water). For this reason the actual $P_i$ concentration in the stroma will be at least 20% higher than the concentration calculated for the sorbitol impermeable $^3H_2O$ space. The amount of $P_i$ taken up into the sorbitol-impermeable $^3H_2O$ space $(P_i)_C$ is

$$(P_i)_C = [P_i]_C \, (S_{H_2O} - S_{Sorb}) \, (\text{nmoles/mg chlorophyll}) \quad (5)$$

If rates of uptake are to be calculated, the time required for the sedimentation has to be added to the incubation time (termination is initiation of centrifugation). Using the Beckman Microfuge at full velocity, the time of sedimentation is about 2 sec with intact chloroplasts.

### Measurement of the Metabolite Levels in the Stroma and the Release of Products during $CO_2$ Fixation by Intact Chloroplasts

Silicone layer filtering centrifugation allows a rapid separation of the chloroplasts from the suspension medium during steady state $CO_2$ fixation. In this way, the efflux of the products of $CO_2$ fixation can be measured with high accuracy and sensitivity. For detection of compounds the chloroplasts are preincubated in the dark for 5–10 min with [$^{32}P$]orthophosphate. Two hundred microliters of the chloroplast suspension containing 20 $\mu g$ of chlorophyll is transferred to a centrifugation tube containing silicone and $HClO_4$. The tube is illuminated for a certain

---

[6] H. W. Heldt, K. Werdan, M. Milovancev, and G. Geller, *Biochim. Biophys. Acta* **314**, 224 (1973).

time, and the incubation terminated by centrifugation. Continuous illumination is maintained until the centrifugation step is completed. Immediately after centrifugation 5 $\mu$l of concentrated $HClO_4$ is added to the top layer. The supernatant and pellet fraction are neutralized with $K_2CO_3$ to pH 6.0, and the acid-soluble extracts are subjected to microscale ion exchange chromatography. For details of the chromatography see Heldt and Klingenberg[7] and Lilley et al.[8] Figure 3 shows ion exchange chromatograms of phosphorylated intermediates in the filtered chloroplasts and the supernatant. This method is very sensitive, chloroplasts equivalent to 5 $\mu$g of chlorophyll are sufficient to obtain a complete analysis as shown in Fig. 3.

### Determination of the pH in the Stroma and in the Thylakoid Space of Intact Chloroplasts

When a dissociable compound is added to a suspension of membrane-surrounded vesicles, the membrane of which is only permeable to uncharged species, the concentration of the uncharged molecule will be equal on both sides. In contrast, the concentration of the corresponding ions will depend on the concentration of protons in both spaces

$$\frac{[H^+]_I}{[H^+]_M} = \frac{[BH^+]_I}{[BH^+]_M} = \frac{[A^-]_M}{[A^-]_I} \qquad (6)$$

where I is internal and M is external medium.

With cationic compounds, the concentration inside will be higher than outside if the internal space is more acidic than the external one and vice versa with anionic substances.

The measurement of the pH in the medium and the concentration ratio between the vesicles and the external medium allows for the calculation of the internal pH. A suitable anionic compound for such pH measurements is 5,5-dimethyloxazolidine-2,4-dione (DMO), and a cationic substance is methylamine. The application of amines for the determination of the pH in the thyalkoid space of broken chloroplasts has been described by Rottenberg et al.[9] and by Gaensslen and McCarty.[10]

For the measurement of the pH in the stroma and the thylakoid space, two parallel samples of 200 $\mu$l chloroplasts (equivalent to 20 $\mu$g chloro-

[7] H. W. Heldt and M. Klingenberg, Vol. 10, p. 482.
[8] R. McC. Lilley, C. J. Chon, A. Mosbach, and H. W. Heldt, Biochim. Biophys. Acta 460, 259 (1977).
[9] H. Rottenberg, T. Grunwald, and M. Avron, Eur. J. Biochem. 25, 54 (1972).
[10] R. E. Gaensslen and R. E. McCarty, Arch. Biochem. Biophys. 147, 55 (1971).

264

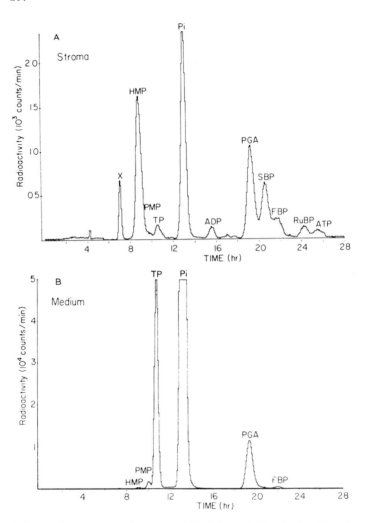

FIG. 3. Ion exchange chromatogram of [32]P-labeled metabolites in the chloroplasts (A) and the supernatant medium (B) during the steady state of $CO_2$ fixation as separated by silicone layer filtering centrifugation. From Lilley *et al.*[8]

phyll) are incubated for 60 sec or longer with 0.04 $\mu$Ci [[14]C]methylamine (final concentration 30 $\mu M$) in one sample and 0.04 $\mu$Ci [14]C-DMO (final concentration 0.5 m$M$) in the other. The very low concentration of methylamine was chosen to prevent any uncoupling by this substance. The incubation is terminated by silicone layer filtering centrifugation. In experiments with illuminated chloroplasts, illumination is continued during centrifugation. The sorbitol- and $^3H_2O$-permeable spaces are determined in parallel experiments. In intact illuminated chloroplasts both methylamine and DMO are accumulated, indicating that there are two different spaces in the sorbitol-impermeable $^3H_2O$ space, one being more acidic and the other more alkaline than the medium. Obviously, these two spaces are identical with the stroma and the thylakoid space. It is possible to calculate the pH in the stroma and in the thylakoid space of intact chloroplasts from measurement of the uptake of $^{14}C$-DMO and [$^{14}C$]methylamine into the sorbitol-impermeable $^3H_2O$ space. This calculation is based on the following three relations

$$[D^-]_U[M^+]_U = [D^-]_S[M^+]_S = [D^-]_T[M^+]_T \tag{7}$$

$$D_C^- = D_S^- + D_T^+ = [D^-]_S V_S^+ [D^-]_T V_T \tag{8}$$

$$M_C^+ = M_S^+ + M_T^+ = [M^+]_S V_S + [M^+]_T V_T \tag{9}$$

where $D_C^-$ and $M_C^+$ are the amount of DMO anion (methylamine cation) in the sorbitol impermeable $^3H_2O$ space (nmoles/mg chlorophyll); $M_T^+$ and $M_S^+$ are the amount of methylamine cation in the thylakoid (stroma) space; $D_T^-$ and $D_S^-$ are the amount of DMO anion in the thylakoid (stroma) space; $[D^-]_U$, $[D^-]_C$, $[D^-]_T$, $[D^-]_S$, $[M^+]_M$, etc., are the corresponding concentrations in the Supernatant (U), etc., $V_C$ is the sorbitol impermeable $^3H_2O$ space; $V_T$ is the thylakoid space; and $V_S$ is the stroma space. Equations (7)–(9) can be combined into a quadratic equation dissolved for $D_S^-$

$$D_S^- = -\frac{a}{2} \pm [(a/2)^2 - b]^{1/2} \tag{10}$$

$$a = \frac{[M^+]_U[D^-]_U(V_T^2 - V_S^2)}{M_C^+} - D_C^- \tag{11}$$

$$b = \frac{D_S^-}{M_C^+}[M^+]_U[D^-]_U V_S^2 \tag{12}$$

From $[D^-]_S$, $[D^-]_U$, and $[M^+]_U$, $M_S^+$ is calculated according to Eqs. (7) and (8). The introduction of this value into Eq. (9) yields

$$[M^+]_T = \frac{1}{V_T}\left(M_C^+ - \frac{[D^-]_U[M^+]_U V_S^2}{D_S^-}\right) \tag{13}$$

The pH in the stroma and in the thylakoid space can be now calculated

$$pH_S = pH_M + \log([D^-]_S/[D^-]_U) \tag{14}$$

$$pH_T = pH_M + \log([M^+]_U/[M^+]_T) \tag{15}$$

Due to the quadratic form of the equation [Eq. (10)], this calculation yields two different solutions. In one solution the thylakoid space is calculated to be more acidic than the stroma. This concurs with earlier investigations with broken chloroplasts showing light-dependent proton transport into the thylakoid space.[11] Therefore, the second solution, in which the stroma space is more acidic than the thylakoid space is disregarded.

Thus, a determination of the pH in the stroma and the thylakoid space requires the measurement of the pH in the medium and of the [14C]methylammonium- and 14C-DMO-anion concentrations in the chloroplasts. In the case of methylamine ($pK = 10.6$) the concentration of the undissociated form is so small at physiological pH values that it can be neglected in the calculation. Therefore, the concentration of [14C]methylammonium anion can be regarded as equal to the concentration of total methylamine + methylammonium anion evaluated from measurement of 14C radioactivity. This is different with DMO (at 4° $pK = 6.64$, at 20° $pK = 6.38$) where a considerable amount may be present in the undissociated form. In order to evaluate the concentration in the supernatant $[D^-]_U$, the concentration of the undissociated DMO $[DH]_U$ has to be subtracted from the concentration of total DMO $[D^{tot}]_U$

$$[D^-]_U = [D^{tot}]_U - [DH]_U \qquad (16)$$

It is a prerequisite of the method for measurement of pH shown here, that the concentration of the undissociated acid in each space is equal. Hence

$$[DH]_U = [DH]_C = [D^{tot}]_U/(1 + 10^{pHM-pK}) \qquad (17)$$

For the evaluation of the concentration of DMO anions in the sorbitol-impermeable $^3H_2O$ space the concentration of undissociated DMO in the supernatant is subtracted from the concentration of total 14C-DMO assayed in the sorbitol-impermeable $^3H_2O$ space.

$$[D^-]_C = [D^{tot}]_C - [DH]_C \qquad (18)$$

The pH determination in the two spaces also requires knowledge of the volumes of the stroma and the thylakoid spaces. The sum of the two spaces is determined as the sorbitol-impermeable $^3H_2O$ space. With intact chloroplasts, the individual size of the two spaces cannot be measured experimentally. Therefore, an estimation of the relative size of the two spaces was carried out by planimetry of electron micrographs from isolated spinach chloroplasts, yielding a mean value of 13% of the sorbitol-

[11] J. Neumann and A. T. Jagendorf, Arch. Biochem. Biophys. 107, 109 (1964).

impermeable $^3H_2O$ space for the thylakoid space.[4] As this value was quite constant in different chloroplast preparations, it has been introduced into all our pH determinations. As shown elsewhere,[6] a deviation in this value of $\pm 50\%$ has no dramatic effect on the results. As a typical result obtained with intact spinach chloroplasts (pH medium 7.6, temperature 20°, with illumination) the pH in the stroma is found to be 8.0 and in the thylakoid space 5.5.[12] The pH gradient across the thylakoid membrane collapses when the chloroplasts are placed in the dark, and the pH in the stroma and the thylakoid space is found to be about 7.0.

If only the pH in the stroma is to be determined, this can be done in a simplified way by measuring the uptake of DMO only. The calculation of the pH in the stroma may then be performed assuming that all the DMO taken up is present in the stroma and none in the thylakoid space. In most cases the error introduced by this simplification is small.

[12] K. Werdan, H. W. Heldt, and M. Milovancev, *Biochim. Biophys. Acta* **396**, 276 (1975).

## [58] Stabilization of Chloroplasts and Subchloroplast Particles

*By* GEORGE C. PAPAGEORGIOU

The technology of the stabilization of soluble enzymes, through chemical manipulations designed to protect them from deactivation, has made exceedingly rapid progress recently.[1,2] In contrast, comparatively little has been achieved with membrane-integrated systems, such as those in the photosynthetic membranes of green plants, in spite of their undisputed technological future as solar energy converters. Because of space limitations, this article describes the attempts to stabilize the structure-dependent functions only of isolated higher plant chloroplasts. Although the very first step of such endeavors is to prepare active chloroplasts, conditions necessary for that are reviewed briefly as they have been discussed in detail earlier in this series.[3-6] Emphasis is placed on chemical manipulations of chloroplast membranes, and in particular on protein

[1] H. H. Weetall, *Anal. Chem.* **46**, 602A (1974).
[2] A. I. Kestner, *Usp. Khim.* **43**, 1480 (1974).
[3] C. A. Price, Vol. 31, p. 501.
[4] T. Galliard, Vol. 31, p. 520.
[5] W. D. Loomis, Vol. 31, p. 600.
[6] P. S. Nobel, Vol. 31, p. 600.

covalent cross-linking techniques, and the effects these have on the functions and the functional life of isolated chloroplasts.

### Factors Contributing to the Inactivation of Isolated Chloroplasts

The inevitable release of the central vacuole sap during cell disruption is probably the most damaging event for chloroplast preparation. The recommendations call for rapid isolation of the organelles, in order to minimize their exposure to the vacuolar sap, and for the fortification of the isolation media with various additives, as shown in Table I.

Aging of isolated chloroplasts during storage, and particularly the inactivation of photosystem II, is attended by the release of lipoprotein particles,[7] and of free fatty acids through the action of lipolytic enzymes.[4,8-10] Chloroplasts release 30% of their phospho- and sulfolipids and 20% of their galactolipids when shocked into hypotonic alkaline media, but this tendency is greatly suppressed by 10 m$M$ $MgCl_2$.[7] *Phaseolus vulgaris* chloroplasts are easily inactivated at pH 6, where galactolipase activity is optimal and survive better at pH 8 where galactolipase is suppressed.[9] In spinach chloroplasts, on the other hand, two galactolipases have been detected: one, with optimum pH 5.5–6.0, releasing fatty acids from digalactosyl diglycerides; and another, with optimum pH 7.5–8.0, releasing fatty acids from monogalactosyl diglycerides.[11] Freshly isolated chloroplasts have 3.1–5.5% of their total fatty acids as free, unesterified molecules. During storage, the proportion of free fatty acids increases (especially the 18:3 linolenic acid), while electron transport across photosystem II decreases.[10]

Free fatty acids suppress the activity of photosystem II (but much less the activity of photosystem I),[8,9] uncouple photophosphorylation, and cause chloroplasts to swell irreversibly.[12] Unsaturated fatty acids (18:3, 16:3) are subject to photoperoxidation, in a process which also bleaches the chlorophylls.[13] Due to the multiplicity of lipid degrading enzymes, only limited protection is afforded by isolating and storing chloroplasts in alkaline media. Further protective measures that have been developed are designed either to suppress the release of fatty acids

---

[7] K.-P. Heise and G. Jacobi, *Z. Naturforsch, Teil C* **28**, 120 (1973).

[8] D. W. Krogmann and A. T. Jagendorf, *Arch. Biochem. Biophys.* **80**, 421 (1959).

[9] R. E. McCarty and A. T. Jagendorf, *Plant Physiol.* **40**, 725 (1965).

[10] G. Constantopoulos and C. N. Kenyon, *Plant Physiol.* **43**, 531 (1968).

[11] R. J. Helmsing, *Biochim. Biophys. Acta* **144**, 470 (1967).

[12] P. A. Siegenthaler, *Biochim. Biophys. Acta* **275**, 182 (1972).

[13] R. L. Heath and L. Packer, *Arch. Biochem. Biophys.* **125**, 189 (1968).

TABLE I
FACTORS WHOSE PRESENCE IN THE ISOLATION AND STORAGE MEDIA IMPROVES THE
ACTIVITIES OF CHLOROPLASTS PREPARATIONS

| Factors | Targets | References |
|---|---|---|
| Isolation medium | | |
| Storage buffering capacity | Neutralization of vacuolar acids | |
| Alkaline pH (7.5–9.0) | Suppression of galactolipases | 4,9–11 |
| BSA | Adsorption of free fatty acids and of their degradation products | 14–17 |
| Antioxidants (thiols, ascorbate, butylated hydroxytoluene) | Inhibition of lipid peroxidations; inhibition of oxidation of phenolics | 5,13,17 |
| KCl, as osmoticum | Loading of thylakoids with suitable counterions for $H^+$ transport | 17 |
| Storage medium | | |
| Low salt, weak buffering capacity | Prevention of membrane protein denaturation | 17 |
| Nonionic osmotica (mannitol, sorbitol, sucrose) | Prevention of membrane protein denaturation; suppression of free fatty acid release | 16–18 |
| High chloroplast density | Preservation of electron transport activities | 15 |
| Alkaline pH, BSA, antioxidants | Targets, as above | — |

or to remove them and their degradation products from the medium (Table I).

Excellent protection is achieved by adsorbing free fatty acids with bovine serum albumin (BSA).[14–17] At room temperature, chloroplasts stored in chloride medium supplemented with 20 mg BSA/ml retained 30% of photosystem II activity for more than 30 hr, whereas at 0° they survived for more than 3 days.[15] Nonionic osmotica (mannitol, sorbitol, sucrose) are superior to electrolytes in storage media, protecting chloroplasts against the release of fatty acids and loss of photoinduced electron transport, proton transport, and phosphorylation.[16–18] Other protective measure include storage of chloroplasts as dense suspensions (2–4

[14] Y. G. Molotkovsky and I. M. Zhestkova, *Biochim. Biophys. Acta* **112**, 170 (1966).
[15] A. R. Wasserman and S. Fleischer, *Biochim. Biophys. Acta* **153**, 154 (1968).
[16] J. F. G. M. Wintermans, P. J. Helmsing, B. J. J. Polman, J. Van Gisbergen, and J. Collard, *Biochim. Biophys. Acta* **189**, 95 (1969).
[17] T. Takaoki, J. Torres-Pereira, and L. Packer, *Biochim. Biophys. Acta* **352**, 260 (1974).
[18] G. Kulandaivelu and D. O. Hall, *Z. Naturforsch. Teil C* **31**, 452 (1976).

mg Chl/ml),[15] freezing in the presence of glycerol in order to prevent ice formation,[15,19] isolation of chloroplasts in KCl medium in order to have thylakoids loaded with suitable counterions for $H^+$ uptake, and presence of thiols in the isolation and storage media for partial suppression (40–50%) of chloroplast galactolipase.[20]

### Cross-linking of Chloroplast Membrane Proteins with Bifunctional Reagents

Enzymes whose tertiary structures are stabilized by means of artificial intra- and intermolecular cross-links, attachment to inert supports, or encapsulation within semipermeable microsacs are, generally, better protected against activation than native enzymes.[1,2] The question is whether we can use the same, or similar, techniques in order to stabilize complex, membrane-integrated systems. Only few attempts have been reported with regard to isolated higher plant chloroplasts, but the limited experience is useful in assessing the applicability of protein cross-linking techniques.

*Glutaraldehyde.* The exact mechanism of protein cross-linking with glutaraldehyde is not yet fully understood.[21,22] Since aldehydes condense with primary amines to form Schiff bases (azomethines), free protein $NH_2$ groups are the expected sites of attack. These groups disappear when soluble proteins,[23,24] and chloroplast membranes[25,26] are reacted with glutaraldehyde. In addition, alkylation of side chain SH, and imidazolyl groups is also possible.[23]

Schiff bases, however, are easily hydrolyzed at low pH, whereas the reaction products of glutaraldehyde with proteins are acid stable. This observation argues, therefore, against the presence of azomethine linkages in glutaraldehyde-cross-linked proteins. Nevertheless, Monsan *et al.*[27] have shown that the Schiff base formed by an $\alpha,\beta$-unsaturated aldehyde (crotonaldehyde) and a primary amine resists acid hydrolysis. When the amine is in excess, a nucleophilic addition to the aldehyde

[19] L. Packer and A. C. Barnard, *Biochim. Biophys. Acta* **126**, 443 (1966).

[20] P. J. Helmsing, *Biochim. Biophys. Acta* **178**, 519 (1969).

[21] K. Peters and F. M. Richards, *Ann. Rev. Biochem.* **46**, 523 (1977).

[22] G. C. Papageorgiou, *in* "Photosynthesis in Relation to Model Systems" (J. Barber, ed.) p. 211. Elsevier, Amsterdam, 1979.

[23] A. F. S. A. Habeeb and R. Hiramoto, *Arch. Biochem. Biophys.* **126**, 16 (1968).

[24] G. Blauer, D. Harmatz, E. Meir, M. K. Swenson, and B. Zvilichovsky, *Biopolymers* **14**, 2585 (1975).

[25] G. C. Papageorgiou and J. Isaakidou, *in* "Bioenergetics of Membranes" (L. Packer, G. C. Papageorgiou, and A. Trebst, eds.), p. 257. Elsevier, Amsterdam, 1977.

[26] J. Isaakidou and G. C. Papageorgiou, *Arch. Biochem. Biophys.* (in press, 1979).

[27] P. Monsan, G. Puzo, H. Mazarguil, *Biochimie* **57**, 1281 (1975).

double bond also occurs, as it has been postulated earlier by Richards and Knowles.[28] The reaction sequence is as follows:

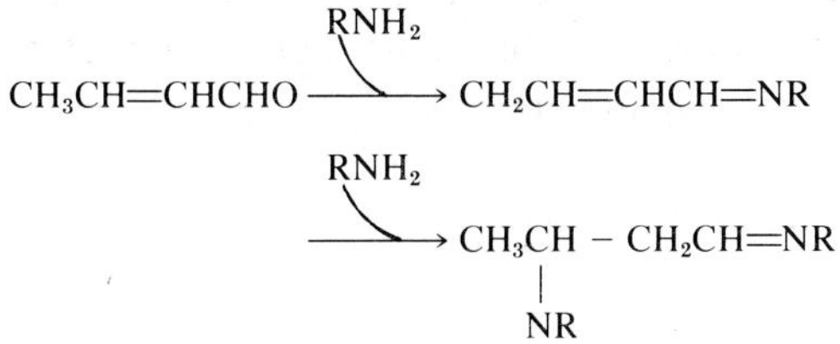

$$CH_3CH{=}CHCHO \xrightarrow{\text{RNH}_2} CH_2CH{=}CHCH{=}NR$$

$$\xrightarrow{\text{RNH}_2} CH_3CH-CH_2CH{=}NR$$
$$\qquad\qquad\qquad\quad \underset{\displaystyle NR}{|}$$

$\alpha,\beta$-Unsaturated polyaldehydes, products of successive aldol condensations, have been detected in aqueous glutaraldehyde solutions by means of ${}^1$H-NMR spectroscopy,[28] but their presence in neutral, or acid commerical solutions has been strongly disputed by others, on the basis of optical and ${}^1$H-NMR spectroscopy[29,30], and ${}^{13}$C-NMR spectroscopy.[31] According to Whipple and Ruta,[31] the main component of aqueous, neutral glutaraldehyde is the hemiacetal monohydrate, which exists in equilibrium with the monomer and the open-chain hydrates. At 23°, the mixture has the following composition.

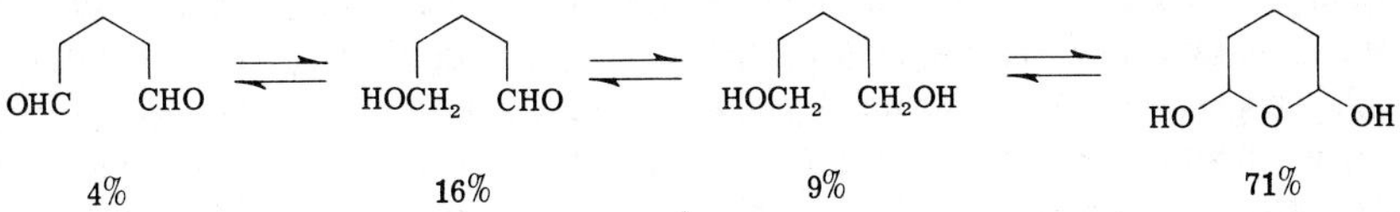

The monomer has been shown to react slowly with soluble proteins and model amino acids to form complex structures, such as those shown below.

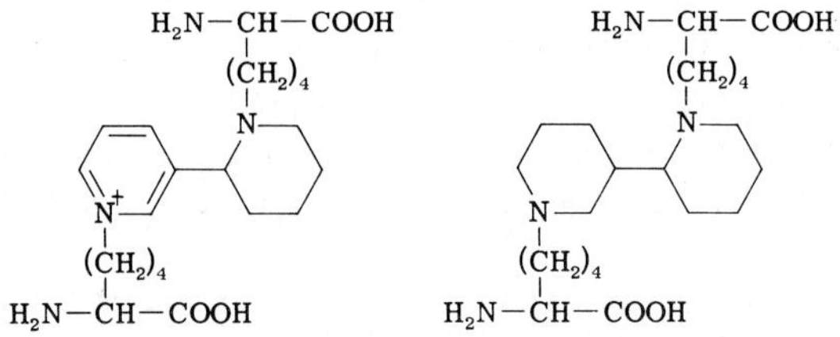

At neutral or acidic pH, the active component of purified glutaraldehyde solutions appears to be the unhydrated monomer, which reacts primarily with lysyl $\epsilon$-NH$_2$ groups to give the structures shown above. Glutaraldehyde polymerizes at slightly alkaline pH (8) to a mixture of

[28] F. M. Richards and J. R. Knowles, *J. Mol. Biol.* **37**, 231 (1968).
[29] P. M. Hardy, A. C. Nicholls, and H. N. Rydon, *Chem. Commun.* 565 (1969).
[30] A. H. Korn, S. H. Feairheller, and E. M. Filachione, *J. Mol. Biol.* **65**, 525 (1972).
[31] E. B. Whipple and M. Ruta, *J. Org. Chem.* **39**, 1666 (1974).

products containing a high proportion of $\alpha,\beta$-unsaturated polyaldehydes. The polymerization is attended by an increase in the absorption at 235 nm. Pure glutaraldehyde absorbs maximally at 280 nm ($\epsilon = 6.5 \ M^{-1}$ cm$^{-1}$, at 23°).[24] The purity of glutaraldehyde solutions is estimated with reference to the absorbance ratio A235/A280, whose acceptable range is between 0.15 and 0.25.[32]

The complexity of the reactions of glutaraldehyde solutions with proteins poses a serious problem, since for reproducible results the reaction conditions should be controlled very carefully. The technical problem is aggravated further by the lack of adequate molecular criteria by which to monitor the progress of the cross-linking reaction. Since glutaraldehyde suppresses the bound positive charge of proteins, electrophoretic mobility cannot be used to estimate the molecular sizes of the products. Approximate indices of cross-linking are the disappearance of free NH$_2$ groups and the loss of osmotic response in the case of chloroplast thylakoids.[22]

Isolated chloroplasts become osmotically inert after treatment with glutaraldehyde.[33-35] For 5 min treatment at 0°, the threshold of inactivation is 30 $\mu$moles glutaraldehyde/mg Chl, but it is lower for longer treatments.[25] These chloroplasts do not transport metal cations actively, and do not phosphorylate, although they preserve some capacity for photoinduced H$^+$ uptake. It is noteworthy that this process is less than 10% inhibited in chloroplasts that have been osmotically inactivated with 120 $\mu$moles glutaraldehyde/mg Chl for 6 min.[35] The inhibition increases to 40% with 30 min treatment,[34] and to more than 90% in heavily fixed chloroplasts (6 mmoles glutaraldehyde/mg Chl for 10 min).[36] Other membrane properties that are inhibited by this cross-linker are the light-induced, light-scattering changes of choroplast suspensions,[36,37] the fast rise of chloroplast fluorescence, and the quenching of chloroplast fluorescence, in the presence of PMS.[38] On the other hand, the fast (ms) light-induced absorbance changes at 518 nm are not affected at all.[37]

Various fractions of residual photosystem II activity have been re-

[32] P. J. Anderson, *J. Histochem. Cytochem.* **15**, 652 (1967).

[33] G. C. Papageorgiou, *Biochim. Biophys. Acta* **461**, 379 (1977).

[34] L. Packer, J. M. Allen, and M. Starks, *Arch. Biochem. Biophys.* **128**, 142 (1968).

[35] J. West and L. Packer, *Bioenergetics,* **1**, 405 (1970).

[36] T. Oku, K. Sugahara, and G. Tomita, *Plant and Cell Physiol.* **14**, 385 (1973).

[37] S. W. Thorne, G. Horvath, A. Kahn, and N. K. Boardman, *Proc. Natl. Acad. Sci. U.S.A.* **72**, 3858 (1975).

[38] B. A. Zilinskas and Govindjee, *Z. Pflanzenphysiol.* **77**, 302 (1976).

TABLE II

REACTION CONDITIONS FOR GLUTARALDEHYDE (GA) FIXATION OF CHLOROPLASTS AND SURVIVING PHOTOSYSTEM II (PS-II) ACTIVITY AS REPORTED BY SELECTED AUTHORS

| Authors | Medium (mM concentrations) | Reaction conditions | | | | Recovered PS-II Activity (%)[a] |
| --- | --- | --- | --- | --- | --- | --- |
| | | GA (%) | GA/Chl | Temperature (°C) | Duration (min) | |
| Park et al.[39] | Leaf infiltration with aqueous GA | 6 | Unspecified | Room | 180 | 25 |
| Hallier and Park[40] | Sucrose 500; Tricine 50; pH 7.4 | 2.5–5 | Unspecified | 0 | 20 | 6 |
| West and Packer[35] | Choline–Cl⁻ 100; Tris-HCl 0.5; pH 7.8 | 0–20 | 0–500[b] | 0 | 6 | 30–24 |
| Oku et al.[36] | Sucrose 400; NaCl 10; Tris-HCl 50; pH 7.8 | 6 | 5400[b] | 0 | 5 | 46 |
| Zilinskas and Govindjee[38] | Phosphate 50; NaCl 10; pH 6.8 | 2 | 120[b] | 0 | 5 | 23 |
| Hardt and Kok[41] | Sucrose 400; NaCl 50; Tris-HCl or Tricine-NaOH 50, pH 7.4 | 5 | 220–300[b] | 0 | 5 | 20–30 |
| Papageorgiou and Isaakidou[26] | Tricine-NaOH 50; pH 7.2 (unstacked thylakoids) | 0–1.5 | 0–150 | 0 | 5 | 70–80 |

[a] Initial rates assayed with ferricyanide, or dichlorophenolindophenol as Hill oxidants, with temperature-shocked chloroplasts.
[b] Estimates.

ported for glutaraldehyde-fixed chloroplasts (see Table II).[25,26,38-43] Apparently, glutaraldehyde has no effect on the $H_2O$ splitting enzyme complex, since $H_2O$ is equally good as a source of electrons as diphenylcarbazide, which couples to photosystem II directly.[26] The most severely affected intermediate of the photosynthetic electron-transport chain is plastocyanin, a copper protein which interacts directly with the reaction center chromophore (P700) of photosystem I. As a result, photosynthetic $O_2$ evolution measured in the presence of cofactors that couple primarily after photosystem I (e.g., ferricyanide) is more severely inhibited in fixed chloroplasts than when measured in the presence of cofactors which couple before photosystem I (i.e., ahead of plastocyanin; e.g., p-phenylenediamine).[41,42]

The available evidence suggests that treatment of chloroplasts with glutaraldehyde engenders two types of inhibition for photosynthetic electron transport. First, a general slow down of electron transport in the segment between photosystems II and I. Second, a more specific inhibition centered on plastocyanin. Both effects are less severe in unstacked thylakoids than in suspensions of stacked thylakoids.[24,43] This explains earlier conflicting results, according to which some authors found a severe inhibition of photosystem I activity in glutaraldehyde-reacted chloroplasts,[25,39,41,42] whereas other reported a partial inhibition only.[36,40]

We turn now to the question as to whether or not cross-linking with glutaraldehyde stabilizes the structure and the function of isolated chloroplasts. The limited relevant information can be summarized as follows. Chloroplasts isolated from glutaraldehyde-infiltrated spinach leaves retain their morphology after extraction with 80 and 100% acetone.[39] Glutaraldehyde-fixed intact chloroplasts lose their membrane envelop when they are osmotically shocked, but they preserve their compact rounded shape, in contrast to the unfixed controls which disintegrate completely. (G. C. Papageorgiou, unpublished results). Following fixation, isolated chloroplasts release less protein when treated with various detergents than do the unfixed controls.[44] Glutaraldehyde affords partial protection to isolated chloroplasts against the inactivation of photosystem II by

[39] R. B. Park, J. Kelly, S. Drury, and K. Sauer, *Proc. Natl. Acad. Sci. U.S.A.* **55**, 1056 (1966).
[40] U. W. Hallier and R. B. Park, *Plant Physiol.* **44**, 554 (1969).
[41] H. Hardt and B. Kok, *Biochim. Biophys. Acta* **449**, 125 (1976).
[42] H. Hardt and B. Kok, *Plant Physiol.* **60**, 225 (1977).
[43] G. C. Papageorgiou and J. Isaakidou, *Z. Pflanzenphysiol.* **89**, 449 (1978).
[44] P. Y. Sane and R. B. Park, *Plant Physiol.* **46**, 852 (1970).

various structure-disruptive treatments, such as trypsin digestion, heating to 50°, and attack by chaotropic agents.[38]

Park *et al.*[39] prepared glutaraldehyde-fixed chloroplasts that could photoreduce dichlorophenolindophenol at approximately one-fourth the rate of the unfixed controls, but the fixed preparations were active even after 24 days of storage at 4° and in darkness, whereas the controls were inactivated within 2–4 days. Hardt and Kok[41] also found that chloroplasts treated with 2.5–5% glutaraldehyde (estimated 60–170 $\mu$moles glutaraldehyde/mg Chl) could photoreduce ferricyanide and *p*-phenylenediamine for more than 30 days when stored at 4° and in darkness. Glutaraldehyde also has a function-stabilizing influence for chloroplasts stored in darkness at more physiological temperatures (17°–20°) but the photosystem II activity survived for 6–8 days only.

These experiments provide no molecular explanation as to why glutaraldehyde prolongs the activity of resting chloroplast preparations. Although they indicate that electron transport operates without mass transport in the membrane, it is still possible that some molecular flexibility is required for such complex processes as transport of reducing equivalents and ions from one side of the thylakoid membrane to the other. Accordingly, glutaraldehyde-immobilized thylakoid membranes may show entirely different stability characteristics at a state of work than at a state of rest. To test this, we assayed ferricyanide-dependent $O_2$ evolution by subjecting the preparation to periodic illumination with saturating white light for 1 min, followed by 1 min of dark rest. Within two such cycles, the activity of the unfixed chloroplasts decreased and became equal to that of the fixed preparations. Thereafter, both preparations were inactivated at the same rate. Similar results were obtained with respect to photosystem I activity, with chloroplasts fixed suboptimally at 10 $\mu$moles glutaraldehyde/mg Chl.[25]

Perhaps, the single most destructive factor for active chloroplasts is the presence of $O_2$. I base this contention on the observation that isolated spinach chloroplasts, which were losing the capacity to photoreduce ferricyanide after a few minutes of saturating illumination at room temperature, could photolyze $H_2O$ for 6.5 hr at 30°–32° in a closed anaerobic system containing an $O_2$ trap (glucose, glucose oxidase) and a peroxide trap (ethanol, catalase).[45]

*Aliphatic Diimido esters.* The properties of these compounds as pro-

[45] I. Fry, G. C. Papageorgiou, E. Tel.-Or, and L. Packer, *Z. Naturforsch. Teil C* **32**, 110 (1977).

tein modifiers have been reviewed in this series.[46,47] In contrast to glutaraldehyde, they react only with $NH_2$ groups, yielding the positively charged amidines as the main product.

$$R_1C(\overset{+}{=}NH_2)OCH_3 + R\overset{+}{N}H_2 \rightarrow R_1C(\overset{+}{=}NH_2)NHR_2 + H^+ + CH_3OH$$

Thus, they offer the advantage both of chemical specificity, and protein charge conservation. Their main shortcomings are the tendency for acid-catalyzed hydrolysis and for the formation of $N$-alkyl imidates at neutral or slightly alkaline pH.[48]

$$R_1C(\overset{+}{=}NH_2)OCH_3 + H_2O \rightarrow R_1COOCH_3 + NH_4^+$$

$$R_1C(\overset{+}{=}NH_2)OCH_3 + R_2\overset{+}{N}H_3 \rightarrow R_1C(\overset{+}{=}NHR_2)OCH_3 + NH_4^+$$

The first side reaction converts diimido esters to monoimido esters (which give monofunctionally substituted products with proteins), while the second reaction permits a single imido ester function to bind two $NH_2$ groups, by a further reaction of the $N$-alkyl imidate.

Diimido esters were introduced to chloroplast studies by Packer and co-workers.[49,50] Isolated spinach chloroplast become osmotically inactive at a concentration of 20–30 $\mu$moles dimethyl suberimidate/mg Chl. In these chloroplasts, 50–55% of the protein $NH_2$ groups have been alkylated, and 45% of detergent-solubilized thylakoid protein cannot enter into polyacrylamide gels that exclude sizes above 300 kilodaltons. Since the size of chloroplast proteins ranges from 10 to 100 kilodaltons, it appears that several subunits have been interconnected. (G. C. Papageorgiou and J. Isaakidou, unpublished results).

Isolated chloroplasts, osmotically immobilized by diimido esters are capable of ferricyanide supported $O_2$ evolution, provided that the fixation reaction is carried out at about 0°, and that no excess cross-linker has been used. Inactive preparations are obtained at higher relative concentrations of the imidates (e.g., 50 $\mu$moles dimethyl suberimidate/mg Chl,[25] or 250 $\mu$moles methyl acetimidate/mg Chl[57]). The inhibition is removed when diphenylcarbazide is used to provide electrons to photosystem II.

[46] M. J. Hunter, and M. L. Ludwig, Vol. 25, p. 585.

[47] F. Wold, Vol. 25, p. 623.

[48] D. T. Browne and S. B. H. Kent, *Biochim. Biophys. Res. Commun.* **67**, 126 (1975).

[49] L. Packer, J. Torres-Pereira, P. Chang, and S. Hansen, *Proc. Int. Congr. Photosynth.* 3rd, 1974, Vol. 2, p. 867 (1974).

[50] G. C. Papageorgiou, G. D. Case, S. Hansen, and L. Packer, *Lawrence Berkeley Laboratory Annual Report,* p. 30 (1974).

This suggests that the $H_2O$-splitting enzyme complex is the site of inactivation for diimido ester-reacted chloroplasts. The probable causes of inactivation are protein cross-linking within the complex, and the combination of alkaline pH and the presence of uncoupling agents in the chloroplast suspension, which is known to inhibit $H_2O$ decomposition.[52] The alkaline pH is required in order to suppress hydrolysis of the imidoester, while the imidoester itself and its hydrolysis products act as uncouplers. In contrast to the specific inhibition of photosystem I activity by glutaraldehyde, diimido esters have no effect at all.[22,25]

During prolonged storage at 0° and in darkness, the electron-transport activity of control and dimethyl suberimidate-treated chloroplasts were found to deteriorate at the same rate. Under continuous illumination, and in the presence of electron-transport cofactors, the imido ester-fixed preparation is inactivated within a few minutes. (G. C. Papageorgiou and J. Isaakidou, unpublished results). Imido ester-treated cells of the cyanobacterium *Anacystis nidulans* irreversibly lose their photosystem II activity, if warmed past the phase transition temperature of the thylakoid membrane lipids.[33]

Table III lists the properties of glutaraldehyde and dimethyl suberimidate fixed chloroplasts.

*Microencapsulation and Entrapment of Chloroplast Fragments.* Microencapsulation of soluble enzymes within semipermeable envelops offers some advantages with regard to stability and handling.[1,2] Microencapsulated chloroplasts have been prepared by Kitajima and Butler[53] by cross-linking protamine and gelatin around them with toluidine isocyanate. The capsules, whose diameter ranged between 10 and 50 $\mu$m, were capable of photosystem I electron transport, but were devoid of photosystem II activity. Longevity tests proved, further, that the activity of the encapsulated particles was preserved better during storage than that of control chloroplasts.

Techniques to entrap chloroplasts within dried poly(vinyl alcohol) films and polyacrylamide gels have been described recently by Ochiai *et al.*[54,55] The poly(vinyl alcohol) is added to the chloroplast suspension (final content 18–20%) and the resulting viscous mixture is spread on a

[51] D. Oliver and A. T. Jagendorf, *J. Biol. Chem.* **251**, 7168 (1976).

[52] E. Harth, S. Reimer, and A. Trebst, *FEBS Lett.* **42**, 165, 1874.

[53] M. Kitajima and W. L. Butler *Plant Physiol.* **57**, 746 (1976).

[54] H. Ochiai, H. Shibata, T. Matsuo, K. Hashimokuchi, and I. Inamura, *Agric. Biol. Chem.* **42**, 683 (1978).

[55] H. Ochiai, H. Shibata, T. Matsuo, K. Hashinokuchi, M. Yukawa, and I. Inamura, *Amino Acid and Nucleic Acid* **37**, 53 (1978).

TABLE III
PROPERTIES OF ISOLATED SPINACH CHLOROPLASTS AFTER REACTION WITH
GLUTARALDEHYDE (GA)[a] AND DIMETHYL SUBERIMIDATE (DMS)[b]

| Property | GA-reacted | | DMS-reacted | |
|---|---|---|---|---|
| | Activity (% of control) | Inactivation threshold[c] | Activity (% of control) | Inactivation threshold |
| Osmotic sensitivity[d] | 30 | 20–30 | 10 | 10–20 |
| Fraction of protein NH$_2$ reacted[e] | 60 | 20–30 | 55 | 30–40 |
| Electron-transport rates | | | | |
|   H$_2$O to ferricyanide | 40 | 40 | 0 | 50 |
|   H$_2$O to $p$-phenylenediamine | 70 | 40 | 0 | 50 |
|   Dichlorophenolindophenol to methylviologen | 8 | 120 | No inhibition | |
| Survival of electron transport activity during dark storage | Positive effect | | No effect | |
| Survival of electron-transport activity during continuous illumination | No effect | | No effect | |

[a] Chloroplast treated with glutaraldehyde for 5 min at 0°C.[25]
[b] Chloroplasts treated with dimethylsuberimidate for 18 hr at 0°C.[25]
[c] The minimum relative concentration of the crosslinker, in $\mu$moles/mg Chl, requires to achieve maximal inactivation.
[d] Measured in terms of light-scattering differences, at 546 nm, between chloroplasts suspended in distilled H$_2$O and in 333 m$M$ NaCl.
[e] Measured according to Isaakidou and Papageorgiou.[26]

glass plate and dried *in vacuo*. Using poly(vinyl alcohol) of a low degree of polymerization (e.g., 550), dried films that can be readily redissolved in buffer are obtained. The reported recovery of activity was 80% for photosystem II and nearly 100% for photosystem I, while the activity loss after 5 weeks of dark storage was about 20%. Insoluble films with smaller fractions of recovered photosystem II and I activities (about 29%) can be prepared by using higher polymers of poly(vinyl alcohols) (e.g., average degree of polymerization 1750). Also, lower fractions of recovered activity (about 11%) were reported for chloroplast preparations entrapped within polyacrylamide gels.

## Concluding Remarks

The stabilization of chloroplast membranes by chemical manipulations, such as protein cross-linking, entrapment, and microencapsulation, encounters unique technical problems that are unknown in the case of soluble proteins. This, as well as the comparatively small research effort expended to the present, accounts for the little progress made in this area. Photosynthetic electron transport offers the advantage that it requires no protein movement for its operation. In principle, therefore, membrane immobilization techniques should not interfere with it. What is required, however, is the standardization and optimization of the technical procedures by which modified photosynthetic membranes with superior functional stability can be prepared.

# [59] H₂ Production by an *in Vitro* Chloroplast, Ferredoxin, Hydrogenase Reconstituted System

## By LESTER PACKER

### Introduction

The manner in which low potential electrons produced by chloroplast photosystems are partitioned between NADPH production, electrons for cyclic photophosphorylation, autoxidation by oxygen of ferredoxin, has been studied in various laboratories; *in vivo* NADPH production successfully competes for low potential electrons.[1] However, in the *in vitro* reconstituted system using washed chloroplasts plus added ferredoxin and hydrogenase[2-5] in which NADPH and the enzymes for NADP reduction are absent, the only reaction which can compete for low potential electrons produced by the photosystems, other than hydrogenase, is the autooxidation by oxygen of ferredoxin as shown in Fig. 1. This side reaction (dashed lines in Fig. 1) (Reaction 1) produced hydrogen peroxide (Reactions 2–5). Assays for hydrogen peroxide production (Reactions 2–6) show that Reactions 2–4 significantly compete for electrons. At higher concentrations of ferredoxin more electrons are diverted to production which can be up to 72% of the $K_3Fe(CN)_6$ Hill reaction rate. Thus photolysis of water by a coupled system composed of chloroplasts, fer-

---

[1] K. Tagawa, H. Y. Tsujimoto, and D. I. Arnon, *Nature (London)* 199, 1247 (1963).

[2] I. Fry, G. Papageorgiou, E. Tel-Or, and L. Packer, *Z. Naturforsch., Teil C* 32, 110 (1977).

[3] D. Hoffman, R. Thauer, and A. Trebst, *Z. Naturforsch., Teil C* 32, 257 (1977).

[4] K. K. Rao, L. Rosa, and D. O. Hall, *Biochem. Biophys. Res. Commun.* 68, 21 (1976).

[5] L. Packer, W. Cullingford, and E. Tel-Or, *Z. Naturforsch., Teil C* 33, (in press).

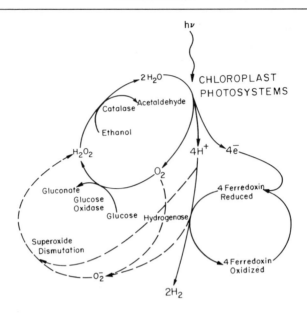

FIG. 1. Hydrogen production by a chloroplast, ferredoxin, hydrogenase system. Reactions shown in solid lines assume that the oxygen trap (glucose + glucose oxidase) functions with 100% efficiency to remove all of the $O_2$ produced by photosystem II coupled to $H_2$ production. In this case 2 $H_2$ would be produced per $O_2$ evolved per $H_2O_2$ and acetaldehyde formed. Reactions shown in dashed lines indicate the competing reaction for low potential electrons brought about by the autoxidation of ferredoxin which generates $O_2^-$ radicals. The occurrence of these reactions has the effect of decreasing the $H_2$ produced and increasing the formation of $H_2O_2$ and acetaldehyde. The extent to which these competing reactions occur has been evaluated (cf. text and Fig. 2).

redoxin, and hydrogenase can function with great efficiency for $H_2$ production but with only a limited lifetime for activity.

## Materials and Methods

Experiments are carried out in 7-ml rubber stoppered vials. Reaction mixtures in a total volume of 2 ml contained 0.1 mg/ml type II spinach chloroplasts, 0.1 mg/ml *Spirulina* ferredoxin (FD), 34 nmoles *Clostridium pasteurianum* hydrogenase, 0.5 mg/ml glucose oxidase, 0.1 mg/ml catalase, 50 m$M$ glucose, 2.5% (v/v) ethanol, 0.5 mg/ml bovine serum albumin, 65 m$M$ HEPES buffer, pH 7.0. Hydrogen production was measured by withdrawing 50 $\mu$l aliquots from the gas phase and injecting into a Varian Model 920 GLC chromatograph with a molecular sieve column (5A mesh 30/60) and thermal conductivity detector.

To estimate the Hill reaction rate, 20–50 $\mu$l additions of a 40 m$M$ solution of potassium ferricyanide $K_3Fe(CN)_6$ were injected into the vials, and the time during which cessation of $H_2$ production occurred was measured. Hydrogen uptake ceases because of the preferential shunting of electrons to ferricyanide.[5] Equal aliquots of $K_3Fe(CN)_6$ are added to duplicate vials as controls.

Acetaldehyde is measured using a modified 2,4-thiobarbituric acid (TBA) assay.[6] The reaction is stopped by adding 0.5 ml trichloroacetic acid to each vial, the contents were centrifuged and a 1 ml aliquot of the supernatant added to 1.4 ml of a TBA solution (.071 gm TBA and 0.7 ml of 1 $M$ NaOH made up to a final volume of 100 ml distilled water) and heated in a boiling water bath for 20 min. The absorbance measurements are made at 499 nm. Under these conditions with both $CH_3CHO$ and glucose present, absorption at 499 nm is proportional to the concentration of $CH_3CHO$.[7]

### Assays

After flushing the reaction mixture with $N_2$ gas, the initial rates of $H_2$ production and acetaldehyde production in the complete coupled system are determined in the absence of glucose oxidase (Fig. 2a) or hydrogenase (Fig. 2b). In the absence of hydrogenase some $CH_3CHO$ formation occurs which ceases after about 1 hr. This indicates that $O_2$ and/or $H_2O_2$ were present in the reaction mixture. Thereafter, acetaldehyde production paralleled $H_2$ production. If glucose oxidase was omitted, $CH_3CHO$ formation continued. This indicates that glucose oxidase was not the only pathway present for formation of $H_2O_2$ from $O_2$. Reduced ferredoxin is known to be autoxidized to form $H_2O_2$, viz.,

$$4\,Fd_{ox} + 2\,H_2O \xrightarrow[\text{Chl}]{h\nu} O_2 + 4\,Fd_{red} + 4\,H^+ \tag{1}$$

$$O_2 + Fd_{red} \longrightarrow O_2^- + Fd_{ox} \tag{2}$$

$$2\,H^+ + O_2^- + Fd_{red} \longrightarrow H_2O_2 + Fd_{ox} \tag{3}$$

or

$$2\,H^+ + 2\,O_2^- \longrightarrow H_2O_2 + O_2 \tag{4}$$

The sum of Eqs. (2) and (3) or twice Eq. (2) plus Eq. (4) is:

$$2\,H^+ + O_2 + 2\,Fd_{red} \longrightarrow H_2O_2 + Fd_{ox} \tag{5}$$

$$H_2O_2 + ethanol \xrightarrow{\text{catalase}} CH_3CHO + 2\,H_2O \tag{6}$$

$$\text{Overall:}\; 2\,\text{Ethanol} + O_2 \longrightarrow 2\,CH_3CHO + 2\,H_2O \tag{7}$$

---

[6] V. S. Waravdekar and D. S. Saslaw, *J. Biol. Chem.* **234**, 1945 (1959).
[7] J. F. Allen, *Biochem. Biophys. Res. Commun.* **66**, 36 (1973).

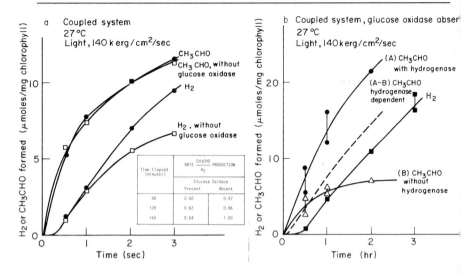

FIG. 2. Hydrogen and $H_2O_2$ production by the coupled system (cf. Packer et al.[5]). (a) Absence of oxygen trap (glucose oxidase). Data shown indicate about 40% of low potential electrons of ferredoxin are autoxidized by $O_2$ to result in $H_2O_2$ and a $CH_3CHO:H_2$ ratio of 0.6 in the presence of the $O_2$ trapping system. (b) Absence of oxygen trap and hydrogenase.

When glucose oxidase was absent (Fig. 2a) $H_2$ production is slightly slower probably due to a smaller pool of reduced ferrdoxin or a slightly higher steady state level of $O_2$ partially inhibiting the enzyme. In the absence of glucose oxidase, with ferredoxin consuming $O_2$, the expected ratio of $CH_3CHO:O_2$ is 2. Therefore, the expected ratio of $CH_3CHO:H_2$ would be 1, which is observed (Fig. 2a and b). When glucose oxidase was present the observed ratio of $CH_3CHO:H_2$ was found to be between 0.6 and 0.7, not 0.5 as predicted by the reactions shown in solid lines only in Fig. 1. This corresponds to 60–80% of the $O_2$ released by water photolysis during $H_2$ production being trapped by glucose oxidase. Other studies[2] show that glucose disappearance (equal to $O_2$ trapped) is also found to be approximately 70% of the value expected according to the reactions in solid lines shown in Fig. 1. The remaining $O_2$ (30%) is trapped by ferredoxin, but since $O_2$ is evolved in the reduction of ferredoxin the percentage of the total $O_2$ production trapped by ferredoxin was about 40%.

The Hill reaction rate under the usual conditions employed (0.1 mg/ml chlorophyll, light intensity of 140 kergs/cm²), as measured by the time required for $K_3Fe(CN)_6$ to be consumed, was 55–60 μmole/hr/mg chlorophyll (Fig. 3). This was accompanied by an initial rate of $H_2$ production

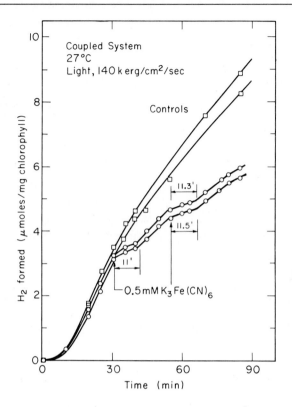

FIG. 3. Method for estimation of the Hill reaction rate in the coupled system for $H_2$ production (cf. Packer et al.[5]).

of 9.8 mmole $H_2$/hr/mg chlorophyll or 33% of the Hill reaction rate. Addition of $K_3Fe(CN)_6$ does not affect the rate of $H_2$ production, but an addition of 1 $\mu$mole of $O_2$ caused an inhibition of $H_2$ production similar to that caused by the addition of $K_3Fe(CN)_6$. This suggests that the oxygen produced by the reduction of the $K_3Fe(CN)_6$ is responsible for the slight inhibition of $H_2$ production after the $K_3Fe(CN)_6$ has been all reduced. Increasing the ferredoxin concentration to 0.55 mg/ml increased the rate of $H_2$ production to 72% of the Hill reaction rate.

Higher rates of $H_2$ production with the reconstituted system can be obtained by optimizing light intensity as by lowering the chlorophyll concentration which yields initial rate of $H_2$ production of approximately 50 $\mu$moles/mg chlorophyll/hr and the total $H_2$ produced over the time interval during which the system retains activity being relatively unchanged.

Summary

The continuous production of molecular hydrogen by an *in vitro* reconstituted system composed of chloroplasts, ferredoxin, and hydrogenase has been studied in several laboratories.[2-5] Such a system can continuously produce $H_2$ by a photosystem II-driven reaction for 6–8 hr when the reaction is carried out in the presence of $O_2$ (glucose and glucose oxidase) and $H_2O_2$ (ethanol and catalase) traps at a rate of 10–25 $\mu$mole $H_2$ produced per milligram of chlorophyll per hour at 25°–29°. This is only about 5–10% of the rate of the Hill reaction found with spinach chloroplasts under similar conditions *in vitro* but with other acceptors.

Acknowledgments

The methods described in this chapter are based upon previous investigations of Ian Fry, George C. Papageorgiou, Elisha Tel-Or, and William Cullingford. The work described here was supported by the United States Department of Energy and NATO.

# [60] Energy-Dependent Reverse Electron Transport in Chloroplasts

*By* YOSEPHA SHAHAK and MORDHAY AVRON

The coupling between light-induced electron transport, the formation of a pH gradient across the thylakoid membrane ($\Delta$pH) and the synthesis of ATP from ADP and $P_i$ has been demonstrated to be a fully reversible process under appropriate experimental conditions. ATP hydrolysis, which is latent in chloroplasts, can be activated by preillumination in the presence of dithiol reagents[1,2] inducing proton uptake[3] from the medium into the inner thylakoid space with the consequent creation of a trans-thylakoid pH gradient in the dark.[2] This light-triggered ATPase further drives the reversal of electron transport toward photosystem II.[4,5] Reverse electron flow can also be driven by an artificially induced pH gradient.[6,7] Both ATP and an artificially induced pH gradient have also

[1] C. Carmeli and M. Avron, Vol. 24 [7], 92.
[2] T. Bakker-Grunwald, *in* "Photosynthesis I" (A. Trebst and M. Avron, eds.), p. 639. Springer Verlag, Berlin and New York, 1977.
[3] C. Carmeli, *FEBS Lett.* **7**, 297 (1970).
[4] K. G. Rienits, H. Hardt, and M. Avron, *FEBS Lett.* **33**, 28 (1973).
[5] K. G. Rientis, H. Hardt, and M. Avron, *Eur. J. Biochem.* **43**, 291 (1973).
[6] Y. Shahak, H. Hardt, and M. Avron, *FEBS Lett.* **54**, 151 (1975).
[7] Y. Shahak, U. Pick, and M. Avron, *in* "Enzymes, Electron Transport Systems" (P. Desnuelle and A. M. Michelson, eds.), p. 305. Elsevier, Amsterdam, 1975.

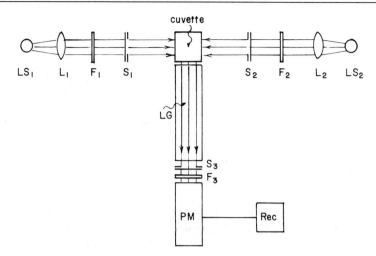

FIG. 1. Schematic description of chlorophyll fluorescence measuring apparatus. LS, Light source; L, lens; F, filter; S, shutter. $LS_1$, a high intensity light source about $10^5$ ergs $cm^{-2}$ $sec^{-1}$ at the surface of the cuvette; $F_1$, 2 cm saturated $CuSO_4$ solution; $LS_2$, low intensity light source, intensity less than 300 ergs $cm^{-2}$ $sec^{-1}$; $F_2$, Corning CS 4-96 and a broad-band (20-nm half-band width) interference filter peaking at 550 nm; LG, light guide (optional). $F_3$, Schott RG 665 and Corning CS 4-77 filters; PM, photomultiplier, S 20 sensitivity; Rec, a fast recorder. After S. Malkin and G. Michaeli, *Photosynth., Two Centuries After Its Discovery By Joseph Priestley, Proc. Int. Congr. Photosynth. Res. 2nd, 1971* p. 149 (1972).

been shown, under appropriate conditions, to induce luminescence from photosystem II.[8,9]

There is a basic difference between ATP and acid–base-driven reactions. The former are steady state reactions which are maintained as long as the ATPase is active,[5,8] while the latter are transient reactions driven by a rapidly formed (and rapidly decaying) transthylakoidal pH gradient.[6,7]

This section describes the experimental techniques developed to measure ATP and acid–base driven reverse reactions including the reverse electron flow induced reduction of Q, the primary acceptor of photosystem II, and oxidation of cytochrome $f$; and the reverse electron flow induced luminescence.

[8] U. Schreiber and M. Avron, *FEBS Lett.* **82**, 159 (1977).
[9] Y. Shahak, Y. Siderer, and M. Avron, *in* "Photosynthetic Organelles, Structure and Function" (S. Miyachi, S. Katoh, Y. Fujita, and K. Shibata, eds.), p. 115. Jpn. Soc. Plant Physiol., Japan, 1977.

## ATP-Driven Reverse Reactions

### Reduction of Q[4,5]

The reduction of Q is measured as the increase in chlorophyll $a$ fluorescence yield[10,11] under a very weak measuring light, which is insufficient to induce significant forward electron flow. Quantitative estimation of the fraction of Q reduced during the reverse reaction can be obtained by independently determining the total amount of Q as the difference between maximal fluorescence yield ($F_{max}$) obtained after the addition of a few crystals of dithionite (which fully reduces Q) and the initial fluorescence yield ($F_o$) obtained upon turning the measuring light on.

*Apparatus.* Q reduction can be measured in any fluorescence apparatus containing a low intensity light source, a photomultiplier sensitive in the red–far red range of the spectrum which can be attached to a recorder or an oscilloscope. In addition, a strong light source is needed for activating the latent ATPase activity.[1,2] A diagram of the apparatus is given in Fig. 1.

*Reagents*

SNT (0.2 $M$ sucrose, 0.1 $M$ NaCl, 0.05 $M$ Tris-HCl pH 7.8)

Tricine, 0.2 $M$, pH 8.0

NaCl, 0.2 $M$

MgCl$_2$, 0.05 $M$

DTT (dithiothreitol), 0.05 $M$ (fresh solution)

PMS (phenazine methosulfate), 10 $\mu M$

ATP, 0.4 $M$, pH 8.0

Chloroplasts, containing about 3 mg chlorophyll per milliliter isolated from spinach

or lettuce leaves in SNT as previously described.[12]

*Reaction Mixture.* The reaction mixture contains in a total volume of 2.0 ml, 0.2 ml of each of the following reagents: Tricine, NaCl, MgCl$_2$, DTT, PMS, and chloroplasts containing about 20 $\mu$g chlorophyll/ml.

*Procedure.* The milliliters of the reaction mixture are placed in a 1-cm$^2$ cuvette. ATPase activity is activated by illumination for 2 min with the strong light source (LS$_1$ in Fig. 1). After turning the activating light off, the shutter of the photomultiplier (S$_3$) is opened and the time course of chlorophyll fluorescence is monitored. The weak measuring light (LS$_2$)

[10] L. N. M. Duysens and H. E. Sweers, *in* "Studies on Microalgae and Photosynthetic Bacteria," p. 353. Tokyo Univ. Press, Tokyo, 1963.

[11] S. Malkin and B. Kok, *Biochim. Biophys. Acta* **126,** 413 (1966).

[12] M. Avron, *Anal. Biochem.* **2,** 535 (1961).

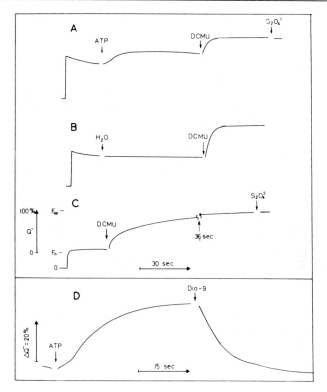

Fig. 2. ATP-induced increase in fluorescence yield. Reaction conditions as described in the text. (A) ATP (1 m$M$) and DCMU (6 $\mu M$) and a few crystals of dithionite were added where indicated. (B) Water in a volume equal to that used for ATP was added where indicated. (C) DTT and PMS omitted. (D) Dio-9, 3.3 $\mu$/ml where indicated. Sensitivity 5 times higher than that employed in (A), (B) and (C). From Rienits et al.[4]

is on all the time. ATP (10 $\mu$l of the stock solution) is added about 30 sec after the illumination, a period during which the high fluorescence yield induced by the strong illumination has essentially decayed. In Fig. 2 the fluorescence increase observed upon addition of ATP and the calibration of the total amount of Q are demonstrated.

*Variations in the Procedure.* (i) Dilution technique: In the procedure described above the compounds present in the light activation stage are also present during the dark ATP-driven reverse reaction. In order to separate between the two stages a dilution technique can be used. For ten times dilution 0.2 ml of the reaction mixture with the chloroplasts (containing 40 $\mu$g chlorophyll) is placed on the bottom of the cuvette and a homemade rod with 45° mirror at its end is put within the cuvette as

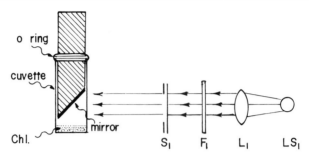

o ring

cuvette

Chl.          mirror          $S_l$     $F_l$     $L_l$          $LS_l$

FIG. 3. Design for preillumination in the dilution technique. See text for details.

described in Fig. 3. After the 2 min preillumination the rod is taken out, 1.8 ml of the reaction mixture (lacking chloroplasts) with the desired compound added or omitted is injected into the cuvette, and the experiment proceeds as described above.

(ii) In experiments using inhibitors such as DCMU [3-(3,4-dichlorophenyl)-1,1-dimethylurea] or NQNO (2-n-nonyl-4-hydroxyquinoline N-oxide) an extremely low measuring light intensity (less than 5 ergs cm$^{-2}$ sec$^{-1}$) is required, since at higher intensities the inhibitors induce by themselves an increase in fluorescence yield (see Fig. 2).

### Oxidation of Cytochrome $f$[5]

ATP-driven cytochrome $f$ oxidation is measured in an Aminco-Chance dual wavelength spectrophotometer as the decrease in the absorption difference between 554 nm and 540 nm, according to Avron and Chance.[13]

*Reaction Mixture.* The reaction mixture is the same as described for ATP-driven Q reduction except for higher chlorophyll concentration around 100 $\mu$g/ml.

*Procedure.* For efficient light activation of the ATPase the sample is initially placed on the bottom of a 10-ml beaker. The 2-ml sample is illuminated for 4 min with a high light intensity, and the activated solution is transferred into the cuvette so that ATP is added 30 sec after preillumination. During measurements the cuvette contents are continuously agitated by a vibrating wire stirrer.

### Simultaneous Measurement of $\Delta pH$ and Q reduction[14,14a]

The pH gradient formed across the thylakoid membrane can be estimated measuring the quenching of the fluorescence of fluorescent amines

[13] M. Avron and B. Chance, *in* "Currents in Photosynthesis" (J. B. Thomas and J. C. Godheer, eds.), p. 455. Donker Publ., Rotterdam, The Netherlands, 1966.

[14] M. Avron and U. Schreiber, *FEBS Lett.* **77**, 1 (1977).

[14a] U. Schreiber and M. Avron, *Biochim. Biophys. Acta* **546**, 436 (1979).

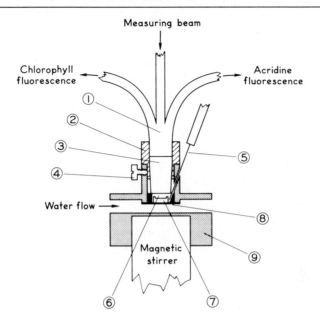

FIG. 4. Measuring device for simultaneously observing ATP-driven changes in chlorophyll and 9-aminoacridine fluorescence. 1, Trifurcated fiber optics: 2, cover of reaction chamber constructed from an external poly (vinyl chloride) ring and an internal transparent lucite rod (3), held in place by a nylon screw (4). 5, Removable syringe for injection of additions; 6, chamber for reaction mixture; 7, rotating magnetic stirring disc, 8, aluminum container which serves also as a heat conductor from the thermostated flowing water; 9, poly (vinyl chloride) body. From Avron and Schreiber.[14]

such as 9-aminoacridine.[15] Simultaneous measurement of 9-aminoacridine and chlorophyll fluorescence changes enables the comparison of ΔpH formation and reverse electron flow which are induced by ATPase activity in the dark.

*Apparatus.* The measuring apparatus constructed by Avron and Schreiber[14] is illustrated in Fig. 4. It employs a bundle of three optical fibers, one of which introduces a weak (about 50 ergs $cm^{-2}$ $sec^{-1}$) blue measuring light to the sample; the second permits a photomultiplier to monitor the blue-green fluorescence of 9-aminoacridine, and the third allows a second photomultiplier to monitor the red fluorescence of chlorophyll *a*. The measuring light (390–440 nm) is filtered through Corning 4-96 and 5-58 filters; the 9-aminoacridine fluorescence measuring photomultiplier is screened by a Corning 4-96 filter and Wratten 58 and 64 filters (transmitting 505–535 nm) and the chlorophyll fluorescence meas-

[15] S. Schuldiner, H. Rottenberg, and M. Avron, *Eur. J. Biochem.* **39** 455 (1973).

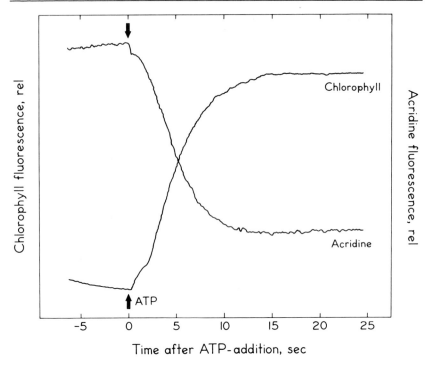

Fig. 5. Simultaneous recording of ATP-driven buildup of ΔpH and reduction of Q. Recorded in the instrument described in Fig. 4, under the conditions described in the text. From Avron and Schreiber.[14]

uring photomultiplier by a Corning 2-64 filter transmitting light longer than 655 nm.

For ATPase activation the fiber bundle is temporarily removed and the cuvette is illuminated with intense heat-filtered white light (about $10^5$ ergs cm$^{-2}$ sec$^{-1}$) from above. The reaction mixture is stirred by magnetic stirrer and is maintained at a constant temperature of 15° by a continuous flow of water from a thermostated bath.

*Reaction Mixture.* The standard reaction mixture is used, except for the addition of 2 $\mu M$ 9-aminoacridine.

*Procedure.* The reaction mixture (about 0.5 ml) is preilluminated for 3 min and monitoring initiated within 5–10 sec after the strong light is turned off. Twenty seconds after the light is off, 3 $\mu$l of 10 m$M$ ATP are injected with a microsyringe, and the two kinds of fluorescence changes are recorded as demonstrated in Fig. 5.

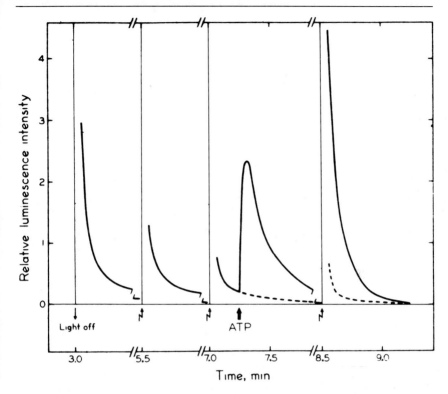

Fig. 6. ATP-driven luminescence. Recorded in the instrument described in Fig. 4 under the conditions described in the text. The light off arrow indicates the end of a 3 min ATPase light-activation period. Flash pairs (6 $\mu$sec each) are applied at the zigzag arrows. The broken lines represent the decay of flash induced luminescence when no ATP is added. Photomultiplier high-voltage turned off during illumination or flashing. From Schreiber and Avron.[8]

## Luminescence[8]

Luminescence is commonly considered to result from the recombination of the oxidized primary electron donor ($Z^+$) and the reduced electron acceptor ($Q^-$) of photosystem II which are produced during illumination.[16,17] Therefore, in addition to the procedure employed thus far for activating the ATPase and producing $Q^-$ via reverse electron flow, a short preilluminating flash is given just prior to the addition of ATP to

[16] J. Lavorel, in "Bioenergetics of Photosynthesis" (Govindjee, ed.), p. 319. Academic Press, New York, 1975.

[17] S. Malkin, in "Photosynthesis I" (A. Trebst and M. Avron, eds.), p. 473. Springer-Verlag, Berlin and New York, 1977.

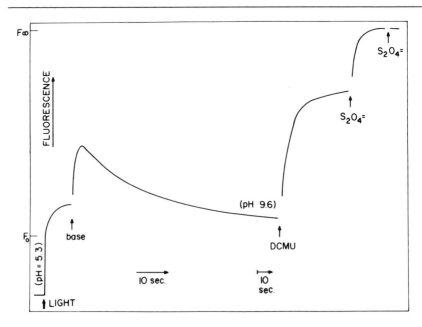

FIG. 7. Acid–base-driven reduction of Q. Conditions as described in the text. DCMU, 1.5 $\mu M$, and a few crystals of dithionite were added where indicated. From Shahak et al.[6]

produce the necessary $Z^+$. The electron in $Q^-$, produced simultaneously by the flash, is rapidly drained into the large secondary acceptor pool (plastoquinone).

*Reaction Mixture.* The same reaction mixture is used, except for the omission of PMS, which accelerates the luminescence decay to the point where no ATP observed effect can be seen (see also Avron and Schreiber[17a]).

*Procedure.* The measuring apparatus described in Fig. 4 was used, utilizing only the chlorophyll fluorescence sensitizing photomultiplier and no measuring light. The reaction mixture (about 0.5 ml) is preilluminated for 3 min with a strong light (about $10^5$ ergs cm$^{-2}$ sec$^{-1}$) to activate the ATPase. After 90 sec in the dark two saturating 6 $\mu$sec flashes (or a 100-msec flash of bright white light) were given and ATP (4 $\mu$l of 0.1 $M$) was added before the flash-induced luminescence has completely decayed. Figure 6 illustrates the ATP-induced luminescence observed. It also illustrates the much larger luminescence observed when the ATPase is active on subsequent flashes. The dashed lines illustrate the luminescence observed in the absence of added ATP.

[17a] M. Avron and U. Schreiber, *Biochim. Biophys. Acta* **546**, 448 (1979).

*Properties*

ATP-driven reverse electron flow is inhibited by energy-transfer inhibitors such as Dio-9 phloridzin or tentoxin; by all uncouplers, such as amines, FCCP (carbonylcyanide-$p$-trifluoromethoxyphenylhydrazone), and gramicidin; and by the electron-transport inhibitors DCMU and NQNO.[4,5] The light-activated state decays within a few minutes in the dark unless ATP is added. The dark reverse electron flow requires the presence of an electron donor such as DTT, ascorbate, or hydroquinone.[5]

## Acid–Base-Driven Reverse Reactions

*Reduction of $Q$[6,7]*

Acid–base-induced Q reduction is measured by the same method as the ATP-driven Q reduction, namely, by the increase in chlorophyll $a$ fluorescence yield in the apparatus described in Fig. 1.

*Reagents*
    Succinate–KOH, 0.03 $M$, pH 5.5
    KCl, 0.3 $M$
    MgCl$_2$, 0.1 $M$
    Tris, 0.2 $M$, pH 10.6
    Chloroplasts suspended in SNT containing about 4 mg chlorophyll/ml

*Reaction Mixture.* The reaction mixture for the acid stage contains in a total volume of 2.0 ml, 0.2 ml of each of the reagents: succinate, KCl, MgCl$_2$, and chloroplasts containing about 40 $\mu$g of chlorophyll, final pH 5.3–5.8

*Procedure.* The reaction mixture is incubated for 1 min in the cuvette at room temperature, followed by illumination from LS$_1$, (see Fig. 1) for 20 sec to reduce the plastoquinone pool. The shutter of the photomultiplier (S$_3$ in Fig. 1) is opened and the fluorescence changes are followed. Transition to the base stage is achieved by injection of a sufficient amount of Tris (about 0.2 ml) to give a final pH of 9.2. The signal observed (Fig. 7) is composed of a fast fluorescence increase followed by a slow decrease which reflects the reoxidation of Q$^-$.

*Luminescence[9]*

Acid–base-induced luminescence has been measured in the apparatus described in Fig. 4 or in the setup described in Fig. 8. The latter is composed of a light source for preillumination, an arrangement permitting injection of base into the cuvette, a photomultiplier sensitive at the red–far red range, an oscilloscope or a fast recorder, and a control unit which

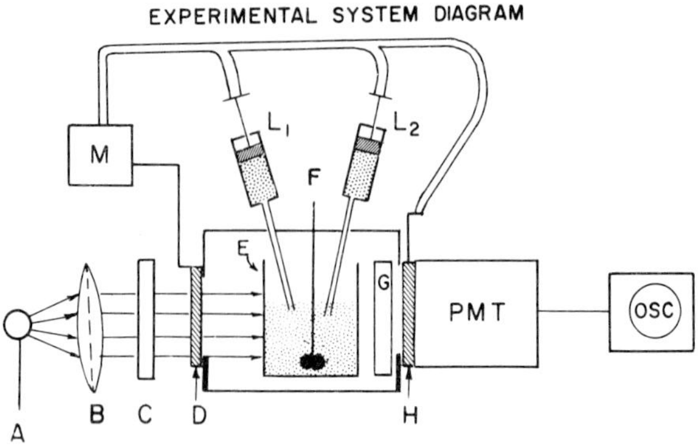

EXPERIMENTAL SYSTEM DIAGRAM

FIG. 8. Experimental arrangement for measuring delayed light emission and acid-base induced luminescence. A, High intensity light source for preillumination. Light intensity is about $10^5$ ergs cm$^{-2}$ sec$^{-1}$ at the surface of the cuvette; B, lens; C, filters: 1 cm saturated CuSO$_4$ solution and a Schott GG 429 filter; D, shutter for preillumination; E, cuvette; F, stirrer; G, Schott RG 665 filter; H, shutter for the emitted light. PMT, photomultiplier with S 20 sensitivity; OSC, oscilloscope or a fast recorder; L$_1$ and L$_2$, pneumatically operated syringes for injection of base and other reagents; M, time control unit starting operation at the closure of shutter D. From Malkin and Hardt.[18]

controls the time between the closure of the illumination shutter, the opening of the photomultiplier shutter, and the injection of the base. The apparatus has been constructed by Malkin and Hardt[18] (Fig. 8).

*Reaction Mixture.* As described for acid–base-induced Q reduction.

*Procedure.* The reaction mixture is incubated for 1 min followed by illumination for 20 sec which induces both charge separation and reduction of plastoquinone. Shutter D (Fig. 8) is closed, shutter H is opened, and the base is injected after a dark period (8–15 sec) during which the native luminescence decays essentially completely. The injection of base gives rise to additional emission of light as demonstrated in Fig. 9.[19,9]

Comparison between acid–base-induced Q reduction and luminescence can be readily made by keeping the reaction mixture for 1 min in the dark after preillumination and firing a single flash (half-life 100 $\mu$sec) a few seconds before the injection of the base to reform charge separation in the luminescence reaction.

*Properties.* Acid–base reverse reactions depend on the plastoquinone

[18] S. Malkin and H. Hardt, *Photosynth., Two Centuries Its Discovery Joseph Priestley, Proc. Int. Congr. Photosynth. Res., 2nd, 1971* p. 253 (1972).

[19] M. Avron, U. Pick, Y. Shahak, and Y. Siderer, *Struct. Biol. Membr., Proc. Nobel Symp., 34th, 1977* p. 25 (1977).

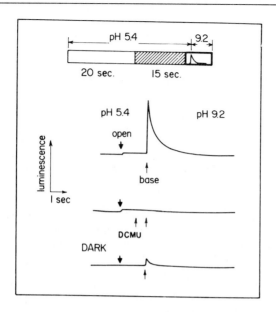

Fig. 9. Proton gradient driven reverse electron flow induced luminescence. Reaction conditions are described in the text. One-tenth milliliter of 0.1 m$M$ DCMU in 10% methanol was injected where indicated. From Avron *et al.*[19]

pool as the electron donor, and thus its oxidation by far-red preillumination or by oxidants (ferricyanide, oxidized DBMIB) inhibit the reaction.[6,7,9] Unlike the acid–base phosphorylation reaction,[20] the reverse electron flow reactions do not require internal buffers in the acid stage.[6] The ATP- or acid–base-induced reverse electron flow are inhibited by electron transfer inhibitors such as DCMU, NQNO and *o*-phenanthroline.[6,9] Energy transfer inhibitors inhibit the ATP-driven reactions but not the acid–base driven reactions. The acid–base-driven Q reduction is rather insensitive to uncouplers except for those which are ADRY (acceleration of the deactivation reactions in the water splitting enzyme Y) agents[21] such as FCCP and SF 6847 (3,5-di-*tert*-butyl-4-hydroxylbenzylidenemalononitrile).[9] However, the acid–base-driven luminescence is inhibited by all uncouplers, with the ADRY agents being somewhat more effective.[9]

[20] S. Schuldiner, *in* "Photosynthesis I" (A. Trebst and M. Avron, eds.), p. 416. Springer-Verlage, Berlin and New York, 1977.
[21] G. Renger, *Biochim. Biophys. Acta* **314**, 320 (1973).

# [61] Single Cell Photosynthesis

*By* JEROME C. SERVAITES and WILLIAM L. OGREN

## Principle

Single leaf cells with high photosynthetic activity are isolated in high yield from thin strips of soybean leaf by stirring the strips in the presence of a macerating enzyme and simultaneously removing the released cells from the site of the stirring action.

## Plant Material

Soybean (*Glycine max* [L.] Merr. cv. Wayne) plants are grown at 35 klux under a 16 hr photoperiod (30° light/20° dark) for 3 weeks, and then transferred to a 6-hr photoperiod. After 1 week of short-day treatment, a mature leaf, usually the third to fifth emergent trifoliate, is shaded with aluminum foil for 1 to 2 days. On the day of cell isolation, the foil is removed, and the leaf is illuminated for 1 hr and detached from the plant. At this time the leaf area is about 0.6 dm², specific leaf weight is about 2 gm fresh weight per dm², and the leaf contains about 4 mg Chl per gram of fresh weight.

A short-day treatment is necessary for high photosynthesis rates in cotton cells,[1] and is considered to be beneficial because it reduces the amount of starch in chloroplasts. Starch granules disrupt chloroplast membranes during preparation, causing a loss of soluble constituents.[2] We also found that a short-day treatment is necessary for high rates of $CO_2$ fixation in soybean cells, and additional activity is obtained by shading. The growth conditions given above are not the only ones which yield cells with high photosynthetic activity, but it is essential that the growth regime used provides low starch levels at the time of cell isolation.

## Cell Isolation

All procedures are carried out at room temperature.
*Infiltration Medium*
　20 m$M$ MES [2-(*N*-morpholino)-ethanesulfonic acid], pH 5.8

---

[1] D. W. Rehfeld and R. G. Jensen, *Plant Physiol.* **52**, 17 (1973).
[2] D. A. Walker, Vol. 23, p. 211.

12.5 m$M$ $K_2SO_4$

2% (w/v) Polyvinylpyrrolidone-40[3,4]

3% (w/v) Macerase (Calbiochem)[5]

The detached leaf is rinsed with distilled water, and the midrib is removed with a razor blade. The leaf sections are placed in a glass petri dish containing 20 ml of infiltration medium and are cut into 1 mm × 1 cm strips. The leaf strips are vacuum-infiltrated for 15 sec with a water aspirator, and the infiltration medium discarded.

*Maceration Medium*

0.3 $M$ Sorbitol

20 m$M$ MES, pH 5.8

12.5 m$M$ $K_2SO_4$

2% (w/v) Polyvinylpyrrolidone-40

3% (w/v) Macerase

The leaf strips and a stirring bar are placed in the maceration chamber of the cell isolation apparatus in Fig. 1. Fifty milliliters of maceration medium are placed in a 50-ml Erlenmeyer flask, which acts as a reservoir and debubbler. The flask is inverted and the medium is circulated to and through the maceration chamber at a rate of 10 ml/min by a tubing pump. When the maceration chamber is filled with medium, the leaf strips are stirred by the magnetic stirrer placed under the maceration chamber assembly. As individual cells are released they are carried through a 420-$\mu$m mesh-opening stainless steel screen, which retains the leaf strips, and collected on a 4.8-cm nylon net[6] (20-$\mu$m mesh-opening) housed in a filter unit.[7] Chloroplasts and cell fragments pass through the nylon net. After 15 min, the pump is turned off, and the net holding the cells is removed from the filter unit.

*Wash Medium*

0.2 $M$ Sorbitol

50 m$M$ Tris-Cl, pH 7.8

5 m$M$ $KNO_3$

2 m$M$ Ca $(NO_3)_2$

1 m$M$ $MgCl_2$

Cells are removed from the nylon net with a gentle stream of wash

---

[3] Obtained from Sigma Chemical Corp., St. Louis, Missouri.

[4] Mention of a trademark or proprietary product does not constitute a guarantee or warranty of the product by the United States Department of Agriculture and does not imply its approval to the exclusion of other products that my also be suitable.

[5] Obtained brom Calbiochem, LaJolla, California.

[6] Nitex nylon, obtained from Tobler, Ernst, and Traber, Inc., New York, New York.

[7] Swinnex-47, obtained from Millipore Corporation, Bedford, Massachusetts.

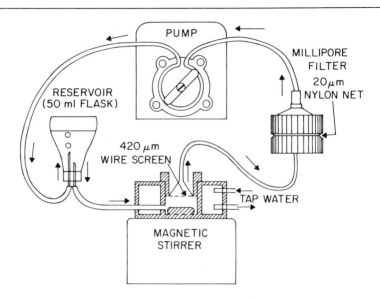

Fig. 1. Design of the cell isolation apparatus. The maceration chamber was constructed from Plexiglas tubing (2.5 cm long × 3.8 cm i.d. × 6 mm thick). The maceration chamber was concentrically surrounded by a second Plexiglas tube (2.5 cm long × 9.4 cm i.d. × 6 mm thick), and both tubes rested on a circular 6-mm thick sheet of Plexiglas. The area between the two pieces of tubing at the top was also covered by a 6-mm piece of Plexiglas. Cool tap water circulated between the two tubes. A Plexiglas plug, machined to fit snugly into the maceration chamber, was covered with a stainless steel net (420 μm mesh opening) and an inverted plastic funnel. When the plug was in place, the size of the maceration chamber was 1.3 cm high × 3.8 cm, a volume of about 14.7 ml. Details of operation are in the text. Reprinted from J. C. Servaites and W. L. Ogren: *Plant Physiol.* **59**, 587 (1977). Copyright (1977) by the American Society of Plant Physiologists. Reproduced by permission of the copyright owner.

medium (20 ml total volume), centrifuged at $100 g$ for 2 min, resuspended in 20 ml wash medium, and centrifuged again. Filtering the preparation through an 80-μm mesh-opening nylon net removes large cell clumps and epidermal cells. The cells are then resuspended in 10 ml of wash medium containing 5 m$M$ dithiothreitol. After resuspension, the cells are placed in a Warburg apparatus and illuminated for 1 hr before assay.

*Comments on the Isolation Procedure*

*Leaf Material.* Up to 3 gm of fresh weight of leaf material can be macerated at one time, yielding 50 to 70% of the cells. Cell yield is reduced when leaf strips greater than 1 mm wide are used.

*Sorbitol.* Commercially available sorbitol frequently contains material which interferes with analysis of the acidic products of photosynthesis. This unidentified material is removed by passing the sorbitol through a column of Dowex-1 acetate.

Takebe *et al.*[8] found that a high (0.6 *M*) sorbitol concentration was essential for the release of cells from tobacco leaves. Thus high sorbitol concentration has been used routinely during cell isolation and assay by subsequent researchers even though severe plasmolysis of the cells occurs.[1] High sorbitol concentration inhibits both cell release and photosynthesis in soybean, and 0.2 to 0.3 *M* sorbitol is optimal.[9] Thus, unlike other species studied, cells can be prepared from soybean without being subjected to plasmolysis.

*Macerating Enzyme.* Before use, Macerase is dissolved in 50 m*M* MES, pH 5.8, and 12.5 m*M* $K_2SO_4$, and precipitated by the addition of 70% ammonium sulfate. The protein is redissolved in fresh buffer and further purified by passage through Sephadex G-25 equilibrated with buffer. This procedure removes unidentified compounds which we found greatly stimulated cell dark respiration. The treatment does not affect cell-releasing activity of the Macerase nor the rate of photosynthesis in the cells released by the purified enzyme. Macerase was routinely used in the original work, but other preparations of macerating enzymes are also suitable, namely, Extractase PC, also called Rohament P (Fermico Biochemics Inc., Elk Grove Village, Illinois) and Pectinol 10-M concentrate (Rohm and Haas Company, Philadelphia, Pennsylvania).

*Maceration.* Stirring action during maceration is essential and cannot be duplicated by shaking. Using the same amount of maceration medium and leaf strips, only 10 to 20% of the cells are released during 2 hr shaking, and the photosynthetic activity of these cells is very low. Rate of cell release, cell yield, and photosynthetic activity are all increased about five times when the released cells are removed from the maceration chamber by the filtering system described in Fig. 1.

*Photosynthesis Assay*

*Assay Medium*
    0.2 *M* Sorbitol
    50 m*M* Tris-Cl, pH 7.8 (prepared and stored $CO_2$-free)
    5 m*M* $KNO_3$
    2 m*M* Ca $(NO_3)_2$

[8] I. Takebe, Y. Otsuki, and S. Aoki, *Plant Cell Physiol.* **9**, 115 (1968).
[9] J. C. Servaites and W. L. Ogren, *Plant Physiol.* **59**, 587 (1977).

1 m$M$ MgCl$_2$

1 $\mu$Ci KH$^{14}$CO$_3$ (concentration as desired, 5 m$M$ for standard assay)

Cells (5 to 20 $\mu$g Chl when determining rates of CO$_2$ fixation, up to 150 $\mu$g Chl when photosynthetic products are to be analyzed)

Because of our interest in photorespiratory metabolism and its relationship to photosynthesis, this procedure includes a step for controlling the O$_2$ concentration in the gas phase over the cells. This step may be omitted if the researcher wants only to determine CO$_2$-saturated photosynthesis rates (5 m$M$ KHCO$_3$ or higher), where the O$_2$ inhibition of photosynthesis is largely overcome. However, when lower bicarbonate concentrations are used, as in a $K_m$ (CO$_2$) determination, for example, it is necessary either to remove O$_2$ from the reaction vessel or to recognize that O$_2$ inhibition of photosynthesis gives rise to high apparent $K_m$ (CO$_2$) values. Control of the O$_2$ concentration is similarly important in the study of chloroplast photosynthesis.

Appropriate quantities of sorbitol, KNO$_3$, Ca(NO$_3$)$_2$, and MgCl$_2$ stock solutions are added to 20 ml reaction vessels. If the amount of $^{14}$CO$_2$ incorporated is to be determined with a scintillation spectrometer, it is convenient to use glass scintillation vials as reaction vessels. The vials are capped with sleeve type rubber stoppers and fitted (outside the center area) with one 16-gauge needle and one 18-gauge needle, and placed in an illuminated photosynthetic Warburg apparatus.

The gas phase is prepared by mixing N$_2$ and O$_2$ from cylinders to the desired O$_2$ concentration. The gas mixture is passed through Ascarite to remove any traces of CO$_2$ and humidified by passing through water. The gas then goes into a manifold and is distributed through rubber tubing to the vials, entering each vial through the 18-gauge needle and exiting through the 16-gauge needle so as not to build up a pressure differential within the vial. After shaking for 15 min, the vials are sealed by pinching off with Day clamps the rubber tubing attached to the syringe needles. The solution in the vials is slightly acidic, facilitating the removal of dissolved CO$_2$. Carbon dioxide-free buffer and bicarbonate are then added by syringe. The reactions are initiated by adding cells by syringe through a 19-gauge needle. The final volume of the assay mixture is 1.0 ml. The triplicate assays are normally terminated after 15 min by the addition of 0.1 ml 6 $N$ acetic acid. In a fume hood, the stoppers are removed, and the samples taken to dryness at 60°.

The contents of the vials are suspended in 0.5 ml water; 10 ml scintillation fluid[10] is added, and the number of disintegrations per minute are determined. To analyze the products of photosynthesis, the contents of

[10] L. Anderson and W. McClure, *Anal. Biochem.* **51,** 173 (1973).

the triplicate reaction vials are pooled and centrifuged to separate the soluble and insoluble fractions. The insoluble fraction, which is mainly starch, is washed with two 1-ml portions of water, and the washes are combined with the soluble fraction. The insoluble fraction must be bleached and solubilized before it can be analyzed. This is done by resuspending the insoluble fraction in the scintillation vial with 0.2 ml water, adding 0.2 ml of 70% perchloric acid and 0.3 ml of 30% $H_2O_2$, sealing the vial, and heating at 70° for 30 min.[11] The soluble fraction contains mainly sugar phosphates, organic acids, amino acids, and sugars.[9] The components of this fraction can be analyzed by any appropriate chromatographic procedure.[12]

*Comments on the Assay Procedure*

*Buffer.* Photosynthesis rates at pH 7.8 are 10 to 20% higher in Tris-Cl buffer than in Tricine [tris(hydroxymethyl)methylglycine], HEPES (*N*-2-hydroxyethylpiperazine-*N'*-2-ethanesulfonic acid), or pyrophosphate. The pH optimum at saturating bicarbonate is 7.8. At subsaturating bicarbonate concentrations, the pH optimum may be more alkaline because of a decreased cell $K_m(CO_2)$ at higher pH.[13]

To prepare $CO_2$-free buffers, degassed distilled water is added to an Erlenmeyer flask and equilibrated to assay temperature in a water bath. Nitrogen- or $CO_2$-free air is continuously bubbled through the water while it is stirred. Dry buffer is added and the pH monitored with a pH meter. The pH is adjusted to the desired level by adding concentrated acid or KOH pellets. The buffer solution is removed by syringe and added to bottles sealed with rubber stoppers and previously flushed with $N_2$. The bottles are stored $CO_2$-free in a desiccator over KOH pellets. Buffer is removed from the bottles and added to the assay vials by syringe.

*Osmoticant.* The optimal sorbitol concentration for assay is 0.2 *M* or less.[9] Sucrose can replace sorbitol and shows a similar concentration effect on photosynthesis rate. The optimal osmoticant concentration for soybean cell photosynthesis is much lower than that for cotton[1] or tobacco.[8] This appears to be a species-dependent parameter, and so if this procedure is to be used with species other than soybean, the optimal osmoticant concentration for isolation and assay must be experimentally determined.

*Stability of Cells.* After isolation, cells are stored at room temperature

---

[11] D. T. Mahin and R. T. Lofberg, *Anal. Biochem.* **16,** 500 (1966).

[12] C. A. Atkins and D. T. Canvin, *Can. J. Bot.* **49,** 1225 (1971).

[13] J. C. Servaites and W. L. Ogren, *Plant Physiol.* **60,** 693 (1977).

in the light and shaken continuously. Photosynthesis rates increase by about 30% over the first 2 hr and then decrease by 10% over the next 3 hr.[9] Thus measurements should not be made immediately after cell isolation. If complicated, multivariate experiments are conducted, the triplicate reaction vessels should not be run simultaneously but randomized over the period of the experiment. Also in lengthy experiments, it is convenient to prepare the gas atmosphere in the vials before beginning the assays. Pregassing the vials permits the assays to be completed in the shortest possible time and minimizes variation in cell activity.

*Illuminance.* Cell photosynthesis is saturated at 6 klux. An illuminance of 11 klux incandescent light is routinely used in our experiments. The cells are stable up to 89 klux; no photoinhibition is observed.[14]

*Length of Assay.* The rate of net $CO_2$ exchange in $C_3$ plants is equal to the rate of $CO_2$ uptake by photosynthesis minus the rate of $CO_2$ evolution by photorespiration. The use of $^{14}CO_2$ in the study of photosynthesis rates in $C_3$ plants is complicated because it takes time for photosynthetic and photorespiratory intermediates to reach a constant specific activity. Initially, $^{14}CO_2$ is fixed in photosynthesis, but the $CO_2$ released by photorespiration is unlabeled, so the rate of photosynthesis as measured by $^{14}CO_2$ uptake is higher than the actual rate of net $CO_2$ fixation.[15] Not until the $^{14}C$ becomes uniformly labeled in the photosynthetic cycle does the rate of $^{14}CO_2$ uptake approximate net photosynthetic $CO_2$ uptake. In soybean cells, this appears to take about 10 min at 0.1 m$M$ $KHCO_3$ and 100% $O_2$.[14] Equilibration occurs more rapidly at higher bicarbonate and lower $O_2$ concentrations.

[14] J. C. Servaites and W. L. Ogren, *Plant Physiol.* **61**, 62 (1977).
[15] L. J. Ludwig and D. T. Canvin, *Plant Physiol.* **48**, 712 (1971).

# [62] Measurement of Phosphorylation Associated with Photosystem I

*By* G. HAUSKA

Formation of ATP in chloroplasts has been primarily studied by measuring partial reactions associated with photosystem I and artificial redox systems, because of their very high turnover. Inhibitors,[1,2] especially

[1] A. Trebst, this volume [65].
[2] S. Izawa, *in* "Photosynthesis I" (A. Trebst and M. Avron, eds.), p. 266. Springer-Verlag, Berlin and New York, 1977.

DCMU,[2a] or isolation procedures[3-7] have been introduced to separate photosystem I reactions and rate limiting electron flow from water through the pool of plastoquinone. Another reason for preference is the relative stability of photosystem I reactions.

In Fig. 1 the possible, artificial electron transport pathways with photosystem I are depicted. The reduction of an electron accepting system ($A_1$) by an electron donating system ($D_1$), usually in the presence of excess ascorbate, has been termed "photoreduction."[8] Three different reaction sites for $D_1$ are possible. In the presence of DCMU the dominant one for most cases is represented by the middle arrow in Fig. 1, i.e., the DBMIB-insensitive, $CN^-$-sensitive pathway involving plastocyanin.[1] If compounds $D_1$ can also be rapidly reduced by photosystem I (dashed arrow in Fig. 1), they are able to catalyze cyclic electron flow, especially in the absence of $A_1$.

Both, photoreduction[9,10] and cyclic electron flow around photosystem I[11-13] are coupled to ATP formation. This phosphorylation shows that one of the two "sites" of energy conservation is associated with photosystem I.[1,14,15] However, certain features of $D_1$ compounds are required (see below).

Measurement of photophosphorylation in general has been described before, in an early volume of this series.[16] Measurement of phosphorylation associated with photosystem II is described in a parallel article of this volume.[17] For two-stage phosphorylation see Volume XXIV of this series.[18]

---

[2a] Abbreviations: DCMU, dichlorophenyl-1,1-dimethylurea; DPIP, 2,6-dichlorophenolindophenol; DAD, 2,3,4,6-tetramethyl-$p$-phenylenediamine; TMPD, $N,N,N',N'$-tetramethyl-$p$-phenylenediamine; PMS; $N$-methylphenazonium methosulfate. Further abbreviations are given in the legend to Fig. 1.

[3] N. K. Boardman, Vol. 23, p. 268.

[4] L. P. Vernon and E. R. Shaw, Vol. 23, p. 277.

[5] G. Jacobi, Vol. 23, p. 289.

[6] K. Shibata, Vol. 23, p. 296.

[7] P. V. Sane, D. J. Goodchild, and R. B. Park, *Biochim. Biophys. Acta* **216**, 162 (1970).

[8] A. Trebst, Vol. 24, p. 146.

[9] M. Losada, F. R. Whatley, and D. I. Arnon, *Nature (London)* **190**, 606 (1961).

[10] A. Trebst and E. Pistorius, *Z. Naturforsch., Teil B* **20**, 143 (1965).

[11] D. I. Arnon, M. B. Allen, and F. R. Whatley, *Nature (London)* **174**, 394 (1954).

[12] A. T. Jagendorf and M. Avron, *J. Biol. Chem.* **231**, 277 (1958).

[13] M. Avron, *Biochim. Biophys. Acta* **40**, 257 (1960).

[14] A. Trebst, *Annu. Rev. Plant Physiol.* **25**, 423 (1974).

[15] G. Hauska and A. Trebst, *Curr. Top. Bioenerg.* **6**, 151 (1977).

[16] F. R. Whatley and D. I. Arnon, Vol. 6, p. 308.

[17] C. Yocum, this volume [55].

[18] A. T. Jagendorf, Vol. 24, p. 103.

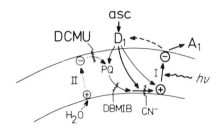

Fig. 1. Electron-transport pathways with photosystem I. The scheme shows the reactions according to the present knowledge of chloroplast membrane topography. The membrane surface exposed to the stroma (suspending medium in isolated chloroplast membranes) is represented by the upper part of the figure. The symbols used are $D_1$ and $A_1$ for electron-donating and accepting redox compounds, respectively; asc for ascorbate; PQ for plastoquinone; DBMIB for 2,5-dibromo-3-methyl-6-isopropyl-$p$-benzoquinone; I and II for photosystem I and II, respectively.

## General Aspects

### Chloroplast Preparations

*In vivo* measurements of cyclic photophosphorylation are considered to be beyond the scope of this article. The reader is referred to the comprehensive review by Gimmler.[19]

*In vitro* measurements may be performed with isolated, envelope-free chloroplasts, or with subchloroplast membrane preparations. Usually it is sufficient to prepare type C chloroplasts,[20] although type B chloroplasts[20,21] might yield somewhat higher rates of phosphorylation. Sucrose[12] (or sorbitol) is the preferential osmoticum in the isolation and suspending medium compared to NaCl,[16] the loss of phosphorylation capacity with time being lower. For surveys of chloroplast preparations consult parallel articles of this volume.[20-22] Chloroplast preparations are difficult to store without loss of photophosphorylation activity. (However, most recently Farkas and Malkin reported that rapid freezing in the presence of 30% ethylene glycol, and storage at $-180°$ preserved the activity in a satisfactory way.[23] Subchloroplast preparations are much

---

[19] H. Gimmler, *in* "Photosynthesis I" (A. Trebst and M. Avron, eds.), p. 448. Springer-Verlag, Berlin and New York.

[20] D. O. Hall, this volume [8].

[21] D. Walker, this volume [9].

[22] P. Böger, this volume [10].

[23] D. L. Farkas and S. Malkin, *Proc. Int. Congr. Photosynth., 4th, 1977* Abstracts, p. 110 (1978).

more stable on storage at low temperature. McCarty has provided a sonication procedure to obtain a preparation with high photophosphorylation activity.[24] Still higher rates have been reported with preparations enriched in photosystem I, using digitonin according to Boardman.[3] To preserve high phosphorylation capacities it was essential to have 0.4 $M$ sucrose present during fragmentation with the detergent.[25,26] Phosphorylating photosystem I preparations may also be obtained by French press treatment. Either very low pressure,[27] or again the presence of an osmoticum such as sucrose[26] is necessary.

*Artificial Redox Systems*

These have been compiled and their properties have been discussed in this volume,[28] and elsewhere[14,15,29] (for a more complete list of references see these summaries).

*Photoreduction.* Electron donors for photosystem I ($D_1$ in Fig. 1) should be oxidizable by the reaction center chlorophyll, P700 ($E_0' = 450$ mV), or by the components between the two photosystems, plastoquinone ($E_0' = 120$ mV), plastocyanin ($E_0' = 380$ mV) and cytochrome $f$ ($E_0' = 360$ mV). They are usually kept reduced by excess ascorbate ($E_0' = 80$ mV), which itself reacts only very slowly with photosystem I. Thus, for thermodynamic reasons, the $E_0'$ of $D_1$ compounds should be within these limits. They should not react directly with $A_1$ compounds. In addition, it has been found that they must be rather lipophilic and must liberate protons upon oxidation, which is in accordance with the present knowledge of the topographical arrangement of photosystem I in the chloroplast membrane, and with the chemiosmostic concept of energy conservation. The oxidizing end of photosystem I is located beyond a lipid barrier with respect to $D_1$ in the suspending medium, as indicated in Fig. 1. These findings and their implications on the sites and the mechanism of energy conservation in chloroplasts have been reviewed in detail.[14,15]

Most commonly used electron donors are DPIP, TMPD, and DAD ($E_0'$ of all three compounds is about 220 mV). The reaction of TMPD is not coupled to ATP formation,[10] which has been attributed to the fact

[24] R. E. McCarty, Vol. 23, p. 302.
[25] G. Hauska, R. E. McCarty, and E. Racker, *Biochim. Biophys. Acta* **197**, 206 (1970).
[26] G. Hauska and P. V. Sane, *Z. Naturforsch., Teil B* **27**, 938 (1972).
[27] C. J. Arntzen, R. A. Dilley, and J. Neumann, *Biochim. Biophys. Acta* **245**, 409 (1971).
[28] S. Izawa, this volume [39].
[29] G. Hauska, *in* "Photosynthesis I" (A. Trebst and M. Avron, eds.), p. 253. Springer-Verlag, Berlin and New York.

that no protons are liberated during its oxidation.[14,15] The reaction with DPIP is coupled.[9] However, the best rates of electron flow coupled to ATP formation are found with DAD.[10,25,30] At high light intensities and with saturating concentrations of DAD (about 2 m$M$), in the presence of a nonlimiting electron accepting system (see below), rates of up to 1000 $\mu$moles ATP formed per milligram of chlorophyll per hour, corresponding to about 2 IEU, can be obtained. The P/e$_2$ ratio is about 0.5 to 0.6 and the $K_m$ for DAD is 0.6 m$M$.[30,30a]

Electron acceptors (A$_1$ in Fig. 1) should be reducible by the reductants of the photosystem I reaction center ($E_0' \sim -550$ mV). To obtain a linear photoreduction without interference from cyclic electron transport, A$_1$ should not be oxidizable and D$_1$ should not be reducible by photosystem I, at least not when both are present. The physiological electron accepting system is NADP$^+$ in the presence of ferredoxin and ferredoxin-NADP$^+$ reductase.[31] In type C chloroplasts, ferredoxin is necessary, and in subchloroplast preparations both proteins have to be added back in saturating amounts. At saturating concentrations of D$_1$ and high light intensity, this system may be rate limiting. The photoreduction of dipyridylium salts does not suffer from this limitation. Methyl viologen is most commonly used, at $10^{-4}$ $M$, which is saturating. Another A$_1$ compound, at least as good as methyl viologen, is anthraquinone 2-sulfonate.

*Cyclic Electron Flow.* Mediators of cyclic electron flow around photosystem I are rapidly oxidized as well as reduced by photosystem I. In principle every D$_1$ compound will also catalyze this reaction in the absence of excess ascorbate and an A$_1$ system. "Cyclic" phosphorylation is observed provided the mediator loses protons upon oxidation. The compound most extensively studied and most commonly used is PMS.[12,13] It catalyzes the highest rates of photosynthetic ATP formation—rates of up to 2500 $\mu$moles ATP formed per milligram of chlorophyll per hour, which equals about 4 IEU, have been reported.[13] Pyocyanine,[25,32] which is a hydroxylated PMS, and DAD[33] in the absence of ascorbate and an A$_1$ system are comparably suitable. The reaction should be run in the absence of oxygen, if interference from autooxidation of either the intrinsic reductants of photosystem I or of the reduced mediators are to be

[30] J. M. Gould, *Biochim. Biophys. Acta* **387**, 135 (1975).

[30a] Another interesting D$_1$ is duroquinol [C. C. White, R. K. Chain, and R. Malkin, *Biochim. Biophys. Acta* **502**, 127 (1978).]. It catalyzes electron transport including the DBMIB-sensitive site (Fig. 1).

[31] B. B. Buchanan and D. I. Arnon, Vol. 23, p. 417.

[32] D. A. Walker and R. Hill, *Plant Physiol.* **34**, 240 (1959).

[33] A. Trebst and E. Pistorius, *Biochim. Biophys. Acta* **131**, 580 (1967).

avoided. This is not necessary if one is just interested in high rates of ATP formation. Reduced PMS and pyocyanine, in contrast to DAD, are highly autoxidizable, so that oxygen will disappear rapidly from the chloroplast suspension during the reaction. PMS,[12] and also pyocyanine, exhibits a concentration optimum between $10^{-5}$ and $10^{-4}$ $M$, while the reaction with DAD, like photoreduction, saturates at about 2 m$M$. This might reflect that cyclic electron flow requires a subtle redox balance to run with optimal rates,[25,34] depending on the redox compound employed. Since PMS and pyocyanine are used in their oxidized forms, electrons are required to start the reaction. In white light PMS, but not pyocyanine, is photoreduced by water or solutes[35] (in the presence of oxygen, pyocyanine is formed[32,35]). In red light electrons may be provided by addition of ascorbate to 2–5 m$M$, but for pyocyanine stronger reductants are required.[25] Alternatively both compounds can be reduced by electron flow from water during a preillumination period before the reaction is started by addition of the phosphorylation substrates and DCMU. DAD is used in its reduced form, and optimal rates are obtained without the complications noted above. On the contrary, addition of another reductant such as ascorbate inhibits cyclic electron flow with DAD by over-reducing the system.[25] In this case the reduction of DAD by photosystem I becomes rate limiting.

*Measurement of Electron Transport.* An estimation of electron transport rates is required if P/$e_2$ ratios are desired.

Spectrophotometric measurement of NADP$^+$ photoreduction has been described in this series.[31] Color formation during oxidation of D$_1$ compounds might interfere.

Photoreduction of methyl viologen or anthraquinone 2-sulfonate is preferentially followed manometrically or with an oxygen electrode,[36] measuring the autooxidation of the reduced compounds. The manometric technique with a photo-Warburg-apparatus allows one to handle several parallel samples conveniently, which is desirable in routine assays, but usually is limited by light intensity. Within the reaction chamber of an oxygen electrode, higher light intensities are more easily reached, and the method is more sensitive, allowing shorter illumination times. Estimation of unambiguous rates of electron transport is complicated by the fact that the $O_2^-$ radical, primarily formed during autoxidation, might oxidize the D$_1$ system before dismutation giving higher rates of oxygen consumption than would correspond to the passage of electrons through the photosystem. Addition of superoxide dismutase overcomes this com-

[34] G. Hauska, S. Reimer, and A. Trebst, *Biochim. Biophys. Acta* **357,** 1 (1974).
[35] A. T. Jagendorf and M. Margulies, *Arch. Biochem. Biophys.* **90,** 184 (1960).
[36] D. C. Fork, Vol. 24, p. 113.

plication.[30,37] Alternatively, the formation of blue, reduced methyl viologen can be measured spectrophotometrically under anaerobic conditions.[38]

Another problem in measuring electron flow through photosystem I
during photoreductions, which is hard to eliminate, is the contribution
from cyclic electron transport,[14] which leads to an overestimation of $P/e_2$
ratios. In photoreduction with DAD, high concentrations of ascorbate
and methyl viologen have been used to attenuate the contribution from
cyclic electron flow as far as possible.[25] Measurement of cyclic electron
transport is more complicated. It requires the assay of the turnover of
some essential component within the cycle, preferentially P700, by more
sophisticated spectroscopic techniques.[39,40] These have not yet been successfully applied to the reaction with PMS, pyocyanine, or DAD.

Collections of $P/e_2$ ratios can be found elsewhere.[1,15,41]

*Reaction Conditions*

*Illuminations.* About $10^6$ ergs/cm²/sec of red light ($>650$ nm) are required to reach saturation of photosystem I in cyclic phosphorylation
with PMS, at low chlorophyll concentration (less than 10 µg/ml).[13] Similarly high light intensities are required in photoreduction with millimolar
concentrations of DAD. Red light should be applied to avoid photochemical reactions of the artificial redox compounds.[35] High light intensities
are relatively easily obtained with single samples by focusing from a slide
projector, or any similar high intensity lamp. This is more difficult if
parallel samples are to be illuminated, e.g., as in a Warburg apparatus
where heat generation is also a problem. In any set up for parallel samples
car should be taken to provide equal intensity of illuminating light. Moving the sample rack back and forth in front of the lamps usually helps.
The time of illumination should be as short as possible (not more than a
few minutes), because of a decrease of phosphorylation activity with
time.[13] Measurement of phosphorylation in discontinuous light, either by
repetitive illumination periods[42,43] or even by single flashes[44] has been

[37] D. S. Ort and S. Izawa, *Plant Physiol.* **53**, 370 (1974).
[38] G. Zweig and M. Avron, *Biochem. Biophys. Res. Commun.* **19**, 397 (1965).
[39] H. J. Rurainski, J. Randles, and G. E. Hoch, *FEBS Lett.* **13**, 98 (1971).
[40] P. C. Maxwell and J. Biggins, *Biochemistry* **15**, 3975 (1975).
[41] D. O. Hall, *in* "The Intact Chloroplast; Structure and Function" (J. Barber, ed.). ASP
Biol. and Med. Press, Amsterdam, 1976.
[42] P. Gräber and H. T. Witt, *Biochim. Biophys. Acta* **423**, 141, (1976).
[43] D. R. Ort and R. A. Dilley, *Biochim. Biophys. Acta* **449**, 95 (1976).
[44] M. Baltscheffsky, *Proc. Int. Congr. Photosynth., 4th, 1977,* p. 17 (1978).

applied to answer questions as to the mechanism and kinetics of energy conversion from light to ATP.

*pH and Temperature.* The pH optimum of photophosphorylation is between 8.0 and 8.5.[13,45,46] Usually the assay is run at room temperature. The temperature coefficient of photophosphorylation seems to be rather low—the rate at 25° is about twice that for 3.5°, but this result[47] might also reflect a limitation by light.

*Ions and Other Solutes.* An optimal structure of the chloroplast membrane system is required for good coupling of phosphorylation to electron transport. This structure is maintained by cations, specifically by $Mg^{2+}$.[48-50] Magnesium (2+) is also a cofactor for the phosphorylation reaction proper, about 5 m$M$ being optimal.[13] High ionic strength and osmolarity are inhibitory, in part by decreasing the diffusion of phosphorylation substrates because of increased viscosity and inaccessible reactive sites in highly condensed chloroplasts.

Tricine is prefered to Tris as buffer; the later acts as an uncoupler.[51] Phosphorylation rates are somewhat higher in the presence of bovine serum albumin.[52] Additions in ethanolic solution (DCMU, redox compounds) should not exceed 1% ethanol in the final reaction mixture.

*Substrates of Phosphorylation.* Optimal concentrations of ADP and phosphate depend on the amount of chlorophyll, on light intensity, and on illumination time. According to Avron,[13] at low chlorophyll concentration, saturation of the phosphorylation rate is reached well below 1 m$M$ ADP but with $P_i$ 10 m$M$ is still not saturating. For the radioactive assay with $^{32}P_i$, 2 m$M$ is the concentration frequently used as a compromise between saturating the rate and specific radioactivity. At high phosphorylation rates, care should be taken to assure that substrates are not consumed during prolonged illumination since an underestimation of the rate would result.

### Assay for ATP Formation

Phosphorylation can be assayed in several ways that have been previously discussed in this series.[53] The oldest, but least sensitive method

[45] M. B. Allen, F. R. Whatley, and D. I. Arnon, *Biochim. Biophys. Acta* **27**, 16 (1958).
[46] U. Pick, H. Rottenberg, and M. Avron, *FEBS Lett.* **48**, 32 (1974).
[47] R. Kraayenhof, M. B. Kartan, and T. Grunwald, *FEBS Lett.* **19**, 5 (1971).
[48] N. Shavit and M. Avron, *Biochim. Biophys. Acta* **131**, 516 (1967).
[49] D. Walz, S. Shuldiner, and M. Avron, *Eur. J. Biochem.* **22**, 439 (1971).
[50] H. Hesse, R. Jank-Ladwig, and H. Strotmann, *Z. Naturforsch, Teil C* **31**, 445 (1976).
[51] N. Good, S. Izawa, and G. Hind, *Curr. Top. Bioenerg.* **1**, 75 (1966).
[52] M. Friedländer and J. Neumann, *Plant Physiol.* **43**, 1249 (1968).
[53] D. Geller, Vol. 24, p. 88.

is the colorimetric test for phosphate. Another approach is to measure the rise in pH, which accompanies the reaction above the second $pK_a$ of the phosphates[53-55].

Enzymatic tests for ATP, either via hexokinase plus glucose-6-phosphate dehydrogenase[53] or via phosphoglycerate kinase plus trisosephosphate dehydrogenase, have also been employed. In this case, adenylate kinase has to be removed from spinach chloroplasts,[12,45,56] otherwise photophosphorylation is overestimated. Adenylate kinase is absent in chloroplasts from some other plants.[13] The sensitivity of the enzymatic test can be substantially increased by fluorimetric measurement of the reduced pyridine nucleotides formed.[53]

For the assay of reasonably high rates of phosphorylation the method of choice is the radiochemical one with $^{32}P_i$ [16,53] (interference from ATP hydrolysis is negligible in chloroplasts[57]). Inorganic phosphates are separated from organic phosphates formed by precipitation as insoluble salts[16,58] or by extraction of the complex formed with molybdate into organic solvents.[13,59] Background radioactivity, mainly due to the presence of pyrophosphate in the $^{32}P_i$ sample, can be decreased either by hydrolysis of the contaminating $PP_i$ before use,[43] by isolating ATP or secondarily formed organic phosphates, by anion exchange chromatography,[53,60,61] or by adsorption to charcoal.[61] The sensitivity of the ATP assay can be increased by about two orders of magnitude using these additional modifications. Radioactivity can be determined by any counting method, most conveniently by Cerenkov radiation with a scintillation counter.[62]

Perhaps the most sensitive test for ATP is measuring the luminescence of luciferin in the presence of firefly luciferase,[63,64] which is suitable for estimation of even the very minute amounts of ATP formed during single, short flashes.[44]

[54] R. A. Dilley, Vol. 24, p. 71.
[55] B. Chance and M. Nishimura, Vol. 10, p. 641.
[56] M. Avron, A. T. Jagendorf, and M. Evans, *Biochim. Biophys. Acta* **26,** 262 (1957).
[57] R. E. McCarty and E. Racker, *J. Biol. Chem.* **243,** 129 (1968).
[58] Y. Sugino and Y. Miyoshi, *J. Biol. Chem.* **239,** 2360 (1964).
[59] O. Lindberg and L. Ernster, *Methods Biochem. Anal.* **3,** 1 (1956).
[60] R. P. Magnusson and R. E. McCarty, *J. Biol. Chem.* **251,** 6874 (1976).
[61] D. J. Smith, B. O. Stokes, and P. D. Boyer, *J. Biol. Chem.* **251,** 4165 (1976).
[62] J. M. Gould, R. Cather, and G. D. Winget, *Anal. Biochem.* **50,** 540 (1972).
[63] L. Strehler, *in* "Methods of Enzymatic Analysis" (H. U. Bergmeyer, ed.), p. 559. Academic Press, New York, 1965.
[64] A. Lundin, A. Rickardson, and A. Thore, *Anal. Biochem.* **75,** 611 (1976).

*Detailed Radiochemical Procedure*

The incubation mixture is the same for phosphorylation coupled to photoreduction or to cyclic electron flow, except that DAD plus methyl viologen are used in the former, and PMS is used in the later case. The stock solutions are pipetted as indicated in Table I, and water is added to obtain the final concentrations. The water should be saturated with air for measurement of photoreduction. For a series of parallel incubations it is advisable to prepare the reaction mixture for all the samples and then distribute it into the individual test tubes of flasks kept in ice before chloroplasts are added. Then the samples are allowed to reach room temperature in the rack of a thermostated water bath (Warburg apparatus). In the case of cyclic phosphorylation, the samples are made anaerobic by flushing with nitrogen or argon (as mentioned above, this is not crucial to obtain high phosphorylation rates, but excludes contribution from noncyclic electron flow). The reaction is usually started by switching on the illuminating red light. Its intensity should be at least $2 \times 10^5$ ergs/cm$^2$/sec. If preconditioning of chloroplasts in the light is required,

TABLE I

Stock Solutions and Reaction Mixture for Photophosphorylations with Photosystem I

| Stock solution | Storage | $\mu$l to be pipetted per ml reaction mixture | Final concentration in reaction mixture |
|---|---|---|---|
| 1 $M$ Tricine–NaOH, pH 8.0 | Refrigerator | 50 | 50 m$M$ |
| 1 $M$ NaCl | Refrigerator | 50 | 50 m$M$ |
| 0.1 $M$ MgCl$_2$ | Refrigerator | 50 | 5 m$M$ |
| 0.1 $M$ ADP, pH 7.0 | Freeze | 30 | 3 m$M$ |
| 0.1 $M$ P$_i$, pH 8.0 | Refrigerator | 20 | 2 m$M$ |
| 0.1 $M$ Na ascorbate | Prepare fresh | 50 | 5 m$M$ |
| 10 m$M$ DAD | Prepare fresh | 100 | 1 m$M$ |
| 10 m$M$ methyl viologen | Refrigerator | 10 | 0.1m$M$ |
| 0.1 m$M$ PMS | Prepare fresh | 50 | 50 $\mu M$ |
| $^{32}$P$_i$ solution | (Shielded by lead) | — | about $10^6$ cpm/ml |
| Defatted bovine serum albumin (50 mg/ml) | Freeze | 20 | 1 mg/ml |
| 1 m$M$ DCMU in ethanol | Refrigerator | 20 | 20 $\mu M$ |
| Chloroplasts in sucrose medium[a] | Prepare fresh | — | 3–10 $\mu$g chlorophyll/ml |

[a] From McCarty.[24]

the reaction can be started by addition of the phosphorylation substrates, either ADP or $^{32}P_i$, or both together. For a series of parallel incubations this can be done from the side arms of appropriate flasks by turning the whole sample rack.

The illumination time should not exceed 2 min (if photoreduction with DAD and methyl viologen is measured with an oxygen electrode, oxygen might be used up in the reaction mixture in less than 2 min, if 10 $\mu$g chlorophyll/ml are present). The reaction is stopped by addition of 40% (w/v) trichloroacetic acid to 2% final concentration (50 $\mu$l/ml reaction mixture). A dark control should be run under identical conditions.

After 10 min the samples are centrifuged for 5 min at 1000 $g$, and 0.4 ml of the supernatant is transferred to a 25 ml test tube containing 1 ml of 5% (w/v) ammonium molybdate and 4 ml 1 $N$ HClO$_4$. After mixing on a Vortex shaker, 5 ml of benzene/isobutanol (1:1) are added, and the phosphomolybdate complex is extracted into the organic phase by mixing the two phases. The upper, organic phase is removed by aspiration via a Pasteur pipette fitted to a trapping flask and a water vacuum pump. Extraction is repeated with 5 ml water-saturated isobutanol, and finally with 5 ml diethyl ether. An aliquot (0.4 ml) of the remaining aqueous solution is either dried on a small aluminum plate and counted in a gas-flow counter, or prepared for Cerenkov counting in a scintillation vial.[62] A standard is prepared by diluting 50 $\mu$l of the reaction mixture, after addition of trichloroacetic acid and centrifugation, to 5 ml; 0.2 ml is used for counting. The factors to calculate total radioactivity from the standard and the radioactivity in organic phosphate from the extracted samples are 530 and 35.4 per ml reaction mixture, respectively. The rate of phosphorylation in reaction mixture, respectively. The rate of phosphorylation in micromoles $P_i$ bound per milligram chlorophyll per hour, is calculated according to the following formula

$$\text{Rate} = \frac{\text{cpm}_{\text{sample}} - \text{cpm}_{\text{dark control}}}{\text{cpm}_{\text{standard}}} \frac{35.4}{530} \frac{60 \times 2 \times 1000}{\text{min(ill)} \times \mu\text{g chlorophyll}}$$

# [63] Quantum Requirements of Photosynthetic Reactions

*By* C. GIERSCH and U. HEBER

## Introduction

The measurement of quantum requirements was discussed earlier in this series.[1] Traditionally, in photosynthesis the quantum yields of oxygen evolution or of $CO_2$ fixation have been measured with intact cells or organisms. Alternatively, the quantum yield of partial reactions of photosynthesis has been determined using chloroplast fragments as experimental material. In both cases, but for different reasons, interpretation of the data has been difficult. In intact cells, dark respiratory processes and photorespiratory glycolate metabolism consume $O_2$ and produce $CO_2$, thereby diminishing photosynthetic gas exchange. Chloroplast fragments are much simpler systems than intact cells, so simple in fact that they cannot perform complex photosynthetic reactions, and activity may easily be lost, resulting in a decrease in quantum yield. Renewed interest in photosynthetic quantum yield measurements came from the availability of chloroplast preparations which can perform photosynthesis at very high rates. In their gas exchange, such chloroplasts are simpler than cells. If properly prepared, they are, for all practical purposes, free of mitochondria and peroxisomes. In the following, the term intact chloroplast refers to this kind of chloroplast preparation.

## Definitions and Conversion Factors

The quantum requirement $r$ of a photochemical reaction is defined as:

$$r = \frac{\text{number of light quanta absorbed}}{\text{number of molecules reacting chemically}}$$

The quantum yield $\Phi$ is the reciprocal of the quantum requirement:

$$\Phi = \frac{\text{number of molecules reacting chemically}}{\text{number of light quanta absorbed}}$$

One mole of quanta is called one Einstein, hence:

$$\Phi = \frac{\text{moles of molecules reacting}}{\text{Einsteins absorbed}}$$

[1] M. Schwartz, Vol. 24, p. 142.

METHODS IN ENZYMOLOGY, VOL. 69

The energy $\epsilon$ of one quantum of wavelength $\lambda$ is given by:

$$\epsilon = h\,\frac{c}{\lambda} = h\nu$$

where $h$ is Planck's constant $= 6.626 \times 10^{-34}$ W sec$^2$, $c$ is the velocity of light $= 2.998 \times 10^8$ m sec$^{-1}$, $\lambda$ is the wavelength in meters, and $\nu$ is frequency (per second). The energy content of one Einstein is given by:

$$E = N\,h\nu = \frac{1.196}{\lambda(\text{nm})}\,10^8 \text{ W sec}$$

where $N(= 6.02 \times 10^{23}$ mole$^{-1}$) is Avogadro's number.

The photon flux $E_\nu$ (Einstein m$^{-2}$ sec$^{-1}$) is obtained from the intensity of the incident light $E_i$ (W m$^{-2}$) by dividing the latter by the energy content of one Einstein:

$$E_\nu\left(\frac{\text{Einstein}}{\text{m}^2\,\text{sec}}\right) = E_i\left(\frac{\text{W}}{\text{m}^2}\right)\frac{1}{N\,h\nu}\left(\frac{\text{Einstein} \times \text{sec}}{\text{W sec}^2}\right)$$

or

$$E_\nu\left(\frac{\text{Einstein}}{\text{m}^2\,\text{sec}}\right) = 8.36 \times \lambda(\text{m})\,E_i(\text{W/m}^2)$$

## Methods

To calculate quantum requirements, it is necessary to know how much light is absorbed by a sample and how much photochemical reaction occurs. Since both the energy of light and light absorption by biological samples differ with wavelength, it is necessary to use fairly monochromatic light. For chloroplast work, light may be considered to be monochromatic if it has a bell-shaped photon flux distribution with half-bandwidths ranging between 5 and 20 nm.[2] Such light can be produced by monochromators and optical filters. Laser light is monochromatic in a much more restricted sense of the word.

*Monochromators.* The limited light output of most commercially available prism or diffraction grating monochromators restricts their use in photosynthetic quantum yield work.

*Optical Filters.* An approximately parallel beam of light having a uniform intensity distribution is preferable for quantum yield work to other optical arrangements. Projectors or microscope lamps equipped with suitable focusing systems may be used as light sources. Since the

[2] K. M. Hartmann, *In* "Biophysik" (W. Hoppe, W. Lohmann, H. Markl, and H. Ziegler eds.), p. 199. Springer-Verlag, Berlin and New York, 1977.

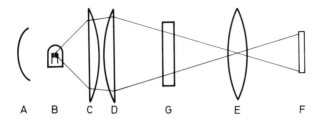

FIG. 1. Schematic diagram of a projection arrangement suitable for producing uniform beams of high intensity. For details see text.

isolation of spectral bands by interference filters drastically reduces light intensity, even these sources will often be found to yield insufficient intensities of monochromatic light. A simple but versatile optical system which is easy to assemble is shown in Fig. 1. Two large condenser lenses (C and D) having a diameter of about 10 cm and a focal length of 15 cm each produce an image of the lamp (B) in the plane of the objective (E). The objective may have a focal length of 20 cm. Lamps with suitable small filaments are the halogen lamps 24 V, 150 or 250 W from Philips or Osram. A parabolic reflector (A) produces another image of the lamp in the plane of the objective E. This reflector should be properly adjusted to produce a second image of the filament above, not on top of the first image. An important advantage of the system depicted in Fig. 1 is that the size of the illuminated sample area and with it the maximal light intensity can be varied by suitably changing the positions of A, C and D, and E. For illuminating the sample F as uniformly as possible, it should be placed in the image plane of the homogeneously illuminated plane surface of lens D. The proper position can be determined by fixing a strip of paper on the plane face of lens D. The sample should be placed where the shaded area in the light beam produced by the strip has sharply defined outlines. The filter arrangement G should include a water filter, a heat absorbing glass filter, and a cutoff filter to produce an optical window which is then perfected by inserting a proper interference filter. Care should be taken that the aperture angle of the incoming beam does not exceed $\pi/12$ rad and that the reflecting surface of the interference filter faces the light source.[3]

*Lasers.* Recently, high energy monochromatic light has become available from lasers. Continuously operating gas lasers are not too expensive, but limited in energy output (some milliwatt) and in the number of avail-

[3] M. Schwartz, Vol. 24, p. 200.

able wavelength bands. Pulsed lasers have high energy output and are the tool of choice for flashlight quantum yield experiments.

*Measurement of Light Intensities.* The intensity of incident light can be measured physically by bolometers and thermopiles and chemically by actinometers. Bolometers have a large receiving surface. Thus, this instrument measures the average light intensity if there are cross-sectional inhomogeneities in the intensity of the incoming beam. The integration is more difficult to obtain by means of a thermopile with a smaller receiving surface, but is not necessary if the beam has a uniform intensity distribution.

The measurement by means of an actinometer is based on constant quantum yields of certain photochemical reactions. Actinometers are secondary standards. They have to be calibrated by other methods. The reduction of $Fe^{3+}$ according to

$$2 Fe^{3+} + C_2O_4{}^{2-} \xrightarrow{h\nu} 2 Fe^{2+} + 2 CO_2$$

is the mechanism of the ferrioxalate actinometer.[4] The reaction is monitored by measuring $Fe^{2+}$ photometrically as the 1,10-phenantroline complex. Details of the preparation of the solution and the quantum yields can be found in the original paper by Hatchard and Parker[4] and in a contribution by Schwartz in Volume XXIV of this series.[1] The quantum yield $\Phi$ of this reaction is about one in the region $405 \leq \lambda \leq 440$ nm.[4] The chlorophyll thiourea actinometer (Gaffron,[5] modified by Warburg and Schocken[6]) shows nearly complete absorption throughout the visible spectrum ($\lambda < 670$ nm). Some discussion of the advantages and drawbacks of the original version[5] and the modifications by Warburg and co-workers[6,7] can be found in a contribution by Kok.[8]

An actinometer employing Reinecke's salt ($K[Cr(NH_3)_2(NCS_4)]$) and operating throughout the visible range has been described by Wegner and Adamson.[9] Its quantum yield is 0.31 (452 nm), 0.29 (520 nm) or 0.27 (750 nm).

*Measurement of Fractional Absorption.* The simplest arrangement is to use an optically thick suspension of chloroplasts. All incident light is

[4] C. G. Hatchard and C. A. Parker, *Proc. R. Soc. London, Ser. A* **235**, 518 (1956).
[5] H. Gaffron, *Ber. Dtsch. Chem. Ges.* **60**, 755 (1927).
[6] O. Warburg and V. Schocken, *Arch. Biochem.* **21**, 363 (1949).
[7] D. Burk and O. Warburg, *Z. Naturforsch., Teil B* **6**, 12 (1951).
[8] B. Kok, *in* "Handbuch der Pflanzenphysiologie" (W. Ruhland, ed.), Vol. 5, Part 1, p. 577. Springer-Verlag, Berlin and New York, 1960.
[9] E. E. Wegner and A. W. Adamson, *J. Am. Chem. Soc.* **88**, 394 (1966).

absorbed; the absorbed light intensity is equal to the incident intensity. Light intensity gradients within the sample are necessarily very large in this method. Measured effects of illumination can, therefore, not be related in a meaningful way to the chlorophyll content of the sample. Moreover, induction phenomena may influence metabolic fluxes in a manner which is difficult to control, since chloroplasts in a stirred sample will alternatively be darkened and illuminated.

Optically thin suspensions avoid these difficulties. For the measurement of absorption, the sample is used as a secondary light source which is excited by the incident beam. The scattered light can be measured by integrating it in an Ulbricht sphere. This is not difficult to construct; cut the hollow sphere of a reasonably large globe into two identical halves (Fig. 2). One of them (A) should receive a central hole (a) to permit the measuring beam to enter. The other one (B) should have a hole (b) for sideways light measurements by a photomultiplier (PM). The latter must be protected from directly scattered or reflected light by a suitable screen (c). An adjustable sample holder between the two halves of the globe brings the sample or a control (white paper) into the light beam. In order to randomize the transmitted and scattered light, the sphere must be provided with a white ("nonabsorbing") diffuse coating on the inside. Suitable is a coat of white paint and an additional coat of MgO from burnt magnesium. The diameter of the sphere should be large as compared with the sample.

Öquist et al.[10] proposed another arrangement for measuring the fractional absorption of intact plants in an Ulbricht sphere. Light enters the sphere from two opposite points and is randomly distributed by two conical screens facing the incoming beam. This avoids mutual shading of parts of the plant which is inevitable when the light beam directly hits the plant. The scattered light is measured as the mean value of eight points equally distributed over the sphere.

In an alternative to the Ulbricht sphere, Noddack and Eichhoff[11] used a hollow ellipsoid with a polished inside surface. The sample is placed in one of its focal points. At the other, a thermopile which is sensitive in all directions monitors the light which is not absorbed. In the ellipsoid photometer, a large proportion of the scattered light is transmitted to the receiver which therefore does not need to be highly sensitive. When an Ulbricht sphere serves for measuring absorption, the percentage of light used for the measurement depends on the ratio of the surface of the

[10] G. Öquist, J.-E. Hällgren, and L. Brunes, *Plant, Cell Environ.* **1**, 21 (1978).
[11] W. Noddack and H. J. Eichhoff, *Z. Phys. Chem., Abt. A* **185**, 241 (1939).

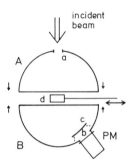

FIG. 2. Experimental setup for measuring the fractional absorption in an integrating Ulbricht sphere. A and B are halves of the sphere; a and b, holes for incident light beam and photomultipler, respectively; c, screen; d, sample; PM, photomultiplier.

sphere to the surface of the receiver.[12] A sensitive measuring device is usually necessary.

### Measurement of Chloroplast Reactions

The sequence of reactions starting from the absorption of quanta and ending with the formation of a product or disappearence of a substrate is usually treated as a "black box." Since most chloroplast reactions are accompanied by oxygen exchange, which can be recorded by an oxygen electrode, the progress of a chloroplast reaction will often be monitored by measuring oxygen uptake or oxygen evolution. The stoichiometry of the reaction must be known, and reaction conditions must be chosen which do not permit competing reactions to assume significant proportions. The problem is exemplified for measurements of the quantum yield of $CO_2$-dependent oxygen production by intact chloroplasts.

Usually, preparations of intact chloroplasts contain a certain percentage of broken chloroplasts devoid of stroma. As these chloroplasts absorb light, but cannot photoreduce $CO_2$, the observed rates of $CO_2$-dependent oxygen evolution must be corrected for broken chloroplasts. The percentage of broken chloroplasts can be determined by measuring the light-dependent reduction of ferricyanide. In contrast to broken chloroplasts, intact chloroplasts cannot photoreduce ferricyanide. First, ferricyanide reduction is measured under saturating light in the untreated preparation. Only the contaminating broken chloroplasts react. In another experiment, all chloroplasts are first osmotically shocked. Subsequently, ferricyanide reduction is measured under the same conditions

[12] B. Kok, in "Handbuch der Pflanzenphysiologie" (W. Ruhland, ed.), Vol. 5, part 1, p. 579. Springer-Verlag, Berlin and New York, 1960.

as in the intact preparation. The ratio between both measurements gives the fraction of broken chloroplasts in the chloroplast suspension.[13,14] A similar method employing ferredoxin and NADP has been published by Robinson and Stocking.[15]

A reaction counteracting $CO_2$-dependent oxygen evolution is oxygen uptake by the Mehler reaction. It occurs in both broken and intact chloroplasts[16] and leads to $H_2O_2$ formation. Hydrogen peroxide inhibits $CO_2$-dependent photosynthesis of intact chloroplasts and, therefore, leads to an increase of the apparent quantum requirement. In the presence of excess catalase, oxygen uptake by the Mehler reaction is balanced by oxygen evolution from $H_2O_2$. Destruction of the latter also relieves inhibition of photosynthesis. As washed preparations of intact chloroplasts do not exhibit significant catalase activity, catalase should always be added.

Another problem in regard to measuring optimal quantum yields during photosynthesis of intact chloroplasts is the consumption of oxygen during glycolate formation in the ribulosebisphosphate oxygenase reaction, which diminishes the oxygen output. Moreover the complete oxidation of sugar phosphates via the oxygenase reaction not only consumes oxygen, but also requires light energy. Since $CO_2$ competitively inhibits glycolate formation,[17] oxygen uptake in the oxygenase reaction can be minimized by increasing the $CO_2/O_2$ ratio in the system. It is preferable to reduce the oxygen pressure rather than to use a high bicarbonate concentration, since the latter is known to decrease the light-induced $\Delta pH$ across the chloroplast envelope by indirect proton transfer.[18] This can deactivate enzymes of the carbon cycle such as fructose bisphosphatase and lead to partial inhibition of photosynthesis. Under such inhibition, the quantum yield is decreased.

## Results

Figure 3[16,19] shows quantum requirements of $CO_2$-dependent, 3-phosphoglycerate-dependent, and oxaloacetate-dependent oxygen evolution by

[13] U. Heber and K. A. Santarius, Z. Naturforsch. Teil B, **25,** 718 (1970).
[14] McC. Lilley, M. P. Fitzgerald, K. G. Rienits, and D. A. Walker, New Phytol. **75,** 1 (1975).
[15] J. M. Robinson and C. R. Stocking, Plant Physiol. **43,** 1967 (1968).
[16] H. Egneus, U. Heber, U. Matthiesen, and M. Kirk, Biochim. Biophys. Acta **408,** 252 (1975).
[17] M. R. Badger, T. J. Andrews, and C. B. Osmond, Proc. Int. Congr. Photosynth., 3rd, 1974, p. 1421 (1975).
[18] U. Heber and P. Purczeld, Proc. Int. Cong. Photosynth., 4th, 1977, p.107 (1978).
[19] U. Heber and M. R. Kirk, Biochim. Biophys. Acta **376,** 136 (1975).

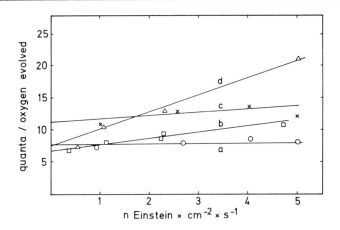

Fig. 3. Quantum requirements of reduction of various substrates by intact chloroplasts as a function of 674-nm light. (a) 3-Phosphoglycerate, (b) $CO_2$, (c) oxaloacetate plus 10 m$M$ $NH_4Cl$, (d) oxaloacetate without uncoupler. Substrate concentrations 2 m$M$ each. Adapted from Egneus et al.[16] and Heber and Kirk.[19]

intact chloroplasts as a function of incident 674 nm light. Both oxygen evolution during the photoreduction of phosphoglycerate and oxaloacetate have a minimal requirement of about 8 quanta per oxygen molecule evolved. This is consistent with a scheme of photosynthetic electron transport involving two photoreactions for the transport of one electron to an acceptor molecule. As the light intensity was increased, only the quantum requirement of oxaloacetate-dependent oxygen evolution increased. As ATP is not consumed during oxaloacetate reduction, this appears to reflect control of electron transport by ATP or by the large transthylakoid proton gradient formed when ATP is not consumed. Indeed, $NH_4Cl$, which is known to decrease the transthylakoid proton gradient, caused a decrease in the quantum requirements of oxaloacetate reduction. Carbon dioxide-dependent oxygen evolution had a distinctly higher quantum requirement than oxygen evolution during the reduction of phosphoglycerate or oxaloacetate. Bioenergetically, $CO_2$ reduction differs from phosphoglycerate reduction only in its higher ATP requirement. It is tempting to speculate that the increased energy requirement for $CO_2$ reduction reflects the need to divert quanta into a reaction synthesizing extra ATP.

Acknowledgments

We wish to thank Dr. J.-E. Hällgren for sending us a copy of his manuscript[10] in advance of publication.

## [64] Light-Emitting Diodes as a Light Source in Photochemical and Photobiological Work

*By* HANS J. RURAINSKI and GERHARD MADER

Numerous investigations in photochemical and photobiological work require excitation of samples by brief flashes of light. The methods used for obtaining such flashes depend largely on the chemical or biological system to be studied and the information desired. Generally, the duration of the exciting pulse should be shorter than the relaxation time of the excited system if a single turnover is to be elicited. If small signals occur, the exciting flash may be applied repetitively and the signals are averaged over many cycles for an improvement of signal-to-noise ratios. In order to avoid overlapping of one turnover with the next, this procedure requires spacing of the flashes with a dark time that is longer than the recovery time for the slowest reaction. Other desirable or necessary qualities of a useful flashing device include rise and decay times that are much shorter than the relaxation time, sufficient light intensity to saturate the system, and a color of the light which is absorbed by the system.

As far as the length of the flash and the dark time between flashes are concerned, flash producing methods are generally limited to a particular range. Mechanical choppers or light shutters, for example, are useful for flash lengths and dark times to approximately 1 msec but generally fail if much shorter pulses are required. At the other extreme, lasers can be pulsed in the nanosecond range, but the same lasers are often of little use if longer pulses are needed. Thus, the experimenter must obtain a flashing device for his particular experiment and has, in general, little variability. Another problem, associated primarily with gas discharge tubes which are useful in the microsecond and millisecond range, is tailing. This means that upon ignition, the light intensity increases or decreases gradually and nonsymmetrically. Also, the ignition pulse may cause electrostatic and electromagnetic disturbances which often require heavy shielding of equipment in the environment.

The following is a description of a simple and inexpensive way to overcome some of these difficulties. Using light-emitting diodes (LED), flashes of arbitrary duration from a few hundred nanoseconds to continuous light can be produced with the same set up. Also the dark time between flashes may be chosen arbitrarily and independently of the flash length. The rise and decay times of LED are in the nanosecond range. Because of their small dimensions, large numbers can be used on a small area, thus compensating for the rather low intensity emitted by a single LED.

## Characteristics of Light-Emitting Diodes

Light-emitting diodes (LED) are semiconductor devices whose active surface of the PN junction emits radiation in the visible or infrared portion of the spectrum. The wavelength of maximum emission depends largely on the type of material used, the concentration of the doping substance and the operating temperature. Numerous LED are commercially available with emission maxima between 560 and 1060 nm and half-bandwidths between 20 and 80 nm. Thus, within these limits the experimenter can choose a diode that is best suited for his purpose. Some arbitrarily selected emitters are listed in Table I.

The intensity of the emission at the semiconductor surface is a function of the current applied to the junction and the ambient temperature. Maximum tolerable currents vary widely among LED from one manufacturer to the next and lie between 20 and 100 mA for continuous radiation. In pulsed operation, the maximally tolerable current and thus the intensity, can be greatly increased to a value which depends on the duration of the pulses and their repetition rate. Values of up to 5 A have been stated for 1 $\mu$sec flashes at a frequency of 1 kHz. The intensity of the emission in the forward direction, i.e., along an axis perpendicular to the emitting surface, is influenced by a plastic body which encloses the semiconductor and which affects the angular distribution of the radiation. Several variations from "point" to fully diffused sources are commercially available. The plastic body also largely determines the physical dimensions of the diodes.

Since the radiation from a single LED is rather low, these devices seem, at first glance, not very useful as actinic light for photochemical or photobiological reactions. However, the entire diode, including electrical leads, is only a few millimeters in size permitting the combination of the output of large numbers on a small area. A possible way to do this is described below.

## Design and Circuitry

In our laboratory, LED arrays have been used as actinic light source in experiments with intact algae and isolated chloroplasts of higher plants. One of the suitable emittors for this purpose is type MV 5152 of Monsanto. Figure 1 depicts the reaction chamber, made of aluminum, whose cover has been removed for viewing. The sample to be measured is

TABLE I

CHARACTERISTICS OF SOME ARBITRARILY SELECTED, COMMERCIALLY AVAILABLE LED[a]

| Characteristic | MV 5254 | LD 56 C | 5082–4658 | CQX 35 | FLV 252 | SG 1004 |
|---|---|---|---|---|---|---|
| Color | Green | Yellow | Red | Red | Far red | Infrared |
| Peak of emission (nm) | 565 | 590 | 635 | 660 | 690 | 940 |
| Half-bandwidth (nm) | 35 | 30 | 40 | 20 | 85 | — |
| Luminous intensity at 10 mA (mcd), continuous operation | 3.0[b] | 16 | 24 | 8.0[b] | 8.0 | 3.0 mW |
| Maximum tolerable current (mA), continuous operation | 35 | 60 | 60 | 50 | 50 | 50 |
| Maximum tolerable current (A) pulsed operation (pulse ≤ 1 μsec) | 5.0 | 0.5 | 1.0 | 1.0 | 1.0 | 1.5 |
| Rise time (nsec) | 100 | 100 | 90 | — | — | — |
| Decay time (nsec) | 100 | 100 | 90 | — | — | — |
| Manufacturer | Monsanto | Siemens | Hewlett-Packard | Tele-funken | Fairchild | RCA |

[a] All data are based on information supplied by the manufacturers and are usually stated for 25°. Most manufacturers, including many not listed here, sell the entire line of LED.
[b] Measured at 20 mA.

FIG. 1. Photograph of the reaction chamber. The cover has been removed. LED, light-emitting diode arrays (24 LED/cm²); HS, heat sink; S, position of the sample cuvette. The dashes indicate the course of the coolant line drilled into the body of the chamber.

contained in a 1-cm³ cuvette which is surrounded on four sides by 1-cm² arrays of the diodes. For spectroscopic measurements, a monochromatic measuring beam crosses the sample through the remaining sides and detects absorption changes caused by the actinic light. If a measuring beam is unnecessary, e.g., when only the "end point" of a reaction is to be determined, the remaining sides of the cuvette are used also for actinic illumination.

In order to increase packing of the LED, their plastic lens has been filed off leaving only enough material around the semiconductors for electrical insulation. Thus, 24 diodes could be placed on an area of 1 cm², raising the total number contributing to the actinic illumination of the sample to 96 or 144, respectively. The arrays were fixed to aluminum heat sinks which facilitate maintenance of a constant temperature. The temperature of the entire reaction chamber including the sample is controlled by pumping a coolant through a channel in the outer rim of the device.

An electrical circuit useful for driving the LED arrays is shown in Fig. 2. In order to take advantage of the rapid rise and decay times of LED, their ability to be pulsed at arbitrary flash lengths from a few hundred nanoseconds to continuous light, and the high current that can be applied under pulsed operation, field effect transistors (MOS-FET,

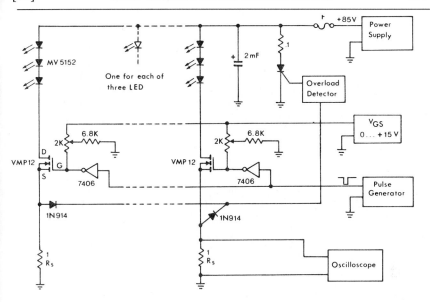

FIG. 2. Schematic of the LED driver and associated auxiliary instrumentation.

Siliconix, type VMP 12) were used as switches. These FET can switch pulses having a maximum current of 3 A and a maximum voltage of 90 V in approximately 5 nsec. Moreover, the low power required to drive the FET greatly simplifies the circuit.

In the present design, three LED in series were connected to a FET which was actuated by a common control voltage ($V_{GS}$) applied at the gate. Trim potentiometers were used to adjust slight differences in drain current among individual specimens. $V_{GS}$ was pulsed with TTL drivers (SN 7406) which in turn were activated by a pulse generator that determined the duration and frequency of the flashes. By varying $V_{GS}$ between 0 and + 15 V, the current through the LED can be regulated from 0 to 2 A. The magnitude of this current is measured as a potential difference across the source resistor $R_s$. A power supply delivered a voltage of 90 V and a current of about 2 A. To obtain the high pulse current required for this design, a capacitor was charged by the power supply. An overload detector incorporated into the circuit disconnects the power supply, thus protecting the LED if overloading should occur accidentally.

Figure 3 shows the shape of the driving and emitted pulses. The rise and decay times of the light are approximately 300 nsec. In the present design, these response times are limited by the TTL driver of the circuit and are somewhat longer than the actual response time of the LED. The spectral quality of the emitted light is shown in Fig. 4. Measurements

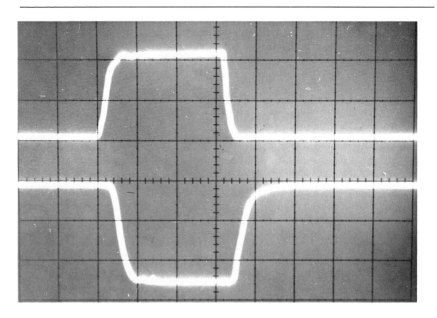

Fig. 3. Switching characteristics of the driver and the LED arrays. Upper trace: electrical pulse of 1 A measured with the circuit shown in Fig. 2. Lower trace: emitted light pulse measured via a photomultiplier. The time scale is 500 nsec/division.

were made with a spectroradiometer (Isco SR). At 25°, the peak of the emission spectrum is located at 635 nm and the half-bandwidth is 42 nm. It is to be noted, that the location of the peak is dependent on the temperature during operation. For the LED used here, a shift toward longer wavelengths of 0.17 nm/°C temperature increase has been stated by the manufacturer. In actual practice, appreciable temperature changes can be expected to occur only under prolonged continuous operation and if no provisions are made for cooling the device.

Figure 5 shows the intensity of a 1-cm² array of LED (i.e., the intensity at each of the faces of the cuvette) as a function of the current applied. With the maximally tolerable current of 35 mA for continuous (dc) operation, the intensity is $2 \times 10^4$ ergs/cm² sec. In pulsed operation the current can be increased to 2.5 A, a limit set by the MOS-FET, resulting in an intensity of approximately $10^6$ ergs/cm² sec.

A few measurements of the ferricyanide Hill reaction demonstrate the use of LED arrays as actinic light in experiments with chloroplasts. At the highest possible current for continuous emission (35 mA), the four

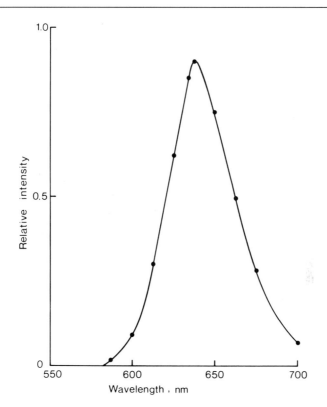

Fig. 4. Spectral distribution of light emitted by the LED (Monsanto, MV 5152). Measurement at ambient room temperature.

arrays, in an arrangement shown in Fig. 1, support a rate of 352 $\mu$mole/ mg chlorophyll hr in the absence and 508 $\mu$mole/mg Chl hr in the presence of an uncoupler (Table II). As the data show, these values approach saturation rates. In pulsed operation, we investigated the turnover of P700 in chloroplasts and intact cells. The pulse width at the intensity maximally tolerable under these circumstances had to be at least 1 msec in order to achieve light saturation. Since P700 shows relaxation times in the microsecond range this flash is too long for single turnover measurements. Moya[1] reported the use of LED as actinic light in a study of fluorescence yields and fluorescence life times in liquid nitrogen. At this temperature, he observed a nearly 10-fold increase in the intensity of the

[1] I. Moya, Dissertation, University of Paris, Centre d'Orsay, 1979.

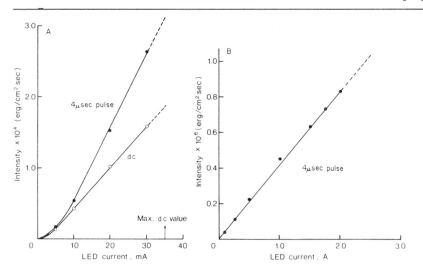

FIG. 5. Intensity of the emission as a function of applied current. Measurements were made with a photomultiplier that had been calibrated against a spectroradiometer (ISCO SR). (A) Open circles, intensity under continuous (dc) operation; closed circles, a 4 $\mu$sec pulse with a dark time of 8 msec. (B) 4 $\mu$sec pulse with a dark time of 8 msec. Pulse lengths were chosen arbitrarily.

TABLE II

FERRICYANIDE REDUCTION RATES IN CONTINUOUS LIGHT OF LED ARRAYS[a]

| | | Rate ($\mu$mole/mg Chl hr) | |
| Current (mA) | Relative intensity | No methylamine | With methylamine |
| --- | --- | --- | --- |
| 10 | 0.23 | 138 | 211 |
| 20 | 0.56 | 259 | 364 |
| 25 | 0.71 | 300 | 433 |
| 30 | 0.87 | 339 | 475 |
| 35 | 1.0 | 352 | 508 |

[a] The relative intensity was calculated from Fig. 5 setting the highest intensity emitted (approximately $2 \times 10^4$ erg cm$^{-2}$ sec$^{-1}$ for each array equal to 1.0). Four arrays were used in the arrangement shown in Fig. 1. The samples contained pea chloroplasts equivalent to 22 $\mu$g chlorophyll/ml 0.8 m$M$ potassium ferricyanide, 5 m$M$ MgCl$_2$, and where indicated, 2.5 m$M$ methylamine. The reaction medium contained 20 m$M$ Tricine, pH 8.0, 0.1 $M$ sucrose, and 20 m$M$ NaCl. Illumination time was 3 min. Measurements at 420 nm using an extinction coefficient of 1 (m$M$ cm.)$^{-1}$

emission and a band shift of the radiation to shorter wavelengths by 20 nm. The LED were modulated up to approx 60 MHz.

## Concluding Remarks

In this report we suggest the use of light-emitting diodes as a useful light source in certain applications of photochemical and photobiological work. We feel that their physical properties as well as ease of operation, low cost, and long life largely offset the low intensity provided by a single LED. As semiconductor technology advances, it can be expected that both the range of wavelengths and the efficiency and intensity of LED will be extended. Also, recently an improved FET (Siliconix, type VN64GA) has been manufactured which can switch 12.5 A with a rise time of 45 nsec. Utilization of this transistor in the circuit of Fig. 2 permits higher driving currents (and, therefore, higher emission intensities of the LED) as well as reduction in the number of FET needed per array.

### Acknowledgments

This work was supported by a grant from the Deutsche Forschungsgemeinschaft. Useful application notes supplied by manufacturers include Hewlett-Packard (AN 945), Monsanto (AN 301, 302, 304, 601, 602, 603), RCA (AN 4741 A), and Telefunken (Optoelectronic Devices).

## [65] Inhibitors in Electron Flow: Tools for the Functional and Structural Localization of Carriers and Energy Conservation Sites

*By* A. TREBST

The inhibition of photosynthetic electron flow in isolated chloroplasts (or bacterial chromatophores) is one of the main tools for the elucidation of the sequence and of the detailed mechanisms of electron carrier function. As discussed in the individual chapters and summarized in Section II some recent insights into photosynthesis are due to studies with inhibitors and accessibility probes, not only into the functional sequence of electron carriers, but also into their structural arrangements which gave evidence for vectorial electron flow), as well as into the mechanism of energy conservation. Inhibitors were instrumental in the final identification of a second coupling site connected with photosystem II, for a concept of artificial energy conservation and for present information on the topology of the membrane.

Inhibitors are indispensible for studies of photosynthesis *in vivo* in intact organisms and for an evaluation of the physiological significance of biochemical observations obtained in *in vitro* systems. Many inhibitors are also of commercial importance as herbicides. In this case, studies on the mode of action and on the relationship of chemical structure to biological activity have stimulated both photosynthesis and herbicide research.

Photosynthetic electron flow from water to NADP, vectorial across the thylakoid membrane, is coupled in two energy conservation sites to ATP formation. The presently available inhibitors of electron flow may be grouped according to five areas of attack on the photosynthetic electron transport system: inhibition of the acceptor and donor sites of photosystem I and II and plastoquinone antagonists (Fig. 1). Some of these inhibition groups were known for two decades, others have been introduced only recently. This article will concentrate on the latter and report only recent findings for the others. This development has been repeatedly reviewed in general[1,2] and also in relation to herbicidal action.[3,4] Very recently Izawa has reported recent progress.[5] Inhibitor studies are closely interrelated with artificial donor or acceptor systems, antibodies and chemical probes as inhibitors, and therefore the relevant chapters by Izawa, Hauska, Yocum, Berzborn and Dilley in this volume should be consulted.

## I. Inhibition of Photosynthetic Electron Flow

### A. Inhibitors of the Acceptor Site of Photosystem I

#### 1. Disulfodisalicylidenepropanediamine (Sulfo DSPD)[5a]

As representative of a group of metal chelating compounds, two sulfosalicylaldehydes connected by a diamine of various chain length,

[1] S. Izawa and N. E. Good, Vol. 24, p. 355.

[2] N. E. Good and S. Izawa, *in* "Metabolic Inhibitors" (R. M. Hochster, M. Kates, and J. H. Quastel, eds.), Vol. 4, p. 179. Academic Press, New York, 1973.

[3] D. E. Moreland, *Prog. Photosynth. Res.* **3,** 1693 (1969).

[4] K. H. Büchel, *Pestic. Sci.* **3,** 89 (1972).

[5] S. Izawa, *in* "Photosynthesis I" (A. Trebst and M. Avron, eds.), p. 266. Springer-Verlag, Berlin and New York, 1977.

[5a] Abbreviations: BSA, bovine serum albumin; DAD, diaminodurene; DBMIB, 2,5-dibromothymoquinone; DCMU, $N_1$-(dichlorophenyl)-$N_3$-dimethylurea; DNP–INT, 2,4-dinitrophenylether of 2-iodo-4-nitro-thymol; DSPD, disalicylidenepropanediamine; EDAC, 1-ethyl-3-(3-dimethylaminopropyl)carbodiimide; PMS, phenazinemethosulfate; PS, photosystem; TMPD, *N*-tetramethyl-*p*-phenylenediamine.

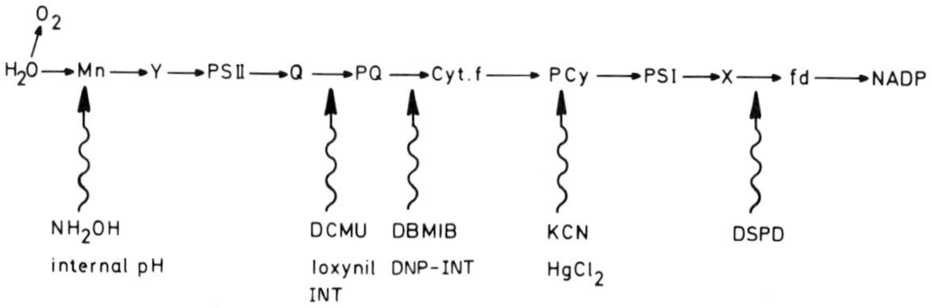

FIG. 1. Five sites of inhibition in the photosynthetic electron transport chain. Mn, manganese in the water splitting complex; Y, primary electron donor of photosystem II; Q, primary acceptor of photosystem II; PQ, plastoquinone; PCy, plastocyanin, X, primary acceptor of photosystem I; fd, ferredoxin; PS, photosystem.

sulfo DSPD was introduced in 1967.[6] Sulfo DSPD at 0.5–1 m$M$ is an inhibitor of photosynthetic ferredoxin reduction (and of Hill reactions dependent on ferredoxin, like NADP, cytochrome $c$ or nitrite reduction

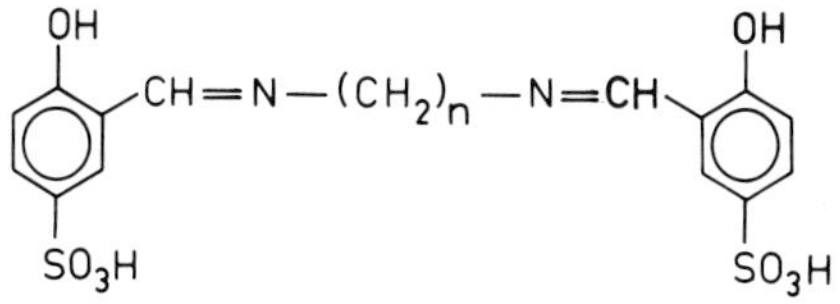

Sulfo – DSPD

n = 3

in isolated chloroplasts.[6] It might actually attack ferredoxin itself.[7] Artificial acceptor systems for photosystem I (like MV or ferricyanide reduction) are not inhibited by sulfo DSPD.[6] Because of its high hydrophilicity, due to the sulfo group, the inhibition is taken as evidence for the accessibility of the endogeneous acceptor side of photosystem I in the thylakoid membrane.

Because of the hydrophilic sulfo group, sulfo DSPD is not useful for studies of photosynthesis in intact systems or organisms. For this the parent compound (DSPD), without the sulfo group, has been used; and it was implied that it acts similarly in intact organisms as does sulfo-DSPD in chloroplasts. DSPD does indeed inhibit $CO_2$-assimilation in

[6] A. Trebst and M. Burba, Z. Pflanzenphysiol. **57**, 419 (1967).
[7] A. Ben-Amotz and M. Avron, Plant Physiol. **49**, 244 (1972).

algae.[8] But recently it was shown that DSPD has a completely different inhibition pattern than sulfo-DSPS.[9] For example DSPD, as against sulfo DSPD, also effects the electron-transport chain in the area of plastocyanin function, and it may also drain off electrons.[9] It is now obvious that DSPD—as against sulfo DSPD—is not a specific antagonist of ferredoxin, and its usefulness for physiological studies of photosynthesis is doubtful.

### 2. Antibodies

Antibodies against the acceptor site of photosystem I were among the first antibodies reported for a chloroplast system. The antibodies against diaphorase and transhydrogenase activity, when discovered in the thylakoid supernatant, actually helped identify the ferredoxin-NADP reductase.[10] Another antibody against the reducing side of photosystem I was prepared later by immunization against whole thylakoid preparations. The preparation contained several antibodies, among them one against the reductase and the coupling factor but also a new antibody which inhibited ferredoxin, but not anthraquinone or ferricyanide reduction.[11] An antigen preparation, neutralizing this antibody, has been obtained from the membrane ($S_{L-eth}$).[12] However, its chemical identity and role in electron flow has not been clarified. Very recently König et al.[13] again reported on the inhibition of the acceptor side of photosystem I by an antibody, this time obtained by immunization against a purified low molecular protein purified from the thylakoid membrane. Very probably this protein and $S_{L-eth}$ are components of the photosystem I complex, as isolated and separated into five bands by Nelson,[14] who prepared antibodies to study their role and respective location in the membrane and at the reaction center. The action of these antibodies against components of photosystem I suggests that the acceptor side of photosystem I (including ferredoxin and reductase) is easily accessible to hydrophilic or high molecular weight compounds and therefore is located on the matrix side of the thylakoid membrane.

[8] H. Gimmler, W. Urbach, W. D. Jeschke, and W. D. Simonis, Z. Pflanzenphysiol. **58,** 353 (1968).

[9] N. Laasch, W. Kaiser, and W. Urbach, Plant Physiol. **63,** 605 (1979).

[10] A. San Pietro, Ann. N.Y. Acad. Sci. **103,** 1093 (1963).

[11] R. Berzborn, W. Menke, A. Trebst, and E. Pistorius, Z. Naturforsch., Teil B **21,** 1057 (1966).

[12] G. Regitz, R. Berzborn, and A. Trebst, Planta **91,** 8 (1970).

[13] F. König, W. Menke, A. Radunz, and G. H. Schmid, Z. Naturforsch., Teil C **32,** 817 (1977).

[14] N. Nelson and B. Notsani, in "Bioenergetics of Membranes" (L. Packer et al., eds.), p. 233. Elsevier, Amsterdam, 1977.

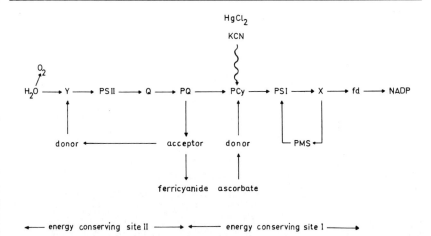

FIG. 2. Photosynthetic reactions in $HgCl_2$ or KCN treated chloroplasts. Inoperative: electron flow to NADP from water as well as from a donor system for photosystem I. Operative: Electron flow from water to an acceptor for photosystem II, coupled via energy conserving site II; photosystem I-driven cyclic electron flow via PMS (via energy conserving site I) or cyclic electron flow driven by photosystem II via a phenylenediamine acting as acceptor and donor (via energy conserving site II).

## B. Inhibitors of Plastocyanin Function

Several new compounds interfering with plastocyanin function or even reacting with plastocyanin directly are now available: $HgCl_2$, KCN, polylysine, antibodies and amphotericin. After inhibition of plastocyanin function, the full activity of photosystem II including electron flow from water to an artificial lipophilic acceptor for photosystem II remains intact (Fig. 2). Photosystem I activity, however, is impaired by plastocyanin inhibitors. That is, noncyclic electron flow is inhibited as are most electron donor systems for photosystem I and cyclic electron flow with the exception of high concentrations of DCPIP or PMS. Cytochrome $f$ photooxidation is also inhibited, but not $P_{700}$ oxidation. In the presence of plastocyanin inhibitors, coupling of the undisturbed electron flow sequence in photosystem II Hill reactions and of PMS cyclic photophosphorylation remains, except when the inhibitor also interferes with the coupling factor (as polylysine does).

### 1. Mercuric Chloride

The inhibitor action of various organic mercurials on the electron flow system is well known, e.g., $p$-mercuric acetate on ferredoxin-NADP reductase (see Honeycutt and Krogmann[15]). Kimimura and Katoh in three papers[16-18] described the effect of $HgCl_2$ on electron flow and localized its inhibition at plastocyanin (Fig. 2). $HgCl_2$ reacts directly with plastocyanin, as later shown by isolation of the inactivated plastocyanin from $HgCl_2$-treated membranes.[19] $HgCl_2$ treatment of chloroplasts did help in establishing photosystem II-driven Hill reactions and its properties.[18] It became clear, for example, that DCPIP is an acceptor of electrons from photosystem II only in fragmented and sonicated chloroplasts but not in intact thylakoid preparations (see Izawa [39] this volume). The concentration of $HgCl_2$ required to inhibit plastocyanin specifically is strictly dependent on and stoichiometric to the chlorophyll concentration; that is, about 10 $\mu M$ for chloroplasts containing 10 $\mu g$ chlorophyll. At higher concentrations of $HgCl_2$, photosystem II activity is also disturbed as shown before for phenylmercuric acid.[15] $HgCl_2$ also affects the coupling system. However, at a carefully chosen minimal concentration of $HgCl_2$ (1 $\mu M/\mu g$ chlorophyll), coupling of photosystem II electron flow but not of photosystem I is observed.[20,21] It seems unjustified to conclude from this that coupling site II is more resistant to $HgCl_2$ (as an energy transfer inhibitor) than coupling site I and by this to imply a heterogeneity of coupling factors for either photosystem I or II.[22]

Recently a combination of $HgCl_2$ and KCN treatment was recommended for plastocyanin inhibition[23] (see below) and used to characterize a photosystem II-dependent cyclic phosphorylation system.

### 2. Potassium Cyanide

Photosynthesis ($CO_2$ assimilation) was long known to be inhibited by KCN, due to sensitivity of the ribulosediphosphate carboxylase. An inhibition of electron flow by high concentrations (10 m$M$ and higher) was reported in 1963 and ascribed to plastocyanin inhibition.[24] This was studied in greater detail much later by Ouitrakul and Izawa,[25] who intro-

[15] R. C. Honeycutt and D. W. Krogmann, *Plant Physiol.* **43**, 376 (1972).

[16] M. Kimimura and S. Katoh, *Biochim. Biophys. Acta* **283**, 279 (1972).

[17] S. Katoh, *Biochim. Biophys. Acta* **283**, 293 (1972).

[18] M. Kimimura and S. Katoh, *Biochim. Biophys. Acta* **325**, 167 (1973).

[19] B. R. Selman, G. C. Johnson, R. T. Giaquinta, and R. A. Dilley, *Bioenergetics* **6**, 221 (1975).

[20] D. A. Bradeen, G. D. Winget, J. M. Gould, and D. R. Ort, *Plant Physiol.* **52**, 680 (1973).

[21] J. M. Gould, *Biochem. Biophys. Res. Commun.* **64**, 673 (1975).

[22] D. A. Bradeen and G. D. Winget, *Biochim. Biophys. Acta* **333**, 331 (1974).

[23] C. F. Yocum and J. A. Guikema, *Plant Physiol.* **59**, 33 (1977).

[24] A. Trebst, *Z. Naturforsch., Teil B* **13**, 817 (1963).

[25] R. Ouitrakul and S. Izawa, *Biochim. Biophys. Acta* **305**, 105 (1973).

duced a gentle treatment of chloroplasts for 30–60 min by still higher concentrations of KCN (pH 7.8, 30 m$M$ KCN + a trace of ferricyanide to keep plastocyanin oxidized). This approach served to clarify the site of inhibition of KCN at plastocyanin function.[26] Furthermore, the coupling of KCN-insensitive photosystem II Hill reactions and thereby the existence of a second coupling site was shown[25] (see Fig. 2). As discussed in the DBMIB system in more detail, this photosystem II electron flow is inhibited by an uncoupler.[26a] Of particular importance is the relative insensitivity to KCN of cyclic photophosphorylation at high concentrations of PMS.[25] This observation was explained by vectorial transmembrane electron transport via proton translocating cofactors and the concept of artificial energy conservation see Section IV,C,1,d.[27,27a] It has been shown that KCN acts directly with plastocyanin.[19,28] Takahashi and Asada reported that NaCN treatment of chloroplast releases copper and manganese.[29]

Recently Yocum[23] reported on a combined treatment of thylakoids with 10 $\mu M$ HgCl$_2$ + 50 m$M$ KCN (at pH 8.0 and 4° for 90 min) to give highly reproducible results on the coupling of photosystem II and of photophosphorylation driven by photosystem II (Fig. 2).

### 3. Polylysine

Polycations, such as polylysines (of different chain length) and histones, were described by Dilley as uncouplers of photophosphorylation.[30] It was reported later that at low salt concentrations and maximum swelling of chloroplasts, it also inhibits electron flow.[31]

Specifically photosystem I activity is inhibited by polylysine (at 1 mg/ml of molecular weight 190,000) or 0.1 mg histone.[31,32] Inhibition is prevented by the presence of high salt (0.05 $M$) and is reversed by addition of plastocyanin, detergent, or even by light-driven electron flow.[31,32] Cytochrome $f$ oxidation by P700 is inhibited, suggesting plastocyanin as

[26] S. Izawa, R. Kraayenhof, E. U. Ruuge, and D. Devault, *Biochim. Biophys. Acta* **314**, 328 (1973).

[26a] J. A. Guikema and C. F. Yocum, *Biochemistry* **15**, 362 (1976).

[27] A. Trebst, *Annu. Rev. Plant Physiol.* **25**, 423 (1974).

[27a] A. Trebst, *in* "Energy Conservation in Biological Membranes" (G. Schäfer and M. Klingenberg, eds.), p. 83. Springer-Verlag, Berlin and Heidelberg, 1978.

[28] S. P. Berg and D. W. Krogmann, *J. Biol. Chem.* **250**, 8957 (1975).

[29] M. Takahashi and K. Asada, *Eur. J. Biochem.* **64**, 445 (1976).

[30] R. A. Dilley, *Biochemistry* **7**, 338 (1968).

[31] J. Brand, T. Baszynski, F. L. Crane, and D. W. Krogmann, *J. Biol. Chem.* **247**, 2814 (1972).

[32] S. Berg, D. Cipollo, B. Armstrong, and D. W. Krogmann, *Biochim. Biophys. Acta* **305**, 372 (1973).

the site of inhibition.[33] The photoreduction of lipophilic acceptors of photosystem II is not inhibited by polylysine and, under careful conditions, ATP formation coupled to it is observed.[34] Because of the implication for plastocyanin location in the membrane, it is important to note that polylysine binds to the surface of the membrane at negatively charged carboxyl groups.[35]

### 4. Antibodies

The use of antibodies against plastocyanin (and cytochrome $f$) is of principal importance for the concept of vectorial electron flow. The early experiments with plastocyanin antibodies suggested its inaccessibility[36] (no inhibition and no agglutination). Sonication of the chloroplast vesicles in the presence of the antibody did show inhibition and thus an inside location for plastocyanin was indicated.[36] As discussed in Section II,A,2, antibody preparations against plastocyanin by Schmid et al.[38] did agglutinate as well as inhibit electron flow to a certain extent, indicating an accessibility of plastocyanin in unstacked thylakoid preparations. Antibodies prepared by Böhme[39] also agglutinated ($\gamma_M$-globulin) and inhibited electron flow up to 40%, but not in all chloroplast preparations. In view of the observation that during preparation of chloroplasts some plastocyanin is always solubilized and becomes attached to the outside, agglutination experiments are less indicative than inhibitors of electron flow. Since only the inside bound plastocyanin allows vectorial (transmembrane) electron flow and proton transport, whereas the outside bound would not, careful experiments on the coupling and stoichiometry of coupled electron flow might resolve the problem of whether only the superimposed artificial and uncoupled electron flow via attached plastocyanin is sensitive to antibodies.

### 5. Amphotericin

Bishop described the inhibition of photosynthetic electron flow by the

---

[33] J. Brand, A. San Pietro, and B. C. Mayne, *Arch. Biochem. Biophys.* **152**, 426 (1972).

[34] D. R. Ort, S. Izawa, N. E. Good, and D. W. Krogmann, *FEBS Lett.* **31**, 119 (1973).

[35] S. Berg, S. Dodge, D. W. Krogmann, and R. A. Dilley, *Plant Physiol.* **58**, 619 (1974).

[36] G. Hauska, R. E. McCarty, R. Berzborn, and E. Racker, *J. Biol. Chem.* **246**, 3524 (1971).

[37] A. Trebst and S. Reimer, *Plant Cell Physiol. Spec. Issue* p. 201 (1977).

[38] G. H. Schmid, A. Radunz, and W. Menke, *Z. Naturforsch., Teil C* **30**, 201 (1975).

[39] H. Böhme, in "Bioenergetics of Membranes" (L. Packer et al., eds.), p. 329. Elsevier, Amsterdam, 1977.

polyene antibiotics amphotericine B[40,40a] and filipin.[41] Photosystem I activity is inhibited by treatment of chloroplasts with 0.5 m$M$ amphothericin for 60 min at 25°, but restored by addition of plastocyanin. Inhibition of amphotericin is related to the presence of sterols (not necessarily present in all plant chloroplasts) and the creation of aqueous pores in the membrane[42] which lead to release of the inside bound plastocyanin.

## 6. DABS

This chemical probe, impermeable through the membrane, does react with plastocyanin[43] among its other inhibitory activities as discussed later (see Section I,E,4; see also Dilley [48], this volume).

## 7. Glutaraldehyde

Glutaraldehyde fixation of chloroplast membranes inhibits electron flow; inhibition was recently localized at the functional site of plastocyanin.[44]

## C. Inhibition of the Oxidizing Site of Plastoquinone

Inhibitors acting on the electron-flow sequence between the oxidation site of plastohydroquinone and plastocyanin reduction turned out to be particularly useful in recent photosynthesis research. Therefore, this group of inhibitors will be discussed in more detail.

An inhibitory class of plastoquinone antagonists was introduced in 1970-1971 with the investigation of halogenated or hydroxylated lipophilic benzoquinones. The dibrominated thymoquinone (DBMIB) established itself as the most useful representative of this group of inhibitors, which has by now been used very extensively in numerous studies. The site of inhibition of DBMIB is at the site of plastoquinone oxidation. This is to say, it is not inhibiting plastoquinone as such, but rather binding to the site on a protein–carrier–complex (Rieske protein?)[44a] at which plastohydroquinone is oxidized. A number of substituted benzoquinones, analogous to DBMIB but recently also certain alkylated dinitroanalines

[40] D. G. Bishop, *Arch. Biochem. Biophys.* **154**, 520 (1973).

[40a] W. G. Nolan and D. G. Bishop, *Arch. Biochem. Biophys.* **190**, 473 (1978).

[41] W. G. Nolan and D. G. Bishop, *Arch. Biochem. Biophys.* **166**, 323 (1975).

[42] A. Finkelstein and R. Holz, *in* "Membranes" (G. Eisenman, ed.), Vol. 2, p. 377. Dekker, New York, 1973.

[43] B. R. Selman, G. L. Johnson, R. A. Dilley, and K. K. Voegeli, *Proc. Int. Congr. Photosynth., 3rd, 1974* p. 897 (1975).

[44] H. Hardt and B. Kok, *Plant Physiol.* **60**, 225 (1977).

[44a] R. Malkin and P. J. Aparacio, *Biochem. Biophys. Res. Commun.* **63**, 1157 (1975).

and certain diphenyl ethers, have been described with inhibitory patterns like DBMIB. In addition EDAC, bathophenanthroline, and valinomycin have been shown to inhibit electron flow between plastoquinone and plastocyanin.

1. DBMIB (Dibromothymoquinone; 2,5-dibromo-3-methyl-6-isopropyl-$p$-benzoquinone).

Dibromothymoquinone

(DBMIB)

*a. General Properties.* One micromolar DBMIB, like DCMU, inhibits all Hill reactions in isolated chloroplast lamellae with electron acceptors more electronegative than about +100 mV [i.e., inhibition of NADP or MV reduction]. However, unlike DCMU, DBMIB does not inhibit oxygen evolution in photoreductions by photosystem II of lipophilic acceptors more electropositive than 100 mV (i.e., photoreduction of a phenylenediimine or benzoquinone). DBMIB, like DCMU, does not inhibit photoreductions by photosystem I at the expense of artificial electron donors, but (unlike DCMU) does inhibit certain cyclic photophosphorylation systems. DBMIB, even at 100 $\mu M$, has no effect on coupled ATP formation per se, it is neither an uncoupler nor energy transfer inhibitor. Reports in the literature[45,46] of such an uncoupler activity of DBMIB misinterpretated the high chemical autoxidizability of DBMIB/ascorbate as a stimulation of electron flow that an uncoupler would cause. It is agreed that DBMIB is a plastoquinone antagonist, and the sensitivity of a photosynthetic system to DBMIB is taken now as evidence for a plastoquinone participation in the system.

*b. Site of Inhibition.* The inhibition of electron flow is at the oxidation of plastoquinol is deduced from the chemical structure of DBMIB, reversal of the inhibition by plastoquinone,[47,48] and the inhibition pattern. There is no influence of DBMIB on Hill reactions by photosystem II,

[45] U. Than Nyunt and J. T. Wiskich, *Plant Cell Physiol.* **14,** 1099 (1973).
[46] G. Ben-Hayyim, Z. Drechsler, and J. Neumann, *Plant Sci. Lett.* **7,** 171 (1976).
[47] H. Böhme, S. Reimer, and A. Trebst, *Z. Naturforsch., Teil B* **26,** 341 (1971).
[48] A. Trebst and S. Reimer, *Biochim. Biophys. Acta* **305,** 129 (1973).

but there is inhibition of cytochrome $f$,[49,49a] plastocyanin, and P700[50,51] reduction; their oxidation is not affected. Haehnel observed no inhibition of plastoquinone reduction by DBMIB by measuring its absorption change at 265 nm, but did observe inhibition of its reoxidation by photosystem I.[51] The influence of DBMIB on the fluorescence induction curve,[52,53] as well as on the oxygen burst,[53] indicates that DBMIB (at 1 $\mu M$) does not disconnect the secondary, a pool (plastoquinone) from photosystem II. Recently, Wood and Bendall[54] described a thylakoid fraction which has a plastoquinol-5-plastocyanin oxidoreductase activity sensitive to DBMIB. From this it follows that the inhibition is between plastohydroquinone oxidation and plastocyanin/cytochrome $f$ reduction. The inhibition of electron flow is due to a DBMIB competition with plastoquinol for the oxidation site at the integral protein complex cytochrome $b_6$/cytochrome $f$/Rieske FeS center.

The number of DBMIB molecules required for inhibition is calculated to be about one per 300 chlorophylls or per one electron-transport chain.[51] DBMIB added in the reduced form as well as in the oxidized form are equally effective,[47,54a] the quinone being quickly reduced by chloroplasts in the light.

   *c. Systems Sensitive to DBMIB.*

   i. Noncyclic and pseudocyclic electron flow from water to photosystem I. Sensitive to 1 $\mu M$ DBMIB are all Hill reactions which proceed via both photosystems I and II with water as the electron donor. For example, 1 $\mu M$ DBMIB inhibits oxygen evolution and the photoreduction of NADP, MV, and anthraquinone sulfonate (whether measured as such or via oxygen uptake in pseudocyclic electron flow). Also, DCPIP and ferricyanide reduction are sensitive to 1 $\mu M$ DBMIB in intact thylakoid vesicles; but they are not inhibited in fragmented systems.[47,54b] The inhibition by DBMIB of DCPIP and ferricyanide reduction therefore depends on the intactness of the vesicle and accessibility of the acceptor site of photosystem II. This result stresses again that DCPIP cannot be generally regarded as a photosystem II electron acceptor since this depends on the kind of system investigated. The DBMIB insensitivity of

[49] H. Böhme and W. Cramer, *FEBS Lett.* **15**, 349 (1971).
[49a] H. Koike, K. Satoh, and S. Katoh, *Plant Cell Physiol.* **19**, 1371 (1978).
[50] W. Ausländer, P. Heathcote, and W. Junge, *FEBS Lett.* **47**, 229 (1974).
[51] W. Haehnel, *in* "Bioenergetics of Membranes" (L. Packer *et al.*, eds.), p. 317. Elsevier, Amsterdam, 1977.
[52] R. Bauer and M. J. G. Wijnands, *Z. Naturforsch., Teil C* **29**, 725 (1974).
[53] Y. de Kouchkovsky, *Biochim. Biophys. Acta* **376**, 259 (1975).
[54] P. M. Wood and D. S. Bendall, *Eur. J. Biochem.* **61**, 337 (1976).
[54a] S. Reimer, K. Link, and A. Trebst, *Z. Naturforsch; Teil C* **34**, 419 (1979).
[54b] A. Trebst, *Proc. Int. Congr. Photosynth.* 3rd, 1974, p. 799 (1975).

DCPIP reduction should be taken as evidence for a photosystem II reduction.

The inhibition of noncyclic and pseudocyclic electron flow from water to PS-I is reversed by a catalytic amount of TMPD.[37,54c] This TMPD bypass is discussed in detail below (Fig. 3).

ii. Cyclic electron flow. Certain cyclic electron-flow systems (involving photosystem I) are sensitive to DBMIB, whereas others are not. The cyclic electron-flow systems (measured as cyclic photophosphorylation) sensitive to 1 $\mu M$ DBMIB are those catalyzed by ferredoxin, menadione, thymoquinone, and duroquinone.[55] Insensitive are those catalyzed by DAD or PMS (see below). The inhibition by DBMIB of ferredoxin, menadione, thymoquinone, and duroquinone systems is reversed by catalytic amounts (0.1 mM) of TMPD.[37,54c,55a] A TMPD bypass restores electron flow as well as coupled ATP formation.[37,54c] This internal TMPD bypass (Fig. 3 and 4) couples the oxidation of the plastohydroquinone onto plastocyanin around the DBMIB block. Proton translocation for coupling in a TMPD bypass in the ferredoxin system is still due to plastohydroquinone; in the other system it probably occurs via the cofactor itself.

iii. Donor system for photosystem I. The usual donor systems DAD, TMPD, or DCPIP, kept reduced by ascorbate, are insensitive to DBMIB (see below). However, as the corresponding cycling system,[55] the durohydroquinone donor system is DBMIB sensitive.[55b] This new duroquinol donor system for photosystem I is very valuable, because it includes the plastohydroquinone oxidation site, thought not necessarily plastoquinone itself.

*d. Systems Insensitive to DBMIB*

i. Photoreductions by Photosystem II. An electropositive electron acceptor may be reduced in a DBMIB-insensitive reaction, the rate depends on the accessibility of the reducing end of photosystem II and the lipophilicity of the acceptor.[47] Ferricyanide is reduced by photosystem II in fragmented thylakoid particles (by sonication or detergents) or *Euglena* chloroplasts (both systems have lost plastocyanin), as is DCPIP.[47]

In intact thylakoid vesicles a lipophilic class II electron acceptor is required for an DBMIB insensitive Hill reaction.[48,56,57] *p*-Phenylenedi-

[54c] A. Trebst and S. Reimer, *Z. Naturforsch., Teil C* **28**, 710 (1973).

[55] G. Hauska, S. Reimer, and A. Trebst, *Biochim. Biophys. Acta* **357**, 1 (1974).

[55a] R. G. Binder and B. R. Selman, *Z. Naturforsch., Teil C* **33**, 261 (1978).

[55b] S. Izawa and R. L. Pan, *Biochem. Biophys. Res. Commun.* **83**, 1171 (1978).

[56] S. Izawa, J. M. Gould, D. R. Ort, P. Felker, and N. E. Good, *Biochim. Biophys. Acta* **305**, 119 (1973).

[57] J. M. Gould and S. Izawa, *Biochim. Biophys. Acta* **314**, 211 (1973).

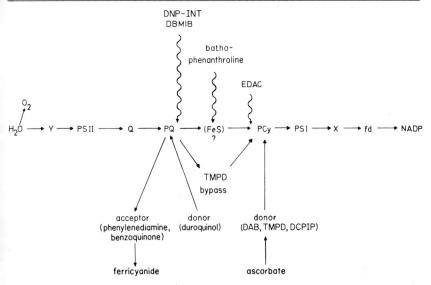

FIG. 3. Inhibition sites between plastoquinone and plastocyanin via a hypothetical nonheme iron. In DBMIB, bathophenanthroline, or EDAC inhibited chloroplasts photosystem II-driven acceptor systems as well as photosystem I-driven donor systems are coupled via energy conserving site II or I, respectively. TMPD bypasses the inhibition site and restores oxygen evolution and electron flow from water to NADP.

amine, various substituted phenylenediamines,[56,57,64] and $p$-benzoquinones[48,58] in catalytic amounts (100 $\mu M$) are used to mediate electron flow from photosystem II to ferricyanide. Instead of ferricyanide, oxygen may be used as terminal electron acceptor for a photosystem II-driven Hill reaction. Methylenedioxydimethyl-$p$-benzoquinone[58] or DBMIB[59] itself (at about 10 to 20 $\mu M$) are good cofactors of such a pseudocyclic electron flow system with oxygen uptake, driven by photosystem II. The reaction is insensitive to DBMIB inhibition, particularly if 100 $\mu M$ manganese ions stimulate the autoxidation of the reduced acceptors.[60] The reduced oxygen species in these autoxidations of acceptors of photosystem II is $H_2O_2$,[61] whereas superoxide radical-anion is formed in the autoxidation of photosystem I acceptors. Silicomolybdate reduction is DBMIB insensitive (as it is also against DCMU).[62] Photoreductions by

[58] A. Trebst, S. Reimer, and F. Dallacker, *Plant Sci. Lett.* **6**, 21 (1976).
[59] J. M. Gould and S. Izawa, *Eur. J. Biochem.* **37**, 185 (1973).
[60] C. D. Miles, *FEBS Lett.* **61**, 251 (1976).
[61] E. F. Elstner and D. Frommeyer, *Z. Naturforsch., Teil C* **33**, 276 (1978).
[62] R. T. Giaquinta, R. A. Dilley, F. L. Crane, and R. Barr, *Biochem. Biophys. Res. Commun.* **59**, 985 (1974).

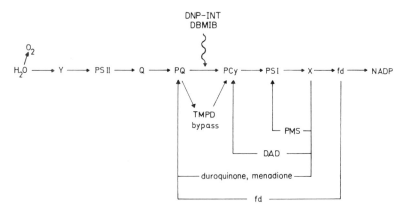

FIG. 4. Cyclic electron flow systems in the presence of DBMIB, coupled to photophosphorylation via energy conserving site I. PMS- and DAD-catalyzed electron flow are not inhibited, whereas menadione- and ferredoxin-catalyzed systems are. A TMPD bypass reverses the inhibition.

photosystem II are coupled to ATP formation with a $P/e_2$ ratio of half of noncyclic electron flow from water to a photosystem I acceptor.[48,56,57] This coupling led to the recognition of a second coupling site in photosynthetic electron flow (Fig. 3).

There are conflicting data on the control of DBMIB-insensitive Hill reactions by coupling conditions. Whereas Gould and Izawa found no stimulation by coupling conditions[57,63] others did.[64,65] Trebst and Reimer[58,64] reported that Hill reactions by photosystem II are inhibited by uncouplers at a pH of 8 and above when phenylenediimines, but not when quinones,[58] are acceptors (also see Cohen et al.[66]). This is explained by an effect of internal pH (equilibrated with the external pH via the uncoupler) on the protonization of the two redox states of the mediator.[67] This in turn is taken as indication that the reducing site for the acceptor reduction by photosystem II is on the inside of the thylakoid membrane. A similar inhibition of photosystem II electron flow by an uncoupler has been reported for $KCN$[26a] or $EDAC$[66] treated chloroplasts.

ii. Cyclic electron flow driven by photosystem II. Recently the properties of a DBMIB-insensitive electron flow system coupled to ATP

[63] J. M. Gould and D. Ort, *Biochim. Biophys. Acta* **325**, 157 (1973).
[64] A. Trebst and S. Reimer, *Biochim. Biophys. Acta* **325**, 546 (1973).
[65] P. Heathcote and D. O. Hall, *Biochem. Biophys. Res. Commun.* **56**, 767 (1974).
[66] W. S. Cohen, D. E. Cohn, and W. Bertsch, *FEBS Lett.* **49**, 350 (1975).
[67] G. Hauska and A. Trebst, *Curr. Top. Bioenerg.* **6**, 151 (1977).

formation and mediated by cofactors shortcircuiting photosystem II has been described.[23]

iii. Photoreductions by photosystem I. Electron donor systems for photoreductions by photosystem I (i.e., DAD, TMPD, or DCPIP + ascorbate as donor and NADP or MV as acceptor) are insensitive to up to 2 $\mu M$ DBMIB[47] (Fig. 3). Above this concentration a 10 to 20% inhibition is observed.[47,68,69] These photoreductions are coupled to ATP formation even in the presence of DBMIB, indicating that plastoquinone does not need participate in coupling of photosystem I (i.e., artificial energy conservation.[27,67]). Donor systems for photosystem I are stimulated by an uncoupler, when a hydroxy-substituted donor is used (such as DCPIPH$_2$), but not in phenylenediamine donor systems with a NH$_2$ group. This effect, opposite to the one in acceptor systems of photosystem II, is also explained by an effect of internal pH on the protonization on the mediator.[67,69]

iv. Cyclic electron flow driven by photosystem I. Cyclic electron flow (Fig. 3) coupled to ATP formation catalyzed by PMS, DAD, or DCPIP are not sensitive to even 100 $\mu M$ DBMIB.[47,55]. As in coupled donor systems for photosystem I, this indicated that plastoquinone is not required for coupling in these systems. The concept of artificial energy conservation was developed to explain this coupling.[27,55,67] According to it, the cofactor itself in the reduced state transports hydrogen ions across the membrane, and a pH gradient is built up when the reduced cofactor is oxidized inside the membrane by the electrogenic photosystem I. The native proton pump associated with plastoquinol oxidation inside is thus no longer required (Fig. 5).

v. TMPD bypass. Compounds, which in the oxidized state are acceptors for photosystem II before DBMIB, are donors for photosystem I after DBMIB in the reduced form if they can bypass the DBMIB block by reconnecting photosystems II and I (Figs. 3 and 4). In such a bypass, oxygen evolution, as well as the reduction of a photosystem I acceptor (such as NADP reduction), is restored.[37,54c] Particularly effective in a bypass is TMPD in catalytic amounts (100 $\mu M$) and in the absence of ascorbate or ferricyanide, which would keep it either reduced or oxidized. The TMPD bypass is coupled to ATP formation with a very high P/e$_2$ ratio.[64,54c,70] This is taken as evidence that it is an internal bypass rather than an external, transmembrane bypass. The TMPD bypass has proven useful also in locating other inhibitor sites[37] (see Sections I, C, 2 and 3).

[68] Y. de Kouchkovsky and F. de Kouchkovsky, *Biochim. Biophys. Acta* **368**, 113 (1974).
[69] G. Hauska, W. Oettmeier, S. Reimer, and A. Trebst, *Z. Naturforsch., Teil C* **30**, 37 (1975).
[70] B. Selman, *J. Bioenerg. Biomembr.* **8**, 143 (1976).

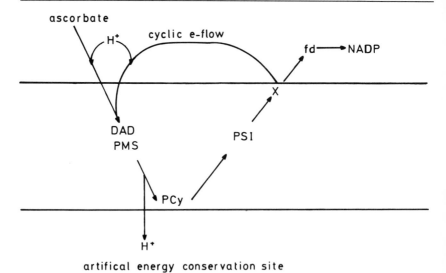

ascorbate

cyclic e-flow

fd———►NADP

DAD
PMS

PSI

X

PCy

H⁺

artifical energy conservation site

FIG. 5. Artificial energy conservation in PMS or DAD catalyzed cyclic electron flow via photosystem I. The native proton translocating site at plastoquinone is inoperative in the presence of DBMIB or other inhibitors between plastoquinone and plastocyanin. It is replaced by a transmembrane proton translocation via the reduced mediator.

*e. Reversal of DBMIB Inhibition by SH Compounds and by BSA.* A TMPD bypass is not a reversal of inhibition because the inhibited step remains inoperative. But thiols (such as mercaptoethanol, cysteine, or DTT) reverse DBMIB inhibition at a ratio of 2 to 4 times thiol/DBMIB via a 1,4-SH addition or nucleophilic attack followed by elimination of a bromine.[71,71a] This reversal of DBMIB in a chloroplast system depends on the pipetting sequence. Thiols immediately inactivate DBMIB, but after a preillumination, DBMIB is in the reduced state and the subsequent addition of a thiol leads to reversal of inhibition only after a few minutes.[54a] Bovine serum albumin (1 mg) reverses the inhibition of DBMIB, independent of its redox state.[54a,71a,77,144]

*f. DBMIB as Quencher of Chlorophyll Fluorescence.* Studies on the induction curve of chlorophyll fluorescence helped elucidate the site of inhibition of DBMIB (see Section I,C,1,b). At higher concentrations (100 μM and more) DBMIB acts also as an effective quencher of the fluores-

---

[71] S. Reimer and A. Trebst, *Z. Naturforsch., Teil C* **31**, 103 (1976).
[71a] W. Oettmeier, *Z. Naturforsch., Teil C* **34**, 242 (1979).

cence of the bulk chlorophyll,[72] as do most quinones.[73] This quenching is effective also at $-190°$ and depends strictly on the oxidized form of DBMIB (in contrast to inhibition of electron flow). This high quenching effect has enabled Butler to develop a general model for the primary photochemistry of photosystem II and its antenna system.[74] The quenching of bacteriochlorophyll fluorescence by DBMIB, seems to be due to replacement of the primary acceptor (a ubiquinone) from the reaction center.[75]

g. *Effect of High Concentrations of DBMIB.* Inhibition of electron flow, as discussed above, requires a DBMIB concentration between 0.2 and 1 $\mu M$, quenching of bulk fluorescence about 10 to 100 $\mu M$. At an intermediate DBMIB concentration, the inhibition pattern of DBMIB on photosynthetic electron flow systems is changed.[52,53,54c] Photosystem II-driven Hill reactions become DBMIB sensitive and, in particularly, the P/e$_2$ ratio falls from 0.5 to about 2 to 0.25.[54c] Also the P/e$_2$ ratio and the rate of electron flow in a TMPD bypass is diminished by 10 $\mu M$ DBMIB.[54c] The change in area above the induction curve of variable fluorescence in the presence of this DBMIB concentration,[52,53] as well as its effect on the oxygen burst, suggest that DBMIB now also affects plastoquinone reduction[76] (Fig. 6). Recently, more results on the second inhibition site of DBMIB were reported.[77,77a] The second site has been attributed to an "intermingling" of DBMIB with the plastoquinone pool[54c] and as an effect on membrane fluidity.[78] Some of the other plastoquinone antagonists[79,80] (see below) usually behave like high concentration of DBMIB.

h. *Relation of Chemical Structure to Inhibitory Potency.* The plastoquinone antagonists synthesized by Folkers *et al.*[79–81] and tested by Bolen *et al.*[81] Arntzen and Neumann *et al.*,[80] and Barr *et al.*[79] (substituted halogenated or hydroxylated benzoquinones with a phytyl or other highly

---

[72] M. Kitajima and W. L. Butler, *Biochim. Biophys. Acta* **376**, 105 (1975).

[73] J. Amesz and D. C. Fork, *Biochim. Biophys. Acta* **143**, 97 (1967).

[74] W. L. Butler, *in* "Photosynthesis I" (A. Trebst and M. Avron, eds.), p. 149. Springer-Verlag, Berlin and New York, 1977.

[75] C. L. Bering and P. A. Loach, *Photochem. Photobiol.* **26**, 607 (1977).

[76] Y. de Kouchkovsky, *Proc. Int. Congr. Photosyn, 3rd, 1974* p. 1081 (1975).

[77] J. A. Guikema and C. F. Yocum, *Arch. Biochem. Biophys.* **189**, 508 (1978).

[77a] A. Stewart and A. W. D. Larkum, *Aust. J. Plant Physiol.* **4**, 253 (1977).

[78] J. N. Siedow, S. C. Huber, and D. E. Moreland, *Biochim. Biophys. Acta* **547**, 282 (1979)

[79] R. Barr, F. L. Crane, G. Beyerm, L. A. Maxwell, and K. Folkers, *Eur. J. Biochem.* **80**, 51 (1977).

[80] C. J. Arntzen, J. Neumann, and R. A. Dilley, *Bioenergetics* **2**, 73 (1971).

[81] J. Bolen, R. Pardini, H. T. Mustafa, K. Folkers, R. A. Dilley, and F. L. Crane, *Proc. Natl. Acad. Sci. U.S.A.* **69**, 3713 (1972).

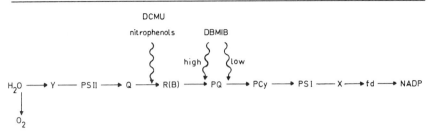

FIG. 6. Inhibition site for DCMU and substituted nitrophenols between Q and B (R) and for high concentrations of DBMIB (10 $\mu M$) between B (R) and plastoquinone, DBMIB at 1 $\mu M$ (= low) inhibits after plastoquinone.

lipophilic long alkyl side chain) show a close chemical relationship to plastoquinone, although they are less specific and less effective than DBMIB. A number of substituted halogenated benzoquinones with only short alkyl side chains and a specificity like DBMIB have been tested for activity as plastoquinone antagonists[82,82a,83] among them butyl substituted mono- or dibromo-, chloro-, or iodo- derivatives. The 2,5-dibromobenzoquinone is an inhibitor, but its effectiveness is enhanced by the introduction of a lipophilic alkyl sidechain of up to four C atoms.[83] As expected iodo-substitution is more effective than bromo or chloro, and two halogens are more effective than one.[82] Also some naturally occurring hydroxylated benzoquinones, helveticone and bovinone, isolated by Steglich, have a DBMIB like inhibitory potency.[54b]

Phenols, with substitution comparable to DBMIB (such as bromonitrothymol) are effective inhibitors of electron flow, but the inhibition site is similar to DCMU (see Section I,D,2). Halogenated naphthoquinones have also been reported to inhibit electron flow,[84] and it has been implied that they might be antagonists of vitamin K.[85] However, from the few structural requirements needed to be effective as a plastoquinone antagonist, the halogenated naphthoquinones might also inhibit plastoquinone function, although with less specificity than DBMIB and possibly at the reducing side.

*i. DBMIB in Bacterial Photosynthesis.* Baltscheffsky[86] reported on an

[82] W. Oettmeier and A. Trebst, *Proc. Int. Congr. Photosynth., 4th., 1977* p. 279 (1978).

[82a] W. Oettmeier, S. Reimer, and K. Link, *Z. Naturforsch., Teil C* **33**, 695 (1978).

[83] A. Trebst, S. Reimer, W. Draber, and H. J. Knops, *Z. Naturforsch. Teil C* **34**, 831 (1979).

[84] H. C. Sikka, R. H. Shimabukuro, and G. Zweig, *Plant Physiol.* **49**, 381 (1972).

[85] H. K. Lichtenthaler and K. Pfister, *in* "Photosynthetic Oxygen Evolution" (H. Metzner, ed.), p. 171. Academic Press, New York, 1978.

[86] M. Baltscheffsky, *Proc. Int. Congr. Photosynth., 3rd, 1974,* p. 799 (1975).

inhibitor effect (at somewhat higher concentrations of DBMIB (5 $\mu M$) than needed to inhibit chloroplasts) on electron flow in *Rhodospirillum rb.* chromatophores, which indicates that DBMIB interferes also with ubiquinone function. At first, it was surprising that DBMIB inhibited cytochrome *c* oxidation, indicating a ubiquinone function between cytochrome *b* and $c_1$. Further experiments by Gromet-Elhanan with DBMIB[87,88] (as well as with a naphthylmercaptobenzoquinone), and evidence from other types of experimentation,[89,90,90a] now indicate that in bacterial photosynthesis a Mitchell Q cycle might be operative. That is, cytochrome *b* is involved in ubiquinone reduction as well as its oxidation. So far evidence for a Q cycle in chloroplast electron flow has not been found (see Section III,A,3).

*j. DBMIB in Respiratory Electron Flow.* Only higher concentrations of DBMIB (about 10 $\mu M$) inhibit also respiratory electron flow in chromatophores[91] and mitochondria.[92] The inhibition site is placed on the oxidation side of ubiquinone. Again the properties of DBMIB inhibition might be suggestive for a Q-cycle as in bacterial photosynthesis.

Recently DBMIB has proved useful in studies on aerobic electron flow in *E. coli.*[93-95] Inhibition is reversed by the addition of ubiquinone.[95] Recent papers report on DBMIB inhibition of respiration in Neurospora[96] and mung bean mitochondria;[78] its multiple nature is attributed to alternate pathways and effect on membrane fluidity.[78]

## 2. Diphenylethers and Dinitroanilines

A new group of compounds inhibiting electron flow, and efficient as DBMIB, has recently been described in the 2,4-dinitrophenyl ethers of

[87] Z. Gromet-Elhanan and H. Gest, *Arch. Microbiol.* **116**, 29 (1978).

[88] Z. Gromet-Elhanan, *Biochem. Biophys. Res. Commun.* **73**, 13 (1976).

[89] A. R. Crofts, D. Crowther, and G. V. Tierney, *in* "Electron Transfer Chains and Oxidative Phosphorylation" (E. Quagliariello, ed.), p. 233. North-Holland Publ., Amsterdam, 1975.

[90] A. Baccarini-Melandri and B. A. Melandri, *in* "Bioenergetics of Membranes" (L. Packer *et al.*, eds.), p. 199. Elsevier, Amsterdam, 1977.

[90a] P. L. Dutton and R. C. Prince, *in* "The Photosynthetic Bacteria" (R. K. Clayton and W. R. Sistrom, eds.), p. 525. Plenum, New York, 1978.

[91] E. H. Evans, and D. A. Gooding, *Arch. Microbiol.* **111**, 171 (1976).

[92] G. Loschen and A. Azzi, *FEBS Lett.* **41**, 115 (1974).

[93] R. K. Poole and B. A. Haddock, *FEBS Lett.* **52**, 13 (1975).

[94] R. L. Houghton, R. J. Fisher, and D. R. Sanadi, *FEBS Lett.* **68**, 000 (1976).

[95] I. L. Sun and F. L. Crane, *Biochem. Biophys. Res. Commun.* **68**, 190 (1976).

[96] A. K. Drabikowska, *Life Science* **21**, 667 (1977).

bromonitrothymol, iodonitrothymol, and ioxynil.[96a,b] At 1 $\mu M$, these compounds inhibit all Hill reactions which require photosystem I but do not inhibit photosystem II-dependent ones (like ferricyanide + phenylenediamine). Donor systems for photosystem I, like DAD, TMPD, or DCPIP, and cyclic systems like PMS, are not inhibited. The durohydroquinone system,[55b] however, is inhibited, the inhibition being reversed by TMPD.[96a] The photosystem II Hill reactions in the presence of DNP-INT are coupled to ATP formation with a $P/e_2$ ratio slightly lower than that obtained in a DBMIB inhibition experiment (with 1 $\mu M$ DBMIB). In fact, the inhibition by the dinitrophenyl ethers appears identical to that by high concentrations of DBMIB (10 $\mu M$), indicating some interference of the compounds not only with a plastohydroquinone oxidizing site, but also with the plastoquinone pool itself. The ferricyanide Hill reaction (in the absence of any lipophilic mediator) is completely blocked by the dinitrophenyl ethers, whereas in the analogous DBMIB experiment there is no complete inhibition and actually a reversal at high concentrations of DBMIB (see above). This points to the advantage of the diphenyl ethers. They have no redox properties, as does DBMIB, and therefore do not act as mediators.

The parent compounds of the dinitrophenyl ethers are the bromo- or iodonitrothymol and ioxynil. As discussed in Section I,D,2, these phenolic compounds are effective DCMU analogs. Dinitrophenylation therefore shifts the point of inhibition.[96b]

Some diphenyl ethers, like nitrofen, are not electron-flow inhibitors but act as energy transfer inhibitors.[96c] Diphenyl ethers are known as

DNP ether of iodonitrothymol          Trifluralin

Inhibitors like DBMIB

[96a] A. Trebst, H. Wietoska, W. Draber and H. J. Knops, Z. Naturforsch., Teil C 33, 919 (1978).

[96b] A. Trebst and W. Draber, in "Advances in Pesticide Science Part 2 (H. Geissbühler, ed.), p. 223. Pergamon Press, Oxford and New York, 1979.

[96c] R. Lambert, K.-J. Kunert, and P. Böger, Pestic. Biochem. Physiol. 11, 267 (1979).

herbicides.[96b,d] Possibly only those with an orthosubstitution may be also photosynthesis inhibitors *in vivo*.[96d] Others are plant hormon antagonists. Robinson *et al.*[96e] have described the herbicide *N*-diisopropyldinitrotrifluoromethylaniline-trifluralin as an inhibitor of photosynthetic electron flow at a concentration of about 10 $\mu M$. Its inhibition pattern is like that of DBMIB. Other herbicidal N-alkylated dinitroanilines are also inhibitors of photosynthetic electron flow.[96b] As with the diphenyl ethers, the herbicidal actions of these compounds *in vivo* does not seem to be due to an inhibition of photosynthesis.

### 3. Bathophenanthroline

This copper and iron chelating agent was shown by Bering *et al.*[97] to inhibit photosynthetic electron flow from photosystem II to I at a concentration of 100 $\mu M$. Bathophenanthroline does not inhibit photosystem II-driven Hill reactions. Bathophenanthroline inhibition might be bypassed by a catalytic amount of TMPD.[37] This inhibition pattern, similar to the one of DBMIB (Fig. 3), places the site of bathophenanthroline inhibition after plastoquinone function and before cytochrome *f*.[49a] From the iron chelating capacity of bathophenanthroline it is speculated[97] that it might act on a nonheme iron protein, functioning between plastoquinone and plastocyanin, as implicated also by electron spin resonance (ESR) studies.[44a]

### 4. EDAC

The use of the water-soluble 1-ethyl-3-(3-dimethylaminopropyl)-carbodiimide as a good electron-flow inhibitor and poor energy-transfer inhibitor (in contrast to DCCD though recently it was shown that DCCD at 50–100 $\mu M$ also blocks electron flow specifically at plastoquinone function.[98a]) was introduced by McCarty.[98] The inhibition of electron flow by 10 $\mu M$ EDAC requires a light pretreatment of the chloroplasts (5 min at pH 8.0, then centrifugation). Photosystem I-dependent reactions, including some donor systems (such as DAD but not TMPD) are then inhibited, but not cyclic photophosphorylation catalyzed by PMS. Photoreductions by photosystem II are not blocked. As discussed for the DBMIB system (see Section I,C,1,d), an uncoupler inhibits this electron

[96d] S. Matzunaka, *in* "Herbicides" (P. C. Kearney and D. D. Kaufmann, eds.), Vol. 2. Marcel Dekker, New York, 1976.

[96e] S. J. Robinson, C. F. Yocum, H. Ikuma, and F. Hayashi, *Plant Physiol.* **60**, 840 (1977).

[97] C. L. Bering, R. A. Dilley, and F. L. Crane, *Biochim. Biophys. Acta* **430**, 327 (1976).

[98] R. E. McCarty, *Arch. Biochem. Biophys.* **161**, 93 (1974).

[98a] P. V. Sane, U. Johanningmeier and A. Trebst, *FEBS Lett.* **108**, 136 (1979).

flow.[66] EDAC does not seem to interfere with plastocyanin directly, and, therefore, a block just before plastocyanin and cytochrome $f$ and after plastoquinone is proposed. This view is strengthened by recent experiments which showed that EDAC inhibition of electron flow from water to photosystem I is reversed by a catalytic amount of TMPD.[37] This system (Fig. 3) described in more detail in Section I,C,1,d,iv, reconnects plastoquinone with plastocyanin in bypassing all inhibitor sites between these two carriers. The TMPD bypass cannot reverse inhibitors of plastocyanin function itself (such as KCN or $HgCl_2$) (see Fig. 3).

### 5. Valinomycin

This $K^+$ ionophore acts on the membrane potential and is also an electron flow inhibitor at 10 $\mu M$ and independent of potassium.[37,99,100] The site of inhibition has been localized by several approaches to be between plastoquinone and plastocyanin.[37,100] Valinomycin does not inhibit photosystem II Hill reactions, and its inhibition of electron flow between photosystem I and II is reversed by a TMPD bypass.[37] Pertinent is a study by Walz on the effect of valinomycin on the aggregation state of a lipid.[101]

### D. Inhibition at the Reducing Side of Plastoquinone

Among this group, represented by DCMU and $o$-phenanthroline, are the first inhibitors of photosynthesis discovered (in 1956 and 1946, respectively) and widely used inhibitors of photosynthetic electron flow which have been of fundamental importance for photosynthesis research.

### 1. DCMU and Related Compounds

DCMU at 1 $\mu M$ and phenanthroline at 100 $\mu M$ inhibit electron flow at the reducing side of photosystem II after the primary acceptor Q and, therefore, all Hill reactions which include the main plastoquinone pool, whether driven by both photosystems or by photosystem II only. As reviewed many times,[1-5] photoreductions by photosystem I at the expense of artificial electron donors as well as cyclic electron flow involving photosystem I are insensitive. Treatments for removal of DCMU sensitivity are discussed below (see Section I,D,5). Reverse electron flow

---

[99] A. Telfer and J. Barber, *Biochim. Biophys. Acta* **333**, 343 (1974).

[100] K. K. Voegeli, D. O'Keefe, J. Whitmarsh, and R. A. Dilley, *Arch. Biochem. Biophys.* **183**, 333 (1977).

[101] D. Walz, *J. Membr. Biol.* **27**, 55 (1976).

from plastoquinone to Q is also DCMU sensitive,[102] giving strong support for the localization of the inhibition site.

Although this site of DCMU inhibition is functionally localized between Q and plastoquinone, its mode of action remains obscure. It is assumed that it binds to a specific peptide bond (see Büchel[4]) of a protein shield covering the reducing side of photosystem II (see model by Renger[103] and ourselves[96b,104]) inducing a conformational change, which in turn affects the redox state of Q. With the evidence for a component between Q and plastoquinone, called B or R and possibly being a plastoquinol anion, the inhibition site of DCMU may have been narrowed to the reaction between Q and R[105,106] (Fig. 6). It has been shown that DCMU[106] as well as o-phenanthroline[107] changes the midpoint potential of R or B, and in this way the equilibrium constant of the reaction between B and Q. The binding studies of DCMU and its analogues and of nitrophenols in normal and trypsinated chloroplasts (see below) indicate now that the DCMU binding protein is part of that protein of the photosystem II complex, which has B (R) as a prosthetic group (here called B-protein) and which catalyzes electron transfer from Q to plastoquinone (Fig. 7).

Also, respiratory electron flow in Saccharomyces in the cytochrome b/c segment is sensitive to DCMU, as it is to 4-hydroxyquinoline-N-oxide(HOQNO).[108] It is interesting to note that photosynthetic electron flow is also inhibited by phenols (see below) and by HOQNO[109] as it is by DCMU.

An effect of DCMU on the donor side of photosystem II and on the S-states of oxygen evolution (as an ADRY reagent) has been pointed out.[110,111]

DCMU is just one of many compounds in a large group of other substituted urea herbicides that inhibit photosynthesis. Numerous compounds (Fig. 8), among them many herbicides, of quite different structures (anilides, biscarbamates, uracils, triazines, triazinones, pyridazi-

[102] K. G. Rienits, H. Hardt, and M. Avron, FEBS Lett. 33, 28 (1973).

[103] G. Renger, Biochim. Biophys. Acta 440, 287 (1976).

[104] See several reports at a conference on "Herbicides in Photosynthesis" at Konstanz 1979, Z. Naturforsch., issue 11, pp. 893–1074.

[105] R. Velthuys and J. Amesz, Biochim. Biophys. Acta 333, 85 (1974).

[106] B. Bouges-Bocuet, Biochim. Biophys. Acta 314, 250 (1973).

[107] D. B. Knaff, Biochim. Biophys. Acta 376, 583 (1975).

[108] B. Convent and M. Briquet, Eur. J. Biochem. 82, 473 (1978).

[109] M. Avron and N. Shavit, Biochim. Biophys. Acta 109, 317 (1965).

[110] A. L. Etienne, Biochim. Biophys. Acta 333, 320 (1974).

[111] B. Bouges-Bocquet, P. Bennoun, and J. Taboury, Biochim. Biophys. Acta 325, 247 (1973).

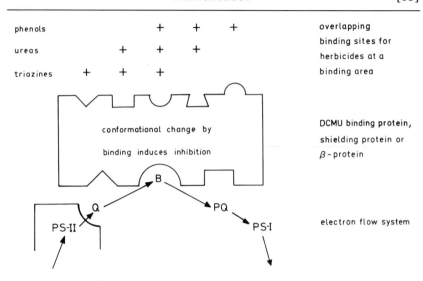

FIG. 7. Schematic representation of a DCMU binding site on the membrane at the acceptor side of photosystem II. That part of photosystem II with B (R) as the prothetic group, which is shielding the primary acceptor Q, carries binding sites for inhibitors like DCMU, triazines and phenoles. It is postulated that a binding area is common to all these inhibitors. They have overlapping though also specific sites for different parts of the molecule in order to accomodate for the resistant species as well as for different chemistry involved in binding. It is assumed that the inhibitor binding induces a conformational change in the B-protein which in turn disturbs its functioning in catalyzing electron flow from Q to PQ.

nones, pyrimidinones, and trifluorobenzimidazoles) inhibit chloroplast electron flow as DCMU does (for reviews, see Moreland[3] and Büchel[4]).

Common to all these compounds is an $sp^2$ hybrid $-\overset{\overset{\displaystyle \|}{}}{C}-NH$, considered to be the basic chemical structural element essential for inhibition and for binding to the membrane (see Büchel[4]). Supposed exceptions to the above element (N-alkylated ureas and ring closed N-acylamides) are also accommodated,[112] if the free electron pair of the nitrogen instead of a NH hydrogen bonding is sufficient for binding.[4]

## 2. Phenols

The inhibition of electron flow in chloroplasts by ioxynil in a manner similar to DCMU is well known.[113,114] Recently halogen-substituted ni-

[112] A. Trebst and E. Harth, Z. Naturforsch., Teil C 29, 232 (1974).
[113] Z. Gromet-Elhanan, Biochem. Biophys. Res. Commun. 30, 28 (1968).
[114] S. Katoh, Plant Cell Physiol. 13, 273 (1972).

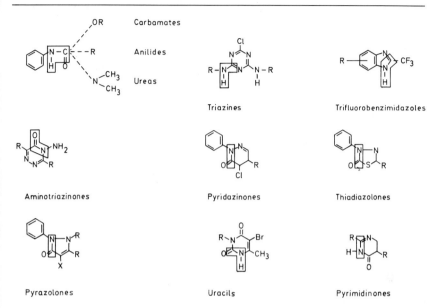

FIG. 8. Analogs of DCMU inhibiting the acceptor side of photosystem II. The basic chemical element essential for inhibition and binding to the membrane is indicated.

trophenols and dinitrophenols with an alkyl substitution have been recognized to be powerful electron flow inhibitors[83,96b,115,115a,116]; 2-bromo-4-nitrothymol or 2,4-dinitrothymol inhibit even at 0.1 $\mu M$.[83,96b] Their well-known uncoupling activity[117] (also of respiratory electron flow) requires higher concentrations. $n$-Butyl-3,5-diiodo-4-hydroxybenzoate,[109] 3,5-diiodosalicylate,[118] and hydroxychinoline $N$-oxide[109] may be grouped among these compounds which do not have the basic chemical element discussed above for DCMU and its analogs (for a detailed discussion see Trebst and Draber.[96b] The comparison and relation of structure to function of about 40 substituted alkyl nitrophenols revealed that the inhibitory potency is proportional to the size of the substituents but not to the p$K$ of the OH group or the lipophilicity of the compound.[83,96b]

[115] J. J. S. van Rensen, W. van der Vet, and W. P. A. van Vliet, *Photochem. Photobiol.* **25**, 579 (1977).

[115a] J. J. S. van Rensen, D. Wong, Govindjee, *Z. Naturforsch., Teil C* **34**, 413 (1978).

[116] E. E. Moreland and J. L. Hilton, *in* "Herbicides" (L. J. Audus, ed.), Vol. 1, p. 493, Academic Press, New York, 1976.

[117] K. S. Siow and A. M. Unrau, *Biochemistry* **7**, 3507 (1968).

[118] P. Homann, *Eur. J. Biochem.* **33**, 247 (1973).

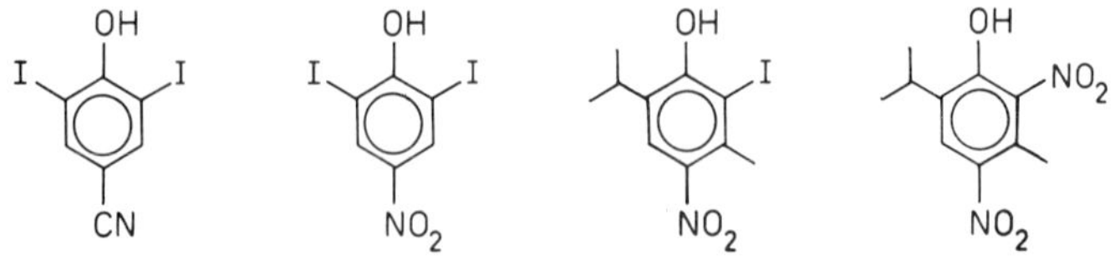

OH      OH      OH      OH

Ioxynil     Diiodonitrophenol     Iodonitrothymol     Dinitrothymol

6,7       6,23       6,7       5,6       $PI_{50}$

Phenol inhibitors like DCMU

The identity of the inhibitor binding area on the B-protein of the thylakoid membrane (but not of specific binding sites for specific substituents in the binding area,[96b] see Fig. 7) for DCMU and its analogs on one side and for bromonitrothymol, ioxynil, and their analogs on the other side has been shown by the replacement technique of Strotmann and Tischer (see Section I,D,4). Ioxynil and bromonitrothymol were shown to replace radioactive labeled metribuzin.[54a] Also trypsin treatment (see below) not only removes the binding efficiency of DCMU, metribuzine and analogs but also of bromonitrothymol.[104] The presence of BSA (1 mg) prevents the inhibition of electron flow by the phenolic compounds but not that by DCMU.[54a]

## 3. Azaphenanthrenes

As the phenols also o-phenanthroline (1,10-diazophenanthrene) does not fit into the scheme of compounds with similar structural elements of the urea herbicides discussed above. Oettmeier and Grewe[119] as well as Satoh[120] have prepared a number of substituted azaphenanthrenes and come to the conclusion that it is not the iron complexing property of o-phenanthroline and other azaphenanthrenes which is responsible for inhibition of chloroplast electron flow. On the other hand, Crane et al. put forward evidence for the importance of the chelating potency for inhibition.[97,121] They also have evidence for shifting inhibition points for o-phenanthroline.[122]

[119] W. Oettmeier and R. Grewe, Z. Naturforsch., Teil C 29, 545 (1974).
[120] S. K. Satoh, Biochim. Biophys. Acta 333, 127 (1974).
[121] R. Barr and F. L. Crane, Plant Physiol. 57, 450 (1976).
[122] U. Banaszak, R. Barr, and F. L. Crane, J. Bioenerg. 8, 83 (1976).

## 4. Binding and Replacement of Analogs

Recently Tischer and Strotmann[123] developed a method for calculating the equilibrium constant for binding by using the distribution of a radioactive labeled inhibitor (atrazine, phenmedipham, or metribuzin were available) between the chloroplast membrane and the supernatant. Furthermore by replacement of a radioactive inhibitor by another unlabeled compound, the binding capacity, and from this, the inhibitory potency and specifity of the compound, can be computed. The replacement techniques permits one to show precisely whether the binding site, and the specific inhibition site, of different compounds is identical or not.

As discussed below (Section I,D,1), this method has led to a much refined view of the structure of the binding protein of DCMU and phenols and its possible identity with the B-protein (Fig. 7).

## 5. DCMU (triazine) Insensitive Reactions

Important recent developments in the biochemistry of the acceptor site of photosystem II came from the attempts to induce DCMU (or other analogs) insensitivity of photosynthetic reactions in chloroplasts.

*a. Resistant Mutants.* Recently, DCMU resistent strains of the green algae Chlamydomonas[124] of Euglena[106,125,125a] and of the blue-green algae Phormidium[126] have been described, as well as triazine-resistent species of Brassica,[127] Amaranthus,[128] Chenopodium,[129] and Senacio.[130] Photosynthesis in the isolated chloroplasts from such species is triazine- but not DCMU-resistent.[128,129] The triazine resistance is shown to be due to an alteration of the thylakoid membrane in that the binding affinity for triazine on the membrane is markedly lower in the mutant, whereas that of DCMU or phenol inhibitors remains the same.[104] This indicates that

[123] W. Tischer and H. Strotmann, *Biochim. Biophys. Acta* **460**, 143 (1977).
[124] J. C. McBride, A. C. McBride, and R. K. Togasaki, *Plant Cell Physiol. Spec. Issue* p. 239 (1977).
[124a] S. Lien, J. C. McBride, A. C. McBride, R. K. Togasaki, and A. San Pietro, *Plant Cell Physiol. Spec. Issue* p. 242 (1977).
[125] R. Galvayrac, J.-L. Bomsel, and D. Laval-Martin, *Plant Physiol.* **63**, 857 (1977).
[125a] D. Laval-Martin, G. Dubertret, and R. Galvayrac, *Plant Sci. Lett.* **10**, 185 (1977).
[126] R. G. Piccioni and D. C. Mauzerall, *Biochim. Biophys. Acta* **504**, 398 (1978).
[127] V. Souza Machado, J. D. Bandeen, G. R. Stephenson, and P. Lavigne *Can. J. Plant Sci.* **58**, 977 (1978).
[128] C. J. Arntzen, C. L. Ditto, and P. E. Brewer, *Proc. Natl. Acad. Sci. USA* **76**, 278 (1979).
[129] V. Souza Machado, C. J. Arntzen, J. D. Bandeen, and G. R. Stephenson, *Weed Science* **26**, 318 (1978).
[130] S. R. Radosevich, K. E. Steinback, and C. J. Arntzen, *Weed Science* **27**, 216 (1979).

the binding area of the B-protein contains a common binding area for all inhibitors but with specific binding sites for DCMU, triazine, and phenols (Fig. 7). The resistent species have lost one or more of such a specific site.

It should be mentioned that a DCMU insensitivity of photosynthesis is also observed in the life cycle of synchronized algae.[131] However, this is very probably due to a change of permeability through outer cell wall. It seems possible that the reported DCMU insensitivity of diphenylcarbazide oxidation in thylakoids from cyanobacteria[131a,b] is due to a chemical artefact.

*b. Trypsin Treatment.* Trypsin treatment of chloroplasts has several effects on photosynthetic electron flow and uncoupled ATP formation (see review by Trebst[27]). It effects NADP reduction at the reductase level, the ATP synthetase, the light-harvesting complex and membrane stacking, and possibly also the donor site of photosystem II. Most relevant for the DCMU inhibition site is the fact that, in addition, trypsin treatment of the thylakoid membrane causes a loss of DCMU sensitivity of photosynthetic ferricyanide reduction. This was first shown by Regitz and Ohad [132,133] for Chlamydomonas chloroplasts and has been taken up recently by Renger.[103,134–136] Incubation of chloroplasts (with 100 μg chlorophyll) and 80 μg trypsin/ml for 10 min in the dark at pH 7.0 (after 10 min addition of trypsin inhibitor) inhibits photosynthetic electron flow from water to any acceptor, except to ferricyanide. This ferricyanide reduction is much less sensitive to DCMU[103,136] in trypsin-treated chloroplasts. Also, the sensitivity toward the other inhibitors of PS-II, like triazine and phenols, is lost.[104,135] The binding sites on the membrane of the inhibitors have been lost by trypsin treatment,[104] because of the destruction of the shielding protein for PS-II or B-protein, which carries the binding sites. This trypsin-sensitive protein has recently been purified.[137] The loss of the B-protein has made the primary acceptor Q accessible to ferricyanide, which therefore is inhibitor insensitive (see reports from Konstant herbicide conference on this problem[104]). Also just

[131] H. Senger, *Plant Cell Physiol., Spec. Issue* p. 229 (1977).
[131a] G. H. Schmid and U. Lehman-Kirk, *Arch. Microbiol.* **115**, 265 (1977).
[131b] G. A. Codd and J. D. Cossar, *Biochim. Biophys. Res. Comm.* **83** 341 (1978).
[132] G. Regitz and I. Ohad, *Proc. Int. Congr. Photosyn., 3rd, 1974* p. 1615 (1975).
[133] G. Regitz and I. Ohad, *J. Biol. Chem.* **251**, 247 (1976).
[134] G. Renger, K. Erixon, G. Doring, and C. Wolff, *Biochim. Biophys. Acta* **440**, 278 (1976).
[135] G. Renger and R. Tiemann, *Biochim. Biophys. Acta* **545**, 316 (1979).
[136] G. Renger, *in* "Bioenergetics of Membranes" (L. Packer *et al.*, eds.), p. 339. Elsevier, Amsterdam, 1977.
[137] E. Croze, M. Kelly, and P. Horton, *FEBS Lett.* **103**, 22 (1979).

low pH changes the accessibility of $Q$ to ferricyanide and the DCMU sensitivity of its reduction.[137a]

Trypsin treatment had been regarded as relatively unspecified and unsuitable for studying the sidedness of the membrane (see review by Trebst[27]). Because the effect of trypsin on oxygen evolution has been overestimated in the earlier experiments, it appears now that, just the opposite, trypsin treatment is an excellent way for accessing the accessibility of components.

The recent preparation of inside-out vesicles from thylakoid membranes[138] permitted reinvestigation of the effect of trypsin[139]: in such inside-out vesicles, trypsin does inactivate oxygen evolution, but not in rightside-out vesicles.[139]

*c. Silicomolybdate.* Girault and Galmiche[140] showed a reproducible DCMU insensitive Hill reaction with silicotungstate as electron acceptor. Crane *et al.*[62,141,142] showed that silicomolybdate is experimentally an even better acceptor of photosystem II, whose reduction by photosystem is only partially sensitive to DCMU (particularly at low pH). Since silicotungstate and silicomolybdate change membrane properties[142a] (also the coupling factor $CF_1$), the reaction rate of their reduction is linear for only one minute. The early, puzzling finding that these DCMU insensitive Hill reactions are not coupled to ATP formation[142] has now been revised[143,144]; the reaction is coupled, as is the DBMIB sensitive Hill reactions, by photosystem II. It appears now that silicomolybdate changes the properties of the B-protein with its DCMU binding site and opens up access of hydrophilic electron acceptors onto the primary acceptor Q comparable to trypsin treatment.

*E. Inhibition and Inactivation of Oxygen Evolution*

1. Tris and NH$_2$OH Treatment

Treatments that inactivate the water splitting system include gentle heating of chloroplasts (55° for 6 min), Tris treatment and hydroxylamine treatment as summarized by Cheniae[145] and Izawa.[5] It was found sub-

[137a] S. Itoh, *Biochim. Biophys Acta* **460** 381 (1977).

[138] B. Andersson and H. E. Akerlund, *Biochim. Biophys. Acta* **503,** 462 (1978).

[139] C. Jansson, B. Andersson, and H. E. Akerlund, *Biochim. Biophys. Acta,* in press.

[140] G. Girault and J. M. Galmiche, *Biochim. Biophys. Acta* **333,** 314 (1974).

[141] R. Barr, F. L. Crane, and R. T. Giaquinta, *Plant Physiol.* **55,** 460 (1975).

[142] R. T. Giaquinta and R. A. Dilley, *Biochim. Biophys. Acta* **387,** 288 (1975).

[142a] G. Ben-Hayyim, and J. Neumann, FEBS Letters **56** 240 (1975).

[143] L. Rose and D. O. Hall, *Biochim. Biophys. Acta* **449,** 23 (1976).

[144] S. P. Berg and A. Izawa, *Biochim. Biophys. Acta* **460,** 206 (1977).

[145] G. M. Cheniae, *Annu. Rev. Plant Physiol.* **21,** 467 (1970).

sequently that during Tris treatment (washing with 0.8 $M$ Tris buffer, pH 8, for 15 min), manganese is released from the membrane into the inside thylakoid space and remains bound unless the chloroplasts are sonicated.[146] This is good supporting evidence for localization of the water splitting system toward the interior space of the membrane. Recently, it was shown that light is required for the inactivation of oxygen evolution by Tris treatment.[147] The treatment of chloroplasts (0.5 mg/ml) with 5 m$M$ hydroxylamine in buffer at pH 7.0 for 15 min at 20°[148] has been particularly advantageous to show the coupling and stoichiometry of ATP formation in artificial donor systems for photosystem II (benzidine compared with tetramethylbenzidine/ascorbate or ferrocyanide, iodine, and others).[148,149] Hydroxylamine treatment is more effective in the absence of chloride ions.[149a]

2. Internal pH

The inhibition of electron flow by high external pH (9–9.3) in the presence of ammonia or methylamine (2 m$M$), is long known[150] and was ascribed to an effect of internal pH (being equilibrated with external pH by uncouplers) on the rate of electron flow.[151] This inhibition by internal pH has recently been found to be due to a specific inactivation of oxygen evolution, because donor systems for photosystem II remain intact at high internal pH.[152–154] For this inactivation procedure (either chloroplasts in HEPES buffer, pH 9.3, and an uncoupler such as 1 $\mu M$ gramicidin or chloroplasts freed of the coupling factor by EDTA treatment) light is required[153] (see, however, Cohn et al.[154]). It was recently specified that it is the $S_2$ state of oxygen evolution that is sensitive to internal pH.[155] Again manganese is released during this treatment as in tris.[155]

The inactivation of the $S_2$ state of oxygen evolution by high internal

[146] R. E. Blankenship and K. Sauer, Biochim. Biophys. Acta 357, 252 (1974).
[147] G. M. Cheniae and I. F. Martin, Biochim. Biophys. Acta 502, 321 (1978).
[148] S. Izawa and D. R. Ort, Biochim. Biophys. Acta 357, 127 (1974).
[149] E. Harth, W. Oettmeier, and A. Trebst, FEBS Lett. 43, 231 (1974).
[149a] P. M. Kelley and S. Izawa, Biochim. Biophys. Acta 502, 198 (1978).
[150] N. E. Good, A. Izawa, and G. Hind, Curr. Top. Bioenerg. 1, 75 (1966).
[151] M. Avron, Photosynth., Two Centuries Its Discovery by Joseph Priestley, Proc. Int. Congr. Photosynth. Res., 2nd, 1971 p. 86 (1972).
[152] E. Harth, S. Reimer, and A. Trebst, FEBS Lett. 42, 165 (1974).
[153] S. Reimer and A. Trebst, Biochem. Physiol. Pflanz. 168, 225 (1975).
[154] D. E. Cohn, W. S. Cohen, and W. Bertsch, Biochim. Biophys. Acta 376, 97 (1975).
[155] J. M. Briantais, C. Vernotte, J. Lavergne, and C. J. Arntzen, Biochim. Biophys. Acta 461, 61 (1977).

pH is in support of the localization of the water splitting reaction at the inside surface of the thylakoid.

## 3. Antibodies

Antibodies against photosystem II particles or components have been prepared[156-159] and shown to inhibit electron flow at the donor side of photosystem II, i.e., inhibit oxygen evolution but not donor systems. Among such inhibitory antibodies are also those directed against lutein and neoxanthin.[160-161] Such an antibody survey of the thylakoid surface was performed in great detail by Menke and his colleagues, who prepared antibodies against a purified proteins from the membrane. An antibody against a protein of molecular weight 11,000[162,163] (also one of 66,000[13]) inhibits at the donor side of photosystem II in unstacked thylakoids with varying degree and effect in different donor systems. The important implication is, of course, that photosystem II and the oxygen evolution system are accessible and therefore located toward the matrix side of the membrane. However, the antibodies are not necessarily directed against the active center, certainly not in the case of the nonfunctional compounds such as lutein. An inhibitory conformational change induced by the antibody on an integral protein on the external side of the membrane could interfere with the functional side on the inside.

## 4. DABS

This nonpermeable probe (see Dilley [48], this volume) interferes also with the water splitting system (reviewed in Giaquinta and Dilley[164]). Its labeling of the membrane as well as its inhibitory potency is enhanced in

[156] J. M. Briantais and M. Picaud, *FEBS Lett.* **20**, 100 (1972).

[157] B. Z. Braun and Govindjee, *FEBS Lett.* **25**, 143 (1972).

[158] F. Koenig, W. Menke, H. Graubner, G. H. Schmid, and A. Radunz, *Z. Naturforsch., Teil B* **27**, 1225 (1972).

[159] B. Zilinskas Braun and Govindjee, *Plant Sci. Lett.* **3**, 219 (1974).

[160] A. Radunz and G. H. Schmid, *Z. Naturforsch., Teil C* **30**, 622 (1975).

[160a] A. Radunz, *Z. Naturforsch., Teil C* **33**, 941 (1978).

[160b] U. Lehmann-Kirk, G. H. Schmid, and A. Radunz, *Z. Naturforsch., Teil C* **34**, 427 (1979).

[161] G. H. Schmid, H. List, and A. Radunz, *Z. Naturforsch., Teil C* **32**, 118 (1977).

[162] G. H. Schmid, W. Menke, F. Koenig, and A. Radunz, *Z. Naturforsch., Teil C* **31**, 304 (1976).

[163] G. H. Schmid, G. Renger, M. Gläser, F. Koenig, A. Radunz, and W. Menke, *Z. Naturforsch., Teil C* **31**, 594 (1976).

[164] R. Giaquinta and R. A. Dilley, *in* "Photosynthesis I" (A. Trebst and M. Avron, eds.), p. 297. Springer-Verlag, Berlin and New York, 1977.

the light, indicative of conformational changes. The reduction of DABS by photosystem I to a more reactive species[165] might explain some of the light-induced effects, but does not change the principal observation that a nonpermeable probe inactivates oxygen evolution. The argument for the action of antibodies above holds here also.

### 5. Heavy Metal Ions

In addition to the inhibition of plastocyanin by mercuric chloride (see Section I,B,1) the influence of copper, lead, and cadmium ions on electron flow and photosystem II in particular has been investigated in view of possible ecological implications.

Copper has long been known to inhibit photosynthesis. The binding of 50 $\mu M$ $Cu^{2+}$ to proteins of the thylakoid membrane[166] seems to affect photosystem II in particular after a light pretreatment.[167] One millimolar lead salts affect photosystem II activities, inhibiting or stimulating depending on pH,[168] its effect sometimes due just to phosphate precipitation. The inhibition by cadmium salts seems to be more specific, 1 to 10 m$M$ cadmium nitrate affecting the donor side of photosystem II.[169-171] Further reports on the inhibition by copper ions of photosystem II have recently appeared.[172-174]

### 6. Unsaturated Fatty Acids

The effect of fatty acids on electron flow (and phosphorylation) is long known, but not in a very specific way. Siegenthaler[175] was able, however, to locate the inhibition by 300 $\mu M$ linolenic acid at the donor side of photosystem II. Its general effect on the functional integrity of the thylakoid membrane was studied by Okamoto et al.[176,177]

[165] W. Lockau and B. R. Selman, Z. Naturforsch., Teil C 31, 48 (1976).
[166] G. Vierke and P. Stuckmeier, Z. Naturforsch., Teil C 32, 605 (1977).
[167] A. Cedeno-Maldonado, J. A. Swader, and R. L. Heath, Plant Physiol. 50, 698 (1972).
[168] M. B. Bazzaz and Govindjee, Environ. Lett. 6, 175 (1974).
[168a] H. Schulze and J. J. Brand, Plant Physiol. 62, 727 (1978).
[169] M. B. Bazzaz and Govindjee, Environ. Lett. 6, 1 (1974).
[170] E. H. Li and C. D. Miles, Plant Sci. Lett. 5, 33 (1975).
[171] M. A. van Duivendijk-Matteoli and G. M. Desmet, Biochim. Biophys. Acta 408, 164 (1975).
[172] Y. Shioi, H. Tamai, and T. Sasa, Plant Cell Physiol. 19, 203 (1978).
[173] E. M. Sorokim, Photosynthetica 8, 221 (1974).
[174] G. Vierke and P. Struckmeier, Z. Naturforsch., Teil C 33, 266 (1978).
[175] P. A. Siegenthaler, FEBS Lett. 39, 337 (1974).
[176] T. Okamoto and S. Katoh, Plant Cell Physiol. 18, 539 (1977).
[177] T. Okamoto, S. Katoh, and S. Murakami, Plant Cell Physiol. 18, 551 (1977).

## 7. Chelators

Barr and Crane[121] describe a number of lipophilic chelators like butanedione derivatives inhibiting at the donor side of photosystem II.

## II. Inhibitors in Recent Developments on the Mechanism of Photosynthesis

Much of present photosynthesis research on the mechanism of energy conservation was stimulated by the Mitchell theory. Actually, photosynthesis research was particularly successful in contributing positive evidence for a chemiosmotic mechanism of coupling of electron flow to ATP formation (for reviews, see Witt,[178,179,179a] Jagendorf,[180,181] Trebst,[27,27a] Hauska and Trebst,[67] McCarty,[182,182a,b] Dilley et al.[183,184,184a] Hall,[185] and Heber[186]). This mechanism requires vectorial electron flow across the membrane, and the coupling sites in the electron flow chain become energy conserving sites consisting of an electron flow loop with an electrogenic step in one and a proton translocating step in the other direction.

### A. Vectorial Electron Flow

A review[27] in 1974 listed the evidence for vectorial electron flow by compiling data for a location of the donor sites toward the inside and a

---

[178] H. T. Witt, *Q. Rev. Biophys.* **4**, 365, 1971.

[179] H. T. Witt, *in* "Bioenergetics of Photosynthesis" (Govindjee, ed.), p. 495. Academic Press, New York 1975.

[179a] H. T. Witt, *Biochim. Biophys. Acta* **505**, 355 (1979).

[180] A. T. Jagendorf, *in* "Bioenergetics of Photosynthesis" (Govindjee, ed.), p. 414. Academic Press, New York, 1975.

[181] A. T. Jagendorf, *in* "Photosynthesis I" (A. Trebst and M. Avron, eds.), p. 307. Springer-Verlag, Berlin and New York, 1977.

[182] R. E. McCarty, *in* "Transport in Plants III" (U. Heber and C. R. Stocking, eds.), p. 347. Springer-Verlag, Berlin and New York, 1976.

[182a] P. C. Hinkle and R. E. McCarty, *Scientific American* **238**, 104 (1978).

[182b] R. E. McCarty, *in* "The Proton and Calcium Pumps" (G. F. Azzone *et al.*, eds.), Vol. 2, p. 65. Biomedical Press, Elsevier, Amsterdam, 1978.

[183] R. A. Dilley and R. T. Giaquinta, *Curr. Top. Membr. Transport* **7**, 49 (1975).

[184] R. A. Dilley, R. T. Giaquinta, L. J. Prochaska, and D. R. Ort, *in* "Water Relations in Membrane Transport in Plants and Animals" (A. M. Jungreis, T. K. Hodges, A. Kleinzeller, and S. G. Schultz, eds.), p. 55. Academic Press, New York, 1977.

[184a] R. A. Dilley and L. J. Prochaska, *in* "The Proton and Calcium Pumps" (G. F. Azzone *et al.* eds.), Vol. 2, p. 45. Biomedical Press, Elsevier, Amsterdam, 1978.

[185] D. O. Hall, *in* "The Intact Chloroplasts" (J. Barber, ed.), p. 135. Elsevier, Amsterdam, 1976.

[186] U. Heber, *J. Bioenerg. Biomembr.* **8**, 157 (1976).

location of the acceptor sites of both photosystems toward the outside of the thylakoid membrane. The evidence, as elucidated by inhibitors of electron flux and by probes for accessibility, did support a scheme of vectorial electron flow, although not without discrepancies. In particular, contradictory data were obtained for the location of the oxygen evolution system and of plastocyanin (cytochrome $f$) on the inside of the vesicle. The present situation is, therefore, reviewed again.

1. Location of the Water Splitting System

The evidence in 1974 against the oxygen evolution system being on the inside came from studies with chemical probes and antibodies, impermeable and therefore inaccessible to components on the inside, but which nevertheless did react (or did agglutinate) with the membrane and did inactivate oxygen evolution. Further evidence for an inactivation of the water splitting system by chemical probes has been provided particularly by Dilley et al.[43,164,183,184] and with antibodies by Menke et al.[158,160–163] However, it became clear also that those inactivation procedures are not necessarily directed against the active site of the oxygen evolution system. It seems quite likely that the large water splitting system (on the inside of the membrane) is intimately connected to photosystem II, which spans the membrane and is in turn connected to the light harvesting protein located toward the outside (as judged from antibody studies by Bose et al.[187]). In this way, the complete system constitutes an integral protein complex. An attack by antibodies or probes on this complex on one side of the membrane will affect the functional activity on the inside. This became particularly obvious when antibodies against lutein or violaxanthin, nonfunctional components of the water splitting system but part of the light harvesting complex, did inhibit oxygen evolution.[160,161] My view is substantiated by present knowledge of structural data on the arrangement of photosystem II particles and the light harvesting system in the membrane.[188,189] Furthermore, evidence for conformational changes in the water splitting system is provided by probes[43,184] and also by inactivation procedures, particularly in the light (see Section V). Preventing these conformational changes by a reaction even at a remote site of the membrane might affect activity. The acces-

[187] S. Bose, J. J. Burke, and C. J. Arntzen, in "Bioenergetics of Membrane" (L. Packer et al., eds.), p. 245. Elsevier, Amsterdam, 1977.

[188] L. A. Staehelin, P. A. Armond, and K. R. Miller, Brookhaven Symp. Biol. 28, 278 (1976).

[189] P. A. Armond, L. A. Staehelin, and C. J. Arntzen, J. Cell Biol. 73, 400 (1977).

sibility of violaxanthin in the membrane to its antibody is, as such, not surprising, because it was already indicated by its deepoxidation in the internal pH space but via an external carrier[190–192] as well as its external formation from zeaxanthin in the so called transmembrane violaxanthin cycle.

On the other hand, more direct data are now available for the inside location of the water splitting system. These are (1) the experiments on the specific inactivation of oxygen evolution (of the $S_2$ state) by internal pH,[152–155,193] (2) the release of manganese into an inside space by Tris treatment,[146] and (3) the liberation of protons of the water splitting reaction toward the inside.[194–196] The model of Lavorel[197] even assumes that the water splitting system is moving on the inside surface, this way connecting several photosystem II units. The dependence of coupling to ATP formation of donor systems for photosystem II on the chemistry of the donor couple (coupled when the donor liberates protons such as benzidine or catechol, but not when only electrons are involved as with tetramethylbenzidine, $I_2$, or ferrocyanide[148,149,198,199]) is an argument for an approach to photosystem II of the donor system (whether water or an artificial donor) from the inside space. This is supported by electron donor studies on the EPR signal IIf.[146] On the other hand, the accessibility of ferrocyanide to the donor site of photosystem II[199] might be used as an argument against it, except that the rate is low.

The recent success in obtaining inside-out vesicles[138] from chloroplast thylakoid membranes now permits the study of the accessibility of those components of the membrane buried inside in rightside-out vesicles. It has been shown that in inside-out vesicles trypsin treatment inactivates the oxygen evolution system,[139] whereas it does not do so in rightside-out vesicles. As already discussed in Section I,D,5, the early experiments on trypsin inactivation of the oxygen evolution system were misleading.

In summary, the evidence for an inside location of the oxygen evolution system has greatly improved.

[190] D. Siefermann and H. Y. Yamamoto, *Biochim. Biophys. Acta* **387**, 149 (1975).

[191] D. Siefermann and H. Y. Yamamoto, *Arch. Biochem. Biophys.* **171**, 70 (1975).

[192] A. Hager, *Planta* **89**, 224 (1969).

[193] G. Renger, M. Gläser, and H. E. Buchwald, *Biochim. Biophys. Acta* **461**, 392 (1976).

[194] W. Junge and W. Ausländer, *Biochim. Biophys. Acta* **333**, 59 (1974).

[195] W. Junge, G. Renger, and W. Ausländer, *FEBS Lett.* **79**, 155 (1977).

[196] C. F. Fowler and B. Kok, *Biochim. Biophys. Acta* **357**, 380 (1974).

[197] J. Lavorel, *FEBS Lett.* **66**, 164 (1976).

[198] A. Trebst, *Proc. Int. Congr. Photosynth., 3rd, 1974* p. 439 (1975).

[199] S. Izawa, D. R. Ort, J. M. Gould, and N. E. Good, *Proc. Int. Congr. Photosynth., 3rd, 1974* p. 449 (1975).

## 2. Location of Plastocyanin

The evidence for the location of plastocyanin (and of cytochrome $f$) on the inside of the thylakoid is largely based on work with antibodies against the two; they inhibit electron flow only when the vesicle is opened up by sonication.[36] The data with impermeable chemical probes (polylysine, DABS, and others) on the other hand, argue for an accessibility of plastocyanin in a hydrophilic cleft on the outside of the membrane.[43,200,200a] Indeed it has been shown now that DABS, CDIS, KCN, and $HgCl_2$, when incubated with the membrane, do react with the copper of bound plastocyanin[19,28,43,200]; this finding dispels the argument that the probes affect plastocyanin function only indirectly. Further, antibody preparations against plastocyanin and cytochrome $f$ were described recently by Schmid et al.[38,201] and by Böhme[39,202,202a] which agglutinate the membrane and inhibit electron flux, although only to 40%. However, there is some inherent discrepancy in the antibody experiments which remains unexplained. The argument that plastocyanin, solubilized during the chloroplast preparation, attaches itself on the outside of the membrane and that this outside bound plastocyanin is reacting with the antibodies has not been fully dispelled. Careful measurements of the coupling of such systems to ATP formation may solve the problem, because only inside bound plastocyanin completes a vectorial loop necessary for proton translocation. Nolan and Bishop[40a] from their amphotericin experiments, also argue for an accessibility of plastocyanin. On the other hand, the properties of donor systems for photosystem I, which also react via plastocyanin or cytochrome $f$ (such as DAD or TMPD) do very much support the notion of a vectorial arrangement of the photosystem I segment, because only hydrogen transporting, lipid-soluble donors produce a pH gradient and coupled ATP formation.[55,67,69,203] Further arguments for a location of plastocyanin on the inside are provided by Bendall,[203a] Lockau[203b] and from work with inside-out thylakoid vesicles.[133,139]

In summary, the evidence for plastocyanin location on the inside of

[200] D. D. Smith, B. R. Selman, K. K. Voegeli, G. Johnson, and R. A. Dilley, Biochim. Biophys. Acta 459, 468 (1977).

[200a] B. R. Selman, D. D. Smith, K. K. Voegeli, G. Johnson, and R. A. Dilley, Proc. Int. Congr. Photosynth. 4th, 1977, p. 499 (1978).

[201] G. H. Schmid, A. Radunz, and W. Menke, Z. Naturforsch., Teil C 32, 271 (1977).

[202] H. Böhme, Eur. J. Biochem. 83, 137 (1978).

[202a] H. Böhme, Eur. J. Biochem. 84, 87 (1978).

[203] G. Hauska, A. Trebst, and W. Draber, Biochim. Biophys. Acta 305, 632 (1973).

[203a] D. S. Bendall, and P. M. Wood, Proc. Int. Congr. Photosynth. 4th 1977, p. 771 (1978).

[203b] W. Lockau, Eur. J. Biochem. 94, 365 (1979).

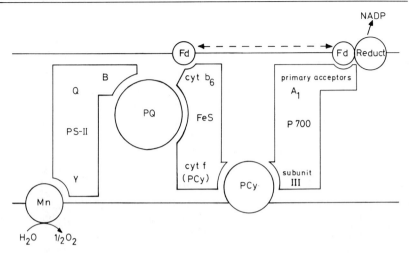

FIG. 9. Components of the photosynthetic electron-flow system from water to NADP, vectorial across the thylakoid membrane. Three integral protein components (PS-II, PS-I, and cytochrome $b_6$, FeS, cytochrome $f$ complex) span the membrane. The oxygen evolution system and plastocyanin are oriented toward the inside, ferredoxin and the ferredoxin–NADP reductase towards the matrix side. Plastoquinone connects the cytochrome complex with the photosystem II complex in noncyclic and with ferrodoxin in the cyclic electron flow pathway. Plastocyanin connects the cytochrome complex with PS-I. Proton channels through the B-protein and the cytochrome $f$ complex complete plastoquinone-driven proton translocation across the membrane in coupling site I.

the membrane remains controversial. In contrast, arguments for the primary donor site of photosystem I (P700) on the inside remain strong.[14,204,205] Therefore, the evidence is still for vectorial electron flow in photosynthesis. The electron-transport system might be drawn in a simple schematic way with three integral proteins (photosystem I and II and the cytochrome $b/f$ complex) spanning the membrane and connected with each other by plastoquinone and plastocyanin (Fig. 9). Proton channels connecting the outside space with plastoquinone through the B-protein and in the cytochrome $f$ complex toward the inside (as indicated by DCCD experiments[98a]) may exist. This would ease the need to postulate redox reactions across the width of the membrane and a location of ceratin carriers strictly on opposite sides of the membrane, if just the photosystems would react across the barrier separating the inside and outside pH spaces.

[204] C. Bengis and N. Nelson, J. Biol. Chem. **250**, 2783 (1975).
[205] W. Junge, Annu. Rev. Plant Physiol. **28**, 503 (1977).

## 3. Mechanism of Plastoquinone Function

The plastoquinone shuttle connects photosystem II and photosystem I and is assumed to carry electrons and hydrogens from the outside to the inside. Not much new direct information on the location of plastoquinone is available. Its reduction by photosystem II and Q is assumed to be on the outside, followed by protonization from the external space[205,206,206a] (see below). Katoh has recently studied the location of Q.[207] The oxidation of plastoquinol on the inside[206a] (see Junge and Ausländer[194]) would lead to the proton gradient responsible for coupling. Although a coupling site between plastoquinone/cytochrome $f$ has been indicated by indirect methods,[208] no direct evidence for plastoquinol movement across the membrane is available. Recently, different models for the plastoquinone shuttle have been reviewed,[67] such as movement of quinone and quinol or of the semiquinone, of a proton pump mechanism via an unidentified carrier, cooperativity of two carriers in reduction, as well as in oxidation of quinone and hydroquinone and the Q cycle. Model liposome systems seem to indicate that a lipophilic hydroquinone may easily connect impermeable redox systems on either side of the membrane (ascorbate to ferricyanide).[209,209a] On the other hand, proteinization of the fully reduced plastoquinol dianion (compound B) on the outside of the membrane, but through a proteinous shield (evidence for a delay in protonization of plastoquinone via a shield is provided by Ausländer and Junge[205,206]) might be all that is necessary to pump protons across a barrier into an inside microspace (see next section). Relevant to the problem of vectorial plastoquinone function is the question at which side of the membrane acceptor systems for photosystem II are reduced. From the influence of internal pH on acceptor systems in DBMIB experiments (see Section I,C,1,d,i) and of the stoichiometry of the TMPD bypass (Section I,C,d,v), it was concluded that it is an inside reduction. On the other side, a crucial test for this assumption is that the stoichiometry of acceptor systems should depend on the proton carrying property of the acceptor, and such experiments failed to support this notion.[70]

[206] W. Ausländer and W. Junge, *Biochim. Biophys. Acta* **357**, 285 (1974).
[206a] R. Tiemann, G. Renger, P. Gräber, and H. T. Witt, *Biochim. Biophys. Acta* **546**, 498 (1979).
[207] S. Katoh, *Plant Cell Physiol.* **18**, 893 (1977).
[208] H. Böhme and W. A. Cramer, *Biochemistry* **11**, 1155 (1972).
[209] G. Hauska, *in* "Bioenergetics of Membranes" (L. Packer *et al.*, eds.), p. 177. Elsevier, Amsterdam, 1977.
[209a] A. Futami, E. Hurt, and G. Hauska, *Biochim. Biophys. Acta* **547**, 583 (1979).

## 4. Additional Electrogenic and/or Proton Translocating Steps

The Mitchell Q cycle[210] proposes an electrogenic electron flow reaction from the inside to the outside, when a cytochrome *b* oxidizes a quinol on the inside, moves an electron back from the inside to the outside (the electrogenic step), and donates the electron back in quinone reduction on the outside. In a Q cycle two protons per one electron are moved in a quinone shuttle. So far no evidence for a Q cycle in chloroplasts has been presented, except perhaps for an additional field measured in intact algae.[211] On the other hand, in bacterial photosynthesis there is now good evidence for a Q cycle[90a] provided at first by DBMIB inhibitor studies (see Section I,C,1,i).

Another proton translocating step might have been indicated by a report on a higher stoichiometry of proton/electron than one in photosystem I reactions.[212,212a] Other measurements, however, indicate that the total stoichiometry of proton/electron in overall photosynthetic electron flow remains at two.[205,213]

## B. Energy Conserving Sites

Vectorial electron flow in the steady state moves 4 protons per 2 electrons and builds up the proton gradient, driving ATP formation. Careful measurements by Pick *et al.*[214] and Portis and McCarty[215,216] strongly support a stoichiometry of 3 protons/ATP in the ATP synthetase reaction and from this an ATP/2e ratio of 1.33 for NADP reduction (also see reviews by Rottenberg[217] and Hall[185]). Two coupling sites (native energy conserving sites or loops in vectorial electron flow) of an ATP/2e ratio of 0.66 each have now been identified along the photosynthetic electron flow system.[27,27a,67,185,198,199] The second site consists of the protons liberated in the water splitting system on the inside and the electrogenic photosystem II across the membrane. The inhibitors between the two photosystems, particularly DBMIB and KCN, were instrumental in

[210] P. Mitchell, *in* "Electron Transfer Chains and Oxidative Phosphorylation" (E. Quagliariello *et al.*, eds.), p. 305. North-Holland Publ., Amsterdam, 1975.
[211] P. Joliot, R. Delosme, and A. Joliot, *Biochim. Biophys. Acta* **459**, 47 (1977).
[212] C. F. Fowler and B. Kok, *Biochim. Biophys. Acta* **423**, 510 (1976).
[212a] R. Velthuys, *Proc. Nat. Acad. Sci. USA* **76**, 2765 (1979).
[213] S. Saphon and A. R. Crofts, *Z. Naturforsch., Teil C* **32**, 810 (1977).
[214] U. Pick, H. Rottenberg, and M. Avron, *FEBS Lett.* **48**, 32 (1974).
[215] R. E. McCarty and A. R. Portis, *Biochemistry* **15**, 5110 (1976).
[216] A. R. Portis and R. E. McCarty, *J. Biol. Chem.* **251**, 1610 (1976).
[217] H. Rottenberg, *in* "Photosynthesis I" (A. Trebst and M. Avron, eds.), p. 338. Springer-Verlag, Berlin and New York, 1977.

identification of coupling site II.[198,199] Furthermore, the concept of an artificial energy conservation site was introduced[27,55,67] when inhibitor studies indicated an artificially induced coupling of a segment of the electron flow system not coupled *in situ* (see Fig. 5, Section I,C,1,d). An artificial site consists of an artificial proton translocation across the membrane via an artificial donor system of photosystem I (see Fig. 5) or photosystem II.

According to the chemiosmotic model, the pH gradient, brought about by vectorial electron flow, is built up across the membrane and is distributed evenly in the inside lumen of the thylakoid. The membrane potential is delocalized over the membrane, and there is just one ATP synthetase system driven by this proton motive force between lumen and external space. For photosynthetic electron flow, this situation is certainly given in acid/base and in postillumination phosphorylation, in artificial proton translocating redox reactions across the membrane in the dark[218] and possibly also in the light in artificial energy conservation with artificial electron donor systems.[67] It might not be true in the transient state, in light flashes and intense light. Studies with the inhibitor $HgCl_2$ seemed to indicate that coupling site I is more sensitive than coupling site II (see Section I,B,1) and therefore a heterogenity of the ATP synthetase.[20-22] Further experiments on a differential response of coupling site I and II are summarized in Dilley and Giaquinta.[183] This interpretation disagrees with the free lateral movement of the coupling factor on the external surface[219,220] and the structural data according to which there is no coupling factor in the partition region.[188,221] Therefore the PMF, generated by electron flux in the stacked area, has to be transported in the inside space to the ATP synthetase on the exposed unstacked region. Furthermore, the sigmoidal response of reconstitution of photophosphorylation upon readdition of coupling factor to EDTA-treated membranes argues for a common pool of energy-rich intermediates available to numerous ATP synthetase systems (see Berzborn and Schröer[222]). Also the energetization of the coupling factor is accomplished equally well by either photosystem II or photosystem I electron flux.[223]

A concept of microspaces[224] has been given some attention recently

[218] B. R. Selman and G. Pscolla, *FEBS Lett.* **61**, 135 (1976).
[219] R. Berzborn, F. Kopp, and K. Mühlethaler, *Z. Naturforsch., Teil C* **29**, 694 (1974).
[220] R. Berzborn, F. Kopp, and K. Mühlethaler, *Proc. Int. Congr. Photosynth., 3rd, 1974* p. 809 (1975).
[221] K. R. Miller and L. A. Staehelin, *J. Cell Biol.* **68**, 30 (1976).
[222] R. J. Berzborn and P. Schröer, *FEBS Lett.* **70**, 271 (1976).
[223] A. E. Grebanier and A. T. Jagendorf, *Biochim. Biophys. Acta* **459**, 1 (1977).
[224] R. P. J. Williams, *Curr. Top. Bioenerg.* **3**, 79 (1969).

in photosynthesis research in spite of the arguments just quoted. Particularly the experiments of Ort *et al.*[225,226] in which ATP formation sets in before the internal thylakoid space seems to be acidified and a pH gradient can be measured, are taken as evidence for special domains in the membrane in which protons are deposited first and which only in steady state are in equilibrium with the internal space. There might be even two different domains for the two coupling sites to account for the $HgCl_2$ experiments mentioned above. Dilley and his colleagues[183,184] favor a model in which protons from the water splitting reaction are released within the membrane. Their data indicate that a conformational change brought about by electron flow from photosystem II to plastoquinone is required for a photosystem II-dependent proton gradient and controls the direction of proton release of the water splitting reaction.[142,227]

The few indications for microspaces and proton channels are refinements of the arrangement of components and mechanisms in vectorial electron flow, which is not necessarily across the complete width of the membrane but across barriers between pH spaces eventually building up in a steady state a delocalized proton potential between the outside and inside lumen (Fig. 9).

[225] D. R. Ort, and R. A. Dilley, *Biochim. Biophys. Acta* **449**, 95 (1976).
[226] D. R. Ort, D. A. Dilley, and N. E. Good, *Biochim. Biophys. Acta* **449**, 108 (1976).
[227] R. T. Giaquinta, D. R. Ort, and R. A. Dilley, *Biochemistry* **14**, 4392 (1975).

# Section V
# Inhibitors

# [66] Delineation of the Mechanism of ATP Synthesis in Chloroplasts: Use of Uncouplers, Energy Transfer Inhibitors, and Modifiers of Coupling Factor 1

*By* Richard E. McCarty

Photophosphorylation in isolated chloroplast inner membranes (thylakoids) and in chromatophores from photosynthetic bacteria is driven by light-dependent electron flow. The intermediate which serves to couple ATP formation to electron transport is probably an electrochemical proton gradient. The inward translocation of protons is associated with electron transport in thylakoids and chromatophores. ATPase complexes, which consist of a hydrophilic coupling factor–ATPase $(CF_1)$[1] and less well-characterized hydrophobic components $(F_0)$, catalyze the synthesis of ATP coupled to the outward flow of protons[2] (Fig. 1).

A large number of reagents and treatments affect photophosphorylation. Inhibitors of electron flow abolish ATP synthesis since they also block the formation of the electrochemical proton gradient, the driving force for ATP synthesis. ATP synthesis may also be uncoupled from electron flow. A reagent which causes a decrease in the phosphorylation efficiency, the ratio of ATP formed per electron pair transferred through the electron transport chain $(P/e_2$ ratio), is classified as an uncoupler. However, this definition of uncouplers is too broad. For example, the $P/e_2$ ratio decreases somewhat at low $P_i$ concentration. Yet, $P_i$ at low concentrations cannot be classified as an uncoupling agent. A more restrictive definition of uncouplers (and treatments which result in uncoupling) must, therefore, be used. In terms of the chemiosmotic[3] hypothesis, an uncoupler increases the permeability of membranes to protons, thereby dissipating the electrochemical proton gradient. Since proton translocation is a direct consequence of electron flow, the proton gradient exerts a back pressure on electron flow. By dissipating the proton gra-

[1] Abbreviations: $CF_1$, coupling factor 1; $F_0$, hydrophobic parts of the ATPase complex; CCP, carbonyl cyanide *m*-chlorophenylhydrazone; DCCD, *N,N'*-dicyclohexylcarbodiimide; Tricine, *N*-Tris(hydroxymethyl)methylglycine; NEM, *N*-ethylmaleimide.
[2] For reviews of photophosphorylation, see A.T. Jagendorf, *in* "Photosynthesis I" (A. Trebst and M. Avron, eds.), p. 307. Springer-Verlag, Berlin and New York, 1977, and R. E. McCarty, *in* "Transport in Plants III" (U. Heber and C. R. Stocking, eds.), p. 347. Springer-Verlag, Berlin and New York, 1976.
[3] P. Mitchell, *Biol. Rev. Cambridge Philos. Soc.* **41**, 445 (1966).

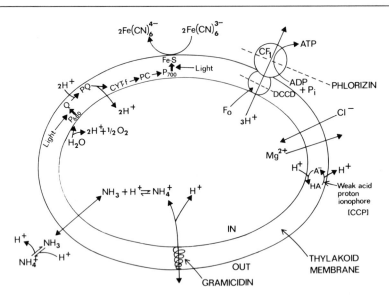

Fig. 1. Sites of action of some uncouplers and inhibitors of photophosphorylation in chloroplast thylakoids. Light-driven electron flow from water to ferricyanide causes the inward translocation of protons. $Cl^-$ and $Mg^{2+}$ move in response to the proton movements. An ATPase complex ($CF_1$–$F_0$) catalyzes ATP formation at the expense of the electrochemical proton gradient generated by electron flow. Phlorizin and $N$-ethylmaleimide (not shown) interact with $CF_1$, whereas DCCD (dicyclohexylcarbodiimide) interacts with a component of $F_0$. Many uncouplers including CCP (carbonylcyanide $m$-chlorophenylhydrazone), are lipid-soluble weak acids which dissipate the proton gradient by carrying protons across the membrane. A dimer of gramicidin molecules forms a water-filled transmembrane channel through which protons, $NH_4^+$, and other small ions can pass. Ammonium ions and other amine cations are accumulated by thylakoids in response to internal acidification.

dient, uncouplers enhance the rate of electron flow while they inhibit phosphorylation. The evidence on the whole supports the concept that uncouplers work by dissipating proton gradients.

Photophosphorylation may also be inhibited by reagents which interfere with the function of the proton translocating ATPase complex ($CF_1$–$F_0$). These reagents have no direct effect on either electron transport or on proton translocation driven by electron flow. Because they interfere with the utilization of the energy of the proton gradient to form ATP, reagents which inhibit phosphorylation in this way are called energy transfer inhibitors. Some inhibitors, for example, dicyclohexylcarbodiimide,[4] of this class appear to interact with the $F_0$ part of the ATPase complex and block proton translocation through the membrane. Other energy transfer inhibitors (phlorizin,[5] for example) interact directly with $CF_1$, the coupling factor–ATPase, to inhibit phosphorylation and associated proton efflux. In addition, a number of reagents which react with functional groups on $CF_1$ under certain conditions have been found to act as energy transfer inhibitors. These reagents have been useful in the study of nucleotide–$CF_1$ interactions as well as in exploring the function of $CF_1$ subunits in the phosphorylation mechanism.

It is not the purpose of this article to provide a list of uncouplers, energy transfer inhibitors, and modifiers of $CF_1$. In a previous article in this series, Izawa and Good[6] presented a thorough survey of inhibitors and uncouplers of photophosphorylation. Instead, I will describe, in practical terms, a few inhibitors that we have found to be most useful.


Uncouplers

Reagents that enhance the permeability of thylakoid membranes to protons uncouple phosphorylation from electron flow. Thus, lipid-soluble weak acids, such as carbonyl cyanide $m$-chlorophenylhydrazone (CCP), which act as proton carriers across both natural and synthetic membranes, are uncouplers (Fig. 1). Unfortunately, many of these proton ionophores will also inhibit electron flow at concentrations only slightly higher than those which cause uncoupling. Some uncouplers are even destroyed by illumination in the presence of thylakoids. Gramicidin is not a proton carrier, but instead, forms an aqueous transmembrane channel through which protons and other small ions can pass. Significantly, it does not inhibit electron flow even at very high concentrations. Moreover, gramicidin is commercially available (Sigma Chemical Co., St. Louis, Missouri) and is inexpensive. Although commercial grades of gramicidin are mixtures of different types of gramicidins, this has little bearing on their use as uncouplers.

The gramicidin(s) are linear pentadecapeptides consisting of alternating D- and L-amino acids. The terminal valine is formylated, and ethanolamine is attached to the terminal carboxyl group through a peptide bond. Note that gramicidin S is a different peptide antibiotic. Gramicidin is sometimes referred to as gramicidin D. Gramicidin A, the most prev-

[4] R. E. McCarty and E. Racker, *J. Biol. Chem.* 242, 3435 (1967).
[5] G. D. Winget, S. Izawa, and N. E. Good, *Biochemistry* 8, 2067 (1969).
[6] Vol. 24 [32].

alent form in commercial gramicidin, has a molecular weight of about 1865. Gramicidins are rather hydrophobic, but may be dissolved in methanol, ethanol, or dimethyl sulfoxide to a concentration of 5 m$M$ or more. Solutions of gramicidin in the millimolar concentration range retain their uncoupling activity for at least several weeks when stored at $-20°$. Methanol, ethanol, or dimethyl sulfoxide at 0.5% have little to no effect on either photophosphorylation or electron flow. Even so, it is good practice to include the same concentration of the solvent used to dissolve the gramicidin in controls. When aliquots of concentrated gramicidin solutions are added to aqueous reaction mixtures, the potency of the gramicidin decreases with time. This phenomenon is likely to be due to the binding of the gramicidin to glass (or polyethylene) surfaces. Thus, the gramicidin should be added either just before or after the thylakoids, so that it may partition into the thylakoid membranes rather than onto glass surfaces.

At saturating light intensities, full uncoupling of photophosphorylation supported by electron flow from water to ferricyanide by thylakoids (50 $\mu$g of chlorophyll per milliliter) occurs at about 0.1 to 0.2 $\mu M$ gramicidin. Higher concentrations are required to inhibit phosphorylation supported by cyclic electron flow. The concentration of gramicidin required to uncouple is dependent on the chlorophyll to gramicidin ratio.

Although gramicidin inhibits photophosphorylation readily, the transmembrane proton concentration gradient ($\Delta$pH) is more resistant. Even with 10 $\mu M$ gramicidin, electron flow from water to ferricyanide can still support a $\Delta$pH of 1.6 units.[7] A $\Delta$pH of this magnitude cannot drive a significant rate of photophosphorylation, but still exerts some back pressure on electron flow. To abolish $\Delta$pH and achieve fully uncoupled rates of electron flow, it is necessary to add NH$_4$Cl at 1 to 5 m$M$ in addition to the gramicidin. Under these conditions, rates of electron flow from water to either ferricyanide or methyl viologen of 1800 $\mu$Eq/hr/mg chlorophyll have been observed.[7,8] Moreover, NH$_4$Cl markedly enhances the ability of low concentrations of gramicidin to inhibit photophosphorylation. In the presence of 1 m$M$ NH$_4$Cl, full uncoupling of phosphorylation from electron flow from water to ferricyanide in chloroplasts (50 $\mu$g chlorophyll per milliliter) was reached at only 10–20 n$M$ gramicidin.[9] These synergistic effects of gramicidin and NH$_4$Cl may be explained by the facts that gramicidin causes the transport of NH$_4^+$ as well as H$^+$ and

[7] A. R. Portis, Jr. and R. E. McCarty, *J. Biol. Chem.* **251**, 1610 (1976).
[8] H. H. Stiehl and H. T. Witt, *Z. Naturforsch., Teil B* **24**, 1588 (1969).
[9] R. E. McCarty, unpublished.

that $NH_4^+$ is accumulated by thylakoids in response to light-driven $H^+$ uptake (see Fig. 1). Assuming a $\Delta pH$ of 2 units, the proton concentration inside thylakoids at an external pH of 8.0 is $10^{-6}$ $M$. In contrast if 0.5 m$M$ $NH_4Cl$ is present, the internal $NH_4^+$ concentration would approach $5 \times 10^{-2}$ $M$, 50,000-fold higher than the internal proton concentration. Thus $NH_4^+$ would be transported at a higher rate than $H^+$.

### Energy Transfer Inhibitors

$N,N'$-Dicyclohexylcarbodiimide[4] (DCCD) and phlorizin[5] (4,4',6'-trihydroxy-2'-glucosidodihydrochalcone) inhibit photophosphorylation, but interact with different parts of the ATPase complex. DCCD binds to a membrane proteolipid which has a molecular weight of 8000.[10] This proteolipid enhances the proton permeability of liposomes which contain the light-dependent bacteriorhodopsin proton pump, and DCCD reverses this effect.[10] In contrast, phlorizin interacts with $CF_1$, since the $Ca^{2+}$-dependent ATPase of the soluble enzyme is inhibited[9] by phlorizin at concentrations which inhibit phosphorylation in thylakoids.

DCCD (206.3 molecular weight) may be purchased from a number of chemical supply houses and dissolves in methanol or ethanol. Concentrated stock solutions (10 to 20 m$M$) lose little inhibitory activity on storage at $-20°$ for several months. DCCD takes time to act. We generally incubate the thylakoids (suspended in 0.4 $M$ sucrose, 0.02 $M$ Tricine–NaOH (pH 8.0), and 0.01 $N$ NaCl) with DCCD for 15 min at $0°$. Photophosphorylation is nearly abolished when thylakoids containing 50 $\mu g$ of chlorophyll are incubated with 50–100 nmole of DCCD in 0.5 ml prior to the assay of photophosphorylation. Although DCCD inhibits coupled electron flow more strongly than that which takes place under nonphosphorylating conditions, DCCD does inhibit uncoupled electron flow to some extent.[4] This result indicates that DCCD probably has a direct effect on electron transport in addition to its effects on photophosphorylation. Triphenyltin chloride, which affects phosphorylation in a manner similar to DCCD, does not inhibit electron flow.[11] Since it promotes a $Cl^-$ for $OH^-$ exchange, triphenyltin chloride cannot be used in $Cl^-$-containing media. Thylakoids from which part of the $CF_1$ has been removed by EDTA extraction become leaky to protons. The incubation of thylakoids with DCCD restores light-induced proton uptake, and, at low concentrations, it also restores some phosphorylation to the $CF_1$-deficient

[10] N. Nelson, *Abstr. Int. Congr. Photosynth., 4th, 1977* p. 269 (1978).
[11] J. M. Gould, *Eur. J. Biochem.* **62**, 567 (1976).

thylakoids.[4] Thus, DCCD probably inhibits photophosphorylation by reacting with a membrane proteolipid to prevent the flow of protons to the $CF_1$.

Several energy transfer inhibitors which interact with $CF_1$ have been described. Phlorizin is, however, the only inhibitor of this type which is available commercially (Sigma Chemical Corporation, St. Louis, Missouri). Before use, phlorizin should be treated with activated charcoal and recrystallized from hot water. One gram of phlorizin (molecular weight 436) dissolves in 22 ml of water at 70° and about 1 liter of water at 22°. Phlorizin is freely soluble in ethanol, and ethanolic solutions appear to be stable for many days at −20°. The ethanol concentration in reaction mixtures should be 0.5% or less.

Since phlorizin acts rapidly, it may be added directly to reaction mixtures. Phlorizin specifically inhibits electron flow which is dependent on phosphorylation.[5] It has no effect on either uncoupled electron flow or on the electron flow under nonphosphorylating conditions. One drawback of the use of phlorizin as an energy transfer inhibitor is that rather high concentrations are required. To fully block electron flow from water to ferricyanide under phosphorylating conditions, 2 m$M$ phlorizin must be added. Photophosphorylation is inhibited about 90% by this concentration of phlorizin. Some compounds closely related to phlorizin, such as 4′-deoxyphlorizin,[5] are nearly 10-fold more effective than phlorizin but are not commercially available.

### Modifiers of Coupling Factor 1

Modification of enzymes with group-specific reagents is used to identify amino acid residues required in some way for the activities catalyzed by the enzymes. Reagents which react primarily with either cysteine sulfhydryl groups or with arginines can act as energy transfer inhibitors. $CF_1$ is the site of action of these reagents.

$N$-Ethylmaleimide (NEM) reacts rapidly with thiols in solution to produce stable thioether linkages.[12] NEM (molecular weight 125.12) may be purchased (Sigma Chemical Co., St. Louis, Missouri) and used without further purification. It is soluble in water up to about 0.2 $M$, and aqueous solutions may be stored at −20° for many weeks with little loss of the NEM. A base-catalyzed hydrolysis of NEM to $N$-ethylmaleamic acid, which reacts only slowly with thiols, occurs. At pH 8.0 and 20°, NEM has a half-life of 220 min, whereas at pH 9.0, the half-life is only 22 min. NEM has an absorption band at 305 nm (extinction coefficient,

---

[12] Vol. 11 [63].

0.620 m$M$ cm$^{-1}$) which disappears when the NEM is either reacted with thiols or converted to the maleamate derivative.

NEM inhibits photophosphorylation only when thylakoids are illuminated in the presence of the reagent prior to assay.[13] The extent of inhibition is very sensitive to the ΔpH across thylakoid membranes, and maximum inhibition (about 60–75%) is observed only at high ΔpH values.[14] Cyclic electron flow supported by the artificial mediator pyocyanine gives higher ΔpH values at high light intensities than that supported by noncyclic electron flow. The following conditions have been found to routinely cause maximum inhibition of photophosphorylation by NEM: thylakoids (50 μg of chlorophyll) are suspended at room temperature in 0.5 ml of a medium which contains 50 m$M$ Tricine–NaOH (pH 8.25), 50 m$M$ NaCl, 5 m$M$ MgCl$_2$, 0.05 m$M$ pyocyanine, and 1 m$M$ NEM. After 90 sec in the light (2.5 × 10$^6$ ergs/cm$^2$/sec) or dark, an amount of either dithiothreitol or β-mercaptoethanol slightly in excess of the NEM is added, and photophosphorylation is assayed with pyocyanine as the mediator. The incubation may be scaled up and the thylakoids washed free of the untreated NEM and pyocyanine so that phosphorylation with other mediators of electron flow may be assayed.

Illumination enhances the incorporation of NEM into the γ subunit of CF$_1$ in thylakoids.[15] This incorporation, which is probably the cause of the inhibition of photophosphorylation, may be determined using $^3$H-NEM (New England Nuclear, Boston, Massachusetts). $^3$H-NEM is somewhat unstable, especially in aqueous solution, and, therefore, the specific radioactivity of the compound should be determined. An aliquot of the $^3$H-NEM solution is subjected to thin layer chromatography on silica gel G with petroleum ether–acetone (9:1, v/v) as the developing solvent. Zones of 0.5 or 1 cm are scraped from the plate and added to scintillation vials containing 9 ml of a scintillation fluid which contains Triton X-100–toluene (1:2, v/v), 0.5% of 2,5-diphenyloxazole, and 0.01% of 1,4 bis[2-(diphenyloxazole)]benzene. The maleamate derivative, the major contaminant, does not migrate in this solvent system, whereas NEM has an $R_f$ of about 0.7. The concentration of NEM may be determined spectrophotometrically, using the difference between the $A_{305}$ of the sample before and after the addition of an excess of dithiothreitol in the calculation.

Since CF$_1$ contains two groups, one on the γ subunit and one on ε

[13] R. E. McCarty, P. R. Pittman, and Y. Tsuchiya, *J. Biol. Chem.* **247**, 3048 (1972).
[14] A. R. Portis, Jr., R. P. Magnusson, and R. E. McCarty, *Biochem. Biophys. Res. Commun.* **64**, 977 (1975).
[15] R. E. McCarty and J. Fagan, *Biochemistry* **9**, 1503 (1973).

which react with NEM in the dark, thylakoids, suspended in 0.4 $M$ sucrose, 0.02 $M$ Tricine–NaOH (pH 8.25), and 0.01 $M$ NaCl, are incubated in the dark for 5 min at 0° in the presence of 2–5 m$M$ NEM. The thylakoids are collected by centrifugation at 10,000 $g$ for 10 min and washed in the buffered sucrose solution. This procedure does not affect either photophosphorylation or the light-dependent inhibition of photophosphorylation by NEM, but increases the light enhancement of the labeling by $^3$H-NEM from 1.5-fold to nearly 20-fold. The reaction with $^3$H-NEM is carried out in the same incubation mixture used for the development of the NEM inhibition except that 25 $\mu$Ci of $^3$H-NEM per milliliter of incubation mixture is also added. Generally, the amount of thylakoids used is 1 to 3 mg. $CF_1$ is then removed from the membrane and purified.[16] $CF_1$ subunits are separated by electrophoresis in the presence of sodium dodecyl sulfate.[17] After fixing and staining the gels, which does not release the $^3$H-NEM from the protein, radioactivity is determined in slices of the gels containing the subunits.[18] Nearly all of the radioactivity is found in the $\gamma$ subunit.

The light-dependent reaction of $CF_1$ in thylakoids with NEM has been exploited as a means to study interactions between membrane-bound $CF_1$ and nucleotides.[19] Moreover, these experiments reinforce the observations that $CF_1$ in thylakoids changes its conformation upon formation of a proton gradient across the membrane. Finally, experiments with NEM and other maleimides suggest that the $\gamma$ subunit plays an essential role in photophosphorylation. More recently, a bifunctional maleimide, $N,N'$-$o$-phenylenedimaleimide, has been shown to cross-link two groups within the $\gamma$ subunit of $CF_1$ in thylakoids.[20] This cross-linking makes thylakoids somewhat leaky to protons.

Recent studies with either purified $CF_1$[21] or thylakoids[22] have suggested that an argininyl residue(s) in $CF_1$ has an essential function in photophosphorylation and in the ATPase activity catalyzed by the enzyme. $\alpha$-Dicarbonyl compounds, such as phenylglyoxal and 2,3-butanedione (Aldrich Chemical Co., Milwaukee, Wisconsin) form complexes with arginyl residues in proteins. The stability of these complexes varies considerably with the buffer used. For example, borate buffer must be used if 2,3-butanedione is the modifying reagent. In contrast, $N$-ethyl-

---

[16] Vol. 23 [49].

[17] Vol. 26 [1].

[18] M. Zaitlin and V. Hariharasubramanian, *Anal. Biochem.* **35**, 296 (1970).

[19] R. P. Magnusson and R. E. McCarty, *J. Biol. Chem.* **250**, 2593 (1975).

[20] M. A. Weiss and R. E. McCarty, *J. Biol. Chem.* **252**, 8007 (1977).

[21] C. S. Andreo and R. H. Vallejos, *FEBS Lett.* **78**, 207 (1977).

[22] R. Schmid, A. T. Jagendorf, and S. Hulkower, *Biochim. Biophys. Acta* **462**, 177 (1977).

morpholine buffer allows the greatest inhibition of photophosphorylation by phenylglyoxal.[22]

Spinach chloroplast thylakoids are incubated with and without the modifying reagents for 20 min at 20° in the dark. Illumination does not enhance the inhibition of photophosphorylation by either 2,3-butanedione or phenylglyoxal.[22] Chloroplast thylakoids are prepared by homogenizing spinach leaves in 0.4 $M$ sorbitol, 0.02 $M$ Tricine–NaOH (pH 7.8), and 0.01 $M$ NaCl and are washed once in 10 m$M$ NaCl[23] If phenylglyoxal (molecular weight of the hydrate 152.15) is used, the thylakoids are resuspended in 0.4 $M$ sorbitol and 0.01 $M$ NaCl to give a chlorophyll concentration of 1 mg/ml. The incubation mixture with phenylglyoxal contains 80 m$M$ $N$-ethylmorpholine-HCl (pH 7.8), 140 m$M$ sorbitol, 5 m$M$ NaCl, 1 m$M$ dithiothreitol, 0.2 mg/ml bovine serum albumin, thylakoids equivalent to 0.33 mg chlorophyll/ml, and 1–20 m$M$ phenylglyoxal. Phenylglyoxal hydrate is dissolved in 125 m$M$ $N$-ethylmorpholine-HCl (pH 7.8). After 20 min at 20°, the mixture is diluted with 7 volumes of 0.4 $M$ sorbitol and 0.01 $M$ NaCl and the thylakoids are collected by centrifugation. The thylakoids are resuspended in half of the volume of the sorbitol–NaCl solution used above and collected by centrifugation. These steps remove $N$-ethylmorpholine which inhibits phosphorylation.

Photophosphorylation and ATPase activity are inhibited about 90% by 20 m$M$ phenylglyoxal in 20 min of incubation. Controls without phenylglyoxal must be run since photophosphorylation is somewhat labile, and 60% or more of the cyclic photophosphorylation capacity may be lost on incubation at 20° even in the absence of the modifiers.

When 2,3-butanedione is used, the thylakoids are resuspended in 20 m$M$ boric acid buffer (pH 7.9), 50 m$M$ choline chloride, 1 m$M$ dithiothreitol, and 0.2 mg/ml bovine serum albumin. 2,3-Butanedione is mixed with 100 m$M$ boric acid buffer (pH 7.9) and added to the thylakoids (0.25 mg chlorophyll/ml) to give final concentrations of 1–20 m$M$. After 20 min at 20°, 10 volumes of the medium (at 0°–4°) used to resuspend the thylakoids are added, and the thylakoids are collected by centrifugation and finally resuspended in the same medium. 2,3-Butanedione at 20 m$M$ inhibits photophosphorylation by 75–80%.

Both phenylglyoxal and 2,3-butanedione appear to act as energy transfer inhibitors. Electron flow and proton uptake are not markedly susceptible to these reagents. In some experiments, however, phenylglyoxal inhibited electron flow through photosystem II. A pronounced inhibition of either the trypsin- or heat-activated $Ca^{2+}$-ATPase of $CF_1$ by phenyl-

[23] Vol. 24 [9].

glyoxal is observed. The $Ca^{2+}$-ATPase of isolated $CF_1$ is inhibited by 2,3-butanedione.[21] The incubation of trypsin-treated $CF_1$[16] with 50 m$M$ 2,3-butanedione for 2 hr at 25° inhibited the ATPase activity of the enzyme by over 80%. The incubations are carried out with 0.2 mg/ml of $CF_1$ in 50 m$M$ boric acid buffer (pH 7.8). Solutions of 2,3-butanedione are freshly prepared in 50 m$M$ boric acid buffer (pH 7.8). NaOH is added to readjust the pH to 7.8 prior to adding the solutions to the enzyme.

The number of arginine residues which react with the modifiers and the sites of the residue(s) with the subunit structure of $CF_1$ remain to be established. It has been tentatively concluded,[21] that modification of only one arginyl residue perhaps at the active site suffices to inhibit the ATPase activity of soluble $CF_1$, but this remains to be proved.

# Section VI
# Nitrogen Fixation

# [67] Techniques for Measurement of Hydrogen Evolution by Nodules

## By F. Joe Hanus, Kevin R. Carter, and Harold J. Evans

Evolution of $H_2$ from legume nodules was first reported by Hoch et al.[1] More recently it was established that $H_2$ was produced from the nitrogenase system. The loss of $H_2$ from nodules has been suggested to be an unnecessary expenditure of cellular energy (in the forms of ATP and reductant) that may result in decreased efficiency of $N_2$ fixation. Hydrogen evolution by nodules of many leguminous plants amounts to an average of about 30% of the total electron flow through the nitrogenase complex. The problems of food production and energy conservation throughout the world have stimulated research activities that hopefully will lead to increases in the efficiency of $N_2$ fixation by legumes. Interest in hydrogen metabolism in $N_2$-fixing organisms therefore has increased.

The evolution of $H_2$ by nodules has been measured by mass spectrometry, gas chromatography, and amperometry. All of these methods, except the amperometric technique, require sampling of the gas mixture over nodules at intervals and subsequent analyses. The amperometric technique allows continuous measurement of $H_2$ evolution by samples of nodules or bacteroids isolated from nodules.

## Mass Spectrometry

Hoch et al.[1] employed mass spectrometry to measure $H_2$ evolution by nodules. The gas mixture above them was sampled and analyzed for masses 2 and 4. In these experiments helium was used as an internal standard. The use of a ratio mass spectrometer for this purpose has been described.[2]

## Gas Chromatography

A gas chromatograph equipped with a thermal conductivity detector has been used for the quantitation of $H_2$ evolved from legume nodules by Dart and Day.[3] They employed a column 6.5 mm diameter by 2 m length of

---

[1] G. E. Hoch, H. N. Little, and R. H. Burris, *Nature (London)* **179**, 430 (1957).
[2] R. H. Burris, Vol. 24, p. 415.
[3] P. J. Dart and J. M. Day, *Plant Soil, Spec. Vol.* p. 167 (1971).

METHODS IN ENZYMOLOGY, VOL. 69

silica gel at room temperature with $N_2$ as the carrier gas. Schubert and Evans[4] used two columns 3.2 mm diameter by 0.84 m of Porapak Q in series with 3.2 mm diameter by 3 m Molecular Sieve 5A in series, with a switching arrangement to facilitate detection of $CO_2$ (which is adsorbed by molecular sieves) as well as $H_2$, $O_2$, and $N_2$. Argon was used as a carrier gas at a very low flow rate and a temperature of 75° in a gas chromatograph equipped with thermistors. The long column and low flow rate of the carrier gas were necessary to separate the $H_2$ peak from the pressure transient peak caused by injection of a relatively large volume of gas on the column. A 6.4-mm diameter by 1.8-m column of Molecular Sieve 5A (40–60 mesh), a flow rate of 30 ml/min and a column temperature of 100° is being used at present. The large column allows the pressure transient peak to appear and dissipate rapidly so that resolution of the $H_2$ peak is possible at a flow rate and temperature that allows good resolution of the $H_2$ peak in less than 1 min. Temperature is not critical to the separation of $H_2$, $O_2$, and $N_2$ on the column of Molecular Sieve 5A as described. Satisfactory separation has been achieved at temperatures ranging from 35° to 150°. At lower temperatures retention times are increased, and broadening of peaks occurs.

Thermistor and filament thermal conductivity detectors are available. The filament detector must be used at reduced filament currents which decreases sensitivity. At high currents, filament oxidation occurs as a result of $O_2$ in the gas samples that are normally analyzed for $H_2$.

Argon, with a thermal conductivity of one-tenth that of $H_2$, is the best readily available carrier gas for use in the detection of $H_2$ with a thermal conductivity detector. Nitrogen, which has a thermal conductivity of one-seventh that of $H_2$, is less expensive than Ar and may be used as a carrier gas with only a slight loss of sensitivity. Since $N_2$ may be present in samples in large quantities and has a longer retention time than either $H_2$ or $O_2$, the use of $N_2$ as the carrier gas eliminates any $N_2$ peaks from the sample thus saving considerable time. When $N_2$ is used as the carrier, 5 nmoles of $H_2$ may be measured reliably with most thermal conductivity detectors. As little as 0.5 nmole of $H_2$ may be detected and estimated.

Gas samples containing $H_2$ may be injected into the gas chromatograph by a gas sampling loop and valve or injected directly into the column by a syringe. The sampling valve has the advantage of greater reproducibility than injection on the column, but requires a large quantity of gas (three to five times the loop volume) in order to completely fill the loop. Experience with direct injection on the column with a syringe has shown that an accuracy of ±10% is easily attained and that ±5% may be achieved with reasonable care. Standard disposable plastic tuberculin

[4] K. R. Schubert and H. J. Evans, *Proc. Natl. Acad. Sci. U.S.A.* **73**, 1207 (1976).

syringes fitted with disposable needles (sizes 25 or 27) have the advantage of low cost and reasonable accuracy when large numbers of samples must be analyzed. Glass syringes are recommended for more precise work. Disposable syringes also may be used for storage of samples prior to analyses. For this purpose, needles are inserted into large rubber stoppers. With plastic syringes the $H_2$ leakage rate for 1 atm of a gas containing $1 \times 10^{-3}$ atm of $H_2$ is approximately 2.5%/hr. With either of the injection techniques the magnitude of the pressure transient increases as the size of the gas sample is increased. A 0.5 ml sample in a 1 ml tuberculin syringe provides a good compromise between accuracy and a reasonable pressure transient. Figure 1 illustrates a chromatogram obtained by use of the technique described above.

The calibration of the thermal conductivity detector output for $H_2$ is conveniently accomplished by diluting appropriate amounts of pure $H_2$ gas with air in bottles or flasks of known volume. Linear detector response should be established in an appropriate range of concentration of $H_2$. For every set of samples analyzed, at least two dilutions of $H_2$ in the range of the concentration of the sample should be analyzed to provide a calibration standard.

Nodule samples are usually taken with a small piece of root attached, to avoid partial nitrogenase inactivation that results from nodule detachment. Serum bottles (21 or 69 ml volume) with rubber serum stoppers are used for nodule samples. Sufficient quantities of nodules should be placed in the bottles to produce adequate $H_2$ for accurate measurement in a reasonable time period. If excess quantities of nodules are placed in the bottles, the $O_2$ concentration may be significantly decreased by respiratory activity during the sampling period.

## Amperometry

The Clark probe for measurement of $O_2$ has long been a standard means of determining partial pressure of $O_2$ in a gas or liquid. The probe consists of a platinum cathode and a silver wire anode immersed in a 50% KCl solution. The electrodes and the KCl are isolated from the external environment by a Teflon membrane that is impermeable to water and ions but permeable to $O_2$. The membrane is held in place by a rubber O ring. For $O_2$ measurement the Pt electrode is polarized at $-0.8$ V with respect to the Ag/AgCl electrode. The reaction at the Pt electrode is

$$O_2 + 4\,e^- + 2\,H_2O \rightarrow 4\,OH^-$$

and at the Ag/AgCl electrode is

$$4\,Ag + 4\,Cl^- \rightarrow 4\,AgCl + 4\,e^-$$

TIME (min)

FIG. 1. Gas chromatograph of a 0.5 ml gas sample containing 0.1% $H_2$, 0.1% $O_2$, and 99.8% $N_2$ (21 nmoles $H_2$, 21 nmoles $O_2$). The sample was analyzed on a Hewlett-Packard Model 5836A gas chromatograph equipped with a 6.4 mm × 1.8 m column of Molecular Sieve 5A (40–60 mesh with fine particles removed using a 60-mesh screen). Conditions of analysis were: column oven temperature, 100°; carrier gas ($N_2$) flow rate, 30 ml/min; attenuation, 8. Retention times were 0.88 min and 1.26 min for $H_2$ and $O_2$ respectively.

The diffusion rate of $O_2$ across the membrane determines the flow rate of electrons (Fig. 2).

The Clark probe (Yellow Springs Instruments, Yellow Springs, Ohio) after a period of preconditioning may also be used to measure $H_2$ either in the gas phase or in solution.[5] This procedure has been used by Jones and Bishop[6] for study of hydrogen exchange in *Anabaena*. With generous help by Professor Jones a procedure for measurement of hydrogen metabolism in legume nodules was initiated by Dr. Karel Schubert in our laboratory.[4]

[5] R. Wang, F. P. Healy, and J. Myers, *Plant Physiol.* **48,** 108 (1971).
[6] L. W. Jones and N. I. Bishop, *Plant Physiol.* **57,** 659 (1976).

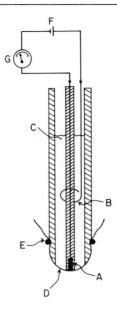

FIG. 2. Schematic representation of a Clark $O_2$ probe. The Pt electrode (A) and the Ag/AgCl electrode (B) are immersed in a 50% KCl solution (C) and isolated from the external environment by a Teflon membrane (D) which is held in place by a rubber O ring (E). A polarizing voltage (F) is applied, and the current flowing through the ammeter (G) is proportional to the partial pressure of $O_2$.

For $H_2$ measurement the polarity of the voltage is reversed so that the Pt electrode is polarized at 0.60 V versus the Ag/AgCl electrode. We have assumed the reaction to be

$$H_2 \xrightarrow{\text{Pt}} 2\,H^+ + 2\,e^-$$

and

$$2\,AgCl + 2\,e^- \xrightarrow{\text{Ag}} 2\,Ag + Cl^-$$

The rate of electron flow is limited by the diffusion rate of $H_2$ through the membrane which is a function of the partial pressure of $H_2$ outside the membrane. A nanoammeter measures the flow of electrons resulting from the oxidation of the $H_2$ and the reduction of the $Ag^+$. The lower limit of reliable detection is on the order of $5 \times 10^{-4}$ atm partial pressure of $H_2$ in a gas mixture.

In order to measure $H_2$ evolution in legume nodules it is necessary to have, in addition to the $H_2$ probe, a means of providing the $+0.6$ V polarizing voltage and a device to measure electrical current in the range of 5 to 100 nA. The probe must be mounted in a gas-tight compartment into which the nodules can be introduced. Also it is desirable, but not essential, to measure $O_2$ consumption by the nodules. In this case a second $O_2$ probe, polarizing voltage supply, and nanoammeter are required, but the nanoammeter for this purpose does not need to measure currents below 100 nA.

Figure 3 is a schematic diagram of an inexpensive device that is similar to the one used in our laboratory to provide the polarizing potential. In addition, a separate adjustment ($R_2$) provides for zeroing or bucking current which produces a current equal in magnitude and opposite in sign to the residual current flowing through the probe when no $H_2$ is present. The switch ($SW_1$) reverses the polarity of the voltage for preconditioning the electrode. No "on/off" switch is needed, since current drain on the batteries is quite low. The probe should be stored with the polarizing voltage "on," otherwise the probe must be resensitized for $H_2$ measurement before use. A potentiometer, $R_1$, is used to adjust the value of the polarizing voltage.

This circuit can also be used to polarize an $O_2$ probe by reversing the polarity and adjusting the voltage to $-0.8$ V. It is desirable to add an "on/off" switch since in the measurement of $O_2$ no sensitization is required and continuous operation in the presence of $O_2$ results in the deposition of excess AgCl on the Ag/AgCl electrode.

In this laboratory we have used Keithley 602 solid state electrometers (Keithley Instruments, Cleveland, Ohio) as current measuring devices and as amplifiers to drive dual pen strip chart recorders. It is possible however to construct inexpensive "op-amp" circuits suitable for driving sensitive input chart recorders.[7] The tracing of $H_2$ evolution from soybean nodules (Fig. 4) was performed using a meter and amplifier constructed for under $30 (and a 5 mV input recorder). A minimal system should be capable of responding to 5–100 nA in two or more sensitivity ranges.

The cuvettes that we have used to hold the electrodes are modifications of the one described by Wang, Healy, and Myers.[5] A glass cuvette constructed to allow insertion of the two electrodes and also to provide for circulation of water from a thermostatically controlled source is desirable. A cuvette constructed of glass has minimum problems of gas adsorption. A cuvette designed by Dr. Karel Schubert was machined from Lucite and has been used in our laboratory, but some problems of $H_2$ adsorption or absorption have been encountered during measurement of $H_2$ in aqueous solutions.

[7] National Semiconductor Corp., "Linear Data Book," p. 213. Natl. Semicond. Corp., 1976.

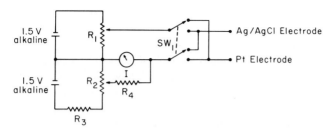

FIG. 3. Schematic diagram of polarizing voltage supply. $R_1$ 10 k$\Omega$ potentiometer to adjust polarizing potential; $R_2$, 1 k$\Omega$ potentiometer to adjust bucking current; $R_3$, 9.1 k$\Omega$ resistor; $R_4$, 510 k$\Omega$ resistor; $SW_1$, DPDT toggle switch for polarity reversal; and I, nanoammeter.

The probe as it comes from the manufacturer is not responsive to low concentrations of $H_2$ and must be sensitized. If the potential is removed from the probe for extended periods of time or if the probe is cleaned with $NH_4OH$ as is suggested for probes used to measure $O_2$, then resensitization will be necessary.

When the probe is new or after cleaning with $NH_4OH$ there is little AgCl on the Ag electrode. Since measurement of $H_2$ requires reduction of $Ag^+$ to Ag it is necessary to operate the probe as an $O_2$ probe for 24 hr to deposit AgCl and the Ag wire. The polarizing voltage is then set to the value for $H_2$ measurement and pure $H_2$ gas introduced by syringe into the chamber. At this time, the probe is only marginally responsive to $H_2$. Each succeeding flushing and $H_2$ treatment is associated with an increase in sensitivity. Although the period of $H_2$ treatment is not critical, periods

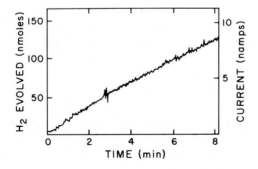

FIG. 4. $H_2$ evolution from soybean nodules inoculated with *Rhizobium japonicum* USDA 16. The measurements were performed using the hydrogen probe as described in the text. Evolution was measured under $N_2$-fixing conditions. The 2.8-ml chamber contained 212 mg of nodules from 32-day-old plants.

of 0.5–1 min seem effective. This procedure should be continued until a current of approximately 5 $\mu$A is obtained with pure $H_2$. This may require several hours.

Membranes must be replaced periodically when appreciable loss of sensitivity is noted. After membrane replacement, the probe exhibits increased sensitivity, noise, and drift. Often it is desirable to avoid changing the membrane and operate with reduced sensitivity to avoid these problems.

To replace the membrane, remove the old membrane and flush out the probe with distilled $H_2O$. After burnishing the platinum electrode with a Kimwipe, flush the probe with 50% saturated KCl solution to which has been added 0.2% Kodak Photo-Flo to improve wetting. Tightly stretch the new membrane over the tip and using the applicator replace the O ring. Handle the membrane only at the edges and avoid touching the tip of the probe.

The probe is stored with the cuvette chamber filled with water. After storing or after extended use, the sensitivity can be increased by drying the chamber and alternating the polarity of the voltage with a frequency of one reversal per second for about 15 sec. This possibly acts to remove interfering contaminants, such as acetylene or carbon monoxide, from the electrodes. The probe is then treated with 1% $H_2$ and flushed with air several times over a 30 min period.

Prior to an assay the electrode is standardized using a 1% $H_2$:99% argon gas mixture. Because of loss of sensitivity over the course of several hours of use, it is advisable to restandardize every 2–3 hrs.

In precise calculations, the volume of the nodules and stems in the chamber cannot be neglected. In practice, approximately 0.3 gm of nodules and stems are placed in a 2.8-ml chamber. The density of the nodule is approximately 1 $gm/cm^3$ thus they would displace approximately 0.3 ml, reducing the effective volume of the gas phase to 2.5 ml. The $H_2$ measurements also have to be corrected to standard temperature and pressure.

Calculations

The following calculations are based on the results presented in Fig. 4.

$C$ = $H_2$ concentration in standard = 1%
$t$ = reaction temperature = 23°
$P$ = atmospheric pressure = 743 mm $H_2$
$I$ = current produced with 1% $H_2$ = 70 nA
$K$ = calibration constant (nmoles/nA)

$dI/dt$ = slope of current versus time = 0.94 nA/min
$V$ = chamber volume = 2.8 ml
$R$ = rate of $H_2$ evolution per gram fresh weight of nodules
$\rho$ = density of nodules = 1 gm/cm$^3$
$Wn$ = weight of nodules = 0.212 gm
$Ws$ = weight of stem = 0.083 gm

Volume of $H_2$ at standard $T$ and $P$ using 1% $H_2$

$$= CV \frac{P}{760 \text{ mm Hg}} \frac{273°}{273° + t}$$

$$= \frac{0.01 \text{ ml } H_2}{\text{ml gas}} 2.8 \text{ ml} \frac{743 \text{ mm Hg}}{760 \text{ mm Hg}} \frac{273°}{273° + 23°}$$

$$= 0.0252 \text{ ml } H_2$$

25.2 $\mu$l $H_2$ (1 $\mu$mole/22.4 $\mu$l) = 1.13 $\mu$moles $H_2$

$$K = \frac{\mu\text{moles } H_2}{I} = \frac{1.13 \ \mu\text{moles}}{70 \text{ nA}} = \frac{16.1 \text{ nmoles}}{\text{nA}}$$

$$R = \frac{K}{Wn} \frac{dI}{dt} = \frac{16.1 \text{ nmoles}}{\text{nA}} \frac{0.94 \text{ nA}}{\text{min}} \frac{1}{0.212 \text{ gm}} 60 \text{ min/hr}$$

$$= \frac{4.283 \ \mu\text{moles } H_2}{\text{g.hr}}$$

Correction factor for volume occupied by stem and nodules:

$$\left( V - \frac{\{Wn + Ws\}/\rho}{V} \right) R = \text{corrected rate}$$

$$\left( 1 - \frac{\{Wn + W\}}{\rho V} \right) R = \left( 1 - \frac{212 \text{ gm} + 0.083 \text{ gm}}{1 \text{ gm/cm}^3 \times 2.8 \text{ ml}} \right)$$

$$R = 0.895 \times 4.283 \ \mu\text{moles } H_2/\text{gm hr}$$

Acknowledgments

The authors wish to express their gratitude to Mrs. Flora Ivers for typing the manuscript. The research program of Dr. H. J. Evans has been supported by the National Science Foundation (PCM 74-17812-A02), the Rockefeller Foundation (GAAS 7628), and the Oregon Agricultural Experiment Station.

## [68] Cultures of *Azospirillum*

### By S. L. ALBRECHT and Y. OKON

The recent isolation of the $N_2$-fixing bacterium *Spirillum lipoferum* (syn. *Azospirillum*)[1] has initiated numerous investigations of this organism. *Spirillum lipoferum* can fix $N_2$ either as a free-living organism or in association with plant roots. This latter finding has generated interest because of its potential contribution to agriculture. For a comprehensive description of $N_2$-fixing bacteria in the rhizosphere, the reader is referred to a review by Döbereiner.[1a]

*Spirillum lipoferum* is probably a ubiquitous soil microorganism, although it is more frequently found in the rhizosphere of plants. Isolations of this organism have been reported from many parts of the world, and the organism seems to be associated with widely diverse plant species.

These bacteria are short, slightly curved gram negative rods, and are motile by means of a single polar flagellum. They display a respiratory metabolism, but some may posess fermentative capability. Colonies of *S. lipoferum* on nitrogen-free, solid agar are small, hard, dull, usually dry, and have entire edges. When grown on nutrient agar or on a medium with combined nitrogen, the colonies are opaque, glistening, raised, and have lobed edges. Older colonies display a characteristic pink pigment on most media.

Isolates of *S. lipoferum* can be separated into two major groups. Group I organisms are generally catalase positive, use sugars poorly if at all, have no apparent vitamin requirement, and generally have the ability to dissimilate nitrate. Group II organisms are usually catalase negative, will grow and fix $N_2$ on sugars forming acid from glucose and ribose, and they have an absolute requirement for biotin. A description of both colonial and cellular morphology, as well as an extensive list of their biochemical characteristics has been compiled by Krieg[2] and Döbereiner and Day.[3]

---

[1] The name *Azospirillum* has been validly published [J. J. Tarrand, N. R. Krieg, and J. Döbereiner, *Can. J. Bacteriol.* **24**, 967 (1978)] and is in common usage at present.

[1a] J. Döbereiner, in "The Biology of Nitrogen Fixation" (A. Quispel, ed.), p. 86. North-Holland Publ., Amsterdam, 1974.

[2] N. R. Krieg, in "Genetic Engineering for Nitrogen Fixation" (A. Hollander, ed.), p. 463. Plenum, New York, 1977.

[3] J. Döbereiner and J. M. Day, in "Proceedings of the First International Symposium of Nitrogen Fixation" (W. E. Newton and C. J. Nyman, eds.), p. 518. Washington State Univ. Press, Pullman, 1974.

Studies of DNA homology[2] suggest that *S. lipoferum* does not properly belong in the genus *Spirillum,* but the organism continues to be used by this name according to the nomenclature and description in ``Bergey's Manual.''[4] Taxonomists are now proposing to call the group I organisms *Azospirillum brasilense* and the group II organisms *Azospirillum lipoferum.*

An extensive collection of *S. lipoferum* strains is maintained by Dr. J. Döbereiner. The origin and a brief description of many of these strains has been published.[5] Döbereiner's strain Sp 7 has been deposited with the American Type Culture Collection as number 29145. The following procedures outline techniques for isolating, culturing and maintaining *S. lipoferum.*

## Cultures

Table I illustrates various applications for *S. lipoferum* cultures and the growth conditions recommended for each specific case. Cells to be use for inoculum in the laboratory or field can be grown under aerobic conditions in $NH_4Cl$-supplemented media or under nitrogen-limiting conditions, respectively. These convenient conditions give the fastest growth rate. Cultures of this type are also useful for producing cells for cytochrome investigations, enzyme extractions or physiological studies. Growth under microaerophilic conditions on a N-free semisolid agar is suggested for the isolation of the organism and for studying the substrates and growth conditions for $N_2$ fixation. This method also may be used to prepare small quantities of $N_2$-fixing cells. The cells can be centrifuged down to a pellet and the semisolid agar easily decanted. It is possible to provide large quantities of $N_2$-fixing cells for studies of cell-free nitrogenase when the organism is grown in liquid culture, with $N_2$ as the sole nitrogen source and at a constant optimal $pO_2$. If a 10% inoculum grown with $NH_4Cl$ under aerobic conditions is added to the liquid medium the bacteria grow very rapidly until they exhaust the small amount of ammonia that was transferred with the ammonia-grown inoculum. There is a very short lag period as the organism adjusts to the shift from the use of combined nitrogen to the use of $N_2$. After this brief transition period, the culture assumes a new and slower rate of growth characteristic of its growth on $N_2$. It is possible to harvest 1.5–2.0 gm of cell paste per liter after 14–16 hr of growth from this type of culture.

---

[4] R. E. Buchanan and N. G. Gibbons, eds., ``Bergey's Manual of Determinative Bacteriology,'' 8th ed. Williams & Wilkins, Baltimore, Maryland, 1974.
[5] C. A. Neyra, J. Döbereiner, R. Lalande, and R. Knowles, *Can. J. Microbiol.* **23,** 300 (1977).

TABLE I

METHODS FOR GROWING *Spirillum lipoferum* FOR SPECIFIC PURPOSES[a]

| Media | Oxygen | Nitrogen source | Carbon source | Replication time | Late exponential phase | | | | Suggested use |
|-------|--------|-----------------|---------------|------------------|------|------|------|------|----------------|
| | | | | | pH | $A_{560}$ | Approximate No. of cell/ml | Yield of cells (gm/liter) | |
| Solid 2% agar | Aerobic (0.2 atm $O_2$) | $NH_4Cl$ | Succinate[b] malate | Colonies formed in 36–48 hr | — | — | — | — | Stock cultures, streaking for purity |
| Solid 2% agar | Aerobic (0.2 atm $O_2$) | Nutrient agar | Nutrient agar | Colonies formed in 24–48 hr | — | — | — | — | Streaking for purity and stock cultures |
| Solid 2% agar | Aerobic[c] (0.2 atm $O_2$) | $N_2$ | Malate succinate | Colonies formed in 48–72 hr | — | — | — | — | Genetic studies, isolation studies |
| Semisolid 0.5% agar | Microaerophilic[d] | $N_2$ | Malate succinate | 20 hr | 7.8 | 0.80 | $6 \times 10^8$ | 1.5 | Cells for nitrogenase isolation, factors affecting growth, and $N_2$ fixation |

| Liquid | Microaerophilic (0.003–0.008 atm O$_2$) | N$_2$ | Malate succinate | 5.5–7 hr | 7.8 | 0.85 | $6.5 \times 10^8$ | 1.5–2.0 | Cells for nitrogenase, batch or continuous culture |
|---|---|---|---|---|---|---|---|---|---|
| Liquid | Aerobic (0.2 atm O$_2$) | NH$_4$Cl | Malate succinate | 1–2 hr | 7.8 | 1.2 | $1 \times 10^9$ | 2.5–3.0 | Inoculum for field studies and cells for biochemical studies |
| Liquid | Aerobic (0.2 atm O$_2$) | Nutrient broth trypticase-soy broth | Nutrient broth trypticase-soy broth | 1–1.5 hr | 7.2 | 1.3 | $1.5 \times 10^9$ | 3.0 | Inoculum, cells for biochemical studies, plant inoculum |

[a] Adapted from a table in an article by Okon, Y., S. L. Albrecht, and R. H. Burris. *Appl. Environ. Microbiol.* **33**, No. 1, 85 (1977).
[b] Or any applicable carbon source.
[c] Conditions within colony may be microaerophillic.
[d] $pO_2$ variable through a gradient in the agar.

*Spirillum lipoferum* can be easily grown in large volume batch cultures. We have grown cultures as small as 3 ml to ones as large as 180 liters. For the large volume cultures it is recommended that at least a 10% by volume inoculum growth in the same medium be used.

*Isolation of Pure Cultures.* The following method for the isolation of *S. lipoferum* is based on a report by Döbereiner and Day.[3] They recognized the poor $O_2$ protection mechanism afforded to the $N_2$-fixing enzyme complex by this organism and devised an effective method for isolating the bacteria. They used enrichment cultures designed to favor the growth and development of *S. lipoferum* by creating a suitable environment that allows the bacteria to find the optimum $O_2$ concentration for $N_2$ fixation. This is accomplished by using a semisolid, N-free medium containing malate and mineral salts. The $N_2$-fixing *S. lipoferum* proliferate under conditions of slow diffusion of $O_2$ and carbon substrate.

Root pieces (3–5 mm) or a soil sample (0.1–0.5 gm) are placed in 10 ml serum bottles containing 3 ml of semisolid (3–5 gm agar/liter) N-free NFb medium (see below) and incubated at 30–35° for 1–4 days. The *Spirillum*, being motile, will migrate toward and proliferate at the region of optimum $O_2$ concentration. As the culture grows, the bacteria, which consist mainly of *Spirillum*, will form a dense white undulating layer or pellicle. This layer is characteristic of this organism's growth in semisolid media. The pellicle, some 2–4 mm under the surface of the agar, will form in 48–96 hr, depending on the quantity of *Spirillum* in the initial inoculum and the incubation temperature. A sterile 0.5% (w/v) alcoholic solution of bromthymol blue dye may be added to the medium. If the dye has been included the culture will then turn from a blue-green color to a brilliant blue as the bacteria grow. The color change, which is indicative of increasing pH, will aid in locating the developing white pellicle which will contrast sharply with the blue color of the medium.

Once the pellicle has developed, the culture may be assayed for nitrogenase activity by sealing the bottle with a serum stopper and injecting 1 ml of acetylene. Cultures with a pellicle invariably will reduce acetylene to ethylene, indicative of nitrogenase activity, within 60 min incubation, provided the pellicle is undisturbed. Ethylene formation can be measured by gas chromatography as described earlier in this series.[6]

To complete this procedure, isolates may be streaked for purity after two or three serial transfers in the semisolid N-free agar. It was found that malate agar plates with 1 gm/liter $NH_4Cl$ or N-free malate agar plates that had been supplemented with 20 mg/liter of yeast extract (Difco) worked well at this stage. Some strains may develop poorly (noticeably

[6] R. H. Burris, Vol. 24, p. 415.

group II) on the NH₄Cl-malate plates; however, addition of 20 mg of yeast extract per liter will ensure their growth. Colonies may develop greenish-blue centers if the bromthymol blue dye is included in the plates. For group I isolates, the pink pigment will develop within 7–9 days; group II pigment formation is somewhat slower. Colonies tentatively identified as *S. lipoferum* are then transferred to nutrient agar or potato agar plates. The potato agar is produced by boiling 200 gm of potatoes in 1 liter of distilled water, removal of the residual potatoes by filtration, and the subsequent addition of 0.25% (w/v) sodium malate, 0.25% (w/v) sucrose, and 2% (w/v) agar. Colonies on nutrient or potato agar develop rapidly. Once the culture is shown to be axenic, the requisite biochemical and physiological tests can be performed.

*Stock Cultures.* Stock cultures are maintained at room temperature on nutrient agar slants in screw-capped tubes; transfer to fresh slants is recommended every 6–9 months. Cultures of *S. lipoferum* have been stored in the cold room or refrigerator (4°) for as long as 12 months and remained viable. The semisolid agar cultures mentioned above may be stored for a short while under sterile mineral oil. There are reports that this organism can be successfully preserved in a lyophilized state or by storage in liquid $N_2$. All stock cultures are tested for purity before use by streaking on nutrient agar and by microscopic examination of a wet mount.

## Growth of Bacteria

*Spirillum lipoferum* displays considerable morphological and metabolic variability depending on the growth media and growth conditions; thus, selection of both is important. The strain of *S. lipoferum* and the ultimate use of the culture will usually dictate the choice of a medium.

*Growth Media. Spirillum lipoferum* is capable of growth on several media. We generally recommend a modification of the Döbereiner and Day (NFb) medium,[3] which provides increased buffering capacity, for most routine work. The basic medium contains the following in grams per liter: $K_2HPO_4$, 6.0; $KH_2PO_4$, 4.0; $MgSO_4 \cdot 7H_2O$, 0.2; NaCl, 0.1; $CaCl_2$, 0.02; $FeCl_3$, 0.01; $NaMoO_4 \cdot 2H_2O$, 0.002. A source of reduced carbon, usually sodium malate or sodium succinate, is added to give a final concentration of 5 gm/liter; NH₄Cl at a concentration of 1.0 gm/liter is generally the choice for a nitrogen source when non-$N_2$-fixing cultures are required. Distilled water should be used in preparing the medium, as tap water may cause serious precipitation problems during medium sterilization. Our experience has shown that doubly glass distilled or deionized water when used with ultra pure chemicals may lead to a microelement insufficiency problem and cultures will fail to grow; addition of trace element solutions or a small quantity of yeast extract (20 mg/liter) will usually eliminate this difficulty. The microelements can be added as the following compounds (mg/liter): $MnSO_4$, 2.1; $H_3BO_3$, 2.8; $Cu(NO_3)_2 \cdot 3H_2O$, 0.04; and $ZnSO_4 \cdot 7H_2O$, 0.24. *Spirillum lipoferum* may also be grown conveniently in nutrient or trypticase soy broth (Difco). It is of interest that all strains tested are resistant to the antibiotic trimethoprim, (A. G. Wood, personal communication). This feature may be useful in their isolation or for use as an antibiotic resistance marker.

*Preparation of Media.* Reagent grade chemicals are used for all these media. Sterilization by autoclaving, except for certain carbon and nitrogen compounds which must be filter sterilized, is performed at 123° (17 psi) for 20 min. All liquid media may be converted to solid or semisolid media by the addition of the requisite amount of agar; 0.3-5.0 gm/liter for semisolid and 20 gm/liter for solid media. In some instances problems of precipitation may occur during autoclaving, these are usually eliminated by autoclaving the phosphates separately and then adding them aseptically after the media have cooled. We have found in certain instances, e.g., growth of large batch cultures, that it is expedient to make up large amounts of the mineral salts mixture ahead of time. All the chemicals are mixed in the dry state, and the mixture is ground in a mortar and pestle to ensure uniform distribution of the salts. This mixture can be stored easily for long periods.

*Inocula.* Inoculum from slants is prepared by adding 10 ml of sterile 50 m$M$ sodium or potassium phosphate buffer, pH 6.8, to a 48-hr-old nutrient agar slant of the culture. However, these bacteria grow so well that cultures may be started by picking a colony from an agar plate or taking a loopful of cells from a slant and transferring it directly to the liquid culture. Such techniques will usually provide a culture of $A_{560}$ = 0.8-1.2 in 16 to 18 hr providing there is a combined nitrogen source in the medium. Growth is generally measured by following changes in absorbance of the culture with time at 560 nm, and linear relationships among absorbance, total nitrogen, and cell numbers have been reported.[7] These cultures can then be used as an inoculum for subsequent cultures or the following types of experiments.

The inoculum for field work can be grown at 30°-36° with vigorous aeration and limiting amounts of combined nitrogen to a density of approximately $7 \times 10^8$ bacteria/ml ($A_{560}$ = 0.9). Growth is usually initiated in this type of culture by the addition of a 2-5% inoculum of a fresh bacterial suspension.

[7] Y. Okon, S. L. Albrecht, and R. H. Burris, *J. Bacteriol.* **127**, 1248 (1976).

Cells intended for physiological or biochemical work are routinely grown in the mineral salts medium containing malate or succinate as the reduced carbon source, nutrient broth, or trypticase soy broth. Growth of small (100–500 ml) $N_2$-fixing cultures for inoculum or other work is conveniently carried out in 1 liter Roux bottles containing 150 ml of the semisolid N-free modified NFb medium.

*Harvesting of Cultures.* The cells can be harvested, usually in log phase, by centrifugation in the cold (4°). Centrifugation for 10 min at 10,000 $g$ is sufficient to pellet the bacteria. The cells can be resuspended and washed several times without damage. However, when harvesting cells actively fixing $N_2$, all operations should be performed anaerobically to avoid damage to the $N_2$-fixing enzyme system. While the choice of the resuspending fluid will depend on the ensuing work, many dispersants have proved satisfactory. They include distilled water, NaCl (0.15–0.75 $M$) solutions, and phosphate buffers (pH 6.8). While yields may vary depending on several factors, it is not uncommon routinely to obtain yields of 2–3 gm of cell paste per liter of culture. Nitrogen fixing cultures will normally yield less than cultures grown on a combined nitrogen source.

### Cell Growth Conditions

Cultural conditions that markedly influence growth patterns and morphology of *S. lipoferum* are considered in the following paragraphs.

*Nitrogen. Spirillum lipoferum* will utilize a wide range of nitrogen sources including $N_2$, $NO_3^-$, yeast extract, peptone, casamino acids, and $NH_4Cl$. All strains tested so far will reduce $NO_3^-$. With ammonium as its N source in liquid culture, *S. lipoferum* shows a generation time of approximately 60 min. It can utilize $N_2$ as a sole N source only if microaerophilic growth conditions are provided and maintained. Addition of a small amount of yeast extract (0.005%) will shorten the lag phase when the organism shifts from combined nitrogen to $N_2$-dependent growth, but will not repress the nitrogenase complex or affect the growth rate. Nitrogenase synthesis and activity is repressed in cultures by the inclusion of significant amounts of combined nitrogen (i.e., 0.25% or more $NH_4Cl$).

There is a marked difference in metabolism depending on whether these organisms are grown on $N_2$ or $NH_4Cl$. When *S. lipoferum* is grown on $N_2$ it will produce high levels of poly-$\beta$-hydroxybutyrate.[8] The $N_2$-grown cells may contain as much as 25% of their dry weight as the

[8] Y. Okon, S. L. Albrecht, and R. H. Burris, *J. Bacteriol.* **128**, 592 (1976).

polymer; ammonia-grown cells may have less than 1%. Cellular morphology mirrors this metabolic difference, as ammonia grown cells tend to be longer, have a smaller diameter, and appear more helical that the $N_2$-grown cells. In addition, the lipid bodies, which may deform the $N_2$-fixing cells, are greatly reduced in number and size when *S. lipoferum* is grown on combined nitrogen.

*Oxygen.* In common with other $N_2$-fixing organisms, the growth rate of *S. lipoferum* is limited by its ability to fix $N_2$; this capacity is markedly influenced by the amount of $O_2$ in its environment. Unlike those $N_2$-fixing species that have evolved various mechanisms that permit nitrogenase activity in the presence of oxygen, *S. lipoferum* possesses no such mechanism; rather it merely ceases to grow on $N_2$ at oxygen pressures any higher than that necessary to accommodate its oxidative phosphorylation activity. Nitrogen-fixation studies[9,10] with oxygenstats have indicated the optimal $O_2$ range for growth is narrow; 0.003 to 0.008 atm. This is a low concentration considering that the organism grows vigorously in air when supplied with $NH_4Cl$. Nitrogen-dependent growth is poor or nonexistent under either limiting or excess $O_2$ pressures. Under these conditions the growth rate will decline and stored carbon will increase.[11] At optimal $pO_2$, batch and chemostat cultures show a growth rate with a doubling time of 5.5–7.0 hr.

*pH and Temperature.* The organism grows best at a pH near neutrality.[6,11] Cultures are commonly started at a pH near 7.0, and they will grow well until a pH of about 7.8 is reached. Above pH 7.8, growth and $N_2$ fixation decline. The range for optimum growth is narrow, 6.8–7.8, and there is little growth below 5.5 or above 8.7. With organic acids as energy sources the pH will rise rapidly. This problem can be reduced by the addition of up to 10 gm/liter phosphate salts to the medium to increase its buffering capacity. Phosphate concentrations in excess of 10 gm/liter were found to be inhibitory. The strains that can grow on sugars also produce acid as they grow. Organic buffers are not recommended for control of pH because of their tendency to interfere with subsequent metabolic studies; however, they will mediate pH effects.

Although the temperature range for growth of *S. lipoferum* ranges from 4° to 45°, fastest growth is observed between 32° and 35°. For most purposes 40° may be considered as a safe maximum temperature. Growth

[9] Y. Okon, J. P. Houchins, S. L. Albrecht, and R. H. Burris, *J. Gen. Microbiol.* **98,** 87 (1977).

[10] M. A. T. Vargas and R. F. Harris, *in* "Limitations and Potentials for Biological Nitrogen Fixation In the Tropics" (J. Döbereiner and A. Hollaender, eds.), p. 373. Plenum, New York, 1978.

[11] J. M. Day and J. Döbereiner, *Soil Biol. & Biochem.* **8,** 45 (1976).

at room temperature is reasonably good. Stagnant liquid cultures will exhibit a positive thermotactic response.

The temperature optimum for nitrogenase activity appears to be between 33° and 40°.[11] Temperatures slightly above 40° provide an initial short-lived increase in nitrogenase activity; as time progresses, however, this high temperature effect becomes inhibitory, and at 45° the temperature is markedly detrimental. Below 33° nitrogenase activity is increasingly depressed as the temperature approaches 10° where the enzyme activity effectively ceases. One-half maximal activity has been established at 18°.

*Carbon Substrates.* Several effective carbon substrates are available for the cultivation of *S. lipoferum*. The organism is commonly grown on organic acids, as many strains (group I) do not utilize sugars well. They grow vigorously on malate, succinate, lactate, and pyruvate.[8,12] While pyruvate may not be convenient to use, the other three are inexpensive, readily available, and stable to autoclaving. Acetate will support moderate growth, but the organism grows poorly with citrate, α-ketoglutarate, oxaloacetate, fumarate, or isocitrate as carbon sources.

Group I isolates will not grow well on sugars, with the exception of galactose. All group II isolates appear to use sugars readily. Glucose, arabinose, galactose, lactose, maltose, rhamnose, xylose, mannose, and glycerol have been reported to support growth to varying degrees.[3,6,11] Ethanol also will support slight growth.

### General Remarks

These methods should be considered as only a starting point, as the investigations into the biology of this organism are just beginning, and in many cases the particular experimental procedure may require some alteration of the medium, physical conditions, or other conditions affecting growth. There is probably no single best way to grow all strains of *S. lipoferum*.

Acknowledgments
    Some of the methods described in this paper were developed with the support of research grants GB 21322 and PCM74-17604 from the National Science Foundation and Public Health Service Grant AI-00848 from the National Institute of Allergy and Infectious Disease. We wish to thank Dr. R. H. Burris and Dr. N. E. R. Campbell for their criticism of the manuscript.

[12] Y. Okon, S. L. Albrecht, and R. H. Burris, *Appl. Environ. Microbiol.* **33**, 85 (1977).

## [69] Asymbiotic Fixation of Dinitrogen by *Rhizobium in Vitro*[1]

*By* W. G. W. KURZ and J. J. CHILD

Studies of the nitrogen fixing symbiosis between legumes and *Rhizobium* have long been restricted by the researcher's inability to obtain fixation by rhizobia separated from its legume host.

Recently several laboratories[1a-5] have demonstrated that the apparent interdependence was not as exclusive as was once thought, and many rhizobial strains can now be induced to fix dinitrogen when cultivated on certain defined media. To obtain fixation three conditions have to be met.

1. A source of fixed nitrogen (e.g., glutamine) has to be a component of the medium.

2. A combination of two carbon sources, a carbohydrate (pentose sugar) and a citric acid cycle intermediate, is needed.

3. A low oxygen tension is essential.

We describe here two techniques which have proved useful for the induction of nitrogenase activity in a range of rhizobial strains growing on (a) solid media and (b) in liquid culture media. Application and modification of those experimental systems should enable biochemists, geneticists and physiologists to evaluate the separate contribution of each partner in the symbiosis.

### Culture on Solid Media

Several methods are available to induce nitrogenase activity in cultured rhizobia growing as colonies on agar.[1a-3] Because of the historical development of the discovery of asymbiotic nitrogen fixation by these organisms,[6,7] most recommended media are modifications of media used

[1] Issued as NRCC No.15961.

[1a] J. D. Pagan, J. J. Child, W. R. Scowcroft, and A. H. Gibson, *Nature (London)* **256**, 406 (1975).

[2] W. G. W. Kurz and T. A. LaRue, *Nature (London)* **256**, 407 (1975).

[3] J. A. McComb, J. Elliot, and M. J. Dilworth, *Nature (London)* **256**, 409 (1975).

[4] J. Tjepkema and H. J. Evans, *Biochem.. Biophys. Res. Commun.* **65**, 625 (1975).

[5] D. L. Keister, *J. Bacteriol.* **123**, 1265 (1975).

[6] J. J. Child, *Nature (London)* **253**, 350 (1975).

[7] W. R. Scowcroft and A. H. Gibson, *Nature (London)* **253**, 351 (1975).

for plant cell tissue culture. These are mixtures of many components. The media need not be complex, however, in view of the successful demonstration of nitrogenase activity in submerged culture in a simple bacterial culture medium.[5]

Surprisingly, considering the known oxygen sensitivity of the nitrogenase enzyme system, activity in agar surface cultures of rhizobia occurs under aerobic conditions, and a normal atmospheric oxygen tension of $p = 0.2$ is found to be optimal for nitrogenase activity.[8] This unexpected requirement for oxygen is explained by the need for suitable atmospheric oxygen tension for growth of the colony to create the necessary microaerophilic physiological condition at some point within the colony allowing fixation to occur. Thus, the specific activities of cultures on solid media likely underestimate the true nitrogenase potential of the rhizobia tested.

Rhizobia are maintained on a medium containing (in gm/liter of water) $K_2HPO_4$, 0.5; $MgSO_4 \cdot 7H_2O$, 0.2; NaCl, 0.1; yeast extract, 0.5; $CaCO_3$, 0.5; and mannitol, 10. Rhizobial strains grown on this medium are washed in water and added to the surface of 10 ml test medium agar slants in a 30 ml test tube plugged with Dispo foam plugs (American Hospital Supply Corp., McGraw Park, Illinois). A test medium which has proved suitable for the induction of nitrogenase activity in a number of rhizobial strains[2] contains (in m$M$) $KH_2PO_4$, 2.2; $CaCl_2 \cdot 2H_2O$, 0.7; KCl, 0.9; $MgSO_4 \cdot 7H_2O$, 0.14; glutamine, 2; myoinositol, 5.6; sodium succinate, 25; L-arabinose, 25; and (in $\mu M$) $MnSO_4 \cdot 4H_2O$, 58; $H_3BO_3$, 82; $ZnSO_4 \cdot 7H_2O$, 3.5; KI, 6; $CuSO_4 \cdot 5H_2O$, 0.8; $Na_2MoO_4 \cdot 2H_2O$, 0.4; $CoCl_2 \cdot 6H_2O$, 0.4; $FeSO_4 \cdot 7H_2O$, 54; $Na_2$-EDTA, 54; thiamin-HCl, 15; nicotinic acid, 41; pyridoxine-HCl, 2.4; and 1% (w/v) agar (Difco Noble) at pH 5.9. After 6–14 days incubation at 25°–30° depending upon the growth rate of the strain investigated, the foam plugs on the test tubes are replaced by serum caps. Repeated measurements may be made on the same sample if the foam plug is inserted in the test tube 1 cm below the rim and a serum cap inserted at the neck. Care should be taken to ensure adequate mixing of the gas phase on either side of the foam plug. After analysis the tube can be evacuated and the original atmosphere restored.

Alternatively the bacterial suspension inoculum may be spread as 1-$cm^2$ areas on test medium agar plates and after 6–14 days incubation agar segments containing the undisturbed bacterial colonies are excised and sealed in vials fitted with serum caps. The atmosphere in the vials is either evacuated and replaced with a gas mixture containing 20% $O_2$–

[8] A. H. Gibson, W. R. Scowcroft, J. J. Child, and J. D. Pagan, *Arch. Microbiol.* **108**, 45 (1976).

10% $C_2H_2$–70% Ar or 3–5% $C_2H_2$ is added to the existing atmosphere. The tubes are incubated at 25°–30° for 2–5 hr and assayed for nitrogenase activity by the acetylene reduction assay technique.[9]

Commercial acetylene is frequently contaminated by trace quantities of ethylene, and care should be taken to ensure that only analytical grade acetylene is used or adequate compensation made for background ethylene. Endogenous ethylene production by the culture can also be estimated from control tubes not exposed to acetylene. Nitrogenase-dependent ethylene production is calculated by subtracting both the endogenous and background levels of ethylene.

### Culture in Liquid Media

Studies of the derepression of nitrogenase synthesis in suspension culture are characterized by the relative simplicity of the media and the absolute requirement for a reduced oxygen tension for activity.[10,11] Thus, it is important to adjust the oxygen concentration of the gas phase to the cell density of the liquid phase in order to obtain high specific rates of acetylene reduction. Higher cell densities require more frequent adjustment of the $O_2$. While reports of successful induction of nitrogenase activity in suspension culture are limited to only a few rhizobial strains, modifications of a basic technique will undoubtedly expand the range of rhizobia capable of fixing atmospheric nitrogen in submerged culture.

Acetylene reduction activities in suspension culture are similar to those reported for isolated bacteroid preparations (20 nm $C_2H_4$/hr/mg protein) and generally much higher than those reported for similar strains under cultivation as colonies on solid media. The requirement for a low level of oxygen in suspension culture is obligatory, and oxygen tensions necessary to obtain nitrogenase activity are within the range of 0.06–0.34% $O_2$.

For general purposes we describe here the experimental system developed by Keister and Ranga Rao.[12] Aerobically grown rhizobia on a medium containing (in m$M$) Na-KPO$_4$, 7.4; CaCl$_2$·2H$_2$O, 0.5; MgSO$_4$, 0.8; sodium gluconate, 22.9; mannitol, 27.4; yeast extract, 0.1% casamino acids, 0.01% and (in $\mu M$) Na$_2$MoO$_4$·2H$_2$O, 11; FeSO$_4$·7H$_2$O, 48; and agar

[9] R. W. F. Hardy, R. C. Burns, and R. D. Holsten, *Soil Biol. & Biochem.* **5**, 47 (1973).
[10] F. J. Bergersen, G. L. Turner, A. H. Gibson, and W. F. Dudran, Biochim. Biophys. Acta **444**, 164 (1976).
[11] D. L. Keister and W. R. Evans, *J. Bacteriol.* **127**, 149 (1976).
[12] D. L. Keister and V. Ranga Rao, *Proc. Int. Symp. Nitrogen Fixation, 2nd, 1976* p. 419, (1977). Academic Press.

1.5% (w/v), pH 7.2, are harvested at log phase of growth, washed and resuspended in an induction medium containing sodium gluconate, 10 m$M$; sodium glutamate, 10 m$M$; $KH_2PO_4$, 3 m$M$; $Na$-$KPO_4$, 10–50 m$M$; zwitterionic buffer, 50 m$M$; $MgSO_4$, 0.4 m$M$; Fe-citrate, 0.1 m$M$; and $Na_2MoO_4 \cdot 2H_2O$, 30 $\mu M$, pH 5.5–7.5. The experiments are carried out in 500 to 1000 ml flat bottom reagent bottles containing 10 to 20 ml of medium and are sealed with rubber sleeve stoppers that are tightly wired shut. By repeated evacuation and filling with argon, the oxygen is removed to a concentration less than 0.005%. The bottles are then filled with a gas mixture of $Ar$-$C_2H_2$-$CO_2$ (89:10:1) and oxygen injected to the appropriate concentration. A concentration of 0.13% $O_2$ in the gas phase was found to give the best specific rates of $C_2H_2$ reduction. The cultures are shaken on rotary shakers (170 rpm) at 27°. The oxygen concentration should be monitored at approximately 12 hr intervals and if necessary adjusted to the desired concentration. Both oxygen and ethylene are determined by gas chromatography.[11]

# [70] Isolation and Characterization of Various Nitrogenases[1]

## By ROBERT R. EADY

Nitrogenase, the enzyme responsible for biological nitrogen fixation, is the subject of intensive international research efforts to establish its structure and mechanism of action. This enzyme, which catalyzes the facile reduction of $N_2$ in a protic environment at ambient temperatures and pressures, is only found in prokaryotes and can be separated into two metalloproteins both of which contain Fe, one of which also contains Mo in addition to Fe.

The increasing use of physical techniques, such as electron paramagnetic resonance (EPR), Mössbauer, and stopped-flow spectroscopy, in these studies has necessitated the development of large-scale purification procedures. The conditions of culture, cell storage and breakage, and extraction procedures can be critical in obtaining active extracts, but a

---

[1] Abbreviations: SDS, sodium dodecyl sulfate; EPR, electron paramagnetic resonance: MES, 2-($N$-morpholino)ethanesulfonic acid; DTNB, 5,5-dithiobis(2-nitrobenzoate); Kp, *Klebsiella pneumoniae;* Av, *Azotobacter vinelandii;* Ac, *Azotobacter chroococcum,* Cp, *Clostridium pasteurianum;* Cv, *Chromatium vinosum;* Bp, *Bacillus polymyxa;* Rj, *Rhizobium japonicum;* Kl, *Rhizobium lupini.*

variety of different methods have been successfully developed. Nitrogenase components are usually purified at room temperature and stored frozen in bead form in liquid nitrogen. Procedures for purification of nitrogenase components from *Azotobacter vinelandii* (Av), *Clostridium pasteurianum* (Cp), and *Rhizobium japonicum* bacteroids (Rj) have been described in Vol. XXIV of this series, and more recently the procedures for *Bacillus polymyxa* (Bp) and *Rhodospirillum rubrum* (Rr) in Vol. 53.

Modifications of the methods described originally have been made which have resulted in increased specific activity, and these will be described here together with procedures for the isolation of nitrogenase components from *Klebsiella pneumoniae* (Kp), *Azotobacter chroococcum* (Ac), and the Mo-Fe protein from *Chromatium vinosum* (Cv). Purification of the unresolved nitrogenase complex of *A. vinelandii* was described in Vol. XXIV and will not be reiterated.

### Nomenclature of the Nitrogenase Components

The nomenclature used by different workers for the component proteins of nitrogenase is confusing: molybdoferredoxin azofermo, component 1, and Mo–Fe protein have been used for the molbydenum- and iron-containing protein; azoferredoxin azoferm, component 2, and Fe protein used for the Fe-containing protein. Here the general terms Mo–Fe protein and Fe protein will be used, and for the proteins from specific organisms the nomenclature of Eady *et al.*[1a] will be used. A capital letter is used to denote the genus and a lower case letter the species from which the protein was isolated. The number 1 refers to the Mo–Fe protein, and the number 2 to the Fe protein. The International Commission on Enzyme Nomenclature has recommended the use of the name nitrogenase as a trivial name for nitric-oxide reductase (EC 1.7.99.2). The use of this number for nitrogenase is to be avoided; the assignment is misleading since these enzymes are different proteins, and the class 99.2 is inappropriate for the reaction catalyzed by nitrogenase.

### General Comments on the Purification and Assay of Nitrogenase

A major problem in the purification of nitrogenase components is their extreme sensitivity to oxygen. Short exposure to air results in irreversible inactivation, the half-life of the Fe protein is typically 45 sec, and that of the Mo–Fe protein 10 min even when purified from aerobic organisms. The successful purification of these proteins can only be

[1a] R. R. Eady, B. E. Smith, K. A. Cook, and J. R. Postgate, *Biochem. J.* **128**, 655 (1972).

achieved if air is rigorously excluded. This is usually done by using buffers sparged with $N_2$ or $H_2$ and which contain approximately 1 m$M$ $Na_2S_2O_4$ to scavenge residual $O_2$. Syringe transfer techniques are used to manipulate buffers and nitrogenase components, and column effluents are collected in flasks flushed with an inert gas. The adequacy of the anaerobic technique employed during a purification can be conveniently checked using papers impregnated with a viologen dye. These are prepared by soaking strips of chromatography paper (about 1 × 15 cm) in 0.25 $M$ Tris-HCl buffer, pH 7.4, containing 1% methyl viologen for 5 min, and then drying. A transient blue color when a few drops of buffer are placed on these papers indicates the presence of residual dithionite.

The reduction of the substrate analog $C_2H_2$ is a sensitive and convenient assay for nitrogenase during purification, and several procedures for this were described in Vol. XXIV of this series. Since purified nitrogenase components have no known intrinsic enzymic activity, the assay of either the Fe protein or the Mo–Fe protein requires that the complementary protein be available. Maximum activity of the Mo–Fe protein is obtained at a 10- to 20-fold molar excess of the Fe protein. The Fe protein has maximum activity at an optimum ratio of the two components; if this ratio is exceeded, inhibition of activity occurs. To obtain reliable estimates of the activity, a number of assays containing a constant amount of Fe protein are assayed with increasing amounts of Mo–Fe protein. In either type of assay specific activity is calculated on the basis of mg of the protein limiting the assay as nmole of product per minute per milligram of protein. The purification procedures described below result in preparations of nitrogenase components which are homogeneous by the standard analytical techniques of protein chemistry. However, the variation in specific activity and metal content of apparently homogeneous preparations of the Mo–Fe protein isolated from a specific organism indicate contamination by inactive or partially active protein. In all probability, the full potential of catalytic activity of either nitrogenase component from any source has yet to be realised.

### General Comments on the Measurement of Protein Concentration Metal and Sulfide Content

$S_2O_4{}^{2-}$ interferes with many of the standard analytical methods used in the characterization of Fe–S proteins, and it is not always practicable to remove it by gel filtration. The high absorption coefficient of $S_2O_4{}^{2-}$ below 330 nm prevents the use of conventional monitoring equipment to determine the protein profile of eluates from columns used in purification. Provided that diluted samples are aerated on a Vortex mixer for 30 sec before analysis, both the Folin and biuret methods are reliable for the estimation nitrogenase proteins. Dry weight measurements have shown that no correction factor is necessary if dried bovine serum albumen is used as a standard.[1a,2]

Fe and Mo estimation by atomic absorption spectroscopy is interfered with by $S_2O_4^{2-}$. In addition, Tris-HCl and the Mo–Fe protein interfere with Mo estimation by this method.[3] Colorimetric estimation of Fe with $\alpha,\alpha'$-bipyridyl and Mo with toluene-3,4-dithiol on wet-ashed samples of protein is the preferred method of analysis.

Sulfide estimation by the methylene blue method is strongly inhibited by $S_2O_4^{2-}$ and by unknown products formed by air oxidation.[4] This interference can be overcome by restricting sample volumes to less than 50 $\mu$l for proteins containing 1 m$M$ $S_2O_4^{2-}$ and by increasing the $FeCl_3$ in the assay mixture to 2.3–4.6 $\mu$mole. The use of various volumes of sample to establish the linearity of color development and checking the recovery of sulphide from ferredoxin from *C. pasteurianum* in the presence of the sample under investigation is recommended.[4]

## Purification Procedures for Nitrogenase Components from Various Organisms

### Clostridium pasteurianum

Two methods for the purification of nitrogenase components are in current use, both of which use extracts prepared by autolysis of dried cells.

Mortenson described a method[5] using protamine sulfate precipitation and gel chromatography which should be consulted for a description of the manipulative procedures involved. The modified method[6] described below is more reproducible and gives a higher yield of Cp1 and Cp2 with increased specific activity. Protamine sulfate is added to the crude extract of 180 gm of dry cells in 50 m$M$ Tris-HCl buffer, pH 8.5. Cp1 and Cp2 are precipitated within the concentration range 1–10% (w/w) of protamine sulfate, related to the initial protein concentration. The precipitated proteins are redissolved by the addition of phosphocellulose (Whatman, P 11) at a fivefold excess over protamine sulfate. All subsequent steps use

[2] M. G. Yates and K. Planque, *Eur. J. Biochem.* **60**, 467 (1975).
[3] H. Dalton, J. A. Morris, M. A. Ward, and L. E. Mortenson, *Biochemistry* **10**, 2066 (1971).
[4] J.-S. Chen and L. E. Mortenson, *Anal. Biochem.* **79**, 157 (1977).
[5] L. E. Mortenson, Vol. 24, p. 446.
[6] W. G. Zumft and L. E. Mortenson, *Eur. J. Biochem.* **35**, 401 (1973).

50 m$M$ Tris-HCl buffer, pH 7.5, containing 0.1 $M$ NaCl and 1 m$M$ Na$_2$S$_2$O$_4$. Cp1 and Cp2 are separated by gel filtration in Sephadex G-100. Two 7.5 × 50 cm columns are used at this stage, Cp1 is eluted as the first dark brown band, and Cp2 as a second yellowish brown band.

*Purification of Cp1.* Cp1 from the Sephadex chromatography step is then concentrated by precipitation with protamine sulfate (0.5–5%, w/w), and when redissolved by phosphocellulose treatment chromatographed on two Sephadex G-200 columns (5 × 50 cm). Although 95% homogeneous at this stage, the specific activity of Cp1 is approximately doubled, and a contaminating "demolybdo" species with an EPR signal at $g = 1.94$ is removed by DEAE chromatography. One gram amounts of Cp1 are applied to a DEAE-cellulose (Whatman, DE 52) column (2.5 × 25 cm) and eluted with a linear NaCl gradient 0.15–0.3 $M$ in 400 ml of Tris-HCl buffer. Cp1 is eluted after a yellowish brown band.

*Purification of Cp2.* Cp2 from the first Sephadex G-100 column is concentrated by precipitation with protamine sulfate (1–4%, w/w) and after being redissolved by phosphocellulose treatment rechromatographed on a Sephadex G-100 column (4 × 90 cm). Finally, Cp2 is applied to a DEAE–cellulose column (2.5 × 25 cm) and eluted with a linear gradient of NaCl from 0.2–0.4 $M$ in 500 ml of Tris-HCl buffer.

### C. pasteurianum—Alternative Procedure

An alternative procedure[7] utilizes polyethylene glycol to precipitate Cp1 and Cp2 from crude extracts and DEAE–cellulose chromatography to separate them. One hundred grams of dried cells are lysed by anaerobic incubation at 30° for 1 hr in 20 m$M$ Tris-HCl buffer, pH 7.4 (15 ml/gm of dry cells), and the crude extract prepared by centrifuging at 27,000 $g$ for 30 min. The pH is then adjusted to 6 by the addition of 0.1 $M$ MES buffer and solid polyethylene glycol 6000 (Union Carbide Corp.) added to 10% (w/v). The precipitate is removed by centrifuging at 27,000 $g$ for 20 min and discarded. Cp1 and Cp2 are precipitated together by the addition of more polyethylene glycol to 30% (w/v) and collected by centrifuging at 27,000 $g$ for 1.5 hr. The precipitate is redissolved in 200 ml of 20 m$M$ Tris-HCl buffer, pH 7.4, containing 10 m$M$ MgCl$_2$, 2 mg of DNase and 4 mg of RNase. Any insoluble material is removed by centrifuging at 15,000 $g$ for 10 min and loading the supernatant onto a DEAE-cellulose (Whatman, DE 52) column (3.5 × 12 cm) equilibrated with 25 m$M$ Tris-HCl buffer, pH 7.4, containing 0.15 $M$ NaCl. Hydrogenase is eluted as a brown band by washing the column with 150 ml of this buffer, 0.25 $M$

[7] M.-Y. W. Tso, T. Ljones, and R. H. Burris, *Biochim. Biophys. Acta* **267**, 600 (1972).

NaCl is used to elute Cp1 as a dark brown band and 0.4 $M$ NaCl to elute Cp2. One millimolar $Na_2S_2O_4$ is included in the buffer used to elute Cp2, which is concentrated approximately threefold using an on-line eluate concentrator (Amincon) and purified immediately as described below.

*Purification of Cp1—Alternative Procedure.* Cp1 from the DEAE-cellulose column is adjusted to pH 6.0 by the addition of 0.1 $M$ MES buffer and solid polyethylene glycol added to 5% (w/v). The precipitate is collected by centrifuging and discarded. Cp1 is then precipitated by the addition of more polyethylene glycol to 14% (w/v) and collected by centrifuging at 20000 $g$ for 30 min. The precipitate is redissolved in 20 ml of 50 m$M$ Tris-HCl buffer, pH 8.0, and chromatographed at 4° on a Sephadex G-200 column (5 × 85 cm) equilibrated with the same buffer. Cp1 is eluted at 850 ml and concentrated to 10–20 mg protein/ml.

*Purification of Cp2—Alternative Procedure.* Cp2 eluted from the DEAE column is immediately further purified to remove contaminating Cp1 by gel filtration on a Sephadex G-100 column (2.5 × 85 cm) equilibrated with 50 m$M$ Tris-HCl buffer, pH 8, containing 1 m$M$ $Na_2S_2O_4$. Cp1 is eluted in the void volume and Cp2 eluted at 180 ml is concentrated on-line to 10–20 mg protein/ml.

*Klebsiella pneumoniae*

In this procedure[1a,8] extracts of *K. pneumoniae* M5a1 are prepared by pressure disruption, Kp1 and Kp2 are purified by DEAE cellulose chromatography and gel filtration. A modification of the original method[1a] resulting in higher yields and specific activity is the use of DEAE–cellulose to concentrate Kp1 in place of the membrane filtration procedure used earlier. All buffers used in the purification contain 0.1 gm/liter of both dithiothreitol and $Na_2S_2O_4$, and, unless stated to the contrary, 25 m$M$ Tris-HCl buffer, pH 7.4, is used. Frozen cells (425 gm) are suspended in 300 ml of buffer and kept on ice. Aliquots (40 ml) are disrupted in a French pressure cell at 15000 psi and a crude extract prepared by centrifuging at 25,000 $g$ for 90 min at 5°. Subsequent steps are carried out at room temperature. An equal volume of a slurry of DEAE–cellulose is added to the crude extract, and after 15 min equilibration the suspension is poured onto a 6.5 × 16 cm column of DEAE–cellulose (Whatman DE 52). After the column has compacted, it is washed with 1 bed volume of buffer. The column is then developed stepwise with 150 ml volumes of 0.15, 0.2, 0.21, 0.22, and 0.25 $M$ NaCl in buffer. Kp1 is eluted as a broad

[8] B. E. Smith, R. N. F. Thorneley, M. G. Yates, R. R. Eady, and J. R. Postgate, *Proc. Int. Symp. Nitrogen Fixation, 1st, 1974* Vol. 1, p. 150 (1976).

dark-brown band after the addition of 0.22 $M$ NaCl. Fractions of 50 ml
are collected, and those with a specific activity greater than 500 nmole
$C_2H_2$ reduced/min/mg of protein are combined for further purification.
Kp2 is eluted with 500 ml of 90 mM-$MgCl_2$.

*Purification of Kp1.* Kp1 eluted from the DEAE column is diluted
twofold with buffer and concentrated by absorption onto a small (4 × 5
cm) column of DEAE–cellulose equilibrated with 25 m$M$ Tris-HCl buffer,
pH 8.7, and eluted with 0.3 $M$ NaCl. Kp1 is obtained at 20–30 mg protein/
ml and 30 ml aliquots chromatographed on a Sephadex G-200 column (5
× 40 cm) equilibrated with 25 m$M$ Tris-HCl buffer, pH 8.7. Kp1 is eluted
as a discrete brown band. Several columns are necessary at this stage to
treat all the Kp1 which is essentially homogenous on disc electrophoresis
at this stage. Kp1 (1.5 gm) is further purified by chromatography on a
DEAE-cellulose column (6.5 × 24 cm) equilibrated with 25 m$M$ Tris-HCl
buffer, pH 8.7, and eluted with a linear gradient, i.e., of 30–90 m$M$
$MgCl_2$. Fractions (30 ml) are collected, and although no clear separation
of Kp1 from other protein is observed, the specific activity is increased
considerably. The distribution of fractions of increased activity within
the salt gradient is rather variable, and it is sometimes necessary to
repeat this step to obtain Kp1 of maximum activity.

*Purification of Kp2.* Kp2 from the first DEAE column is diluted
fourfold and absorbed onto a small (2.5 × 4 cm) column of DEAE-
cellulose and eluted with 90 m$M$ $MgCl_2$. Aliquots (30 ml) are chromatographed on Sephadex G-200 equilibrated with buffer containing 50 m$M$
$MgCl_2$. Contaminating Kp1 is eluted in the void volume, and Kp2 is
eluted as a discrete brownish yellow band. It is necessary to reconcen-
trate and repeat this step to obtain Kp2 with maximum activity which for
unknown reasons is higher than that obtained originally. It has been
shown that this last step will remove $O_2$-inactivated Kp2 from the native
protein.[1a]

*Azotobacter vinelandii*

Conditions for growth of *A. vinelandii* OP (ATCC 13705), the preparation of crude extracts by pressure disruption, and the purification and
crystallization of Av1 were described[9] in Volume XXIV. The alternative
procedure[10] described below allows homogeneous preparations of both
Av1 and Av2 to be obtained, and Av1 to be crystallized reproducibly.
Frozen cells are resuspended in 25 m$M$ Tris-HCl buffer, pH 7.4, and

[9] R. C. Burns and R. W. F. Hardy, Vol. 24, p. 480.
[10] V. K. Shah and W. J. Brill, *Biochim. Biophys. Acta* **305**, 455 (1973).

disrupted by osmotic shock treatment with 3 $M$ glycerol. Crude extract containing 2.5–3 gm of protein is loaded onto a DEAE-cellulose (Whatman DE 52) column (2.5 × 17 cm) equilibrated with 25 m$M$ Tris-HCl buffer, pH 7.4, containing 0.1 gm/liter of $Na_2S_2O_4$. This buffer is used throughout the purification procedure unless stated to the contrary. The column is successively eluted with 2 bed volumes of buffer, 3 bed volumes of 0.1 $M$ NaCl, 1 bed volume of 0.25 $M$ NaCl to elute Av1, and 1 bed volume of 0.5 $M$ NaCl to elute Av2. Both components are eluted in a volume of approximately 20 ml.

*Purification of Av1.* Av1 eluted from the DEAE column in 0.25 $M$ NaCl is then heat-treated at 52° for 10 min and denatured protein removed by centrifuging at 20,000 $g$ for 10 min. The supernatant, containing Av1, is concentrated threefold by ultrafiltration using an XM-50 membrane (Amicon), diluted with 14 ml of buffer, and rechromatographed on a DEAE-cellulose column (2.5 × 17 cm) equilibrated with buffer containing 0.15 $M$ NaCl and 3 gm/liter $Na_2S_2O_4$. The column is washed with 2–3 bed volumes of this buffer, and then Av1 is eluted with 0.25 $M$ NaCl in a volume of 18.5 ml. Av1 is then concentrated to 4 ml by ultrafiltration, diluted sixfold with buffer, and reconcentrated. Needle-shaped crystals of Av1 begin to form during the reconcentration. The contents of the ultrafiltration cell are transferred to a serum-capped centrifuge tube and kept at 38° for 1 hr before collection of the crystals by centrifuging at 20,000 $g$ for 10 min. The supernatant is then removed and the pellet resuspended in 2–3 volumes of 25 m$M$ Tris-HCl buffer, pH 7.4, containing 42 m$M$ NaCl and centrifuged for 20 min. The pellet of Av1 crystals is redissolved in 3 ml of 25 m$M$ Tris-HCl buffer, pH 7.4, containing 0.25 $M$ NaCl. A small amount of insoluble white amorphous material is removed by centrifuging. Additional crystals of Av1 can be obtained by reconcentration of the mother liquor and repetition of the process described above. Av1 can be obtained crystalline after the heat-treatment step, but lower yields result. The crystallization procedure described here is more reproducible[10,11] than the original dilution method.[9]

*Purification of Av2.* Av2 eluted from the DEAE column in 0.5 $M$ NaCl is concentrated to 9 ml by ultrafiltration using a UM20E membrane (Amicon). After a twofold dilution with buffer, it is rechromatographed on a DEAE-cellulose column (2.5 × 18 cm) equilibrated with buffer containing 0.25 $M$ NaCl. The column is washed with 2 bed volumes of this buffer before Av2 is eluted with 0.35 $M$NaCl. Quantities (10–20 mg) of Av2 are further purified by preparative gel electrophoresis using a Fractophorator (Buchler Instruments, Fort Lee, New Jersey) and a

[11] W. A. Bulen, *Proc. Int. Symp. Nitrogen Fixation, 1st, 1974* Vol. 1, p. 177 (1976).

pulsed power supply (Ortec 4100, Ortec Inc., Oak Ridge, Tennessee). A 1-cm polyacrylamide stacking gel (6%, w/v) and a 4-cm separating gel (8%, w/v) are used with water cooling. The electrode buffers are 65 m$M$ Tris borate, pH 9, containing 0.3 gm/liter $Na_2S_2O_4$, and the eluting buffer used is 65 m$M$ Tris-HCl, pH 7.4, containing 0.1 gm/liter each of $Na_2S_2O_4$ and dithiothreitol. The gel is prerun for 4–6 hr at 5 mA (75 V at 100 pulses/sec and a discharge capacitance of 1 $\mu$F). Electrophoresis of Av2 is under these conditions for 10–12 hr when the current is increased to 20 mA (150V at 200 pulses/sec). Av2 is eluted in 1.5–2 ml of buffer.

*Azotobacter chroococcum*

In this procedure[2] extracts of *A. chroococcum* (NCIB 8003) are prepared by pressure disruption. Ac1 and Ac2 are separated by DEAE chromatography and purified by gel filtration. Freshly harvested cells (1 kg) are suspended in 500 ml of 75 m$M$ Tris-HCl buffer, pH 7.8, containing 0.1 gm/liter dithiothreitol and kept on ice. The periodic addition of 40% KOH is necessary to maintain the pH of the suspension which is treated in a French pressure cell at 3000 psi in 40 ml aliquots. After disruption the pH is adjusted to 7.4 and a crude extract prepared by centrifuging at 40,000 $g$ for 30 min under $N_2$ at 5°. Subsequent purification is carried out at room temperature in buffers containing 0.1 g/l of both dithiothreitol and $Na_2S_2O_4$. Catalytic amounts of RNase and DNase are then added to the crude extract which is incubated under $N_2$ for 30 min before an equal volume of a slurry of DEAE-cellulose in 25 m$M$ Tris-HCl buffer, pH 7.4, is added. After a further 1 hr incubation the suspension is poured onto a column (10 × 20 cm) of DEAE–cellulose (Whatman DE 52). After the slurry has settled the column is washed with 2 liters of 25 m$M$ Tris HCl buffer, pH 7.4, before Ac1 and Ac2 are eluted together with 90 m$M$ $MgCl_2$. This fraction is then either dialyzed overnight or diluted sixfold with 25 m$M$ Tris-HCl buffer, pH 7.4, before rechromatography on a DEAE–cellulose column (4 × 25 cm). Two columns are used at this stage, and both are eluted with volumes of 100 ml of buffer and then NaCl solutions from 0.1–0.25 $M$ NaCl in 0.05 $M$ steps. Ac1 is eluted in 0.15 $M$, or more usually 0.2 $M$ NaCl. Ac2 is eluted with 90 m$M$ $MgCl_2$ in a volume of 300 ml.

*Purification of Ac1.* Ac1 eluted from the second DEAE–cellulose column by 0.2 $M$ NaCl (total volume 290 ml) is further purified by gel filtration of 40 ml aliquots on a Sephadex G-200 column (5 × 40 cm) equilibrated with 25 m$M$ Tris-HCl buffer, pH 8, containing 0.1 $M$ $MgCl_2$. Ac1 eluted from this column is then concentrated on a small (3 × 5 cm) column of DEAE–cellulose and eluted with 90 m$M$ $MgCl_2$ in 25 m$M$ Tris-

HCl buffer, pH 7.4. The concentrated material is then chromatographed on a Sephadex G-100 column equilibrated with 25 m$M$ Tris-HCl buffer, pH 7.4, containing 50 m$M$ $MgCl_2$. To obtain homogeneous preparations of Ac1, it is usually necessary to repeat the Sephadex G-200 chromatography step.

*Purification of Ac2.* Aliquots (30 ml) of Ac2 eluted in 90 m$M$ $MgCl_2$ from the second DEAE column are chromatographed on a Sephadex G-100 column (5 × 40 cm) equilibrated with 25 m$M$ Tris-HCl buffer, pH 7.4, containing 50 m$M$ $MgCl_2$. This is repeated up to three times to obtain homogeneous Ac2. A DEAE–cellulose column (2 × 5 cm) equilibrated with 25 m$M$ Tris-HCl buffer, pH 7.4, is used to concentrate Ac2 (eluted with 90 m$M$ $MgCl_2$) before Sephadex chromatography at each stage.

*Chromatium vinosum*

Of the nitrogenase components of this organism, only Cv1 has been purified and partially characterized.[12] *Chromatium* strain D is grown under an atmosphere of 5% $CO_2$ in $N_2$ on a modified Pfennig's medium. Cultures from a 20% inoculum are harvested after approximately 72 hr when levels of acetylene reduction are high. Extracts are prepared by pressure disruption 400 gm of the cell paste are suspended anaerobically in 3 volumes of 20 m$M$ Tris-HCl buffer, pH 8.5, and disrupted at 10,000 psi in a homogenizer. Then 1 $M$ $MgCl_2$ is added to a final concentration of 20 m$M$ based on the volume of buffer added to the cell paste, and the suspension is centrifuged for 30 min at 23,000 $g$. Additional 1 $M$ $MgCl_2$ is then added to give a final concentration of 20% based on the volume of the supernatant. The pH is then adjusted to 7.2 with 1 $M$Tris, and $Na_2S_2O_4$ (0.5 mg/ml) is added. Photosynthetic lamellae are then precipitated by the addition of 50% (w/v) polyethylene glycol 6000 to a final concentration of 8% (v/v) and removed by centrifuging at 23,000 $g$ for 30 min. Protamine sulfate (50 mg/ml) is added to the supernatant to precipitate nitrogenase. During this process the nitrogenase activity of the supernatant is monitored to establish the concentration range over which precipitation occurs—usually 5–15%. Precipitated material is collected by centrifuging at 23,000 $g$ for 10 min and resolubilized by stirring for 30 min with a sevenfold excess of cellulose phosphate in 20 m$M$ Tris-HCl buffer, pH 8. The supernatant is then diluted to 500 ml with 20 m$M$ Tris HCl buffer, pH 7.5, containing 0.1 $M$ $MgCl_2$. Cv1 and Cv2 are then separated by chromatography on a DEAE-cellulose column (4.5 × 25 cm) equilibrated with the same buffer. Cv1 is eluted with 200 ml of buffer

---

[12] M. W. C. Evans, A. Telfer, and R. V. Smith, *Biochim. Biophys. Acta* **310**, 344 (1973).

containing 0.25 $M$ NaCl and Cv2 with 0.6 $M$ NaCl. Cv1 is well resolved at this stage but Cv2 contains residual Cv1 and ferredoxin.

*Purification of Cv1.* Cv1 from the DEAE-cellulose column is concentrated by ultrafiltration using an XM-100 membrane (Amicon) and chromatographed on a Sephadex G-200 column (4.5 × 40 cm) equilibrated with 40 m$M$ Tris-HCl, buffer, pH 8, containing 0.1 $M$ NaCl. Complete purification requires that this step is repeated.


### Criteria of Purity

The most widely used test of the purity of nitrogenase components are disc electrophoresis of the native proteins under anaerobic conditions and of the denatured proteins in the presence of SDS. The Mo–Fe proteins when purified by the methods described above are often homogeneous by these criteria before the maximum specific activity has been obtained, suggesting a considerable degree of contamination with inactive protein. The successful removal of a 'demolybdo'' species of the Mo–Fe proteins is indicated[6] by the absence of an EPR signal of the $g$ = 1.94 type in the $Na_2S_2O_4$-reduced protein at temperatures below 30° K. Although the correlation between metal content and specific activity is not direct,[13] in general the Mo and Fe content increase with specific activity to 2 Mo atoms and 24–32 Fe atoms/mole in those of the highest activity. The yields and specific activities of nitrogenase components purified by the procedures described above are shown in Table I.


### Properties of the Mo–Fe Proteins

The tendency of these proteins to aggregate (with no loss in activity) on storage in liquid nitrogen, and to dissociate at high dilutions undoubtedly account for the range of values that appeared for the molecular weight in the earlier literature. Within the last five years a number of different techniques have been used to determine molecular weights which now, in most instances, fall in the range of 200,000–235,000 (see Table II). Because of the comparative ease of maintaining anaerobic conditions, gel filtration has been the method most widely used.

The subunit structure of purified Mo–Fe proteins determined by electrophoresis in polyacrylamide gels containing SDS[14] is consistent with

---

[13] W. H. Orme-Johnson and L. C. Davis, *in* ''Iron Sulfur Proteins'' (W. Lovenberg, ed.), Vol. 3, p. 15, Academic Press, New York, 1977.

[14] K. Weber and M. Osborn, *J. Biol. Chem.* **244**, 4406 (1969).

TABLE I
YIELD AND SPECIFIC ACTIVITIES OF MO–FE PROTEIN AND FE PROTEIN PURIFIED FROM
DIFFERENT ORGANISMS

| Organism | Mo–Fe protein | | Fe protein | |
|---|---|---|---|---|
| | Specific activity (nmole product/min/mg) | Yield (gm) | Specific activity (nmole product/min/mg) | Yield (gm) |
| C. pasteurianum | >2000 | 1–2 | >2200 | 0.5[a] |
| | 2500 | 0.24 | 3100 | 0.18[b] |
| K. pneumoniae | 2250 | 0.8 | 1200 | 0.128[c,d] |
| A. vinelandii | 1638 | 0.048 | 1815 | 0.031[e] |
| | 1444 | 0.378 | — | —[f] |
| A. chroococcum | 2072 | 1 | 2100 | 0.73[g] |
| C. vinosum | 1600 | 0.5 | — | —[h] |

[a] W. G. Zumft and L. E. Mortenson, Eur. J. Biochem. 35, 401 (1973).
[b] M.-Y. W. Tso, T. Ljones, and R. H. Burris, Biochim. Biophys. Acta 267, 600 (1972).
[c] R. R. Eady, B. E. Smith, K. A. Cook, and J. R. Postgate, Biochem. J. 128, 655 (1972).
[d] B. E. Smith, R. N. F. Thorneley, M. G. Yates, R. R. Eady, and J. R. Postgate, Proc. Int. Symp. Nitrogen Fixation, 1974 1st Vol. 1, p. 150 (1976).
[e] V. K. Shah and W. J. Brill, Biochim. Biophys. Acta 305, 445 (1973).
[f] R. C. Burns and R. W. F. Hardy, Vol. 24, p. 480.
[g] M. G. Yates and K. Planque, Eur. J. Biochem. 60, 467 (1975).
[h] M. C. W. Evans, A. Telfer, and R. V. Smith, Biochim. Biophys. Acta 310, 344 (1973).

the native proteins being tetrameric with a molecular weight near 220,000, but are often contradictory as to the number of types of subunit.[15] Both one or two types of subunit present in equal numbers were reported for various Mo–Fe proteins, and for Av1, conflicting data appeared from different laboratories. It has recently been shown[16] that the number of subunit bands obtained with Kp1, Ac1, Av1, and Rj1 using this method depends on the commercial source of the SDS used in the electrode buffers. One or two discrete well-resolved bands are obtained with different SDS brands (Fig. 1), emphasizing the unreliability of this method for determining the subunit structure of these proteins. The molecular weights near 50,000 and 60,000 obtained for the subunits of Cp1 agree well with the values of 52,500 and 62,000 from gel filtration in Sepharose 6B equilibrated with 6 $M$ guanidine HCl. The subunits of Cp1, Kp1, and

[15] R. R. Eady and B. E. Smith, in "Dinitrogen Fixation" (R. W. F. Hardy, F. Bottomley, and R. C. Burns, ed.), Vol. 2, p. 399 (Wiley (Interscience), New York, 1979.
[16] C. Kennedy, R. R. Eady, E. Kondorosi, and D. Klavans Rekosh, Biochem. J. 155, 383 (1976).

TABLE II

MOLECULAR WEIGHT OF PURIFIED MO–FE PROTEINS ISOLATED FROM VARIOUS SOURCES

| | C. pasteurianum | K. pneumoniae[e] | A. chroococcum[f] | A. vinelandii | R. japonicum[l] bacteroids | R. lupini[n] bacteroids |
|---|---|---|---|---|---|---|
| Gel filtration | 200,000[a] 210,000[b] | 220,000 | 216,000 | 216,000[g] 221,000[h] | 180,000[m] | 194,000 |
| Ultracentrifugation | 168,000[d] | 200,400 | — | 270,000[i] 245,000[j] | 202,000 197,000 | — |
| Glycerol gradient centrifugation | 201,000[b] | | | | | |
| Disc electrophoresis | | 217,000 | 230,000 | | | |
| Sum of subunit molecular weights | 220,000[a] | 221,800 | 220,000 | 280,000[k] 232,000[h] | 210,000 | 228,000 |
| Multiple of minimal molecular weight from amino acid composition | 221,800[c] | 229,000 | 227,000 | | | |
| Osmometry | | | | 229,000[h] | | — |
| Electron microscopy | — | — | — | 264,000[k] | — | — |
| Average | 203,000 | 219,000 | 223,000 | 234,000 | 197,000 | 211,000 |

[a] T. C. Huang, W. G. Zumft, and L. E. Mortenson, J. Bacteriol. 113, 884 (1973).
[b] M.-Y. W. Tso, Arch. Microbiol. 99, 71 (1974).
[c] J. S. Chen, J. S. Multani, and L. E. Mortenson, Biochim. Biophys. Acta 310, 51 (1973).
[d] H. Dalton, J. A. Morris, M. A. Ward, and L. E. Mortenson, Biochemistry 10, 2066 (1971).
[e] R. R. Eady, B. E. Smith, K. A. Cook, and J. R. Postgate, Biochem. J. 128, 655 (1972).
[f] M. G. Yates and K. Planque, Eur. J. Biochem. 60, 467 (1975).
[g] D. Kleiner and C. H. Chen, Arch. Microbiol. 98, 100 (1974).
[h] W. A. Bulen, Proc. Int. Symp. Nitrogen Fixation, 1st, 1974 Vol. 1, p. 177 (1976).
[i] R. C. Burns, R. D. Holsten, and R. W. F. Hardy, Biochem. Biophys. Res. Commun. 39, 90 (1970).
[j] R. H. Swisher, M. L. Landt, and F. J. Reithel, Biochem. Biophys. Res. Commun. 66, 1476 (1975).
[k] J. T. Stasny, R. C. Burns, B. D. Korant, and R. W. F. Hardy, J. Cell Biol. 60, 311 (1974).
[l] D. W. Israel, R. L. Howard, H. J. Evans, and S. A. Russell, J. Biol. Chem. 249, 500 (1974).
[m] F. J. Bergersen and G. L. Turner, Biochim. Biophys. Acta 214, 28 (1970).
[n] M. J. Whiting and M. J. Dilworth, Biochim. Biophys. Acta 371, 337 (1974).

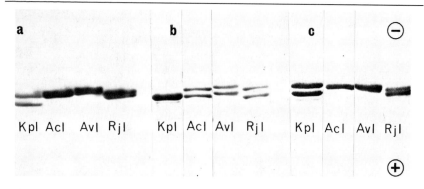

Fig. 1. Banding pattern of various Mo–Fe proteins on 10% acrylamide gels using three different brands of SDS. The gels and sample buffer contained Serva brand SDS, the electrode buffer contained SDS from (a) Schwartz, (b) Serva, or (c) Koch-Light. [After C. Kennedy, R. R. Eady, E. Kondorosi, and D. Klavans Rekosh, *Biochem. J.* **155**, 383 (1976).]

Av1 have been isolated. Those of Cp1 by repetitive gel filtration in Sephadex G-100 equilibrated with 20 m$M$ HCl and 0.1% $\beta$-mercaptoethanol,[17] Kp1 by preparative SDS electrophoresis,[16] and Av1 by chromatography on carboxymethyl cellulose equilibrated with 7 $M$ urea at pH 4.6.[18] Unfortunately these conditions all result in the loss of Mo and Fe from the subunits.[17] The isolation and properties of a low molecular weight $O_2$-sensitive Mo-containing cofactor from Mo–Fe proteins is described by Shah (this volume, Article [72]).

The nonidentity of the subunits of Kp1 has been established by comparison of peptide maps of tryptic digests of the separated subunits. The amino acid compositions of subunits of a given Mo–Fe protein are very similar, and it is difficult to distinguish between them on this basis.[16] The biochemical data are consistent with the Mo–Fe proteins being tetramers of the $\alpha_2\beta_2$ type. Support for this formulation is provided by a genetical analysis of the *nif* genes of *K. pneumoniae* where two separate genes determine the synthesis of Kp1 subunit polypeptides.[19]

The Mo, Fe, and acid-labile sulfide content of several Mo–Fe proteins are shown in Table III. The correlation between metal content and activity is not direct[13]; presumably loss of activity can occur without loss of metal. Preparations of Mo–Fe proteins with the highest specific activity

---

[17] T. C. Huang, W. G. Zumft, and L. E. Mortenson, *J. Bacteriol.* **113**, 884 (1973).
[18] R. H. Swisher, M. L. Landt, and R. J. Reithel, *Biochem. J.* **163**, 427 (1977).
[19] R. A. Dixon, C. Kennedy, A. Kondorosi, V. Krishnapillai, and M. J. Merrick, *Molec. Gen. Genet.* **157**, 189 (1977).

TABLE III
MOLYBENUM, IRON AND ACID-LABILE SULFIDE CONTENT OF HIGHLY PURIFIED MO–FE
PROTEINS[a]

| Protein | Specific activity (nmole product/min/mg) | Mo (gm atom/mole) | Fe (gm atom/mole) | $S^{2-}$ (gm atom/mole) |
|---|---|---|---|---|
| Cp1 | 1100 | 1.38 | 19.3 | 22[b] |
| | 1200 | 1.3 | 20.5 | —[c] |
| | 2250 | 2 | 24 | 24[d] |
| | 2500 | 1–1.5 | 12–18 | 8–15[e] |
| Kp1 | 1200 | 1 | 17.5 ± 0.7 | 16.7 ± 1[f] |
| | 2150 | 2.01 ± 0.3 | 32.5 ± 3 | —[g] |
| Av1 | 1100 | 1.54 | 24 | 20[h] |
| | 1400 | 1.6 | 28–31 | 21–33[i] |
| | 1700 | 1.88 | 24 | 22[j] |
| Ac1 | 2000 | 1.9 ± 0.3 | 22 ± 2 | 22 ± 2[k] |
| Rj1 | 1000 | 1.3 | 28.8 | 26.2[h] |
| Cv1 | 1600 | 1.4 | 19 | 15[m] |

[a] Data from refs b, h, and l have recalculated on the basis of a MW of 220,000.
[b] H. Dalton, J. A. Morris, M. A. Ward, and L. E. Mortenson, *Biochemistry* **10**, 2066 (1971).
[c] J. P. Vandecasteele and R. H. Burris, *J. Bacteriol.* **101**, 794 (1970).
[d] T. C. Huang, W. G. Zumft, and L. E. Mortenson, *J. Bacteriol.* **113**, 884 (1973).
[e] M.-Y. W. Tso, *Arch. Microbiol.* **99**, 71 (1974).
[f] R. R. Eady, B. E. Smith, K. A. Cook, and J. R. Postgate, *Biochem. J.* **128**, 655 (1972).
[g] B. E. Smith, R. N. F. Thorneley, M. G. Yates, R. R. Eady, and J. R. Postgate, *Proc. Int. Symp. Nitrogen Fixation, 1st, 1974* Vol. 1, p. 150 (1976).
[h] D. Kleiner and C. H. Chen, *Arch. Microbiol.* **98**, 100 (1974).
[i] R. C. Burns and R. W. F. Hardy, Vol. 24, p. 480.
[j] W. A. Bulen, *Proc. Int. Symp. Nitrogen Fixation, 1st, 1974* Vol. 1, p. 177 (1976).
[k] M. G. Yates and K. Planque, *Eur. J. Biochem.* **60**, 467 (1975).
[l] D. W. Israel, R. L. Howard, H. J. Evans, and S. A. Russell, *J. Biol. Chem.* **249**, 500 (1974).
[m] M. C. W. Evans, A. Telfer, and R. V. Smith, *Biochim. Biophys. Acta* **310**, 344 (1973).

contain 2 Mo and 22–32 Fe atoms and an approximately equivalent number of acid-labile sulfide atoms per molecule. Mössbauer and EPR studies often require the use of protein enriched with $^{57}$Fe or $^{95}$Mo. The successful exchange of these metals from Mo–Fe protein has not been reported, and isotopic substitution is achieved by growing the organism on purified medium enriched in the relevant metal isotope.

The amino acid composition of purified Mo–Fe proteins are very similar.[15] All the common amino acids are present and acidic residues at 19.4–21.6% are approximately twice as abundant as basic residues. C-terminal amino acids are alanine and leucine in Cp1[20] and serine in Av1,[18] and the N-terminal residues in Rl1 are serine.[21]

Concentrated solutions of Mo–Fe protein are dark brown but the visible absorption spectra are very broad. Features at 525 and 557 nm in some crystalline preparations of Av1 have been attributed to cytochrome contamination.[10] Av1 is the only Mo–Fe protein which has been crystallized. The procedure is based on the insolubility of this protein at low salt concentrations at neutral pH[22]—a property which is also used in its purification.[9,10] Two types of crystals have been obtained, neither of which is a suitable form for X-ray structure analysis. When the ionic strength is lowered by the rapid dilution of concentrated protein solutions with buffer, small crystals (1–4 × 30–60 $\mu$m) are formed,[9,10] the optimum concentration of NaCl for crystallization is 40–80 m$M$. Membrane concentration of solutions to saturation results in the formation of aggregated forms[11] (1–3 mm × 40 $\mu$m). Crystallization is not a good criterion of purity[10] and Kp1, Cp1, and Ac1 do not crystallize under these conditions.

The only spectroscopic technique successfully used to probe the environment of Mo in the Mo–Fe protein is the recently developed technique of X-ray absorption edge spectroscopy. Comparison of the K X-ray absorption edge of Cp1 with a range of Mo-containing model compounds enable a coordination charge of 2.3 ± .03 to be calculated.[23] The assignment of Mo in Cp1 to a pair of antiferromagnetically coupled Mo(V) atoms has been criticized,[24] and more recent experiments with $S_2O_4^{2-}$-reduced Cp1 are consistent with the ligand environments of the Mo involving sulfur atoms.[25]

At temperatures below 30°K the $S_2O_4^{2-}$-reduced proteins exhibit a characteristic EPR spectrum with $g$ values near 4.3, 3.7, and 2.01 (see Table IV and Fig. 2). This type of signal is not shown by any other class of FeS protein and has been assigned to transitions in the $S = \frac{1}{2}$ ground state of spin $S = \frac{3}{2}$ system.[26–28] Kp1 and Av1 substituted with [57]Fe show

[20] J.-S. Chen, J. S. Multani, and L. E. Mortenson, *Biochim. Biophys. Acta* **310**, 51 (1973).

[21] M. J. Whiting and M. J. Dilworth, *Biochim. Biophys. Acta* **371**, 337 (1974).

[22] R. C. Burns, R. D. Holsten, and R. W. F. Hardy, *Biochem. Biophys. Res. Commun.* **39**, 90 (1970).

[23] S. P. Cramer, R. K. Eccles, F. W. Kutzler, K. O. Hodgson, and L. E. Mortenson, *J. Am. Chem. Soc.* **98**, 1287 (1976).

[24] B. E. Smith, *J. Less-Common Met.* **54**, 465 (1977).

[25] S. P. Cramer, K. O. Hodgson, W. O. Gillum, and L. E. Mortenson, *J. Amer. Chem. Soc.* **100**, 4630 (1978).

[26] B. E. Smith, D. J. Lowe, and R. C. Bray, *Biochem. J.* **130**, 641 (1972).

[27] G. Palmer, J.-S. Multani, W. G. Zumft, and L. E. Mortenson, *Arch. Biochem. Biophys.* **153**, 325 (1972).

[28] E. Munck, H. Rhodes, W. H. Orme-Johnson, L. C. Davis, W. J. Brill, and V. K. Shah, *Biochim. Biophys. Acta* **400**, 32 (1975).

TABLE IV

EPR Signals Associated with $S_2O_4^{2-}$-Reduced Mo–Fe Protein and Signals Observed under Different Conditions in the Steady State

| | | Turnover under | | |
| | $S_2O_4^{2-}$-reduced | Ar or $N_2$ | low CO (0.001 atm) | high CO (0.1 atm) |
|---|---|---|---|---|
| Kp1 | 4.27 3.73 2.018 high pH[a] | 2.139 2.001 1.977[h] | | |
| | 4.32 3.63 2.009 low pH[a] | 2.092 1.974 1.933[h] | 2.073 1.969 1.927[h] | 2.17 2.06 2.06[h] |
| | | 2.125 2.000 2.00[h,i] | | |
| | | 5.7 5.4[h] | | |
| Acl | 4.29 3.65 2.013[b] | 2.14 2.001 1.976[j] | 2.08 1.97 1.92[j] | 2.16 2.07 2.04[j] |
| Av1 | 4.30 3.67 2.01[c] | 2.09 2.01 1.98[e] | 2.01 1.98 1.93[e] | 2.17 2.08 2.05[e] |
| Cp1 | 4.27 3.78 2.01[d] | 2.13 1.99 1.98[e] | 2.07 1.975 1.92[e] | 2.15 2.07 2.04[e] |
| Bp1 | 4.37 3.53 2.01[k] | — | — | — |
| Rr1 | 4.34 3.65 2.01[e] | — | 2.08 1.975 1.93[e] | 2.17 2.09 2.05[e] |
| Rj1 | 4.17 3.73 2.03[f] | 2.14 2.01 1.98[e] | — | 2.18 2.11 2.06[e] |
| Cv1 | 4.3 3.68 2.01[g] | — | — | — |

[a] B. E. Smith, D. J. Lowe, and R. C. Bray, *Biochem. J.* **135**, 331 (1973).
[b] M. G. Yates and K. Planque, *Eur. J. Biochem.* **60**, 467 (1975).
[c] R. C. Burns, R. D. Holsten, and R. W. F. Hardy, *Biochem. Biophys. Res. Commun.* **39**, 90 (1970).
[d] G. Palmer, J. S. Multani, W. C. Cretney, W. G. Zumft, and L. E. Mortenson, *Arch. Biochem. Biophys.* **153**, 325 (1972).
[e] L. C. Davis, M. T. Henzl, R. H. Burris, and W. H. Orme-Johnson, *Biochemistry* (in press).
[f] H. J. Evans, P. E. Bishop, and D. Israel, *Proc. Int. Symp. Nitrogen Fixation, 1st 1974* Vol. 1, p. 234 (1976).
[g] M. C. W. Evans, A. Telfer, and R. V. Smith, *Biochim. Biophys. Acta* **310**, 344 (1973).
[h] D. J. Lowe, R. R. Eady, and R. N. F. Thorneley, *Biochem. J.* **173**, 277, (1978).
[i] Only observed in the presence of added ethylene.
[j] M. G. Yates and D. J. Lowe, *FEBS Lett.* **72**, 121 (1976).
[k] W. H. Orme-Johnson and L. C. Davis, in "Iron–Sulphur Proteins" (W. Lovenberg, ed.), Vol. 3, p. 15. Academic Press, New York, 1977.

about a 5-G linewidth broadening which is not observed when the proteins are substituted with $^{95}$Mo.[26,27] Correlation with Mössbauer data for these proteins (see below) indicate that this signal is associated with Fe atoms, probably in two $Fe_4S_4$ clusters. The $g$ values and linewidth of the EPR spectrum of Kp1 are pH dependent[29] and the interconversion of the high and low pH forms is associated with a $pK_a$ of 8.7 at 0°. This equilibrium is perturbed by the binding of $C_2H_2$ in favor of the high pH form. Oxidation of Kp1, Cv1, and Cp1 with the redox dye Lauth's violet bleaches

[29] B. E. Smith, D. J. Lowe, and R. C. Bray, *Biochem. J.* **135**, 331 (1973).

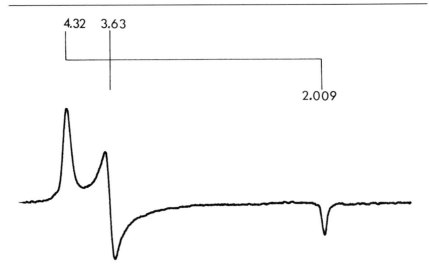

FIG. 2. EPR spectrum of $S_2O_4{}^{2-}$-reduced Kp1 at pH 6.8. At this pH only the low pH form is observed. The spectrum was recorded at 12°K with a microwave frequency of 9.16 GHz and a microwave power of 150 mW. Kp1 was at 12.3 mg protein/ml. [M. J. O'Donnell, B. E. Smith, and D. J. Lowe, unpublished.]

the EPR signal without loss of activity.[12,26,27] Oxidation with air, or excess $K_3[Fe(CN)_6]$ results in the loss of activity and the appearance of a complex series of signals near $g = 2$.

During turnover, i.e., in the presence of Fe protein, MgATP and $S_2O_4{}^{2-}$, the EPR spectrum of the $S_2O_4{}^{2-}$-reduced Mo–Fe protein is 90% bleached. Correlation with Mössbauer data (see below) indicates that this is due to a further reduction, not oxidation, of this EPR center.[13,15] In the steady-state EPR-active intermediates present in low concentration with signals comparable to those observed in simpler Fe–S proteins have been reported (see Table IV). The intensities of these signals are perturbed by $C_2H_2$, $C_2H_4$, the product of $C_2H_2$ reduction, and the inhibitor CO.[30,31] EPR signals which correspond to the −1 and −3 oxidation of $Fe_4S_4$ clusters are observed (see Table IV). The signals observed under these conditions presumably arise from the redistribution of electrons within the redox centers of the Mo–Fe protein so that EPR active intermediates accumulate to a detectable level. They provide evidence for the binding of substrate product and inhibitor to the Mo–Fe protein, although

[30] D. J. Lowe, R. R. Eady, and R. N. F. Thorneley, *Biochem. J.* **173,** 277 (1978).
[31] R. H. Burris and W. H. Orme-Johnson, *Proc. Int. Symp. Nitrogen Fixation, 1st, 1974* Vol. 1, p. 248 (1976).

the lack of linewidth broadening observed with $^{12}C$ compounds[30,32] does not support direct binding to the $Fe_4S_4$ clusters as occurs with hydrogenase.[33]

The Mössbauer spectra of $^{57}Fe$-enriched Kp1[34] and Av1[28] are very similar despite the difference in metal content of the proteins used in these studies. The low temperature spectra of the $S_2O_4^{2-}$-reduced proteins are complex and consist of three overlapping quadrupole doublets and a component with paramagnetic hyperfine structure. At higher temperatures an additional quadrupole doublet appears, derived from the paramagnetic component at lower temperatures (Fig. 3). The spectral parameters indicate that Fe atoms are in three major environments. In both proteins approximately 55% of the Fe atoms are ferrous; 2 Fe atoms in Kp1 and 3–4 Fe atoms in Av1 have been assigned to spin-coupled high-spin ferrous atoms. The remainder of the ferrous Fe is low spin: 8 Fe atoms in Kp1 and 8–10 Fe atoms in Av1. The doublet which appears on raising the temperature has been correlated with EPR active center of the $S_2O_4^{2-}$-reduced proteins and is associated with 8 Fe atoms in Kp1 and 8–10 atoms in Av1. The Mössbauer parameters of this species at 77° K ($\Delta E = 0.71$, $\delta = 0.37$) are similar to those of the $Fe_4S_4$ cluster of oxidized HiPiP. During turnover this is the only component of the spectrum which changes; and a new species with parameters close to those of reduced HiPiP ($\Delta E$ 1.12, $\delta = 0.42$) appears. These data allowed the assignment of the bleaching of the EPR signal of the $S_2O_4^{2-}$-reduced protein which occurs under these conditions to a "superreduction," rather than an oxidation.[34] Kp1 oxidized with the redox dye Lauth's violet retains full activity but the overlapping quadrupole doublets observed at 77°K in the $S_2O_4^{2-}$-reduced protein are replaced by a single doublet ($\Delta E = 0.75$, $\delta = 0.37$). Mo–Fe proteins isolated from different sources show a considerable variation in redox potential and in the number of electrons involved in these processes.[13,15] The midpoint potential of Cp1 and Cv1 has been measured by the EPR potentiometric technique. In this method[35] the potential is poised by the addition of $S_2O_4^{2-}$-or $K_3[Fe(CN)_6]$ to the protein plus redox mediator under anaerobic conditions and the system allowed to equilibrate for 10 min. The potential is then recorded and a sample removed for the measurement of the intensity of the $g = 3.7$ feature of the EPR spectrum. One electron

[32] L. C. Davis, M. T. Henzl, R. H. Burris, and W. H. Orme-Johnson, *Biochemistry* (in press).

[33] D. L. Erbes, R. H. Burris, and W. H. Orme-Johnson, *Proc. Natl. Acad. Sci. U.S.A.* **72**, 4792 (1975).

[34] B. E. Smith and G. Lang, *Biochem. J.* **137**, 169 (1974).

[35] P. L. Dutton, *Biochim. Biophys. Acta* **226**, 63 (1971).

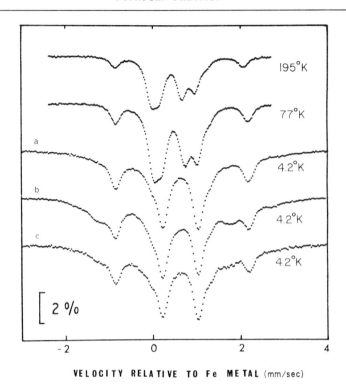

VELOCITY RELATIVE TO Fe METAL (mm/sec)

FIG. 3. Mössbauer spectrum of $S_2O_4^{2-}$-reduced Kpl enriched with $^{57}$Fe. (a) 4.2° K, zero magnetic field. (b) 4.2° K with a 100 G magnetic field parallel to the γ-ray beam. (c) 4.2° K with a 550 G magnetic field perpendicular to the γ-ray beam. [After B. E. Smith and G. Lang, *Biochem. J.* **137**, 169 (1974).]

processes were observed for Cp1 which gave a single $E_{m,7.5}$ of $-20$ mV,[36] and for Cv1 where two centers with $E_{m,7.5}$ of $-60$ mV and $-260$ mV were observed.[37] At lower potentials, a 30% decrease in the intensity of the EPR signal of Cv1 occurred with an $E_{m,7.5}$ of $-460$ mV.[38]. The value for Cp1 is higher than that obtained by the dye-equilibration method where a four electron process was associated with an $E_{m,7.5}$ of $-70$ mV.[39] Electrolytic reduction of Av1 oxidized with methylene blue gives two

[36] W. G. Zumft, L. E. Mortenson, and G. Palmer, *Eur. J. Biochem.* **46**, 525 (1974).
[37] S. L. Albrecht and M. C. W. Evans, *Biochem. Biophys. Res. Commun.* **55**, 1009 (1973).
[38] M. C. W. Evans and S. L. Albrecht, *Biochem. Biophys. Res. Commun.* **61**, 1187 (1974).
[39] M. Walker and L. E. Mortenson, *Biochem. Biophys. Res. Commun.* **54**, 669 (1973).

TABLE V

MOLECULAR WEIGHT AND SUBUNIT COMPOSITION OF Fe PROTEINS

| | Cp2 | Kp2 | Av2 | Ac2 | Rl2 |
|---|---|---|---|---|---|
| Gel filtration | 55,000[a] 56,000[b] | 62,000[c] | 64,000[d] 66,500[e] | 63,000[g] | 65,000[h] |
| Sedimentation velocity | — | 68,000[c] | — | — | — |
| Amino acid composition | — | 67,800[c] | — | 67,000[g] | — |
| Sum of subunit MW | 55,000[a] | 69,200[c] | 66,000[d] 68,000[e] | 61,000[g] | 64,000[h] |
| Subunit composition | one type | one type | one type | one type | one type |
| Subunit MW | 27,500[a,b] | 34,600[c] | 33,000[d,f] | 30,800[g] | 32,000[h] |

[a] G. Nakos and L. E. Mortenson, *Biochemistry* **10**, 455 (1971).

[b] M.-Y. W. Tso, *Arch. Microbiol.* **99**, 71 (1974).

[c] R. R. Eady, B. E. Smith, K. A. Cook, and J. R. Postgate, *Biochem. J.* **128**, 655 (1972).

[d] D. Kleiner and C. H. Chen, *Arch. Microbiol.* **98**, 100 (1974).

[e] W. A. Bulen, *Proc. Int. Symp. Nitrogen Fixation, 1st, 1974* Vol. 1, p. 177 (1976).

[f] R. H. Swisher, M. Landt, and F. J. Reithel, *Biochem. Biophys. Res. Commun.* **66**, 1476 (1975).

[g] M. G. Yates and K. Planque, *Eur. J. Biochem.* **60**, 467 (1975).

[h] M. J. Whiting and M. J. Dilworth, *Biochim. Biophys. Acta* **371**, 337 (1974).

reduction waves at $-320$ and $-450$ mV, both associated with a six electron reduction.[40]

## Properties of the Fe Proteins

The most useful criterion of purity for these proteins is electrophoresis in 8–10% polyacrylamide gels containing SDS. Interpretation of electrophoresis patterns obtained under nondenaturing conditions is complicated by the formation of size and charge isomers unless extreme precautions are taken to exclude oxygen. Oxygen damage results in a characteristic banded pattern on the gel and makes the identification of contaminating proteins difficult. The Fe proteins are unstable to prolonged storage except at liquid nitrogen temperatures. There is some variability in stability to storage at temperatures near 0° Kp2 and Ac2 are stable but some

[40] G. D. Watt and W. A. Bulen, *Proc. Int. Symp. Nitrogen Fixation, 1st 1974* Vol. 1, p. 248 (1976).

TABLE VI
IRON AND SULFIDE CONTENT OF VARIOUS HIGHLY PURIFIED Fe PROTEINS

| | Fe content (gm atom/mole) | $S^{2-}$ content (gm atom/mole) | Specific activity (nmole $C_2H_2$ reduced min/mg/protein) | MW |
|---|---|---|---|---|
| Ac2 | $4^a$ | 3.9 | 2000 | 65,400 |
| Rl2 | $3.1^b$ | — | 434 | 65,000 |
| Kp2 | $4^c$ | 3.8 | 980 | 66,800 |
| Cp2 | $4^d$ | 4 | 2708 | 55,000 |
| | $3-4^e$ | 3-4 | 3100 | |

[a] M. G. Yates and K. Planque, *Eur. J. Biochem.* **60**, 467 (1975).
[b] M. J. Whiting and M. J. Dilworth, *Biochim. Biophys. Acta* **371**, 337 (1974).
[c] R. R. Eady, B. E. Smith, K. A. Cook, and J. R. Postgate, *Biochem. J.* **128**, 655 (1972).
[d] G. Nakos and L. E. Mortenson, *Biochemistry* **10**, 455 (1971).
[e] M.-Y. W. Tso, *Arch. Microbiol.* **99**, 71 (1974).

preparations of Av2 and Cp2 rapidly lose activity.[15] The extreme oxygen sensitivity of these proteins has been referred to in the context of the problems it poses to their purification; 45 sec is a typical half-life for the rate inactivation by air; oxygen damage is accompanied by the oxidation of sulfhydryl groups and loss of tertiary structure. ATP has been shown to enhance the sensitivity of Kp2 and Ac2 to oxygen.[41,42]

The native molecular weights of highly purified Fe proteins from various organisms are shown in Table V. Cp2, with a MW of 55,000, is smaller than other Fe proteins investigated, which have average molecular weights in the range 62,000–66,000. The subunit structures of these proteins have been investigated by SDS electrophoresis, and in all cases a single band was observed. Ultracentrifuge data for Av2 in 8 $M$ urea are consistent with a single species of MW 30,000. These data are summarized in Table V and are consistent with the Fe proteins being dimers comprised of identical subunits. In the case of Cp2, this assignment is supported by amino acid sequence studies.[43]

The amino acid compositions of Cp2, Ac2, Kp2, and Rl2 are similar[15]; tryptophan is absent, and, as with the Mo–Fe proteins, acidic residues

[41] R. R. Eady, B. E. Smith, R. N. F. Thorneley, M. G. Yates, and J. R. Postgate, *in* "Nitrogen Fixation by Free-living Microorganisms" (W. P. D. Stewart, ed.), p. 377. Cambridge Univ. Press, London and New York, 1975.
[42] M. G. Yates, *FEBS Lett.* **8**, 281 (1970).
[43] M. Tanaka, M. Haniu, K. T. Yasunobu, and L. E. Mortenson, *in* "Iron and Copper Proteins" (K. T. Yasunobu, H. F. Mower, and O. Hayaishi, eds.), p. 83, Plenum, New York, 1976.

(about 20%) are about twice as abundant as basic residues. The amino acid sequence of Cp2 has been determined. Unlike the ferredoxins, the 6 cysteine residues that each subunit of this protein contains are not clustered but are widely distributed among the 273 residues of the polypeptide chain.[43] The Fe proteins contain approximately 4 gm atoms each of Fe and $S^{2-}$ per molecule (see Table VI). Cluster extrusion techniques using thiolate ligands have recently been used to establish that these are present in a single $Fe_4S_4$ cluster in Cp2[44,45] (this volume, [71]). How the single $Fe_4S_4$ cluster is distributed between the two identical subunits is not known, but from the distribution of cysteine residues in the protein it has been suggested that it bridges the two subunits.

At temperatures below 30°K $S_2O_4^{2-}$-reduced Fe proteins have a rhombic EPR spectrum with $g_{av} = 1.94$ similar to those of reduced ferredoxins (see Table VII). The intensity of this signal depends on the source the protein was isolated from, but always integrates to less than 1 electron/mole. Oxidation of Cp2[27] and Ac2[2] with a tenfold excess of PMS bleaches the EPR signal without significant loss of activity. Stopped-flow studies have shown that on rereduction of oxidized Ac2 with $SO_2^-$· (derived from the dissociation of $S_2O_4^{2-}$), 1 electron is rapidly ($K > 10^8$ $M^{-1}$ sec$^{-1}$) taken up and the EPR signal fully restored.[46] Spectral measurements indicate that a second electron is subsequently taken up in three slow phases without further change in the EPR spectrum. This behavior is difficult to reconcile with a redox process involving a single $Fe_4S_4$ cluster that this protein is thought to contain. During turnover the EPR signal is 90% bleached indicating that the protein is predominantly in the oxidized form.

The Mössbauer spectrum of $^{57}Fe$ substituted $S_2O_4^{2-}$-reduced Kp2[34] at 195°K is a symmetrical doublet ($\delta$ 0.52 mm/sec, $\Delta E$ 1.05 mm/sec) which at 4.2°K broadens into a multiplet indicating magnetic character (Fig. 4). At 77° K a minor species at +1.3 mm/sec is apparent in the spectrum, indicating some differences in the environment of the Fe atoms. These spectra are very similar to those of reduced ferredoxins which contain $Fe_4S_4$ clusters[47] and are strikingly different from reduced ferredoxins

[44] W. H. Orme-Johnson, L. C. Davis, M. T. Henzl, B. A. Averill, N. R. Orme-Johnson, E. Munck, and R. Zimmerman, in "Recent Developments in Nitrogen Fixation" (W. Newton, J. R. Postgate, and C. Rodriguez Barrueco, eds.), p. 131. Academic Press, New York, 1977.

[45] W. O. Gillum, L. E. Mortenson, J.-S. Chen, and R. H. Holm, *J. Am. Chem. Soc.* **99,** 584 (1977).

[46] M. G. Yates, R. N. F. Thorneley, and D. J. Lowe, *FEBS Lett.* **60,** 89 (1975).

[47] R. N. Mullinger, R. Cammack, K. K. Rao, D. O. Hall, D. P. E. Dickson, C. E. Johnson, J. D. Rush, and A. Simopulos, *Biochem. J.* **151,** 75 (1975).

TABLE VII

EPR SIGNALS OF HIGHLY PURIFIED Fe PROTEINS

| | | g Values | | Integrated intensity (spin/mole) | Specific activity (nmole $C_2H_2$ reduced min/mg protein) |
|---|---|---|---|---|---|
| Kp2 | 2.053 | 1.942 | 1.865 | 0.45 ± 0.07 | 600–1500[a] |
| Ac2 | 2.05 | 1.94 | 1.87 | 0.17[b]–0.24[c] | 2000[b] |
| Av2 | 2.05 | 1.94 | 1.88 | — | 1815[d] |
| Cp2 | 2.06 | 1.94 | 1.88 | 0.25 | 2400[f] |
| | 2.06 | 1.94 | 1.88 | 0.79 | 1200[e] |
| | 2.05 | 1.94 | 1.87 | 0.2 | 2200[g] |

[a] B. E. Smith, D. J. Lowe, and R. C. Bray, *Biochem. J.* **135**, 331 (1973).

[b] M. G. Yates and D. J. Lowe, *Eur. J. Biochem.* **60**, 467 (1975).

[c] M. G. Yates, R. N. F. Thorneley, and D. J. Lowe, *FEBS Lett.* **60**, 89 (1975).

[d] V. K. Shah and W. J. Brill, *Biochim. Biophys. Acta* **305**, 445 (1973).

[e] W. H. Orme-Johnson, W. D. Hamilton, T. Ljones, M.-Y. W. Tso, R. H. Burris, V. K. Shah, and W. J. Brill, *Proc. Natl. Acad. Sci. U.S.A.* **69**, 3142 (1972).

[f] W. H. Orme-Johnson, L. C. Davis, M. T. Henzl, B. A. Averill, N. R. Orme-Johnson, E. Munck, and R. Zimmerman, *in* "Recent Developments in Nitrogen Fixation" (W. Newton, J. R. Postgate, and C. Rodriguez Barrueco, eds.), p. 131. Academic Press, New York, 1977.

[g] G. Palmer, J. S. Multani, W. C. Cretney, W. G. Zumft, and L. E. Mortenson, *Arch. Biochem. Biophys.* **153**, 325 (1972).

containing $Fe_2S_2$ clusters,[48] where the two types of Fe atoms can be distinguished in the Mössbauer spectra.

The optical spectra of Kp2, Cp2, Ac2, and Rj2 consist of a broad absorption devoid of any major features.[15] The increase in absorption near 425 nm on oxidation has been used in presteady state studies on the electron transfer between the Fe protein and Mo–Fe protein[49] and the reduction of oxidized Fe protein by $SO_2^-$.[46,50] The increase in absorbance of Cp2 on oxidation by Cp1 *plus* ATP in the presence of limiting $Na_2S_2O_4$ parallels the decrease in intensity of the EPR signal of Cp2.[51]

The circular dichroism spectra of Kp2 and Cp2 do not show any transitions in the visible region,[1a,20] in contrast to the ferredoxins. The α-helix content of both proteins decreases markedly on $O_2$ inactivation. The redox potential of Cp2 has been determined using both optical and

[48] W. R. Dunham, A. J. Bearden, I. T. Salmeen, G. Palmer, R. H. Sands, W. H. Orme-Johnson, and H. Beinert, *Biochim. Biophys. Acta* **253**, 134 (1971).

[49] R. N. F. Thorneley, *Biochem. J.* **145**, 391 (1975).

[50] R. N. F. Thorneley, M. G. Yates, and D. J. Lowe, *Biochem. J.* **155**, 137 (1976).

[51] T. Ljones, *Biochim. Biophys. Acta* **321**, 103 (1973).

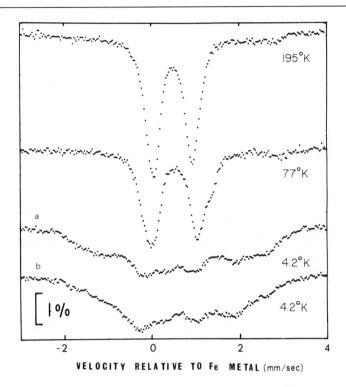

FIG. 4. Mössbauer spectrum of $S_2O_4{}^{2-}$-reduced Kp2 enriched with $^{57}$Fe. (a) 4.2° K with a 100 G magnetic field parallel to the $\gamma$-ray beam. (b) 4.2°K with a 550 G magnetic field perpendicular to the $\gamma$-ray beam. [After B. E. Smith and G. Lang, *Biochem. J.* **137**, 169 (1974).]

EPR potentiometric techniques. In the optical method the $S_2O_4{}^{2-}$-reduced protein is titrated with various oxidized redox dyes, and after equilibration the potential determined by spectral measurements of the dye which can be correlated with the change in protein absorbance. This technique gives a midpoint potential $E_{m,7.5} = -240$ mV, and an extrapolated value of 1.4–2 electrons to reduce fully oxidized Cp2.[39] The number of electrons required for this process is in good agreement with the value of 2 obtained[46] for the reduction of PMS-oxidized Ac2 by $S_2O_4{}^{2-}$. The redox properties of Cp2 have been investigated using the EPR potentiometric technique and the $H^+/H_2$ hydrogenase couple[31] or $S_2O_4{}^{2-}$ and $K_3Fe(CN)_6$ to poise the potential.[36] From the variation in intensity of the EPR signal, values of $E_{m,7.5}$ of $-294 \pm 20$ mV and $-270$ mV were obtained. As discussed below, a decrease in potential to $-400$ or $-490$ mV was observed[31,36] in the presence of MgATP and to $-380$ mV in the presence of MgADP. Binding of ATP to Cp2, Kp2, and Ac2 has been demonstrated by gel filtration in columns equilibrated with $^{14}C$-labeled nucleotide.[15] More recently[52] gel equilibration studies have given quantitative data for Cp2. MgATP binds to two sites with a $K_d = 16.7 \, \mu M$; MgADP inhibits the binding of MgATP to one of these sites and increases the affinity of the uninhibited site for MgATP approximately twofold. MgADP binds with a $K_d = 5.2 \, \mu M$. The MgATP-induced change in the EPR signal from rhombic to axial symmetry has also been used to monitor the binding of MgATP to Cp2 and Kp2. The data are consistent with tight binding of 2 molecules of ATP to Cp2 but looser binding to Kp2 with $K_d = 0.4 \, mM$. Problems associated with the interpretation of binding as monitored by EPR changes are discussed in Orme-Johnson and Davis[13] and Eady and Smith.[15] The change in conformation caused by MgATP binding is reflected in a large increase in titer and reactivity of sulphydryl groups in Kp2 to DTNB[53] and Cp2 by the increased reactivity of the Fe atoms with $\alpha,\alpha'$-bipyridyl.[54] MgADP does not alter the latter rate but increases sulfhydryl group reactivity, although to a lesser extent than MgATP. The midpoint potential of Cp2 ($-294 \pm 20$ mV) is decreased to $-400 \pm 20$ mV by binding of MgATP. MgADP has a similar effect. The changes in conformation which accompany the decrease in redox potential are the consequence of binding not hydrolysis of ATP.[31] The 110 mV decrease in potential has been attributed[36] to the 70-fold tighter binding of MgATP to the oxidized Fe protein with $K_d = 0.3 \, \mu M$.

[52] M.-Y. W. Tso and R. H. Burris, *Biochim. Biophys. Acta* **309**, 263 (1973).
[53] R. N. F. Thorneley and R. R. Eady, *Biochem. J.* **133**, 405 (1974).
[54] G. A. Walker and L. E. Mortenson, *Biochem. Biophys. Res. Commun.* **53**, 904 (1973).

# [71] Quantitative and Qualitative Characterization of the Iron–Sulfur Centers of Proteins

By L. E. MORTENSON[1] and W. O. GILLUM

## I. Introduction[2–7]

There are three well-defined iron–sulfur centers known to be present as active site chromophores of various enzymes and electron-transferring iron–sulfur proteins. These are the $Fe_4S_4^*(Cys)_4$ center found in the so-called "bacterial" 4 Fe and 8 Fe type ferredoxins (Fd), the $Fe_2S_2^*(Cys)_4$ center found in the "plant" 2 Fe type ferredoxins, and the $Fe(Cys)_4$ center found in rubredoxin.[8] In addition, there appears to be at least one more type of center, that found in the MoFe protein of nitrogenase. The

[1] The work of L. E. Mortenson was supported by AI04865-18 from the National Institutes of Health and PCM77-2465 from the National SCIENCE Foundation.

[2] The process of removing iron–sulfur sites from intact proteins, as described in this review, is called "extrusion," core extrusion or active site extrusion. Displacement is the analogous term used in Orme-Johnson et al.[3,4] Ligand exchange specifically describes the replacement of one ligand for another on the irons of the active site and does not depend on the full set of extrusion conditions. Ligand exchange is described in Holm et al.[5,6] and is often used synonymously with extrusion, core extrusion or active site extrusion. $Fe_4S_4^*(Cys)_4$, $Fe_2S_2^*(Cys)_4$, and $Fe(Cys)_4$ are the shorthand chemical formulas for the known iron–sulfur organizational types in proteins and enzymes, where Cys is cysteine and $*$ is $S^{2-}$. They are referred to as active sites or centers. The first two are called iron–sulfur clusters where their "cores" are $Fe_4S_4^*$ and $Fe_2S_2^*$, respectively. The single Fe is nominally the "core" of the $Fe(Cys)_4$ center. These centers are termed tetramers, dimers, and monomers where Cys is viewed as a generalized thiol $(Rs^-)$.[5–7] The total oxidation state of a center is defined by the sum of the formal oxidation states of all the irons, sulfides, and terminal thiols in a particular site. $1-$, $2-$, and $3-$ are the physiologically usable oxidation levels known; however, $4-$ is theoretically available to $Fe_4S_4^*(SR)_4$ and $Fe_2S_2(SR)_4$ centers.

[3] B. A. Averill, J. R. Bale, and W. H. Orme-Johnson, *J. Am. Chem. Soc.* **100**, 3034 (1978).

[4] W. H. Orme-Johnson and R. H. Holm, Vol. 53, p. 268.

[5] L. Que, Jr., M. A. Bobrik, J. A. Ibers, and R. H. Holm, *J. Am. Chem. Soc.* **96**, 4168 (1974).

[6] J. J. Mayerle, S. E. Denmark, B. V. DePamphilis, J. A. Ibers, and R. H. Holm, *J. Amer. Chem. Soc.* **97**, 1032 (1975).

[7] R. W. Lane, J. A. Ibers, R. B. Frankel, G. C. Papaefthymion, and R. H. Holm, *J. Amer. Chem. Soc.* **99**, 84 (1977).

[8] R. H. Holm, *in* "Biological Aspects of Inorganic Chemistry" (D. Dolphin, ed.), p. 71. Wiley, New York, 1977.

latter has not been completely characterized but has a reported composition of $Mo_1Fe_8S_6$.[9]

In addition to its presence in the electron-transferring ferredoxins, the $Fe_4S_4^*(Cys)_4$ center has been found as the sole chromophore and active center of both clostridial hydrogenase and the Fe protein of nitrogenase.[10] There exists a more complicated class of metalloproteins that contains other types of chromophores in addition to the iron–sulfur centers. This group includes xanthine oxidase, which contains $Fe_2S_2^*(Cys)_4$ centers[11] as well as a Mo center and a flavin component, and trimethylamine dehydrogenase, which contains a flavin and a $Fe_4S_4^*(Cys)_4$ center.[12] Finally, there is the class comprised of the MoFe proteins of the nitrogenase systems which are thought to contain more than one type of iron–sulfur centers, i.e., the $Fe_2S_2^*(Cys)_4$ and $Fe_4S_4^*(Cys)_4$ cores as well as another with the stoichiometry $Mo_1Fe_8S_6$.[9] Recent structural evidence using XAS (X-ray absorption spectroscopy) has shown the Mo to be in an environment closely associated with both iron and sulfur indicating a metallic center previously uncharacterized.[13]

Although electron paramagnetic resonance (EPR) spectra of these iron–sulfur proteins are indicative of the type of centers present, use of this method to identify and quantitate the iron–sulfur centers is difficult since the EPR spectra of all iron–sulfur proteins of even a single type are not the same, and other paramagnetic centers interfere.[14] In addition, in some iron–sulfur proteins, even under the best experimental conditions, not all the protein is in the state that has the known EPR signal. Therefore, quantitation of the EPR signal would not give an accurate value for the numbers and kinds of centers in the protein. In this article we describe those methods found suitable for measuring the type and amount of iron–sulfur centers found in various proteins and the conditions necessary for quantitation of these centers.

In general, the method requires (a) treatment of the protein with an organic solvent to partially denature the protein, which allows the iron–sulfur center to be accessible to ligand exchange and produces a less

[9] V. K. Shah and W. H. Brill, *Proc. Natl. Acad. Sci. U.S.A.* **74**, 3249 (1977).

[10] W. O. Gillum, L. E. Mortensen, J.-S. Chen, and R. H. Holm, *J. Amer. Chem. Soc.* **99**, 584 (1977).

[11] G. Wong, D. Kurtz, R. H. Holm, L. E. Mortenson, and R. G Upchurch, *J. Amer. Chem. Soc.* **101**, 3078 (1979).

[12] C. O. Hill, D. J. Steenkamp, R. H. Holm, and T. P. Singer, *Proc. Natl. Acad. Sci. U.S.A.* **74**, 547 (1977).

[13] S. P. Cramer, K. O. Hodgson, W. O. Gillum, and L. E. Mortenson, *J. Amer. Chem. Soc.* **100**, 3398 (1978).

[14] W. H. Orme-Johnson, *Annu. Rev. Biochem.* **42**, 159 (1977).

aqueous environment to stabilize the extruded thiol–FeS complex; (b) conversion to a stable core oxidation state (usually the 2– total oxidation state of the center); (c) the extrusion of the FeS center from the uncharacterized protein by use of an excess of a thiol reagent where the interaction between known FeS centers, and this thiol ligand is thoroughly characterized; and (d) the identification of the extruded FeS center complexed to the known thiol by spectrophotometric or other means.[10,15]

## II. Quantitative Identification of Iron–Sulfur Centers

### A. Materials

Hexamethylphosphoramide (HMPA)[16] and benzenethiol (PhSH) are fractionally distilled and stored under dinitrogen. HMPA may be frozen to further retard decomposition. $(Ph_4As)_2[Fe_4S_4(SPh)_4]$, $(Et_4N)_2$ $[Fe_2S_2(SPh)_4]$, and $o$-xylyl-$\alpha,\alpha'$-dithiol are synthesized by procedures described elsewhere.[5,6] The core extrusion or ligand-exchange technique requires a vacuum manifold capable of reducing oxygen concentrations to the nanomolar range. The entire reaction may be completed and analyzed in quartz semimicrocuvettes attached via a stainless steel needle through an appropriate rubber septum. For a description of the use of alternative thiols, solvent mixtures and reaction vessels see Averill et al.[3]

Protein samples used must be purified to the highest degree based on accepted criteria for purity and activity. Otherwise, the quantitative and even the qualitative nature of the iron–sulfur centers will remain equivocal.

### B. Methods[10,15]

Stock protein solutions should be placed in a buffer of appropriate pH (7–9) that is compatible with the solubility of the organic solvent system. Tris-Cl 50 m$M$ at pH 8.5–9.0 has proved amenable to 80% HMPA–20% Tris solutions over a range of protein molecular weights up to 400,000. Stock solutions should be adjusted to yield about 30–45 $\mu M$ iron–sulfur cluster in the final solution if a 1.0 cm cell is to be used. The 1.0 cm semimicrocuvettes are not a limitation, but merely a convenient volume (about 1.6 ml).

[15] L. Que, Jr., R. H. Holm, and L. E. Mortenson, J. Amer. Chem. Soc. 97, 463 (1975).
[16] Attention is drawn to the possible carcinogenic hazard of this compound; see, for example Chem. & Eng. News 54(5), 3 (1976); T. Fujinaga, K. Izutsu, and S. Sakura, Pure Appl. Chem. 44, 117 (1975).

In a typical experiment 0.2 ml stock protein solution is introduced anaerobically into a cuvette (containing an anaerobic atmosphere) through a rubber septum secured by tape. The septum should be tight so that the removal of needles will not dislodge it. After addition of the protein, the system is degassed six to ten times by alternately cycling between about 100 $\mu$m pressure and deoxygenated argon at 1 atm. Precaution should be taken to prevent excessive foaming during these evacuation cycles. All syringes should be flushed with anaerobic buffer or with argon before using to prevent the introduction of $O_2$ when accepting and transferring samples. The protein solution in the cuvette is cooled in an ice bath and kept on the anaerobic vacuum system during addition of the denaturing aprotic HMPA. At this point 0.8 ml of $O_2$-free and cooled HMPA is transferred from a gas-tight syringe to the cuvette. This addition should be slow since rapid mixing can allow the high heat of solution from HMPA/$H_2O$ to destroy the protein and hence its FeS center. After an additional three to four evacuation and gassing cycles, the cuvette is detached from the vacuum system, mixed slowly, and brought to room temperature over a 5 to 10 min period. A spectrum of the mixture is then taken between 700 and 300 nm which, when compared to an aqueous sample of the protein, will show a shift in its absorption maxima characteristic of unfolding of the protein (Fig. 1). To complete the dissociation of the iron–sulfur cluster(s) from the protein, a 100-fold molar excess of benzenethiol over the iron in the sample is added anaerobically by syringe. The final concentrations of Fe and thiol in the cuvette correspond to 120–180 $\mu M$ and 12–18 m$M$, respectively. Spectra are taken periodically until ligand exchange is complete as signaled by the development of a stable spectrum of the new iron–sulfur complex based on the well-characterized synthetic thiol derivatives (Fig. 1). The reaction for C. pasteurianum ferredoxin containing two $Fe_4S_4^*(Cys)_4$ centers is complete within 30 min, while spinach ferredoxin requires about 90 min at the stated thiol concentrations to yield its one $Fe_2S_2^*$ core. An alternative method is to add the thiol and the HMPA together, which will allow minimal decomposition of the FeS protein while the extrusion proceeds. Since the thiol is a good reducing agent, the addition of the combined reagents is a precautionary procedure in the case of very $O_2$-sensitive proteins; however, most systems tested did not require this extra protection, and solutions usually were stable for >24 hr.

Quantitation of this technique depends on the absorption spectra and extinction coefficients of well-characterized synthetic iron–sulfur compounds that are identical to the end product chromophores produced by the extrusion reactions. The nonpresumptive procedure is to synthesize $(Ph_4As)_2[Fe_4S_4(SPh)_4]$ and $(Et_4N)_2[Fe_2S_2(SPh)_4]$, place them in the appro-

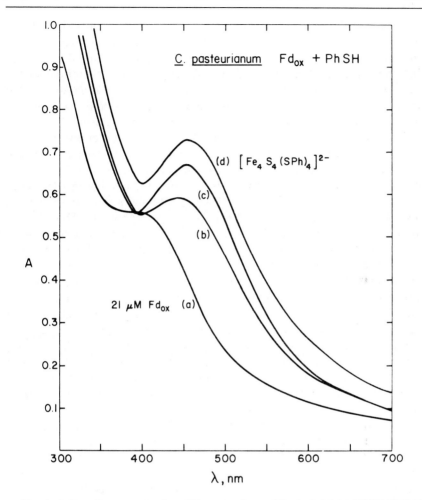

FIG. 1. Active site core extrusion of *C. pasteurianum* $Fd_{ox}$ in 4:1 (v/v) HMPA/$H_2O$ (50 m$M$ Tris-Cl, pH 8.5), $l = 1$ cm, 25° (a) 21 $\mu M$ $Fd_{ox}$, (b) 1 ml of solution (a) + 0.2 $\mu$l of PhSH 6 min after mixing; (c) solution (b) after 54 min; (d) solution (c) + 0.8 $\mu$l of PhSH 4 min after mixing. Spectrum (d) was constant with time. Ox, oxidized or the (2−) state for the proteins in this and in Fig. 2.

priate extrusion media, and calculate the extinction coefficients. This further allows calibration of the spectrophotometer with literature values. However, it is reasonably acceptable to apply the known literature values directly, for example, $[Fe_4S_4(SPh)_4]^{2-}$, $\epsilon_{458} = 17,200$ $M^{-1}cm^{-1}$ and $[Fe_2S_2(SPh)_4]^{2-}$; $\epsilon_{482} = 11,900$ $M^{-1}cm^{-1}$. The latter constants refer to the concentration of FeS cluster, but extinction coefficients on a per iron basis are also used. The ratio of the concentration of extruded cluster to that of the original protein is the number ($n$) of clusters per individual protein. To verify the absorption maxima, background spectra should be examined. For example, mersalyl ([3-[[2-(carboxymethoxy)benzoyl ]amino]-2-methoxypropyl]hydroxymercucy monosodium salt) treatment, which destroys iron–sulfur centers, allows one to measure absorption of the protein without the center, and a correction for apoprotein absorption can be made. No significant absorption was observed in the visible region for the apoprotein of *C. pasteurianum* ferredoxin. Absorbancies of all reagents (e.g., thiols and/or oxidants) should also be subtracted from the measured spectrum before calculating concentrations of the extruded FeS centers. In many cases this will be insignificant and in general less than 4% of the total absorbance. After utilizing these corrections, the authors have found only a 5% error in the extrusion technique with perhaps the exception of the very air sensitive iron protein of the nitrogenase system from *C. pasteurianum* (<10%).

## III. Variations of the General Method

Although the stated method can be used over a large concentration range for the prototype (*C. pasteurianum* ferredoxin) and is easily accomplished, the identification and quantitation of the iron–sulfur centers of many iron–sulfur proteins demands additional observations and techniques. These specific methods are discussed in terms of the known iron-sulfur types.

### A. $Fe_4S_4^*$ Proteins

1. In some proteins, the iron–sulfur centers are inaccessible to the extruding thiol even after the "denaturing" solvent is added. For example, when a 100-fold excess of benzenethiol was added to solutions of the high potential protein from *Chromatium* and *B. stearothermophilus* ferredoxin, the familiar spectrum of $[Fe_4S_4(SPh)_4]^{2-}$ began to develop, but these extrusion reactions came to equilibrium without accounting for all the iron of the proteins in the extruded products. To extrude the centers fully, a mole ratio of PhSH/Fe of 1000:1 rather that of 100:1 was required. Band shape, absorbance maxima, and recovery of the total iron present in the product are criteria that show analytical completion of the reaction. In addition, absorbance of the benzenethiol may contribute significantly at very high concentrations; therefore, its absorption must be subtracted from the spectrum obtained.

2. Since the conclusion as to how many iron–sulfur centers are present depends upon the production of $[Fe_4S_4(SPh)_4]^{2-}$ (i.e., the 2− total oxi-

dation state of the analog), the redox nature of the original iron–sulfur protein to be extruded must be considered. It is known (or inferred) that the oxidation states of the holoproteins are equivalent to the $1-$, $2-$, and $3-$ total oxidation states of the analogs.[8,17] Therefore, unknown mixtures of chromophores could exist which might be impossible to analyze in the form of the holoprotein isolated. If proteins containing only $Fe_4S_4{}^*$ cores are considered, there are four possibilities for the initial protein oxidation state. (a) The oxidized state of many ferredoxins (e.g., *C. pasteurianum*) contains the $(2-)$ tetramer $(Fe_4S_4{}^*)$ as isolated, and their centers are easily extruded as such. (b) The oxidized state of the high potential iron-sulfur protein (HIPIP), equivalent to a $(1-)$ tetramer, may be present. For this core state, the use of the general method for extrusion has been shown to destroy the center and yield purple solutions with spectra resembling that of $FeCl_3$ plus thiol.[10] To prevent this, the protein is first reduced with dithionite to the $(2-)$ state before extrusion or, alternatively, one may add the thiol to the organic phase before it is applied to the oxidized $(1-)$ protein.[3] For the latter, reduction of the iron–sulfur center by thiol from the $(1-)$ to the $(2-)$ state apparently occurs concomitant with ligand exchange, and its cubane structure is maintained. (c) A third possibility is that the $Fe_4S_4{}^*$ core present in a protein is equivalent to $[Fe_4S_4(SPh)_4]^{3-}$. The extrusion analog of such a protein would be expected to yield a spectrum equivalent to the $3-$ analog. Experimentally it has been found, however, that a solution of analytically pure $[Fe_4S_4(SPh)_4]^{3-}$ autoxidizes in HMPA to give an equilibrium mixture (based on spectral analysis) of $\sim 90\%$ $[Fe_4S_4(SPh)_4]^{2-}$ and presumably $\sim 10\%$ of the $3-$ species. Addition of a solution of ferricyanide converts the remaining 10% of the $3-$ species to the $2-$ oxidation state. Extrusion in HMPA of the $3-$ form of *C. pasteurianum* Fd showed a similar conversion to the $2-$ form; although, in this case $\sim 20-30\%$ of the sites required further oxidation. (d) The final case involves the presence of more than one oxidation state in a protein. An example of this is hydrogenase from *C. pasteurianum* which contains a mixture of oxidation states of the tetramer (see discussion in Gillum *et al.*[10]).

From the above discussion a procedural sequence for extrusion of $Fe_4S_4{}^*$ cores emerges. First, apply the general method to the resting state of the isolated protein, then to any reduced cores $(3-)$ that are present add an oxidant (i.e., ferricyanide) until the spectrum becomes invariant. If the protein iron is not fully recovered in the analog clusters produced, then increase the thiol concentrations in a repeat experiment. If full

---

[17] R. H. Holm and J. A. Ibers, *in* "Iron-Sulfur Proteins" (W. Lovenberg, ed.), Vol. 3, p. 205. Academic Press, New York, 1977.

recovery still is not obtained, add the thiol to the organic phase initially so that any $1-$ clusters will be reduced to $2-$.

3. Another complication can arise when iron is present that is not associated with the clusters or concerned with activity present in the extrusion mixture. Such iron could come from protein decomposition during the procedure or be present as an impurity in the preparation. Such adventitious iron can, like mixtures of ferric chloride and benzenethiol, develop violet-colored complexes with absorption maxima at 550–560 nm and a broad tail that extends into the infrared.[10] It is difficult to correct for the absorption of these complexes since they decrease with time. Obviously, the presence of such complexes would increase absorbance at 452 nm and cause quantitative inaccuracies. In addition, the lack of knowledge of their presence would lead to the erroneous conclusion that iron–sulfur *dimers* exist in such preparations (see below). Since the tetramer spectrum drops rapidly from its maximum at 452 nm toward the infrared (see Fig. 1), we have used the absorbance ratio $A_{458}/A_{550}$ as a criterion for an acceptable extrusion reaction of a $Fe_4S_4{}^*$ protein site. Although analytical samples of $Fe_4S_4(SPh)_4]^{2-}$ gave an average value for this ratio of 2.18, extrusion of proteins containing $Fe_4S_4{}^*$ cores only reached a limit of 2.02, even though all iron was accounted for. A criterion for an acceptable extrusion is suggested to be $A_{458}/A_{550} > 1.8$. Values of 1.9 have consistently been observed when good anaerobic technique was exercised.

4. The reaction mixture for extrusion of iron–sulfur centers has a safeguard in that the high concentration of thiol used protects against oxidation and also can reduce tetramers from the $1-$ to the $2-$ oxidation state. Furthermore, a semiquantitative feature of the use of 80% HMPA/ 20% $H_2O$ is that all proteins containing at least one tetrameric cluster in the $2-$ total oxidation state show a well-defined peak or shoulder at 400–410 nm in the solvent mixture before addition of thiol. The band is either red-shifted from the intact protein peak or resolved from a nearly featureless spectra as in the cases of the Fe protein of nitrogenase and hydrogenase from *C. pasteurianum*. This observation is quite useful in initial judgements as to the choice of the extrusion procedure for unknown FeS proteins. It is unlikely that the 1 Fe and 2 Fe type iron–sulfur proteins would give the latter spectral shift,[6,7,10] but it is obvious that the presence of a large number of $Fe(Cys)_4$ and $Fe_2S_2{}^*(Cys)_4$ centers relative to tetrameric sites or of reduced $Fe_4S_4{}^*$ cores could make this approach equivocal.

## B. $Fe_2S_2{}^*$ Proteins

1. Most of the conclusions found for tetrameric sites in the preceding section apply to the $Fe_2S_2{}^*$ cores as well. A significant difference lies in

the reduced state of the 2 Fe ferredoxins. Their oxidation state analogue, $Fe_2S_2(SPh)_4^{3-}$, has not been isolated since spontaneous dimer $\rightarrow$ tetramer conversion occurs as seen below [Eq. (1)].

$$2\ [Fe_2S_2(SPh)_4]^{3-} \rightarrow [Fe_4S_4(SPh)_4]^{2-} + 4\ PhS^- \tag{1}$$

This rapid conversion,[6,18] if it occurred, could lead to misinterpretation of core type. One $Fe_2S_2^*$ protein, reduced spinach ferredoxin, has been quantitatively extruded (with our general method) as $Fe_2S_2^*(SPh)_4^{2-}$ when the reduction was accomplished either before or after introduction of HMPA. This is not an absolute control, however, since a protein with two reduced dimeric sites in close proximity might yield tetramer during extrusion before the ligand exchange could be completed. Since several other $Fe_2S_2^*$ proteins have been identified using very similar methods,[3,4] the interconversion does not seem to be a problem under the conditions used.

2. There exists a second route to tetramer from dimer as shown:

$$2\ Fe_2S_2(SPh)_4^{2-} \rightarrow Fe_4S_4(SPh)_4^{2-} + PhSSPh + 2\ PhS^- \tag{2}$$

This conversion occurs at low pH and low thiol concentration.[8,18] In this case, high pH and excess thiol in the extrusion medium would protect experimental integrity by favoring the back reaction of Eq. (2). A pure example of $(Et_4N)[Fe_2S_2(SPh)_4]$ showed no spectral change over 44 hr under our standard reaction conditions, and the absorption spectrum of the final extrusion mixture with spinach ferredoxin has remained unchanged for >24 hr.

The reverse of Eq. (2) has not been observed by us. However, Averill *et al.*[3] have reported its occurrence under certain conditions and obviously such conditions must be avoided.

3. Extraneous iron could be even a greater problem for quantitation of $Fe_2S_2^*$ species than for their $Fe_4S_4^*$ relatives, since simple iron salts plus benzenethiol yield products with spectra very similar to $[Fe_2S_2(SPh)_4]^{2-}$ in the 300–560 nm range. Further, the unstable nature of the impurity lends a continuously varying spectrum which obscures the dimeric absorption. This feature can affect the value of the $A_{458}/A_{550}$ ratio (~1.2 for spinach ferredoxin) and remove its credibility as a criterion for impurity. It is unfortunate that the simple iron species formed from adventitious iron is too transitory to quantitate; however, an unusually high absorbance at 700 nm relative to that seen for quantitative $Fe_2S_2^*$ extrusions may be taken as evidence of impurity.

It should be noted that the $o$-xylyl-$\alpha,\alpha'$-dithiol derivative of $Fe_2S_2^*$ or

[18] J. Cambray, R. W. Lane, A. G. Wedd, R. Johnson, and R. H. Holm, *Inorg. Chem.* **16,** 256 (1977).

$Fe_4S_4^*$ centers gives quite characteristic spectra in dimethyl sulfoxide (DMSO), and this reagent has been exploited by other workers.[3,4,15] This ligand does not seem to give well resolved spectra in HMPA.[3]

## C. 1 Fe–0 S Proteins

It is obvious from the above discussions that benzenethiol is not the reagent of choice for ligand exchange of a protein with only one iron and no inorganic sulfide. o-Xylyl-α,α'-dithiol, on the other hand, seems a logical choice, since it forms a stable analog with a simple Fe atom.[7] The successful quantitative removal of the single iron site from oxidized C. pasteurianum rubredoxin was accomplished with the dithiol in 4:1 (v/v) DMS/$H_2O$.[7] Other dithiols have been examined with somewhat less success.[3]

The formation of dimers and tetramers in the presence of o-xylyl-α,α'-dithiol will interfere with the detection of rubredoxin-like centers and vice versa. Comparison of intensities of published spectra[3,7,15] and correction for these interferences may be the only solution to the question. The problem of the conversion of monomer → dimer/tetramer should not be encountered unless sulfide is available in the total extrusion solution.[18] The reverse conversion, however, would contribute to the problem of extraneous iron when extruding $Fe_2S_2^*$ and $Fe_4S_4^*$ centers.

## D. Proteins with Other Chromophores

The methods described above have proved ideal for those proteins, such as the ferredoxins, hydrogenase, and the Fe protein of nitrogenase, that are soluble in the aprotic solvent used (HMPA) and that have only one type of chromophore. There are, however, many proteins that are more complex and possess other chromophores in addition to a single type of FeS center. An example is trimethylamine dehydrogenase[12] which has a flavin component in addition to a $Fe_4S_4^*(Cys)_4$ center. To circumvent this problem the absorption of the other chromophore (flavin), in the same spectral region as the PhSH derivative, was substracted from the overall spectrum of the extruded core. The amount of absorption to subtract was determined by treating a similar sample of protein in the extrusion solvent with the organic mercurial, mersalyl, which destroyed the iron–sulfur center but left the other chromophore intact. To further separate the absorption of the two chromophores the extruding reagent used was p-methoxybenzenethiol since the absorption spectrum of its $Fe_4S_4^*(SR)_4$ analog is red-shifted.

An additional FeS center has recently been described but not fully characterized, i.e., the so called "FeMo" cofactor isolated from the MoFe protein of nitrogenase.[9] Attempts to exchange the chromophore of this cofactor by ligand exchange with thiols has not yielded characterizable spectra. However, the EPR spectrum of the cofactor is sharpened in the presence of excess PhSH indicating a change in ligand environment of the iron.[19] Methods are now being developed for the characterization of this interesting center.[13,]

### E. $Fe_2S_2^*$ and $Fe_4S_4^*$ Cores in the Same Protein

There is a possibility that both $Fe_2S_2^*$ and $Fe_4S_4^*$ cores can exist in a single protein. This might be true of the MoFe protein of nitrogenase.[11,20] For proteins that only contain such mixed centers and no other iron centers or other interfering chromophores, the general method could be used. This can be seen in Fig. 2 where various mixtures of an 8-Fe Fd and a 2-Fe Fd were subjected to extrusion by PhSH in HMPA and the resulting spectra recorded. The theoretical absorption spectrum of the mixture after extrusion (calculated from the extinction coefficients of the respective extruded ferredoxins) are in good agreement with the experimentally obtained spectra. Thus, the use of simultaneous equations at two or more wavelengths, one could calculate the numbers of each center present.

If there are other centers present as well as $Fe_2S_2^*$ and $Fe_4S_4^*$, then either the "EPR method" of Orme-Johnson (Section V,A) or the fluorine NMR (nuclear magnetic resonance) method of Wong, Kurtz, Holm, and Mortenson (Section V,B) would be the method of choice. Specific details of the former method will be published in the near future.

### F. Additional Considerations

The preceding discussions are meant to be general guidelines for a set of specific reactions known to occur quantitatively with the restrictions outlined. Detailed accounts of these limitations relative to the analog compounds have been compiled by Holm[8] and others.[3,10] For appraisal of alternative reaction conditions see Averill et al.[3] and Orme-Johnson and Holm.[4]

### IV. Summary of Precautions[21]

In the preceding descriptions of the extrusion method, certain experimental conditions must be controlled to insure a quantitative and qualitative procedure. The most important guidelines to observe are (a) dis-

[19] J. Rawlings, V. K. Shaw, J. R. Chisnell, W. J. Brill, R. Zimmerman, E. Münck, and W. H. Orme-Johnson, J. Biol. Chem. 253, 1001 (1978).
[20] E. Münck and W. H. Orme-Johnson, personal communication.
[21] Figures 1 and 2 and section IV are taken from Gillum et al.[10]

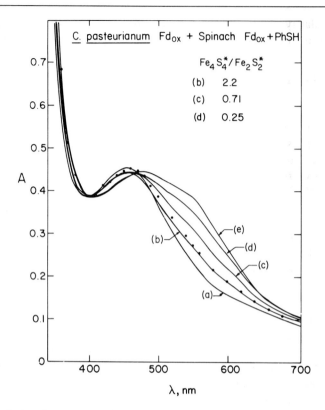

FIG. 2. Active site core extrusions of mixtures of *C. pasteurianum* $Fd_{ox}$ (8-Fe $Fd_{ox}$) and spinach $Fd_{ox}$ (2-Fe $Fd_{ox}$); experimental conditions are described in Gillum *et al.*[10] Spectra of the following initial protein solutions 90 min after addition of benzenethiol are shown: (a) 13 $\mu M$ 8-Fe $Fd_{ox}$; (b) 9.6 $\mu M$ 8-Fe $Fd_{ox}$ + 8.7 $\mu M$ 2-Fe $Fd_{ox}$; (c) 6.4 $\mu M$ 8-Fe $Fd_{ox}$ + 18 $\mu M$ 2-Fe $Fd_{ox}$; (d) 3.2 $\mu M$ 8-Fe $Fd_{ox}$ + 26 $\mu M$ 2-Fe $Fd_{ox}$; (e) 35 $\mu M$ 2-Fe $Fd_{ox}$. The solid points represent the reconstructed curve for solution (b) [0.75(a) + 0.25(e)]. Initial core mole ratios are indicated.

ruption of protein tertiary structure in a solvent medium in which the protein is soluble and stable; (b) anaerobicity at all stages of the reaction due to the oxygen sensitivity of analogs and proteins in aqueous–non-aqueous solutions; (c) employment of soluble thiols whose analog chromophores are substantially red-shifted compared to protein absorption; (d) addition of sufficient thiol and, where appropriate, oxidant to promote quantitative extrusion in a conveniently short time period; (e) adjustment of protein oxidation level, aqueous component pH and thiol concentration so as to prevent spontaneous dimer → tetramer conversion; and (f) minimization of extraneous iron content in protein preparations. The last is particularly important if 2 Fe and 4 Fe sites separately or as mixtures are to be quantitated.

## V. Development of Other Methods

### A. EPR

A clever extension of the general method has been devised by Bale et al.[22] In this method the proteins to be analyzed are unfolded in solvent in the presence of thiol just as described for the general method. The displaced core is in turn displaced from the "displacing" thiol by apoproteins of 2 Fe and 4 Fe ferredoxins. Since the apoproteins of these ferredoxins are very specific for their respective cores, only the correct 2 or 4 Fe cores will be accepted when a mixture of cores is presented to them. Thus if a protein having either a single type of core or a mixture of cores in addition to interfering chromophores is subjected to this method, either 2 Fe or 4 Fe ferredoxins are produced. The latter are then identified and quantitated by the use of EPR procedures, i.e., measurement of the height of spectra at various $g$ values compared to the holoferredoxins of the apoproteins used. Thus, another advantage to this method is the fact that optically resolved solutions are not required.

It is essential to assure system equilibrium after each step of the procedure and to provide an excess of the apoprotein (about 10:1) over the original protein to be identified. A recovery of about 90% of the cores has been obtained.

### B. NMR[6]

Another method of identifying iron–sulfur clusters has been developed using fluorine NMR. This approach takes advantage of the contact shifted fluorine resonances under the influence of different magnetic environments (i.e., iron–sulfur cores). The fluorinated dimeric and tetrameric species produced are separated by about 2.6 ppm when $p$-trifluoromethylbenzenethiol is used as the extruding reagent. In the case of all known centers, a signal enhancement proportional to the 24 fluorines is realized.

The general extrusion reaction is performed in the NMR tube with the exception that deuterated solvents are used. The sample is subjected to 1–5 hr of Fourier transform NMR irradiation, and the signals are integrated. To the final extrusion solution a known amount of iron–sulfur analog cluster made with the fluorinated thiol is added, and data are again averaged for the same amount of time under identical conditions. The

[22] J. R. Bale, B. A. Averill, and W. H. Orme-Johnson, (submitted for publication).

concentration of the unknown protein solution can be calculated from these data and compared to the iron present in the intact protein.

Because of the signal-to-noise ratio of the NMR technique, the optical method has the greater sensitivity. However, the NMR procedure has the advantage of not being affected by other prosthetic groups (e.g., flavins and sirohemes) and promises to be a generally applied extrusion procedure. With this technique one should be able to identify and quantitate unequivocally the type or types of iron–sulfur centers in proteins containing more than one center, even in the presence of other chromophores that interfere with the optical method. This includes such proteins as the MoFe protein of nitrogenase, xanthine oxidase, nitrate reductase, and sulfite reductase.

# [72] Isolation of the Iron–Molybdenum Cofactor from Nitrogenase

By VINOD K. SHAH

Activation of inactive nitrate reductase in extracts of *Neurospora crassa* mutant strain Nit-1 by acid-treated molybdoenzymes[1,2] offered a new concept of a cofactor common to different molybdoenzymes. When extracts of mutant strains of *Azotobacter vinelandii* and *Klebsiella pneumoniae* defective in component I (Fe–Mo protein) of nitrogenase were screened for[3-5] activation by acid-treated component I, strains analogous to Nit-1 were found. Acid-treated component I activated inactive component I in extracts of *A. vinelandii* mutant strain UW45, *K. pneumoniae* mutant strain UN106, and wild-type *A. vinelandii* that was derepressed for nitrogenase synthesis in tungsten-containing medium.[4-6] All efforts to isolate the activating factor by using conventional biochemical methods were unsuccessful. A unique method for the isolation of an iron–molyb-

[1] P. A. Ketchum, H. Y. Cambier, W. A. Frazier, III, C. H. Madansky, and A. Nason, *Proc. Natl. Acad. Sci. U.S.A.* **66,** 1016 (1970).
[2] A. Nason, K. Y. Lee, S. S. Pan, P. A. Ketchum, A. Lamberti, and J. DeVries, *Proc. Natl. Acad. Sci.* **68,** 3242 (1971).
[3] V. K. Shah, L. C. Davis, J. K. Gordon, W. H. Orme-Johnson, and W. J. Brill, *Biochem. Biophys. Acta* **292,** 246 (1973).
[4] H. H. Nagatani, V. K. Shah, and W. J. Brill, *J. Bacteriol.* **120,** 697 (1974).
[5] R. T. St. John, H. M. Johnston, C. Seidman, D. Garfinkel, J. K. Gordon, V. K. Shah, and W. J. Brill, *J. Bacteriol.* **121,** 759 (1975).
[6] H. H. Nagatani and W. J. Brill, *Biochim. Biophys. Acta* **362,** 160 (1974).

METHODS IN ENZYMOLOGY, VOL. 69

denum cofactor (FeMo-Co) described here yields concentrated stable preparations suitable for electron paramagnetic resonance and Mössbauer spectroscopy.

This method can be used for the isolation of FeMo-Co from component I from a variety of nitrogen-fixing organisms. Molybdenum cofactors (Mo-Co) from other molybdoenzymes also can be isolated by this technique.[7]

### Methods

*Growth of the Organism.* *Azotobacter vinelandii* mutant strain UW45, which does not fix $N_2$, was grown in modified Burk's medium[8] containing $100\mu g$ N/ml as ammonium acetate at 30°. Small-scale cultures were grown in 200 ml of medium in 1 liter baffle flasks on a rotary shaker. Large-scale cultures were grown in 15 liters of medium in a 5 gallon carboy with vigorous aeration. Cells were harvested at 3 hr after exhaustion of $NH_4^+$, and the cell paste was stored in a tightly closed container at $-20°$ until use. Nitrogenase is fully derepressed during this 3 hr cultivation in N-free medium.[3]

The wild-type organism used was *A. vinelandii* OP which was grown in vigorously aerated modified Burk's N-free medium in a 150-liter pilot plant fermentor.[9] Cells were harvested in a Sharples centrifuge, while the culture was still in logarithmic growth phase. The cell paste was stored in tightly closed containers at $-20°$ and samples were used as needed.

*Preparation of Extracts.* Frozen cells were thawed and suspended in 4 volume of 4 $M$ glycerol in 0.025 $M$ Tris-HCl buffer (pH 7.4) for 30 min followed by centrifugation at 12,000 $g$ for 10 min at 0°–4°. After the glycerol was decanted, the pellet was loosened and the cells were lysed by vigorous shaking with 4 to 5 volumes of 0.025 $M$ Tris-HCl buffer (pH 7.4) containing 0.3 mg sodium dithionite per milliliter. Deoxyribonuclease I (5–10 $\mu g$/ml) was added and the contents were flushed with $N_2$ gas. The lysate was centrifuged at 12,000 $g$ for 30 min. All buffers were thoroughly sparged with $N_2$ before use.[10]

*Purification and Crystallization of Component I.* All operations were carried out under a $N_2$ or $H_2$ atmosphere at room temperature. The reagents used were sparged with $N_2$ and contained 0.3 mg/ml sodium dithionite. Chromatography on DEAE–cellulose was performed by a

---

[7] P. T. Pienkos, V. K. Shah, and W. J. Brill, *Proc. Natl. Acad. Sci. U.S.A.* **74**, 5468 (1977).

[8] G. W. Strandberg and P. W. Wilson, *Can. J. Microbiol.* **14**, 25 (1968).

[9] V. K. Shah and W. J. Brill, *Biochim. Biophys. Acta* **305**, 445 (1973).

[10] V. K. Shah, L. C. Davis, and W. J. Brill, *Biochim. Biophys. Acta* **256**, 498 (1972).

method similar to that described by Bulen and LeComte[11] with the modifications as indicated.[9] A Whatman DE52 (microgranular) column (2.5 × 17 cm) was reduced with 400 ml of 0.025 $M$ Tris-HCl buffer pH 7.4), and 100–150 ml of crude extract (about 3 gm of protein) was applied to the column with intermittent washing with the same buffer. The column was eluted successively with 2 bed volumes of 0.025 $M$ Tris-HCl buffer (pH 7.4) and 3 bed volumes of 0.1 $M$ and 1 bed volume of 0.25 $M$ NaCl in 0.025 $M$ Tris-HCl buffer (pH 7.4). Component I, appearing as a dark brown band in the 0.25 $M$ NaCl fraction, was collected in a 60-ml serum bottle fitted with a serum stopper and kept anaerobic by flushing with $N_2$ or $H_2$. More than 90% of component I was eluted in about 20 ml. This component I-containing fraction was heat-treated at 52° for 5 min under $H_2$ with agitation followed by rapid cooling and centrifugation at 12,000 $g$ for 10 min at 0°–4°. Component I was concentrated in an ultrafiltration cell with an Amicon XM100A membrane and then diluted with 5 volumes of 0.025 $M$ Tris-HCl buffer (pH 7.4) to bring the NaCl concentration to 0.042 $M$ and concentrated again. The NaCl concentration for optimal crystallization is 0.042 $M$ NaCl in 0.025 $M$ Tris-HCl buffer pH 7.4. This can also be achieved by repeated dilution of concentrated component I with 0.042 $M$ NaCl in 0.025 $M$ Tris-HCl (pH 7.4) following concentration. Component I, which begins to crystallize in the ultrafiltration cell, is transferred to a serum-capped centrifuge tube. The ultrafiltration cell is washed with a minimum volume of 0.042 $M$ NaCl in 0.025 $M$ Tris-HCl (pH 7.4) and the washing was combined with component I. The tube containing concentrated component I was kept in a water bath at 38° for 1 hr for maximum crystallization. The preparation was centrifuged at 20,000 $g$ for 10 min at 30°. The supernatant solution was removed and saved in a $O_2$-free serum-stoppered bottle. The dark brown pellet was washed twice by suspending it in 4–5 volume of 0.042 $M$ NaCl in 0.025 $M$ Tris-HCl (pH 7.4) followed by centrifugation at 20,000 $g$ for 30 min at approximately 30°. The component I pellet was dissolved in 0.25 $M$ NaCl in 0.025 $M$ Tris-HCl (pH 7.4) and centrifuged at 12,000 $g$ for 10 min at 0°–4° to remove a small amount of white amorphous material. Component I is distributed in 1–2 ml aliquots in serum-stoppered 5 ml vials flushed with $H_2$ and stored at −20°. The purification procedure[9] for obtaining crystalline component I is summarized in Table I. A small additional crop of crystalline component I can be obtained by subjecting the supernatant solution obtained by crystallization to a DEAE-cellulose column that had been washed with 300 ml of 0.15 $M$ NaCl in 0.025 $M$ Tris-HCl buffer (pH 7.4) and then repeating the technique.

[11] W. A. Bulen and J. R. LeComte, *Proc. Natl. Acad. Sci. U.S.A.* **56**, 979 (1966).

TABLE I

PURIFICATION OF NITROGENASE COMPONENT I OF *A. vinelandii*

| Fraction | Volume (ml) | Total activity (units × $10^{-4}$) | Total protein (mg) | Specific activity[a] | Recovery (%) |
|---|---|---|---|---|---|
| Crude extract | 200 | 24.0 | 5120 | 47 | 100 |
| DEAE-cellulose column fraction | 44 | 22.0 | 1126 | 195 | 92 |
| 52° supernatant | 42 | 22.0 | 792 | 278 | 92 |
| Crystallized | 5.5 | 16.9 | 103 | 1641 | 70 |

[a] Measured in nmoles ethylene formed per minute per milligram of protein. Activity of component I (in all fractions including crude extract) is obtained by performing the assays in presence of excess of component II.

*Isolation of the Iron–Molybdenum Cofactor.*[12] All operations were carried out under a $H_2$ atmosphere at 0°–4°. The reagents used were deoxygenated on a gassing manifold by evacuating and flushing with deoxygenated $H_2$ with constant mixing. These deoxygenated reagents contained 1.2 m$M$ sodium dithionite, added as a 0.1 $M$ aqueous solution (in 0.013 $N$ NaOH) just before use. The syringe-needle assembly used for each step was flushed with $H_2$ before use.

Approximately 75 mg of crystalline component I in 4 ml of 0.25 $M$ NaCl in 0.025 $M$ Tris-HCl buffer (pH 7.4), was diluted with 8 ml of water in a glass centrifuge tube containing four glass beads (4 mm size). To this diluted solution of component I, citric acid solution was added to a final concentration of 15 m$M$, mixed thoroughly, and allowed to stand for 3 min. Disodium hydrogen phosphate solution was added to a final concentration of 25 m$M$; the contents were mixed thoroughly and allowed to stand for 25–30 min for the complete precipitation of component I. The citrate/phosphate-treated component I was centrifuged at 8000 $g$ for 10 min, and the supernatant solution was removed with a 4 inch long hypodermic needle on a syringe. This citrate/phosphate supernatant solution contains only traces of FeMo-Co activity, but approximately 50% of the iron present in component I and can be discarded. The citrate/ phosphate-treated pellet of component I was washed twice by thoroughly suspending it in 8 ml of dimethylformamide, followed by centrifugation at 8000 $g$ for 10 min. This dimethylformamide supernatant solution contains less than 1% of the total FeMo-Co activity and can be discarded. The dimethylformamide-washed pellet was eluted three times by sus-

[12] V. K. Shah and W. J. Brill, *Proc. Natl. Acad. Sci. U.S.A.* **74**, 3249 (1977).

TABLE II
ISOLATION OF FeMo-Co FROM NITROGENASE COMPONENT I

| A. vinelandii fraction | Volume (ml) | Total activity (units) | Total Fe (nmole) | Total Mo (nmole) | Activity/ Mo ratio | Fe/Mo ratio | Yield (%) |
|---|---|---|---|---|---|---|---|
| Crystalline component I | 5 | 121,100 | 7273 | 441 | 275 | 16.5 | 100 |
| Citrate/phosphate supernatant solution | 17.2 | 113 | 3614 | 1.3 | — | — | — |
| DMF supernatant solution[a] | 16.4 | 1,702 | 88 | 13.4 | — | — | — |
| NMF supernatant solution (FEMo-Co)[b] | 12.4 | 106,300 | 3074 | 391 | 272 | 7.9 | 88 |
| Residual pellet | 5 | 4,125 | 188 | 31.2 | — | — | — |

[a] Values are for the pooled two $N,N$-dimethylformamide supernatant solutions.
[b] Values are for the pooled three $N$-methylformamide supernatant solutions.

pending it in 4 ml of $N$-methylformamide containing 5 m$M$ $Na_2HPO_4$ (added as a 0.2 $M$ aqueous solution) for 30 min, followed by centrifugation at 8000 $g$ for 10 min. The three $N$-methylformamide supernatants contain, respectively, 70, 15, and 5% of the FeMo-Co activity and molybdenum present in component I. Representative data[12] for the distribution of FeMo-Co activity and of molybdenum and iron in different fractions are shown in Table II.

*Further purification of FeMo-Co.* The FeMo-Co (about 90,000 units) was applied to an anaerobic Sephadex G-100 column (77 × 1 cm) in $N$-methylformamide to remove any contaminating Fe centers, Mo, Fe, or denatured component I protein that might be present. The column was equilibrated and eluted at a flow rate of 12 ml/hr with $N$-methylformamide containing 5 m$M$ $Na_2HPO_4$ and 1.2 m$M$ $Na_2S_2O_4$. Fractions were collected anaerobically and analyzed for FeMo-Co activity, Mo, and Fe. The FeMo-Co has a distinct brown color which was followed visually on the column. The results of a representative column are presented in Fig. 1. The FeMo-Co activity was eluted as a single peak, and the recovery of FeMo-Co activity, Mo, and Fe in this peak was 97–98%. The distribution of Fe and Mo coincides with FeMo-Co activity in different fractions, suggesting that the Fe is an integral part of the FeMo-Co. The Fe/Mo ratio in the FeMo-Co is 8:1.[12]

*Activation of Inactive Nitrogenase.* The acetylene reduction assays for nitrogenase activity were performed under $H_2$ atmosphere in 9 ml serum bottles, and all the reagents were thoroughly evacuated and flushed with $H_2$ before use. Cell-free extracts (6–7 mg of protein) of A. vinelandii

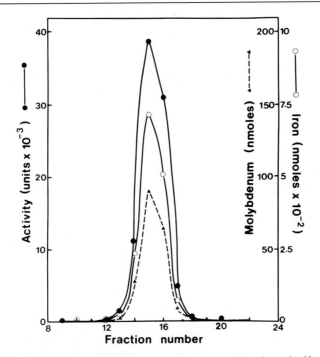

FIG. 1. Elution profile of FeMo-Co from a Sephadex G-100 column in $N$-methylform-amide. Dextran blue, FeMo-Co, and FeCl$_3$ are eluted at about 20, 35, and 50 ml, respectively. ●, Activity; ○, iron; ▲, molybdenum.

mutant strain UW45 or *K. pneumoniae* mutant strain UN106 derepressed for nitrogenase synthesis in a medium containing molybdate, or the wild-type *A. vinelandii* derepressed in a medium containing tungstate instead of molybdate, were incubated with varying amounts of FeMo-Co for 30 min at room temperature. After this preincubation, 0.8 ml of reaction mixture (made anaerobically) containing 2.5 μmoles ATP, 30 μmoles creatine phosphate, 0.2 mg creatine phosphokinase, 5 μmoles MgCl$_2$, 15 μmoles Tris-HCl buffer (pH 7.4), and 20 μmoles Na$_2$S$_2$O$_4$ was added. The reaction bottles were brought to 1 atm by pricking the serum stopper with a 22-gauge hypodermic needle. Acetylene (0.5 ml) was injected in each assay bottle, which was incubated at 30° for 15 min in a water-bath shaker. The reaction was terminated by injecting 0.1 ml of 30% (w/v) trichloroacetic acid. Ethylene formed was measured after 30 min with a Packard 407 gas chromatograph with Porapak N column.[4,10]

*Definition of Specific Activity.* Specific activity is expressed as the nanomoles of ethylene formed per minute per nanomoles of molybdenum.

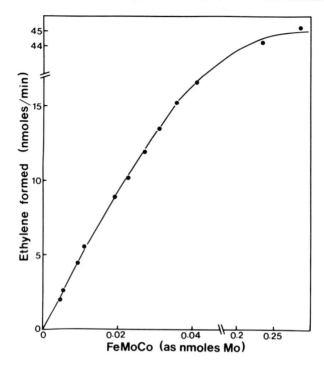

FIG. 2. FeMo-Co-dependent activation of inactive component I in extracts of mutant strain UW45.

One unit of FeMo-Co is defined as the amount required to produce 1 nmole of ethylene per minute under the reconstitution conditions used.

*Properties.* The FeMo-Co contains 8 iron atoms and 6 acid-labile sulfides per molybdenum atom. The specific activity of FeMo-Co is 272 (nmoles of ethylene formed/min/nmole of Mo). There is better than 98% reconstitution between FeMo-Co and inactive component I in *A. vinelandii* mutant strain UW45. Activation of inactive component I in extracts of mutant strain UW45 was dependent on the FeMo-Co concentration (Fig. 2) and followed saturation kinetics. Similar results were obtained when extracts of *K. pneumoniae* mutant strain UN106 or wild-type *A. vinelandii* that was derepressed for nitrogenase synthesis in tungsten-containing medium, was titrated with FeMo-Co. Per unit of FeMo-Co, activation of inactive component I in extracts of wild-type *A. vinelandii* derepressed for nitrogenase synthesis in tungsten-containing medium was about 20% lower than that obtained with strain UW45. This lower activation might be due to the competition by a corresponding tungsten-containing moiety with the FeMo-Co.[12]

TABLE III
Stability of FeMo-Co

| Conditions | Activating ability[a] (%) |
|---|---|
| Condition A[b] | |
| Aerobic for 1 min at 0° | 0 |
| Aerobic for 1 min at 25° | 0 |
| Anaerobic for 10 days at −20° | 100 |
| Anaerobic for 2 days at 0° | 99.5 |
| Anaerobic for 24 hr at 25° | 99 |
| Condition B[c] | |
| Aerobic for 1 min at 0° | 0 |
| Anaerobic for 18 hr at −20° | 81.1 |
| Anaerobic for 4 hr 0° | 56.2 |
| Anaerobic for 4 hr at 25° | 19.6 |
| Condition C[d] | |
| Aerobic for 1 min at 0° | 0 |
| Anaerobic for 18 hr at −20° | 71.2 |
| Anaerobic for 2 hr at 0° | 53.6 |
| Anaerobic for 2 hr at 25° | 16.3 |

[a] Based on activation of inactive component I *in vitro* in extracts of strain UW45. Activating ability of FeMo-Co in $N$-methylformamide assayed at 0 time is considered as 100%.
[b] FeMo-Co in $N$-methylformamide.
[c] FeMo-Co in 25 m$M$ Tris-HCl, pH 7.4, containing 1.2 m$M$ dithionite.
[d] FeMo-Co in citrate/phosphate buffer, pH 5.8 (ionic strength 0.45 $\mu$g), containing 1.2 m$M$ dithionite.

The FeMo-Co isolated from nitrogenase component I of *Clostridium pasteurianum, Rhodospirillum rubrum, Bacillus polymyxa,* and *K. pneumoniae* also activates inactive component I in the extracts of mutant strain UW45.[12] These observations show that FeMo-Co from anaerobic, facultative, and photosynthetic $N_2$-fixing organisms can activate component I in the obligate aerobe, *A. vinelandii.* Component I from *C. pasteurianum* does not cross-complement with component II from *A. vinelandii.*[13,14] However, the FeMo-Co from *C. pasteurianum* component

[13] R. W. Detroy, D. F. Witz, R. A. Parejko, and P. W. Wilson, *Proc. Natl. Acad. Sci. U.S.A.* **61,** 537 (1968).
[14] D. W. Emerich and R. H. Burris, *Proc. Natl. Acad. Sci. U.S.A.* **73,** 4369 (1976).

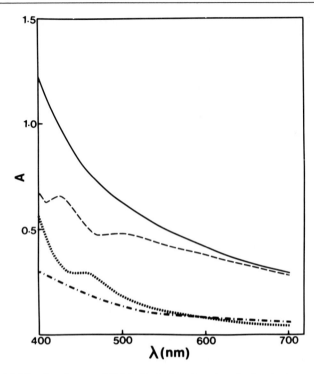

Fig. 3. Visible absorbance of FeMo-Co before (——) and after (····) exposure to air. Difference (– – –) spectra of the preceding two spectra. Effect of 2 equivalents of sodium mersalyl/Fe (–.–.–) on the visible absorbance spectrum of FeMo-Co. The FeMo-Co used for these studies had 26,500 units of activity per milliliter.

I fully activated inactive component I of *A. vinelandii* mutant strain UW45. It seems that FeMo-Co in various $N_2$-fixing organisms is very similar. The Fe/Mo ratio in FeMo-Co from *C. pasteurianum* component I was also 8:1. The FeMo-Co from *C. pasteurianum* also activated inactive component I in extracts of *K. pneumoniae* mutant strain UN106 or wild-type *A. vinelandii* that is derepressed for nitrogenase synthesis in tungsten-containing medium.

The FeMo-Co is extremely sensitive to $O_2$. Brief exposure to air completely abolished the activating ability of FeMo-Co (Table III). The storage temperature had no effect on the activating ability of FeMo-Co (in *N*-methylformamide) as long as it was kept anaerobic. The FeMo-Co is unstable in aqueous medium even under strictly anaerobic conditions, and the storage temperature has a pronounced effect on the stability of FeMo-Co in aqueous media. Nitrogenase component I is stable, whereas

FeMo-Co is unstable in an aqueous environment. This suggests that the component I protein core may be responsible for maintaining an aprotic environment at the active site of the FeMo-Co in component I.[12]

The visible spectrum of FeMo-Co has no distinctive features (Fig. 3). There appears to be a steady decline in absorbance from about 400 to 700 nm. Upon progressive oxidation of the FeMo-Co, the absorbance decreases, suggesting destruction of the chromophore. The destroyed chromophore is represented by the difference spectrum of FeMo-Co before and after exposure to air (Fig. 3). Addition of sodium mersalyl to FeMo-Co resulted in a substantial reduction in visible absorbance.

The electron paramagnetic resonance (EPR) spectrum of FeMo-Co (with two $g$ values near $g = 4$ and one near $g = 2$) represent fundamentally the same type of structure as observed in component I; the broad features and $g$ shifts reflect differences in the local environment.[15] The FeMo-Co from component I of *A. vinelandii, C. pasteurianum* and *K. pneumoniae* give identical EPR signals. Combined Mössbauer and EPR studies[16] have shown four spectroscopic classes of iron centers in component I. One of these centers, labeled $M_{EPR}$ (about 40% of the total Fe), undergoes reversible redox reactions during catalytic turnover of the enzyme system,[16] and this center is recovered in excellent yields in FeMo-Co. On brief exposure to air, FeMo-Co loses the EPR signal with concomitant loss of activating ability.[12,15]

[15] J. Rawlings, V. K. Shah, J. R. Chisnell, W. J. Brill, R. Zimmerman, E. Münck, and W. H. Orme-Johnson, *J. Biol. Chem.* **253**, 1001 (1978).
[16] E. Münck, H. Rhodes, W. H. Orme-Johnson, L. C. Davis, W. J. Brill, and V. K. Shah, *Biochim. Biophys. Acta* **400**, 32 (1975).

# [73] Heterocyst Isolation

*By* PETER FAY

## Introduction

Filamentous blue-green algae (cyanobacteria), members of the families Nostocaceae, Rivulariaceae, Scytonemataceae, and Stigonemataceae, when grown in a medium lacking combined nitrogen produce distinct cells called heterocysts by the transformation of a small proportion (about 5%) of the normal vegetative cells.[1,2] Heterocysts display char-

[1] G. E. Fogg, *New Phytol.* **43**, 164 (1944).
[2] G. E. Fogg, *Ann. Bot. (London)* [N. S.] **13**, 241 (1949).

METHODS IN ENZYMOLOGY, VOL. 69

acteristic structural and functional properties.[3] They appear under the light microscope as round yellowish-green cells, with a relatively homogenous content, surrounded by a thick secondary envelope. Heterocysts lack a functional photosystem II,[4,5] fix no $CO_2$,[6,7] and evolve no $O_2$ when illuminated,[8,9] though retain an active photosystem I capable of photophosphorylation.[10,11] They also possess an elevated respiratory metabolism[6,12-14] and can carry out oxidative phosphorylation.[10,11] The biochemical transformation of heterocysts is apparently essential in the performance of their principal function of atmospheric $N_2$ fixation.[15,16]

The once enigmatic heterocysts[3] have of late been the subject of intense studies due to their association with the fundamental biological process of $N_2$ fixation, and because they provide a model system for the study of a one-dimensional pattern of cell differentiation.[17,18] In order to distinguish between the physiological and biochemical properties of heterocysts and vegetative cells, it is essential to effect the isolation and complete separation of the two cell types. Further, the structural and functional integrity which heterocysts possessed within the intact filament must be preserved.

A method for the isolation of heterocysts, devised first in our laboratory,[6] was based on the selective mechanical disruption of vegetative cells in a French pressure cell. It was observed during the preparation of cell-free extracts from blue-green algae that heterocysts resisted breakage at a pressure (77 MN m[-2]) which caused the disruption of all vegetative cells.[19] Repeated differential centrifugation of this material resulted in a

[3] P. Fay, in "The Biology of Blue-Green Algae" (N. G. Carr and B. A. Whitton, eds.), p. 238. Blackwell, Oxford, 1973.

[4] M. Donze, J. Haveman, and P. Schiereck, Biochim. Biophys. Acta 256, 157 (1972).

[5] S. M. Klein, J. M. Jaynes, and L. P. Vernon, Proc. Int. Congr. Photosynth. Vol. 1, p. 703 (1975).

[6] P. Fay and A. E. Walsby, Nature (London) 209, 94 (1966).

[7] C. P. Wolk, J. Bacteriol. 96, 2138 (1968).

[8] S. Bradley and N. G. Carr, J. Gen. Microbiol. 68, xiii (1971).

[9] E. Tel-Or and W. D. P. Stewart, Nature (London) 258, 715 (1975).

[10] W. E. Scott and P. Fay, Br. Phycol. J. 7, 283 (1972).

[11] E. Tel-Or and W. D. P. Stewart, Biochim. Biophys. Acta 423, 189 (1976).

[12] P. Fay and S. A. Kulasooriya, Arch. Mikrobiol. 87, 341 (1972).

[13] F. Winkenbach and C. P. Wolk, Plant Physiol. 52, 480 (1973).

[14] R. B. Peterson and R. H. Burris, Arch. Microbiol. 108, 35 (1976).

[15] P. Fay, W. D. P. Stewart, A. E. Walsby, and G. E. Fogg, Nature (London) 220, 810 (1968).

[16] W. D. P. Stewart, A. Haystead, and H. W. Pearson, Nature (London) 224, 226 (1969).

[17] C. P. Wolk, Proc. Natl. Acad. Sci. U.S.A. 57, 1246 (1967).

[18] M. Wilcox, Nature (London) 228, 686 (1970).

[19] R. M. Cox, P. Fay, and G. E. Fogg, Biochim. Biophys. Acta 88, 208 (1964).

purified heterocyst preparation. When the metabolic activities of such isolated heterocysts were examined, surprisingly, no significant $^{15}N_2$ incorporation was measured, though vigorous $^{15}N_2$ fixation had taken place in the intact filament.[6] With the introduction of anaerobic techniques and the provision of essential requirements for the nitrogenase reaction (ATP-generating system and electron donor), it was possible to demonstrate nitrogenase activity in heterocysts isolated either in the French pressure cell or by sonication,[16,20] but levels of enzyme activity were much lower than those in whole filaments. This discrepancy, and the contraversy of whether or not vegetative cells of heterocystous algae can also synthesize nitrogenase,[21] called for an ultrastructural examination of heterocysts isolated by various methods.[22] The isolation of structurally intact and functionally competent heterocysts was thought essential in order to establish whether these cells are the sole or primary sites of $N_2$ fixation in these organisms.[15]

### Culture Methods

Although heterocystous species represent almost half of the known taxa of blue-green algae (708 out of 1650 listed by Geitler[23]), relatively few have been isolated in axenic culture.[24] In most studies on isolated heterocysts *Anabaena cylindrica* was used as the source organism but other species may be equally suitable.[5,25]

Information on culture media, apparatus and growth conditions required for the culture of blue-green algae can be found in a number of recent review articles and books.[26,27] The composition of a medium suitable for the growth of $N_2$-fixing species is given in Table I.

The use of an exponentially growing (log phase) culture is preferred for the isolation of heterocysts, partly because such cultures are usually free from extracellular mucilage which hinders the purification procedure, but also because at this stage cultures do not produce akinetes (spores)

[20] C. P. Wolk, *Ann. N.Y. Acad. Sci.* **175**, 641 (1970).
[21] R. V. Smith and M. C. W. Evans, *Nature (London)* **225**, 1253 (1970).
[22] P. Fay and N. J. Lang, *Proc. R. Soc. London Ser. B.* **178**, 185 (1971).
[23] L. Geitler, *in* "Kryptogamenflora von Deutschland, Österreich und der Schweiz" (L. Rabenhorst, ed.), Vol. 14. Akad. Verlagsges., Leipzig, 1932.
[24] E. A. George, "List of Strains." Culture Centre of Algae and Protozoa, Cambridge, 1976.
[25] H. Fleming and R. Haselkorn, *Proc. Natl. Acad. Sci. U.S.A.* **70**, 2727 (1973).
[26] N. G. Carr, *in* "Methods in Microbiology" (J. R. Norris and D. W. Ribbons, eds.), Vol. 3B, p. 53. Academic Press, New York, 1969.
[27] G. E. Fogg, W. D. P. Stewart, P. Fay, and A. E. Walsby, "The Blue-Green Algae." Academic Press, New York, 1973.

TABLE I
MEDIUM FOR THE CULTURE OF BLUE-GREEN ALGAE[a]

| Compound | Concentration ($\mu M$) | Compound | Concentration ($\mu M$) |
|---|---|---|---|
| NaCl | 4000 | $MnSo_4 \cdot 4H_2O$ | 9.2 |
| $MgSO_4 \cdot 7H_2O$ | 1000 | $MoO_3$ | 1.0 |
| $K_2HPO_4$[b] | 1000 | $ZnSO_4 \cdot 7H_2O$ | 0.77 |
| $CaCl_2 \cdot 6H_2O$ | 500 | $CuSO_4 \cdot 5H_2O$ | 0.32 |
| Fe–EDTA | 72 | $Co(NO_3)_2 \cdot 6H_2O$ | 0.17 |
| $H_3BO_3$ | 46 | pH 7.6 | |

[a] Modified after M. B. Allen and D. I. Arnon, *Plant Physiol.* **30**, 366 (1955).
[b] Autoclaved separately.

which would contaminate the heterocyst preparation; spores like heterocysts survive the isolation procedure.[6] For similar reasons continuous
cultures provide suitable material for heterocyst isolation.[27,28]

An excellent way to produce uniform and active heterocystous material is by controlled induction of heterocyst differentiation.[29] The algal
material is grown initially in the presence of ammonium-nitrogen which
suppresses heterocyst formation.[2] When such nondifferentiated material
is transferred into a medium free from combined nitrogen, and thus the
cellular C:N ratio rises above a critical level, heterocyst differentiation
(and nitrogenase synthesis) begins and proceeds more or less synchron-
ously, reaching a steady value after about 40 hr.[29]

## Methods Used for Heterocyst Isolation

The heterocystous filament of a blue-green alga is an organized
multicellular structure in which the individual cells are part of a higher
structural and functional unit. This view is corroborated by the presence
of intercellular connections between adjacent cells of the filament,[30,31] by
the transport of metabolites along the filament,[7] and by the presence in
the filament of gradients of metabolic activity.[12] More specifically, it was
shown that nitrogenase activity in the heterocyst is dependent on the
association of the latter with vegetative cells for the maintenance of

[28] P. Fay and S. A. Kulasooriya, *Br. Phycol. J.* **8**, 51 (1973).
[29] S. A. Kulasooriya, N. J. Lang, and P. Fay, *Proc. R. Soc. London Ser. B* **181**, 199 (1972).
[30] H. Drawert and I. Metzner, *Ber. Dtsch. Bot. Ges.* **69**, 291 (1956).
[31] H. S. Pankratz and C. G. Bowen, *Am. J. Bot.* **50**, 387 (1963).

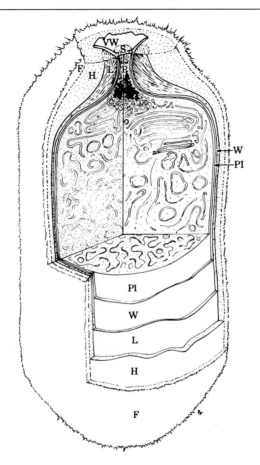

FIG. 1. Three dimensional view of a heterocyst sectioned in three planes and illustrating the construction of surface layers. Pl, plasmalemma; W, cell wall; L, laminated layer; H, homogeneous layer; and F, fibrous layer of the envelope; P, pore channel; S, septum; VW, portion of the vegetative cell wall. (From Lang and Fay.[32])

reducing conditions[12] and to enable the interchange of materials required or produced during $N_2$ fixation.[7,16] Hence separation of the heterocyst from the rest of the filament will inevitably affect the function and the survival of the isolated cell.

The cell wall of blue-green algae is a multilayered structure composed of four distinct layers incorporating a mucopolymer component similar to that found in the wall of gram-negative bacteria.[27] The cross wall (septum) in the filamentous forms, which develops during cell division as an annular ingrowth, is mostly composed of the two inner wall layers

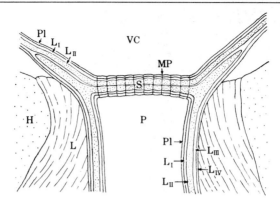

FIG. 2. Diagrammatic section through the septum between heterocyst and adjacent vegetative cell. P, pore channel; Pl, plasmalemma; $L_I$ to $L_{IV}$, successive layers of the cell wall; L, laminated layer; H, homogeneous layer of the envelope; S, septum; MP, microplasmodesmata; VC, vegetative cell. (From Lang and Fay.[32])

only. During heterocyst differentiation, an additional envelope is laid down outside the cell wall which itself is constructed of three layers, an outermost fibrous, a homogenous middle, and a laminated dense inner layer (Figs. 1 and 3).[32] The envelope is particularly well developed about the pore channel. The contact between heterocyst and vegetative cell is restricted to a small septum at the terminal end of the pore channel which apparently contains the fine intercellular connections (microplasmodesmata), similar to those present in the crosswall between two vegetative cells (Figs. 2 and 3).[32]

The elaborate heterocyst envelope, which probably has an important but yet unknown physiological function (however, cf. Winkenbach et al.[33] and Granhall[34]), also appears to provide protection against mechanical and chemical damage. The relative resistance of heterocysts to physical forces and degrading enzymes, which resides within the heterocyst envelope, is the basis to all current methods used for heterocyst isolation.

### Differential Destruction of Vegetative Cells by Osmotic Shock

This is one of the simplest methods used to obtain extracts from microorganisms, it involves exposing the cells to sudden changes of the osmotic pressure. A concentrated suspension of the algal material is

[32] N. J. Lang and P. Fay, Proc. R. Soc. London Ser. B **178,** 193 (1971).
[33] F. Winkenbach, C. P. Wolk, and M. Jost, Planta **107,** 69 (1972).
[34] U. Granhall, Physiol. Plant. **38,** 208 (1976).

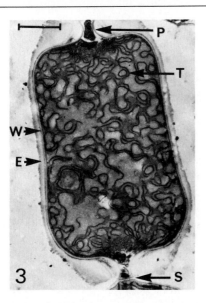

FIG. 3. Median section through a heterocyst of an untreated filament. E, envelope; W, cell wall; T, thylakoids; P, pore channel; S, septum displaying microplasmodesmata. Scale, 1 μm. (From Fay and Lang[22].)

mixed with a hypertonic solution (of glycerol or sucrose), and the filaments are left in the mixture for about 10 min. The viscous suspension is then withdrawn into a syringe and ejected into a hypotonic solution (buffer or medium). The treatment first causes water to leave the cytoplasm and the cells to shrink and later effects a rapid intake of water into the cells. The sudden osmotic change generates an excessive pressure capable of disrupting the vegetative cell walls and plasma membranes.

The heterocyst protoplasts, including the plasma membrane and cell wall, contract in the hypertonic solution but the thick envelope retains its shape. When these cells are viewed in the light microscope they have the appearance of being plasmolyzed.[22] In the hypotonic solution cells rapidly become swollen but in the heterocysts swelling is restricted by the robust envelope, and these cells appear intact under the light microscope. However, electron microscope examination shows that, in fact, the heterocyst wall is ripped open at the pore region, the septum is ruptured, and the plasma membrane fragmented (Fig. 4).[22] A considerable loss of granular cell material may occur. Evidently, the sudden increase in osmotic pressure caused the disruption of heterocyst wall and membrane before swelling could be controlled by the tough envelope.

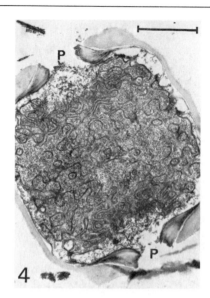

FIG. 4. Median section through a heterocyst isolated after osmotic shock treatment. The pore channel (P) is forced apart and the septum ruptured. Note severe loss of granular cytoplasmic material. Scale, 1 μm. (From Fay and Lang.[22])

## Differential Disruption of Vegetative Cells in a French Pressure Cell

When heterocystous algal material concentrated by centrifugation is placed in a pressure cell and forced through the needle valve at a pressure of about 58 MN m$^{-2}$ (570 atm or 8400 psi), the shearing forces generated selectivity disrupt the vegetative cells while heterocysts seem unimpaired when viewed in the light microscope.[6] Protection against the physical forces is apparently due to the resistance afforded by the multilayered envelope. However, when heterocysts isolated in this way are inspected using the electron microscope, considerable damage is revealed (Fig. 5).[22] The pore channel is ruptured, the heterocyst broken open, and the cell contents partially lost.

## Differential Disruption of Vegetative Cells by Sonication

Sonication is used as another method for the selective physical destruction of vegetative cells.[7,14,16] High frequency (400–600 kHz) ultrasonic pressure waves are believed to produce localized regions of high vacuum (cavitation) in the liquid, in which the alga is suspended, followed by powerful collapse of the minute gas bubbles which will destroy any vegetative cell at the vicinity of such implosions. Again, according to

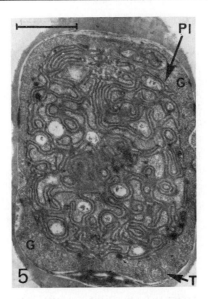

FIG. 5. Heterocyst isolated by the French press technique. The plasmalemma (Pl) is ruptured; granular cytoplasmic matrix (G) and fragmented thylakoids (T) have leaked out from the protoplast. Scale, 1 $\mu$m. (From Fay and Lang.[22])

electron microscope evidence, heterocysts could suffer extensive damage from the powerful sound waves (Fig. 6), though this damage may not be detected by light microscopy.[7,14]

*Selective Enzymic Destruction of Vegetative Cells*

Incubation of heterocystous filaments in a buffered isotonic solution containing egg white lysozyme and EDTA at 35° may result within 1 hr in extensive lysis of vegetative cells while heterocysts remain unchanged.[22]

Lysozyme is routinely used to disrupt bacterial cells through its action on the murein sacculus.[35] The enzyme splits the glycosidic bond between N-acetylglucosamine and N-acetylmuramic acid which leads to the breakdown of wall structure. Addition of a chelating agent (EDTA) promotes the action of lysozyme apparently by the removal of divalent cations from the lipopolysaccharide layer of the cell wall, thereby reducing the ability of the wall to withstand osmotic stress.[36] Blue-green algae

[35] R. Repaske, *Biochim. Biophys. Acta* **22**, 189 (1956).
[36] R. Repaske, *Biochim. Biophys. Acta* **30**, 225 (1958).

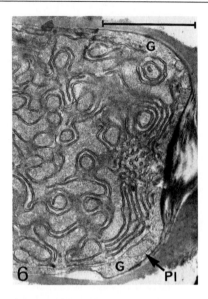

FIG. 6. Polar region of heterocyst isolated by sonication. The plasmalemma (Pl) is ruptured, allowing leakage of granular cytoplasmic matrix (G). Scale, 1 μm. (From Fay and Lang.[22])

like bacteria are susceptible to lysozyme action,[37] though the speed of enzymic degradation varies greatly according to species, age of culture, and the nature of sheath surrounding the filament.

Electron microscopic inspection of heterocysts isolated by lysozyme action shows that the triple-layered heterocyst envelope is unaffected,[22,34] and this apparently provides protection to the wall beneath. The septum between heterocyst and vegetative cell, though thought to contain murein, appears to remain intact, and thus the internal structure of the heterocyst is unchanged. The integrity of isolated heterocysts is confirmed by the finding that intracellular protein composition and pattern are not affected by lysozyme treatment.[25]

The conclusion is clearly that methods involving powerful mechanical forces to disrupt the vegetative cells inevitably injure heterocysts. Only heterocysts isolated after selective enzymic destruction of the vegetative cells preserve an integrity comparable with those in untreated filaments. Though the lysozyme treatment is in general preferable, other methods may be favoured in cases when loss of soluble or fine granular heterocyst content is unimportant, e.g., in the isolation of the heterocyst envelope.[33]

[37] B. D. Vance and H. B. Ward, J. Physiol. 5, 1 (1969).

## The Lysozyme Procedure

The current lysozyme procedure for heterocyst isolation, originally described by Fay and Lang[22] and modified by Fleming and Haselkorn,[25] Bradley and Carr[38] and Tel-Or and Stewart,[9] has been developed to meet three basic requirements: ultrastructural integrity, a high degree of purity and high yield of heterocystous material.

*Isolation*. The algal material, grown as described above, is harvested by low speed centrifugation (1000–2000 $g$, 20 min), washed free from supernatant medium and resuspended at a density of about 25 mg dry alga/ml of an isotonic buffer containing $10^{-3}$ $M$ Tris-HCl (pH 7.6), 2 × $10^{-3}$ $M$ EDTA, and 0.5 $M$ mannitol. An alternative buffer solution is composed of 30 m$M$ HEPES ($N$-2-hydroxyethyl-piperazine-$N'$-ethane sulfonic acid), 30 m$M$ PIPES (piperazine-$N,N'$-bis-2-ethane sulfonic acid), 1.0 m$M$ MgCl$_2$, and 0.35 $M$ mannitol, pH 7.2[39] The use of an isotonic isolation medium is essential to protect the fragile heterocyst septum against osmotic damage following lysis of the adjacent vegetative cells.

The temperature of the algal suspension is raised to 35° in a water bath before lysozyme is added at a concentration of 1.0 mg/ml, and the reaction mixture incubated at this temperature with shaking (80 oscillations/min) for a period preferably not longer than 1 hr. Progress of the reaction is followed by frequent microscopic examination of the mixture using phase contrast illumination. In certain cases, several hours long incubation is necessary for the complete lysis of vegetative cells.

*Purification*. When the lysis of vegetative cells is complete, the heterocysts are freed from the vegetative cell debris by extrusion of the lysate through a French pressure cell[38] or Yeda press[9] at a pressure not exceeding 200 kN m$^{-2}$ (30 psi or 2 atm), or simply by drawing the suspension through a Pasteur pipette.[25] The lysate is then diluted 20-fold with mannitol–Tris buffer (without lysozyme and EDTA), and the detached heterocysts are collected by centrifugation at 1200 $g$ at 4° for 15 min. The residue is resuspended in the isotonic buffer and recentrifuged at 750 $g$ at 4° for 10 min. This procedure is repeated two (or more) times until the vegetative cell debris is eliminated. This method of differential centrifugation provides a heterocyst preparation with a purity of 95% or more.

The purified preparation is best maintained in the isotonic buffer on ice for further experimentation. If transfer into hypotonic medium is

[38] S. Bradley and N. G. Carr, *J. Gen. Microbiol.* **96**, 175 (1976).
[39] R. B. Peterson and R. H. Burris, Arch. Microbiol. **116**, 125 (1978).

required, this should be done by the gradual reduction of the tonicity of the medium to avoid cell rupture.

The procedure described for heterocyst isolation may require further modification to preserve the integrity of certain metabolic systems or the activity of enzymes. To protect nitrogenase, the heterocysts must be isolated and purified under strictly anaerobic conditions. The concentrated algal suspension is placed in the dark and deoxygenated by bubbling through the suspension $O_2$-free $N_2$, $H_2$, or an inert gas for about 10 min. All solutions and reagents are treated similarly and kept under anaerobic gas cover. Addition of sodium dithionite (0.5 m$M$) to the suspending medium will provide further protection to the oxygen-sensitive nitrogenase by scavenging traces of $O_2$.

Cell-free extracts from isolated heterocysts can be obtained by rupturing the cells in a French pressure cell at a pressure exceeding 110 MN m$^{-2}$ (15,000 psi or 1000 atm) or by any other mechanical device applying similar forces to break the heterocyst envelope. The crude extract is centrifuged at 5000 $g$ for 20 min to remove envelope and wall debris. Further centrifugation of the supernatant at 35,000 $g$ for 30 min will effect the separation of particulate (membrane) and soluble fractions.

# [74] Leghemogloblins

## By M. J. DILWORTH

### Introduction

First recognized by Kubo[1] in 1939, the $O_2$-binding hemoproteins of legume root nodules have been variously described as leghemoglobins or legoglobins. They occur in the infected cells of legume root nodules with an overall concentration of 1–3 mg leghemoglobin (Lb) per gram fresh weight, depending on the plant species. Their distribution within the infected cell is still controversial; evidence from autoradiography[2] and diaminobenzidine staining,[3,4] suggesting that they are restricted to the membrane envelope around the bacteroids, while ferritin-labeled antibody techniques suggest a location in the plant cytoplasm.[5]

[1] H. Kubo, *Acta Phytochim.* **11**, 195 (1939).
[2] M. J. Dilworth and D. K. Kidby, *Exp. Cell Res.* **49**, 148 (1968).
[3] F. J. Bergersen and D. J. Goodchild, *Aust. J. Biol. Sci.* **26**, 741 (1973).
[4] G. Truchet, *C. R. Hebd. Seances Acad. Sci.* **274**, 1290 (1972).
[5] D. P. S. Verma and A. K. Bal, *Proc. Natl. Acad. Sci. U.S.A.* **73**, 3843 (1976).

Heme synthesis for Lb appears to be a bacteroid property,[6,7] though the regulation to match heme and apoprotein synthesis is unknown. Heme–protein association in the laboratory occurs spontaneously under physiological conditions.[8]

Genetically, leghemoglobins are coded for by the plant DNA, as shown by production of dissimilar leghemoglobins by dissimilar plants infected with the same *Rhizobium*.[9,10] The mRNA for soybean Lb *a* and *c* has been isolated, shown to have a poly(A)segment,[11] and translated with either soybean nodule ribosomes or a wheat germ system.[5] Inhibitor experiments are consistent with leghemoglobin synthesis on plant cytoplasmic ribosomes,[12] specifically free polysomes,[5] while pulse-chase experiments indicate that leghemoglobins are relatively stable during the active $N_2$-fixing life of the nodule.[12]

Leghemoglobin synthesis begins very shortly after nodule initiation, usually before nitrogenase synthesis can be demonstrated.[3,13] It is restricted to infected cells of the nodule. In legumes with meristematic nodules (lupin, broad bean, etc.) leghemoglobin synthesis continues in proportion to nodule growth; in nodules lacking meristems (soybean, kidney bean) leghemoglobin synthesis occurs only over a brief period of nodule formation, and the concentration remains more or less constant until nodule senescence. It is obviously important to choose an optimum time for nodule harvest which maximizes both nodule and leghemoglobin yield. Generally, plants actively forming seed will show rapid loss of leghemoglobin from their nodules; the early stages of flowering are usually the latest convenient times for nodule harvest.

Individual legume species usually yield more than one leghemoglobin component from their nodules; in soybean at least, these are different gene products with different sequences[14,15] and not artifacts of isolation. Microheterogeneity within a component has been noted for soybean leghemoglobins $a$[14] and $c_2$,[15] broad bean Lb I,[16] and kidney bean PLb$a$.[17]

[6] J. A. Cutting and H. M. Schulman, *Biochim. Biophys. Acta* **192**, 486 (1969).

[7] C. A. Godfrey and M. J. Dilworth, *J. Gen. Microbiol.* **69**, 385 (1971).

[8] N. Ellfolk and G. Sievers, *Acta Chem. Scand.* **19**, 2409 (1965).

[9] M. J. Dilworth, *Biochim. Biophys. Acta* **184**, 432 (1969).

[10] J. A. Cutting and H. M. Schulman, *Biochim. Biophys. Acta* **229**, 58 (1971).

[11] D. P. S. Verma, D. T. Nash, and H. M. Schulman, *Nature (London)* **251**, 74 (1974).

[12] D. R. Coventry and M. J. Dilworth, *Biochim. Biophys. Acta* **447**, (1976).

[13] M. J. Dilworth and C. A. Appleby, *in* "Dinitrogen Fixation" (R. C. Burns, ed.), Vol. 1, Part 2. Wiley, New York (in press).

[14] N. Ellfolk and G. Sievers, *Acta Chem. Scand.* **25**, 3532 (1971).

[15] J. G. R. Hurrell and S. J. Leach, *FEBS Lett.* **80**, 23 (1977).

[16] M. Richardson, M. J. Dilworth, and M. J. Scawen, *FEBS Lett.* **51**, 33 (1975).

[17] P. Lehtovaara and N. Ellfolk, *FEBS Lett.* **43**, 239 (1974); *Eur. J. Biochem.* **54**, 577 (1975).

Leghemoglobin function within nodules has been elucidated only recently. Its functions appear to be twofold—facilitation of $O_2$ diffusion across the nodule into the $N_2$-fixing bacteroids to support oxidative phosphorylation and ensuring that the $O_2$ concentration delivered to the bacteroids is buffered at a value which gives efficient oxidative phosphorylation without damaging nitrogenase function.[18,19]

## Plant Growth

Seeds of the legume required should be inoculated with an effective *Rhizobium* strain for the particular species. Plants can be grown in the glasshouse, in sand–vermiculite mixtures, with daily application of a dilute nutrient solution containing 1 m$M$ $CaCl_2$, 0.5 m$M$ $KH_2PO_4$, 0.25 m$M$ $MgSO_4$, 0.25 m$M$ $K_2SO_4$, 10 $\mu M$ ferric citrate (or ferric-EDTA), 2 $\mu M$, $H_3BO_3$, 1 $\mu M$ $MnSO_4$, 0.5 $\mu M$ $ZnSO_4$, 0.2 $\mu M$ $CuSO_4$, 0.1 $\mu M$ $Na_2MoO_4$, and 0.1 $\mu M$ $CoSO_4$, pH 5.5.

Alternatively, field plantings can be used, preferably in light-textured soils to facilitate nodule recovery. A suitable fertilizer dressing is 11 gm/m² of a mixture of commercial superphosphate, rock phosphate, KCl, and $MgSO_4 \cdot 7H_2O$ (2:4:2:1 by weight) and 0.4 gm/m² of a mixture of $CuSO_4 \cdot 5H_2O$, $ZnSO_4 \cdot 7H_2O$, $Na_2MoO_4$, and $MnSO_4 \cdot 4H_2O$ (2:2:1:2 by weight).[20]

## Estimation

Purified preparations of leghemoglobin are most conveniently assayed by the pyridine hemochromogen method.[21] Equal volumes of 4.2 $M$ pyridine in 0.2 $M$ NaOH and leghemoglobin solution are mixed and the resulting hemochrome reduced with a few crystals of sodium dithionite. Absorbance at 556 nm is measured against a reagent blank, and converted to leghemoglobin concentration using $\epsilon_{mM}^{556 \ nm} = 34.6$.

$$\text{Lb concentration (mg/cm}^3) = A_{556 \ nm} \times \frac{2}{34,600} \times MW$$

In less purified preparations, one cuvette of the leghemoglobin–alkaline pyridine reagent is oxidized with a minimal amount of $K_3Fe(CN)_6$

[18] C. A. Appleby, G. L. Turner, and P. K. MacNicol, *Biochim. Biophys. Acta* **387**, 461 (1975).

[19] F. J. Bergersen and G. L. Turner, *J. Gen. Microbiol.* **91**, 345 (1975).

[20] M. J. Dilworth and D. C. Williams, *J. Gen. Microbiol.* **48**, 31 (1967).

[21] K. G. Paul, H. Theorell, Å. Åkeson, *Acta Chem. Scand.* **7**, 1284 (1953).

and another reduced as above with dithionite. The value of $A_{556nm}$ *minus* $A_{539nm}$ is determined and used to calculate leghemoglobin concentration from $\Delta\epsilon_{mM} = 23.4$.[22]

## Isolation

*Soybean Leghemoglobin.*[23] Nodules are harvested 25–28 days after appearance and homogenized in 4 volumes of cold (4°) 0.1 $M$ phosphate buffer, pH 6.8. The addition of polyvinylpyrrolidone (Polyclar AT, General Aniline Corp.) in a ratio of 0.3 gm per gram nodules has been recommended to remove polyphenols.[24] The homogenate is clarified by centrifugation (10,000 $g$, 20 min) and fractionated with solid ammonium sulfate between 55 and 80% saturation, with the pH maintained at 6.8.[25] The precipitate is dissolved in a minimum volume of 0.1 $M$ Tris-HCl buffer, pH 7.7, containing 0.1 m$M$ EDTA. After dialysis for a minimum of 5 hr against the same buffer, the solution is centrifuged (100,000 $g$, 30 min) and the supernatant concentrated by pressure filtration through a Diaflo UM10 membrane (Amicon Corp., Cambridge, Massachusetts).

The next steps combine oxidation of the Lb components to remove ferrous oxyLb species and simplify the subsequent chromatogram, and gel filtration at alkaline pH to remove bound nicotinate[26] from the ferric Lb. Fifteen milliliters of 0.1 $M$ Tris-HCl buffer, pH 9.2, at 2° are added to 50 ml of 2.7 m$M$ Lb (from 767 gm nodules) at the same temperature. Solid $K_3Fe(CN)_6$ (176 mg, 4 Eq) is then added, and the solution stirred until $O_2$ evolution ceases. The mixture is then added to a Sephadex G-15 (Pharmacia, Sweden) column (55 × 5 cm diameter) equilibrated with 0.1 $M$ Tris-HCl buffer, pH 9.2 at 2°, at 2 cm³/min, and eluted at the same rate with the same buffer. Three colored bands are eluted—red ferric Lb, yellow ferricyanide plus nicotinate, and a purple band, in that order. Lb fractions are collected and pooled, 140 ml containing 116 μmoles Lb. If desired, the absorbance ratio 560 nm/620 nm in 0.1 $M$ MES buffer, pH 5.2, can be used to estimate whether the ferric Lb nicotinate complex is still present; the ratio should be 1.7 if complete removal of nicotinate has been achieved.[23]

[22] F. J. Bergersen, G. L. Turner, and C. A. Appleby, *Biochim. Biophys. Acta* **292**, 271 (1973).
[23] C. A. Appleby, N. A. Nicola, J. G. R. Hurrell, and S. J. Leach, *Biochemistry* **14**, 4444 (1975).
[24] C. A. Appleby, *Biochim. Biophys. Acta* **188**, 222 (1969).
[25] J. B. Wittenberg, F. J. Bergersen, C. A. Appleby, and G. L. Turner, *J. Biol. Chem.* **249**, 4057 (1974).
[26] C. A. Appleby, B. A. Wittenberg, and J. B. Wittenberg, *J. Biol. Chem.* **248**, 3183 (1973).

The Lb is concentrated to about 20 ml over a Diaflo UM10 membrane, and dialyzed for 36 hr against two changes of 40 ml of 10 m$M$ sodium acetate buffer, pH 5.2. DEAE–cellulose (Whatman DE-52, Reeve Angel & Co., London) is equilibrated with 0.5 $M$ sodium acetate buffer, pH 5.2, and washed with 80 ml distilled water so that the effluent conductivity and pH approximate those of 10 m$M$ acetate buffer, pH 5.2. A column is poured (25 × 5 cm diameter) and washed with 30 ml of the 10 m$M$ acetate buffer, pH 5.2. The dialyzed Lb is diluted to 50 ml with 10 m$M$ acetate buffer, pH 5.2, before loading onto the column. The column is eluted at 2 cm³/min with a linear gradient generated from 40 ml of 10 m$M$ and 40 ml of 100 m$M$ sodium acetate buffer, pH 5.2. The elution diagram is illustrated in Fig. 1.

Further separation of Lb $c_1$ and $c_2$ can be achieved by concentrating the appropriate peaks, dialyzing them against dilute acetate buffer, and rechromatographing them on columns of DEAE-cellulose (36 × 1.5 cm diameter) developed with a linear gradient generated from 200 ml 20 m$M$ and 200 ml 40 m$M$ sodium acetate buffers, pH 5.2.

*Broad Bean Leghemoglobin.*[16] Nodules (120 gm) are harvested at about 8 weeks, washed and homogenized under $N_2$ in 4 volumes of 0.1 $M$ potassium phosphate buffer, pH 7.6, containing 2.5 m$M$ EDTA to inhibit polyphenol oxidase activity. The homogenate is centrifuged under $N_2$, and the Lb fractionated between 48 and 80% saturation with solid ammonium sulfate at 4°. The precipitate is dissolved in a minimum volume of 0.1 $M$ potassium phosphate buffer, pH 7.6, and dialyzed overnight with stirring against 40 ml distilled water.

After centrifugation (30,000 $g$, 20 min) the Lb is absorbed onto a column (10 × 2.2 cm diameter) of DEAE–cellulose (Whatman DE-52) equilibrated with 3 m$M$ phosphate buffer, pH 6.8. Nonadsorbed material, including interfering carbohydrates which make the solutions extremely viscous, is washed through the column with 50 ml of 3 m$M$ potassium phosphate buffer, pH 6.8, before the Lb is eluted with 50 m$M$ phosphate buffer, pH 6.8.[27] The Lb is dialyzed free of phosphate against distilled water, and concentrated over a Diaflo UM10 membrane before being chromatographed on a column (48 × 2.2 cm diameter) of DEAE-cellulose equilibrated with 13 m$M$ sodium acetate buffer, pH 5.2. Elution is at 0.75 ml/min with the same buffer; green Lb I (the acetate complex of ferric Lb) elutes at about 0.65 ml. A similar procedure can be used for snake bean Lb I, which elutes at about 10 ml.

The pH of the Lb peak is adjusted to 7.0 with 1 $M$ Tris-HCl buffer, pH 8.0, and the volume reduced to 3–5 ml by pressure filtration on a

[27] W. J. Broughton and M. J. Dilworth, *Biochem. J.* **125**, 1075 (1971).

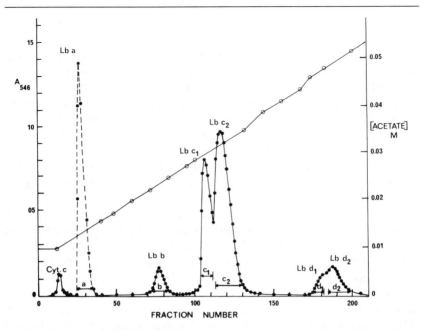

FIG. 1. Elution profile for soybean ferric Lb components on DEAE-cellulose (acetate) columns of pH 5.2. (From Appleby *et al.*[23])

Diaflo UM10 membrane. The Lb is applied to a column (90 × 2.2 cm diameter) of Sephadex G-75 (Pharmacia, Sweden) in water, and eluted with water at 0.75 ml/min, fractions being monitored at 280 and 405 nm to ensure removal of heme-free protein peaks.

*Kidney Bean Leghemoglobin.*[28] Nodules (200 gm) stored at −16° are extracted in a mortar with ice-cold 20 m$M$ phosphate buffer, pH 6.8, and the extract centrifuged. Lb is fractionated out of solution between 55 and 80% saturation with solid ammonium sulfate with the pH maintained between 6.5 and 7.0 with ammonium hydroxide addition. The precipitate is dissolved in 20 m$M$ phosphate buffer, pH 6.8, dialyzed against distilled water, and concentrated by pressure filtration over a Diaflo UM10 membrane.

Leghemoglobin is chromatographed on a column (36 × 2.5 cm diameter) of DEAE-Sephadex A-25 (Pharmacia, Sweden) equilibrated in 15 m$M$ acetate buffer, pH 5.6. A convex concentration gradient generated by connecting a closed reservoir containing 250 ml 15 m$M$ acetate buffer,

[28] P. Lehtovaara and N. Ellfolk, *Acta Chem. Scand.* **29**, 56 (1975).

pH 5.6, with a supply of 52 m*M* acetate buffer of the same pH is used to elute PLb*a*. When the reservoir is then connected to a supply of 100 m*M* acetate buffer, pH 5.6, PLb*b* is eluted. PLb*b* is thought to be a deamidation product of PLb*a*.

*Yellow Lupin Leghemoglobin.*[29] The use of fresh nodules is recommended for lupin Lb preparation; extended storage of nodules at −20° results in poor yields of Lb. Nodules (120 gm) are extracted as for soybean nodules, and the same ammonium sulfate fractionation and dialysis followed.[9] The dialyzed Lb is oxidized with ferricyanide; since lupin oxyLb oxidizes very slowly or not at all at pH 8.6, a pH of between 5.2 and 6.8 and a temperature of 20° are required for satisfactory oxidation. The ferric Lb is desalted through an alkaline Sephadex G-25 column to remove nicotinate and concentrated by adsorption to a small column of DEAE-cellulose (phosphate form). After washing with 10 m*M* phosphate buffer, pH 6.96, the ferric Lb is eluted with 50 m*M* phosphate buffer of the same pH, concentrated, and equilibrated with 50 m*M* Tris-HCl buffer, pH 8.1. The Lb components are separated on a column of DEAE–Sephadex A-50, through which a pH gradient from 8.1 to 7.1 in Tris-HCl buffer is run.

An alternative procedure[30] first separates three different ferric Lb components and two ferrous oxyLb components on DEAE-cellulose (acetate) columns, using stepwise elution with 20, 40, and 80 m*M* ammonium acetate, pH 7.0. After ferricyanide oxidation of the ferrous oxyLb components to their corresponding ferric Lb forms, final purification was achieved by rechromatography on DEAE–cellulose using 20, 35, or 60 m*M* ammonium acetate, pH 7.0, to elute the different components.

Since more extract details of the procedures have not been published, only the outlines are given.

### Storage

Storage at −196° (liquid $N_2$) is recommended, particularly to preserve tertiary structure.[31]

[29] Ts. A. Egorov, M. U. Feigina, V. K. Kazakov, M. I. Shakhnarokov, S. I. Mitaleva, and U. A. Ovchinnikov, *Bioorg. Khim.* **2**, 125 (1976).
[30] I. P. Kuranova, A. I. Grebenko, N. V. Konareva, I. F. Syromyatnikova, and V. V. Barynin, *Biokhimiya* **41**, 1603 (1976).
[31] J. G. R. Hurrell, N. A. Nicola, W. J. Broughton, M. J. Dilworth, E. Minasian, and S. J. Leach, *Eur. J. Biochem.* **66**, 389 (1976).

## Cell-Free Synthesis

Soybean nodule polysomes have been prepared and shown to incorporate [4,5-³H]leucine into a fraction precipitating with anti-Lb antisera when appropriately incubated with postribosomal wheat supernatant and cofactors.[11] The antigen–antibody complex, solubilized in SDS, coelectrophoresed with authentic Lb. Subsequent work[5] indicates that free polysomes rather than membrane-bound polysomes are involved with Lb synthesis.

Poly(A)-rich RNA was extracted from the polysomes with chloroform–phenol and fractionated on oligo(dT)-cellulose[5]; the fractions giving greatest incorporation into anti-Lb antisera precipitable material with a wheat germ cell-free protein-synthesizing system corresponded to 9 S and 12 S RNA.[11] Such RNA was shown to contain poly(A) segments by hybridization with ³H-poly(U).[11] Although the ratio of Lbc to Lba is usually 2 to 5,[32] most label incorporated in the wheat germ system given poly(A)-containing RNA was into Lba.[11] Heme was without effect on the synthesis of Lb.

## Properties

The physical, chemical and spectroscopic properties of the leghemoglobins have been summarized in two recent reviews.[13,33]

*Physical.* Primary amino acid sequences are known for soybean Lba[14] and Lbc₂,[15] broad bean Lb I,[16] kidney bean PLba[17], and yellow lupin Lb I.[29] Molecular weights range from about 15,800 to 17,200. Glycine is the N-terminal amino acid for all the Lb's except soybean Lba,[15] snake bean Lb I,[13] and lupin Lb III,[30] which have valine, valine, and alanine, respectively. The sequences show very high variability between plant species; none of them contain cysteine residues, and the methionine contents are either low or zero. Measured p$I$ values are all acidic; 4.4 and 4.7 for soybean Lbs $a$ and $c$,[34] 5.1 for lupin Lb I,[35] and 4.7 for kidney bean PLba.[28]

The heme moiety is in all cases protoheme (IX[28,35,36]; the binding to

[32] W. H. Fuchsman, C. R. Barton, M. M. Stein, J. T. Thompson, and R. M. Willett, *Biochem. Biophys. Res. Commun.* **68**, 387 (1976).

[33] C. A. Appleby, *in* "The Biology of Nitrogen Fixation" (A. Quispel, ed.), p. 522. North-Holland Publ., Amsterdam, 1974.

[34] N. Ellfolk, *Acta Chem. Scand.* **14**, 1819 (1960).

[35] W. J. Broughton, M. J. Dilworth, and C. A. Godfrey, *Biochem. J.* **127**, 309 (1972).

[36] N. Ellfolk and G. Sievers, *Acta Chem. Scand.* **19**, 268 (1965).

soybean apoLb$c$ appears stronger than to apoLb$a$, and both Lb's lose heme to the apoprotein of horseradish peroxidase.[37]

Leghemoglobins crystallize in space group C2 with four asymmetric units per cell. The crystal dimensions are listed in the tabulation below.

| | $a$ (nm) | $b$ (nm) | $c$ (nm) | $\beta$ |
|---|---|---|---|---|
| soybean Lb$a$[38] | 5.73 | 4.42 | 4.69 | 90°31' |
| lupin Lb[39] | 9.295 | 3.831 | 5.215 | 98°50' |

In kidney bean PLb$a$[17] and in soybean Lb$a$,[40] the pattern of helix-promoting and helix-breaking residues favors similar folding to that in myoglobins, and the formation of a similar hydrophobic pocket around the heme possibly more accessible to ligands than in other globins. X-ray structural analysis of lupin Lb I at 0.5 nm resolution[41] supports the general pattern of chain folding being very similar to that in myoglobin.

*Chemical.* Heme binding to apoLb involves one of the propionic acid residues.[42] Four of the heme iron coordination sites are filled by planar, pyrrole nitrogen atoms; in ferric Lb, the fifth (proximal) link to heme is an imidazole ring nitrogen atom,[43] almost certainly histidine-97 (numbering as in the lupin Lb sequence[29]). At alkaline pH, the sixth ligand is hydroxyl, while at neutral or acid pH it is probably water.[44]

Ferric Lb is a thermal equilibrium mixture of high- and low-spin forms[45] with about 60% high-spin component at 20° and pH 5.3. The sixth heme ligand appears to be water in the high-spin ferric Lb and the imidazole of histidine-63 in the low-spin component.[43]

"Thermal equilibrium" ferric Lb reverts to a high-spin ferrous myoglobin type structure when reduced with dithionite,[46] while fluoride rapidly produces a single high-spin ferric Lb fluoride,[44] indicating that the ferric

[37] N. Ellfolk, U. Pertila, and G. Sievers, *Acta Chem. Scand.* **27**, 3601 (1973).
[38] W. A. Hendrickson and L. E. Love, personal communication.
[39] E. G. Arutunyan, V. N. Satsev, G. Ya. Zhiznevskaya, and L. I. Borodenko, *Kristallografya* **16**, 237 (1971).
[40] N. Ellfolk, *Endeavour* **21**, 139 (1972).
[41] B. K. Vainshtein, E. G. Arutunyan, I. P. Kuranova, V. V. Borisov, N. I. Sosfenov, A. G. Pavlovskii, A. I. Grebenko, and N. V. Konareva, *Nature (London)* **254**, 163 (1975).
[42] N. Ellfolk and G. Sievers, *Acta Chem. Scand.* **19**, 2409 (1965).
[43] C. A. Appleby, W. E. Blumberg, J. Peisach, B. A. Wittenberg, and J. B. Wittenberg, *J. Biol. Chem.* **251**, 6090 (1976).
[44] N. Ellfolk, *Acta Chem. Scand.* **15**, 975 (1961).
[45] A. Ehrenberg and N. Ellfolk, *Acta Chem. Scand.* **17**, S343 (1963).
[46] C. A. Appleby, *Biochim. Biophys. Acta* **189**, 267 (1969).

TABLE I

ABSORPTION MAXIMA OF SOME LEGHEMOGLOBIN LIGAND COMPLEXES[a,b]

| | | Band position (nm) | | | | |
|---|---|---|---|---|---|---|
| Complex | pH | Change transfer | $\alpha$ | $\beta$ | Change transfer | $\gamma$ |
| Ferric leghemoglobin | | | | | | |
| Hydroxide | 10 | | 572 | 541 | | 412 |
| Water | 5–7 | 625 | 560(sh) | 530(sh) | 496 | 403.5 |
| Acetate | 4.5–5.5 | 618–622 | 569 | 530 | 495 | — |
| Fluoride | 5–7 | 605 | 555(sh) | 525(sh) | 486 | — |
| Cyanide | 5–7 | | 538–540 | | | 415–417 |
| Nicotinate | 5–7 | | 557 | 528 | | 407.5 |
| Ferrous leghemoglobin | | | | | | |
| No ligand | 5–8 | | 555–556 | | | 427 |
| Oxygen | 5–8 | | 574–575 | 540–541 | | 411–412 |
| Carbon monoxide | 7 | | 562–563 | 538 | | 417 |
| Nitric oxide | | | 570 | 545 | | — |
| Nicotinate | 5–7 | | 554 | 525 | | 418.5 |

[a] Modified from Dilworth and Appleby.[13]

[b] A dash indicates the presence of a band whose position is not accurately recorded; a blank means that the band is absent or insignificant.

Lb is not a stable hemochrome. It, therefore, appears that the thermal equilibrium ferric Lb has its sixth coordination position relatively open to exogenous ligands.

The redox potentials for unligated soybean Lb[47] and lupin Lb[48] are close to +220 mV at pH 7, reaching +270 mV between pH 5 and 6.

The absorption maxima for Lb and some of its derivatives are summarized in Table I (modified from Dilworth and Appleby[13]) for soybean; values for lupin Lb are similar.[49]

Two important complexing agents for Lb are acetate and nicotinate. The green ferric Lb acetate has charge transfer bands near 620 and 495 nm[45]; it is a high-spin form almost unaffected by temperature. It has one molecule of acetate per molecule of ferric Lb; similar complexes are formed with propionate, butyrate, or valerate.[44] Acetate appears to be the species bound rather than un-ionized acetic acid, and binding occurs

[47] R. W. Henderson and C. A. Appleby, *Biochim. Biophys. Acta* **283**, 187 (1974).

[48] B. P. Atanasov and Ya. V. Peive, *Dokl. Akad. Nauk SSSR* **216**, 1161 (1974).

[49] Ya. V. Peive, B. P. Atanasov, G. Ya. Zhizenvskaya, and     Krasnobaeva, *Dokl. Akad. Nauk SSSR* **202**, 965 (1972).

within the heme pocket[43] and not at an allosteric site. The complex is either not reducible with dithionite or very much more slowly reduced than free ferric Lb.[50] No ferrous Lb acetate complex is formed.[50]

During chromatography of crude lupin or soybean Lb preparations on DEAE-cellulose (acetate) columns, different coloured (red-brown and green) ferric Lb bands can be seen. The red-brown band of ferric soybean Lb$a$ is a ferric hemochrome[46]; the ligand is nicotinate,[51] with the un-ionized ring nitrogen possibly ligated to heme iron and the ionized carboxyl making a stabilizing linkage to a protonated residue in the hydrophobic pocket.[26,51] Nicotinate also forms a ferrous Lb hemochrome,[26] which is unreactive with CO. It lowers the affinity of Lb for $O_2$; at low (2 torr) pressure, nicotinate will deoxygenate ferrous oxyLb,[51] while at higher $O_2$ pressure $O_2$ may displace nicotinate.[13] Ferric Lb nicotinate lacks the charge transfer band at 620 nm, and the absence of this band in nodule spectra cannot therefore be taken as evidence for absence of ferric Lb species in nodules.

### Gas Binding

*Oxygen.* The $O_2$ affinity of Lb is very high; $p_{1/2}(O_2)$ for soybean Lb is 90 n$M$ dissolved $O_2$ at pH 7 and 20°,[52] and the oxygenation is hyperbolic as expected for a monomeric globin without heme–heme interaction.[53,54] Different components of soybean Lb appear to have slightly different $O_2$ affinities.[52]

The high affinity is due to a very fast loading reaction; the $O_2$ "on" constant ($k'$) is 1.1 to 1.8 × $10^8$ $M^{-1}$ sec$^{-1}$ at neutral pH,[54,55] and the $O_2$ "off" constant ($k$) between 4.4 and 11 sec$^{-1}$ for soybean Lb.[54,55] The equilibrium association constant $K$ ($k'/k$) must lie between 1.4 and 2.7 × $10^7$ $M^{-1}$; the dissociation constant must be about $5 × 10^{-8}$ $M$ or 50 n$M$ dissolved $O_2$, in reasonable agreement with direct equilibrium measurements. The kinetic parameters are important in allowing Lb to function in facilitated $O_2$ diffusion in the legume root nodule.

*Carbon monoxide.* CO also binds very strongly to Lb; the partition constant, $M$, which measures the relative affinities of CO and $O_2$ for ferrous Lb is 63 for soybean Lb$a$.[54] Direct measurement shows that 0.89

[50] B. A. Wittenberg, J. B. Wittenberg, and C. A. Appleby, *J. Biol. Chem.* **248**, 3178 (1973).
[51] C. A. Appleby, B. A. Wittenberg, and J. B. Wittenberg, *Proc. Natl. Acad. Sci. U.S.A.* **70**, 564 (1973).
[52] C. A. Appleby, *Biochim. Biophys. Acta* **60**, 226 (1962).
[53] J. Behlke, G. Sievers, and N. Ellfolk, *Acta Chem. Scand.* **25**, 746 (1971).
[54] T. Imamura, A. Riggs, and Q. H. Gibson, *J. Biol. Chem.* **247**, 521 (1972).
[55] J. B. Wittenberg, C. A. Appleby and B. A. Wittenberg, *J. Biol. Chem.* **247**, 527 (1972).

n$M$ CO is sufficient to produce 50% conversion of ferrous oxyLb to ferrous carboxyLb, while the "off" constant for CO is 0.012 sec$^{-1}$.[54]

*Dinitrogen.* Spectrophotometric evidence for formation of an Lb–N$_2$ complex[56] must be rejected since the spectra have been shown to result from a mixture of ferrous Lb and ferrocytochrome $c$.[57] Evidence for N$_2$ binding to Lb at a "secondary gasation site" has been presented.[58]

[56] K. Abel, N. Bauer, and J. T. Spence, *Arch. Biochem. Biophys.* **100**, 339 (1963).
[57] C. A. Appleby, *Biochim. Biophys. Acta* **180**, 202 (1969).
[58] G. J. Ewing and L. G. Ionescu, *J. Phys. Chem.* **74**, 2341 (1970).

# Addendum

# Addendum to Article [20]

By KERRY T. YASUNOBU and MASARU TANAKA

Since this article was written, some additional papers have appeared in which additional ferredoxin sequences are reported. These are according to their Fe and the S contents: (1) the 2Fe–2S type—*Equisetum telmateia* (horsetail) ferredoxins I and II,[1] *Equisetum arvense* (horsetail) ferredoxins I and II,[2] *Phytolacca americana* (pokeweed) ferredoxins I and II,[3] *Porphyra umbilicalis* (red algal) ferredoxin,[4] *Aphanothece sacrum* (blue green algal) ferredoxin II,[5] *Halobacterium halobium* (halobacterial) ferredoxin,[6] and the *Cyanidium caldarium* (acido-thermal algal) ferredoxin[7] and (2) the 8Fe–8S type—the *Chlorobium thiosulfatophilum*, strain Tassajara (photosynthetic green sulfur bacterial) ferredoxin[8]. The phylogenetic data obtained from the sequences of the 2Fe–2S chloroplast ferredoxins are discussed in the paper by Wakabayashi *et al.*[3] In addition, the preliminary structural data from the crystal X-ray diffraction studies for the *Spirulina platensis*[9] and the *Aphanothece sacrum*[10] 2Fe–2S ferredoxins and the *Azotobacter vinelandii* 4Fe–4S ferredoxin[11] have now been reported.

[1] T. Hase, K. Wada, and H. Matsubara, *J. Biochem. (Tokyo)* **82,** 267 (1977).

[2] T. Hase, K. Wada, and H. Matsubara, *J. Biochem. (Tokyo)* **82,** 277 (1977).

[3] S. Wakabayashi, T. Hase, K. Wada, H. Matsubara, K. Suzuki, and S. Takaichi, *J. Biochem. (Tokyo)* **83,** 1305 (1978).

[4] P. W. Andrew, L. J. Rogers, D. Boulter, and G. Haslett, *Eur. J. Biochem.* **69,** 243 (1976).

[5] T. Hase, S. Wakabayashi, K. Wada, and H. Matsubara, *J. Biochem (Tokyo)* **83,** 761 (1978).

[6] T. Hase, S. Wakabayashi, H. Matsubara, L. Kerscher, D. Oesterhelt, K. K. Rao, and D. O. Hall, *J. Biochem. (Tokyo)* **83,** 1657 (1978).

[7] T. Hase, S. Wakabayashi, K. Wada, H. Matsubara, F. Jüttner, K. K. Rao, I. Fry, and D. O. Hall, *FEBS Letters* **96,** 41 (1978).

[8] T. Hase, S. Wakabayashi, H. Matsubara, M. C. W. Evans, and J. V. Jennings, *J. Biochem. (Tokyo)* **83,** 1321 (1978).

[9] T. Tsukihara, K. Fukuyama, H. Tahara, Y. Katsube, Y. Matsuura, N. Tanaka, M. Kakudo, K. Wada, and H. Matsubara, *J. Biochem. (Tokyo)* **84,** 1645 (1978).

[10] A. Kunita, M. Koshibe, Y. Nishikawa, K. Fukuyama, T. Tsukihara, Y. Katsube, Y. Matsuura, N. Tanaka, M. Kakudo, T. Hase, and H. Matsubara, *J. Biochem. (Tokyo)* **84,** 989 (1978).

[11] C. D. Stout, *J. Biol. Chem.* **254,** 3598 (1979).

# Author Index

Numbers in parentheses are reference numbers and indicate that an author's work is referred to although the name is not cited in the text.

## H

# J

Jackson, J. B., 319
Jackson, R. C., 392
Jacobi, G., 614, 649
Jacobson, E. L., 470
Jacobson, M. K., 470
Jagendorf, A. T., 86, 116, 224, 250, 538, 505, 612, 614, 615(9), 622(51), 623, 649, 650(12), 652(12), 653, 654(35), 656, 707, 713, 717, 724, 725(22)
Jahn, T. L., 25
James, D. M., 258, 263(20), 264(20), 266(20), 267(20), 269(20)
James, T. L., 404
Janiszowska, W., 381
Jank-Ladwig, R., 519, 655
Janossy, A., 594
Jansson, C., 703, 709(139)
Jarvis, P. G., 453, 454(5), 456(5)
Jaynes, J. M., 802, 803(5)
Jensen, L. H., 229, 402
Jensen, R. A., 45
Jensen, R. G., 95, 100, 123, 642, 645(1), 647(1)
Jeschke, W. D., 678
Jewess, P. J., 461
Johansson, B. C., 314, 315, 316
Johnson, C. E., 231, 773
Johnson, G. L., 680, 681, 683, 708(43), 709, 710(43)
Johnson, P. W., 231
Johnson, R., 787, 788(18)
Johnston, H. M., 49(15), 51, 790
Jolchine, G., 156(15, 18, 19, 21), 157, 158, 162(19)
Joliot, A., 471, 472, 712
Joliot, P., 14, 206, 215(28), 216(28), 349, 351(1), 471, 472, 712
Jolly, S. A., 270, 275(5), 277(5), 278(5)
Jonasson, K., 70(51), 71, 75(51)
Jones, E. A., 223
Jones, H. G., 17
Jones, J. B., 236
Jones, L. L., 591
Jones, L. W., 409, 474, 732
Jones, R. L., 70, 72(18)
Jordan, E., 446
Jost, M., 806, 810(33)

Joy, K. W., 256, 260(8), 262(8), 264
Joyard, J., 290, 292, 294(8), 297, 299(5), 301(5)
Junge, W., 355, 493, 602, 603, 604, 685, 709, 710, 711, 712(205)
Jungermann, K., 393
Jupp, A., 26

# K

Kabat, E. A., 492, 498(1), 501(1)
Kagamiyama, H., 238
Kagawa, Y., 503
Kahn, A., 618
Kahn, J. S., 345
Kaiser, W., 678
Kalberer, P. P., 382, 383(2), 384(1)
Kaltschmidt, E., 373
Kamen, M. D., 156(15), 157, 158(15), 237
Kamin, H., 257, 258, 261, 263(18), 264(13, 18), 265, 266, 267, 268(18), 269, 547
Kan, K.-S., 151, 153(3), 154(3), 344
Kanai, R., 70, 72(12), 82
Kane, J. F., 45
Kaney, A. R., 40, 45, 47(11)
Kao, K. N., 69, 70, 71, 72(20), 74(39), 75(54), 77(4), 80, 84
Kapadia, G., 434
Kaplan, D., 280
Kaplan, N. O., 465, 466(1)
Kaplan, S., 35(19), 36, 168, 170
Karlish, S. J. D., 584
Karlsson, R., 403
Karnovsky, M. J., 527
Karny, O., 306, 307(9), 309(9)
Kartan, M. B., 655
Kartha, K. K., 70, 71, 74(42), 75(54)
Kasprzyk, Z., 381
Katan, M. B., 516
Kates, M., 294
Katoh, S., 119, 121, 149, 205, 218(25), 223, 286, 352, 353(35, 36), 355(36), 417, 418(16), 504, 680, 699, 706, 711
Katz, J. J., 396, 404
Kaufman, P., 79
Kaufmann, R., 595
Kaur-Sawhney, R., 70, 72(14), 80(14), 81
Kawai, T., 5
Kawashima, N., 331

Sun, A. S. K., 475
Sun, I. L., 694
Sussdorf, D. H., 496, 498(28)
Suzuki, K., 230
Suzuki, R., 109
Swader. J. A., 706
Swan, A. G., 563
Swan, M., 465, 467(11)
Swanson, C. A., 561, 562(1), 571(1)
Swartz, H. M., 239
Sweeney, W. V., 231
Sweers, H. E., 14, 632
Swenson, M. K., 616, 618(24), 620(24)
Swisher, R. H., 765, 766, 768(18), 773
Sybesma, C., 156(26), 157, 158(26)
Syromyatnikova, I. F., 818, 819(30)

## T

Taboury, J., 698
Tachiki, K. H., 379
Taenami, S., 232
Tagasaki, R. K., 409
Tagawa, K., 229, 233, 241, 247(5), 248, 250, 625
Tahara, H., 229
Taiz, L., 70, 72(18)
Takahashi, M., 351, 681
Takamiya, A., 377, 480
Takamiya, K., 336, 337
Takaoki, T., 615
Takebe, I., 70, 79(32), 80(32), 645, 647(8)
Talberg, A. B., 588, 594(15), 596(15)
Talukdar, S., 377
Tamai, H., 706
Tamminga, J. J., 289
Tamura, G., 260, 262, 263(23), 264(23)
Tanaka, M., 235, 236, 237, 238, 401, 402(41), 772, 773(43)
Tanaka, N., 229, 236
Tanford, C., 364, 443
Tanner, C. B., 457
Tarrand, J. J., 740
Tardieu, A., 529
Tatsube, Y., 229
Taylor, W. E., 230
Telfer, A., 475, 603, 696, 762, 764, 767, 769, 770(12)

Tel-Or, E., 106, 107(6), 108(6), 110, 232, 352, 355(29), 359, 621, 625, 627(5), 628(5), 629(5), 630(2, 5), 802, 811
Terdro, S., 237
Terry, N., 459
Tevini, M., 379
Than Nyunt, U., 684
Thauer, R., 393, 625, 630(3)
Thayer, W. S., 515
Theorell, H., 814
Thimann, K. V., 70, 72(24), 73(24), 82(24)
Thomas, G., 381
Thompson, J. T., 819
Thompson, K. H., 339, 341(16)
Thore, A., 656
Thornber, J. P., 26, 142, 144(4, 5), 145(3, 5), 146(1, 5), 147, 148(1, 4, 5), 149, 150, 151, 152, 153(1a, 3), 154, 156, 157, 159, 168, 172, 174, 175, 176(4), 177, 178, 287, 344, 359, 366, 370, 379, 434, 443(4), 481
Thorne, S. W., 92, 123, 618
Thorneley, R. N. F., 758, 764, 767, 769, 770, 771(30), 774, 775, 776, 777(46), 778
Threlfall, D. R., 375, 376, 380, 381
Tiede, D. M., 156(35), 157, 159, 174, 176(4), 177(4), 178
Tiemann, R., 702
Tierney, G. V., 693
Tillberg, J. E., 545
Tillberg, O., 403
Ting, I. P., 70(60), 71, 76(60), 104
Tió, M. A., 453
Tischer, W., 701
Tischler, L. R., 271
Tisue, T., 396, 397(22)
Tiwari, D. N., 40, 44(20)
Tjepkema, J., 750
Tobler, U., 115
Togasaki, R. K., 10, 112, 114(25, 26), 701
Tolbert, N. E., 270, 271, 275(5), 277(5), 278(5), 333, 334
Tollin, G., 399, 400, 401, 403, 404, 405, 406(45, 58, 59)
Tomita, G., 352, 355(23), 360(23), 618, 619(36), 620(36)
Tonn, S. J., 167, 169, 172
Torbjörnsson, L., 403
Torres-Pereira, J., 615, 622
Totten, R. E., 70

# Subject Index

## A

A-III particle, properties of, 135–141

A 23187, chloroplast ion fluxes and, 599, 603

Absorption spectra
  of Fe proteins, 776
  of Mo–Fe proteins, 766

*Acer*, cultures, protoplasts from, 75

*Acetabularia mediterranea*, chloroplast material from, 120

Acetaldehyde, assay of, 627

Acetate
  leghemoglobin and, 821–823
  plastoquinone biosynthesis and, 381

Acetic anhydride, incorporation into chloroplasts, 507, 509, 510

Acetimidate, chloroplast membrane cross-linking and, 622

Acetylene, nitrogenase assay and, 744, 752, 753, 796–797

Acid–base-driven reverse electron transport
  luminescence and, 639–640
  properties of, 640–641
  reduction of Q and, 639

Acrylamide, limiting concentrations, choice of, 441

Actidione, mutant selection by, 10

Action spectra, observation of, 474

Acylases, phosphatidic acid biosynthesis and, 299, 301

Adenosine diphosphate
  CF$_1$ $\alpha$- and $\beta$-subunits and, 306
  Fe proteins and, 778
  membrane-bound, phosphorylation of, 324–325

Adenosine-5'-(1-thiodiphosphate), coupling factor 1 and, 483

Adenosine triphosphatase
  activation of, 630
  bacterial coupling factor and, 315–316, 320–321
  chloroplast envelope membranes and, 125
  fluorescamine and, 519
  tonoplast, release of, 78

Adenosine triphosphate
  binding to Fe proteins, 778
  formation, assay of, 655–658
  [$\gamma$-$^{32}$P]-labeled, preparation of, 303
  protons and, 712
  reverse electron transport and
    luminescence, 637–638
    oxidation of cytochrome $f$, 634
    properties of, 639
    reduction of Q, 632–634
    simultaneous measurement of $\Delta$pH and Q reduction, 634–636

Adenylate kinase, photophosphorylation and, 656

Adrenal glands, iron–sulfur protein of, 230, 238

Affinity chromatography, of bacterial coupling factor, 318

Agar, washing of, 39

*Agmenellum quadruplicatum*
  mutagenesis in, 41
  MNNG treatment of, 42
  mutant types, 43–45

Alfalfa, ferredoxin of, 232
  amino acid sequence, 237, 238

Algae
  isolation of ferredoxin from
    assays, 232–233
    extraction, 233
    purification, 233–234
  photorespiration assay in, 460–461
  plastoquinones in, 377, 380

Aliphatic diimido esters, chloroplast membrane cross-linking and, 621–623

Alkaline phosphatase, phosphatidic acid and, 301

*Amaranthus*, triazine-resistant mutants, 702

Amido black, gel staining and, 369

Amines, stimulation of proton uptake by
  calculation, 546
  materials and methods, 545–546
  principle, 545
  validity and drawbacks of method, 546

Amino acid(s)
  composition of bacteriochlorophyll $a$-protein and, 339, 340

redox properties, 199
spectrophotometric, 198–199
stability, 198
purification of, 196–198
Cytochrome $b$-559
assay of, 182
properties
chemical properties and molecular weight, 195–196
purity, 191
redox, 193–195
spectrophotometric, 192–193
stability, 191–192
purification, 189–191
rate of reduction, 203
Cytochrome $c$
ferredoxin–NADP oxidoreductase assay and, 251
flavodoxin and, 406
reduced, preparation of, 224
reduction
ferredoxin assay and, 233
inhibition of, 677
Cytochrome $f$
assay of, 181–182
donors of electrons to, 427–428
midpoint oxidation–reduction potential, 219, 221
oxidation
ATP-driven reverse electron transport and, 634
polylysine and, 681–682
P700 and, 281
properties
autoreduction, 188–189
chemical, 186, 188
purity, 186
spectrophotometric, 186, 187
purification of, 183–184
procedure, 184–186
reduction
DBMIB and, 685
rate equation for, 207–215, 216
time course of, 203, 204

### D

*Datura* pollen, protoplasts from, 73
DEAE–BioGel A, subchloroplast fragments and, 131

DEAE–cellulose
bacterial coupling factor and, 318, 319, 320
bacteriochlorophyll $a$ and, 337
cytochrome $b$-559 preparation and, 190
cytochrome $f$ preparation and, 184–185
δ- and ε-subunit of $CF_1$ and, 308
ferredoxin isolation and, 231, 234
ferredoxin–NADP oxidoreductase purification and, 252
ferredoxin–nitrite reductase purification and, 259
ferredoxin–thioredoxin reductase and, 391
flavodoxin and, 395, 396, 397, 398
leghemoglobin and, 816, 817, 818, 822
NDP–NTP kinase and, 345–346
nitrogenase and, 757, 758, 759, 760, 761, 762
nitrogenase component I and, 793–794
plastocyanin preparation and, 226
reaction center preparation and, 160, 173, 174–175
thioredoxin preparation and, 387, 388–389
DEAE–Sephadex
bacterial coupling factor and, 315
$CF_1$ depleted of δ-subunit and, 307
$CF_1$ preparation and, 305
ferredoxin–nitrite reductase purification and, 259
leghemoglobin and, 818
plastocyanin purification and, 227
Delayed light emission, salt stimulation of, chloroplast ion fluxes and, 599–602
Densitometric tracing, of slab gels, 440
4'-Deoxyphlorizin, potency of, 722
Deoxyribonucleic acid homology, of *Spirillum lipoferum,* 741
*Desulfovibrio,* flavodoxin and, 393
*Desulfovibrio desulfuricans,* ferredoxin of, 231
*Desulfovibrio gigas,* ferredoxin of, 231
amino acid sequence, 235, 236
*Desulfovibrio vulgaris,* flavodoxin of, 400, 401, 402, 406
Detergents, reaction center preparation and, 156–157
Detergent fractionation, of thylakoid fragments, 131
Dextran sulfate, protoplast stability and, 80

oxidation–reduction titration of iron–sulfur centers and, 243

photophosphorylation and, 581, 582

purification of, 580

2,6-Dimethyl-*p*-benzoquinone, as electron acceptor, 422

Dimethyl-3,3′-dithiobispropionimidate, coupling factor and, 505

Dimethylmethylenedioxy-*p*-benzoquinone, as electron acceptor, 422

5,5-Dimethyloxazolidine-2,4-dione, pH measurement and, 609–613

*N,N*-Dimethyl-*p*-phenylenediamine, as electron acceptor, 421

2,5-Dimethylquinone, as electron acceptor, 422

Dimethyl suberimidate
chloroplast membrane cross-linking and, 622

coupling factor and, 505

3-(4,5-Dimethylthiazolyl-2)-2,5-diphenyltetrazolium bromide, pyridine nucleotide assay and, 465, 468, 469

Dinitroanilines, plastoquinone oxidizing site and, 695

2,4-Dinitrothymol, inhibition by, 700

1,5-Diphenylcarbazide
as electron donor, 433

inhibition of electron transport and, 504

Diphenylcarbazone, quantum yield measurement and, 481

Diphenyl ethers, plastoquinone oxidizing site and, 694–695

1,6-Diphenyl-1,3,5-hexatriene, as membrane probe, 514

Diquat, mutant selection by, 11–12

Disalicylidene propanediamine, inhibition by, 677–678

Disc gel electrophoresis, *see also* Electrophoresis
cytochrome $b_6$ and, 197–198

of cytochrome *f*, 185

for separation of cytochromes *f* and *b*-559, 190–191

Dissociation constant
calculation of, 489

of CF1-FTP complex, measurement of, 487–488

Dissociation rate constant
calculation of, 491

of CF1–FTP complex, measurement of, 488

Disulfide bonds, polypeptide electrophoresis and, 364, 365, 372

Disulfodisalicylidenepropanediamine, photosystem I acceptor site and, 676–678

5,5-Dithiobis(2-nitrobenzoate)
Fe proteins and, 778

nitrite reductase and, 264

2,2-Dithio-bis-5-nitropyridine, phosphorylation and, 504–505

Dithionite
active site core extrusion and, 783

cytochrome assays and, 181, 182

ferredoxin–nitrite reductase assay and, 256

flavodoxins and, 405–406

interference with analysis of nitrogenase, 755–756

iron–sulfur center reduction and, 241, 243, 246

reaction center isolation and, 175

Dithiothreitol
cytochrome assay and, 182

thioredoxin and, 382

Dodecylbenzenesulfonate, bacteriochlorophyll *a*-protein and, 338

Dopamine, oxygen uptake and, 430

Dowex-1 acetate, glycolate isolation and, 463

Dowex 1-chloride
labeled ATP preparation and, 303

nitrate reductase and, 271

Dowex 1-X8, nucleotide separation on, 324

Driselase, source of, 78

*Dunaliella parva,* preparation of chloroplast material from, 115–116

*Dunaliella tertiolecta,* nitrite reductase of, 262, 263, 264

Duroquinone, oxidation–reduction titration of iron–sulfur centers and, 243

# E

EDTA
chloroplast-bound manganese and, 351–352

chloroplast membrane polypeptides and, 373

## J

## K

## L